HANDBOOK OF
CHEMICAL HEALTH AND SAFETY

HANDBOOK OF
CHEMICAL HEALTH AND SAFETY

ROBERT J. ALAIMO

Editor

American Chemical Society
Washington, D.C.

2001

OXFORD

UNIVERSITY PRESS

Oxford   New York

Athens   Auckland   Bangkok   Bogotá   Buenos Aires   Calcutta
Cape Town   Chennai   Dar es Salaam   Delhi   Florence   Hong Kong   Istanbul
Karachi   Kuala Lumpur   Madrid   Melbourne   Mexico City   Mumbai
Nairobi   Paris   São Paulo   Shanghai   Singapore   Taipei   Tokyo   Toronto   Warsaw

and associated companies in
Berlin   Ibadan

Developed and distributed in partnership by the
American Chemical Society and Oxford University Press

Published by Oxford University Press, Inc.
198 Madison Avenue, New York, New York 10016

Oxford is a registered trademark of Oxford University Press.

Library of Congress Cataloging-in-Publication Data
Handbook of chemical health and safety / Robert J. Alaimo, editor.
    p.   cm.
  Includes bibliographic references and index.
  ISBN 0-8412-3670-4
    1. Hazardous substances—Safety measures—Handbooks, manuals, etc.
    I. Alaimo, Robert J., 1940–
T55.3.H3 H344   2001
660′.2804—dc21   99-52958

1 3 5 7 9 8 6 4 2

Printed in the United States of America
on acid-free paper

Disclaimer

The Handbook of Chemical Health and Safety is intended for use by chemists, chemical engineers, and health and safety professionals. The material contained in this handbook has been compiled from sources believed to be reliable and to have expertise in the topic. This handbook is intended to serve as a starting point for chemical health and safety practices and does not purport to cover all related issues, specify minimum legal standards, or represent the policy of the American Chemical Society (ACS).

No warranty, guarantee, or representation is made by ACS as to the accuracy or sufficiency of the information contained herein, and ACS and its members assume no liability or responsibility in connection herewith. Users of this book should consult and comply with pertinent local, state, and federal laws and should consult legal counsel if there are any questions or concerns about the applicable laws, safety issues, and chemicals set forth herein.

Contents

Preface   xi

Contributors   xiii

Abbreviations and Acronyms   xix

SECTION I. SAFETY MANAGEMENT AND POLICY
Robert J. Alaimo, *Section Editor*

1. Chemical Laboratory Safety Training   3
   Kenneth P. Fivizzani

2. Working Alone   9
   W. Carl Gottschall

3. Personal Hygiene   11
   Ruth A. Hathaway

4. Drug Enforcement Agency Chemical Regulations   15
   Robert J. Alaimo

5. Laboratory Security   21
   Richard L. Fleming

6. Addressing the Human Dynamics of Health and Safety   25
   E. Scott Geller

7. Effective Management of Contractor/Visitor Safety   32
   Richard G. Raymer

8. The Chemical Hygiene Plan   38
   George H. Wahl, Jr.

9. Chemical Safety Information on the Internet   45
   Ralph B. Stuart III

10. Understanding the Toxic Substances Control Act: Compliance and Reporting Requirements   53
    Fred Hoerger
    Robert Hagerman

SECTION II. RISK MANAGEMENT
Lawrence Doemeny
Elizabeth K. Weisburger
Section Editors

11. Introduction to Risk Assessment   63
    Leslie Stayner

12. Communicating Risk   67
    Richard J. Hackman
    Kimberlee Vollbrecht
    Leslie M. Yee

13. Standard Operating Procedures   72
    Jay A. Young

14. Occupational Exposure Limits   75
    Edward V. Sargent
    Bruce D. Naumann

15. Industrial Hygiene Exposure Assessment—Data Collection and Managment   81
    Paul Hewett

16. Industrial Hygiene Exposure Assessment—Data Analysis and Interpretation   102
    Paul Hewett

17. Reproductive Hazards in the Workplace   130
    Elizabeth Anne Jennison

18. Epidemiology   136
    Neely Kazerouni
    Shelia Hoar Zahm
    Ellen F. Heineman

19. Carcinogenesis   141
    Elizabeth K. Weisburger

20. Toxicology   148
    Elizabeth K. Weisburger

21. Process Safety Reviews   157
    Dennis C. Hendershot

22. Safe Entry into Confined Spaces   166
    Laura L. Hodson

23. The Control of Hazardous Energy
    (Lockout/Tagout)   172
    David J. Van Horn

SECTION III.   EMERGENCY MANAGEMENT
    Russell Phifer, *Section Editor*

24. Emergency Response Planning and Training   181
    Louis N. Molino, Sr.

25. Emergency Equipment   186
    John C. Bronaugh

26. Emergency Evacuation/Shelter-in-Place
    Plans   194
    John J. McNamara

27. Flood Contingency Plans   198
    Ruth A. Hathaway

28. Seismic Safety   201
    Stanley H. Pine

29. Accident/Incident Investigation   205
    Stephen Sichak

30. Chemical First Aid   208
    John R. McInerney

SECTION IV.   LABORATORY EQUIPMENT
    Douglas B. Walters, *Section Editor*

31. Specialized Instrumentation and Monitors   215
    Dan Agne
    Sal Agnello

32. Inert Atmospheres Work   221
    Gerald W. Boicourt

33. Preventive Maintenance   234
    Jerry R. Hines

34. Health and Safety in the Microscale Chemistry
    Laboratory   239
    Zvi Szafran
    Mono M. Singh
    Ronald M. Pike

35. Laboratory Scale-Up and Pilot Plant
    Operations   249
    Amy L. Romanowski

36. Reducing Electrostatic Hazards Associated with
    Chemical Processing Operations   256
    Vahid Ebadat
    James C. Mulligan

37. Centrifuge Safety   265
    Martha A. McRae

38. Safe Use of Laboratory Glassware   272
    Robert J. Alaimo

39. Pressure/Vacuum Containing Systems and
    Equipment   277
    Kenneth K. Miles

40. Laboratory Ovens and Furnaces   282
    Robert J. Alaimo
    Stephen A. Szabo

41. Refrigerator, Freezer, and Cold Room Use in
    Chemical Laboratories   289
    Robert J. Alaimo

42. Specialized Laboratory Containment-Control
    Hoods   292
    Lou DiBerardinis

43. Laboratory Chemical Hoods   299
    Lou DiBerardinis

44. Biological Safety Cabinets   307
    Raymond W. Hackney, Jr.

SECTION V.   CHEMICAL MANAGEMENT
    Jay A. Young, *Section Editor*

45. Material Safety Data Sheets   317
    Robert J. Alaimo
    Lynne A. Walton

46. Personal Protective Equipment   322
    S. Z. Mansdorf

47. Incompatibles   338
    Leslie Bretherick

48. Corrosives and Irritants   343
    Jay A. Young

Contents   ix

49. Hazardous Catalysts in the Laboratory   345
    Francis P. Daly

50. Flammables and Combustibles   347
    James G. Gallup

51. Identifying Oxidizing and Reducing Agents   358
    Rudy Gerlach

52. Peroxidizable Organic Chemicals   361
    Richard J. Kelly

53. Disposal of Shock- and Water-Sensitive,
    Pyrophoric, and Explosive Materials   371
    George C. Walton

54. Compressed Gases   377
    George Whitmyre

55. Hydrogenations   383
    J. M. Lambert, Jr.
    W. S. Hamel

56. Chemical Inventory Control and Methods   391
    Sharon E. Stasko

57. Chemical Storage   397
    Lyle H. Phifer

58. Cryogenic Safety   401
    George Whitmyre

59. Explosive and Reactive Chemicals   404
    Irv Kraut

SECTION VI.   RADIOLOGICAL AND BIOLOGICAL
SAFETY
    Nelson Couch, *Section Editor*

60. Non-Ionizing Radiation   413
    Joseph M. Greco

61. Non-Ionizing Radiation: Radio-Frequency
    and Microwave Radiation   421
    R. Timothy Hitchcock

62. Ionizing Radiation: Radiation Safety Program
    Elements   428
    Philip E. Hamrick

63. Ionizing Radiation: Fundamentals   436
    Daniel D. Sprau
    Philip E. Hamrick
    Frederick L. Van Swearingen

64. Radiation Emergency Response,
    Decontamination, PPE   446
    Bob Wilson

65. Dosimetry   452
    Carmine M. Plott

66. Lasers   458
    Wesley J. Marshall

67. Biological Safety: Program Elements   465
    Debra L. Hunt

68. Research Animal Biosafety   471
    Scott E. Merkle

69. Biological Safety: Emergency Response and
    Decontamination Procedures   476
    Frederick L. Van Swearingen
    Antony R. Shoaf

SECTION VII.   LABORATORY DESIGN
    Stephen Szabo, *Section Editor*

70. Isolation Technology   487
    Dennis Eagleson

71. Lab Design/Radiosynthesis Lab Design   495
    Nelson W. Couch
    John J. Nicholson
    Shimoga R. Prakash

72. High-Pressure Test Cells and Barriers   501
    George W. Moncrief

73. Clean Rooms for Semiconductors   508
    John A. Mosovsky

74. Ergonomics in Design   513
    Laura S. Nystrom

75. Ergonomic Factors in Laboratory Design   521
    Joshua O. Kerst

76. Design Criteria for New Laboratories and
    Renovations   528
    Janet Baum

77. Moving and Decommissioning Laboratories   554
  Richard Fink

78. General Ventilation Design and Control
  Systems   560
  John P. Martin

79. Elevators, Stairs, Ramps, and Step Stools   566
  Philbert R. Romero

SECTION VIII.   ENVIRONMENTAL MANAGEMENT
  Russell Phifer, *Section Editor*

80. The Disposal of Chemical Wastes   573
  Russell W. Phifer

81. Biological Waste Management   598
  Robert Emery
  Wayne R. Thomann

82. Safe Handling of Biohazardous Materials
  for Transport   603
  John H. Keene

83. Radioactive Waste Management and
  Transportation   608
  Robert E. Uhorchak

84. Waste Minimization: General   615
  Cynthia Klein-Banai

85. Waste Minimization: Laboratories   622
  Cynthia Klein-Banai

86. Environmental Controls and Liabilities   626
  Russell W. Phifer
  James Harless

  Index   631

Preface

This Handbook was planned to be the most complete and comprehensive resource in the field of chemical health and safety. It will be the single most useful text on issues related to chemical safety. It is intended to replace many texts on the subject and be the first place any chemist, chemical hygiene officer, or safety specialist will look for answers. This reference was prepared by subject matter experts and will be the most up-to-date treatment of the subject available. It is the broadest and most complete resource available in the field.

This valuable reference will be the starting point for any chemical safety study, providing complete information and guidance. It will provide the reader with the tools necessary to create and maintain a safe work environment. It will contain the answers to both simple and complex questions related to workplace and environmental laws and regulations. It will be the most useful training tool available to the safety professional to ensure compliance in the ever-changing regulatory world.

This Handbook provides information on proper chemical equipment handling. Guidance on purchasing, storage, use, and disposal is presented in easy-to-understand terms. It includes useful examples of each of these items. The safe use of specialized equipment is outlined and the most current technology described.

Emergency preparedness plans are presented, enabling facilities to preplan for most emergency situations. Spill prevention and cleanup plans are included, providing useful information when accidents occur.

The Handbook will describe methods to help establish a safety culture in the workspace. It will be a how-to guide to starting and maintaining a successful safety program.

In addition to the traditional topics covered in a chemical safety text, the Handbook includes industrial hygiene issues, radiation safety practices, safe laboratory and facility design criteria, biosafety issues, ergonomics, and much more. The authors selected as contributors are experts in their fields and this ensures the Handbook to be the most authoritative and complete available today.

The development and the completion of this Handbook could not have happened without the hard work and dedication of three individuals, W. Thayer Long, Margaret Brown, and Mildred Hazard. Thayer served as the American Chemical Society's initial project gatekeeper and kept the project moving forward. Margaret picked up where Thayer left off and diligently tied all the loose ends together. She, almost single-handedly, turned a collection of manuscripts into a book. Mildred, in her role as my Administrative Assistant, coordinated, collated, and collected the manuscripts. The Editor acknowledges and appreciates the assistance of these three individuals.

This book is dedicated to my parents, Samuel R. and Michaline C. Alaimo, in their memory, and to Cheri, Christi, Rob, and my wife Barbara.

Robert J. Alaimo

Contributors

Dan Agne
Simplex Time Recorder Company
100 Simplex Drive
Westminster, MA 01441-0001

Sal Agnello
FieldTech Inc.
626 E. Wisconsin Avenue
Milwaukee, WI 53202

Robert J. Alaimo
Procter & Gamble Pharmaceuticals
Route 320, Woods Corners
Norwich, NY 13815

Janet Baum
Health, Education + Research Associates, Inc.
7470-A Delmar Boulevard
St. Louis, MO 63130

Gerald W. Boicourt
BASF Corp.
1609 Biddle Avenue
Wyandotte, MI 48192-3729

Leslie Bretherick
Woodhayes, West Road
Bridport, Dorset,
DT6 6AE, England

John C. Bronaugh
Haws Corp.
1455 Kleppe Lane
Sparks, NV 89431-6467

Nelson W. Couch
Triangle Health and Safety, Inc.
112 Bounty Lane
Durham, NC 27713

Francis P. Daly
Apyron Technologies
4030-F Pleasantdale Road
Atlanta, GA 30340

Lou DiBerardinis
Massachusetts Institute of Technology
77 Massachusetts Avenue, Building 20C-204
Cambridge, MA 02139

Laurence Doemeny, Senior Scientist
NIOSH
Centers for Disease Control and Prevention
4676 Columbia Parkway
Cincinnati, OH 45226-1998

Dennis Eagleson
The Baker Company
P.O. Drawer E
Sanford, ME 04073

Vahid Ebadat
Chilworth Technology
11 Deer Park Drive
Monmouth Junction, NJ 08852

Robert Emery
Environmental Health and Safety
Houston Health Science Center
University of Texas
P.O. Box 20036
Houston, TX 77225

Richard Fink
Massachusetts Institute of Technology
77 Massachusetts Avenue, Building 56-255
Cambridge, MA 02139-4307

Kenneth P. Fivizzani
Nalco Chemical Company
One Nalco Center
Naperville, IL 60563-1198

Richard L. Fleming
Procter & Gamble Pharmaceuticals
Route 320, Woods Corners
Norwich, NY 13815

James G. Gallup
Rolf Jensen and Associates, Inc.
549 W. Randolph Street, 5th Floor
Chicago, IL 60661

E. Scott Geller
Department of Psychology
Virginia Polytechnic Institute
Blacksburg, VA 24061-0436

Rudy Gerlach
164 W. High Street
New Concord, OH 43762

W. Carl Gottschall
17 Bradbury Lane
Littleton, CO 80120-4162

Joseph M. Greco
Radiation/Laser Safety Officer
Eastman Kodak Company
Health, Safety and Environment; B-320
Rochester, NY 14652-6261

Richard J. Hackman
Procter & Gamble
6110 Center Hill Center
Cincinnati, OH 45224

Raymond W. Hackney, Jr.
Health and Safety Office
University of North Carolina at Chapel Hill
212 Finley Golf Course Road
Chapel Hill, NC 27514

Robert Hagerman
6233 Forest Haven Drive South
Glen Arbor, MI 49636

W. S. Hamel
Parr Instrument Company
211 53rd Street
Moline, IL 61265

Philip E. Hamrick
National Institutes of Health
National Institute of Environmental Health Sciences
P.O. Box 12233
Research Triangle Park, NC 27709

James Harless
Techna Corporation
44808 Helm Street
Plymouth, MI 48170

Ruth A. Hathaway
Hathaway Consulting
1810 Georgia Street
Cape Girardeau, MO 63701

Ellen F. Heineman
National Cancer Institute
6120 Executive Boulevard
Rockville, MD 20852

Dennis C. Hendershot
Rohm & Haas Co., Engineering Division
P.O. Box 584
Rt. 413 and State Road
Bristol, PA 19007

Paul Hewett
NIOSH/DRDS
1095 Willowdale Road
Morgantown, WV 26505-2888

Jerry R. Hines
Westhollow Technical Center
Shell Development Company
P.O. Box 1380
Houston, TX 77251-1380

R. Timothy Hitchcock
1111 Yorkshire Drive
Cary, NC 27511

Laura L. Hodson
Research Triangle Institute
Safety and Occupational Health, Building 5
P.O. Box 12194
Research Triangle Park, NC 27709

Fred Hoerger
2813 Schade West Drive
Midland, MI 48640

Debra L. Hunt
Duke University Medical Center
Box 3149
Durham, NC 27710

Elizabeth Anne Jennison
Solutia Inc.
575 Maryville Centre Drive
St. Louis, MO 63141

Neely Kazerouni
Division of Cancer Epidemiology and Genetics
National Cancer Institute
6120 Executive Boulevard, Room 7031
Rockville, MD 20852

John H. Keene
Department of Preventive Medicine
Medical College of Virginia
Virginia Commonwealth University
Richmond, VA 23284

Richard J. Kelly
Lawrence Livermore National Lab
Hazards Control Department, MS L-143
P.O. Box 808
Livermore, CA 94551

Joshua O. Kerst
Humantech, Inc.
900 Victors Way, Suite 220
Ann Arbor, MI 48108

Cynthia Klein-Banai
University of Illinois at Chicago
Health and Safety Section, M/C 645
Environmental Health and Safety Office
1140 S. Paulina Street
Chicago, IL 60612-7217

Irv Kraut
Ace-Tech, Inc.
875 North Rt. 25
Aurora, IL 60505

J. M. Lambert, Jr.
Parr Instrument Company
211 53rd Street
Moline, IL 61265

S. Z. Mansdorf
L'Oreal
1, Avenue Eugene Schueller
93601 Aulnay sous Bois, CEDEX
France

Wesley Marshall
U.S. Army Center for Health Promotion
   and Preventive Medicine
5158 Blackhawk Road
Aberdeen Proving Ground, MD 21010-5403

John P. Martin
Earl Walls & Associates, Inc.
5348 Carroll Canyon Road
San Diego, CA 92121-1797

John R. McInerney
Dyncorp of Colorado, Inc.
Building 122, RFETS
P.O. Box 464
Golden, CO 80404-0464

John J. McNamara
Chester County Emergency Services
Government Services Building
601 Westtown Road
West Chester, PA 19382-4558

Martha A. McRae
Desert Research Institute
2215 Raggio Parkway
Reno, NV 89512-1095

Scott E. Merkle
National Institute of Environmental Health Sciences
National Institutes of Health
P.O. Box 12233
Research Triangle Park, NC 27709

Kenneth K. Miles
1613 Paisano NE
Albuquerque, NM 87112

Louis N. Molino, Sr.
33 E. King's Highway,
Mt. Ephraim, NJ 08059

George W. Moncrief
1542 Pennsbury Drive
West Chester, PA 19382-7752

John A. Mosovsky
Lucent Technologies
9999 Hamilton Blvd.
Breinigsville, PA 18031

James C. Mulligan
Chilworth Technology
11 Deer Park Drive
Monmouth Junction, NJ 08852

Bruce D. Naumann
Merck & Co.
1 Merck Drive
WS 2F-45
Whitehouse Station, NJ 08889-0100

John J. Nicholson
DuPont Pharmaceuticals Company
500 S. Ridgeway Avenue
Glenolden, PA 19036

Laura S. Nystrom
DuPont SHE Excellence Center
D-6098
1007 Market Street, N-2521
Wilmington, DE 19898-0001

Lyle H. Phifer
Chem Service, Inc.
P.O. Box 599
West Chester, PA 19381-0599

Russell W. Phifer
WC Environmental, LLC.
P.O. Box 1718
439 South Bolmar Street
West Chester, PA 19380

Ronald M. Pike
363 East 1st Avenue
Salt Lake City, UT 84103-2613

Stanley H. Pine
Chemistry Department
California State University
5151 State University Drive
Los Angeles, CA 90032

Carmine M. Plott
400-Y Park Ridge Lane
Winston-Salem, NC 27104

Shimoga R. Prakash
DuPont Pharmaceuticals Company
P.O. Box 30
109 Elkton Road
Newark, DE 19714

Richard G. Raymer
Procter & Gamble
Sharon Woods Technical Center
11510 Reed Hartman Highway
Cincinnati, OH 45241

Amy L. Romanowski
UOP Research Center
50 East Algonquin Road
Des Plaines, IL 60019

Philbert R. Romero
Los Alamos National Laboratory
P.O. Box 1663
Mail Stop K-403
Los Alamos, NM 87544

Edward V. Sargent
Merck & Co.
1 Merck Drive
WS 2F-45
Whitehouse Station, NJ 08889-0100

Antony R. Shoaf
Shoaf Scientific Consultants, Inc.
2386 Horseshoe Neck Road
Lexington, NC 27295

Stephen Sichak
318 N. Russell
Mt. Prospect, IL 60056-2447

Mono M. Singh
National Microscale Chemistry Center
Merrimack College
North Andover, MA 01845

Daniel D. Sprau
Department of Environmental Health Sciences, Safety,
    and Technology
School of Industry and Technology
East Carolina University
Greenville, NC 27858

Sharon E. Stasko
Vertere, Inc.
225 Chapman Street
Providence, RI 02905

Leslie Stayner
Risk Evaluation Branch
NIOSH, C15
4676 Columbia Parkway
Cincinnati, OH 45226

Ralph B. Stuart III
University of Vermont
Environmental Safety Facility
655 D Spear Street
Burlington, VT 05405

Stephen A. Szabo
2323 Turner
Ponca City, OK 74604

Zvi Szafran
Vice President Academic Affairs
New England College
7 Main Street
Henniker, NH 03242

Wayne R. Thomann
Duke University Medical Center
P.O. Box 3149, DUMC
Durham, NC 27710

Robert E. Uhorchak
Research Triangle Institute
3040 Cornwallis Road
P.O. Box 12194
Research Triangle Park, NC 27709-2194

David J. Van Horn
Van Horn & Associates
7 Fields Landing
Rehoboth Beach, DE 19971-4144

Frederick Van Swearingen
Department of Physical Sciences
Winston-Salem State University
601 Martin Luther King, Jr. Drive
Winston-Salem, NC 27110

Kimberlee Vollbrecht
Procter & Gamble Company
One Procter & Gamble Plaza
Cincinnati, OH 45202

George H. Wahl, Jr.
Department of Chemistry, Box 8204
North Carolina State University
Raleigh, NC 27695-8204

Douglas B. Walters
KCP, Inc.
6807 Breezewood Road
Raleigh, NC 27607

George C. Walton
Reactives Management Corporation
1025 Executive Boulevard, Suite 101
Chesapeake, VA 23327

Lynne A. Walton
Procter & Gamble Company
8700 Mason-Montgomery Road
Mason, OH 45040

Elizabeth K. Weisburger
9301-A Wescott Place
Rockville, MD 20850-3463

George Whitmyre
Chemical Engineering
University of Delaware
Colburn Lab
150 Academy Street
Newark, DE 19716

Bob Wilson
University of Kentucky
Environmental Health and Safety
Radiation Safety Office
102 Animal Pathology Building
Lexington, KY 40546-0076

Leslie M. Yee
Procter & Gamble Company
Two Procter & Gamble Plaza
Cincinnati, OH 45202

Jay A. Young
12916 Allerton Lane
Silver Spring, MD 20904

Sheila Hoar Zahm
National Cancer Institute
6120 Executive Boulevard
Rockvile, MD 20852

Abbreviations and Acronyms

These abbreviations and acronyms are used so frequently throughout this book that they are defined here only and not in the text.

ACGIH	American Conference of Governmental Industrial Hygienists	FDA	Food and Drug Administration
ACS	American Chemical Society	FR	*Federal Register*
AIHA	American Industrial Hygiene Association	MSDS	Material Safety Data Sheet
ANSI	American National Standards Institute	NCI	National Cancer Institute
ASTM	American Society for Testing and Materials	NFPA	National Fire Protection Association
CDC	Centers for Disease Control and Prevention	NIH	National Institutes of Health
		NIEHS	National Institute of Environmental Health Sciences
CFR	Code of Federal Regulations	NIOSH	National Institute for Occupational Safety and Health
DHHS	Department of Health and Human Services	NRC	Nuclear Regulatory Commission
DOT	Department of Transportation	NTP	National Toxicology Program
EPA	Environmental Protection Agency	OSHA	Occupational Safety and Heatlh Administration

SECTION I

SAFETY MANAGEMENT AND POLICY

ROBERT J. ALAIMO

SECTION EDITOR

1

Chemical Laboratory Safety Training

KENNETH P. FIVIZZANI

The role of training in a safety program cannot be over-emphasized. Good safety training is informative, useful, memorable, and designed for the specific audience. Safe work habits are not intuitive; they are learned by both training and experience. If training is perceived as an opportunity to educate rather than just to comply with regulations and policies, the result will justify the commitment of time and resources.

Managers and faculty members should take an active role in safety training, emphasizing the top priority of a safe work environment. These same key people must enforce safety policies consistently and personally follow every rule and policy. All chemists who serve an administrative, managerial, or teaching function must convey both correct information and a proactive attitude about safety in the research/teaching laboratory. Safety training is not an afterthought or add-on but an essential part of any curriculum in laboratory research.

Training Topics

Emergency Response

"How do I call for help in an emergency?" The answers to this question are the content of this section. Common types of emergencies that should have a planned response include fire, hazardous chemical spill or vapor release, medical emergency, electrical power (lab hoods) failure, and severe weather. For each type of emergency, an employee must know how to summon help and what response is required of all employees during the emergency. Section III of this handbook contains specific information regarding emergency response planning.

Personal Protective Equipment (PPE)

Employees must be given appropriate PPE and be trained as to how to use that equipment correctly (see Fig. 1.1).

Safety glasses and goggles, gloves, lab coats, safety shoes, hard hats, earplugs, and respirators might be required in laboratories, plants, or pilot plants. The procurement process should be straightforward so that PPE is considered by the staff to be readily available. Some employees can be trained to carry out job hazard analyses to determine what PPE is required for specific lab operations. Training must be reinforced by supervisors, teachers, and research colleagues on an ongoing basis. As a corollary, research personnel must be vigilant in refusing to allow any individual to be photographed in a laboratory environment without appropriate PPE.

Other Safety Equipment

Everyone working in a laboratory must know the location of eyewashes and safety showers. These items are rarely used but are crucial in certain critical situations. Lab hoods are safety equipment (not storage cabinets); training is required to learn how to use hoods properly. If employees or students are required to use fire extinguishers in their jobs, they must be trained in the proper use of extinguishers. Fire departments recommend that extinguisher training include practice using the extinguisher to put out fires.

OSHA Regulations

There are published lists of OSHA regulations that apply to research and teaching laboratories. For example, see pages 199–201 of *Prudent Practices in the Laboratory: Handling and Disposal of Chemicals* (National Academy Press; Washington, DC, 1995). The most significant regulations for lab operations, including support personnel, are the Hazard Communication Standard, 29 *CFR* 1910.1200, and Occupational Exposures to Hazardous Chemicals in Laboratories (commonly known as the Lab Standard), 29 *CFR* 1910.1450. Section (f) of the Lab

FIGURE 1.1 Misunderstood safety training. (Reproduced with permission from Shoe cartoon. Copyright Tribune Media Services, Inc.)

Standard specifies the required employee training. All lab employees must be informed about the contents of these two standards as well as the contents and location of the Chemical Hygiene Plan for the lab. Employees who use respirators must also understand and comply with the regulation of respirators, 29 *CFR* 1910.134. Note that there may also be state laws that apply to laboratory operations.

Local Policies and Procedures

Every laboratory has specific safety policies and procedures. The most basic policies must be explained to everyone working in the lab. Administrative procedures may be important only to supervisors or instructors. What are the initial safety training requirements before an individual is approved to work in the lab? Are there any absolute (sometimes called *cardinal*) rules, the violation of which will result in expulsion from the lab and/or termination of employment? How and when must accidents and injuries be reported? How does one properly dispose of lab wastes?

Specialized Training

Some employees will require specific training because of their work assignments. Use of highly toxic or extremely hazardous materials often will require training that covers the hazards and safe use of those materials. Some laboratories require such training, for example, for any material appearing in 29 *CFR* 1910, subpart Z, OSHA's Toxic and Hazardous Substances. Use of each of these materials is governed by a specific OSHA regulation. The training required for approved use of a hazardous material may vary depending on the formal education and work history of an individual. For example, employees having a bachelor's degree in chemistry usually have an awareness of general hazard categories and which types of chemicals would be in each classification. Experienced chemists understand that strong acids and bases are corrosive and that many organic materials are flammable or combustible. It is recommended that every employee review the hazards and properties of highly toxic and extremely reactive substances before using them.

Staff members who use ultraviolet light, lasers, X-rays, or radioactive isotopes may need specialized training in these areas. OSHA's Bloodborne Pathogens regulation, 29 *CFR* 1910.1030, covers those who come in contact with blood or blood products. The regulation details what information must be covered in the annual training required for those exposed to bloodborne pathogens.

Members of an Emergency Response Team (ERT) who respond to hazardous spills or leaks must be trained according to OSHA's Hazardous Waste Operations and Emergency Response (HAZWOPER) regulation, 29 *CFR* 1910.120. Members of a medical or first aid team should be trained in general first aid, cardiopulmonary resuscitation (CPR), and bloodborne pathogens.

Methods of Training

For each of the topics mentioned above, what is the most effective way to convey the needed information to those who need it? All training should be presented at a level appropriate for the intended audience. Different audiences will require different approaches. Chemists may have a good working knowledge of the technical content of a safety training program; clerical support staff or maintenance personnel will need to learn the fundamentals of most technical safety topics. Remember your audiences when considering the purchase of commercial training materials. Try to involve line management or faculty members in the training. Testing provides a way to monitor employee comprehension and retention of the information; testing also documents compliance with training requirements.

Videos

Safety training videos are a common part of many laboratory safety programs. These videos are available for a large number of topics; they are convenient and easy to use. Videos can be especially effective in demonstrating unsafe conditions or reenacting accidents and incidents.

What factors should you look at when you evaluate safety videos? You need to consider content, realism and credibility, minor flaws in the presentation, and any subtle messages conveyed by the actors' depiction of laboratory activity. Always preview a video before you purchase it, and remember the different audiences that will see the video as part of their training. The following checklist (see Fig. 1.2) will help in the evaluation of videos.

Formal Presentations

Another common training method is for an individual to make an oral presentation covering the topic of interest. Advantages here include the fact that training can be customized and the cost of training materials is minimal. The instructor must have adequate understanding of the topic and should be an effective communicator.[2] Well-designed overheads or slides will improve the clarity of the presentation as well as ensure that the essential points are discussed. Guest speakers, such as a member of the local fire department, an industrial hygienist, or an occupational health nurse or physician, should be considered for specialized topics. There should be adequate time for the audience to ask questions of the speaker.

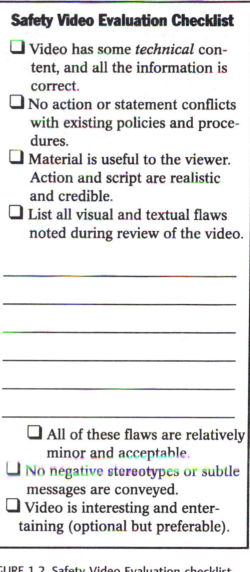

Safety Video Evaluation Checklist

☐ Video has some *technical* content, and all the information is correct.

☐ No action or statement conflicts with existing policies and procedures.

☐ Material is useful to the viewer. Action and script are realistic and credible.

☐ List all visual and textual flaws noted during review of the video.

_____

_____

_____

_____

_____

_____

☐ All of these flaws are relatively minor and acceptable.

☐ No negative stereotypes or subtle messages are conveyed.

☐ Video is interesting and entertaining (optional but preferable).

FIGURE 1.2 Safety Video Evaluation checklist. (From Ref. 1. Copyright 1996 American Chemical Society.)

Interactive Training

Recently, computer-based interactive training programs have been developed.[3,4] These programs can be put on a personal computer and used by the employee or student at his/her convenience. This method provides the greatest flexibility in terms of pacing the instruction and scheduling the training. A test at the end of the program measures comprehension and can be used to determine if additional or refresher training is needed. Finding an interactive program that is appropriate for use by all lab employees can be a challenge. If the program has a highly technical orientation, those with less formal education and training may be overwhelmed. A program that focuses on the fundamental concepts and issues may be perceived as boring or a waste of time by the technically educated or experienced employees or students.

Hands-on Training

Whenever a training participant learns by actually doing some activity, there is a high probability that the information will be remembered. Some safety topics can only be done properly by active participation. Fire extinguisher training and CPR are examples in which a hands-on approach is essential. Safe driving and first aid are examples of topics in which hands-on experience is very helpful, but classroom instruction is also valuable. Hands-on training usually requires more preparation, and the number of participants in any session may be limited.

Documentation

All safety training sessions must be documented. In addition to providing evidence of who received what training, the importance of the training is emphasized. There are four essential components of good safety training documentation: (1) the date the training actually occurred (an announcement/invitation memo does not document that the training ever took place); (2) a description of the subject and the training activity, including the instructor; (3) a list of individuals who completed the training; and (4) the signature of an appropriate individual verifying that all the information contained in the documentation is correct.

New Lab Employees and Transferees

The Laboratory Administration (managers or faculty) must establish the minimum initial safety training that is required of anyone who works in the lab. This initial training must be completed before a new employee can do any type of work in the lab. As far as visiting or touring the lab, the new employee (or job applicant) can be considered as a visitor (see below).

The initial training could involve reading a safety or procedures manual, attending an initial safety training presentation, meeting with a safety professional for a discussion of general lab safety policies, or some other appropriate safety activity. The employee's supervisor should be actively involved in this initial training, at least to the point of quizzing the employee for compre-

hension and addressing any questions the new employee may have.

Within a reasonable period of time, the new employee must receive all safety training required by law, specifically training on the Hazard Communication Standard, the Lab Standard, and the Chemical Hygiene Plan. As specific job assignments are made, the employee must receive the appropriate safety training before starting that particular assignment.

Transferees from other departments in your organization may be excused from repeating any safety training that they have already received within a reasonable period of time. If a transferee's most recent job did not involve laboratory work, it is recommended that he/she complete all lab safety training as if he/she were a new employee.

Inexperienced Lab Employees

Some technically trained employees will come to your organization with no prior lab experience other than lab coursework, in which every detail of the experiments has been worked out by the instructor or lab manual author. These new employees need to learn everything about designing and carrying out laboratory work. They need to learn basic lab techniques, such as selecting the proper type and size of glassware, methods of heating reactions, use of condensers and other means of cooling experimental systems, transfer of chemicals, and how to work with pressurized and vacuum systems. All of these skills have a safety component as well. An advantage of training a true lab novice is that there are no unsafe habits to correct. Inexperienced lab workers must have someone with whom they feel comfortable asking basic questions about lab operations and safety. They must understand that safety is always a top priority and that working safely is the rule and not the exception.

Employees with Lab Experience

New employees with previous lab experience may have developed a cavalier or indifferent attitude about safety. They need to be convinced of the high expectation that is placed on lab employees to work safely. This approach may be different from what they experienced with a former employer (see Fig. 1.3). Even the experienced new employees should complete the minimum safety training required of all lab employees. Depending on the situation, they may be excused from specialized training that they may have received elsewhere.

New Ph.D. scientists can be the most difficult to convince of the value of good safety practices, depending on their specific research background. Having just com-

DILBERT

FIGURE 1.3 The evolution of safety training. (DILBERT reprinted by permission of United Features Syndicate, Inc.)

pleted their dissertation, they tend to be quite confident of their own knowledge and capabilities. They may possess the naive sense of invincibility that is typical of many young people. They have spent four to six years engaged in active research at a large university, where safety may not have been a major priority. Although they rarely say it, one senses that they are thinking "we never bothered with that in grad school." These new researchers need factual information to bring them to the realization that everyone needs to learn more about safety. Take advantage of their current expertise and ask for their input when developing safety procedures and policies.

Students

In academic laboratories,[5] most of the workers are students. Students have a different perspective and different motivations from individuals employed by a company or school. Their legal status is different from that of lab employees.

Nonemployee Students

Typical high school and undergraduate college students are not classified as employees according to OSHA. Thus, the legal requirements covering employees in OSHA regulations do not apply to students. However, school administration and faculty have a professional and moral obligation to provide a safe work environment for students. All teachers of chemistry should include the appropriate chemical health and safety information as part of their instruction. Everyone who teaches or trains lab workers, whether in academia or industry, is obliged to teach them how to work safely.

Teaching Assistants/Graduate Students

Teaching Assistants (TAs), whether undergraduate or graduate students, are paid by the college or university. They are employees and must comply with all government regulations that pertain to employees. Most science graduate students are either TAs or Research Assistants, both paid positions, so they are also employees. In addition, TAs are analogous to first-line supervisors in industry in that they provide safety training and are responsible for the students' safety in the labs. Thus, TAs must be trained both as lab employees and as safety trainers for their students.

Summer Students (Industry)

Many companies employ summer students to substitute for vacationing employees. Often these summer students are children of employees. Although they may be required to have some science courses, their lab experience is basically a summer job, and they have little expectation of acquiring knowledge or experience that will be useful in later life. Because of their lack of lab experience and relatively short period of employment, summer students may require special training to cover the safety fundamentals that new employees would learn over the first few months of employment.

Co-op Students (Industry)

Co-op students are interested in learning about labwork as part of a career in science or engineering. As such, they have more of a vested interest in learning safe work habits. They too, however, are relatively inexperienced and will need special safety training. As part of their overall co-op experience, they should eventually assume more responsibility for ensuring a safe work environment in their lab area.

Nontechnical Support Staff

Examples of nontechnical support staff include secretaries, maintenance personnel, mail clerks, library and computer professionals, and stockroom and lab dishwashing aides. If their job requires them to enter the labs, then they should receive some basic safety training, including required PPE and emergency response. In a very general sense, they need to know where the more hazardous areas of the lab are located. The concept of "designated area" should be explained to them, noting that these are not dangerous zones but simply areas where hazardous materials are used and stored. Because of their limited knowledge of chemistry, nontechnical employees are usually very accommodating when it comes to complying with safety policies and procedures in the labs.

Part-Time and/or Temporary Employees

With respect to safety training, these employees should be treated exactly the same as full-time permanent employees. For those working in the lab, the minimum initial safety training and all training required by government regulation must be completed. If the temporary employee is inexperienced, he/she could be asked to attend the special safety training programs that the co-op and summer students attend. If the employee does not work directly in the lab, safety training should be the same as for the nontechnical support staff. A good lab safety program includes *all* employees, technical and nontechnical, full and part time, permanent and temporary.

Custodial Staff

Custodial staff members are a subset of nontechnical support staff, but one aspect of their job presents a unique concern among nontechnical employees. Although these employees have very limited technical education or experience, they enter the labs regularly, perhaps even daily, and they are often present after the technical staff has gone home at the end of the workday. Custodial staff may be outside contractors, and they may have limited reading or speaking skills in the vernacular language of the laboratory. Safety training in this case concentrates on how a laboratory has different hazards from the office or classroom areas of a facility. Specifically, these employees must learn not to touch or bump into lab equipment; not to turn off water or electric power without prior notification of lab personnel; and if a lab chemical is spilled or leaks, to call for trained emergency responders or technical expertise before attempting to clean up the spill. Custodial workers may not understand the need for PPE in the labs, so supervisors will have to monitor compliance with the appropriate safety rules. In addition, the lab staff may have to take certain precautions for the safety of custodial workers, such as segregating hazardous (glassware, sharp objects) and nonhazardous trash or removing objects from floors that are scheduled to be cleaned.

Contractors

Outside contractors present a challenge to any organization's safety program and safety performance. Contract employees come and go, frequently changing jobs or work areas. It is not always clear who supervises these employees. Because they work in so many different settings, they may be oblivious to the surrounding activity in a lab.

Contractors who work in the lab must receive the training required of all lab employees, for example, training on the Hazard Communication and Lab Standards. But who provides this training? The school or company where the lab is located is not the employer of the contract employee. Is the owner or manager of the contracting company (the employer) capable of doing lab safety training? One might presume that whoever trains the lab employees could also train the contractors as well, but this approach may have legal complications. If you tell a contract employee that he/she has a legal right to medical consultation if exposed to excessive levels of a hazardous chemical, could it be inferred that your organization will provide and pay for that service? Who ensures that each contract employee has the appropriate hazard communication training for the particular job being done that day? In labs with very good safety records, it is not unusual to observe that contract employees have the most frequent and most serious accidents.

Visitors

Most labs welcome visitors, at least during the normal workday hours. Many laboratories, however, do not have any rules or even guidelines for visitors. It is a good idea, whenever feasible, to require visitors to sign in and sign out of lab areas. They may be asked to wear visitor name tags to clearly identify them. VISITORS MUST ALWAYS COMPLY WITH SAFETY RULES THAT APPLY TO THEM IN THE LABORATORY. The most common occasions for infractions of safety rules are the requirement to wear eye protection in the lab and the prohibition against eating, drinking, or smoking in laboratory areas.

Other questions must be considered in developing a visitors safety policy. Can visitors wander through the labs alone? Are there any restrictions on children or pregnant women visiting (any developmental or reproductive hazards)? At a minimum, lab workers must be trained to take ownership of their labs and prevent others (visitors, colleagues, managers, and faculty) from committing unsafe acts or exposing themselves to unsafe conditions.

References

1. Fivizzani, K. P. *Chem. Health Saf.* **1996**, *3*(2), 33–35.
2. Barker, T. *Safety + Health* **1996**, *153*(6), 54–58.
3. Burke, A. *Industrial Safety and Hygiene News* **1997**, *31*(11), 31–32.
4. Gordon, J.; Hequet, M. *Training* **1997**, *34*(3), 24–31.
5. American Chemical Society. *Safety in Academic Chemistry Laboratories*, 6th Ed.; Washington, DC, 1995; see page 41 for a discussion of safety training.

2

Working Alone

W. CARL GOTTSCHALL

Working Alone in the Chemical Workplace

DON'T!
Rather than writing a one word chapter and slighting the topic, what follows is intended to convey some of the reasons why all experts concur in saying: *Do not work alone with chemicals.*

Indeed, most chemical workplaces have a policy which states: "Working with chemicals is absolutely forbidden unless there are at least two people present," or words to that effect. There are good reasons, morally, legally, and financially, for such a policy wherever and whenever possible. In a sense, the admonition may seem like a paraphrase of the Murphy's Law, "If something can go wrong, it will," approach to everything, but when considering the health and safety, perhaps even the life of an individual, long odds become less relevant.

Working with chemicals in a laboratory or elsewhere without anyone else present has been and likely will remain a topic of sometimes heated discussion. Just don't do it is the best, most universal advice. The buddy system's advantages are legion, and generally that approach is required everywhere, at all times when working with hazardous chemicals and associated equipment and apparatus.

Accidents and Incapacitations

Slips, trips, and falls are the most common accidents and obviously occur in chemical workplaces as well as anywhere. A fall in a chemical workplace, however, has the potential to involve hazardous chemicals or equipment not associated with any usual activity of the individual. Should the fall be incapacitating, the perhaps hours-long absence of another person and hence the lack of assistance could cause serious complications. A seizure, stroke, or other "natural" event is bad enough when assistance is readily available, but when one is isolated, it is far worse.

Another example of the need for a buddy in laboratory situations is the accidental splash of a chemical that causes burning pain in the eyes; reflexively, the eyes close and hence even a nearby eye wash station might not be reached quickly enough if at all. Even when reached, proper utilization is often difficult or nearly impossible without the aid of another individual.

As the potential danger or hazard of the operation(s) being performed increases, the need for additional safety precautions or human interfaces increases. Distilling some corrosive substance such as an acid would require more precautions than distilling brine; working with a toxic substance that presents adverse impacts at levels below the level of human detection capabilities requires even greater precautions. The factors and possibilities to consider can rapidly become enormous and demonstrate the wisdom of having a knowledgeable buddy present at all times.

The American Chemical Society (ACS) clearly and unambiguously recommends: "Never perform any work when alone in the chemical workplace or laboratory."[1] And in another publication: "Never perform any work when alone in a chemical laboratory. At least two people must be present."[2]

Unattended Operations

Beyond the aspect of human involvement alone, one should also address the question of unattended reactions or equipment left to operate without a human presence. Again, the ACS speaks to this area of concern:

Unattended Operation of Equipment
Reactions that are left to run unattended overnight or at other times are prime sources for fires, spills, and explosions. Do not let equipment such as power stirrers, hot plates, heating mantles, and water condensers run overnight without fail-safe provisions. . . .
Check unattended reactions periodically. Always leave a note plainly posted with a phone number where you . . . can be reached in case of emergency. Remember that in the middle of the night, emergency personnel are entirely dependent on accurate instructions and information.[2]

Although the individual setting up the experiment may not be subject to any personal adverse result should an accident occur, one must also recognize that security personnel or other nearby individuals might be. Accordingly, it is recommended that the lights be left on in the workplace to draw attention to some activity therein. It is mandatory that appropriate signs with sufficient information be left in prominent places. Obviously one should also contact any individuals potentially impacted by some malfunction.

It is not sufficient to consider only potential personnel injuries, although that area should be addressed first. Bear physical plant ramifications in mind also. The individual's experiment or even his/her laboratory or the entire building could be destroyed. Destruction of records, papers, or theses in progress, and loss of specially synthesized materials or expensive equipment could cause irreparable damage. Potential involvement of explosives, or the release of toxic gases or carcinogens from other

nearby laboratories that become subsequently involved should indicate why responsible authorities do not readily permit unattended processes.

It Is Your Decision

The preceding sections indicate only some of the things that can go wrong. The historical record of accidents that have happened in isolation could fill a book with scary scenarios. Nonetheless, there are exceptional situations which could arise or might be mandated by the press of other matters that require that an individual work alone. While not recommending nor even condoning such activity, it is appropriate to recognize that it does occur. Under such circumstances, steps should be undertaken both to minimize the possibility of an accident and to mitigate the results, should one occur.

Obviously, each chemical workplace is unique and it is not possible here to list all of the precautions that should be taken when circumstances require that a person must work alone. But at the minimum, the precautions should include provisions that enable the isolated worker, even if incapacitated, to call for help, and further, they should include provisions for frequent, regular visits by a second person who would be able to provide competent assistance in the event of an accident. Finally, it would be prudent to notify a friend or family member that one is going to be working alone until some set time, so that failure to return would prompt a follow-up.

If the need to work alone is a frequent situation, it is imperative to have a program in place to assure the

FIGURE 2.1 Man-Down Transmitter. (Photo courtesy of Transcience.com, Stamford, CT.)

safety of the worker. If security guards are present in the facility, ask that they stop by during their security rounds to check on the worker. Periodic phone calls to or from the individual would be another way of keeping in touch. The work area could also be equipped with an emergency panic button that would alert security personnel or emergency services that an emergency situation exists. One of the best options is a personal alarm system. Personal alarm systems are available that will send out an audible alarm when activated, followed by a radio transmitted alarm to a central station. One type of unit is referred to as a Man-Down Transmitter and is small enough to be worn on a belt (see Fig. 2.1). The unit has the capability of having the alarm set off either manually with the push of a button or by a built-in tilt sensor. The tilt sensor will activate the unit after a 10-second delay followed by a 10-second audible alarm. If the unit is not reset at that time it will initiate the radio alarm transmission as well as the audible alarm alerting emergency personnel. An example of such a system is available from Transcience.com.[3]

References

1. American Chemical Society, *Chemical Safety Manual for Small Businesses*, 2nd ed.; Washington, DC, 1992.
2. American Chemical Society, *Safety in Academic Chemical Laboratories*, 6th ed.; Washington, DC, 1995.
3. Transcience.com, 11 Ryan Street, Stamford, CT 06907; (800) 243-3494; info@transcience.com.

3

Personal Hygiene

RUTH A. HATHAWAY

Personal Hygiene

How can you tell that a person is a chemist? They wash their hands before using the bathroom. This is an old joke among various laboratory personnel. It may sound odd, but it should be part of the personal hygiene habits of everyone who works with chemicals. There are many dos and don'ts that warrant consideration. This chapter will explore some of these.

Look at your workplace setting. Examine it closely. Are gloves worn? What type of eye protection is used? Is special clothing such as lab coats worn? Why are these precautions taken? If just one of these items is required, then personal hygiene must be considered. We wear protective equipment to minimize exposure. Unless one is completely encapsulated prior to and during the exposure window, one must assume one has been exposed to the chemicals that exist in the area. This means that vapors and/or particulates have settled on and infiltrated the clothing one wears and one's skin and hair. Below is a head-to-toe review.

Head

Hair can become a liability. Any hair that obscures vision must be restrained. One cannot work safely when one cannot see clearly all the area within one's vision. If hair is worn long, it must be secured so that it does not come in contact with anything one is working with, such as chemicals, flames, or heated surfaces when one tilts or moves the head. This precaution will aid in avoiding undue exposure and potentially severe injury. Never run your fingers through your hair or scratch your scalp with unwashed hands. This only exposes more area to the chemicals in the work area.

Cosmetics should never be applied while in an area where chemicals are used, handled, or stored, nor should they be worn in a thick layer when working in these areas. I have seen people wear their makeup so thick that if they should be splashed with a chemical, they would never sense it. A thick layer of makeup hinders one in discerning that there is a problem, such as a splash on the face. It may even cause one to scratch one's face because the makeup is reacting with a chemical with which one is working. If makeup must be worn, then do so sparingly. Never rub or touch the face with unwashed hands. This only enhances exposure.

Contact lenses should only be worn when permitted (see chapter 48 on corrosives and irritants) and even then only with additional eye protection where chemicals are used, handled, or stored. Never remove, clean, or replace contact lenses in the workplace. When removing contact lenses, one exposes the eyes to direct chemical exposure from the fingers. If one cleans and replaces the contact lenses, one can trap chemicals between the eye and the lens. If an adjustment or removal is necessary, do so outside of the workplace. Wash hands thoroughly prior to removing or adjusting the lenses.

Safety goggles must be kept clean and in good working order; that is, they must not be broken, bent, or scratched. If they are dirty, they will obscure the work area. One may not even be able to tell if splashing has occurred due to spots already on the safety goggles. Always have safety goggles in place before entering, while occupying, and when exiting any area where chemicals are used, handled, or stored. Do not adjust or clean safety goggles in locations where the inside surface may be inadvertently exposed to additional chemicals. Any protective eye wear that is worn must be impact- and chemical-resistant. It must meet ANSI standards as required by 29 *CFR* 1910.133. Some plastic lenses may "dissolve" or be permanently etched by organic compounds. If you are unsure about your lenses, check with your chemical hygiene officer, safety officer, supplier, or optometrist.

Earrings can present a problem and should not be worn in areas where chemicals are used, handled, or stored. Nonpierced earrings can fall off, perhaps into the solution one is working with. Even pierced earrings can lose their fastener and fall off. Long and/or dangling earrings can interfere with one's work by continually moving around the neck area. One may inadvertently touch them, move them, or rub the neck area, causing additional contact exposure. They may even become tangled with hair, creating a nuisance. These earrings can even interfere with the proper fit of some protective equipment.

A person working in areas where chemicals are used, handled, or stored really needs to think seriously about body piercing. With piercing of eyebrows, eyelids, noses, lips, and tongues being "in fashion," these fashion accessories can impede the proper fit of protective equipment. All of them add an additional route for exposure to the body.

Almost everyone experiences the occasional cold or allergy with a runny nose. Try to avoid using handkerchiefs in areas where chemicals are used, handled, or stored. If possible, leave the area, wash the hands thoroughly, and then take care of the problem nose. Otherwise, use care so as to avoid transferring chemicals directly to sensitive nasal tissue. Avoid storing handkerchiefs or facial tissue where chemicals are used, handled, or stored as they will absorb the chemical vapors and particulates present in the workplace.

Headphones should not be worn unless required. They impair one's capacity to hear an alarm or distress call. This can include not only fire alarms but also the boiling over of a heated liquid, or the cry of a co-worker for help.

Eating and drinking in areas where chemicals are used, handled, or stored should never be permitted. This includes chewing gum, candy, and tobacco products. One should assume that whatever one is working with will be and has been absorbed by the food and/or drink one is consuming. Likewise, these items must never be stored with hazardous chemicals, on a shelf, in a drawer, or in a refrigerator. Glassware intended for use with chemicals should never be used to store food. Glassware, like gloves, absorbs whatever it comes in contact with, and therefore can deposit it back onto the food and into the drink.

Do not pipet anything by mouth. Similarly, great care must be exercised when "smelling" chemicals. Gently waft the odor to the nose rather than bringing the nose to the material. The sense of smell must be protected as it is one of our defenses that alert us to a dangerous situation.

Hands

Always wash hands before touching nonexposed areas. This can be prior to using a pencil on the secretary's desk or the phone in the main office, and (especially) prior to using the rest room. Items outside of the workplace are outside the workplace for a reason. Therefore, care must be taken not to transport contaminants to the outside. Even if gloves are worn, the hands should be thoroughly washed after the gloves are removed. All gloves have a finite life. Eventually, chemicals are absorbed into the glove material, and little by little they diffuse to the inside of the glove. (See Chapter 46 for more information on gloves.) If the policy states that gloves should only be removed outside of the working area, thoroughly wash the gloves (while still on the hands) before leaving so as not to add additional contamination to the door knob/handle when exiting. This will also prevent unnecessary exposure to the outside area.

It is advisable not to wear rings in areas where chemicals are used, handled, or stored. Rings with stones can reduce the lifetime of gloves by causing stress marks and/or tearing. The skin under rings is much more sensitive than any other area on the hands. When chemicals come in contact with this area, exposure is quicker and more intense. Exposure is longer in duration when the chemical (including vapors) becomes trapped between the ring and the skin.

Regularly clean under one's nails. One never knows what might be trapped there. Do not bite one's nails. This transfers whatever chemical is on the fingers and under the nails into the mouth, thereby causing ingestion of the chemical.

Fingernail polish should not be worn where chemicals are used, handled, or stored for two reasons. First, most organic solvents will remove part or all of the polish. Since it usually is worn for fashion purposes, this makes it unfashionable. Second, overexposure to some chemicals, for example, cyanide and selenium compounds, is often first detected by the discoloration of the tissue under the nails.

Do not use hand lotions where chemicals are used, handled, or stored. They can aid in the absorption of chemicals through the intact skin. Sometimes dryness and chafing of the skin are due to exposure and/or an allergic reaction to a chemical. Hand lotion may only compound the problem rather than remedy it.

Clothing

Clothing that is worn should be loose fitting and modest. The more skin covered and the less skin susceptible to exposure, the better. In an emergency, one's life may depend on the speed with which one can shed that clothing. For example, snap fasteners on lab coats are preferable to buttons, especially buttons with tight-fitting holes. Do not wear tight-fitting clothing when working with chemicals. On the other hand, do not wear extremely loose or baggy clothing that may be caught on glassware and/or equipment and thereby create a hazardous situation or accident. The workplace is not the location for a fashion statement.

Workplace apparel, such as lab coats, pants, jackets, and other protective clothing should be stored outside of the workplace in a restricted storage area. This will prevent their exposure to chemicals in the workplace. These items are for the workplace and should be worn only in the area intended. They should not be worn while eating lunch or when running out for a short break. They should be removed, stored properly, and the hands thoroughly washed before exiting. Pockets in this type of clothing are handy to store items in. However, they also serve as a collection site for stray particulate matter and chemicals. Therefore, they must be considered contaminated. With this in mind, do not store gum, cough drops, facial tissue, or spare safety goggles in the pockets.

One should never wash work apparel with nonwork apparel. By washing all of one's laundry together, all the wash may be affected. If one's workplace does not provide a laundry service, then the chemist must be careful in this regard. Wash work clothing separately from other clothing. If one has been using chemicals that may pose a health risk, it is advisable to run an empty cycle through the washer in order to minimize the risk as much as possible. Some chemicals emit an irritating odor that becomes attached to the clothing. This clothing may need to be washed more than once just to reduce the odor. If possible, one should have two types of clothing—work clothing and street clothing. It is advisable to change prior to leaving work. If this is not possible, then one should change immediately upon reaching home, so as not to transfer any chemical and/or residue to household furniture. Keep these garments clean. The more soiled the garment, the sloppier the chemistry!

Body

Jewelry, such as wrist watches and bracelets, really has no place where chemicals are used, handled, or stored. While most of us cannot live without a watch, watches pose the same problem as rings. Vapors and/or chemicals can become trapped between the jewelry and the skin. Because the skin under the jewelry is usually more sensitive, the chemical/vapor will be absorbed faster. The exterior surfaces of watches with plastic cases or backings have a tendency to dissolve when they come into contact with some organic solvents. When this happens, the watch may actually fuse to the arm hair. This can cause a fair amount of pain when hair is torn off the arm as the watch is removed. Necklaces should never be worn in the workplace. They can dangle and become entangled with articles of equipment. Other than as a fashion accessory, they offer no asset to one's work attire and can cause a serious accident.

Practice cleanliness. Do not wear soiled clothing; sloppiness breeds sloppiness. Avoid heavy use of colognes and perfumes as their strong odor could obscure the odor of a hazardous vapor or gas. Just as soiled clothing retains odors and chemical residue, so does skin and hair that was not covered or protected. Therefore, it is advisable to shower and wash one's hair thoroughly after leaving the workplace. Until the chemical is washed off, exposure to that chemical continues. Since the long-term effects of many chemicals are unknown, one should not take unnecessary risks. If one lives with small children, this is important since young children are more vulnerable than adults when exposed to chemicals. Keep workplace clothing away from small children. This also holds true for pets.

Feet

Wear sensible shoes that provide the feet with proper support. The shoes should be nonporous and should completely enclose the feet. Cloth and canvas shoes should never be worn where chemicals are used, handled, or stored as they may allow liquids and vapors to pass through and/or they may absorb liquids they come in contact with. Sandals are strictly taboo. If the shoes do not enclose the feet, the feet can be exposed to not only chemicals but to sharp objects such as broken glass. Wear shoes with good tread so as to provide protection against slipping. Leather is an excellent absorber of organic solvents; wear rubber or other specified footwear in workplaces where this would be a hazard. Shoes worn in the workplace should be worn only at the workplace. Unless there is a method to completely clean them before leaving, leave them there. Otherwise, one can track and transfer chemicals everywhere one walks. The goal is to avoid the transfer of chemicals to another location where they should not be present.

Other

Equipment for work with chemicals should stay in the workplace. This means that the pencil and/or pen that one uses in the workplace should remain in the workplace. It has been exposed to the chemicals in that area. Think of it as contaminated. With this in mind, do not put the pencil in one's mouth even if only to chew on it while thinking. Whatever is on it will be in one's mouth.

Do not bring outside items into the workplace that are not meant to be in the workplace. They may be mistaken as misplaced and thus be removed and returned without proper cleaning.

Respect others. Work clothing and personal protective equipment should be assigned to one person and one person only. This includes the outer office and secretary's office. If one must use items such as a pencil or the telephone, be careful and think beforehand. Make sure the hands have been washed thoroughly. Make every attempt not to bring chemicals from the workplace into these rooms. With visitors having access to these areas, it is our responsibility to protect them and other co-workers from undue chemical exposure.

Children and pets should not be brought into a workplace where chemicals are used, handled, or stored. Both have a tendency to wander around and explore and do not understand or know about the hazards that may confront them. Even if they do not "get into" anything, they are exposed to the chemicals and vapors present.

Be aware of precautions for medications that you may be taking. Those that cause drowsiness or impair vision mean that you must exercise all the more care while working. It is wise to inform at least one co-worker of medication that you may be taking so that if an emergency situation arises involving you, emergency personnel can be informed. Do not store in or take the medications into the workplace.

When good hygiene is not practiced, work performance becomes sloppy. Our work is a reflection of how we view ourselves and the importance we place on our own health. Thus, it is imperative to know the chemicals that we are working with and their hazards, to know the limitations and the warning signs before anything should happen, both planned or unplanned, and the necessary first aid procedures if all else fails.

References

American Chemical Society. *Safety in Academic Chemistry Laboratories*, 6th ed.; Washington, DC, 1995.

Clayton, G. D.; Clayton, F. E., Eds. *Patty's Industrial Hygiene and Toxicology*, 4th Ed.; Wiley & Sons: New York, 1994.

National Academy of Sciences. *Prudent Practices in the Laboratory: Handling and Disposal of Chemicals*; Washington, DC, 1995.

Young, J. A., Ed.; *Improving Safety in the Chemical Laboratory: A Practical Guide*, 2nd ed.; Wiley & Sons: New York, 1991.

4

Drug Enforcement Agency Chemical Regulations

ROBERT J. ALAIMO

The Drug Enforcement Agency (DEA) regulates a class of chemicals that require special licensing, ordering procedures, storage requirements, handling, and use and disposal practices. Historically these materials were defined as controlled substances because they were regulated by various federal agencies and may possess the potential for abuse. The Controlled Substances Act (CSA) of 1970 revamped, replaced, or amended various federal laws related to the control of drugs and chemical substances with abuse potential. The new regulation established a five-schedule control plan for the management or control of the substances previously regulated under a number of the old regulations. The Bureau of Narcotics and Dangerous Drugs (BNDD) was given the authority to regulate the legitimate use of controlled substances by manufacturers, researchers, and others, and the Food and Drug Administration (FDA) was given the responsibility for determining the scientific basis for scheduling. The Drug Enforcement Administration (DEA) was created in 1973 by the consolidation of a number of other federal agencies with drug control responsibilities. The requirements of the CSA are described in the Code of Federal Regulations (CFR) Volume 21 Part 1300 to end.[1]

Since 1973, a number of significant changes and modifications to the drug control laws have taken place. In 1978 psychotropic drugs were added to the CSA. In 1984 the authority for expedited scheduling of chemical substances was added to the CSA. Additional responsibilities were added to the DEA's role over the next few years. The CSA was amended in 1988 to regulate a number of specific chemical substances that were being diverted from legitimate uses to illegal production of controlled substances. This law, known as the Chemical Diversion & Trafficking Act (CDTA) of 1988, dealt with the distribution of 12 precursor and 8 essential chemicals. Control of these specific chemicals was further expanded with the implementation of the Domestic Chemical Diversion Control Act (DCDCA) of 1993. This law required registration of manufacturers, distributors, importers, or exporters of the chemicals now referred to as List I and List II chemicals.

The DEA is the sole federal agency for the regulation of this special class of drugs and chemicals, but not the only regulatory body involved with control. The individual states also have various agencies, working in conjunction with the DEA with the responsibility for enforcement of controlled substance regulations. Discussion of specific state requirements is outside the scope of this chapter.

Regulations and Registrations

The Controlled Substances Act (CSA) requires that every person who manufactures, distributes, or dispenses any controlled substances or proposes to engage in the manufacture, distribution, or dispensing of any controlled substance shall obtain annually a registration unless exempted by law. The regulation specifies a separate registration for each independent activity involving controlled substances. The activities of primary focus in this section are related to the use of controlled substances in research or chemical analysis. Research use of controlled substances may be limited to certain drug schedules, but also can include research-related manufacturing activities involving those schedules. A registrant with a research registration can also conduct chemical analysis, and import and distribute those controlled substances listed in the schedules in which he is authorized to conduct research. The registrant possessing a registration to conduct chemical analysis with controlled substances shall be allowed to manufacture and import substances for analytical purposes.

A separate registration is required for each principal place of business. In general, this means an industry or institutional laboratory in one location possessing a researcher's license may not use controlled substances under that registration at a related laboratory at another address. A company or academic institution with several laboratories in the same city may be required to have a separate license for each location at which they intend to use controlled substances. The registrations are renewed annually and specific fees are associated with each type of registration.

Generally, a controlled substance officer (CSO) is responsible for the administration of the site's controlled substance program. This individual will have the responsibility for day-to-day operations of the program.

Schedules of Controlled Substances

The CSA classifies each of the regulated controlled substances into one of five schedules or classes based on the material's medical use, potential for abuse, safety, or potential for dependence. The Act provides the authority to add substances or remove substances from a schedule or change the schedule of a substance from one schedule to another. The usual symbolic representation for controlled substance schedules is a capital C followed by a roman numeral. A schedule I material is shown as (C I), a schedule II substance is shown as (C II), and so forth. The basis for placement of a material into each of the five schedules is described in "A Security Outline of the Controlled Substances Act of 1970" prepared by the Office of Diversion Control (U.S. Government Printing Office: 1991-298-665 [40675]) and is listed as follows.

Schedule I (C I)

The controlled substances in this schedule are those that have no accepted medical use in the United States, are not accepted as safe for use under medical supervision, and have a high abuse potential. Examples of substances in this schedule are heroin, LSD, peyote, methaqualone, and certain fentanyl analogs.

Schedule II (C II)

The controlled substances in this schedule have a high abuse potential with severe psychological or physical dependence liability, but have an accepted medical use in the United States. The C II controlled substances consist of certain narcotic, stimulant, and depressant drugs. Examples of C II controlled substances are opium, morphine, codeine, cocaine, methylphenidate, and pentobarbital.

Schedule III (C III)

The substances included in this schedule are listed based on the criteria that they possess abuse potential less than the substances in schedules I and II and a currently accepted medical use in the United States, with potential for abuse that may lead to low to moderate physical or psychological dependence. Examples of substances included in this class are certain stimulants and depressants, preparations containing narcotic drugs in limited quantities such as codeine products, derivatives of barbituric acid, nalorphine, and anabolic steroids.

Schedule IV (C IV)

The substances in this schedule have an abuse potential and dependence liability less than those listed in schedule III. Abuse of these substances may lead to low physical or pyschological dependence when compared to agents in C III. Chemicals in this schedule have an accepted medical use in the United States. Examples of agents in this class include barbital, phenobarbital, chloral hydrate, flenfluramine, pentazocine, and certain depressants, stimulants, and minor tranquilizers.

Schedule V (C V)

The substances in this schedule have a low potential for abuse relative to substances controlled in schedule IV and have a currently accepted medical use in the United States. Physical and/or psychological dependence is low compared to C IV substances. The agents in this class are buprenorphine, certain drugs containing limited quantities of narcotics such as elixers, and certain cough syrups and the stimulant pyrovalerone.

Controlled Substance Code Number

Controlled substances have been assigned an "Administration-Controlled Substance Code Number" for the purposes of identification of the substances or class on Certificates of Registration. These numbers are required on applications for importing procurements or manufacturing quotas. Applicants for import and export permits must include these numbers on the application form. Additional details of this requirement are contained in 21 *CFR* 1308.03. Box 1 contains a number of examples of some of the more common controlled substances and their Administration-Controlled Substance Code Number.

Controlled Substance Analogs

Controlled substance analogs are defined as materials that have chemical structures similar to a substance controlled in schedules I or II or have a central nervous system effect similar to substances in C I or C II. These substances, if not already controlled in another schedule,

```
┌─────────────────────────────────────────────────────────────┐
│                          Box 1                              │
│            Examples of DEA Administration-Controlled        │
│                   Substance Code Numbers                    │
│                                                             │
│          Schedule I                    Schedule II          │
│   Codeine-N-Oxide      9053     Codeine            9050     │
│   Dihydromorphine      9145     Morphine           9300     │
│   Methaqualone         2565     Thebaine           9333     │
│   Heroin               9200     Pentobarbital      2271     │
│                                                             │
│          Schedule III                  Schedule IV          │
│   Lysergic Acid        7300     Dextropropoxyphene  9278    │
│   Nalorphine           9400     Barbital            2145    │
│   Anabolic Steroids    4000     Fenfluramine        1670    │
│                                                             │
│                      Schedule V                             │
│            Buprenorphine            9064                    │
│            Pyrovalerone             1485                    │
└─────────────────────────────────────────────────────────────┘
```

approved as a drug, or undergoing review by the U.S. FDA, are automatically controlled in schedule I.

Security Requirements

The security requirements for controlled substances in schedules I–V are described in detail in 21 *CFR* 1301. 71-76 and in *A Security Outline of the Controlled Substances Act of 1970*.[2] The principal intent of these security requirements is to deter or prevent the theft or diversion of controlled substances. The regulation is very subjective and leaves much room for interpretation. Approval of the security system and the security procedures is required prior to the issuance of the registration. The physical security controls for drugs in the various schedules is evaluated by the DEA, and a number of factors are considered in determining the adequacy of control. Among the factors considered are:

- the type of activity involved, such as chemical analysis, compounding, or bulk manufacture
- the type and form of the controlled substances, such as solids or liquids, finished dose forms or pure or mixed agents
- the quantity of controlled substances normally handled
- the location of the premises, such as high- versus low-crime areas or urban versus rural
- type of building construction
- the type of safes, vaults, or cages used for storage
- the type of closures used on the doors, vaults, and gates such as combination, key, or padlocks

- the control of the keys and locks, such as number of keys and other control procedures
- alarm systems available and how the alarm is transmitted, tested, and maintained
- the perimeter control of the facility, such as fences, security guards, and control of unsupervised public access
- supervision and identification of employees and workplace accountability
- procedures for the control of the movement of visitors and guests in controlled areas
- the availability of local police and security forces and their adequacy of size, training, and response time
- the adequacy of internal controls for monitoring the controlled substances

The DEA is required to review the security system and programs of certain applicants prior to approval of the registration. The substances in schedules I and II have much stricter security controls than do the substances in schedules III–V. The differences in physical security controls are described in detail in 21 *CFR* 1301.72–76.

The general security requirements for the management of List I chemicals are described in 21 *CFR* 1309.71 and require that all applicants and registrants must provide effective controls and procedures to guard against theft and diversion. Specifically, guidance on the storage and control of access to the chemicals is provided in this part. Particular attention is required in the storage of and access to these materials. The chemicals in List I must be stored in containers sealed to prevent tampering or in areas under physical control or supervision of designated individuals. The effectiveness of the security procedures can be assessed by the DEA.

Ordering Controlled Substances

The procedures for ordering of controlled substances in schedules I and II are described in 21 *CFR* 1305. Chemical catalogs identify those materials that are controlled substances and suppliers will not accept orders for those materials without proper certification of DEA registration. This can include a copy of the current DEA registration. When ordering controlled substances in schedules I and II, a special DEA-supplied order form called DEA Form 222 must be used. These forms are supplied in "books" of either 7 or 14 forms. The order forms are in triplicate with a carbon paper separator. The first page of the triplicate is white and is marked "SUPPLIER'S COPY 1"; the second page is green and is marked "DEA COPY 2"; the third page is blue and is marked "PURCHASER'S COPY 3." A sample of a completed order form is shown as Figure 4.1.

Following the completion of the order form, the purchaser sends Copy 1 and Copy 2, with the carbon paper intact, to the supplier and retains Copy 3 for his/her files. The order form must be signed by an authorized person, usually the individual who has signed the application or who has received power of attorney from the registrant. The orders must be filled within 60 days of order placement or the form becomes invalid and the purchase canceled. The order forms must be free of erasures, alterations, or mistakes or the suppliers are obligated to refuse the order form and return it to the purchaser. Upon completion of the order execution the supplier completes the supplier's portion of the form and sends the DEA Copy 2 to the DEA Division Office at the end of the month in which the order was filled. In order to purchase materials in schedule I and II, the registrant must be specifically licensed to receive these materials and have the appropriately authorized DEA Form 222.

Controlled substances in schedules III–V can be obtained without the use of the special DEA Form 222. However, the purchaser must be registered with the DEA to receive materials in the particular schedule. Suppliers usually require a copy of the purchaser's current DEA registration prior to filling any order for a controlled substance.

Laboratory Use of Controlled Substances

The laboratory use of controlled substances usually involves the issuance of a research or analytical registration from the DEA. Since these operations generally require the use of small quantities of controlled substances, the secured storage arrangements required by the regulation can usually be fulfilled by use of a narcotic safe that meets the DEA's specifications. Security in the laboratory itself can be fulfilled by storage of the controlled substances under lock and key in an approved narcotic cabinet. During actual laboratory experimental use, the materials must be kept under the physical control of the researcher. When the area is vacated and the materials cannot be conveniently returned to the secure storage cabinet, the area itself must be secured. The researcher is responsible and accountable, through experimental records, for all the materials released for work. The inventory control of the laboratory use of the controlled substances must be accurate and kept up-to-date. Certain exemptions on record keeping are allowed for researchers operating under FDA-approved and -monitored Investigational New Drug (IND) activities. The name, address, and registration number of the establishment maintaining the records must be provided to the DEA. The details of the exemption are outlined in 21 *CFR* 1304.03.

Reporting Loss, Theft, or Breakage

The controlled substance registrant is required to have in place a procedure to report the theft, loss, or breakage involving controlled substances. It is the responsibility of the registrant to notify the appropriate DEA office of any substantial loss of controlled substances. DEA Form 106 is to be used when reporting the loss of controlled substances. State regulations may also come into play when the loss or theft of controlled substances is involved. A necessary part of an overall employee security program is the reporting of employee theft or diversion of controlled substances from the workplace. Employees who have knowledge of any controlled substance theft from the workplace have an obligation to report the pilferage to their security officials. Failure to report theft or diversion should be considered a factor in an employee's continued employment in the controlled substance area.

If during the course of normal work involving controlled substances, accidental breakage or spillage of containers occurs, a report should be made to the controlled substance officer. The contents of the spill should collected, inventoried against the use record, and returned to the controlled substance officer for destruction. The contents should never be discarded in the normal laboratory waste containers.

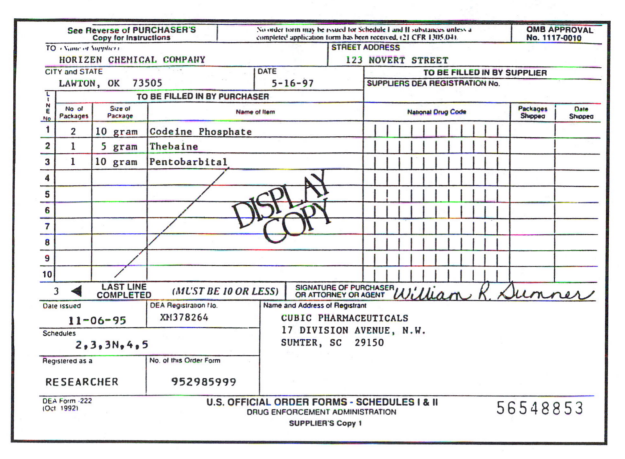

FIGURE 4.1 DEA Form 222.

Disposal

The disposal of controlled substances that was mentioned briefly in the preceeding section are covered in greater detail in this section. The procedure for the destruction and disposal of controlled substances is covered in 21 *CFR* 1307.21–22. The general procedure as described in the regulation requires that the registrant contact the DEA regional office requesting the forms needed to report the disposal. The request should include a cover letter advising the DEA that these materials are no longer needed and the registrant wants to have them destroyed. The DEA can authorize the disposal of the controlled substances through one of the following:

- by transfer to another registered individual who is authorized to possess the substance
- by delivery to the local DEA office
- by destruction in the presence of a DEA agent or other authorized person
- by such other means as deemed appropriate by the DEA that will ensure that the substances do not become available to unauthorized persons

The DEA has also indicated that these procedures shall not be construed as affecting or altering those disposal procedures adopted by any state. The individual states may have special disposal options available to the registrant that would be appropriate.

Record Keeping

The DEA's record keeping obligations for controlled substance registrant are covered under 21 *CFR* 1304. The handling of controlled substances requires that accurate and complete records be maintained. These records must be made available for inspection by the DEA during routine inspections. The requirements imposed on the registrant include:

- preparing and maintaining the required records
- the filing of the appropriate reports with the DEA
- the taking of a physical inventory of controlled substances at specified time intervals

---

Box 2
List I Chemicals

1. Anthranilic acid
2. Benzyl cyanide
3. Ephedrine
4. Ergonovine
5. Ergotamine
6. N-Acetylanthranilic acid
7. Norpseudoephedrine
8. Phenylacetic acid
9. Phenylpropanolamine
10. Piperidine
11. Pseudoephedrine
12. 3,4-Methylenedioxy-phenyl-2-propanone
13. Methylamine
14. Ethylamine
15. Propionic anhydride
16. Isosafrole
17. Safrole
18. Piperonal
19. N-Methylephedrine
20. N-Methylpseudoephedrine
21. Hydroiodic acid
22. Benzaldehyde
23. Nitroethane

There are certain exemptions to the record keeping requirements for specific classes of registrants, and these are outlined in 21 *CFR* 1304.03. All registrants are required to maintain an inventory that is an accurate and complete record of all controlled substances in the registrant's possession. The inventory requirements for each class of registrant are discussed in detail in 21 *CFR* 1304.11–19. Disposal records are another record required of all registrants.

All records, including inventories, disposal, and purchase orders required under 21 *CFR* 1304 are required to be kept for a minimum of 2 years from the date of the record. These records must be available for inspection and copying by the DEA or their authorized agents. State requirements may vary, and some may require as much as a 5-year retention time for certain records.

List I and List II Chemicals

The Chemical Diversion & Trafficking Act of 1988 and the Domestic Chemical Diversion Control Act of 1993

placed under federal control certain chemicals that were being diverted for the illegal production of controlled substances. The 1988 and 1993 Acts defined the chemicals subject to control and these are described in 21 *CFR* 1309, 1310, and 1313. The List I parent chemicals are identified in Part 1310.02(a) (see Box 2). The law also covers esters, salts, and optical isomers of many of the chemicals listed. List II chemicals are described in Part 13 10.02(b) (see Box 3).

Manufacturers, distributors, importers, and exporters of the chemicals in List I are expected to:

- register with the DEA
- comply with certain security requirements
- keep records and make reports to the DEA of transactions
- provide notice of import or export to the DEA

When the List I chemicals are imported, exported, or distributed, the DEA evaluates:

- building access and security systems
- hiring policy
- ordering procedures

Box 3
List II Chemicals

1. Acetic anhydride
2. Acetone
3. Benzyl chloride
4. Ethyl ether
5. Potassium permanganate
6. 2-Butanone (methyl ethyl ketone)
7. Toluene
8. Hydrochloric acid
9. Sulfuric acid
10. Methyl isobutyl ketone

- receiving procedures
- shipping procedures
- export/import procedures
- storage locations

The handling of the List II chemicals does not require registration with the DEA; however, the record keeping requirement is still in effect. An explanation of the reporting requirements and the threshold quantities for the various chemicals in List I and List II is contained in 21 *CFR* 1310.04. A person who acquires a listed chemical for internal use or for end use in a product is not required to be registered or maintain records. In most cases, this will exempt laboratory operations utilizing List I and List II chemicals unless they are imported, exported, or distributed.

A useful reference to controlled substances regulations is the *Controlled Substances Handbook*, published quarterly by Government Information Services, Arlington, VA 22203, and available by subscription.

References

1. *Code of Federal Regulations*; Volume 21, Part 1300 to End, U.S. Government Printing Office: Washington, DC, 1998.
2. *A Security Outline of the Controlled Substances Act of 1970*; U.S. Government Printing Office: Washington, DC, 1991.

5

Laboratory Security

RICHARD L. FLEMING

The ultimate goal of an ideal security program is to eliminate or reduce security-related threats to a company. Its goal should be to prevent losses from occurring wherever possible and to limit unpreventative losses through an effective application of:

- deterrence and prevention
- detection
- response
- training and education

An assets protection program should be developed and implemented to control potential losses to people, assets, and proprietary information. A Security Contact/Manager should be appointed for responsibility of overall security. Security guidelines should be written (tailored to the specific business and site layouts). The basic goals of any security program include:

- protecting people
- protecting intangible assets
- preventing unauthorized access
- protecting company and employee personal tangible property

This can be accomplished through the utilization of security personnel, electronic, physical equipment, and an ongoing training awareness program. The security program should create secured zones within a facility by designating security perimeters at which intrusion protection, intrusion detection, and access control can be exercised effectively. This should channel the flow of personnel and equipment through established entrances/exits, which should be kept to a minimum to best control them.

Security Program/Purpose

The first step is to identify a Security Contact/Manager or team responsible for establishing a Security Program. The plan and procedures are to address the company's specific needs, ensuring that all aspects of an adequate security program are addressed. In most instances, the basic responsibility for security within an organization is a line function, with the Security Contact/Manager leading efforts in security-related matters within the organization and providing training of departmental Security Contacts to coordinate auditing of overall security procedures. The program should include establishing a system assuring immediate reporting of and procedures to handle theft and frauds of company property, unusual uses or disappearances of company money, malicious damage or vandalism, unauthorized disclosure of sensitive information, threats against the company or employees, trespass on company property, possession or use of illegal drugs/substances on company property, and civil disturbances.

Access Control

Controlling access to a facility is the single most important component of good security to protect our employees, company assets, and company information. Controlling and limiting entrances/exits for employees, contractors, and visitors is essential. An employee ID badge system with cardreaders and a sign in/out contractor and visitor log sheet must be maintained and include special badges to identify visitors (to be worn while on company premises). It is recommended that core business hours be established for employees, contractors, and visitors, restricting access outside of the core hours. In a laboratory environment where safety and confidentiality of one's work is a must, access by outsiders should be restricted to business purposes only and be controlled at all times. Visitors within the workplace are always to be escorted by an employee.

Security Personnel

Security guards should be assigned at the facility, twenty-four hours a day, seven days a week, stationed at entrances/exits to monitor and control employees, contractors, and visitors. Only authorized individuals are to be given access and only authorized company assets are allowed to be removed from the premises. Processes should be established to easily control the illegal removal of assets. Routine security guard facility tours must be established for after-hours checks that perimeter doors are locked and that specific critical equipment is working, in the event of chemical spills or fire and to ensure that pipes and so forth are not leaking. Any of these incidents may pose a threat to personal safety, cause damage to property, and disrupt the business. In the laboratory environment the tour should include scheduled walks through the labs to check special after-hours experiments for detection of any unusual odors or spills. Usually security personnel monitor the facilities' fire system(s) and initiate emergency evacuations. Prior to assignment all security personnel are to be trained in Facility GMP and Safety requirements. Contract security personnel are often used in this capacity. However, they historically are low paid with a high turnover. Many of the activities previously suggested may be better performed by company employees.

Security Equipment

This equipment is available from various security equipment vendors for ID badging, cardreader access control, keys/cores, and closed circuit video transmission (CCVT). ID badging systems range from a simple camera and laminating machine to various types of computer-generated systems. Door Access Cardreaders (see Fig. 5.1) range from a single stand-alone cardreader to multiple cardreaders and alarm-computer systems that can be programmed to allow or limit an individual access to specific doors and times. These systems can also be tied into fire/pull station alarms alerting security personnel to respond to an emergency. Special key/cores are available to companies with exclusive key ways to assure no duplication. CCVT consists of cameras, monitors for surveillance of entrances, exits, and key areas of the facility (see Fig. 5.2). This system usually includes video recorders that continually videotape. These tapes are then kept for a specific period of time and can be reviewed, if a problem occurs.

Protection of Assets

Inventory of company assets (with serial numbers where applicable) should be maintained. A paper trail process should be developed to authorize removal to control or dispose of assets. Authorized paperwork must be provided to security upon removal of assets from the premises. Logging of all incoming and outgoing trucks and random truck inspections are deterrents to unauthorized removal of assets.

Protection of Information

Confidential paper files should be controlled at all times and kept locked in a file cabinet. Confidential paper files that are being discarded should always be shredded, using a strip or cross-cut paper shredder available through most office equipment suppliers. Computer files should also be protected at all times through password protection of individual computers and mainframes. Personal computers should always be turned off when an individual is out of the lab/office and passwords should never be shared with others. Diskettes should be locked up at all times when not being used.

Security Contacts and Security Audits

Depending on the size of the company and the number of functions/departments, Security Contacts should be established for each group. The role of the Security Contact is to function as a liaison between security and functional/department employees, to share security information/expectations, and to provide security training; to be a central contact for individuals to report security concerns and incidents; and to develop periodic security audits of after-hours security checks of laboratories/offices. A typical security audit consists of checking for

FIGURE 5.1 Photo of a typical entrance with a cardreader to control access to the building.

FIGURE 5.2 Photo of security work station with CCVT monitors.

OFF SHIFT AUDITS		YES	NO	DETAILS FOR "NO"
1.	Office Doors Locked			
2.	Furniture Locked (Desks, Files, Bookcases, Etc.)			
3.	Computers Off/Locked/Password Protected			
4.	Clean Desk Policy Followed			
	a. Paper/Manuals Put Away/Secured			
	b. Computer Discs Put Away/Secured			
5.	Mail Area Secure			
6.	Laboratory Secured			
7.	Copier Equipment Locked/Secured	NA	NA	
8.	Fax Equipment Locked/Secured	NA	NA	
9.	Tools/Equipment Locked/Secured			
DAILY OPERATIONS AUDIT		**YES**	**NO**	**DETAILS FOR "NO"**
1.	ID Badge Worn/Visible			
2.	Visitors Escorted by Employee			
3.	Pass Out Slips Used Properly			
4.	Sensitive Information is Shredded			
5.	Keys/Combination Locks are Controlled			
6.	Confidential/Restricted Stamp Used Properly			
7.	No Sensitive Information is Posted in Public Areas			
	- Break Areas			
	- Accessible Operating Areas			
8.	Contractors/Truckers Use Assigned Break Areas/Restrooms			
Comments		Laboratories Offices		

FIGURE 5.3 Security audit checklist.

clean desks (files, lab notebooks, internal telephone books, etc., are not left out); that desks, file cabinets, and designated areas are secured by being locked after normal work hours, and that computers and fax and copy machines are shut off.

Security Audit Checklist

To assist individuals in performing security audits, you may want to develop a security audit checklist to ensure uniformity in the audits (see Fig. 5.3).

Workplace Violence

In today's society, the safety and security of employees is a major concern of everyone. Workplace-violence incidents have become the norm in both large cities and small communities alike. We too often read in the newspaper or hear on the news of such terrible events. Companies should have preventative measures, as well as a plan to react to a workplace-violence situation. Many companies today have a workplace-violence policy out-lining employee expectations as well as making clear that violation of the policy will lead to disciplinary action and that the company will take affirmative steps to maintain a safe workplace. Company nurses, HR managers, security, and supervisors should be knowledgable of the symptoms of potentially disgruntled employees who, under stress, may react violently to others at the workplace. Employee-awareness training should be established and employees encouraged to report co-workers' unusual behaviors. It is also suggested that a Workplace Violence Resource Team be established to become knowledgeable on the subject and to evaluate individuals with symptoms or a history of violence that could result in potential workplace violence. This evaluation should include steps to get these individuals help through an Employee Assistance Program as a preventative measure.

Civil Disturbances

Employee injury, business disruption, and possible destruction of company property are possibilities whenever a civil disturbance occurs. Protesters/pickets cause disruption to the business and can often resort to violence and property damage. Therefore, it is important that procedures be established to handle and respond to such an event. You should know the rights of the protesters. For example, they can legally assemble peacefully, with or without signs, and walk within so many feet of the center of the highway at the entrance(s) of a facility. Protesters cannot assemble and stand without moving, obstruct people or vehicles from entering/exiting the facilities, trespass on private property, do harm or threaten employees or visitors, or damage or deface company or private property. It is suggested you work with your local authorities to define your state's laws on this subject. You should involve them in your plan and request that they respond in the event of such a disturbance. Often the authorities will immediately respond to the site where the situation is occurring and explain to the protesters what they can and cannot do legally.

6

Addressing the Human Dynamics of Health and Safety

E. SCOTT GELLER

The human dynamics of occupational safety and health has become an increasingly popular topic at national and regional safety conferences. Safety and health professionals realize that reducing injuries below current plateaus requires more attention to psychological factors. The traditional three Es of safety—Engineering, Education, and Enforcement—have made their mark. Now it's time to focus on the human dynamics of injury prevention—the psychology of health and safety.

Engineers have designed safer environments, including personal protective equipment; and top-down policies and regulations have been imposed to get organizations to apply the engineering safeguards. First, relevant groups and individuals are educated about the safe way to handle a process, whether at work, at home, or traveling in between. Then, if the safety education is not followed, enforcement strategies are implemented when possible to motivate compliance. Thus, failure to follow prescribed safety rules at work or on the road can result in costly penalties to organizations, groups, or individuals. This justifies the common industrial slogan, "Safety is a condition of employment."

The three Es have reduced the frequency and severity of injuries in the workplace, at home, and on the road. Regarding road safety, for example, it has been estimated that the earliest automobile safety standards saved at least 28,000 American lives by 1974.[1] And countless more lives have been saved by increases in the use of vehicle safety belts and child safety seats, primarily as a result of state laws passed in the 1980s. Similarly, the use of machine guards and personal protective equipment (PPE) in industrial settings has prevented numerous work injuries.

Many more lives would be saved and injuries prevented if more people used available PPE and took prescribed safety precautions. The current rate of vehicle safety belt use in the United States, for example, is about 67%.[2] While this is a dramatic improvement from the 15% usage levels prior to state belt-use laws and accompanying education campaigns, there is still much room for improvement. Consider that most of the riskiest drivers still don't buckle up.[3,4,5] Consider also that many employees put themselves at risk daily because they do not comply with safety rules. And such at-risk behavior is even greater on the home front. How many people do you know, for example, who mow their lawns while wearing safety glasses, ear plugs, and steel-toed shoes?

Safety and health professionals recognize the need to get more people to follow safe work practices. Unsafe or at-risk behavior is a prime cause of most injuries. Thus, when more people decrease their at-risk work practices and follow safe procedures, industrial injuries will be reduced. And if more people would take their safe routines beyond the workplace, unintentional injuries and fatalities from vehicle crashes and residential mishaps would be substantially reduced. This requires a fourth E—Empowerment.

Empowerment for Health and Safety

Empowerment implies personal commitment and involvement in promoting health and safety. When people feel empowered they value a particular goal and believe they can achieve it. People who feel empowered to reach safety-related goals follow safe work practices because they subscribe to the safety mission (e.g., to achieve an injury-free environment) and believe their efforts contribute to that mission. They use PPE consistently because they hold safety as a value, not because of a top-down rule or regulation. As a result, they practice safe behaviors at home (where enforcement is impossible).

Feeling empowered for health and safety is certainly an ideal state to achieve. A workforce empowered for health and safety goes beyond the call of duty to achieve daily goals that contribute to the shared vision of an injury-free workplace. How can we get ourselves to such an empowered state, and how can we help others do the same? To enable ourselves and others to feel empowered for health and safety, we need to understand some basic principles of human behavior and attitude. These human dynamics explain why the traditional safety E's have not been sufficient to empower people to actively care for their health and safety. They also imply guidelines for increasing people's involvement in safety and health campaigns so that they feel empowered rather than perform merely to fulfill a condition of employment.

Behavior-Based versus Person-Based

Dealing with the human dynamics of safety and health essentially boils down to improving people's attitudes and behaviors regarding a particular campaign or process designed to prevent illness or injury. Some consultants or trainers advocate a direct focus on attitudes (or values), whereas others target behaviors directly. Safety professionals need to understand the difference between these approaches and learn how to integrate both in their attempts to address the human dynamics of health and safety. Research (not common sense) provides direction for how to use both of these approaches to improve health and safety, and these guidelines are presented in this chapter.

The Behavior-Based Approach

Changing people from the "outside in" is referred to as a behavior-based approach. With this approach, intervention strategies target behaviors directly in an attempt to encourage people to improve their attitude. Behavior-change programs can take many forms, but essential to most behavior-based intervention is the identification of one or more critical behaviors to increase or decrease, a method of tracking change in a target behavior, and a system for recognizing or rewarding progress toward achieving a behavior-change goal.

Proponents of behavior-focused intervention can refer to a vast research literature that shows the cost-effective impact of this approach to improving behavior in numerous areas, including industrial safety.[6,7,8] However, most of the improvement documented in these studies was relatively short term, and few of these studies evaluated changes in people's attitudes or feeling states as a function of the behavior-change intervention. Indeed, the most frequent criticism of the behavior-based approach to safety is that it only influences short-term and small-scale improvement.[9,10,11] If you want to influence long-term change in people, you need to get inside them and influence their attitudes and values. This implies a person-based approach, whereby people are changed from the "inside out."

The Person-Based Approach

Addressing attitudes or values directly sounds like a reasonable and logical way to change people, and in fact this is the approach used most often to deal with the human dynamics of safety and health. This person-based approach is reflected in the traditional educational approach to safety (as discussed above), and appears regularly in the corrective action plans of industrial injury investigations, with proposals like "the employee shall be retrained." The ambiguity of an education-focused action plan, however, illustrates a major limitation of this approach to improve the human dynamics of health and safety.

In order to apply person-based approaches to improve human potential, clinical psychologists receive specialized therapy or counseling training for four years or more, as well as a year-long internship. Such intensive professional training is necessary because tapping into an individual's perceptions, attitudes, or values is a demanding and complex process. In addition, these inside dynamics of people are extremely difficult to measure reliably, making it hard to assess therapeutic progress and obtain straightforward feedback regarding one's counseling skills. As a result, the person-based therapy process can be very time-consuming, involving numerous one-on-one sessions between professional therapist and client.

Therefore, it is understandable that advocates of the person-based approach to safety and health promotion have had little to advise regarding a specific strategy or technique.[10,11,12] The most that they propose is group-awareness training, or education sessions. They offer minimal advice on how to conduct such sessions in order to maximize impact. This is partly because there is little empirical evidence on the best way to educate, besides the popular slogan: *Tell them and they'll forget; demonstrate and they'll remember; involve them and they'll understand.*

It certainly seems intuitive that the impact of education increases directly with the students' active participation. This relationship is illustrated daily through personal experience. We learn more from situations that provoke our active involvement. In other words, when we react overtly to information presented, we are more likely to internalize the lesson and buy-in to its meaning and ramifications. Thus, our behavior influences our thinking and perhaps our attitude. This is, of course, the basic premise of the behavior-based approach, and a rationale for applying behavior-based techniques (as discussed below) to increase the person-based influence of education.

An Integrated Perspective

When people choose to act in certain ways, they typically adjust their self-talk and mental attitude to be consistent with their actions; and when people alter their self-talk, attitudes, or values, relevant behaviors change as a result. Thus, both behavior-based and person-based approaches to changing human dynamics can influence both attitudes and behaviors, either directly or indirectly. However, a key in the first sentence of this paragraph is "choose to act." If people believe their behavior is totally a function of outside contingencies (as when they wear PPE or drive the speed limit only to avoid a negative consequence), they feel no obligation to internalize the rule or alter relevant attitudes. They can still maintain the belief that PPE is unnecessary and that exceeding the posted speed limit is a desirable risk. And without relevant adjustment in person factors, PPE will not be used when mowing one's lawn, and speed limits will be exceeded when traffic conditions allow it and the perceived probability of getting caught is low. This principle of perceived choice or personal control has critical implications for the design of intervention programs to motivate behavior change, and is addressed below.

Trying Both Approaches

Most parents, teachers, first-line supervisors, and safety captains have used both person-based and behavior-based approaches when attempting to influence other peoples' knowledge, behaviors, attitudes, or values. When we lecture, counsel, or coach others in a one-on-one or group situation, we are essentially using a person-based approach; and when we recognize, correct, or punish others for what they do, we are operating from a behavior-based perspective.[13] Of course, we are not always effective with our person-based or behavior-based techniques. Often we don't apply the right principles to the process, or we are not skillful at implementing person-based or behavior-based procedures. Contrary to some opinions,[10] it takes more than common sense to influence the human dynamics of health and safety.

As mentioned above, clinical psychologists require years of education, practice, and professional feedback to address such internal dimensions of people as attitudes, values, or emotions. On the other hand, the behavior-based approach was designed for implementation by individuals with minimal professional training.[14] It was founded on the notion that human dynamics need to be addressed in the situations where beneficial change is needed (such as the home, school, rehabilitation institute, or workplace). Thus, it is necessary to teach the managers or leaders in these settings (e.g., parents, teachers, supervisors, or co-workers) the behavior-change techniques most likely to be effective under the circumstances.

Because the behavior-based approach is straightforward, objective, and relatively easy to administer, it has been popular in many situations. And because intervention progress can be readily monitored by regular observation of target behaviors, program administrators can receive objective feedback regarding the impact of their efforts. Such feedback motivates continued application of techniques that work, and refinement of procedures that do not produce expected outcomes.

Developing Internal Motivation

Ways to increase the beneficial influence of behavior-change techniques are often inspired by a consideration of person-based concepts. For example, the limited impact of powerful external contingencies, as discussed above, implicates the need to consider internal person factors when designing motivational processes. More specifically, external consequences (whether rewards or penalties) should not provide complete justification for the desired behavior, if you want to cultivate the kind of internal motivation that can lead to attitude change and value formation.[15,16,17] This critical point has been supported by a substantial body of research, instigated in part by a pioneering study conducted more than 30 years ago.[18]

The researchers used a mild or severe threat to prevent young boys, aged 7 to 9, from playing with an attractive toy. For Mild Threat, the boys were told merely that "It is wrong to play with the robot." In contrast, the boys in the Severe Threat condition were told, "It is wrong to play with the robot. If you play with the robot, I'll be very angry and will have to do something about it." After giving a mild or severe threat, the researcher left the room in which four toys were available for the subject to play with, including the robot. Researchers watched the boys from behind a one-way mirror to observe the impact of the threat. Both threats worked. Only one of the 22 boys in each condition touched the forbidden toy.

About six weeks after this session, a young woman returned to the boys' school and took them out of class one-at-a-time to participate in a different experiment. She did not mention the earlier session, but instructed each boy to take a drawing test. While scoring the test, she told the boy he could play with any toy in the room. The same five toys from the prior session were available, including the robot. Of those boys from the prior Severe Threat condition, 17 (or 77%) played with the robot. But only 7 (33%) of the boys from the Mild Threat condition played with the previously forbidden toy. Apparently, more boys given the mild threat developed an internal rationale (or justification) for avoiding the robot, and as a result avoided this toy even after the external control was removed. Thus, the long-term impact of a behavior-based intervention can be increased by considering internal, person-based factors, especially the perception of choice.

Perceiving Choice

As noted above, when people perceive that their behavior is determined by personal choice (at least to some degree), they feel obligated to change their internal feelings or attitude to be consistent with the behavior. In the experiment described above, for example, several boys in the Mild Threat condition probably felt their lack of play with the forbidden toy was partly their own choice, since the threat was not sufficient to justify their avoidance behavior. As a result they developed an internal attitude consistent with avoiding a certain toy and playing with others. Consequently, the more personal choice associated with an intervention program, the more likely will behavior change be accompanied with internal thinking or feeling states that support the behavior when the external controls are no longer available.

This is a prime reason for using positive rather than negative consequences to motivate behavior change. When people perform to achieve positive or rewarding consequences, they perceive that they have choice (or personal control).[19] They choose to earn the reward or to not get involved. On the other hand, when people be-

have in a certain way to avoid negative or unpleasant consequences (as when locking out a power source only to avoid losing their job), they do not believe they have much choice. In fact, they feel controlled. Under such circumstances they are not likely to develop internal support for their behavior.

While perceived choice and a positive attitude are more apt to accompany reward than penalty contingencies, it is probably not advisable to elimate all punishment contingencies from the workplace, community, or home. Recall that most boys in the experiment described above complied with the experimenter's instruction, regardless of the severity of the threat. That's a critically important point. External contingencies must be sufficient to get the desired behavior started, accompanied with appropriate education and training, but they should not be more powerful than necessary. And this includes rewards as well as penalties. We don't want people complying with safety rules only to gain a reward or avoid a penalty. We need to consider inside (person-based) factors when designing programs to influence outside behaviors. The next section describes a process that focuses on behavior change in a way that also facilitates the development of internal attitudinal support.

The DO IT Process

This process addresses the human dynamics of health and safety by focusing directly on behavior with a consideration of person-based factors. The DO IT process is founded on research in behavioral science that the most effective way to motivate people to go beyond the call of duty for health and safety is to involve them in developing and administering the process. For people to actively care for safety over the long term, they need to believe in the process and feel good about implementing it themselves.

The DO IT process puts people in control of improving behaviors and thereby preventing injuries. It offers a tool for solving the behavioral aspects of safety problems. It provides objective data for exploring why certain at-risk behaviors are occurring and for evaluating the impact of interventions designed to increase safe behavior or decrease at-risk behavior. If an intervention does not produce the desired effect, it is either refined or a completely different behavior-change approach is tried. As depicted in Figure 6.1, each letter of DO IT reflects one of the steps in this four-stage continuous improvement process.

"D" for Define

The process begins by defining certain behaviors to work with. These are the targets of the behavior-change pro-

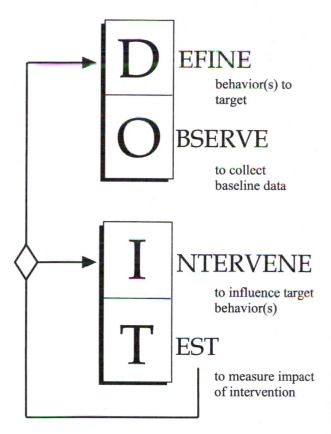

FIGURE 6.1 Safety improves with continuous cycling through a four-step process.

cess and are at-risk behaviors that need to decrease, or safe behaviors that need to occur more often. An at-risk target is:

- a behavior that has led to a serious injury or fatality
- a behavior that could lead to a serious injury or fatality
- a behavior that has led to several minor injuries or "near misses"
- a behavior that could potentially lead to a serious injury or several minor injuries because many people perform it

Avoiding at-risk behaviors often requires certain safe behaviors, and therefore safe targets might be behaviors that substitute for particular at-risk behaviors. On the other hand, a safe target behavior can be defined independently of an associated at-risk behavior. The definition of a safe target might be as basic as using certain PPE or "walking within pedestrian walkways," or the safe target could be a process requiring a particular sequence of safe behaviors, as when lifting a heavy object or when locking out energy sources appropriately.

Deriving a precise definition of a DO IT target is facilitated with the development of a checklist that can be used to evaluate whether the target behavior or process is being performed safely. And developing such behavioral definitions can be an invaluable learning experience. After viewing a videotape on hand safety, for example, a team of workers could develop a checklist of safe versus at-risk use of hands in their own work areas. This enables a translation of general motivational information into a specific tool that can be used to protect each other's hands. When people get involved in deriving a behavioral checklist, they own a training process that can improve human dynamics on both the outside (behaviors) and inside (feelings and attitudes). Even more improvement in behavior-based and person-based factors occurs when people use their behavioral definitions to perform the next stage of DO IT.

"O" for Observe

When people observe each other for certain safe or at-risk behaviors, they are "actively caring" in the best sense of the word. They realize everyone performs at-risk behaviors, sometimes without even knowing it. Thus, the observation stage cannot be fault-finding, but rather is a fact-finding process facilitating the discovery of behaviors and conditions that need to be changed or continued (supported) in order to prevent injuries.

There are many ways to complete the four stages of DO IT, and having people choose the specific procedures for their work areas not only enables appropriate customization but also increases individual motivation, commitment, and ownership. This is one way of incorporating person-based factors into a behavior-based safety process. Regarding the observation process, teams of workers need to decide:

- What kind of checklist will be used during observation?
- Who will conduct the behavioral observations?
- How often will the observations be conducted?
- How will data from the checklist be summarized and interpreted?
- How will people be informed of results from the observation process?

Helpful guidance related to answering these and other procedural questions for the DO IT process can be found in several sources,[20,21,22] and a videotape training series complete with workbooks and a facilitator's guide is available.[23] There is not one generic set of procedures for every situation, and the customization and refinement of a process for a particular setting should never stop. Participants need to make regular changes in their observation protocol in order to keep the process evergreen and maximally effective. It is often advantageous to begin with a limited number of behaviors and a relatively simple checklist. This reduces the possibility that some people will feel overwhelmed at the start, enables the broadest range of voluntary participation, and provides numerous opportunities to successively improve the process by expanding its coverage of both behaviors and work areas.

"I" for Intervene

During this stage, interventions are designed and implemented in an attempt to increase safe behaviors or decrease at-risk behaviors. Interventions target behaviors, not the people who are behaving at-risk. The names of those observed are never included on the behavioral checklist. Therefore, intervention means changing external conditions of the system in order to make safe behavior more likely than at-risk behavior. When designing interventions, the ABC model or three-term contingency should be considered, with A for activator, B for behavior, and C for consequence. Activators direct behavior (as when the ringing of a telephone or door bell signals certain behaviors from residents), and consequences motivate behavior (as when residents answer or do not answer the telephone or door depending on current motives or expectations developed from prior experience at telephone or door answering).

The most motivating consequences are soon, certain, and sizable. And, positive consequences are preferable to negative consequences. Although negative consequences can be as effective as positive consequences in motivating behavior, positive consequences enable greater perception of personal control, choice, and freedom (as discussed above). As a result, positive consequences can benefit internal person-based factors (such as attitude and value). This is a special value of incentive/reward programs when they are designed from a behavior-based perspective.[24]

The process of observing and recording frequency of safe and at-risk behavior on a checklist provides an opportunity to give individuals and groups valuable feedback. When the results of a behavioral observation are shown to individuals or groups, they receive the kind of information that enables practice to improve performance. Without feedback, practice only makes permanence. Feedback is necessary for practice to make perfect. There is considerable research evidence that providing workers with feedback regarding their safe and at-risk behaviors is a very cost-effective intervention approach for improving safety performance.[25,26,27] Figure 6.2 offers specific

guidelines for giving rewarding and correcting behavioral feedback.

In addition to behavioral feedback, researchers have found a number of intervention strategies to be effective at increasing safe work practices. These include worker-designed safety slogans, "near miss" and corrective action reporting, safe behavior promise cards, individual and group goal setting, actively caring thank-you cards, safety coaching, as well as incentive/reward programs for individuals or groups. Space limitations prevent description of these various interventions, but informative discussions are available for the safety professional[13,20,22] as well as a videotape training series.[23] The following points implicate important guidelines for the design of effective behavior change interventions:

- Activators are usually easier and less expensive to implement than consequences, but are typically less effective.
- Activator signs should specify the desired behavior and be placed in areas where the behavior is needed.
- The target audience should be empowered to create activator messages.
- Activator messages should be varied frequently and implicate consequences.
- Goals should be specific, motivational, achievable, relevant, and trackable (SMART) and lead to positive consequences.
- Incentive/reward programs should focus on process activities (as in safe behaviors) rather than outcome (as in total recordable injury rate).
- Everyone who meets certain behavioral criteria should earn a safety reward.
- Safety rewards should include a safety message that is displayable (as in a safety slogan on a coffee mug, hat, or jacket).
- Contests should not reward one individual or group at the expense of another.
- Tangible rewards are for symbolic value only, and should not be viewed as a payoff.
- Progress toward achieving a group safety reward should be systematically monitored and publicly posted.
- Recognition for individual safety achievement should be personal, genuine, private, and frequent.
- Safety celebrations for injury-reduction milestones should be unannounced, show top-down support, and facilitate discussion of the processes used to achieve the outcome.
- Substitute an "employee of the *moment*" paradigm for "employee of the *month*."

Give it One-on-One Make it Timely Identify the Observed Behavior	
Rewarding Safe Behavior	**Correcting At-Risk Behavior**
Give personal praise and thanks Listen to reaction Reaffirm approval and offer encouragement	Specify the desired behavior Listen to reaction Solicit potential solution Request commitment for change Express concern and caring

FIGURE 6.2 Guidelines for giving feedback to reward safe behavior and correct at-risk behavior.

"T" for Test

The test phase of DO IT provides work teams with the information they need to refine or replace a behavior-change intervention, and thereby improve the process. If observations indicate significant improvement in the target has not occurred, the work team analyzes and discusses the situation, and refines the intervention or chooses another intervention approach. On the other hand, if the target reaches the desired frequency level, the participants can turn their attention to another set of behaviors. They might add new critical behaviors to their checklist, thus expanding the domain of their behavioral observations. They might design a new intervention to focus only on the new behaviors.

The test phase of DO IT can obviously lead to a variety of possible follow-up activities. The key is for the work group to decide from their objective data what to do next. Every time the participants evaluate an intervention approach, they learn more about how to improve safety performance. They have essentially become behavioral scientists, using the DO IT process to (a) diagnose a human dynamics problem; (b) monitor the impact of a behavior-change intervention; and (c) refine interventions for continuous improvement. The Test phase provides the motivating consequences to support this learning process and keep the participants involved. They have data to reassure themselves that they are in control of safety. They have reason to believe they can contribute regularly to the ultimate purpose—an injury free work environment. They are empowered to actively care for health and safety.

Conclusion

As reviewed here, behavior-based psychology holds the tools for increasing safe behavior and reducing at-risk behavior throughout a work culture. However, it is not enough to have techniques to conduct behavior-based observation and feedback, behavior-based incentive/reward programs, behavior-based coaching, and behavior-based injury investigation. People need to perceive sufficient personal control, ownership, empowerment, and interdependency in order to use these behavior-based tools. For long-term improvement, safety-related rules and contingencies need to be internalized as self-discipline. This requires integrating person-based psychology with a behavior-based approach.

The general procedures reviewed here need to be customized for a particular work culture. The journey toward such discovery and specification will not be without bumpy roads, missed turns, and inconvenient detours. The basic principles covered here can be used as a rough map for involving people in achieving continuous improvement in safety and health. Be prepared, however, to blaze new trails and traverse uneven terrain. And don't forget to stop along the way to recognize and celebrate your small-win accomplishments. This provides the fuel to continue your journey to the ideal destination—a Total Safety Culture.

References

1. Guarnieri, M. J. *Safety Research* **1992**, *23*, 151–158.
2. Novak, L. National Highway Traffic Safety Administration, 1997, personal communication.
3. Evans, L.; Wasielewski, P.; von Buseck, C. R. *Human Factors* **1982**, *24*, 41–48.
4. Wagenaar, A. C. Report UMTRI-84-2, 1984.
5. Waller, J. A. *Annu. Rev. Public Health* **1987**, *8*, 21–49.
6. Geller, E. S. *The Psychology of Safety*; Chilton Book Company: Radnor, PA, 1996.
7. Krause, T. R.; Hidley, J. H.; Hodson, S. J. *The Behavior-based Safety Process*; Van Nostrand Reinhold: New York, 1990.
8. Petersen, D. *Safe Behavior Reinforcement*; Aloray, Inc: Goshen, NY, 1989.
9. Bradford, D. M.; Ryan, R. F. *Professional Safety* **1996**, *41*(12), 34–36.
10. Eckenfelder, D. *Values-driven Safety*; Government Institutes, Inc.: Rockville, MD, 1996.
11. Smith, T. A. *Professional Safety* **1995**, *40*(2), 18–23.
12. Topf, M. D.; Petrino, R. A. *Professional Safety* **1995**, *40*(12), 24–27.
13. Geller, E. S. In *Essentials of Safety and Health Management*; Lack, R. W., Ed.; CRC Press/Lewis Publishers: Boca Raton, FL, 1996.
14. Ullman, L. P.; Krasner, L., Eds. *Case Studies in Behavior Modification*; Holt, Rinehart, & Winston: New York, 1965.
15. Cooper, J.; Scher, S. J. In *The Psychology of Persuasion*; Brock, T.; Shavitt, S., Eds. Freedman: San Francisco, CA, 1990.
16. Goethals, G. R.; Cooper, J.; Naficy, A. *J. Per. Soc. Psychol.* **1979**, *37*, 1179–1185.
17. Lepper, M.; Green, D., Eds. *The Hidden Cost of Reward*; Erlbaum: Hillsdale, NJ, 1978.
18. Freedman, J. L. *J. Exp. Soc. Psychol.* **1965**, *1*, 145–155.
19. Skinner, B. F. *Beyond Freedom and Dignity*; New York: Alfred A. Knopf, 1971.
20. Geller, E. S. *The Psychology of Safety*; Chilton Book Company: Radnor, PA, 1996.
21. Krause, T. R.; Hidley, J. H.; Hodson, J. J. *The Behavior-Based Safety Process*; Van Nostrand Reinhold: New York, 1990.
22. McSween, T. E. *The Value-Based Safety Process*; New York: Van Nostrand Reinhold, 1995.
23. *Actively Caring for Safety*; Tel-A-Train, Inc.: Chattanooga, TN, 1994.
24. Geller, E. S. *Professional Safety* **1996**, *41*(10), 34–39.
25. Chokar, J. S.; Wallin, J. A. *J. Appl. Psychol.* **1984**, *69*, 524–530.
26. Geller, E. S. *J. Organ. Beh. Man.* **1980**, *2*(3), 229–240.
27. Sulzer-Azanoff, B.; De Santameria, M. C. *J. Appl. Behav. Anal.* **1980**, *13*, 287–295.

7

Effective Management of Contractor/Visitor Safety

RICHARD G. RAYMER

Many companies have corporate values and policies that clearly express a commitment to a safe and healthy workplace for employees. In fact, many companies are establishing long-term goals to have injury-free or even incident-free workplaces. In the last few years, these values, policies, and goals have been extended to not only include a company's employees but to also include every contract employee and visitor to come on to that company's site. The principle is that any incident, any injury or illness has precisely the same human cost as another. This is true whether the incident involves a full-time company employee, a person on-site to do construction or maintenance work, a temporary agency employee to assist with administrative tasks, a driver delivering supplies, or a visitor to the site. The starting point for making meaningful change is the concept that senior management must set a clear expectation that it is not acceptable to have one level of safety for company employees and a different level for others who come on-site.

For senior management to insist on one level of safety for everyone coming on-site, there must be a sufficient degree of dissatisfaction with the status quo. The health and safety leaders must help management recognize the problem and understand the internal and external forces driving this change. Some of the internal forces that companies using contractors recognize are:

- a significant increase in the use of temporary agency or contract employees for tasks ranging from cleanup to packing product, running processes, doing construction, and performing maintenance

- more utilization of service providers
- incredible increases in capital construction
- contract and temporary employees, oftentimes inexperienced or lacking knowledge, working alongside company employees and creating hazards to everyone in the area

These internal forces, usually caused by companies redeploying their workforce or making other work system changes, result in more contract employees on sites doing more tasks that were previously done by experienced company employees.

Along with these internal forces, outlined here are several external forces that can influence companies to take a more active role in the health and safety of noncompany employees.

- There have been several highly publicized, large-scale incidents, particularly in the process industry. Companies are recognizing their vulnerability from poor contractor safety management and loss of public trust.
- Many contract companies do not have the resources to provide a level of employee safety equivalent to the host company.
- The courts are holding companies/owners liable for injury and illness to contract employees. Rulings have held that owners must oversee the work of contractors and should know what their contractors are doing; that owners are expected to be aware of the hazards associated with the work being done; that the hiring of a contractor without checking its previous safety performance is foolhardy.
- Regulatory agencies in all parts of the world are citing the employer/owner who actually creates the hazard and who has responsibility for correcting the hazard whether or not their own employees are exposed.
- In the United States, OSHA intensified its focus on contractor safety in the 1992 Process Safety Management (PSM) of Highly Hazardous Chemicals Standard, 29 *CFR* 1910.119(h).

The safety professional analyzing the status of these and other internal and external forces within his/her company can provide the needed perspective to senior management to recognize the problems, become dissatisfied with the status quo, and provide the needed support to begin implementing an effective contractor/visitor health and safety process. In addition, many companies are motivated to become more active in contractor/visitor safety because accident data clearly indicate that a problem exists. In fact, companies leading the way in contractor/visitor safety have some of the most respected employee safety records in their industry. This causes a significant gap in performance when comparing company results to contractor safety results and the needed dissatisfaction with the status quo to take action.

Contractors

The simplest definition of a contract employee is a person coming on-site under a contractual agreement or a purchase order. This is different from an individual coming on-site as a visitor or business invitee (probably not under any contract agreement) such as package delivery people, sales people, or a vendor to look at a job for bid. There are three major categories of contract employees, each with some unique characteristics:

1. construction and maintenance
2. temporary services
3. service providers

Construction and Maintenance Contractors

Typically, construction and maintenance contractors are firms that provide a workforce to a company to build, revise, renovate, repair, or maintain equipment or facilities under a contract or purchase order. Usually there is an established job or construction site involving multiple crafts and a general contractor utilizing several subcontractors or suppliers. The company has a construction, project, or maintenance manager responsible for the work being performed by the construction or maintenance contractors.

This is typically a very diverse workforce and is coupled with several other conditions like incomplete structural connections, temporary facilities, tight work areas, varying work surface conditions, an ever-changing work site, and the transient nature of the workers. This situation results in the most hazardous of work environments. Recent National Safety Council reports indicate that, on average, 2100 deaths and 205,000 disabling injuries occur each year in the U.S. construction industry. This accounts for approximately 20% of work-related fatalities in a market segment that employs only 5% of the national labor force.

Temporary Services

Temporary services are firms that provide individuals to a company usually to supplement an existing workforce.

Workers are assigned to accomplish specific tasks or to produce a specified end product. The type of work being done by these individuals includes administrative, secretarial, packing, warehousing, loading, emergency clean-up, and other short-term work. At times, teams of people are provided to operate an entire process. The company has a line or staff manager responsible for the work being performed by the temporary service firm.

Temporary service firms tend to have a transient workforce and perform low- to medium-hazard tasks. The types of injury and illness that result are laceration, strain, sprain, contusion, and musculoskeletal disorder.

Service Providers

Service providers are firms that provide an individual or a small workforce to do specific tasks. The type of work being done by these individuals ranges from servicing/maintaining office equipment to pickup/delivery of uniforms, operating food services, filling vending machines, doing janitorial tasks, moving dropped trailers, providing facilities and grounds maintenance, providing security, and so forth. The company has a facilities services or maintenance manager responsible for the work that is performed by the service provider.

The tasks tend to be low to medium hazard but people may come on-site unaware of required safe work practices or be placed in an unexpected hazardous situation, such as a service person working on energized office or security equipment, a guard confronted with a workplace violence situation, plowing snow on an unfamiliar road, and so forth.

Components for a Successful Contractor Health and Safety Process

This section discusses the components that are common to companies that have implemented a contractor health and safety process. As the descriptions of the types of contract providers indicate, contract work has different degrees of complexity and exposure to hazards. There are several components that should be part of a successful contractor health and safety process. The likelihood of an incident occurring and the predicted severity of that incident determine the level of detail needed. The hazards present on a construction site are much greater than those faced by a small number of people brought in on a one-time basis to repack product. The attention to detail in each component must be appropriate for the complexity and hazard level of the situation.

Management commitment and support is the first component of any contractor/visitor health and safety program. As discussed above, senior management must insist on one level of safety for everyone who comes on a site. This commitment is essential to the development and implementation of the other components of the program. In addition, expectations, performance measures, tracking, and an accountability system must be established.

Contractor selection process is the second component of a successful process. Contractor selection is based in part on the contractor's safety performance and demonstrated ability to effectively manage safety. A prequalification process that defines minimum expectations of a contractor and any subcontractor must be utilized. Typical expectations of the selection process are:

- demonstrated safety performance in the top 50% of contractors doing similar work using workers compensation experience modification rate, total incident rate, and/or lost/restricted workday rate
- written health and safety program
- people assigned as responsible and accountable for safety
- safety orientation and training methods for all contract employees
- hazard recognition and control program
- injury/illness investigation and follow-up process
- regulatory awareness and compliance program
- company performance reports
- documented substance abuse policy including pre-employment screening

The prequalification process helps contractors improve their safety focus and creates a pool of potentially qualified contractors to bid on jobs.

Additional contractor requirements must be identified for high-risk work. A prebid meeting may be held to include health and safety performance and review of any high-risk requirements. The high-risk requirements along with the minimum expectations in the prequalification process are built into the contract specifications.

Following the bid review process, a contractor is selected to perform the work or provide a service. A contract is signed and/or a purchase order issued which includes the health and safety requirements. This should include wording that indicates that failure to meet health and safety requirements is grounds for termination of the contract.

The third component is prejob planning and expectations exchange. This process is intended to develop a spirit of cooperation and partnership between the company and the contractor. It also provides contractor supervision and safety supervisors an orientation and training in the company's safety values, policies, and procedures. The company person responsible for the work provides the contractor management with:

- risk assessments and job hazard analysis for operations applicable to the contractor's work
- information on chemicals/hazards on the site
- information and training for special situations where the company has special expertise and understanding, e.g., unique raw materials used in a process
- general site safe practices, e.g., emergency action and response plans
- site or company permit systems and procedures
- the company's approach to safety management and job risk assessment to serve as a model
- information on joint job/work site inspections to identify unsafe behaviors and hazards
- the procedure for the contractor to report measures of performance periodically and how this information will be utilized

In addition to providing this information and training, the company representative reinforces the expectations that the contractor and any subcontractors used by the general contractor will:

- meet laws and regulations
- implement or maintain a comprehensive health and safety management system
- report performance results versus expectations and develop action plans to close any gaps
- follow the company's and site's general safe practices and permit-system requirements
- perform job hazard analysis (JHA) for jobs without JHAs
- correct imminent danger conditions immediately and, if required when other hazards are identified, stop work until proper safety precautions are taken

The last component is a performance evaluation system. Stated differently, this is the accountability and feedback process. Performance measures are provided by the contractor to the company manager responsible and held accountable for the safety performance of the work being done by the contractor. These measures are also summarized for all contractors on a site and provided to senior management.

The manager responsible for the work being done by the contractor uses the performance information to evaluate the contractor. This evaluation is not one-way. The contractor participates and shares fully in the evaluation process.

An additional component that some companies have included is owner-controlled or owner-provided insurance. This is usually done with major alliance contractors and results in improved injury reporting, treatment of injuries by on-site providers or designated health care

facilities, incident investigation/analysis, and cost control/savings.

In summary, the components of an effective contractor health and safety process are:

- management commitment and support
- contractor selection process
- prejob planning and expectation exchange
- performance evaluation system
- owner-provided insurance, in some instances

The degree of detail used to implement each of these components is dependent on the complexity of the work being done and the degree of exposure to hazards. This is partly indicated by the type of contractor required to perform the work.

Developing and Implementing the Process

To implement a contractor/visitor health and safety process, most companies form a cross-functional team with representatives from the major operating groups, regions of the world, and appropriate corporate staff functions (safety, construction/engineering, legal, purchasing, human resources, and security). This approach ensures that the process has broad input, support, buy-in, and "line ownership." The initial work of the team is to create a contractor/visitor health and safety process and determine the performance measures to be used.

The team or a subteam should benchmark with other companies in its industry and with construction contractors. Benchmarking can be done in several ways including gathering information through trade associations, contacting other safety professionals, executing information searches through a library or the Internet, and attending seminars on the topic. This will provide a "search and reapply" basis for the team or subteam to draft a process. The process or approach is then reviewed by the people in the organization directly involved in implementation to initially gather input and, eventually, achieve consensus.

Once the process is defined, several approaches are available for implementation. Examples of how some major companies have implemented a contractor/visitor health and safety process include:

- developing individuals at each company location as highly trained process owners to coordinate activity to ensure the safety and health of all contractors and visitors
- utilizing a combination of company resources, contracted on-site safety consultants, and contractor safety supervisors to implement the process
- developing a "best approach" that each company location can use as a guidance document to de-

velop the specific program elements to meet defined company requirements

An active communication network is established during implementation to share learnings and, following initial implementation, to promote continuous improvement.

Visitors

Definition

A visitor or business invitee is a person coming to a site for a specific reason (probably not under contract) and for a relatively short time. Examples include company employees not permanently assigned to a site, employee family members, delivery drivers, sales people/vendors, meeting attendees, catering services, tour groups, or individuals providing isolated maintenance services such as sewer cleaning, utility or security equipment repair, and so forth.

Components for a Successful Visitor/Business Invitee Health and Safety Process

This section discusses the components that companies use to protect the safety of visitors/business invitees. The need is to give people basic information for their safety with the expectation that the instruction is followed. This provides a minimum level of training on the site's safe practices and procedures.

Typically the process involves security personnel presenting a visitor/business invitee with a pocket-size card, booklet, or piece of paper providing information on:

- the company's safety policy and expectations
- security information
- safe practices including:
  - personal protection equipment
  - use of walkways, truck routes, and other traffic information
  - smoking
  - personal hygiene or quality assurance requirements, e.g., hair and beard covering, etc.
  - hazardous materials
- work that requires special procedures (e.g., energy isolation) or permits (e.g., confined-space entry)
- emergency alarm and evacuation procedures
- reporting of injury/illness while on the site
- safe areas during unloading or other work
- the expectation that they stay with their host or escort while visiting the site

A signature of the visitor/business invitee entering a site should be required to indicate he/she understands the information. Individuals issuing the information to a visitor/business invitee must be knowledgeable and able to answer questions that might arise.

Most companies require a visitor/business invitee to be escorted or directed to wait in a safe area. An escort is particularly important if an individual is working in a remote area of a site, or in areas where equipment is operating or hazardous materials are present.

In summary, a minimum level of training is provided to protect the safety of visitors/business invitees. This is done by providing information when entering the site, requiring a signature of the visitor/business invitee to indicate the information is understood, and providing an escort or safe area to wait in while on the site.

Performance Measures

Process Measures

After a company has defined the components of its contractor/visitor health and safety process, measurable criteria indicating success should be defined. For example, if visitor/business invitee signatures are required to indicate he/she understood the health and safety information provided, the visitor log should match the signature sheets. Measurable criteria are defined for each component and then some type of rating system is used to determine the level of implementation. This can be as simple as a 0 for nothing has been done to a 10 for fully implemented and working. Numbers between 0 and 10 can be defined to indicate intermediate levels of implementation. An average of the sum of the individual component ratings would provide an overall process-implementation measure. The overall rating number can be utilized to establish interim goals and as an ongoing expectation of performance. If appropriate, the overall rating can be supplemented with the requirement that some individual components must always be at a certain rating.

Ratings can be determined by company and/or contractor representatives doing a self-assessment of the contractor/visitor health and safety process. The ratings from site to site can be calibrated by external assessments. These assessments provide a means to focus resources and put the process in place independent of injury/illness occurrence. However, injury/illness or other outcome measures are important to ensure that the process is right and achieving the expected results.

Outcome Measures

Most companies that have implemented contractor/visitor health and safety programs require contract companies to report, at least once a quarter, OSHA or equivalent recordable injury/illness cases with and without lost/restricted days, the number of lost/restricted days, and the number of employee exposure hours. This is used to calculate lost time accident, OSHA recordable, and lost workday rates.

Other measures can be added to be more proactive in preventing incidents before injury/illness occur. Some of these are:

- targeted risk areas (e.g., elevated work or electrical exposure) using a behavior observation process to measure the number of people working safely versus the total number of people working in the risk area
- near-miss incidents where injury/illness did not occur but could have
- work-site audits to measure compliance with key regulatory requirements and orderliness/cleanliness standards (e.g., spills, slip/trip hazards, etc.)
- training completed and/or safety meetings conducted

Utilizing the Measures

Whatever process or outcome measures are used, the evaluation of performance should not be one-way. Contractors must participate and share fully with the company manager responsible for the work in the evaluation. Gaps between performance and expectations must be worked on to improve both the process and the results. Performance evaluation results are fed back to the prequalification process and poor-performing contractors are dropped from the qualified list.

A combination of incident rates and process measures should be reported to senior management. This is needed to ensure upper management's continued support and commitment to contractor/visitor safety.

Challenges in Implementing a Global Standard

Implementing a contractor/visitor health and safety process in all regions of the world brings with it some unique challenges. Some examples of the challenges that companies have faced include:

- cultures where reporting injury/illness is not considered honorable
- countries that do not allow drug testing as a condition of employment
- areas of the world that allow child labor to do construction work or assembly

- different measures from OSHA's to determine the level of severity of an injury
- places where wearing shoes and shirts is an initial effort before personal protection equipment can even be considered
- problems with people selling personal protection equipment because of its value on the "black market"
- people driving powered industrial trucks who have never operated a motorized vehicle

These and other issues must be identified and creative solutions or compromises developed in the process, for example, not requiring drug testing in countries where it is not allowed. An example of a creative solution to personal protection equipment (PPE) being sold was to give each contract employee an allowance for PPE. If they kept their PPE through the entire job, the remaining allowance was given in cash. This kept people in their PPE and was less expensive than reissuing equipment.

Persistence is imperative. The challenge of implementing a contractor/visitor health and safety process around the world reinforces the importance of senior management insisting on one level of safety for all people coming on a site.

Conclusions and Summary

Companies are shifting their vision of an injury-free or incident-free workplace to not only include employees but contractors and visitors. In fact, there is a shift from maintaining an arms-reach relationship to it being allowable to become involved in contractor safety issues to it becoming expected of companies.

It is also becoming increasingly clear that ensuring the safety of all people is not only morally right but fosters public trust and is cost-effective. Companies that have implemented a contractor/visitor health and safety process report reductions in incident rates by 50% in two to three years. They also report significant cost reductions through owner-controlled or owner-provided insurance.

It is an ambitious undertaking for a company to expect equivalent health and safety performance from its contractors and visitors. It means putting in place an entire health and safety process without having direct control over the process. It requires the full support and commitment of senior management. It cannot be done without line ownership, and it has unique challenges when expanded to all operations around the world. Companies, however, are finding that it eliminates double standards and heightens the sense of importance of health and safety for its own employees.

References

E. I. du Pont de Nemours and Company. Submittal to the Business Roundtable 1995 Owner Safety Awards.

The Eli Lilly and Company Construction Contractor Safety Program. Submittal to the Business Roundtable's 1994 Construction Industry Safety Excellence Awards Program.

Hislop, Richard. *Key Elements of a Successful Construction Safety Program*; National Safety Council, 10/31/96.

Monsanto Contractor/Guest Environmental, Safety & Health Process. Submittal to the Business Roundtable's 1994 Construction Industry Safety Excellence Awards Program.

Occupational Health & Safety Reporter. *OSHA Obligations Remain Despite Use of Contingent Workers Attorney Says*, BNA, 8/14/96.

OSHA Instruction CPL 2.103.

8

The Chemical Hygiene Plan

GEORGE H. WAHL, JR.

In every laboratory, there should be a document that describes the recommended ways of carrying out typical procedures, the housekeeping rules that all employees must follow, how to order new chemicals, and how to perform just about every laboratory activity. In most labs, this document is the Chemical Hygiene Plan or CHP.*

Beginning in 1991, the U.S. Occupational Safety and Health Administration (OSHA) began enforcing the Laboratory Standard.[1] This regulation requires among other things the development of a Chemical Hygiene Plan (CHP) specific to the facility. Very wide latitude is given; the goal is to have a document that is truly "user friendly" and that provides good practical advice on most areas of interest to a person working with chemicals. Unlike many earlier OSHA standards that specified the appropriate way to perform a task, the Lab Standard is "performance based." That is, the specification of the best way to achieve maximum protection from hazardous chemicals in a given laboratory is left up to the Chemical Hygiene Officer, often the author of the CHP, to determine, specify, and ensure compliance.

Background

OSHA—Worker's RTK/Haz Comm

One of the first efforts of OSHA in the area of informing workers about the hazards presented by chemicals was the Hazard Communication Standard (29 *CFR* 1910. 1200), also known as the "Worker's Right to Know" (RTK). (This standard should not be confused with the EPA regulation "Community Right to Know." That rule requires chemical facilities to reveal the identity and quantities of chemicals they emit into the environment so that persons living or working in the surrounding area will be informed.)

This Haz Comm Standard was designed for the routine and predictable use of hazardous chemicals and provides a rather rigid set of requirements on employers. Among other directives, it requires the employer:

1. to determine which of the materials at the job site are hazardous
2. to maintain an easily accessible inventory of all these hazardous materials and provide Material Safety Data Sheets (MSDSs) for all of them
3. to ensure that all containers of hazardous materials are properly labeled with the identity of the material, appropriate hazard warnings, and the name and address of the material's manufacturer, supplier, or importer

Haz Comm also requires that when an employee is first hired or moved to another work location, or a new hazardous material is introduced into the work area, specific training be provided. This training must include an explanation of the Haz Comm Standard; identification of those hazardous materials present along with their potential health effects and their location at the job site; the location and availability of the local written hazard communication program and MSDSs; a description of available personal protective equipment and safe work practices; procedures to detect and measure hazardous

---

*Facilities such as industrial quality control laboratories, medical laboratories that use only simple test kits, and many pilot plants would not be covered and must therefore comply completely with Haz Comm (29 *CFR* 1910. 1200).

material concentrations in the workplace; and an explanation of the labeling system in use.

Labels

The two items of primary concern in the OSHA Haz Comm standard are labels and Material Safety Data Sheets. Each facility must provide its workers with adequate training so that they will be able to recognize a hazardous chemical by reading its label. Chemical labels provide a wealth of information. Not only must a label contain the name of the chemical, but it must also describe the principal hazard posed by the chemical, appropriate personal protective equipment (PPE) to be worn when using it, and actions to be taken in case the chemical is spilled.

A modern chemical label will also contain the chemical's Chemical Abstracts Service Number. This CAS Number is a unique descriptor of the material. It is very useful in computer searching for additional information about the chemical, and it is often required when transporting and disposing of a chemical. Using it when ordering a chemical will help to ensure that you receive exactly what you ordered.

Labels from the same supplier will also contain a color coding that will make it very simple to store chemicals by reactivity class. This will help to prevent inadvertent mixing of incompatible chemicals in case bottles break on a shelf or fall off a shelf. Unfortunately, there is not yet a uniform color coding system that guarantees that a chemical from supplier X will have the same storage color coding as the same chemical from supplier Y.

Whenever labeling a new container of a chemical, you must take responsibility and be sure that anyone else in the laboratory will have a clear idea of what is in the container. As a minimum, the label should contain the name of the chemical, the principal hazard it poses, and the date it was acquired (see Box 1).

Box 1

Look around your lab and inspect the various commercial chemical labels. Make sure you can find the CAS Number, the principal hazard posed by the chemical, the recommended PPE for using that chemical, and the storage code. (This would also be a great time to decide whether or not you really need to keep all of the chemicals you find on your shelves!)

Box 2

All MSDSs must supply physical and chemical properties of the material; toxicity data and information on health hazards; storage, handling, and disposal information; and emergency and first aid procedures.

Material Safety Data Sheet (MSDS)

Chemical manufacturers (and distributors) must assemble a wide variety of information on their products. This information is assembled into a form called a Material Safety Data Sheet (or MSDS). This MSDS is then sent along with the chemical to provide the user with up-to-date information. In the laboratory, everyone working with chemicals must have ready access to the MSDS for any chemical they are using.

The exact format and quality of an MSDS will vary from one manufacturer to the next, but they all contain the same categories of information.

The MSDS, typically 2–12 or more pages, is among the best sources of information to use when determining the hazards and risks of using a chemical. However, they often exaggerate the dangers posed by the chemical and are rarely easy to read (see Box 2).

Other Sources of Chemical Hazard Information

Most facilities will have a variety of other information resources. You may find a chemical information shelf in or near the lab, an institutional safety manual, one individual who is both very knowledgeable and also easily approachable, a good library, an environmental health and safety center, and increasingly, on-line resources.

The field of computer-accessible information is growing rapidly. Many important resource books and MSDSs are now available on CD-ROMs. Access to the Internet opens up an essentially unlimited array of information. A very active discussion group, SAFETY, is maintained at the University of Vermont and is available worldwide for free (<rstuart@esf.uvm.edu>). SAFETY also maintains an archive of past discussions that may be easily searched. Numerous other sites provide easy access to many MSDSs, regulations (e.g., at <http://www.osha.gov> all OSHA regulations may be searched), different Chemical Hygiene Plans, and just about any safety information imaginable. These on-line sources are also sure to grow rapidly as computer access becomes universally available. The home page of the ACS Division of Chemical Health and Safety (<http://chas.cehs.siu.edu>) through its *hot buttons* "CH&S Netways" and "Links to other related sites" provides up-to-date links to most of the information resources of interest to the safety and health professional.

The Laboratory Standard

The Laboratory Standard ("Occupational Exposure to Hazardous Chemicals in the Laboratory"; 29 *CFR* 1910. 1450) became effective May 1, 1990, and compliance was required as of January 31, 1991. Since 1991, U.S. laboratories in which hazardous chemicals are used or stored (with a few exceptions) have been required to have a written Chemical Hygiene Plan (CHP). The purpose of the CHP is for a facility to have a set of *locally* developed procedures and specific work practices that are necessary to ensure that employees are protected from health hazards associated with hazardous chemicals with which they work. Taking into account local conditions, the CHP describes how the "prudent" person would handle chemicals.

Relation to Other OSHA Standards

The Lab Standard was designed to supersede other health-based OSHA standards that require rather specific and nonflexible compliance, such as the Haz Comm Standard described above. However, other OSHA rules on topics not directly related to this standard remain in force. For example, the importance of the General Duty clause (29 *CFR* 1910 Sec 5), which requires the employer to provide a safe workplace and the employee to comply with all workplace rules remains undiminished and is often the most cited of OSHA violations. OSHA standards designed to minimize eye or skin contact remain in force, as do those that describe Permissible Exposure Limits (PELs) or "action levels."

Worker Protection

When deciding if a chemical should be used, due attention should be given to the hazards it presents, such as flammability, toxicity, or explosivity. Whenever possible, it is wise to choose a less rather than more hazardous substance.

Once it has been determined that a specific chemical is needed, the provision of appropriate means to protect the employee must be considered. A laboratory worker may be protected from exposure to hazardous chemicals in one of three ways, in order of importance: Engineer-

---

Box 3

A laboratory worker may be protected from exposure to hazardous chemicals in one of three ways: Engineering Controls; Administrative Controls; or Personal Protective Equipment.

---

ing Controls; Administrative Controls; or Personal Protective Equipment (PPE).

OSHA strongly recommends that the hierarchy of protective measures be as given above. "Engineering Controls" must always be the first line of defense. They are additions to the permanent construction of the laboratory, such as additional ventilation, automatic shutoff valves, sprinkler systems, and so forth, and as such are the most reliable.

The other two choices rely on compliance by the employee and are therefore more problematic and less reliable. "Administrative Controls," such as rules for the proper care and use of a hazardous chemical, are an important part of workplace practice and add to the education, as well as the protection, of the employee.

Only when adequate protection is not guaranteed by either Engineering or Administrative Controls should PPE be required. PPE rules might be general, such as "chemical splash goggles must be worn at all times in the laboratory," or specific for various chemicals or various operations (see Box 3).

Unlike many regulations that specify measurable things that must be done, the Lab Standard relies on locally developed administrative controls or workplace rules to ensure the maintenance of a safe environment. The CHP will describe the rules that apply to specific activities.

Performance-Based Standard—Flexible

The Lab Standard also permits different workplace rules in different locations. This encourages the development of those rules that are most appropriate for each location. One size need not fit all! (See Box 4.)

Box 4

You design how you will minimize your own exposure to chemical hazards using generally accepted good laboratory practices: the practices that the prudent person would choose.

Chemical Hygiene Plan

Employer–Employee Responsibilities

In any workplace in which hazardous chemicals are used on a laboratory scale, it is the clear responsibility of each employer to produce a Chemical Hygiene Plan (CHP) suitable for the materials, processes, and facilities employed. This CHP must be capable of protecting employees from health hazards associated with the hazardous chemicals used in that laboratory.

It is also the employer's responsibility to make this CHP readily available to all employees, to train these employees according to the CHP, and to review the CHP at least annually, or when there is a significant change in the materials, processes, or facilities used.

The employee must follow the CHP in all procedures to which it pertains, and must advise the employer whenever the CHP needs updating or correction.

In addition to the many safety resources mentioned above, there are also a variety of tools that make preparation of a CHP a relatively simple task. The American Chemical Society published "Developing a Chemical Hygiene Plan"[2] in 1990, and later "Developing a High School Chemical Hygiene Plan." The latter is available in hard copy as well as in a choice of computer-readable formats. Although directed at the high school market, it is very easily adaptable to produce just about any CHP. Many similar commercial offerings are also available. Whenever a CHP from a similar facility is available it should be consulted as well.

Standard sources of safety information (such as "SAFETY in Academic Chemistry Laboratories"[3] and "Prudent Practices"[4]) should be referenced in the CHP, used as the source for much of the CHP information (as appropriate), and made available to employees along with the CHP.

As a "performance-based" standard, the actual format and design of the CHP is not specifically outlined. Rather, broad suggestions are provided in Appendix A (of the Standard). Thus, two CHPs from different facilities may bear very little resemblance. However, if they contain the required eight elements (Box 5) they may both be adequate and in accord with the Lab Standard. The eight elements need not be separate sections of the CHP. Rather, they are points against which to measure the adequacy of a plan.

Although not specifically required by the Lab Standard, the development of a CHP should begin with the preparation and review of a complete *chemical inventory*. With this information in hand, the Chemical Hygiene Officer (CHO), or other author of the CHP, may determine what hazardous chemicals, if any, are present in the facility. These are the materials on which the CHP must focus to minimize exposure of the laboratory workers. This inventory also provides a unique opportunity for downsizing the chemical holdings, thereby removing some sources of exposure and also limiting possible future legal liability. When made readily available to those working in the lab, this inventory will also greatly diminish the chances of ordering chemicals that are already on hand.

A typical CHP will have a cover page that accurately and concisely describes the area governed, names the person(s) responsible, provides emergency phone numbers, gives the effective date of the document, and briefly outlines the material to be found in the CHP.

The second page should be an accurate table of contents with page numbers, to make the document as user friendly as possible. The actual contents and the order in which they appear are totally the responsibility of the person preparing the CHP. However, every CHP must cover the eight required elements described below (listed in Box 5).

Box 5

The Required* Elements of any Chemical Hygiene Plan (CHP) are:

1. Standard Operating Procedures (SOPs) for Handling Toxic Chemicals
2. Criteria to Be Used for Implementation of Measures to Reduce Exposures
3. Requirements That Hoods and other Protective Equipment Shall Be Functioning Properly
4. Employee Information and Training (including emergency procedures)
5. Requirements for Prior Approval of Laboratory Activities
6. Medical Consultations and Medical Examinations
7. Designation of Responsible Personnel
8. Special Precautions for Work with Particularly Hazardous Substances

*The Laboratory Standard also contains Appendix A composed of nonmandatory procedures for compliance based on recommendations from the original "Prudent Practices." In this nonmandatory section there is a list of eleven typical components of a CHP provided as guidance to employers in the development of a plan. They must still cover the eight major points above.

Standard Operating Procedures (SOPs)

One of the largest sections of any CHP is usually the one that describes the Standard Operating Procedures, or SOPs, that will be followed in the facility. These will usually include the following.

Personal Protective Equipment (PPE)

These SOPs will almost always describe the appropriate Personal Protective Equipment that is to be used under a variety of circumstances.

Eye Protection In particular, unambiguous guidance on choosing the type of eye protection to be used should be included. The type of PPE of course should match the hazard presented. Whenever there is danger of splashes of hazardous chemicals such as corrosive acids or bases, or flammable material such as most organic solvents, chemical splash goggles are clearly required. Chemical splash goggles come in a variety of sizes, designs, and colors. They all are meant to fit snugly around the forehead, temples and upper cheeks, and nose. They do not allow any direct pathway from the outside to the eye. They provide only indirect ventilation. That is, there are no simple holes in the goggles to allow direct passage of air in and out. Rather, any openings to the outside direct the incoming air away from the eyes to prevent a direct splash to the eyes.

Because of this indirect ventilation, some designs of chemical splash goggles soon become uncomfortably hot and will fog on the inside due to the buildup of perspiration. This can usually be alleviated by the choice of a different design of chemical splash goggle, and/or by the use of an antifog lens coating.

Make sure that the chemical splash goggles assigned to you fit well and do not fog on prolonged use. If they do not fit, they do not provide adequate protection. If they fog, they themselves present a hazard to your safe work in the lab.

Work Practices

Most safety experts will agree that the principal cause of laboratory accidents is poor housekeeping. When a chemical is spilled on a clear and clean bench, it is usually a simple cleanup problem. However, if the same chemical spill occurs on a dirty, cluttered bench top, the possibilities for unexpected reactions are magnified. Also, in the rush to clean up the mess, glassware may be accidentally knocked over, causing other problems.

If the material spilled is an aqueous solution and there are loose, frayed electric cords on the bench top, the chances for electrocution, or at least loss of power, are great.

Working alone is another dangerous practice. The CHP should address this, describe clearly the institution's policy on the subject, and perhaps offer alternative work practices.

Personal Hygiene Practices

Common but important practices, such as washing hands whenever leaving the lab, not wearing the lab coat outside the lab, proper procedures for removing gloves, and so forth, should be spelled out in reasonable detail.

Procedures for Ordering, Storage, and Disposal of Chemicals

Classification of Chemicals There are literally millions of chemicals. How do we possibly deal with them? We classify them according to important properties, or according to structural types, or which elements they contain, or in many other ways.

From a safety point of view, it is important to be able to classify chemicals according to the hazards they pose. Lab workers must be particularly well versed in this classification system. They should be able to determine the hazard presented by reading labels, or MSDSs, or by consulting other local sources, and they must understand the symptoms of overexposure.

The CHP should mention the existence of the Chemical Inventory of the laboratory, clearly describe how to consult it, and how to add or delete chemicals from it. How is a chemical ordered? Certainly, the first step should always be to consult the inventory. If the chemical is already available on-site, much time is saved. More importantly, quantities on hand are thereby minimized, lessening the danger of hazardous spills, and decreasing the cost of ultimate disposal.

Who may order a chemical? The laboratory has a well-developed procedure for ordering new chemicals. It should be clearly outlined so that every person working in the lab can understand it.

The procedures to be followed for proper chemical storage, including segregation by reactivity class, use of secondary containment, and entry into the laboratory Chemical Inventory, must be described.

Chemical disposal involves considerable personal and environmental risks and can involve the facility in expensive fines when done improperly. The CHP must define what can and what cannot be done in this area. Very often there will be designated personnel who alone will handle disposal. The methods for contacting the disposal group should be clearly delineated.

Criteria to Be Used for Implementation of Measures to Reduce Exposures

One of the great innovations of the CHP is the specific inclusion of a requirement that employees be aware of the signs and symptoms of overexposure. The plan should also outline the choices available to reduce exposure to chemicals. These choices might include the use of hoods or glove boxes; alternate work practices; alternate chemicals; and Personal Protective Equipment.

Requirements That Hoods and Other Protective Equipment Shall Be Functioning Properly

Just about every laboratory has at least one hood or other device for minimizing exposure to chemicals. The CHP must contain a description of what constitutes acceptable performance, how that condition may be measured, and what to do when the device is not functioning properly. Although this subject involves complicated engineering, it must be described so that any person working in the laboratory will understand it.

Employee Information and Training (Including emergency procedures)

To ensure that everyone concerned is given the proper training and also has the opportunity of contributing to future refinements of the CHP, the document must clearly describe when training will take place and what will be covered.

In addition to a review of the CHP, the training must include a description of the Lab Standard and a review of the MSDSs and labels used in the facility. Ideally, there should also be some measure of the effectiveness of the training, such as a quiz or discussion session that tests the employees' comprehension of the material.

Emergency Procedures

Another excellent result of the production of a CHP is the ready access to a description of recommended procedures to follow in case of a variety of emergency situations such as spills, cuts, and fires. By deciding beforehand how to deal with a potential accident, some of the panic that is often associated with emergency situations is removed.

Evacuation

Under what circumstances will an evacuation of the facility be declared? Who may do it? How is it done? Where do we go after we leave the facility? What do we do there?

All of these questions should be answered by the CHP. They should be clearly written, and each employee should review them, and perhaps practice them at least annually or whenever the employee changes job location or begins a significantly new job.

Requirements for Prior Approval of Laboratory Activities

The CHP describes the typical operations and procedures followed in the lab. However, the nature of lab work frequently involves nonstandard activities. The CHP must unambiguously describe the types of activities that will require special approval before commencing. The procedures to follow when requesting prior approval and the person(s) who must be consulted will be clearly delineated. Situations that might trigger a request for prior approval might include a scale-up of a reaction by a specified amount; introduction of a new piece of equipment; significant changes in reaction temperature or pressure; or use of a new category of chemical. The procedures for requesting prior approval need to be clearly delineated, perhaps by describing a prior approval form that will be used at the facility.

Medical Consultations and Medical Examinations

Whenever an employee develops signs or symptoms associated with a hazardous chemical with which he/she has been working; or when exposure monitoring reveals an exposure level routinely above the action level (or PEL) of an OSHA regulated substance; or when an event such as a spill, leak, or explosion makes the likelihood of a hazardous exposure real, the employee is entitled to medical consultation and/or examination at the employer's expense.

The CHP should describe the specific local procedures to be followed, the information provided to the physician, and the information that will be provided to the employee.

Designation of Responsible Personnel

As a minimum, the CHP must identify the Chemical Hygiene Officer, as well as the reporting structure of the

CHO. It may also identify a Chemical Hygiene Committee and describe its duties and meeting times.

Special Precautions for Work with Particularly Hazardous Substances

The CHP should clearly delineate any special procedures to be followed when working with carcinogens, reproductive toxins, substances that have a high degree of acute toxicity, and chemicals of unknown toxicity. An initial section should provide simple rules for determining if a chemical belongs in one of these categories. This might include the definitions of these materials as given in the Lab Standard.

For maximum protection of employee health, a "Designated Area" might be described as the only place in which such materials may be used or stored. A designated area may be an entire laboratory, a portion of a lab, or just a specific laboratory fume hood or glove box. Any designated area must be clearly labeled and access should be restricted to those persons who have been trained to work there. Well-defined procedures for working in a designated area should also be included in this section.

The Appendix to the Lab Standard

As stated above, the Appendix to the Standard contains an amazing array of specific "Prudent Practices" that might be appropriate to your CHP. The hallmarks of these "prudent" practices are (1) minimize all chemical exposures; (2) avoid underestimation of risk; (3) provide adequate ventilation; (4) institute a chemical hygiene program; and (5) observe Permissible Exposure Limits and Threshold Limit Values. (See Chapter 14.)

Consult the Appendix, read it carefully, and decide which segments would be suitable for inclusion in your CHP.

The Lab Standard

A copy of the Laboratory Standard (29 *CFR* 1910.1450) should be kept with the CHP so that all employees may consult it at any time that the CHP is available. Initial training must insure that all employees know of the existence, location, and provisions of this rule.

As with any regulation, the Laboratory Standard is not "easy reading." However, with a little guidance in which the Background, Mandatory, Nonmandatory, and Reference Sections are dissected and explained, it can become a useful reservoir of information, and even of good advice to the serious laboratory professional.

Box 6
The CHP provides a unique venue to demonstrate that "SAFETY is everybody's business!"

Conclusion

The ideal Chemical Hygiene Plan is never truly complete. The annual review will usually uncover areas that need improvement, expansion, or deletion. As chemical procedures are perfected, alternate, less hazardous means of accomplishing current laboratory goals will be found and substituted for earlier procedures. Through the regular involvement of all employees in training, as well as in the annual evaluation of the CHP, everyone will develop an increased sense of "ownership" and a stronger safety culture will result (see Box 6).

A CHP is a dynamic document. Ideally, it is not a hardbacked, bound document to be kept on the shelf "just in case." In large organizations, it is very unlikely that any one CHP will suffice for the great variety of operations carried out by diverse staff in different locations. Rather, it should be a "performance-based" document, tuned to the needs of a given laboratory, area, or operation.

One approach is to have a menu of elements available electronically from which a well-designed CHP could be prepared for a specific lab. Another nearby lab performing different functions might have a significantly different CHP developed from the same menu. Each of these CHPs would then be the training instrument for the personnel in the area it covers; and would be the principal source of chemical safety information for those same people.

References

1. *Fed. Regist.*; 1990, 55 (21), (29 *CFR* 1910.1450).
2. Young, J. A.; Kingsley, W. K.; Wahl, Jr., G. H. *Developing a Chemical Hygiene Plan*; American Chemical Society: Washington, DC, 1990; ISBN 0-8412-1876-5.
3. American Chemical Society. *SAFETY in Academic Chemistry Laboratories*, 6th ed.; Washington, DC, 1995; ISBN 0-8412-3259-8 (one free copy may be obtained by calling 1-800-227-5558).
4. *Prudent Practices in the Laboratory: Handling and Disposal of Chemicals*; National Academy Press: Washington, DC, 1995; ISBN 0-309-05229-7. In addition to being a definitive and up-to-date source of safety information, it also includes eighty-eight "Laboratory Chemical Safety Summaries" (LCSSs). These LCSSs are an attempt to prepare MSDS-like sheets in a consistent format and focus on the information of most importance to the laboratory worker.

9

Chemical Safety Information on the Internet

RALPH B. STUART III

The Internet started off as the domain of computer hobbyists and university research scientists in the early part of the 1990s, but since has broadened into a communication tool important to safety professionals as well as the general public. However, making the best use of this resource takes information and experience, as any other form of professional communication requires. This chapter provides some basic information about the Internet and describes some of the resources on the Internet that are valuable for health and safety purposes. Remember, the Internet is still rapidly growing and changing, so additional resources are likely to be available beyond those listed here.

What Is the Internet?

There has been a lot of discussion of the Internet and the World Wide Web in the popular media, much of it sensational and at least somewhat misleading. Although the use of Internet involves many details, for the purposes of this chapter, it can be described with a few relatively simple concepts.

The Physical Internet

Physically, the Internet is a group of computers and computer networks that are physically connected and speak the same language. The physical connections can be through commercial phone lines, over high-speed cables specifically designed for computer communication, or through mobile means such as wireless connections. The significance of these physical connections is that they permit the use of software protocols that allow a wide variety of computers and other electronic equipment to exchange data. Because of these shared protocols, an Internet connection can be made with devices as diverse as personal computers, mainframe computers, and video cameras. This creates a highly versatile medium, which a wide variety of people can use and actively participate in. For these reasons, the Internet has become the method of choice for public access to electronic information.

Information Functions of the Internet

While the physical and technical basis for exchanging files between computers is interesting to some, what good is it? There are three specific functions that the Internet serves for users of safety information. First, the Internet provides the ability to access a wide variety of technical information that is available in the form of electronic data files on the World Wide Web (WWW). Second, the Internet makes possible discussions among people who are interested in similar issues. Third, the Internet presents new avenues for distributing health and safety information to the audiences that safety professionals serve.

The Culture of the Internet

It is important to remember that the Internet is not controlled by a single organization or authority. The Internet protocol itself is a language that computers and computer networks speak. Consequently, the Internet has the organizational nature of any language: there are many local authorities, but no global power. This decentralized nature results in a culture on the Internet that is diverse and disorganized at the casual user level. However, the Internet is also inventive. There are many different ways of getting a particular thing done on the Internet. If you find yourself wondering if you can use the Internet in a particular way, it is likely that other people are wondering the same thing and are working on it themselves. This results in the ever increasing number of tools available to use on the Net and many pleasant surprises.

Using Internet Information

The Internet contains a continuously growing body of information. However, the types of information that are available on the Net are fairly simple, and the information tools that have been developed to collect data are consistent in their approach to the task. These factors, plus the speed and low cost of searching the Internet, make it a useful tool for professional communication, including research and networking. This section describes

Internet information tools in a general way. Remember that the tools that are available to you will depend on the type of computer you are using and how you are connected to the Internet. As with everything else on the Internet, the details of these tools change regularly, so be prepared for surprises, usually pleasant.

What Is Internet Information?

There are two basic types of information available on the Internet. The first is the collection of file libraries on the World Wide Web. The second type of information is informal discussions between people of similar interests and of varying expertise. Which type of information will be more valuable to you will depend on your particular needs.

One of the main attractions of the Internet is the availability of public file libraries. These are put together for various purposes by academic institutions, commercial and noncommercial organizations, and government agencies. These libraries have gradually become interconnected to form the World Wide Web. You visit these libraries by using software tools called Web browsers.

The other major type of information on the Internet is interactive discussions organized around particular subjects. These take place in discussion groups, which operate through either e-mail lists or newsgroups. The information in these discussions is less formal than file libraries on the web, and thus it is a different sort of information than Web sites contain. This distinction is further explained in the research section of this chapter.

Internet Information Tools

While all Internet information is the same in the sense that it is made up of electronic bits of data, the interpretation of those bits is what provides the informational value. Much of the work of interpreting the bits is done by computer software called information tools. The major tools are generically described here.

E-mail

Electronic mail (e-mail) is one of the few Internet tools that conveys its function in its name. E-mail is the ability to exchange notes with other people connected to the computer network and is the most commonly used feature of the Internet. To use e-mail, you need an e-mail address and e-mail software. These are usually provided by the same source as your Internet connection. As a medium, e-mail shares many features with familiar means of communication such as telephones and face-to-face conversations. However, it has attributes that make it

distinct from these media. In particular, it is much faster and usually more convenient than paper mail for exchanging information. It is important to understand these differences in order to use e-mail effectively.

E-mail Lists

While e-mail conversations with other individuals can be useful, the major difference between e-mail and other communication methods is the e-mail mailing list. These lists are run by automated software that receives a note and then redistributes it to a list of interested people. The lists provide kinds of interaction that cannot be achieved over the phone or at conferences and provide for sharing information and expertise in a uniquely convenient and timely way. However, they do require some practice to be used effectively.

Joining an e-mail list is straightforward. You send the appropriate "subscribe" command to the address of the mailing list management program that runs the list. However, choosing which lists to subscribe to takes practice. It is difficult to participate effectively in a busy e-mail list on an occasional basis. The speed of e-mail discussions is such that they can begin and end within 48 hours. Tracking discussions and moving them into directions that you find useful means reading list traffic at least every other day. Browsing a list on a weekly basis will enable you to keep track of what information is available in the list's archives, but it is unlikely that people will respond to messages about issues that came up two weeks ago, unless a new angle is given to the question. Less active lists do not require as much effort to follow, but they are less likely to generate useful discussions.

Web Browsers

Web browsers are the software your computer uses to collect files from Web sites on the Internet. This software is designed around the ability to place text and graphics on the computer screen. Recently, however, effort has been put into making Web browsers and Web sites more interactive, so that you can run programs on Web sites that customize information to your needs. This is likely to make using the Web more complicated than a simple point-and-click experience, but that is the essential style of most Web sites.

Web Directories

Exploring the Internet has been made much easier in the last few years by the development of Web directories. These are Web sites that list a large number of general interest Internet sites, organized into subject categories. One of the most popular of these directories is Yahoo (http://www.yahoo.com). Yahoo's home page lists a variety of subject areas. Clicking on one of these titles leads to subcategories and then to a list of Web sites that provide information about these subject areas. By using Yahoo as a starting point, you can get a sense of the capabilities of the Web and the type of information available there.

Search Engines

Usually, one is looking for information on a specific subject. Using a subject-based directory such as Yahoo is not the most efficient way of approaching this sort of search on the Internet. Rather, you want to be able to specify particular words that you want to find and have the computer find files that contain those words for you. This requires the use of a Web search engine. One of the most powerful search engines currently available on the Net is Google found at http://www.google.com/.

Conducting Web Searches

The difference between search engines and Web directories is best described by example. A search for the words "hazardous materials" with Yahoo produced 32 hits. Google produced about 100,000. Yahoo found sites that specifically listed "hazardous materials" as keywords associated with their site—primarily commercial producers of hazardous materials handling equipment and training courses. Google found many other files containing the words "hazardous materials," including job descriptions, safety plans, and information about specific chemicals.

Refining the Search

As the "hazardous materials" example demonstrates, simply putting in the first words that occur to you to search on can be rather inefficient. Fortunately, Internet search engines provide ways of refining your search so that it can be more selective and the results more useful. While the precise format of these refinements varies from search engine to search engine, the concepts they use are similar.

The first step in refining your searches involves phrasing your question in such a way that you are clear what answer you need. This is often rather easy (does OSHA have any regulations that specifically cover the use of this chemical?), but at other times this can be more difficult (e.g., is this workplace situation a confined space?).

If you are having trouble coming up with a question that describes your need, it may be better to think up the name of a magazine article that would be just what you need. A four- or five-word phrase is a good place to start your search. It is important to think about possible other meanings for the words you select. For example, "safety" may refer to chemical concerns in your mind, while it refers to law enforcement issues in many other people's minds. An ambiguous word such as this is usually a poor choice to search for.

The Internet as a Research Tool

Although there is a lot of safety information on the Internet, searching for a specific piece of information on the Internet can be a frustrating experience. This is because information on the Internet is available in a variety of formats and is of varying quality. To use the Internet successfully for research purposes, a plan for your search is necessary. While it is not always a sure thing that you will find exactly what you are looking for on the Internet, it is very likely that you will find something that is helpful.

This section first describes some of the tools that you can use to become familiar with the types and amount of information available on the Net. Then it describes a strategy for answering specific questions that arise in your daily work.

Finding Information on the Internet

One of the biggest challenges facing health and safety professionals is finding information when they need it. While a well-stocked paper library can go a long way toward answering this need, changing regulations and new uses for hazardous chemicals make paper an unreliable and often expensive medium for researching the latest aspects of hazardous material use. Happily, the Internet has developed into an important research tool over the last few years and now provides a legitimate alternative to an extensive library of books, manuals, and regulations.

Just as being familiar with the contents of a paper library makes searching for information there easier, being familiar with how the Internet works makes searching for particular information easier. So it is a good idea to spend some time building your Net-surfing skills before you need to answer a particular question.

An Internet Search Strategy

In order to conduct research on the Internet efficiently, it is important to have a search strategy in mind while you are looking. A typical plan may be as follows.

Refine the Question

Consider whether the Internet is the best place to look for the information you are interested in. Information that does not change often or is of wide application is probably available on paper, in a library. On the other hand, for very specific or very new information, the Internet can be an unmatched resource.

Decide What Kind of Information You Want

Looking for a specific piece of data (e.g., the flash point of acetone) is different from looking for a technical interpretation of that data (use of acetone requires adequate ventilation due to its flammability), which is different from looking for a rule of thumb (use acetone in a fume hood if you're using more than 500 milliliters). These different types of information will be found with different strategies in different places on the Net.

Formal Databases   For specific pieces of data, formal databases are the best places to look. There is a variety of such sources, such as MSDS collections and databases containing government regulations. These databases are usually indexed to allow for keyword searches. Selecting keywords carefully will make your search more efficient.

Professional Interpretations   For technical interpretations of raw data, the best places to look are in collections of policies and procedures that are available on line. Such collections are usually associated with Web sites that companies and institutions put on-line for the convenience of their employees or customers. A good way to find these documents is to use the Internet indexes that are available. For this search, use keywords that apply as specifically as possible to your item of interest. Be prepared for many "false hits"—returns that are of no interest—for these searches. However, you are also likely to find several useful unexpected sources of information.

Informal Information   Because much safety knowledge requires technical expertise to apply appropriately, it is unlikely to be found in the formal information sources on the Net. However, the Internet has many informal information collections available in the archives of electronic mailing lists that operate on the Net. These are the first places to check for this type of information. The SAFETY archives (http://list.uvm.edu/archives/safety.html) are a good place to start such a search. Even if you do not find the information you are after there, you may find a reference to another Internet resource that has the information you are looking for.

Select Keywords for Searching

The result of refining your question should be a set of keywords that you want to search for. You will use these keywords in performing searches at various Web sites that are likely to contain appropriate information. For example, if you are simply after the flashpoint of acetone, "flashpoint" and "acetone" are appropriate keywords. Whereas if you are concerned about ventilation requirements for using acetone, "flashpoint" is not likely to be helpful and "flammable liquid" may be a reasonable substitute for "acetone."

Keywords need to be as specific as possible while allowing for variations in terminology that are likely to arise. Using keywords such as "safety" or "health" are likely to produce too many sites for most purposes. Most Web site indexes allow you to use logical connectors such as "and," "or," and "not" when conducting your search. This can help you refine your keyword search until you have about 20–40 hits. Lists of hits longer than that are probably too long to effectively search and an indication that your keyword strategy should be refined.

Select a Web Site to Start Your Search

Once you have a good idea of what kind of information you need to answer your question and what keywords are likely to be associated with that information, you are ready to start searching the file libraries on the Net. Start with a hot list of Web sites that you are familiar with.

It is best to test out your initial set of keywords by using it with one search engine, with the idea of seeing how many useful responses it produces before using other search engines. In this phase of your research, it is better to start with a subject-focused index.

For example, at http://list.uvm.edu/archives/safety. html, there is an index to the SAFETY e-mail list archives. This database consists of over 80,000 e-mail messages discussing a wide variety of safety issues. As well as providing access to the discussions by SAFETY participants, this database provides a good way to refine your search strategy in a limited universe of subjects. The advantage of using this database is that the keywords are likely to be used in the same context in which you are thinking of them, so the results are likely to be germane to your question.

By searching in this database, you can see if the phrase you chose is commonly used by other safety professionals to describe the situation you are thinking of. If the results are not related to your specific concerns, you can change the words you are using until the results are more appropriate. Once the keywords are refined in this search, they are likely to be more effective when searching larger databases such as Yahoo and AltaVista.

In general, more than 20 hits in a search indicate that the search needs to be further refined. Various search engines provide different ways of manipulating the keywords you have decided to use. For example, large databases usually require that all of the keywords you enter be included in the file for it to be considered a relevant hit. This is because the wide variety of subjects they cover creates many hits on an individual word. Most Internet search engines require you to use an alternate screen to conduct a nonstandard search. The details of composing these searches are usually found on those screens.

Ask a Discussion Group

If your search of the file libraries fails to produce the information you are after, or you are looking for more informal information than is available at Web sites, it is time to post a request for information to an appropriate e-mail list or a newsgroup. To increase your chances of success when you ask a question of a list, be sure to follow netiquette guidelines appropriate to that group.

First, check the archives of the group's discussion to see if it is the right group to ask the question of and to be sure that it is not a question that has been asked and answered repeatedly. When framing the question, be as specific as possible in asking the question, so that those who read it can determine what type of answers are appropriate (i.e., general pointers to the professional literature v. specific interpretations of your information).

Check Your Information!

Always be sure to confirm information you've gotten from the Net before you act on it. Remember that the information available on the Internet was written based on someone else's assumptions, in ignorance of the details of your particular situation. There may be specific, critical differences between the situation that you face and that of the person writing the information. The effort involved in confirming net information may range from asking yourself "Does this make sense?" to checking a paper reference source, to consulting with a professional with more expertise than yours.

Networking on the Network

In addition to the ability to access file libraries for research purposes, the Internet provides significant opportunities for professional networking. As is the case with any professional network, the Internet can help you to learn more about the basic technical issues of your field; get tips about approaching specific problems; be aware of new issues developing in the field; find prospective partners or consultants; and celebrate (or commiserate) with others in similar circumstances. The advantage of the Internet for networking activities is that it provides a convenient way to have ongoing discussions with geographically dispersed colleagues. These discussions can take place either in a group or individually. This section describes some of the considerations involved in using the Internet for this purpose.

General Considerations

As the Internet has grown in popularity, the time required to remain current with it has increased as well. While the technical details of using e-mail and Web sites have simplified significantly in the 1990s, the task of wading through all the possibly relevant information sources has become more complex. It is important that you have a clear idea of what your goals are for using the Internet; otherwise, you may find that you can devote a lot of time to using it without much payback.

Using the Internet to network with other people with similar interests can minimize this learning curve. The benefits of networking on the Internet are that it is a low-budget, low-travel way to be involved in your professional community. Productive professional relationships can be developed with a wide range of people without face-to-face meetings. These relationships usually start in discussion areas such as e-mail lists or newsgroups. They often develop into private correspondence that is able to be more speculative than public discussions can be.

The primary costs of developing a network of professional contacts over the Internet are time, patience, and a network connection. Fortunately, a powerful Internet connection is not required, as most networking happens via e-mail with low graphical content and small files. A good place to find more information about the process of networking and how it is affected by the Internet is *The Network Observer*, which is written by Phil Agre and available on-line.

Finding Discussion Groups

There are thousands of discussion groups operating on the Internet, either through e-mail or on newsgroups. Some are formally organized with charters (such as the SAFETY list described in Appendix 1); others are simply collections of e-mail addresses held together by someone's personal e-mail software. Finding discussions that are of most interest to you can be a bit of a challenge. However, there are several good places to check.

The first place to check is trade magazines and professional journals. "What's New on the Internet?" is a favorite topic for articles. These articles usually include both a list of Web sites relevant to the profession and discussion groups. It is important to note what type of forum is being discussed when reading these lists; using a newsgroup is significantly different from using an e-mail list from the computer point of view.

Another place to look for relevant discussions are the various "lists of lists" that exist on the Internet. One example of such a list is available at Tile.Net (http://www.tile.net). This site includes descriptions of a wide variety of mailing lists, organized by name, subject area, or location of the host machine. For example, a search on the word "safety" found a variety of lists, including several local safety discussions, a radiation safety list, and one which includes food safety as part of its discussion.

Remember that formal descriptions of discussion groups can often be significantly different from the actual subjects talked about within the group. You can search the actual text of many discussions within their archives to determine their actual focus.

Providing Safety Information on the Internet

In addition to providing a new information resource for safety professionals, the Internet is becoming the site of routine interactions both in the workplace and for the general public. This provides new opportunities for safety professionals to get out their messages, whether these are aimed at changing workers' behavior to improve workplace safety, advertising the availability of their products or services, or giving the public access to government information. In many cases, the unique advantages of the electronic information delivery (e.g., flexibility in delivery, asynchronous delivery of information, interesting interactive graphical presentations) make it ideal for providing safety information to at least some of the audiences that the safety professional serves.

This section provides an introduction to some of the conceptual issues involved in using the Internet for this purpose. It focuses on the process of developing, maintaining, and promoting a Web site; however, remember that there are other ways of using the Internet to sell safety besides Web sites. An active e-mail presence or pointing people to information that other people have provided on the Internet may be as effective and a more efficient way of serving the audience of interest.

Planning Your Web Site

If you want your organization to have a presence on the Internet, there are a number of decisions that you need to make before you start. Why does your organization want to be on the Internet? Is your main purpose to provide access to information that you would provide for free by other means? For example, many government organizations are setting up Web servers to provide access to the full text of publications that they now make available on paper. Their aim is to provide improved access to their public information and, not incidentally, to reduce the cost of reproduction and shipping.

Do you want to promote your organization's products and services? If so, you should plan on providing more information to your customers than you do on paper. As well as providing in-depth product and service information, you should be prepared to provide new information that is not likely to be found elsewhere. If you are hoping to attract new customers via the Internet, remember that Internet users expect to get something useful for free, even from commercial sites.

Do you want to provide safety training and information electronically? This requires careful planning about how the electronic information will fit into the culture of your organization. Remember that the Internet will be only one part of the "Information Ecosystem" that your audience is exposed to.

Selecting Information to Put on the Net

Once you have defined your goals, the first issue is what type of information you want to include and in what form you want to distribute it. These choices require balancing several issues, including the types of information you are trying to deliver (e.g., material safety data sheets, contact information for help in resolving safety problems, or training materials); the ability of your audiences to access the Internet; and the level of documentation you require to demonstrate regulatory compliance. Providing safety information to a group of workers provides a good example of these trade-offs.

The advantage of using a Web site for safety training is that it is possible to include a large amount of information there, most of which has already been generated for other purposes. It is relatively easy to move training handouts and overheads from paper and transparencies to a form accessible at your Web site. People in your audience can then access these materials at their convenience or use them to review the information you have already presented. At the same time, you can leverage the many other electronic resources that are available on the Internet to provide background information or more details (e.g., the text of the OSHA regulation or a standard operating procedure from another institution) on the issue.

The disadvantage of this approach is that a Web site must be carefully planned to make the information you have put there easy to find. For example, although safety professionals are used to working with programs on a regulatory issue basis (e.g., bloodborne pathogens separately from chemical hygiene issues), this is probably a confusing way to organize information for the average worker who is looking for an answer to the question, "How do I clean up this stuff I just spilled?"

Despite the challenges of providing safety information over the Internet (or an institutional Intranet), the potential advantages are attractive enough that many people are doing this.

Making a Web Site Effective

Development of a health and safety Web site presents the same challenge as the development of other safety information: How do you gain and hold the attention of the people who should have the information you have to share?

Three factors that promote this are be fun, be useful, and change over time. Fortunately, Web sites lend themselves to these criteria. As you browse the Web, you will find many examples of Web sites that are some or all of these. Unfortunately, these Web sites require quite a bit of effort to develop and maintain. It is important to keep the ambition of your Web site within the scope of the resources you have to devote to it.

Designing the Web site as a whole can be a major challenge. There are a variety of guides available on the Internet that highlight some of the issues involved. A Web search on usability will quickly identify several.

Appendix 1: Introduction to the SAFETY Mailing List

What Is SAFETY?

SAFETY is an electronic mailing list that started in 1989. People can send e-mail to the list and it will be redistributed to the list subscribers. Discussions on SAFETY involve environmental and occupational health and safety issues, although a wide range of subjects and participants is encouraged. There are currently more than 80,000 messages in the SAFETY archives. A keyword index to these archives, daily digests of the discussions, and other useful SAFETY-related files can be found at http://list.uvm.edu/archives/safety.html. To subscribe to the list, send: SUB SAFETY your name to LISTSERV @LIST.UVM.EDU.

Typical issues discussed on safety include chemical safety issues, indoor air quality, interpretation of safety standards, hazardous waste disposal, safety management, and electronic resources on these topics. A breakdown of over 11,000 messages sent to SAFETY in its first four and a half years showed the following percentages in these subject categories:

Categories	% of Total Messages
Chemicals	16
Miscellaneous	12
Fire/General safety	9
Safety management	8.5
Waste management	8
Regulations	7.5
Electronic resources	7
Lab safety	6
Biosafety	4
Indoor air quality	3
Emergency response	3.5
Fume hoods/Ventilation	3.5
Organizations	2
Personal protective equipment	2
Radiation safety	2
Jobs	1

These numbers have changed somewhat in the past few years, but are a reasonable estimate of distribution of interests among SAFETY subscribers.

More specifically, typical subjects that have been discussed on SAFETY include (in alphabetical order):

1. Airborne lead concentrations in firing ranges
2. Carpeting and indoor air quality
3. Contiguous properties and EPA ID numbers
4. Disposal of fluorescent lights/ballasts
5. Eyewash stations—Installation & maintenance
6. Fire evacuation programs and fire drills
7. Fume hoods—Installation & maintenance
8. HF burns and first aid
9. Heavy metal hazards associated with museum specimens
10. How do I search the SAFETY archives?
11. Indoor air quality management issues
12. MSDSs on the Internet
13. Perchloric acid hood evaluation and decontamination
14. Photocopier emissions
15. Picric acid disposal
16. Residential carbon monoxide monitors
17. Safe liquid nitrogen handling procedures

18. Safety departments in universities and colleges
19. Safety glasses/goggles in laboratories
20. Safety information on the Internet
21. Safety showers—Installation & maintenance
22. Sources of good safety videotapes
23. Survey of academic hazardous waste programs
24. The use of disclaimers in e-mail
25. Use of contacts in chemical environments
26. Where to go to study safety/industrial hygiene/ occupational health

Appendix 2: Major Health and Safety Web Sites

Note: This list is not meant to be all-inclusive. Rather, it is designed to provide a sampling of Web sites from government, academic, commercial, and other sources that represents the kinds of information that can be found on the Web both within the United States and internationally.

Agency for Toxic Substances and Disease Registry
http://atsdr1.atsdr.cdc.gov:8080/atsdrhome.html

American Board of Industrial Hygiene
http://www.abih.org

American College of Occupational and Environmental Medicine
http://www.acoem.org

American Conference of Government Industrial Hygienists
http://www.acgih.org

Biosafety Resource Page
http://www.orcbs.msu.edu/absa/resource.html

Canadian Centre for Occupational Health and Safety
http://www.ccohs.ca

Center for Safety in the Arts
http://artswire.org:70/1/csa

Centers for Disease Control and Prevention (CDC)
http://www.cdc.gov

ChemFinder
http://chemfinder.camsoft.com

Consumer Product Safety Commission (CPSC)
http://www.cpsc.gov

Cornell Ergonomics Web
http://ergo.human.cornell.edu

Department of Energy Technical Information Service
http://dewey.tis.eh.doe.gov

Division of Chemical Health and Safety, American Chemical Society
http://chas.cehs.siu.edu

Documentation for IDLH Concentrations
http://www.cdc.gov/niosh/idlh-1.html

Duke University Occupational and Environmental Medicine
http://occ-env-med.mc.duke.edu/oem

Environmental Chemicals Data and Information Network
http://ulisse.etoit.eudra.org/Ecdin/Ecdin.html

Environmental Protection Agency
http://www.epa.gov

ErgoWeb
http://www.ergoWeb.com

European Agency for Safety and Health at Work
http://www.eu-osha.es

Hazardous Substance Release/Health Effects Database
http://atsdr1.atsdr.cdc.gov:8080/hazdat.html

Health and Safety Promotion in the European Union
http://www.hsa.ie/hspro/index1.html

Health Canada, Health Information Network
http://www.hwc.ca/links/healthcan/hinfo_e.htm

Howard Hughes Medical Institute Lab Safety
http://www.hhml.org/science/labsafe

Health Canada, Laboratory Biosafety Guidelines
http://hwcWeb.hwc.ca/hpb/lcdc/bmb/biosafty/ index.html

ILO Database on International Labour Standards
http://ilolex.ilo.ch:1567/public/english/50normes/ infleg/iloeng/index.htm

International Agency for Research on Cancer (IARC)
http://www.iarc.fr

International Occupational Safety & Health Information Centre
http://turva.me.tut.fi/cis/home.html

Manitoba Workplace Safety & Health Division
http://www.gov.mb.ca/labour/safety/index.html

MEDLINE database
http://www.nlm.nih.gov/databases/freemedl.html

National Institute of Environmental Health Sciences
http://www.niehs.nih.gov

National Institute for Occupational Safety and Health
http://www.cdc.gov/niosh/critdoc2.html

Occupational Safety and Health Administration
http://www.osha.gov

Oklahoma State University Online Training Modules
http://www.pp.okstate.edu/ehs/modules/home.htm

Radiation and Health Physics Home Page
http://www.sph.umich.edu/group/eih/UMSCHPS/

SafetyLine, Australia
http://www.wt.com.au/safetyline

Seton Identification Products
http://www.seton.com

Typing Injury FAQ
http://www.cs.princeton.edu/~dwallach/tifaq/

University of Edinburgh, Health Environment and Work
http://www.med.ed.ac.uk

Vermont SIRI MSDS Collection
http://siri.org

World Health Organization
http://www.who.ch

Young Worker Awareness, Workplace Health and
  Safety Agency, Ontario, Canada
http://www.whsa.on.ca

10

Understanding the Toxic Substances Control Act

Compliance and Reporting Requirements

FRED HOERGER

ROBERT HAGERMAN

The Toxic Substances Control Act (TSCA)[1] was signed into law on October 11, 1976, by President Gerald Ford. A wide variety of regulations and ongoing programs aimed at preventing or anticipating potential unreasonable risks are authorized by this law. The Environmental Protection Agency (EPA) is responsible for maintaining an inventory listing of all chemicals in commerce and for reviewing new chemicals proposed for manufacture. The agency has broad authority to collect information on the health and environmental effects of chemicals and certain aspects of their commercial distribution, and can require testing of chemicals to ascertain potential health and environmental effects.

Reporting requirements under TRI, the Toxic Release Inventory,[2] and certain programs aimed at pollution prevention are closely related to the purpose and programs of TSCA. These, along with the TSCA requirements, are administered at EPA in the Office of Pollution Prevention and Toxics. This office is under the supervision of the Assistant Administrator for Prevention, Pesticides, and Toxic Substances. This chapter summarizes the current status of the TSCA and related programs, but the reader is cautioned that the programs are ongoing and that additional chemical-specific requirements are likely from time to time.

Purpose of TSCA and Reporting Policy

Some background on the purpose and Congressional intent regarding TSCA is helpful in understanding the re-porting policies that have evolved over the past 20 years. The original purpose of TSCA was to provide a broad range of authority to identify and control those toxic substances that pose or may pose an unreasonable risk to human health or the environment. Identification of unreasonable risks, either real or potential, requires information. Thus, there are a wide range of reporting requirements, both general and chemical specific, that manufacturers, processors, and importers need to comply with. Some of these flow from self-implementing requirements, such as reporting substantial risk information in a timely manner; others flow from very chemical-specific regulations, such as to perform and report on specific tests on a specified chemical.

TSCA was drafted with the intent that its use be coordinated with the authorities of other laws. Again, although subject to interpretation, the Occupational Safety and Health Administration (OSHA) and environmental-release laws take precedence over TSCA. However, TSCA is the primary law with broad testing and information collection authority over chemicals. As indicated in Box 1, pesticides, drugs, and food additives are excluded from TSCA requirements, although chemical intermediates used in their manufacture are subject to TSCA. Genetically engineered substances not subject to Federal Insecticide, Fungicide and Rodenticide Act (FIFRA) or Food and Drug Administration (FDA) authority fall within the scope of TSCA.

Generally, it can be said that the statutory authority vested in TSCA has been used largely for information

---

Box 1

Exclusions to the definition of "chemical substance" and thus to the scope of TSCA:

- *Pesticides* when regulated under Federal Insecticide, Fungidice and Rodenticide Act (FIFRA)
- *Foods and food additives* when regulated under Food, Drug and Cosmetic (FD&C Act)
- *Drugs, medical devices, and cosmetics* when regulated under FD&C Act
- *Tobacco and tobacco products*
- *Radioactive materials or by-products* regulated under Atomic Energy Act of 1954
- *Certain munitions* subject to taxation under the Internal Revenue Code of 1954

---

generation and collection with less emphasis on regulation requiring controls. Thus, compliance requirements have been largely on reporting information. However, this information generation and review has resulted in many voluntary actions to reduce risks. Some industry observers have felt that the law has strongly promoted corporate policies for evaluating and reducing risks.

TSCA is intended to apply to a broad range of chemical substances but not to duplicate the provisions of other laws. Thus, it is critical to understand certain key definitions and exclusions.

A "chemical substance" is defined in section 3 of the statute as any organic or inorganic substance of a particular chemical identity including any combination of such substances that occur in nature or are the result of a chemical reaction, and any element or uncombined radical. The definition excludes mixtures, but not the components of mixtures.

TSCA authorities may be applied to those commercial entities that conduct any of the following five activities:

- manufacture (to produce, manufacture, or import into United States customs territory)
- processing (preparation of a chemical substance or mixture after its manufacture for distribution in commerce as a chemical substance or as a component of an article)
- distribution in commerce (applies to chemical substances, mixtures, or articles containing a chemical substance or mixture, and means to sell in commerce; to introduce or deliver for introduction into commerce; or to hold after introduction into commerce)

- use (undefined in the statute)
- disposal (undefined)

Another important definitional provision of TSCA (section 26 (c)) authorizes the EPA to take action with respect to categories of chemical substances or mixtures in the same manner as to individual chemical substances.

The Chemical Inventory and New Chemicals

A key benchmark in complying with TSCA is the inventory of existing chemicals.[3] This list was first compiled in the late 1970s and was a listing of all chemicals manufactured in or imported into the United States at that time. As additional chemicals have been introduced into commerce, they have been added to the inventory. Today the inventory consists of about 70,000 substances. About 55,000 of these commercial chemicals are either polymers or substances produced in low volumes (less than 10,000 pounds per year).[4]

The published inventory lists substances by name according to a complex nomenclature system established by Chemical Abstracts Services (CAS), under contract and supervision of the EPA. Each name is tied to a unique number known as the CAS number. A public inventory list is available, both electronically and in hard copy. The specific chemical identities of a number of substances have been classified by manufacturers as confidential business information and are thus not on the public inventory.

A prospective manufacturer or importer of a chemical may be uncertain as to whether the chemical is listed on the complete inventory. Upon filing of a bona fide letter of intent to manufacture or import, the EPA will search the confidential inventory and, if the substance is already listed on the inventory, disclose the assigned CAS number and correct nomenclature of the substance to the proposed manufacturer or importer.

Premanufacture Notification

Substances not listed on the complete, or so-called Master Inventory, and proposed for manufacture or import are classified as new substances. A manufacturer or importer must submit a Premanufacture Notification (PMN), to the EPA at least 90 days before manufacture is legally permissible. The notification must include chemical identity, physical and chemical properties, and all health and safety data available on the compound.

During the 90-day review period (which may be extended under some circumstances), the EPA determines whether the new substance is likely to pose an unreasonable risk. The EPA review is usually carried out by a group of technical experts, who examine the submitted data and consider structure activity relationships and the likely potential use and exposure patterns.

If the review warrants, EPA may require generation of test data or place limitations or prohibitions upon manufacture or use. In 1995, more than 2300 PMNs were submitted. The EPA required that notifiers on 44 chemicals generate test data before production volume would exceed specified limits. In many instances proposed manufacturers entered into consent agreements with the EPA to control workplace exposure below certain limits, to label certain hazards, to conduct product stewardship programs, or to limit the type of use of the product. The EPA has issued significant new use regulations (SNURs), which require submission of a new PMN, if manufacture is to result in uses other than those proposed in the original PMN. Sixty-four SNURs were issued in 1995.[4]

If EPA review has not posed the need for restrictions, manufacture or importation may be initiated after the 90-day PMN review period. A Notice of Commencement of Manufacture (NOC) must be filed with the EPA within 30 days of actual manufacture or importation. Once the NOC is filed, the substance is placed on the inventory as an existing chemical. Once listed, other entities may manufacture or import the substance without filing a PMN.

A prescribed form[5] and filing fee are necessary for making the PMN notification. The form may be submitted on paper (EPA Form No. 7710-25 or equivalent EPA-approved format) or electronically. Unless all required information is included in the form, the EPA treats the submission as deficient and the 90-day review period does not begin until a complete notification is received.

Exemptions

Certain types of polymers described at 40 *CFR* 723.250 are exempt from some PMN requirements. Manufacturers or importers of polymers that meet the complex criteria to qualify need only to submit a "postcard" notification identifying the manufacturer and the number of polymeric substances manufactured in the previous year. In addition, the manufacturer must maintain extensive records demonstrating compliance with the exemption requirements.

"Small quantities" that are necessary for research and development purposes are exempt from PMN requirements. Substances may be manufactured for test marketing if a simplified notification procedure is followed and the scope of the test marketing is specifically defined.

There are also partial exemptions for substances to be produced only in small quantity, and for those with low environmental release and low human exposure.[6]

Compliance

Compliance with the new chemical provisions of the law involves many subtle, but critical, aspects. Complex chemical or polymer structures may make it difficult to determine whether a substance is on the inventory or is entitled to exemption status. Process changes may result in formation of products or substances that are not listed in the inventory. Some substances are listed on the inventory as products of specific reactants, reaction ratios, or process conditions, even though changes from the specified reactants, reactant ratios, or conditions may give rise to an identical chemical substance; in these cases a PMN would be required. These are just a few of the many complications that can arise if detailed and expert consideration is not given to the compliance aspects of the distinction between new and existing chemicals. The EPA imposes large monetary fines for failure to submit a PMN prior to manufacture; the fines can be assessed on the basis of the number of days of manufacture without having made notification (potentially accumulating at up to $27,500 per day).

Reporting of Chemical Information

Section 8(a) of TSCA requires manufacturers and processors to report information about their activities which is reasonably required for the purposes of the Act. Information to prepare the first Chemical Inventory was reported by manufacturers and importers complying with a rule issued under this section. The initial rule also required the reporting of the quantity of each chemical and the location of its manufacture. Every four years, manufacturers and importers update information about the quantity manufactured or imported and the location of these activities according to the requirements of the Inventory Update Reporting rule.[7]

From time to time, basic exposure-related information must be reported by manufacturers and importers under provisions of the Preliminary Assessment Information Reporting rule.[8] In one instance, much more detailed information was collected from manufacturers and importers under the Comprehensive Assessment Information Reporting rule.[9] These rules have generally been used to support EPA evaluations concerning the need to test chemicals suggested by the Interagency Testing Committee or other EPA programs. The two rules are "model" rules: they were promulgated only once, but are amended as needed by addition of chemicals for reporting of information.

The EPA has also issued a few chemical-specific rules for substances of concern to the agency that are not presently produced in or imported into the United States. These rules require advance notification to the EPA of an intent to manufacture or import and other information specific to each substance.[10]

Notification of Substantial Risk Information

Manufacturers, processors, or distributors of a substance or mixture must immediately notify the EPA if they obtain "information that reasonably supports the conclusion that such substance or mixture presents a substantial risk of injury to health or the environment" unless the information is "already known" to the agency. This requirement is in Section 8(e) of the statute; therefore, no regulation is necessary for its enforcement. However, the language is very much subject to interpretation. EPA issued a formal guidance on notification[11] in 1977 and revised it further[12] in 1991 as part of a compliance audit program of industrywide scope. In this latter guidance, the EPA withdrew prior interpretive guidance with respect to the definition of reportable nonemergency incidents of environmental contamination. The EPA intended to issue revised guidance for such reporting in 1997.

It is important to note that the EPA has interpreted the notification requirement to apply to natural persons as well as juridical persons; thus, individual employees in for-profit entities are liable for compliance. However, this personal liability does not apply if a company establishes and operates a system that allows employees to inform company management of potential substantial risk information. Management then transmits reportable information to the EPA, thus minimizing the need for numerous duplicative reports, since many individuals may become aware of information at about the same time.

Compliance involves critical consideration of toxicity information, such as indicators of potential carcinogenicity, reproductive or other chronic effects, and nonemergency releases to the environment. It appears that information to be reported involves the severity of potential toxic or ecological effects with little consideration of the degree of potential exposure.

Reporting of Significant Adverse Reactions

Manufacturers, importers, processors, and distributors of chemical substances and mixtures are required by Section 8(c) of TSCA to keep records of oral or written reports they may receive which allege that substances or mixtures are the cause of significant adverse reations.[13] The records must be kept for 30 years for a workplace incident, otherwise for 5 years. Significant adverse reactions are those that may indicate a substantial impairment of normal activities or long-lasting or irreversible damage to health or the environment.

Allegations could be attributable to a company product, a raw material, process operations, emissions, effluents, or discharges. Thus, compliance requires a referral network for those most likely to receive relevant reports. This network could include, for example, occupational health professionals, field marketing personnel, legal department (worker's compensation or liability claims), security (neighborhood complaints), and first-line supervision (worker complaints). The referral process may include screening to determine whether a given report must be retained in the company files. Compliance also involves training employees about their right to report allegations and the process for making reports.

Reporting of Health and Safety Studies

Manufacturers and processors of chemicals specified in lists at 40 *CFR* 716.120 are required to submit to the EPA copies of health and safety studies of listed chemicals or mixtures in their possession.[14] Lists of studies in progress and a list of studies of which they are aware must also be supplied. The EPA must also be notified of new studies initiated during a 10-year period after listing, and copies of such studies must be submitted upon completion. As Congress directed, the regulatory definition of health and safety study is quite broad, encompassing almost any study (formal or not) containing data bearing on the effect of a listed chemical on health or the environment, although there are several exemptions. For example, monitoring to determine compliance with workplace exposure limits is not considered a reportable study unless the results of several such monitoring measurements are aggregated and analyzed.

Compliance with the submission requirements when a chemical is added to the list involves a fairly straightforward search of appropriate company files and screening of potential submissions in accord with the criteria of the regulation. Ongoing compliance during the 10 years after listing may be more difficult, requiring either affirmative systems to trigger reporting when a study is initiated or completed, or ongoing awareness by those who may initiate submissions.

The agency usually asks for information on production volume and related information (under the provision of Section 8(a) of TSCA) at the time it requests reporting of health and safety studies.

Global Commerce in Chemicals

Reporting of Exports

Section 12(b) of TSCA requires exporters of certain chemicals to notify EPA of their activities. The chemicals and mixtures triggering export notification are those which are subject to one of the following:

- testing requirements under Sections 4 or 5
- manufacturing or other limitations under Sections 5 or 6
- judicial review under Section 7 (imminent hazard proceedings)

Reporting is required regardless of the reason for export or even if the chemical is present in small concentrations. Exporters must notify the EPA the first time they ship a chemical to any given country. The EPA in turn notifies the embassy of the destination country. The regulation[15] established no exemptions (e.g., for small quantities, low concentrations, impurities, etc.). Thus, the EPA has received many "trivial notices." The EPA has announced its intention to substantially modify the regulation.

If a company has a system with which it can identify all intended export shipments in a timely manner, compliance is relatively simple. Those with many exports emanating from diverse locations would likely require a computer-based system to ensure adequate compliance.

Certification of Imports

Importers of chemicals (but not of articles) must certify either that each chemical in a shipment is in compliance with TSCA and its regulations, or that it is not subject to TSCA.[16]

Compliance systems will usually involve the materials handling function (e.g., purchasing) in the case of imports of relatively large quantities of chemicals, and sensitization of those who may initiate ad hoc imports (e.g., of research and development samples) to the need for certification. Certifications are made on import documents (bill of lading or commercial invoice) or by letter accompanying the shipment. Small shipments that bypass formal customs inspection (e.g., mailed packages) may be certified directly to the EPA by timely letter.

Testing of Chemical Substances

Section 4 of TSCA contains provisions for the EPA to require manufacturers to test substances or mixtures for potential health and/or environmental effects. Over the years testing methodology has been defined for a wide variety of effects and more efficient means of administering and complying with testing programs have evolved.

The manufacturer's testing obligations and requirements flow from the EPA Master Testing List.[17] The master testing list is dynamic in nature. In 1995 it included some 500 specific chemicals and 10 categories of substances. These were drawn from a subset of the chemical inventory of three to four thousand high-volume chemicals and another 11,000 substances of relatively high toxicity or exposure potential.[4] The master testing list has been developed and is periodically updated by a series of internal EPA reviews, by recommendations from an interagency committee, and from public input. As a general, but not binding, strategy, the EPA attempts to collect and review existing health and safety studies prior to imposing testing requirements.

TSCA testing programs are implemented in three ways.

1. Voluntary testing programs are initiated by a company or a consortium of companies, usually within a trade association. Recent examples include a product stewardship/testing program on siloxanes, and exposure assessment of formaldehyde in newly constructed single-family housing.

2. Another approach involves enforceable consent orders, whereby companies agree with the EPA to the specific scope and nature of testing; the agreement is subject to penalties if not consummated. An order initiated in 1995 involves industry obtaining data on dermal absorption rates for about 80 chemicals. Three other enforceable consent orders related to toxicity tests on specific chemicals.[4]

3. The EPA also uses formal rulemaking to generate new information. The affected manufacturers must generate specific data by prescribed tests within specified times. As examples of rulemaking, the EPA in 1995 issued a rule requiring testing of nine substances for developmental and reproductive effects[18] and another rule for testing 21 substances[19] that have been identified as hazardous air pollutants under the Clean Air Act.

It can be seen that compliance with TSCA testing requirements is complex. Typical decisions involve integrating company priorities for information generation with EPA priorities, assessing the feasibility of mounting a joint program with other producers, and estimating the relative efficiency of voluntary efforts versus compliance with a future consent order or regulation. In addition, one needs to consider whether there is a reasonable trade-off of less testing for greater risk-management practices that substantially reduce human exposure and/or environmental-release potentials.

International programs on testing supplement TSCA testing. There are several globally oriented industry associations involved in testing, such as the International Isocyanate Producers. The EPA, the OECD (Organization for Economic Cooperation and Development), and industry groups have embarked on a Screening Information Data Set program. The objective is for industry in the various OECD nations to share the burden in developing basic information on some 100 chemicals produced in high volume. The screening data will provide the basis for assessing need and scope of further testing.

Regulation of Hazardous Chemical Substances and Mixtures

The wide variety of possible restrictions on manufacture and use of chemical substances are detailed in Box 2. Such restrictions are authorized under Section 6 of TSCA in order to prevent potential unreasonable risks of injury to human health and the environment.

Regulations are to be based upon consideration of potential injury, benefits of the substance, and economic consequences. In practice, Section 6 rulemaking has been used infrequently, for example, to regulate the disposal of a dioxin-containing waste and to prohibit the use of hexavalent chromium compounds as inhibitors in cooling towers for comfort cooling systems. Over the years, EPA strategy on toxics, industry initiatives, and application of other environmental laws appear to have evolved into more efficient approaches than the formal and extensive rulemaking provisions of Section 6.

Box 2

EPA may issue regulations that:

- prohibit manufacture, processing, or distribution (MPD)
- limit the amount of MPD
- limit uses or concentrations in a particular use
- provide for warning labels
- require maintenance of process and compliance records
- prohibit the manner or method of use
- prohibit or prescribe the manner of disposal
- require notices of unreasonable risks
- require reports on quality control procedures

Risk Management of Existing Chemicals: Strategy and Actions Flowing from Testing and Reporting

The extensive testing, substantial risk, and other reporting activities under TSCA provide much information to the EPA and the public. The EPA utilizes a two-stage review process to target potential concerns to a smaller group of substances, uses, and technologies to which the agency will focus its risk-management (RM) thrusts. The RM1 process involves screening chemicals to identify those that are likely to be of greatest concern. A much smaller number of chemicals are evaluated in the RM2 phase with the objective of determining options for risk-reduction actions.

The RM process and other evaluations have led to a variety of approaches for controlling and reducing risks. These include community empowerment, government partnerships, voluntary agreements with industry, product stewardship, information dissemination, pollution prevention, and goal setting. Currently, the EPA has ongoing programs of these types involving benzidine dyes, paint strippers, cleaning products purchased by the General Services Administration, land application of pulp and paper mill sludge, and the development of exposure-limit guidelines.[4]

Issues That May Shape Future Requirements

Confidential Business Information

Confidentiality of trade secret information, such as identity in formulations, efficient manufacturing processes, and end uses has given many manufacturers and processors competitive advantages in the marketplace. TSCA, and the implementing regulations, have provided for confidential business information (CBI) protection if the reporting entity designates the portion of reported information that is confidential, states the reasons for the claim, and certifies that the information has not and will not be divulged to outside parties. The EPA maintains the information in secured areas inaccessible to the public.

Over the years, there has been criticism of the CBI policies with the perceptions that some companies are hiding adverse effects information, that needed information is inaccessible to state regulatory and health officials, and that various publics could initiate protection strategies if information were available. It can be anticipated that CBI issues may influence TSCA policy on reporting and the public availability of information.

Chemical Use Inventory

Legislation in 1989 created the Toxics Release Inventory (TRI) and resulted in annual reporting of emissions for several hundreds of compounds. The results have generally been regarded as a highly effective environmental program. Corporate attention to reducing emissions and discharges has increased significantly with reductions of more than 50% in 5 years in many instances. The EPA, with considerable support from environmentalist groups, has been considering extending the TRI concept by using TSCA to collect and disseminate use information, frequently referenced as a Chemical Use Inventory. For example, the Inventory Update Rule under Section 8(a) of TSCA might be expanded to collect information on the internal disposition of substances by manufacturers (mass balance reporting) and the functional use of substances. Although the future of such proposals is uncertain, it appears that the extent and nature of future reporting will continue to undergo evolutionary change.

The Unreasonable Risk Standard

TSCA, in contrast to many other environmental laws, does not have many mandated lists of chemicals for action or deadlines for action. Instead, the EPA has considerable latitude in setting its priorities, and regulatory actions are based on whether they meet a standard of reducing potential unreasonable risks. This standard is consistent with many of the concepts embodied in regulatory reform. However, the standard is controversial, with TSCA critics feeling that too much administrative effort is required to impose controls. These factors, along with the evolution of the pollution prevention concept, have led to a significantly new public policy that is complementary to TSCA.

This new policy involves voluntary actions:

- unilaterally by industry, for example, product stewardship programs and life cycle analysis of products
- cooperative programs between EPA and industry as described earlier
- advisory committees working with companies on local facility concerns

Inclusion of TSCA-like Requirements in Other Laws

TSCA has not been substantially amended or reauthorized in its now 20-year history. However, many other environmental laws have been enacted or reauthorized during this period. Some issues that might have been addressed under TSCA have been included in these other laws. For example, the phase-out of hard chlorofluorocarbons was included in Clean Air Act amendments. More recently, requirements for evaluating the possible estrogenic effects of chemicals were included in revisions of the pesticide and clean water statutes.

Expertise and Organization Required for Compliance

TSCA's broad requirements extend to many firms who do not consider themselves chemical manufacturers or processors. Any for-profit entity that uses chemicals should provide itself a source of expertise in TSCA statutory and regulatory requirements to determine any compliance needs. Such expertise must also include a thorough understanding of both the processes and products (or services) of the firm. In small companies, legal counsel may work closely with a technical expert. In larger companies, the expertise may flow from a team of experts in diverse disciplines, including, for example, legal, process engineering, occupational health, computer systems, marketing, research and development, and distribution. In any event, it will be necessary to develop some level of cross-functional expertise and communication.

Development and maintenance of effective compliance systems involve an organizational approach that assigns explicit "ownership" of specific compliance requirements to appropriate functions or individuals. The "owner" then calls on other functions as needed. For example, it would be appropriate to assign accountability for ensuring timely submission of PMN notices to individual research personnel, who could then draw on the expertise of a regulatory specialist or attorney to help in preparation of the PMN. On the other hand, the regulatory specialist would bear responsibility for tracking testing or reporting regulations and determining applicability to company products with the aid of product experts or computerized data bases.

The EPA maintains an Office of Technical Assistance for TSCA. Personnel in this office can provide information on TSCA programs, are able to answer questions on the complex compliance aspects, and can direct individuals to specific experts or sectors in the agency. The office maintains a phone hot line for queries (202-554-1404).

Public Participation—An Opportunity for Stakeholders

TSCA has many provisions for stakeholder participation in public policy. All proposals for a regulatory requirement are published in the *Federal Register* with an opportunity for public comment in writing, and sometimes also in a public hearing. These are usually preceded by an advanced Notice of Proposed Rulemaking. In addition, the EPA often convenes informal workshops or conferences on potential regulatory issues.

The Office of Technical Assistance mentioned above is a source of information about pending or future actions, proposals, and public meetings. Much of the future activity is also noted in the Regulatory Calendar published periodically in the *Federal Register.*

References

1. Toxic Substances Control Act; Public Law 96–409, Oct. 11, 1976.
2. Toxic Release Inventory; 40 *CFR* 372, 1988. See also Public Law 99–499, 1986.
3. Reporting for the Initial Inventory; 40 *CFR* 710, 1977.
4. *Annual Report of the Office of Pollution Prevention and Toxics, FY 1995*; Publication No. EPA 745-R-96-005; EPA: Washington, DC, 1996.
5. Premanufacture Notification; Premanufacture Notice Requirements and Review Procedures; 40 *CFR* 720, Appendix, 1983.
6. (Exemptions for) Chemical Substances Manufactured in Quantities of 10,000 Kilograms or Less per Year and Chemicals with Low Environmental Release and Human Exposures; 40 *CFR* 723.50. 1995.
7. Partial Updating of TSCA Inventory Database; Production and Site Reports; 40 *CFR* 710, 1986.
8. Chemical Information Rules: Manufacturers Reporting; Preliminary Assessment Information Reporting (PAIR); 40 *CFR* 712, 1982.
9. Comprehensive Assessment Information Rule (CAIR); 40 *CFR* 704, 1988.
10. Chemical-Specific Reporting and Recordkeeping Rules; 40 *CFR* 704, Subpart B, 1984, 1988.
11. (Interpretive Guidance for) Notification of Substantial Risk Under Section 8(e); 43 *FR* 11109, 1978.
12. Notice of Availability of Revised Guidance; 56 *FR* 28458, 1991.
13. Records and Reports that Chemical Substances Cause Significant Adverse Reactions to Health or the Environment, Recordkeeping and Reporting Procedures; 40 *CFR* 717, 1983.
14. Health and Safety Data Reporting; Submission of Lists and Copies; 40 *CFR* 716, 1983.
15. Chemical Imports and Exports; Notices of Export Under Section 12(b); 40 *CFR* 707, Subpart D, 1980.
16. Imports and Exports; General Import Requirements and Restrictions; 40 *CFR* 707, Subpart B, 1983. 19 *CFR* 12.118–12.127.28, 1983.
17. EPA Master Testing List; 61 *FR* 65936, 1996.
18. Proposed Multi-Substance Rule for Testing for Developmental & Reproductive Toxicity and Neurotoxicity; 56 *FR* 9092; 56 *FR* 9103, 1991.
19. Proposed Hazardous Air Pollutant Test Rule; 61 *FR* 33178, 1996.

SECTION II

RISK MANAGEMENT

LAWRENCE DOEMENY

ELIZABETH K. WEISBURGER

SECTION EDITORS

11

Introduction to Risk Assessment

LESLIE STAYNER

In a sense, risk assessment has existed as long as humans have been faced with decisions about what action to take when confronted with a hazardous situation. However, risk assessment has only been recognized as a formal scientific discipline within the past few decades. The growth of risk assessment has been fueled by societal concerns about the myriad potential environmental hazards that have been identified by environmental scientists. Numerous reports of hazards related to exposures at work, pesticides in food, air and water pollution, and hazardous waste sites have created serious public concerns about possible threats to people's health. Confronted with these hazards the public has increasingly demanded that scientists not only identify hazards, but also help them to place these hazards in proper perspective.

Risk assessments have become a requirement for standards setting by all United States regulatory agencies, including the Occupational Safety and Health Administration (OSHA) and the Environmental Protection Agency (EPA), as a result of court decisions and legislation. In a 1980 landmark case, the Supreme Court decision on benzene (*Industrial Union* v. *American Petroleum*; 100 S.C.T. 2844, 1980) made the assessment of risk a requirement for setting occupational health standards. The Court remanded the OSHA standard for benzene on the basis that OSHA had inadequately demonstrated that lowering the standard would significantly reduce the risk of harm. In this decision the Court stated:

> Some risks are plainly acceptable and others are plainly unacceptable. If for example the odds are one in a billion that a person will die of cancer by taking a drink of chlorinated water, the risk clearly could not be considered significant. On the other hand, if the odds are one in a thousand that regular inhalation of gasoline vapors that are 2% benzene will be fatal a reasonable person might well consider the risk significant and take the appropriate steps to decrease or eliminate it.

This statement has been taken by OSHA as guidance for determining when a risk is significant enough to warrant regulatory action. Requirements for risk assessment have been legislatively mandated in enabling legislation for the EPA's Clean Water and Clean Air Acts.

In response to the regulatory needs, methods for risk assessment have been rapidly evolving during the past few years. This chapter presents a brief overview of the current concepts and methods used in risk assessment. Several books and exhaustive reviews have been written on these issues and interested readers are referred to these sources for additional information (NAS, 1983; NAS, 1995; CCERP, 1985; OSTP, 1985; EPA, 1987; Paustenbach, 1989).

Definition of the Risk Assessment Process

In 1983 the National Academy of Sciences (NAS) published a landmark report that attempted to define the scope of risk assessment in the federal government. The NAS report developed a useful paradigm for viewing the risk assessment process, which is summarized in Figure 11.1. This model draws a sharp distinction between risk assessment and risk management. According to the NAS report, risk assessment involves "the use of the factual base to define the health effects of exposure of individuals or populations to hazardous materials and situations," whereas risk management involves "the process of weighing policy alternatives and selecting the most appropriate regulatory action, integrating the results of risk assessment with engineering data and social, economic, and political concerns to reach a decision." The

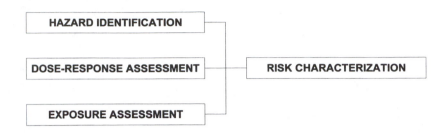

FIGURE 11.1 Model of the risk assessment process.

NAS report highlighted the need for a clear organizational separation between the functions of risk assessment and risk management activities.

The NAS report divided the risk assessment process into four distinct elements:

- *Hazard Identification*—The qualitative evaluation of the adverse health effects of a substance in animals or in humans.
- *Dose-Response Assessment*—The process of estimating the relationship between the dose of a substance(s) and the incidence of an adverse health effect.
- *Exposure Assessment*—The evaluation of the types (routes and media), magnitudes, time, and duration of actual or anticipated exposures and of doses, when known; and, when appropriate, the number of persons who are likely to be exposed.
- *Risk Characterization*—The process of estimating the probable incidence of an adverse health effect to humans under various conditions of exposure, including a description of the uncertainties involved.

Risk characterization is dependent on information derived from hazard identification, exposure, and dose-response assessments. These three elements combined are often referred to as quantitative risk assessment (QRA). The hazard identification element has also been frequently referred to as qualitative risk assessment. Approaches for addressing each of the components of the risk assessment process for occupational and environmental health and safety hazards are described in the following sections.

Hazard Identification

The determination of whether exposure to a chemical, biologic or physical agent, psychologic stressor or other hazard poses a risk to human health is generally a complex process. In the past, occupational hazards have largely been identified through direct observations of ad-

verse events such as scrotal cancer among chimney sweeps (Pott, 1775) and liver cancer among workers exposed to vinyl chloride (Waxweiler et al., 1976). Ideally, potential hazards should be identified before agents are introduced into the environment or workplace through the use of experimental tests such as acute toxicity, mutagenicity, carcinogenicity, and adverse reproductive effects for chemical exposures. Unfortunately, thorough testing has not been performed for many of the hazards currently found in the environment and workplace.

Dose-Response Assessment

Perhaps the most controversial component in the QRA process is the selection of a mathematical model that quantitatively describes the relationship between the dose (or exposure) and the probability of an adverse health effect. Methods for dose-response modeling may be broadly divided into those that have been used for carcinogenic and noncarcinogenic hazards. They may be further divided into biologically based and empirically (or statistically) based modeling approaches. At this time, biologically based models have been solely developed for estimating risks associated with exposures to carcinogens. The most widely used biologically based models have been based upon the Armitage and Doll multistage theory of carcinogenesis (Armitage & Doll, 1961). This theory suggests that for a cell to become cancerous it must progress through a number of irreversible changes (or stages) that must take place in a certain order. It is also assumed that each stage may be represented by a Poisson process, and that dose rate would be linearly related to the transition rates governing one or more of the stages.

Several mathematical formulations of the multistage model have been developed including models in which the tumor response is treated as a quantal (all or none) response (Guess & Crump, 1976) or as a time-to-tumor response (Crump & Howe, 1984). A "linearized" procedure based on the quantal multistage model has been used extensively by the EPA (Anderson, 1983). In this procedure, risk estimates are derived from the upper 95% confidence interval of the linear parameter (q^{1*}) of the multistage model. The linearized multistage procedure was designed to provide conservative (i.e., protective of public health) estimates of risk.

"Two-stage" models have been proposed as an alternative to the multistage model for carcinogenic risk assessments (Moolgavkar & Knudson, 1981). These models attempt to incorporate information on the kinetics of cell growth and differentiation in addition to allowing for two mutational events. Exposure rate may be modeled as having an effect on either the rate of the mutational events, or on the rate of normal cell growth, or on the rate of proliferation of initiated cells. While these models offer the potential for incorporating additional biologic data (and hence improvement in the risk-assessment process), their use requires additional information that may not be available (e.g., growth rates of normal and initiated cells).

Finally, empirical (i.e., statistical) models may be used for modeling dose-response relationships for both cancer and noncancer health effects. While these models are descriptive in nature and may lack a biologic basis, they are sufficiently flexible to allow for a wide range of possible dose-response curves. The Cox proportional hazards (Cox, 1972) and Poisson regression (Frome et al., 1973) are empirical models that have been frequently used for the analysis of epidemiologic mortality studies. The probit and logit models are empirical models that have been used for modeling noncancer health effects from toxicologic studies. The shape of the probit and logit models is sigmoidal and the risk estimates derived from them drop off rapidly at low doses. These models have been referred to as tolerance distribution models, and may be viewed as consistent with the existence of a distribution of individual thresholds in the population (Van Ryzin & Rai, 1980).

The EPA (1996) has recently proposed an alternative method for dose-response evaluation for carcinogens in which a best-fitting empirical model is fitted to the data and used to estimate the effect level corresponding to a 10% effect level (ED_{10}) or a lower level of risk if the data permit it. A straight line would be drawn from the ED_{10} to the origin to estimate the probability of carcinogenic effects when the mode of action of a carcinogen was considered to lead to linearity at low doses. If nonlinearity was assumed, then a "margin of exposure analysis" would be used rather than estimating the probability of effects at low doses. A margin of exposure analysis simply contrasts the level of current exposures to the ED_{10}.

Noncarcinogenic Hazards

The uncertainty (or safety) factor technique is a semiquantitative method that has a long history of use for establishing exposure limits particularly for noncarcinogens (Dourson & Stara, 1983). This method is based on the concept of a threshold level of exposure below which adverse effects are not likely to occur. The threshold level is thought to exist somewhere between the no observed adverse effect level (NOAEL) and the lowest observed adverse effect level (LOAEL) from a toxicologic or epidemiologic study.

The NOAEL or the LOAEL is divided by uncertainty factors to derive a "safe" level of exposure. To account for human intraspecies variability to the adverse effects of a chemical, an uncertainty factor of 10 is generally used to derive exposure limits based on valid studies of chronic exposure. An uncertainty factor of 100 is generally used to derive exposure limits from studies of chronic exposure of experimental animals to account for uncertainties related to both intraspecies and interspecies variability. An additional uncertainty factor of up to 10 may be applied if the data are from studies of subchronic exposures in animals or humans, or if the exposure limit is based on a LOAEL rather than NOAEL to reflect the uncertainties introduced by these factors. Finally, a modifying factor (MF) of ≤10 may be applied based on professional judgment that scientific uncertainties exist in the database that are not reflected by any of the other uncertainty factors described above.

The benchmark dose (BD) has been proposed as an alternative to the uncertainty factor approach for deriving exposure limits (Gaylor, 1983; Crump, 1984). The BD is defined as the 95% lower confidence interval on the exposure producing some low level of risk such as 0.01 or 0.1 derived from a statistical model. Unlike the uncertainty factor approach, the BD is appropriately influenced by both the sample size and the shape of the exposure-response curve. It has been proposed that uncertainty factors may also be applied to the BD to yield a desired level of low risk (Gaylor, 1983). In this respect, the BD approach may be thought of as equivalent to linear low-dose extrapolation from the bottom range of the data. The EPA (1995) has developed proposed guidelines for the evaluation of neurologic effects that utilize the benchmark approach in addition to the uncertainty factor method.

Exposure Assessment

The level of exposure to a hazard is obviously a crucial determinant of risk. No risk exists in the absence of exposure. In order to fully characterize risk, information is needed on the distribution of exposures in the target population. Such information is unfortunately often lacking or inadequate for environmental hazards and particularly for occupational hazards. In the past, in order to overcome this lack of data, risk assessors have been forced to make assumptions about both the level and duration of exposure to a hazard. In order to err on the side of protecting the public's health, government agencies have frequently used worst-case scenarios to estimate exposures for risk assessments. For example, the EPA might estimate exposures to air pollution for an individual who resided for a lifetime on the fence of an industrial facility. OSHA frequently estimates risk for workers who are exposed for a "working lifetime" (e.g., 45 years) at the current or proposed regulatory standard.

Increasingly, risk assessors have attempted to make more realistic assumptions about exposures and to evaluate the distribution of exposures in the population rather than relying on worst-case exposure scenarios. Monte Carlo methods (e.g., see Finley & Paustenbach, 1994) have been recently introduced for developing more realistic estimates of the distribution of exposures in environmental settings.

Risk Characterization

Risk characterization is the final and perhaps the most crucial step of the risk-assessment process. The characterization of risk essentially involves providing the risk managers and public with the necessary and appropriate information for making informed decisions.

Risk characterization was the subject of a recent report produced by a panel convened by the National Research Council (NRC, 1996). This report offered the following new definition of risk characterization:

> Risk characterization is the synthesis and summary of information about a potentially hazardous situation that addresses the needs and interests of decision makers and of interested and affected parties. Risk characterization is a prelude to decision making and depends on an iterative, analytic-deliberative process.

It is important to recognize that a complete risk characterization should include, in addition to the risk estimates themselves, a clear description of the assumptions made in the analysis and the uncertainties underlying the risk estimates. When possible uncertainty analyses should be conducted to estimate the potential impact of assumptions on the risk estimates.

Conclusion

This chapter has briefly summarized current concepts and practices used in assessing occupational and environmental health risks. It should be recognized that risk assessment is a new and evolving science. Methods for conducting risk assessment are still in an embryonic stage and undergoing rapid development. Extrapolations that must be made from high to low doses and from animal to man introduce considerable uncertainties into the risk-assessment process. These uncertainties need to be clearly identified and discussed as a part of any well conducted risk assessment. Despite these uncertainties, risk-assessment methods provide a coherent and scientific basis for developing environmental and occupational health standards.

As a recent NAS review of risk-assessment methodologies emphasized (NAS, 1996), it is probably best to view risk assessment as an iterative process. In this process risk assessments are conducted, needed research is identified and performed to improve the risk assessment, and so forth, until further improvements in the risk assessment would not significantly alter the risk estimates.

References

Anderson, E. L. *Risk Anal.* **1983**, *3*, 277–295.

Armitage, P.; Doll, R. Stochastic Models for Carcinogenesis. In *The Fourth Berkeley Symposium on Math Stat and Prob.*; University of California Press: Berkeley, 1961.

CCERP. *Risk Assessment and Risk Management of Toxic Substances, A Report to the Secretary of Health and Human Services (DHHS) from the Executive Committee DHHS Committee to Coordinate Environmental and Related Programs (CCERP)*; 631–008/244497DC, U.S. Govt. Printing Office: Washington, DC, 1995.

Cox, D. R. *J. Royal Stat. Society*; Series B 34 **1972**, 187–202.

Crump, K. S.; Howe, R. B. *Risk Anal.* **1984**, *4*(3), 163–176.

Crump, K. S. *Fundam. Appl. Toxicol.* **1984**, *4*, 854–871.

Dourson, M. L.; Stara, J. F. *Regul. Toxicol. Pharmacol.* **1983**, *3*, 224–238.

EPA. *The Risk Assessment Guidelines of 1986*; EPA/600/8-87/045, Washington DC, 1987.

EPA. Proposed Guidelines for Neurotoxicity Risk Assessment. *Fed. Regist.* **1995**, *60*, (192), 52031–52056.

EPA. Proposed Guidelines for Carcinogen Risk Assessment. *Fed. Regist.* **1996**, *61* (79), 17959–18001.

Finley, B.; Paustenbach, D. *Risk Anal.* **1994**, *14*(1), 53–73.

Frome, E. L; Kutner, M. K.; Beauchamp, J. J. *J. Am. Stat. Assoc.* **1973**, *106*, 418–432.

Gaylor, D. W. The Use of Safety Factors for Controlling Risks. *J. of Tox. Env. Health* **1983**, *11*, 329–336.

Guess, H.; Crump, K. *Math. Biosci.* **1976**, *32*, 15–36.

Moolgavkar, S. H.; Knudson, A. G. *J. Natl. Cancer Inst.* **1981**, *65*, 559–569.

NAS. *Risk Assessment in the Federal Government: Managing the Process*; National Academy Press: Washington, DC, 1983.

NAS. *Understanding Risk: Informing Decisions in a Democratic Society*; National Academy Press: Washington, DC, 1996.

NAS. *Science and Judgement in Risk Assessment*; National Academy Press: Washington, DC, 1996.

OSTP. Chemical Carcinogens: A Review of the Science and Its Associated Principles. *Fed. Regist.* **1985**, *50*(5), 10372–10442.

Paustenbach, D. J. *The Risk Assessment of Environmental Hazards: A Textbook of Case Studies*; Wiley & Sons: New York, 1989.

Potts, P. Cancer Scroti. In *Chirurgical Observations*; Hawes, Clarke and Collins: London, 1775; pp. 63–68.

Van Ryzin and Rai. In *The Scientific Basis of Toxicity Assessment*, Witeschi, H., Ed.; Elsevier/North Holland Biomedical Press: Amsterdam, 1979; pp. 273–290.

Waxweiler, R. J.; Stringer, W.; Wagoner, J. K. et al. *Ann. NY Acad. Sci.* **1976**, *271*, 40–48.

12

Communicating Risk

RICHARD J. HACKMAN

KIMBERLEE VOLLBRECHT

LESLIE M. YEE

As scientists, engineers and health professionals, we are often expected to communicate information of critical importance to others. Examples include (1) new data on the potential health effects of a manufactured material; (2) the results of air sampling in the workplace for a chemical known to cause sensitization; (3) a report on the presence of cancer-causing materials emitted from a local factory; and (4) a serious medical diagnosis. In each of these situations, the technical professional must communicate to an audience of one or more individuals who will very likely be concerned, anxious, or even angry. The ability to reach these audiences effectively and deliver your information has become a critical skill in the effective management of risks.

While no formal definition of risk communication has been universally accepted, Vince Covello has described this process as "the purposeful exchange of information about the existence, nature, form, severity, or acceptability of risks." We also think of risk communication as the way by which individuals convey information to potentially concerned audiences about topics related to their safety, health, and well-being. Covello, Paul Slovic, and Peter Sandman are considered to be among the foremost leaders in the research of risk communication. Many of the tools offered in this chapter have been developed from the foundation of their teachings.

Risk Communication Principles

Covello has observed that the following three key equations are the basis for effective risk communication:

1. Perceptions = Realities
2. Communication = Skill
3. Goal = Trust and Credibility

The first equation states that perceptions are realities; what is perceived as real is real in its consequences. Fundamental to effective risk communication is in-depth understanding of the knowledge, attitudes and perceptions of the target audience.

The second equation states that communication is a skill. Effective risk communication is not something that comes naturally. Instead, it is a product of knowledge, preparation, training and practice.

The third equation states that the goal of the risk communication effort is first and foremost to establish trust and credibility. Facts will not be accepted if trust has not been built. Education is possible only after trust and credibility have been established. This last equation is the most important of all.[1]

Why is the establishment of trust and credibility the first goal of risk communications? Earning trust and credibility is an essential first step to achieving the overall goal of risk communications, that is, providing the audience with a *true perception of the risk*. This may involve reducing their concern regarding a risk if their concern level is unreasonably high. An example would be reducing the concern of workers removing lead-based paint, by explaining that such work can indeed be *safely* done if specific work practices, personal protective equipment, and appropriate medical monitoring are applied.

Conversely, effective risk communications may also involve *raising* the level of concern regarding risk if the audience's concern level is unreasonably low. An example would be raising the level of concern of an audience for preventing motor vehicle crashes in order to encourage them to more consistently use defensive driving techniques and vehicle safety belts.

Please note that much of the research on risk communication has been performed in North American and Western European societies. While we believe the majority of the tools for effective risk communication will work in most geographies, there are obviously cultural differences that need to be carefully evaluated in every situation.

Steps in Effective Risk Communication

Identify the Audience and Evaluate Their Needs

The first step in the development of a risk communication plan is to identify and characterize your potential audience and evaluate relevant needs. Your audience must be considered broadly. They may be co-workers, community members, consumers, regulatory agency officials, public interest groups, contractors, the media, or anyone else affected by your operations, products, or services.

In order to best evaluate the needs of your audience, you must first understand the background of the audience. From this, you can then work to gain insight into the underlying issues your audience cares about the most. Understanding and meeting your audience's concerns will serve as the basis for the entire communication process.

If the communication is to a single individual or small group, meeting the person(s) involved and establishing a relationship is preferred. When communicating to a larger audience, it is important to know the demographics of the group. From this, you can explore more fully the nature of their concerns. It is also helpful to become familiar with the internal politics of the group and determine the key leaders and influencers.

Understand the Nature of the Communication

The next step in the plan is to understand more fully the nature of the communication. Important questions to ask include:

- Will the communication be to small or large groups?
- Will the communication be to members of your own organization or to those outside?
- Is the information to be communicated good news, bad news, or something in between?
- Is the information expected, or will it come as a surprise?
- Do you have an existing relationship with the audience? If so, has it been good?
- What other concerns will the audience have?
- Will the setting be formal or informal?

It is important to realize that survey after survey has indicated the most trusted individuals for communicating health-related information are independent health professionals (e.g., nurses, physicians, pharmacists, industrial hygienists, and others). University scientists and other neutral professional organizations are also highly respected sources of this information. The one area in which university scientists are generally not considered as trusted is biotechnology (e.g., genetic engineering), where academia is perceived to be part of the problem rather than part of the solution. In general, industry and government representatives are not well trusted. Interestingly, company nurses and physicians are generally trusted; however, the companies for which they work are not trusted. This occurs because nurses and physicians are perceived to have their own professional codes of ethics which supersede any allegiances to their employers.

When considering the nature of the communication, it is also important to consider the perspectives the audience may have already developed. Research performed by Slovic and Fischoff[2] have indicated that the public evaluates risk and perceives a higher level of danger according to the following factors:

Perceived Lower Risk	Perceived Higher Risk
Voluntary	Involuntary
Familiar	Unfamiliar
Controllable	Uncontrollable
Controlled by self	Controlled by others
Fair	Unfair
Not memorable	Memorable
Not dreaded	Dreaded
Chronic	Acute
Diffuse in time and space	Focused in time and space
Not fatal	Fatal
Immediate consequences	Delayed consequences
Natural	Artificial
Exposure to adults	Exposure to children
Detectable	Undetectable

Understanding that audiences generally perceive risk in this way will enable you to better understand the nature of the communication setting you will encounter.

Set Clear Communications Objectives

As you begin to plan your communication strategy, it is critical to establish objectives. In doing this, you determine the results you hope to achieve. Working from these desired results, you can identify barriers, formulate action steps and develop your overall plan to meet them. The objectives you establish become the road map for future actions. These objectives can be varied, but may include:

- audience understanding of key test results
- community support for a project
- audience awareness of a new perspective
- agreement to meet again to resolve certain issues
- no negative publicity
- agreement to proceed with an important new product or process
- alignment on a course of action

If you will be sharing information that is unfavorable, keep in mind that the audience will be judging you for intent. Often, they will be trying to categorize you into one of three scenarios:

1. Rogue scenario—a bad organization does bad deeds
2. Stupid scenario—a good organization unknowingly commits a blunder
3. Tragic scenario—bad things happen to good people and good organizations

As a minimum, one of your communication objectives should be that your audience views your organization as positive in its intentions.

Determine the Message, Messenger, and Setting

Having completed the above steps, the next actions focus on the development of the message, determining who will deliver the message, and where the communication will take place.

The message must first demonstrate the speaker's genuine care for the needs of the audience by directly addressing their concerns. This is crucial in order to develop a relationship of trust and credibility. Dr. Covello has demonstrated that four factors are critical to this process:

1. caring and empathy
2. competence and expertise
3. honesty and openness
4. dedication and commitment

Of these four factors, caring and empathy are believed to be the largest contributors (by a large margin) to the development of trust and credibility. The audience will perceive caring and empathy through a combination of message content, tone of voice, and nonverbal cues such as body language and appearance. The audience will quickly form an opinion as to whether the speaker can be trusted in a relatively short period of time—sometimes as short as 30 seconds. The other three factors are also critical to supporting a relationship built on trust. Development of the message must contain elements of these four concepts.

Communicators must carefully select words and concepts for the message content. For example, words and phrases that communicate competence and expertise within a technical audience, may unintentionally raise concerns or fears in the general public. The technical content of the communication should be adjusted to the familiarity level of the audience. Whenever possible, advance planning for crisis communication should include a review of relevant materials for public perceptions and concerns. Learning about the most effective use of language in communicating with the public should be incorporated into day-to-day communication habits where possible.

The selection of the person(s) who will do the communicating should be given careful consideration. The communicators need to be viewed as credible sources of information by the intended audience. (Ideally, at least one speaker will have established a reputation as a credible spokesperson based on prior relationships with members of the audience.) Female speakers are often viewed as caring and empathetic from the onset, and therefore need to supply more facts and figures to be perceived as competent. Male speakers, however, should emphasize caring, empathy, and dedication rather than competence. Audiences often expect males to be competent but uncaring. In all cases, the speakers must genuinely care about the topic to be communicated, and be responsive to the needs of the audience.

If the topic to be communicated is controversial or the information is likely to be questioned by the audience, the use of an independent expert resource should be considered. Since university scientists are often believed to be more credible and impartial sources of risk-related information, thought should be given to using these or other outside organizations to support or verify the information to be presented. If their conclusions are different from those you have developed, you will very likely need to reconsider your message. Again, where issues and communication needs can be identified ahead of time, research on the availability of locally credible, reassuring spokespeople could prove valuable.

The setting for the communication also needs to be evaluated closely. If the communication is to a single individual or small group, the location for the discussion should be private and free from distractions or interruptions. This will allow meaningful conversation and provide an atmosphere of confidentiality, if needed. If a larger group setting is anticipated, adequate preparation is essential. The meeting room should be open and pleasant. The acoustics should be evaluated so microphones can be arranged both for the speaker and the audience if necessary. All barriers (e.g., podium, tables) between the speaker and the audience should be removed. The seating should be arranged so all attendees feel a part of the discussion. The speaker should not be elevated above the audience on a platform or podium.

Whenever possible, it is recommended that holding more frequent smaller meetings is preferred to holding a few larger meetings. The small-group atmosphere allows for more effective dialogue between the audience and the speaker, and facilitates development of ongoing trusting relationships. A chartpad should also be present so that issues or concerns raised by the audience can be captured for future resolution.

Preparation for Communication

As you prepare for your communication, the following concepts become important:

- Prepare early and begin the communication process with a concerned audience well before a crisis develops. This may even involve sharing work in process or incomplete data, particularly with key local thought leaders. By doing this, you help build a relationship with the audience based on mutual trust and establish a network of credible individuals who can reinforce your good intentions. This also helps to avoid having to communicate surprises or other unanticipated information. Working toward an atmosphere of ongoing and regular communication should be the goal.

- Tap into the informal networks of communication among opinion-formers and other persons who can help you best understand the concerns of the audience. This will allow you to be more sensitive to the needs of the audience and prepare information to meet their needs.

- Practice your presentation before a small, unbiased audience. This will help you become even more familiar with the information to be presented, and may cause likely questions or concerns to surface. It will also give you an indication as to whether your choice of language is likely to enable you to communicate clearly and effectively with the audience.

- Anticipate tough questions. This is especially true if you are communicating information that is controversial or unpopular. Your preparations for the communication session should include developing a list of likely questions as well as "worst-case" questions that may arise from the audience. Plan on having to respond to these questions during the discussions.

Covello and Allen have offered the following "Seven Cardinal Rules for Risk Communication"[3]:

1. Accept and involve the public as a legitimate partner.
2. Plan carefully and evaluate your efforts.
3. Listen to the public's specific concerns.
4. Be honest, frank, and open.
5. Coordinate and collaborate with other credible sources.
6. Meet the needs of the media.
7. Speak clearly and with compassion.

Delivery of Presentation

Having considered the preceding information, you are now ready to deliver your communication. A recommended presentation structure[4] would include:

1. Begin by acknowledging the concerns of the audience and describe your interest in their issues. Offering a personal story or other anecdote that builds common ground with the audience is often a good way to begin. It is important to show you genuinely care about the topic to be discussed. If you do not feel this, someone else more directly associated with the issue should be the communicator. (Caution: do not assume you have a complete understanding of the audience's concerns. Investigate these thoroughly!)

2. Provide a summary statement or conclusion. This should be short and given in positive language.

3. State the supporting facts for your summary. Where possible, cite an independent third-party source to support your conclusion. Try to present at least two other facts that strengthen your position. Illustrating these facts with a story or other narrative may be helpful. Then bridge back to the conclusion.

4. Repeat the summary or conclusion.

5. Indicate what further actions you are going to take to address the audience's concerns—including how you will keep them involved and informed.

Following this general format will allow the audience to hear the information in a logical sequence. Questions, answers, and further dialogue on a particular topic can then move forward.

When communicating with the audience, it is imperative that your nonverbal communication be positive. This means that all barriers between the audience and the speaker(s) have been removed. It also means the speaker(s) make good eye contact with the audience and present themselves in an open, honest manner. The audience will make a judgment as to whether the speaker(s) are trusted and credible both on what is said, and how it is presented.

Other recommendations for effective risk communication include:

- Use language and terms familiar and comfortable to the audience. Avoid technical jargon and limit the use of acronyms.
- State your position in a straightforward, positive way. Do not attack the credibility of organizations or individuals who are unfavorable to your position.
- Make sure risk comparisons are correct, easily understood, and fit the situation being discussed. Risk comparisons should only be used to offer perspective—not as absolute fact. They should be intuitively acceptable to the audience. Examples may include comparison of risk for the same setting but at different times, or comparison to a risk that is more familiar to the audience.
- Provide information on the steps you have accomplished to reduce the risk being discussed. Never speculate or offer worst-case scenarios.
- Discuss the risk communication topic in a serious, honest, caring manner. Never joke about the concerns expressed by the audience.
- Listen to questions intently and offer your best information. If you do not know an answer, indicate this, and promise an answer by a specific time. Deliver on your promise.

Communication Follow-Up

After your risk communication session has occurred, the actions you take in follow-up will be critical. If specific measures have been agreed upon with your audience, it is essential that completion of these measures be prompt. Interim progress reports also need to be provided to reassure your audience you are working diligently on their items of concern. In many situations, follow-up communications/meetings are recommended to help build the relationship with the audience, and to bring closure to the follow-up action items.

The audience from your risk communication session will be watching intently to see that you complete any agreed-upon actions. Implement them swiftly, and you will gain credibility and move a large step forward to building a positive, trusting relationship in the future.

Honesty and Integrity

As you prepare for your risk communication efforts, it is paramount to recognize *your ethical responsibility to communicate the truth*. The recommendations presented here and in other publications on risk communication offer tools to help you present your information more effectively to a concerned audience. You have an ethical obligation to make sure your presentation materials and statements are accurate and correct. You will be viewed by your audience as a spokesperson for your organization. You will be held accountable for the information you provide. It must be true and accurate. The basis for any relationship built on trust is honesty and integrity. The same needs to be true for all your risk communication endeavors.

Conclusions and Key Learnings

The information we have presented here reflects our experiences in communicating with our employees, community members, and consumers. As discussed above, this chapter also reflects many of the principles developed by Covello, Sandman, and Slovic, who are the real pioneers of risk communication. We have learned that no two communication situations are the same. Methods that work well in one setting may not work as well in another. One theme, however, seems consistent: the communicator must have a genuine concern for the affected individuals. They must be willing to investigate the cause of the audience's concerns, and honestly present the information developed. Doing this with empathy and caring will serve as the basis for a successful interaction. This can be a difficult challenge, given the desire of most organizations for a concise and brief presentation of materials.

As we have worked through our risk communication challenges, we have found the following key learnings have served us well:

1. Be honest and sincere in your interactions.
2. Listen to the concerns of the audience and implement action plans to address them.
3. If you will have repeated interactions with the audience, the focus of the initial meetings should be on building the relationship, rather than presenting data.
4. You should be perceived as caring individuals rather than as instruments of a larger organization.
5. Work with small groups whenever possible and encourage their input. Small group settings are

more personal for both the presenter(s) and the audience—thus allowing for better interaction.

6. Overprepare for your presentations and be ready to answer tough questions.

7. Whenever possible, request a list of questions or concerns in advance of the meeting. This will allow you to be better prepared to respond to audience needs.

We hope you find the information in this chapter useful, and we wish you success in your future risk communications.

References

1. Covello, V. T. Risk Communication and Occupational Medicine. *J. Occup. Med.* **1993**, 35(1).
2. Fischoff, B.; Lichtenstein, S.; Slovic, P.; Keeney, D. *Acceptable Risk;* Cambridge University Press: Cambridge, MA, 1981.
3. Covello, V.; Allen, F. *Seven Cardinal Rules of Risk Communication;* U.S. Environmental Protection Agency, Office of Policy Analysis: Washington, DC, 1988.
4. Lum, M.; Tinker, T. *A Primer on Health Risk Communication Principles and Practices;* U.S. Dept. of Health and Human Services, Public Health Service, Agency for Toxic Substances and Disease Registry, Division of Health Education: Atlanta, GA, 1994.

13

Standard Operating Procedures

JAY A. YOUNG

The OSHA regulation, "Occupational Exposure to Hazardous Chemicals in Laboratories," 29 *CFR* 1910.1450, also known as the "Laboratory Standard," requires that the Chemical Hygiene Plan for a laboratory include eight elements that "indicate [the] specific measures that the employer will take to ensure laboratory employee protection." The first-named element is "Standard operating procedures relevant to safety and health considerations to be followed when laboratory work involves the use of hazardous chemicals."

SOPs Defined

A standard operating procedure (SOP) is a written set of steps to be followed by workers performing a routine task. Typically, an SOP details an operation in terms of the steps to be taken by a worker in the order in which they should be accomplished. As is evident from the above, an SOP that meets the requirements of the Laboratory Standard must include safety and health considerations as an integral part of the set of steps to be followed.

Routine tasks that are within the purview of an SOP are not restricted to those procedures that are carried out routinely on a daily basis. Also included are routine tasks such as the emergency shutdown of a laboratory operation, failure of equipment, steps to be taken in the event of a power outage, and the repair of an apparatus or device.

Usually, SOPs are prepared by persons other than the worker who performs the task and thus may manifest a lack of understanding of aspects such as the details of the workers' workspace, the workers' need for access to supplies for the task, or (sometimes) the fact that a worker has only two hands. Although the steps may seem logical to the writer of the SOP, the writer may never have examined the workplace nor observed the task itself. In some instances the writing of the SOP is even carried out with no consultations with the worker who is to perform the task. Even worse, in some instances the task of writing an SOP is done by outside sources, or by part-time help or even interns with no experience at all.

An SOP is a complex document. It must deal not only with the steps required in an operation, but also with the nature of the task, as well as the ability of the worker to accomplish that task.

An SOP should be as much a training document as a tool. The training of the worker should be in accordance with the task at hand. It should reckon with the workers' ability to follow instructions; it should consider the education and knowledge, and the skill and experience, of the worker. It should consider the risks of each step in the task and explain the associated hazards and precautions.

Writing and Revising the SOP Drafts

Although a single person might well write a first, rough, draft outline for an SOP, the following drafts and revisions thereto should be prepared by at least three persons working together: the writer of the rough draft, a worker who anticipates that he or she will be one of those assigned to carry out the procedure, and a safety and health professional. As an initial step in preparing an SOP, the worker (who preferably is familiar with the task or with a similar task) demonstrates the way in which he or she would accomplish one or more steps of the procedure. The safety professional notes the accompanying hazards and the engineering and administrative controls, the protective equipment, and the precautions that should be employed. The writer outlines the steps taken with regard to their efficiency, their fitness within the workspace environment, and the equipment to be used.

Once this study is complete, the three persons can formulate a jointly-written draft SOP by considering the following items[1]:

- each step to be taken and in what order
- the hazards associated with each step
  the precautions to be taken
- the potential harm to the environment
  the procedures to minimize or eliminate such harm
- other factors to be considered
  nature of the workplace
  waste handling
  community interests and concerns

Preparing the Final Draft

Review the SOP draft to be sure that these matters are included:

I. Initial Details
   A. Title (and number) of the SOP. The title should be descriptive, preferably with an "action word" (usually a verb) as the initial word and some reference to doing it safely or properly. For example, Using an IR spectrometer properly, Transferring concentrated acid safely.
   B. Effective date
   C. Revision date (if appropriate)
   D. Author(s)
II. Purpose and Scope of the SOP
   A. State the purpose and scope using specific terms and details. For example, "This SOP has been written for the safe use and handling of solutions of HCN in glacial acetic acid in hood number 6 in the R and D laboratory."
   B. Describe the specific operations covered by this document.
   C. Specifically identify the workers covered by this document.
III. Description of the Task
   A. Location(s); where the work is to be done.
   B. Number of people required.
   C. Skill levels of each person or group of persons.
   D. Equipment required. Identify each item that is needed. Thus, do not state "glass blowing equipment"; rather, list each piece: "torch and tubing," "oxygen cylinder and regulator," "graphite rod," "tungsten probe," and so on.
   E. Safety considerations. Warnings and precautions should be prominently displayed, for example, in bold print or in italic font. Do not rely on differently colored ink; some workers may be color-blind. It may be best to place all such notices together at the beginning of the SOP or it may be thought best to spread them out, placing them at those places in the SOP where they apply, or better yet to do both. Whatever the choice that is made, all SOPs should be consistent. The wording should be brief, emphatic, to the point, stating plainly the nature of the hazard(s) and the precautionary measure(s) to be followed.

      Suitable examples of hazard and precautionary wording can be found on the labels for hazardous chemicals. Be wary of the use of icons only with no words associated as safety or warning symbols; experience has shown that such nonword symbolic "icons" are often and easily misinterpreted. The ancient Egyptians used icons that became unintelligible, even to the Egyptians themselves, in a short time. If icons with or without words must be used, their use and meaning should be consistent in all SOPs, all signage, all posters, all painted surfaces, wherever they appear throughout the employer's establishment, without exception.
   F. Personal Protective Equipment required. Again, specify each item.
   G. Emergency equipment and action required. In addition to the list of each equipment item, including spill cleanup materials and hazardous waste containers, be sure all persons know whom to call and where to go if evacuation is necessary.
   H. Finished product or result expected.
IV. Definitions of All Terms and Acronyms, Specifica-

tions of Pertinent Regulations. A list may be inevitably lengthy; if so, arrange in logical order, not necessarily alphabetically, and position the list in an easily locatable place, for example, at the beginning or end of the SOP or on a uniquely colored or otherwise remarkable page.

V. Training

    A. A description of the training to be received by all employees who are effected by the SOP.

        1. List the behavioral requirements. For example, "The employee should be able to *calculate* the concentration of _____." "... to *operate* the P____ E____ model no. XYZ gas chromatograph." "... to *repair* the carnsnort *using* the spangler and neswang." That is, avoid phrases such as "The employee should know how to _____" or "... should be aware of _____."; instead, describe the behavior that the employee would exhibit if he or she did know how to _____ or was aware of _____.

        2. Describe how the effectiveness of the training is to be evaluated, for example, by *observing* the behavior of the trainee when carrying out the procedure; does the behavior show that the employee does (or does not) know ..., is (or is not) aware of.... Also, evaluate the training by asking questions such as "Why not do it this way?" "What would happen if ...?" "How can you be sure that ...?"

    B. The recommended frequency of refresher training, for example, on a yearly or other basis.

VI. Length

    A. Is the SOP too long, or too brief? If judged lengthy, can or should it be divided into two or more SOPs? Or, should it be subdivided into parts such as an "Introduction," "Setting up," "Preparing the _____," "Finishing steps," and "Wrap-up." If it appears to be too short, does it cover the topic completely? Does it fit the scope that is stated in the SOP?

    B. Is it clearly written, understandable? Is it suitable for the audience of workers who will follow its guidelines?

Revising an Accepted SOP

SOPs are not written in stone; they should be revised whenever the need arises. Such a case might be when the educational or experience background of the workers changes sufficiently to make a revision desirable, for example, if newly employed workers are more familiar with a non-English language. Or, when new information suggests that there is a different, and better, way to proceed, for example, when a new analytical procedure becomes available.

Certainly the revision of an SOP already in use is mandatory when the investigation of an accident indicates that the cause involves a step in the SOP or a misunderstanding of the wording of a sentence or phrase in an SOP.

An SOP should also be revised when, after a close call, an almost-accident, the investigation shows that the near-accident could have been avoided if the SOP had been differently worded, or the steps performed in a different order.

And, an SOP should never even be used if, when given to a worker to carry out for the first time, it is evident that the worker does not understand what is to be done. Such an SOP should be not only be revised; one should start again, from scratch.

Reference

1. Kingsley, W. K. *Chem. Health and Safety* **1998**, 5(4), 28–31.

14

Occupational Exposure Limits

EDWARD V. SARGENT
BRUCE D. NAUMANN

The earliest information on the health effects of chemical exposures came from observational evaluations of working conditions, illness investigations, acute animal toxicity experiments, and crude clinical studies. For many of these investigations exposure data were lacking. Advances in the science and use of risk assessment and toxicology, coupled with improvements in industrial hygiene sampling and analysis, have provided a better understanding of the health effects of chemical exposure.

The essential cornerstone of any occupational health and safety program is the establishment of workplace or occupational exposure limits (OELs). Over the last 100 years, a decline in chemical exposures in the workplace has paralleled the development and use of standards for acceptable concentrations of contaminants in workplace air. The availability of occupational exposure limits has allowed industrial hygienists to identify and correct unacceptable exposures.

The purpose of this chapter is to develop an understanding of the importance of, and basis for, occupational exposure limits. Current sources of occupational exposure limits will be reviewed. Since occupational exposure limits are not available for the majority of chemicals handled in the workplace, particularly in the laboratory setting, techniques for deriving occupational exposure limits and classifying materials into hazard/safe-handling categories will be outlined.

Current Workplace Exposure Limits

ACGIH TLVs

The American Conference of Governmental Industrial Hygienists (ACGIH) has been instrumental in developing occupational health standards for the workplace and has provided air quality threshold limits for substances since 1950. Originally, the limits were termed maximum allowable concentrations (MACs), which were defined as the amount of airborne contaminant in the form of gas, vapor, fumes, or dust that can be tolerated by man with no bodily discomfort or impairment of bodily functions, either immediately or after years of exposure. After a few years MACs were renamed, and are still called Threshold Limit Values, or TLVs. This change in names from MAC to TLV reflected the evolution in the philosophy of airborne exposure limits. MACs are based on the premise that workers should not be subjected to any additional toxicity above that level, whereas threshold limits are an average allowable air concentration.

All occupational exposure limits including Threshold Limit Values are intended for use in the practice of industrial hygiene. Without them, the industrial hygienist is not easily able to evaluate and control potential health hazards present in the workplace. To assist the industrial hygienist in this responsibility and to serve as a convenient reference, the ACGIH publishes *Threshold Limit Values for Chemical Substances and Physical Agents and Biological Exposure Indices*,[1] commonly called the *TLV Booklet*. The *TLV Booklet* is revised annually and is regarded as an industry standard.

Threshold Limit Values have been further defined and separated into three different categories: a time-weighted average (TLV-TWA), a short-term exposure level (TLV-STEL), and a ceiling concentration (TLV-C). The time-weighted average concentration is appropriate for a normal 8-hour workday and 40-hour workweek and is an airborne level to which nearly all workers may be repeatedly exposed, day after day, without adverse effect. The short-term exposure limit is the average concentration to which workers can be exposed continuously for a short period of time without suffering from irritation, irreversible tissue damage, or narcosis of sufficient degree to increase the likelihood of accidental injury, impair self-rescue, or materially reduce work efficiency. It is intended as a supplement to the TLV-TWA where there are recognized acute effects in addition to toxic effects of a chronic nature, although STELS can be assigned to TLVs without 8-hour TWAs. The STEL is defined as a 15-minute time-weighted average exposure that should not be exceeded at any time during a workday, even if the 8-hour TWA is below the TLV. Exposures at the STEL should not be longer than 15 minutes and should not be repeated more than four times per day. There should be at least 60 minutes between successive exposures to allow for recovery from effect. The ceiling limit is a concentration that should not be exceeded during any part of the work day. The definition of the TLV-C is consistent with the original maximum allowable concentration.

A wide variety of reported adverse effects have been listed in the documentation of TLVs.[2] In an effort to provide the industrial hygienist with as much usable information as possible on each substance, the ACGIH summarizes adverse effects due to exposure and the basis for establishing the exposure limit. By far, the most frequent adverse effect which serves as the basis for a TLV is irritation (Table 14.1). Acute or subchronic organ-specific (systemic) toxicity accounts for the next largest percentage of compounds, with an approximately equal amount of data derived from animal and human studies. A large number of compounds have TLVs based on neurotoxicity; however, many are cholinesterase-inhibiting insecticides. Relatively few TLVs are based on sensitization, carcinogencity, or reproductive toxicity.

Threshold limits are based on the concept that exposure to chemicals results in a dose-dependent response. For most chemicals a no-observed-effect level (NOEL) can be derived from available studies. TLVs do not, however, provide universal protection to all individuals who might be exposed to a compound. Employees who might be considered more sensitive include individuals with hypersensitivities, idiosyncratic reactions, and pregnancy. Often such cases are dealt with by using administrative controls. Special consideration is also given to dermally absorbed compounds, mixtures, and unusual work schedules. A "skin notation" is reserved for compounds with potential significant contribution to exposure by the cutaneous route.

The prevailing opinion on mutagens and genotoxic carcinogens is that thresholds do not exist. The TLVs for these compounds are often based on an endpoint other than carcinogenicity. However, the TLV Committee further classifies carcinogens into several categories: A-1 for Human Carcinogens, A-2 for Suspect Human Carcinogens, A-3 for Animal Carcinogens, A-4 for Not Classifiable as a Human Carcinogen, and A-5 for Not Suspected as a Human Carcinogen. For many chemicals identified as carcinogens, it is recommended that exposures be maintained as low as reasonably achievable, even though it is assigned a TLV.

OSHA Permissible Exposure Limits

With the passage of the Occupational Safety and Health Act of 1970, the Occupational Safety and Health Administration (OSHA) and the National Institute of Occupational Safety and Health (NIOSH) were formed. In order to enforce air standards within the workplace, OSHA adopted the 1968 ACGIH TLV List and the American National Standards Institute (ANSI) air standards to promulgate the first official Permissible Exposure Limits (PELs) which were legally binding. Unfortunately, OSHA is not able to revise the list without separate rulemaking and the list has become outdated. NIOSH, in an effort to update the PELs, developed criteria documents and recommended exposure limits (RELs) for some 70 substances. While OSHA has not adopted the RELs into law, they continue to provide a comprehensive analysis of occupational hazards.

On January 19, 1989, OSHA published a proposal to revise its existing standards with the Air Contaminants Standard, 29 *CFR* 1910.1000. The amendments lowered PELs for 212 compounds, established PELs for 164 substances not currently regulated, and left the remaining PELs unchanged. This was done primarily by adopting the 1987–1988 ACGIH TLVs, although several NIOSH RELs were also included. Unfortunately the air contaminants rule was vacated by the courts, leaving the PELs promulgated in 1970 in force. OSHA continues work on establishing mechanisms to update PELs on a more frequent basis.

TABLE 14.1 Basis for ACGIH TLVs

Basis for TLV	Number of Compounds
Irritation	404
CNS effects	160
Respiratory effects	117
Liver effects	115
Blood effects	85
Kidney effects	62
Skin effects	50
Cancer	49
Sensitization	37
Cholinergic effects	33
Cardiovascular effects	32
Anoxia (cellular)	31
Reproductive effects	30
Ocular effects	22
Pneumoconiosis	20
Asphyxiation	17
Nervous system effects	16
Gastrointestinal effects	15
Pulmonary function changes	12
Bone effects	7
Metal fume fever	5
Metabolic disorders	4
Thyroid effects	3
Hearing loss	2

Adapted from the 2000 TLV CD-ROM, "TLVs and Other Occupational Exposure Values." (Personal Communication: Kim Stewart, ACGIH.)

AIHA-WEELs

The American Industrial Hygiene Association established the Workplace Environmental Exposure Level (WEEL) Committee in 1976 to establish occupational exposure limits for chemical agents for which no alternative exposure guide exists, for example, for materials that do not already have a TLV or PEL. As of 2000, 93 WEEL Guides have been published.[3] These documents and a WEEL booklet, similar to the publications prepared by ACGIH for TLVs, are published by AIHA.

Worldwide Occupational Exposure Limits

Many countries throughout the world have established occupational exposure limit programs.[4] In some instances OELs are not given any legal status, while in others they are promulgated and enforced in legislation. For the most part, the OELs adopted are based on the TLV concept developed by ACGIH although in some countries MACs are still used. In many countries the ACGIH TLVs were adopted directly and in their entirety; in others only certain TLVs were selected based on nationwide application. In still others, TLVs were adjusted for country-wide differences in workweek duration, such as the 48-hour workweek in Argentina.

Published OELs, whether TLVs, PELs, WEELs, or other limits, are typically only available for relatively common laboratory reagents and solvents and commercial products. However, there is a need to provide guidance on how to ensure that exposures to all chemicals handled in laboratories, pilot plant operations, and manufacturing areas, including isolated synthetic intermediates, are maintained below acceptable levels. The following sections discuss methods for determining acceptable levels of exposure for chemicals and the practices required to achieve these levels in the workplace.

Methods for Establishing Occupational Exposure Limits

OELs have been established for only 1700 of the nearly 100,000 chemicals estimated to be in commerce; thus, it is often necessary to establish internal OELs for compounds handled in the workplace. Exposure limits may not have been established because there is (1) a low perception of hazard for the chemical, (2) lack of toxicity data on which to base an exposure limit, or (3) a limited number of exposed employees. Each of these reasons is applicable to the handling of chemicals in the laboratory. Often chemicals are handled in relatively small quantities by more highly trained individuals under tighter engineering controls. Also, compounds under development are more likely handled in the laboratory when available toxicity data on which to base an exposure limit are limited.

The most common method of determining an OEL is to identify a threshold dose (no-effect level) and apply an uncertainty or safety factor to it. The concept of exposure limits is based on the principle that exposure to a chemical agent may be permitted up to some tolerance limit greater than zero. This assumes a nonlinear dose response relationship and allows for the estimation of a threshold dose, usually expressed as the no-observed-adverse-effect level (NOAEL).

In order to facilitate the derivation of a numerical occupational exposure limit (OEL), the equation shown below can be used:[5,6]

$$\text{OEL (mg/m}^3) = \frac{\text{NOAEL (mg/kg)} \times \text{BW (kg)}}{\text{UF} \times \text{MF} \times V \text{ (m}^3) \times \alpha \times S}$$

where the NOAEL is corrected for body weight (BW) and divided by a composite uncertainty factor (UF), a modifying factor (MF) which accounts for professional judgment, and the volume air breathed in an 8-hour work day (10 m^3). The OEL may also be adjusted for bioavailability (α) and for compounds not rapidly cleared from the body that result in higher steady state (S) plasma concentrations with repeated exposures compared to single exposures.

Uncertainty or safety factors are applied to no-effect levels in order to ensure that the OEL is adequately protective of the exposed population. UFs typically range from 1 to 1000 depending on how well defined the no-effect level is, the seriousness of the endpoint, and the degree of variability in the response.[7] Recent attempts have been made to more accurately quantify each source of uncertainty and provide the scientific rationale for "suggested" values to use for each uncertainty factor.[6,8] In some cases it may also be necessary to estimate a NOAEL from a lowest-observed adverse effect level (LOAEL) by applying an additional adjustment factor. Depending on the endpoint, size of dose groups, spacing of dose levels, and slope of the dose response curve, the LOAEL-to-NOAEL adjustment factor should almost always be less than five in a well-designed study[9] and frequently less than three.[6,7,10] Additional uncertainty factors are applied when adjusting for use of subchronic animal data rather than chronic data as the basis for the OEL. Available data indicate that an adjustment factor significantly less than 10 would also be adequate for subchronic to chronic extrapolation.[6,7,10–13]

Data are often used from experiments using non-occupationally relevant routes of exposure to establish OELs. Workers are generally exposed via the inhalation and dermal routes, and to some lesser extent, from the oral route. If the amount of deposition and absorption of a compound in the respiratory tract is unknown it is conservatively assumed to be 100%. It is possible to adjust the OEL to account for differences in bioavailability when they can be quantified.

Professional judgment often plays an important role in making adjustments to default uncertainty factors. A modifying factor can be used to adjust OELs in order to account for (1) the complexity of information used to derive OELs, (2) the number of assumptions and sources of uncertainty inherent to the OEL setting process, (3) the lack of a prescriptive process for establishing OELs, and (4) the myriad work environment factors that directly affect the frequency and duration for exposure.

Use of Data in Establishing Exposure Limits

The availability of data for deriving an OEL will vary greatly from substance to substance. In instances where the compound is of commercial interest and enough compound exists, sufficient animal testing data may be obtained on which to base an OEL. This is particularly true for laboratory chemicals. For some compounds this may not be possible although there are several approaches that can be used to estimate an OEL and define safe handling practices. One approach is to establish whether qualitative or quantitative/activity relationships exist with other, better studied compounds. This assumes that the homologous, unstudied chemical causes the same effect as a reference chemical for which there are sufficient data to establish an OEL. This method was used by the ACGIH to establish TLVs for a homologous series of alkylamines.

Another approach is to estimate an OEL conservatively on the basis of limited data. For many compounds, only data on physical properties and acute toxicity are available. Physical properties can influence the potential for exposure and subsequent effects by determining the possibility for entry, mode of entry, reactivity with tissue, and degree of absorption, distribution, and elimination, but do not allow one to readily predict all types of toxicity. Acute lethality data are not the best data to use in setting exposure limits, but do permit an evaluation of the relative toxicity of chemicals. However, since these tests are required in order to determine shipping classification, they are often the only toxicity data available. Information on irritation potential can be very useful in setting exposure limits because this often represents the most sensitive endpoint. An animal test for evaluating sensory irritation has shown good correlation with TLVs and is predictive of what would be acceptable in humans.[14]

Sensitization is not a practical endpoint on which to base exposure limits. This is because animal sensitization studies are not always predictive of potential human effects, particularly in defining the dose-response kinetics necessary for setting exposure limits. Thresholds for both induction and elicitation need to be determined and both are difficult to define in standard sensitization assays. Since sensitizers and allergens usually affect only a portion of an exposed population, it is often practical to remove sensitized workers from the workplace or minimize exposure by use of engineering controls or personal protection.

Long-term or lifetime studies are the best source of data to establish exposure limits, particularly if the exposure limits are for protection from chronic effects of chemical exposure. The most useful animal tests are repeat-dose studies in which a NOAEL or LOAEL is identified. It is likely that such studies are performed using a non-occupational route of exposure such as gavage. If so, the results will have to be extrapolated to the inhalation route by quantifying differences in bioavailability in order to be relevant. Chronic studies are not available for the majority of chemicals in commerce. Due to cost and time constraints, it is unlikely that such test results will be available for noncommercially viable chemicals, such as isolated intermediates and low-volume laboratory reagents. In such cases, NOAELs or LOAELs from shorter-term studies, such as 2-week, 14-week, or 6-month studies, may be used. Mutagenicity data are often obtained early in the development of a chemical product. The Ames microbial mutagenesis assay is often one of the first tests performed on a newly developed chemical. For certain types of chemicals, such as new pharmaceuticals, a positive Ames test may preclude development. Other *in vitro* tests including mammalian cell gene mutation assays, sister-chromatid exchange, unscheduled DNA synthesis, DNA alkaline elution, and *in vitro* cytogenetics studies have also been widely used. Most mutagenicity tests are sensitive and predictive of carcinogenic potential. Tests are often combined using a tiered scheme in order to increase their predictability using a weight-of-evidence evaluation. Mutagenicity data without further studies to determine *in vivo* biological relevance, that is, carcinogenicity or reproductive toxicity, cannot be used to quantify risk because dose-response information is lacking. Exposure limits, therefore, cannot be determined solely from these data.

Performance-Based Occupational Exposure Limits

For some compounds it may be impossible to establish a numerical exposure limit. Compounds handled early in

TABLE 14.2 Performance-Based Category Enrollment Criteria

Enrollment Criteria	Performance-Based Category				
	1	2	3	4	5
Potency (mg/day)	>100	>10-100	0.1-10	<0.1	<0.1
Severity of acute (life-threatening) effects	low	low/mod	moderate	mod/high	high
Acute warning symptoms	good	fair	fair/poor	poor	none
Onset of warning symptoms	immediate	immediate	may be delayed	delayed	none
Medically treatable	yes	yes	yes	yes	yes/no
Need for medical intervention	not required	not required	may be required	may be required immediately	required immediately
Acute toxicity	slightly toxic	moderately toxic	highly toxic	extremely toxic	super toxic
Sensitization	not a sensitizer	mild sensitizer	moderate sensitizer	strong sensitizer	extreme sensitizer
Likelihood of chronic effects (e.g., cancer, repro, systemic)	unlikely	unlikely	possible	probable	known
Severity of chronic (life-shortening) effects	none	none	slight	moderate	severe
Cumulative effects	none	none	low	moderate	high
Reversibility	reversible	reversible	may not be reversible	may not be reversible	irreversible
Alteration of quality of life (disability)	no	no	yes/no	yes	yes

Reprinted with permission from Reference 15. Copyright, 1996, *Am. Ind. Hyg. Assoc. J.*

the research and development process often have not been tested but yet are being handled in the workplace. Isolated intermediates in a multistep chemical synthesis may also be handled without the benefit of toxicity data. In the pharmaceutical industry, potent receptor-mediated drugs have been developed for which it is not possible to identify a no-effect level on which to base an OEL. For other classes of compounds such as carcinogens, mutagens, and sensitizers, establishment of a numerical limit can be equally difficult.

The inability to quantify numerical limits has led to the development of alternative methods for controlling exposures to acceptable levels. A performance-based approach for chemicals was developed from the philosophy and practices used by research facilities that handle pathogenic microorganisms.[15] For human pathogens, control strategies developed by the Centers for Disease Control and Prevention (CDC) are specified for each of four biosafety levels that apply to organisms of increasing pathogenicity.[16] In a similar fashion, chemicals can be assigned to one of five hazard categories that correspond to specific strategies known to control exposures to acceptable levels (for that class of chemicals) based on previous experience. Assignments are made following a thorough review of all available data and consideration of a number of enrollment criteria. One example of a series of performance-based enrollment criteria and subsequent categories is shown in Table 14.2. In this scheme, the enrollment criteria include consideration of

not only the types of toxicity (e.g., cancer, sensitization), but also the potency, severity, reversibility, and the need for medical treatment.

Enrollment criteria for performance-based control categories corresponds to the definitions for particularly and extremely hazardous chemicals found in OSHA's Laboratory Standard and nonmandatory appendices (24 CFR 1910.1450). In a performance-based system, each of the categories correspond to predefined strategies which provide the necessary degree of control to protect employees and the environment. As shown in Table 14.3, these containment strategies can range from conventional good manufacturing or laboratory practices to

TABLE 14.3 Performance-Based Containment Categories

Performance-Based Category	Containment Level
1	Good laboratory or manufacturing practice
2	Good laboratory or manufacturing practice with more stringent controls
3	Essentially no open handling (closed systems recommended)
4	No open handling (closed systems required)
5	No manual operations/human intervention (robotics/remote operations recommended)

Reprinted with permission from Reference 15. Copyright, 1996, *Am. Ind. Hyg. Assoc. J.*

"no open handling" and the use of isolation technologies and robotics. To further implement performance-based exposure limits, unit operations and design matrices need to be developed to translate these assignments into specific recommendations and standard operating procedures that, when followed effectively, achieve the necessary level of control. Matrices should be developed for all unit operations involving the chemical in question.[15] Matrices for general design concepts and for laboratory operations, including research and quality control laboratories, can also be developed. These control strategies are consistent with OSHA's recommendations on specifying designated areas, use of containment devices, and special procedures for particularly hazardous materials (i.e., highly toxic materials, reproductive toxins, and carcinogens).

Documentation and Communication of Exposure Limits

All exposure limits, whether numerical or performance-based, should be documented in the form of a monograph which provides a summary of the underlying data, the rationale for its use in calculating a limit, and all sources of uncertainty. Complete documentation gives line managers and safety and industrial hygiene professionals important information on which to base decisions regarding exposure controls and to communicate them effectively.

When established, occupational exposure limits are required for inclusion in material safety data sheets (MSDSs) prepared for each compound. If a numerical OEL is derived, the hazards identified in the OEL monograph should also be reported in the MSDS in addition to the limit. Since the monograph serves as a mechanism to identify data gaps it should be reviewed on a regular basis and the OEL should be revised to reflect any significant new information. This is particularly important with compounds for which OELs are established early in development. If a performance-based system is used, the performance-based category should be reported and the appropriate elements of the unit operations matrices and SOPs should be incorporated into the appropriate sections of the MSDS.

Establishment of OELs requires close communication between safety professionals, engineers, industrial hygienists, occupational toxicologists, physicians, and line management. To implement these recommendations properly, all stakeholders must have a full understanding of the basis for the OEL so that they can design appropriate control strategies, evaluate handling operations, and communicate accurately to employees in reports and training programs. Industrial hygienists who will be sampling for the presence of the chemical in the workplace must have valid and sensitive analytical methods. Occupational physicians on the other hand need to fully understand the toxicological and pharmacological basis of an OEL in order to evaluate employees medically who may have been exposed to the chemical in question.

References

1. ACGIH. *Threshold Limit Values (TLVs®) for Chemical Substances and Physical Agents and Biological Exposure Indices (BEIs®)*; American Conference of Governmental Industrial Hygienists: Cincinnati, OH, 2000.
2. ACGIH. *Documentation of the Threshold Limit Values and Biological Exposure Indices*; American Conference of Governmental Industrial Hygienists: Cincinnati, OH, 1996.
3. AIHA. *The AIHA 2000 Emergency Response Planning Guidelines and Workplace Environment Exposure Level Guides Handbook*; American Industrial Hygiene Association: Fairfax, VA, 2000.
4. AIHA. *Occupational Exposure Limits-Worldwide*; American Industrial Hygiene Association: Fairfax, VA, 1987.
5. Sargent, E. V.; Kirk, G. D. *Am. Ind. Hyg. Assoc. J.* **1988**, 49, 309–313.
6. Naumann, B. D.; Weideman, P. A. *Hum. Ecol. Risk Assmt.* **1995**, 1, 590–613.
7. Dourson, M. L.; Stara, J. F. *Regul. Toxicol. and Pharmacol.* **1983**, 3, 224–238.
8. Renwick, A. G. *Food Addit. and Contam.* **1993**, 10, 275–305.
9. Weil, C. S. *Toxicol. Appl. Pharmacol.* **1972**, 21, 454–463.
10. Kadry, A. M.; Skowronski, G. A.; Abdel-Rahman, M. S. *J. Toxicol. Env. Health* **1995**, 44, 83–95.
11. Lewis, S. C.; Nessel, C. S. *Toxicologist* **1994**, 14, 401.
12. Weil, C. S.; McCollister, D. D. *Agr. Food Chem.* **1963**, 11, 86–491.
13. Woutersen, R. A.; Til, H. P; Feron, V. J. *J. Appl. Toxicol.* **1984**, 4, 277–280.
14. Alarie, Y.; Neilson, G. D. *Toxicol. Appl. Pharmacol.* **1982**, 65, 459–477.
15. Naumann, B. D.; Sargent, E. V.; Starkman, B. S.; Fraser, W. J.; Becker, G. T.; Kirk, G. D. *Am. Ind. Hyg. Assoc. J.* **1996**, 57, 33–42.
16. CDC. *Biosafety in Microbiological and Biomedical Laboratories*, 3rd ed.; Richmond, J. Y., McKinney, R. W., Eds; U.S. Government Printing Office: Washington, DC, 1993.

15

Industrial Hygiene Exposure Assessment—Data Collection and Management

PAUL HEWETT

Exposure to toxic materials always entails some level of risk. This risk reflects both the inherent toxicity of a substance and the frequency, duration, and severity of exposure. Risk management refers to the process of assessing and, if necessary, reducing exposure and therefore risk for exposed individuals. An exposure assessment is an essential component of risk management for determining a course of action. The actual measurement of current exposures to gases, vapors, or particulates may not be required, as there are qualitative and semiquantitative exposure assessments. It is often the case, however, that measurements are necessary for initial or baseline evaluations. Furthermore, periodic sampling and occasional audits are necessary for validating earlier assessments and for evaluating exposure trends in the work environment. Consequently, an industrial hygienist is often faced with questions regarding the collection, analysis, interpretation, and management of occupational exposure data.

The purpose of this chapter is to suggest appropriate questions and provide reasonable answers regarding the *collection* and *management* of occupational exposure data. Chapter 16 (Hewett[1]) covers data *analysis* and *interpretation*. Both chapters presume some familiarity with exposure limits and exposure-measuring instrumentation. Furthermore, quantitative exposure assessment is only one component of a "comprehensive exposure assessment" program.[2,3] Readers should consult the reading list at the end of this chapter for more information regarding comprehensive exposure-assessment programs and broader discussions regarding industrial hygiene, instrumentation, and statistics.

An additional purpose of this chapter is to provide guidance for developing what could be called a philosophy for occupational exposure management. Because exposure-monitoring programs must be designed and tailored for a wide variety of work environments, it is critical that we first adopt reasonably consistent interpretations of the occupational exposure limits and agree on the goal of an effective exposure-monitoring program.

Occupational Exposure Management

Occupational exposure management refers to risk management in the workplace; that is, the process of assessing and controlling risks associated with exposure to toxic chemicals, physical agents (e.g., radiation, heat, noise, vibration), biological agents (e.g., bacteria, fungal spores, and other biologic aerosols), and ergonomic hazards. Exposure management incorporates the traditional industrial hygiene functions of *hazard recognition*, *hazard evaluation*, and *hazard control* (Table 15.1) and requires the coordinated activities of plant management, medical professionals, toxicologists, control technology and process engineers, and safety professionals. However, many of the responsibilities of exposure management are assigned to an industrial hygienist; that is, a professional "qualified by education, training, and experience to anticipate, recognize, evaluate, and develop controls for occupational health risks."[4] The end result of effective exposure management is an adequately controlled "exposure profile"—or distribution of exposures—for each employee.

Hazard Recognition

Hazard recognition is the first step in the process of exposure (risk) management. In principle, it consists of a three-part basic characterization: (a) characterize or describe the work environment; (b) assemble information regarding toxicology and applicable OELs; and (c) define initial or tentative exposure groups. However, exposure management may proceed from the identification of a single, predominant toxic substance, followed by the identification of all exposed employees. Or it might begin by first grouping workers by similarity of process, job or task, area, and controls, and then proceeding to a comprehensive assessment of the work environment, which includes an inventory of all potentially toxic substances, for each exposure group.

TABLE 15.1 Occupational Exposure Management = Risk Management

Risk Management Step	Risk Management Action
Hazard recognition	Characterize or describe the work environment for each group
	Assemble information on toxicology and applicable OELs
	Define initial or tentative exposure groups*
Hazard evaluation	Collect/model exposure data, then . . .
	Analyze exposure data, then . . .
	Interpret exposure data
	Manage the exposure database
Hazard control	Substitute less toxic/hazardous materials, and/or . . .
	Enclose process or worker, and/or . . .
	Install/modify general or local ventilation, and/or . . .
	Modify work practices, and/or . . .
	Implement administrative controls,** and/or . . .
	Require interim personal protective equipment

*The exposure group definitions and potential for new or additional exposures should be reassessed any time there are significant changes in the process, production rate, ventilation controls, assigned tasks, or work practices or when new workers are introduced.

**Many OSHA 6(b) standards forbid the rotation of workers through high-exposure areas.

Workplace Characterization

A basic characterization or description of the workplace is needed for each exposure group.[2,5,6] This characterization should be documented and include (a) a description of the workplace; (b) a description and review of the production processes, work patterns, emission sources, and existing controls (engineering, administrative, and personal protective); (c) a list of the job descriptions and the tasks associated with each job; (d) an inventory of the chemical, physical, biological, and ergonomic hazards associated with each job or task; (e) the number of workers per shift, by job title, and an evaluation of any real or potential differences between shifts; and (f) shift length and recovery time information (necessary for adjusting exposure limits for nontraditional work shifts, discussed below).

Toxicology

Information on health effects can be obtained from material safety data sheets, chemical suppliers, standard references on occupational toxicology, trade or professional organizations, and federal agencies such as OSHA and NIOSH. Applicable federal, authoritative, or corporate OELs should be identified and the relevant OEL documentation reviewed to determine the reasons and rationale for setting the OEL. For example, one should consult the ACGIH TLV documentation when using any ACGIH TLV, or the Federal Register preamble for each OSHA 6(b) Permissible Exposure Limit (PEL).*

Exposure Groups

Exposure groups have gone by several names: "homogeneous risk group,"[7] "exposure zone,"[8] "homogeneous exposure group,"[5,6] "uniform exposure group,"[9,10] and "similar exposure group,"[2] among other terms. In this chapter the term "exposure group" will be used to refer to any logical grouping, based on either observation or any objective methodology, that is expected to result in a reasonable degree of homogeneity with respect to the conditions of exposure (e.g., similarity of process, toxic substance, jobs/tasks, and controls). It is possible for an exposure group to consist of a single employee who is engaged in unique or distinctly different activities.

While the "exposure group" will be our basic unit for aggregating workers, it must be recognized that we are not interested in controlling the average risk in each exposure group; we are interested in controlling the risk for each and every member of the exposure group. The exposure group concept is used simply because most employers lack the resources to routinely monitor the exposures of *each* employee.

Ideally, all exposure groups should be perfectly homogeneous; that is, workers within each exposure group should perform identical tasks using identical work practices and be subject to identical controls for identical periods of time. If this were so, measurements collected from any worker could be used to evaluate the exposure profiles of all workers within the group. However, in reality all exposure groups are heterogeneous with respect to the above factors, but to a greater or lesser degree, depending on the inherent exposure variability for a particular work environment and workforce, and the skills and experience of the industrial hygienist when establishing initial exposure groups.

*OSHA 6(b) PELs are those that have been promulgated since 1970. These are more *complete* standards than the Table Z1, Z2, and Z3 standards in the sense that they include specific requirements for exposure monitoring, medical surveillance, hierarchy of controls, use of personal protective equipment, and so on. These additional requirements, when implemented, further reduce or manage risk. The *Federal Register* preamble justifying a 6(b) PEL should be reviewed. The complete text for many 6(b) PELs can be found on the OSHA Internet home page.

Hazard Evaluation

Once suitable and sufficient background information has been assembled it is necessary to determine if the substance in question represents a *hazard* to the employees, given the conditions of use and the frequency, duration, and severity of exposure. This hazard evaluation, or exposure assessment, comes in three varieties: qualitative, semiquantitative, and quantitative. A quantitative exposure assessment refers to the collection of current exposure measurements and is warranted whenever information regarding exposures is missing or uncertain. Qualitative and semiquantitative exposure assessments are used to determine the need for a quantitative exposure assessment by addressing the question: "Are significant exposures likely to occur under the expected conditions of use?"[11]

A qualitative assessment might involve the determination that the substance in question is present in insignificant quantities or that the operation or process is totally enclosed with an extremely low probability of inadvertent release, even during maintenance activities.* For example, according to the OSHA benzene standard[12] products containing less than 0.1% of benzene are exempt from regulation.

A semiquantitative exposure assessment utilizes "objective" exposure data. For example, in the OSHA cadmium standard, objective data are defined as "information demonstrating that a particular product or material containing cadmium or a specific process, operation, or activity involving cadmium cannot release dust or fumes in concentrations at or above the action level [i.e., half of the Time-Weighted Average (TWA) PEL] even under the worst-case release conditions."[13] Objective data may consist of historical exposure data (previous data from the same work environment and usually not more than one year old), analogous exposure data (data from similar processes or operations), or predictions from exposure modeling[2] (statistical models or physical/chemical models). Any of these exposure assessments, but particularly quantitative exposure assessments, can be divided into four stages: data collection, analysis, interpretation, and management.

Data Collection

A sampling strategy should be devised before actual data collection begins. The sampling strategy indicates the type of survey (baseline survey, surveillance, audit, or other) and the procedure for selecting workers to be monitored.

---

*Routine and periodic maintenance activities often result in high exposures.

Data Analysis

If sufficient data are collected during a baseline evaluation or subsequent reevaluations, then summary statistics and compliance statistics can be calculated. The data should first be evaluated to determine if the lognormal distribution assumption applies. Generally six or more measurements are required before stable statistics can be calculated.[2,5] Data analysis procedures are presented in Chapter 16.

Data Interpretation

A written "decision logic" is necessary for determining if a particular set of exposure data indicates that the work environment is "acceptable" or "unacceptable," or if more information/data are needed. A decision logic may consist of simple decision rules or a formal statistical test. Decision rules are useful when only a limited number of measurements are available. Formal statistical tests are usually applied when sufficient data are available, usually six or more measurements. Either way, the goal is to determine whether or not the "exposure profile" for each employee in the exposure group is acceptable. Data interpretation procedures are presented in Chapter 16.

Data Management

Although mentioned last, the identification of the relevant descriptors of exposure data and the development of a data management system should come early in the hazard evaluation process. There are many potential users of and uses for exposure data.[14,15] Exposure data may be used for determining compliance with existing federal regulations, for assessing the status of existing exposure controls, and later for estimating cumulative exposures in an epidemiological study. The data may used in the future by researchers or the designers of other facilities. There is growing concern that industrial hygienists collect and safeguard for future use not only the exposure measurements, but also comprehensive descriptive information regarding the work environment and workforce. The resulting "occupational exposure databases" could then be used to evaluate the efficacy of different types of controls and provide accurate industrywide exposure data for trade organizations and standards-setting organizations.

Hazard Control

Once it is determined that the exposure profile for an individual worker or exposure group is unacceptable, steps must be taken to reduce exposures. A written "hazard control" plan should be developed, maintained, and continuously updated. If feasible, exposures should be reduced through substitution of less toxic substances, engineering controls (e.g., process enclosure, and local and general ventilation), work practice modification, or, as a last resort, through the use of personal protective devices. Often all that is required is the fine tuning or modification of existing controls. The evaluation of individual work practices and analysis of the task components of a job often leads to ways of substantially reducing exposure. In any case, additional measurements are usually warranted in order to verify the need for additional controls or to evaluate the effectiveness of any intervention. Burton[16] provides excellent overviews of the topic of hazard control.

Comments

Exposure Management Is a Long-Term Responsibility

Exposure management does not end until the substance in question is no longer used. Processes change, controls deteriorate, and new workers are introduced, so there is always a need for periodic reassessments, resampling, and internal audits. Every exposure management program should incorporate a "continuous improvement" concept.[2,17,18] For example, after initial or baseline exposure assessments where the focus is on exposure groups, industrial hygienists then focus on evaluating and controlling exposures during individual tasks. The expectation is that by periodically auditing, evaluating, and controlling task-based exposures, along with periodic evaluations of full-shift exposures, exposure groups tend to become more homogeneous and exposures in general tend to decline. Furthermore, as reviewed by Hewett,[19] most authorities recommend that every overexposure be evaluated to determine if the work environment has deteriorated.

Documentation of the Absence of Exposure

Because employees and local communities are increasingly concerned with emissions of presumptively toxic materials both within a facility and into the general environment, it is often important to accurately document the *absence* of exposure, or the fact that exposures are *minimal*.[2] For such data to be convincing for risk-communication purposes, they must be collected using a thoughtfully designed exposure-monitoring program, similar to those designed for employees known to be significantly exposed.

Biological Monitoring

For many substances Biological Exposure Indices (BEIs) have been developed.[20] These provide an additional means for assessing worker exposure. The documentation for each BEI should be consulted for guidance regarding the comparison and interpretation of biological measurements.[21]

Dermal Absorption

For numerous substances, principally organic and organo-metallic chemicals, skin contact and absorption represents an important route of exposure. Many of the substances in the ACGIH TLV booklet[20] have a "skin" designation indicating that skin absorption can be significant. Because significant skin contact can invalidate a favorable assessment of airborne exposure, the potential for skin contact should be evaluated, along with the potential for exposure by inhalation. (See Ref. 2 for an introduction to assessing exposure by dermal absorption.)

The Role of Judgment and Experience

Because the number of exposure measurements is often too small to permit conclusive determinations, and most work environments are rarely stable (i.e., exposures change due to season changes, controls becoming less effective, and production-level changes), there is always a role for experience and judgment when exposure measurements are interpreted. There is no substitute for a sound knowledge of the process and good observational skills. The employer or representative should correlate observations of employee work practices and knowledge of the process parameters (production rate, substance, rate of use, ventilation) with exposure measurements.

The periodic classification of each work environment as acceptable or unacceptable requires numerous judgments. Judgment is important for defining initial and revised exposure-group definitions, for determining when to resample, for ranking and prioritizing exposure groups for evaluation, for determining if an operation is reasonably stable or dynamic and subject to change, and so on. The accuracy of these judgments can be expected to improve as one gains experience relating actual measurements with observations. The collection and analysis of exposure data should be used to verify or validate one's judgment, or when one is simply uncertain regarding the magnitude of exposures.[2]

There will always be uncertainty in the estimation of an exposure profile for an individual or exposure group, particularly if historical or surrogate exposure data are used. The confidence interval calculations described in

Hewett[1] (Chapter 16) can be used to quantify the uncertainty regarding our estimates of the parameters of this exposure profile. Furthermore, there will always be uncertainty in the level of protection offered by any OEL.[2] For example, legal OELs are often dated and more protective OELs have since been recommended by other organizations. Or an internal or corporate OEL may be based on "no observed adverse effect" data and the level of long-term risk is simply unknown. In these and similar instances one is advised to be conservative (i.e., tend towards overprotection) when interpreting exposure data or adopt an interim or working OEL that is a fraction of the legal OEL.[2]

Occupational Exposure Limits (OELs)

In order to design an ethical and defensible exposure-monitoring program it is necessary to assign a statistical interpretation to an occupational exposure limit (or OEL). In this chapter the various types of OELs are assigned statistical *interpretations*. Each interpretation is designed to be consistent with the *definition* assigned to the OEL by the sponsoring organization (e.g., Occupational Safety and Health Administration [OSHA], American Conference of Governmental Industrial Hygienists [ACGIH], and National Institute for Occupational Safety and Health [NIOSH]). Other interpretations are possible. However, as noted by Roach[22] in 1967, "[it] is important that hygienic standards should not be given widely different interpretations."

Unless less toxic substitutes can be found or processes totally enclosed, an OEL is needed for determining whether a work environment is currently acceptable or unacceptable.[2,5,23] A workplace is judged to be acceptable if the exposure profile for each employee is sufficiently controlled.

Components of an OEL

OELs can be thought of as "legal," "authoritative," "internal," or "working." Legal, or regulatory, OELs are those set and enforced by state or federal agencies, such as OSHA and MSHA. Authoritative OELs are recommended by organizations such as the ACGIH and AIHA, and federal agencies such as NIOSH. Companies often devise internal, or corporate, OELs for substances for which there are no legal or authoritative OELs,[23] or when the legal or authoritative OEL is dated. In the absence of a legal, authoritative, or corporate OEL, the industrial hygienist should devise a working or provisional OEL, to be used until a corporate or other OEL becomes available.[2,23]

Each OEL consists of three components: (1) a concentration, (2) an averaging time, and (3) a target.[19] (Legal OELs often include additional requirements for exposure monitoring, medical monitoring, respiratory protection, and/or exposure controls.) The "concentration" refers to the obvious numerical value and units of the OEL. The "averaging time" refers to the period for which an average exposure is estimated. The appropriate averaging time is set by the originator of the OEL (e.g., OSHA, ACGIH, NIOSH).

Ideally, the target (or focus) of all legal and most authoritative OELs is the *individual worker*. Our goal is to protect *each* individual worker. However, due to limited resources we often focus on exposure groups rather than individuals. Consequently, our *conclusions* regarding the exposure group must be reasonably *predictive* of the exposures experienced by each member of the group.

An OEL of some sort is necessary for the evaluation of data. Since only a relative handful of the tens of thousands of substances and mixtures encountered in industrial operations have OELs, many corporations find themselves compelled by both ethical and liability considerations to develop corporate occupational exposure limits. According to Paustenbach[23] companies should accept three propositions: (1) OELs—legal, authoritative, internal, or working OELs—are needed whenever employees are exposed to toxic agents; (2) the company should fully document the rationale for establishing a corporate OEL, and (3) corporate or provisional OELs should be set even if adequate toxicological and epidemiological data are not available. Once a corporate or provisional OEL is set by the corporate risk assessors, the plant risk manager or industrial hygienist should treat it like any legal or authoritative OEL.

Exposure Measurements and Exposure Profiles

The term "exposure measurement" refers to a single estimate of the average exposure across the averaging time specified by the OEL. For example, a typical exposure measurement is an estimate of the average exposure across a single shift. Such a measurement would be compared to a TWA OEL. However, often the full-shift measurement is itself calculated using several partial-shift measurements. Calculation procedures for estimating full-shift, time-weighted average (TWA) exposures can be found in nearly any industrial hygiene reference and the ACGIH TLV booklet.[20]

Though not often done, it is technically possible to measure the full-shift TWA exposure of a single worker for each of the approximately 250 working days per

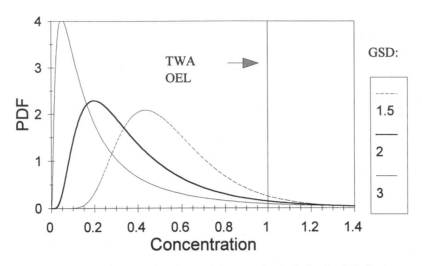

FIGURE 15.1 "Indirect control" model. Hypothetical single-shift limit, or TWA OEL, is set at 1. The exceedance fraction is fixed at 0.05 for each distribution. (PDF = probability density function.)

GSD	Mean	% > OEL
1.5	0.56	5 %
2.0	0.41	5 %
3.0	0.30	5 %

year. These measurements could then be plotted in a histogram, the shape of which would be an estimate of the true exposure distribution for that worker during that year. The term "exposure profile"[2,5] is used to refer to this distribution. Most industrial hygienists find that the lognormal distribution is an adequate model for describing such exposure profiles and for predicting future exposures.[10] As depicted in Figures 15.1 and 15.2, lognormal distributions tend to be skewed, with more measurements toward the low end and a long tail extending toward the higher exposures.

The exposure profile of an exposure group is a composite of the individual exposure profiles of group members. The lognormal distribution can often be used to describe this exposure profile as well.

Acceptable Exposure Profiles

Because exposures derive from continuous distributions with a zero lower boundary and a nearly unlimited upper boundary, there will always be some finite probability that a random exposure will exceed the applicable OEL. An "acceptable" exposure profile is usually one where such "over-exposures" occur infrequently. However, there are few OELs where an acceptable exposure profile has been defined in rigorous statistical terms. Therefore it is necessary to assign a practical, or work-

ing, *statistical interpretation* to each OEL. Table 15.2 contains statistical interpretations as recommended by several organizations or authorities. Starting with these definitions, we can then proceed to define both acceptable and unacceptable exposure profiles.

The exposure parameter most often mentioned in the industrial hygiene literature for rating the acceptability of an exposure profile is the 95th percentile exposure. There is general agreement that an exposure profile is "acceptable" if the *true* 95th percentile is equal to or less than the OEL.[2,7,8] A variation on this theme is to calculate the fraction of overexposures, or exceedance fraction.* If the true exceedance fraction is less than or equal to 5%, then the 95th percentile exposure is also less than or equal to the OEL. Exposure profiles considered "acceptable" according to this exposure control model are depicted in Figure 15.1

Rapid acting substances are typically assigned ceiling limits. These OELs are designed to limit exposures as measured over a few seconds (e.g., when measured with a direct reading instrument) to a few minutes. The ACGIH defines a TLV-Ceiling as a value that "should not

---

*The exceedance fraction relates to a specific OEL, whereas the 95th percentile can be compared to any OEL. Consequently, the 95th percentile is a more useful statistic where there are several applicable OELs (such as an OSHA PEL, NIOSH REL, and ACGIH TLV).

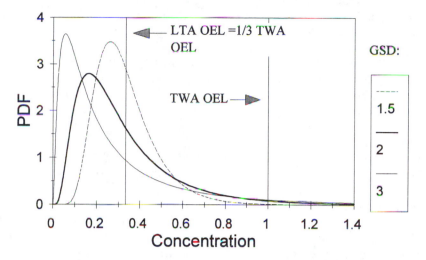

FIGURE 15.2 "Direct control" model. Hypothetical single-shift limit, or TWA OEL, is set at 1. The LTA (long-term) OEL is fixed at 1/3 the single shift OEL, per the recommendation of the AIHA (Ref. 5). (PDF = probability density function.)

GSD	Mean	% > OEL
1.5	0.333	0.2 %
2.0	0.333	2.7 %
3.0	0.333	6.1 %

be exceeded during any part of the working exposure."[20] In practical terms, if the true 99th percentile instantaneous or short-term exposure is less than the ceiling OEL, then one could reasonably conclude that the within-shift exposure profile was controlled during *that particular shift*.

An evaluation scheme based on the 95th percentile exposure is well suited to acute effect substances that may result in effects after roughly 15 minutes to several hours. Such substances will have either a short-term exposure limit (STEL) OEL or a TWA OEL (depending on the rapidity of action). For these substances it is important to control the dose across a short period of time.

Average exposure across longer periods is less relevant because it is assumed that accumulation does not occur and that there is complete recovery before the next shift. If the true 95th percentile short-term exposure is less than the STEL OEL, then one could reasonably conclude that the within-shift exposure profile was controlled during *that particular shift*.

For chronic disease agents it is important to limit the cumulative dose acquired by the employee. Perhaps the most relevant exposure parameter is the long-term average exposure, which suggests that an exposure profile can be deemed "acceptable" if the true long-term average exposure is less than a long-term average OEL, or

TABLE 15.2 Working Statistical Definitions for OELs

Type of OEL	Working Statistical Definition	Supporting References
Ceiling	99th percentile instantaneous exposure or short-term (i.e., less than 15-minute) exposure within each shift	
STEL	95th percentile 15-minute exposure within each shift	2, 7, 8, 9, 20
TWA	95th percentile full-shift exposure	2, 6, 7, 8, 24, 31
LTA*	(a) 10%–25% of the TWA OEL	28, 29
	(b) 33% of the TWA OEL	2, 5

*The averaging time should not be more than one year for most work environments and no more than two years for stable work environments.

LTA OEL. However, few legal or authoritative LTA OELs have been developed.[24]* Instead, single-shift TWA OELs have been used to indirectly limit cumulative dose for chronic disease agents to presumably acceptable levels.[19] Properly implemented, single-shift limits will reduce the true long-term average exposure to a fraction of the TWA OEL. For example, Figure 15.1 depicts several typical lognormal distributions where the 95th percentile is equal to the TWA OEL. Using this "indirect" control model, the mean value of each distribution is approximately half or less of the TWA OEL.

In principle, a LTA OEL is useful for gauging whether or not the true average exposure of each employee is actually being limited to a reasonable value.[19] OSHA[12,25] has stated that in a controlled work environment the average exposure should be "well below" the OSHA PEL. How much less? The AIHA[2,5] suggested that in the absence of a legal or authoritative LTA OEL it is reasonable to set an LTA OEL for chronic disease agents at one-third of the single-shift TWA OEL. Others have suggested no more than half[26] or no more than one-tenth to one-fourth[27,28] of the single-shift limit. In the absence of a legal or authoritative LTA OEL, it is reasonable to use the AIHA recommendation of one-third the single-shift limit (or TWA OEL). Furthermore, as a general principle, at no time should the "measured" long-term average exceed one-half of the single-shift limit. Exposure profiles considered "acceptable" according to this "direct" control model are depicted in Figure 15.2.

Few so-called chronic disease agents solely produce chronic effects. For example, crystalline silica, usually considered a chronic toxicant, can also produce acute and accelerated silicosis, depending on the dose rate (frequency, duration, and severity of exposure).[29] Cadmium at lower levels is considered a chronic toxicant producing lung cancer and kidney disease, but excessive exposure within a single shift can cause life-threatening pneumonitis and pulmonary edema. For these and similar substances, it would be a mistake to assume that dose rate can be ignored. In other words, both the dose rate and cumulative dose are important, suggesting that *both* a single-shift OEL and long-term average OEL are necessary for properly judging an exposure profile. A single-shift limit would function as basically a "single-shift excursion limit" for a LTA OEL. The single-shift limit would be used to evaluate whether or not the exposure profile is *currently* controlled, by comparing each measurement, or the 95th percentile (or upper confidence limit on the

95th percentile) of a set of measurements, to the TWA OEL. The LTA OEL limit would be used to determine if, over each observation interval of 6 months to a year, the long-term mean of the exposure profile is adequately controlled. In practice, however, virtually all OELs are single-shift or short-term (e.g., 15-minute) limits.

Unacceptable Exposure Profiles

According to NIOSH, "no more than 5% of an employee's true daily exposure averages should exceed the standard."[7] The CEN (Comité Européen de Normalisation) recommends that corrective action take place if the exceedance fraction exceeds 5%.[6] The AIHA recommends that exposures be controlled to the point that only a small fraction exceeds the limit; for example, "no more than 5% of the exposures exceed the [OEL]."[5] Similar recommendations have been made by others (see Hewett[19] and Still and Wells[30]). Consequently, for the purposes of this chapter an unacceptable exposure profile is one where the probability of overexposure, or exceeding the OEL, is greater than 5%.

One should not infer from this discussion that exposures above an OEL can be ignored when the 95th percentile exposure is less than the OEL. As discussed later, standard industrial hygiene practice is to *investigate* each exposure above an OEL.[2,7,19,31] A measured overexposure is a signal that the work environment may have changed and is no longer acceptable.

Rating the Degree of Exposure Control

It is often useful, for risk communication purposes, to be able to state that exposures for an acceptable exposure group are "highly controlled," "well-controlled," or "controlled." It is also useful to be able to describe unacceptable exposure groups as "poorly-controlled" or "uncontrolled." In the absence of any official definitions for these terms, the "exposure categories" listed in Table 15.3 may be used. The "rule-of-thumb" descriptions for the categories were adapted from the AIHA monograph[5] on exposure assessment. The recommended statistical interpretations were designed to be consistent with the concepts of acceptable and unacceptable exposure profiles developed earlier in this chapter. For example, if the point estimate of the 95th percentile is less than half of the TWA OEL, then you would be justified in stating that exposures "appear to be well-controlled." If the 95% upper confidence limit (UCL) for the 95th percentile exposure is less than half of the TWA OEL, then you would be justified in stating that exposures "are well-controlled (with at least a 95% confidence)." The data analysis techniques discussed in Hewett[1] (Chapter 16) can be

---

*In many European nations the long-term OEL for vinyl chloride monomer is 3 ppm, while the single-shift TWA OEL is 7 ppm. Exposures should be controlled so that the long-term, annual limit is not exceeded.

TABLE 15.3 Exposure Category Rating Scheme

Exposure Category	Rule-of-Thumb Description*	Qualitative Description	Recommended Statistical Interpretation	Notes
Highly controlled	Employees have little or no inhalation contact	Exposures infrequently exceed 10% of the OEL	$P(c > 0.1 \cdot OEL) \leq 0.05$	1,2,3
Well controlled	Employees have frequent contact at low concentrations and rare contact at high concentrations	Exposures infrequently exceed 50% of the OEL and rarely exceed the OEL	$P(c > 0.5 \cdot OEL) \leq 0.05$ $P(c > OEL) \leq 0.01$	4,5,6 7
Controlled	Employees have frequent contact at low concentrations and infrequent contact at high concentrations	Exposures infrequently exceed the OEL	$P(c > OEL) \leq 0.05$	8
Poorly controlled	Employees often have contact at high concentrations	Exposures frequently exceed the OEL	$P(c > OEL) > 0.05$	9
Uncontrolled	Employees frequently have contact at high concentrations	A large percentage of the exposures exceed the OEL	$P(c > OEL) \gg 0.05$	10,11

Note. c refers to an 8-hour TWA exposure concentration or a short-term exposure concentration. OEL refers to the TWA OEL or STEL OEL. The term exposures should be read as exposures of each employee.
*The rule-of-thumb descriptions were adapted from the AIHA (1991).
 1. Infrequently refers to an event that occurs no more than 5% of the time.
 2. $P(c > 0.1 OEL) \leq 0.05$ is read as the probability that an exposure (c) is greater than one-tenth the OEL and is less than or equal to 0.05.
 3. Alternative statistical definition: 95th percentile $\leq 0.1 \cdot OEL$.
 4. High concentrations are defined as concentrations that exceed the TWA OEL or STEL OEL.
 5. Rarely refers to an event that occurs no more than 1% of the time.
 6. Alternative statistical interpretation: 95th percentile $\leq 0.5 \cdot OEL$.
 7. Alternative statistical interpretation: 99th percentile $\leq OEL$.
 8. Alternative statistical interpretation: 95th percentile $\leq OEL$.
 9. Alternative statistical interpretation: 95th percentile $> OEL$.
10. Alternative statistical interpretation: 95th percentile $\gg OEL$.
11. Exposures are considered largely uncontrolled if the point estimate of the exceedance fraction is much greater than 0.05. For example, a point estimate of an exceedance fraction of 0.25 would clearly indicate that exposures are uncontrolled.

used to determine if an exposure profile meets the criteria for a "highly-controlled," "well-controlled," or "controlled" rating.

Comments

What Does It Mean to Be "In Compliance"?

The term "in compliance" has sometimes been defined to narrowly refer to the situation where each of a small number of measurements is less than the applicable OEL, federal or otherwise. Tuggle[32] showed how unacceptable work environments can often be declared "in compliance" when the sample size is small. For this reason professional industrial hygienists[2,5,6,8,30] advocate the collection of sufficient measurements to evaluate exposure profiles. Consequently, readers are encouraged to equate the concept "in compliance" with the concept, previously discussed, of an "acceptable exposure profile" for each worker. An "acceptable exposure profile" indicates that the individual long-term mean exposure, single-shift excursions above the OEL, and the probability of a citation are controlled to arguably acceptable values.[19]

Other Interpretations

It should be noted that the interpretations presented here regarding the ACGIH TLVs and OSHA PELs for chronic disease agents are not shared by all.[9,10] There are those who believe that these exposure limits should be interpreted as upper control limits to the long-term average exposure. Hewett[24] argues that there is abundant evidence that the ACGIH TLVs and OSHA PELs for chronic disease agents were and are intended to be interpreted as upper control limits for each single shift, and not upper limits for long-term, lifetime average exposure as some maintain. However, the documentation for a particular OEL should be consulted to determine the proper interpretation before designing an exposure monitoring program.

Exposure Monitoring Programs

Once it is determined that a quantitative exposure assessment is necessary, a written, documented exposure-monitoring program should be developed.[2,5] An exposure-monitoring program should cover the four hazard evaluation components: (a) data collection, (b) data analysis, (c) data interpretation, and (d) data management. First, it is necessary to clearly establish the goal for an exposure-monitoring program, regardless of its actual design.

Goal of an Effective Exposure-Monitoring Program

It is important that the collection and interpretation of exposure measurements result in the *correct* classification of the work environment for each exposure group, and not grossly underestimate or overestimate the exposures of any employee within each exposure group. Consequently, the goal of an effective exposure-monitoring program is to *periodically* obtain *sufficient*, *valid* and *representative* exposure measurements so that the *work environment* for each *worker* is *reliably classified* as either *acceptable* or *unacceptable*. Each italic term is explained below.

"Periodically"

Since few industrial processes remain constant for extended periods (months to years) exposure profiles should be expected to change over time.[33] For example, processes change, production levels vary, ventilation systems degrade, new workers are introduced, and tasks and work practices vary. Exposure sampling should occur frequently enough that a significant and deleterious change in either the contaminant generation process or the efficacy of the exposure controls is not permitted to persist for long.

"Sufficient"

The industrial hygienist should collect a sufficient number of measurements so that uncontrolled work environments are reliably detected. A critical attribute of a dataset, or collection of exposure measurements, is its predictive value for each member of the exposure group. A single exposure measurement that is under the OEL has limited predictive value when attempting to demonstrate that a work environment is (and is likely to remain) acceptable.* On the other hand, a single overexposure occurring by itself, or in a small dataset, is highly suggestive that exposures may be poorly controlled or uncontrolled.[19] A collection or sample of exposure measurements has better predictive value.

"Valid and Representative"

A valid exposure measurement is one that is collected using a reasonably accurate sampling and analytical method and at a location that permits a reasonable estimate of the selected employee's personal exposure. When characterizing the exposure profile of an exposure group, measurements should be representative in the sense that (a) production levels, environmental controls, and work practices were not manipulated or optimized for the benefit of the survey; (b) the work environment is reasonably stable within each observation interval; and (c) measurements were collected in a random fashion (i.e., the sample days were randomly selected).

Practically speaking, however, nonrepresentative sampling has its uses. It is both reasonable and efficient to collect measurements solely on days of expected maximal exposure and/or solely from employees known or suspected to routinely experience the greatest exposures.† It is also reasonable in many industrial environments to collect measurements in campaign fashion (i.e., on consecutive days), rather than in a strictly random fashion, because in most cases there is little serial correlation between measurements.[34] When evaluating exposures relative to a short-term OEL or ceiling limit OEL, the usual strategy is to purposefully sample during periods that are representative of peak or maximum probable exposures.[2,34]

"Work Environment"

We are interested in rating the quality of the work environment for each exposure group. The work environment can be defined as the physical space in which the members of the exposure group spends the majority of their time, or it can be more abstract and refer to a specific combination of exposure agent, shift, job, task, and work area.

---

*A single exposure measurement that is extremely low, e.g., less than 10% of the TWA PEL or STEL, can have considerable predictive value *provided* the day, employee, shift, and task (when comparing short term exposures to the STEL) represents a worst case exposure.

†The term "representative employee" is often used to by OSHA to refer to the maximum-risk employee.

"Worker"

Both common sense and federal law[35,36] dictate that each worker should expect a work environment devoid of unreasonable risks. While we recognize that our goal is to protect *each* individual worker, limited resources compel industrial hygienists to (a) aggregate workers into exposure groups; (b) determine which exposure groups warrant priority attention[2,5]; and (c) evaluate the "exposure profile" of each exposure group in order of priority.[2,5,8] Consequently, our data-collection strategies, and data analysis and interpretation procedures must be designed so that our *conclusions* regarding the exposure group are reasonably *predictive* of the exposures experienced by each member of the group.

"Reliably Classified"

A properly designed exposure-monitoring program will have a high probability of classifying a work environment as acceptable when the exposure profile is truly acceptable (according to some definition of acceptable), and a high probability of classifying the work environment as unacceptable when the exposure profile is clearly unacceptable. It may be difficult to design a program that maximizes both probabilities, in which case the program design should focus on maximizing the latter probability. This is discussed in more detail in the section on program performance characteristics below.

"Acceptable or Unacceptable"

Exposures (i.e., the exposure profile) for an exposure group are classified as either *acceptable* or *unacceptable*. Acceptable exposures can be described qualitatively as either minimal, well controlled, or controlled. Unacceptable exposures are either poorly controlled or uncontrolled (Table 15.3). The determination, after a baseline survey, that the exposure profile is acceptable for an exposure group *does not* automatically imply that the exposure profiles for all group members are also acceptable. As discussed below, periodic follow-up surveys and random sampling of all group members is commonly used to verify that maximum-risk employees have not been overlooked. This, combined with task-analysis and control of within-shift peak exposures, tends to ensure that the assessment of the group exposure profile is sufficiently predictive of the exposure experiences of all group members.

Prioritization

The number of unique exposure groups—or agent/process/job/area combinations—can be considerable for most workplaces. However, in instances of mixed exposures it is often permissible to select one or more *index* compounds for measurement and control. Exposures to other less toxic, less hazardous, or difficult-to-measure substances are likewise reduced and controlled.

Several schemes have been advanced for prioritizing between exposure groups. In 1991 the AIHA Exposure Assessment Strategies Committee[5] recommended the use of a "health-effect rating" and an "exposure rating" (see also Rock[31]). More recently the AIHA introduced the concept of "critical exposure group" to refer to exposure groups that should be evaluated first.[2] Similar schemes have been found useful by many corporations and have become an integral part of what the AIHA[2] refers to as "comprehensive exposure assessment."

Critical Concepts

Observation Intervals

"Observation interval," or "survey period,"[30] refers to the length of time until the next re-sampling survey. It is necessary that during this time the exposure profile remain reasonably stationary. Otherwise, the results of the last survey will have little to no predictive value. The observation interval may vary with exposure group, depending on the nature of the process, cycles and trends in production, the rate of worker turnover, and the maintenance and upkeep of existing controls. The observation interval may be as short as a few days or weeks for highly dynamic work environments, and as long as a year or more for well-controlled, highly stable work environments. For acutely toxic substances the observation interval can be as short as a single day; in other words, exposure monitoring takes place daily.

A key feature of the observation interval concept is that measurements collected during the *current* observation interval are usually the most important for predicting future exposures. Current high exposures should not be combined with measurements from the previous interval in order to produce an "acceptable" exposure profile. The current high exposures indicate that the exposure profile may no longer be adequately controlled and that future exposures may be excessive.

Program Performance Characteristics

The problems of exposure management are similar to those in maintaining quality control in a manufacturing process. The ability of a quality control procedure to detect *acceptable* and *unacceptable* manufactured products or lots is described by the procedure's Performance or Operating Characteristic curve.[37] It is possible to determine the performance characteristics for any exposure-monitoring program. For example, Tuggle[32] presented the performance curve for the NIOSH sampling strategy and decision logic. He showed that the probability of incorrectly concluding that the work environment is acceptable was quite high, even when the exceedance fraction is clearly unacceptable, that is, 0.25 or greater. Since the NIOSH strategy and decision logic has been incorporated into numerous OSHA 6(b) single-substance standards, this suggests that an exposure-monitoring program based on such minimalistic requirements will often misclassify work environments as acceptable when in fact the true exceedance fraction exceeds 0.05.

Computer simulation may be required to determine the operating characteristic curve for most exposure monitoring programs and so is not often done. However, it is important to recognize that for any exposure-monitoring program there are two "risks" that can be minimized: the "employer's risk" and the "employees' risk." Employer's risk refers to the probability of *incorrectly* concluding that the work environment for an exposure group is unacceptable, when indeed the true exposure profile is acceptable. Employees' risk refers to the situation where the work environment has been *incorrectly* judged to be acceptable, when in fact the true exposure profile is unacceptable. Note that the Employees' Risk *does not necessarily correspond to the risk of developing disease*. It refers to the fact that it is the employees that are affected when an unacceptable level of overexposures remains undetected for the observation interval. In traditional terms, employer's risk corresponds to the α-error (Type I error) for the exposure monitoring program, while employees' risk refers to the β error (Type II error).

The conscientious employer will always design an exposure-monitoring program that focuses on minimizing the employees' risk. This is because any decision that results from the analysis of a finite set of exposure data is extrapolated in two ways. First, the decision is extrapolated to all unmeasured employees in the exposure group. If the decision is wrong, it can potentially affect a great many employees. Second, the decision is extrapolated into the future for the entire exposure group for the entire duration of the observation interval, that is, until the collection of the next set of exposure measurements. If the decision is wrong, it will affect the entire exposure group for what may be an extended interval.

It is entirely logical to design an exposure-monitoring program that is concerned with minimizing employees' risk and tolerating a rather large employer's risk. This is because the employer's risk does not correspond to the probability of implementing expensive, potentially unneeded controls. Employers usually *verify* the presence of an unacceptable exposure profile with one or more follow-up surveys before implementing expensive controls.

Exposure Histories

Over time an "exposure history" will develop for each exposure group. This exposure history can be used to detect long-term trends or to evaluate cyclic patterns of exposure associated with production trends or seasonal variations. Exposure histories also allow you to "calibrate" your industrial hygiene judgment regarding the stability or variability associated with various processes, operations, or tasks. Furthermore, consistently low measurements from year to year, even when only a few measurements are collected per year, can provide convincing evidence that exposures were, and are likely to continue to be, "acceptable." Periodic task analysis should also help improve group homogeneity, so that the exposure experience of the group is more predictive of each group member.

Data Collection

Data collection requires a sampling strategy that specifies the type of survey and the process for selecting whom to monitor within each group, how many measurements to collect, and how often or under what conditions this process is repeated. The sampling strategy should specify the approximate frequency of periodic reassessments and the conditions that will trigger a special reassessment. Periodic surveys have a dual purpose. First, resampling is essential for verifying that the earlier assessment was correct. Second, resampling is often the only means of detecting a trend toward increasing exposures. There are several types of sampling strategies and at least five types of exposure-monitoring surveys.

Sampling Strategies

There are four generic sampling strategies: individual based, maximum-exposed employee based, high-risk activity or task based, and exposure group based. Hybrid combinations are always possible and, in fact, commonplace. Sampling strategies based on the exposure group concept are prominently featured in the recommendations of authoritative organizations, such as the AIHA[2,5] and CEN.[8]

Individual Based

Ideally, the exposures experienced by each employee should be regularly estimated, preferably by numerous exposure measurements collected either in campaign fashion (i.e., within a short period of time or during consecutive shifts) or across several months. For some occupations or work environments, where there are only a few workers and exposures range from significant (e.g., greater than 10% of the OEL) to "poorly controlled" (see Table 15.3), it may be necessary to periodically monitor 100% of the workers. Each employee basically constitutes a single exposure group.

Strategies that involve the regular monitoring of all exposed employees are, for practical reasons, not often implemented. However, it is important to recognize that regardless of the type of strategy adopted out of expediency or due to limited resources, the goal of an exposure-monitoring program is to make accurate *decisions* regarding the exposure profile of *each* employee.

Maximum-Risk Employee Based

Strategies that focus on the maximum-risk employees (MREs) in each exposure group are based on the idea that if the MRE exposures are judged acceptable, then it is logical to assume that the *all workers within the group are adequately protected*. For example, in the mid-1970s NIOSH[7] developed a sampling strategy and decision logic that featured (a) the selection of the "maximum risk employee" (MRE), or the "employee [per exposure group] presumed to have the highest exposure risk"; and (b) the collection of one or a few exposure measurements. It was designed to impose a "minimum burden to the employer [i.e., risk manager] while providing adequate protection to the exposed employees," but had several weaknesses. For example, the ability of industrial hygienists to reliably select one or more maximum-risk employees from an exposure group has been questioned by several researchers. Furthermore, the NIOSH strategy will not reliably detect unacceptable work environments, even when the true exceedance fraction greatly exceeds 0.05.[32] Consequently, one should view the NIOSH scheme as the basis for a minimalistic exposure-monitoring program that is best suited for *auditing* work environments where exposures were previously determined, by a comprehensive exposure assessment, to be controlled, well-controlled, or minimal.

Slightly modified versions of this strategy were incorporated into OSHA's single substance 6(b) standards (e.g., lead, benzene, asbestos, formaldehyde, among others). OSHA also recognized that such a strategy represented a "token" commitment that will not accurately classify all work environments.[34] Nonetheless, for initial evaluations or where resources are limited or re-sampling intervals are broad, the MRE concept is still recom-

mended and commonly used by industrial hygienists as a means of efficiently determining the acceptability of the work environment for the members of an exposure group.

High-Risk Activity or Task Based

Investigators are increasingly interested in determining which task (i.e., component of a job) or work practice contributes most to the worker's overall exposure.[38–42] Once a particular task or work practice is identified as the primary contributor to exposure, it is often possible to substantially reduce the daily average exposure by adding task-specific controls and/or through the modification of individual or group work practices.

Exposure Group Based

A common exposure sampling strategy involves the concept of "homogeneous exposure groups"[5] or "similar exposure group."[2] Basically, workers are aggregated on the basis of work similarity, exposure agent(s), environment (workplace, process, task, and controls) similarity, and identifiability.[6] One or more measurements are collected from each of *n* randomly selected workers per exposure group. This strategy was designed for efficiency; that is, a decision is reached for each exposure group with a limited number of measurements. The measurements obtained from the group are felt to characterize the work performed by each group member and therefore can be extrapolated to all members of the exposure group, measured or not. Such a strategy is implicitly based on the concept that effective occupational exposure management for the exposure group results in effect exposure management for each member of the group.

In 1991 the AIHA Exposure Assessment Strategies Committee presented the "homogeneous exposure group" (HEG) concept.[5] In theory, workers in an HEG have "identical probabilities of exposure to a single environmental agent." However, the committee recognized that exposures on any single day will vary from worker to worker. Furthermore, in practice, the individual exposure profiles are expected to be similar, but not identical. For an initial or baseline evaluation the industrial hygienist should randomly select 6–10 workers per HEG and collect 6–10 measurements over a relatively short period of time. The industrial hygienist then analyzes the data and decides, using a combination of "statistical analysis and professional judgment," whether or not the "exposures demonstrate an acceptable work environment."* Exposures for an HEG are usually deemed acceptable if it is highly likely that 95% of the measurements are less than the OEL (determined using upper tolerance limits). The Committee now uses the phrase "similar exposure group,"[2] but the general concept remains the same.

The European standard for exposure assessment adopted by the CEN[6] is also based on the HEG concept. The CEN acknowledges that within an HEG, exposures are subject to both "random and systematic" variation and provides a "rule of thumb" for assessing group homogeneity.† This standard recommends simple decision rules for classifying *each* exposure measurement collected from an HEG. However, if six or more measurements are randomly collected then one can use statistics to estimate the probability of overexposure for individuals within the HEG. The CEN suggests that if this probability is less than 0.1% and the work environment is reasonably stable, then exposure monitoring can be reduced or eliminated until a significant change occurs. If this probability exceeds 5%, then corrective action should take place. Otherwise, periodic monitoring should be used to confirm that the point estimate of the probability of overexposure remains less than 5%.

Exposure group–based strategies are best suited to exposure groups that are reasonably homogeneous; that is, there are only minor systematic differences between the individual exposure distributions of the group members. If the exposure group is heterogeneous and there are large systematic differences between individuals, then such a strategy may miss group members that are routinely overexposed.[6] Several researchers have shown that exposure groups often have a great deal of between-worker variability.[43,44] Consequently, this assumption may not be valid without an analysis of objective data. For this reason, occasional random sampling of all members of an exposure group is advised in order to assess the degree of group heterogeneity. In addition, as noted previously, task and work practice analysis is expected to decrease between-worker differences in exposure, thus making the group experience more predictive of individual experiences.‡

---

*The Committee allows that the industrial hygienist may select the "most exposed worker" when determining whether or not an HEG is in compliance with a government standard.

†[I]f an individual exposure is less than half or greater than twice the arithmetic mean [for the HEG], the relevant work factors should be closely reexamined to determine whether the assumption of homogeneity was correct."

‡Such data would include repeat measurements randomly collected from a random sample of workers within each exposure group, which then could be analyzed using ANOVA techniques (see Ref. 10).

Sampling Surveys

For each of the above sampling strategies an individual survey will tend to fall into one of the following categories: (1) baseline, (2) surveillance, (3) audit, (4) research (epidemiological), and (5) diagnostic. For baseline, surveillance, and audit surveys, the goal of the survey may be to merely demonstrate "compliance" with the minimalistic requirements of federal or state regulations. Or the goal of the survey may be to accurately characterize the exposure profile of each exposure group or for the maximum risk employees within each group. Harris[45] and the AIHA[2,5] provided expanded discussions regarding these surveys. Harris in particular recommends several "levels" of effort where measurements are collected to satisfy both compliance and research (risk-assessment) needs.

Baseline

A baseline or initial-exposure sampling survey is intended to collect sufficient exposure measurements to accurately characterize and judge the exposure profile of an exposure group. Consequently, one should be concerned with forming reasonably accurate exposure groups and with collecting sufficient measurements per group. For example, the AIHA[2,5] recommends that at least 6–10 measurements are required for a baseline evaluation. In principle, *a new baseline survey* is required whenever there are changes that have the potential to significantly alter the exposure profile of the exposure group. These changes include production level changes; seasonal effects; introduction of new workers; deterioration of existing controls; changes in job descriptions, tasks, and/or work practices, to name a few. For example, the effectiveness of general dilution ventilation and local exhaust controls can vary substantially between summer and winter months.

Surveillance

The routine monitoring that occurs after a baseline survey can be considered surveillance sampling. Sufficient measurements should be collected in a timely fashion so that trends are identified and the initial exposure rating of the exposure group is validated. The frequency of monitoring depends on the stability of the process and the degree of existing control. Routine quantitative exposure surveys or exposure surveillance monitoring may not be the best choice in those situations where exposures are just marginally controlled (i.e., point estimate of the exceedance fraction is less than but near 0.05). In such situations, it is entirely possible that the cost of regular monitoring and surveillance activities approaches or exceeds the cost of effective controls.[2]

Audit

An audit survey, as distinguished from a surveillance survey, is conducted by outside investigators, such as corporate level inspectors or compliance officers from a regulatory agency. It is intended to quickly check for evidence that the exposure profile for an exposure group is unacceptable. This is often accomplished by selecting and monitoring one or more maximum-exposure employees or tasks. If the exposure of this employee is above the OEL, then it is reasonable to assume that this and other employees may be routinely overexposed. The company is then obligated to initiate a more comprehensive evaluation, such as is done during baseline surveys, and take remedial action if warranted.

Research (Risk Assessment)

Many companies participate in cohort studies or recognize that well-developed exposure databases are necessary for future risk assessments.[45] Since a goal of these studies is to determine the exposure-response relationship for a particular substance, it is necessary to routinely and accurately characterize the exposure profile for each exposure group—regardless of exposure level—included in the study cohort during each observation period of the study. A common observation period is a year interval.

Diagnostic

A diagnostic survey is designed to locate the source of exposure, identify the task or activity that contributes most to exposure, and test the efficacy of a control. In general, short-term measurements or direct reading instruments are used and the time and location are often deliberately selected in order to measure maximum within-shift exposures. Such measurements should not be used when estimating the exposure profile for an exposure group, but can be used for prioritizing between groups when allocating limited industrial hygiene resources.

Data Collection Issues

In 1973 Hosey[46] suggested that industrial hygienists address five questions before collecting measurements that are reasonably representative of worker exposures.

- Where to sample?
- Whom to sample?
- How long to sample?
- How many measurements to collect?
- What shift to sample?

At that time industrial hygienists typically collected short-term, or "grab sample" measurements, even when evaluating full-shift exposures. Since then instrumentation has improved, permitting the collection of full-shift exposure measurements using personal, battery-powered sampling pumps or direct-reading devices with data-logging capabilities. Consequently, our answers to the above questions may differ somewhat from those first proposed by Hosey. Furthermore, we can add several additional questions to the list:

- What season to collect measurements?
- When to resample?
- When to reduce sampling?

Sample Location

When measuring exposures for comparison with a TWA, STEL, or Ceiling OEL, the usual procedure is to place a reasonably accurate exposure-measuring device on the worker, within the worker's breathing zone.* There are instances where area sampling locations are reasonably predictive of individual worker exposures; however, before using general area exposure measurements to assess the quality of risk management for individual workers you should first determine the degree of correlation between personal and area measurements.

Worker Selection

Random selection is recommended when the goal is to characterize the exposure profile of an exposure group.[2,5,8] In reality, true "statistically based" random selection is seldom practiced. The phrase "representative employee" is often encountered in OSHA 6(b) (single substance) regulations to refer to the employee expected to have the highest exposure. Although some authorities question whether or not an industrial hygienist can reliably select this "representative employee," also called the maximum-risk employee, the selection of the employee "closest to the source of the hazardous material being

---

*The breathing zone has been variously defined as a sphere or hemisphere with a six inch to two foot radius about a worker's head. The concept behind the various definitions is that an individual worker's exposure is best estimated by a sampling device placed as close to the worker's mouth and nose as is reasonable and safe. Typically, a sampling device is clipped to the worker's left or right lapel.

generated"[7] is an often used technique to efficiently evaluate a work environment.

Measurement Averaging Time

The averaging time for the majority of TWA OELs is 8 hours and 15 minutes or less for short-term exposure limits and ceiling OELs. In order to combine measurements for data analysis it is essential that the averaging time, sampling methodology, and sampling strategies be similar for each measurement in the dataset.[7] Unless the averaging times are nearly identical, the sample geometric mean and geometric standard deviation (discussed in Chapter 16) may be misleading and lead to incorrect decisions. For example, it would be improper to mix 15-minute STEL and full-shift TWA measurements. Furthermore, measurements collected for substantially different purposes usually should not be combined. For example, measurements collected from randomly selected employees should not be combined with measurements collected from one or more purposefully selected maximum-risk employees.

Sample Size

How many measurements are necessary to adequately characterize an exposure profile? There is no easy answer. Strict use of rigorous sample size formulae often leads to sample size estimates—*per individual worker or exposure group*—well beyond the reach of most exposure monitoring programs.[48] Several consecutive, partial-shift "industrial hygiene samples" may be required just to generate a single full-shift TWA measurement. For a LTA OEL, where the relevant averaging time is no more than one year, *n* full-shift TWA measurements are necessary before one can calculate a single "measurement" of long-term exposure.[19,47]

For TWA OELs and baseline or initial-exposure surveys there are several *rules-of-thumb*. For example, the CEN[6] recommended a minimum of six measurements before using "statistical principles" to evaluate the exposure profile of an exposure group. The AIHA[2,5] recommended 6–10 measurements for baseline evaluations, but cautioned that the accurate estimation of the true 95th percentile may require a much larger sampling commitment. Ayer[49] recommended that "the number of samples [i.e., measurements] in a sampling round equal the square root of number of workers." For surveillance or periodic monitoring Roach[50] recommended that "1 in 25 or 1 in 50" shifts be monitored when exposures exceed 10% of the OEL. Hewett and Ganser[51] suggested a method for relating the sample size to the desired width of the confidence interval around the point estimate of the exceedance fraction.

In summary, there is no easy answer or magical sample size formula applicable to all situations. The above cited *rules-of-thumb* provide general guidance. If exposures are expected to be minimal, then it is reasonable to limit the sample size and devote resources to more problematic exposure groups. Marginally controlled work environments should receive more attention. Work environments known to be poorly controlled or uncontrolled are unlikely to benefit from additional sampling. Controls should be implemented and consideration given to interim worker protection through the use of personal protective equipment.

Shift Selection

Frequently there is more than one shift and the work on each shift is functionally identical. However, one should not automatically assume that the two shifts constitute a single exposure group. OSHA's 6(b) standards generally require that measurements be taken on each shift:

> Representative samples for each job classification in each work area shall be taken for each shift unless the employer can document with objective data that exposure levels for a given job classification are equivalent for different work shifts.[34]

This regulation constitutes generally sound advice. For example, Rock[31] described a situation where there were significant differences in the exposure profiles between day and night shift painters. Measurements collected exclusively from the day shift would have severely underestimated night shift exposures.

Time of Year

It is generally accepted that indoor exposures tend to be greater during the colder months. Doors are generally closed and ventilation systems lose effectiveness. As with shift, one should not assume that measurements collected during one season are representative of all seasons. When in doubt, baseline exposure monitoring should occur several times during the first year, until an exposure history is developed.

Frequency of Re-sampling

Roach[50] recommended that "an appraisal should be made each month of one or more employee's exposure over one shift" for each exposure group where measurements are routinely between 10% and the OEL. Corn[52] recommended resampling frequencies ranging between 3 and 12 months, depending on past results, confidence in the engineering controls, and the toxicity of the substance. The AIHA[2,5] recommended periodic reevaluations at least annually for each exposure group.

Resampling should also take place anytime there is a change in process, tasks, production levels, and ventilation controls that may result in new or additional exposures. Furthermore, "immediate" reevaluation is warranted if there is an employee complaint, a process change, and any "real or suspected occupational illness."[2,5] Reevaluation may also be necessary when significant new toxicological data or regulatory changes occur.

The CEN[6] recommended resampling even when a baseline survey shows that exposures are below the OEL:

> [S]ubsequent measurements at appropriate intervals should, if necessary, be taken to ensure that the situation continues to prevail. The nearer the concentration recorded comes to the limit value, the more frequently measurements should be taken.

The CEN then provided two examples of decision schemes for determining the number of measurements and sampling frequency for an exposure group that has already undergone a comprehensive exposure assessment.

As a general principle a follow-up survey should be conducted after every initial or baseline exposure assessment to verify that the initial or baseline exposure assessment was correct. Each exposure group should be reevaluated at least once per year, and more often if the work environment is particularly variable. This reevaluation may involve simply updating the background information for this exposure group. However, a new baseline survey is warranted if there are indications that the process has changed, controls have deteriorated or are no longer effective, or that work practices or job tasks have changed.

Reduction of Sampling

Roach[50] recommended that the frequency of resampling be reduced to once every 2 months when exposures are consistently controlled to below one-third of the OEL. Exposures consistently below one tenth of the OEL require only an occasional evaluation, and then focused on checking the ventilation system and detecting "process leaks."

OSHA 6(b) standards often specify the conditions for reducing the sampling burden. For example[34]:

> The employer may discontinue periodic monitoring if results from two consecutive sampling periods at least 7 days apart show that employee exposure is below the action level [i.e., 50% of the standard] and the STEL.

According to the CEN,[6] if the exceedance fraction is less than 0.1% and the work environment is reasonably stable, then exposure monitoring can be reduced or eliminated until a significant change occurs.

Comments

Regulatory Compliance Sampling Strategies

In the United States most federal regulations require that employers collect only one or a few measurements per exposure group, and usually from one or more maximum-risk employees (when they can be reliably identified). These measurements are not required to be evaluated and interpreted using statistical techniques such as described here, but are merely compared to the legal OEL. Such minimalistic strategies will reliably render the correct decision only when exposures are already "minimal," "well-controlled," or grossly "uncontrolled."[19,32] Industrial hygienists should be aware that the strict, uncritical application of these minimalistic mandatory requirements can often lead to the incorrect conclusion that exposures are acceptable. Consequently, there is considerable room for voluntary improvement and enhancement, as OSHA intimated in Appendix B of the 1992 final standard for formaldehyde.[53] Exposure profile analysis, as advocated by the AIHA,[2,5] the CEN,[6] and numerous authorities is preferred (see related discussion in Hewett[19]).

Data Management

Each datum generated by an exposure-monitoring program should be documented and stored in a database of some sort, along with the descriptive information necessary for giving meaning to that number. Industrial hygienists recognize that there is considerable potential associated with maintaining databases of exposure data and with sharing these data with trade organizations, academia, and federal agencies.[54] The many potential uses include[55]:

- assessing compliance with applicable OELs
- assessment of control measures: exposure data can be used to monitor the continued effectiveness of existing control measures or for developing cost-effective controls

- risk assessment: exposure databases are critical to exposure-response epidemiological studies
- regulatory risk management: exposure databases can be used to develop "better informed" regulatory policies and guidance
- risk communication: such databases can be used to effectively and accurately communicate risk evaluations to employers, employees, and regulatory agencies

One of the tasks facing those developing company, corporate, industry or trade organization, or national and international exposure databases is the identification of the "essential data elements." Table 15.4 contains the major categories of essential data elements as suggested by a joint ACGIH-AIHA Task Group on Occupational Exposure Databases[56] and a European working group.[57] Both articles include an overview of the relevant literature and provide recommendations for developing an effective exposure data management program.

The common feature of these proposals is the concept of "accountability." Each risk manager (e.g., industrial hygienist) should (a) continuously document all critical decisions, such as the criteria for defining and refining exposure groups; and (b) maintain relevant exposure information for use by risk managers and/or researchers in the years to follow.

Along similar lines, the AIHA Exposure Assessment Strategies Committee[2] recommends that exposure data be managed using six relational databases:

1. workplace data—workplace description, process flowcharts, floor plans, etc.
2. environmental agent data—an inventory of the chemical, physical, and biological agents present

3. similar exposure group data—basis for establishing each exposure group
4. worker data—data for linking specific workers to each exposure group for each observation period
5. exposure assessment data—relevant data regarding each exposure assessment to include the interpretation of the data (i.e., acceptable, unacceptable, or uncertain)
6. monitoring data—the data relevant to each exposure measurement (e.g., date collected, location, sampled worker)

Setting up such database systems is not a trivial task and sometimes takes years. But once in place such systems permit the analysis of exposure trends, efficient targeting of resources, effective communication of exposure analyzes to workers and management, documentation of remedial actions, and documentation of effective exposure control.

Additional Issues

Nontraditional Work Schedules

Brief and Scala[57] observed that the TLVs were designed for a traditional 8-hour work shift and 40-hour workweek. They introduced a conservative method for *reducing* the TLVs to reflect a "novel" work schedule. Extended work shifts and/or more than 40 hours of exposure per week reduce the "recovery" time for each worker and "stretch the reliability and even viability of the data base for the TLV."[58] The ACGIH[59] recommends that, among other models, the "Brief and Scala model" be used as guidance for reducing the TLV when there is a nontraditional work schedule. OSHA specified a simpler scheme for the 1978 lead[60] and the 1994 cadmium[61] PELs.

Sampling and Analytical Error versus Environmental Variability

All sampling and analytical methods yield estimates of exposure that vary somewhat from the true exposure at the location of the sampling device. The overall accuracy of a sampling and analytical method depends on the imprecision and bias* of the method. NIOSH[62] considers a method sufficiently accurate if, after accounting for bias and imprecision, it will yield an estimate that is within

TABLE 15.4 Suggested Major Categories for the Essential Data Elements in an Occupational Exposure Database

ACGIH-AIHA Task Group[57]	Rajan et al.[56]
Facility/site	Premises
Survey tracking (e.g., original report)	Workplace
Work area	Worker activity
Employee information	Product
Process and operation	Chemical agent
Chemical agent information	Exposure modifiers
Exposure modifiers	Measurement strategy
Sample information	Measurement procedure
Sampling device information	Results
Engineering controls	References (e.g., original
Personal protective equipment	report)
Results: chemical, noise	

Note. The references describe the data elements associated with each category.

---

*Sampling and analytical methods for organic chemicals often have less than 100% recovery of the substance, resulting in a negative bias in the calculated exposures.

plus or minus 25% of the true concentration (within a range of one-half to twice the PEL) 95% of the time.

Exposure measurements collected from day to day often vary considerably, reflecting the variability in the process parameters, production rate, variations in the daily operation of the exposure controls, individual work practices, and so on. The normal random errors in the sampling and analytical method will contribute somewhat to the day to day (between-day) variability in exposures. However, this contribution is usually inconsequential.[63] In other words, if an exposure exceeds the TWA PEL or STEL it is more likely due to work-related factors than to errors arising from the sampling and analytical method. For this reason, the conscientious employer will *investigate* any overexposure for *assignable* causes that possibly can be easily rectified.

Suggested Reading

Comprehensive Exposure Assessment

CEN (Comité Européen de Normalisation). *Workplace atmospheres—Guidance for the assessment of exposure by inhalation of chemical agents for comparison with limit values and measurement strategy.* European Standard EN 689, effective no later than Aug. 1995 (English version), Feb. 1995.

Mulhausen, J., Damiano, J., Eds. *A Strategy for Assessing and Managing Occupational Exposures*, 2nd ed.; American Industrial Hygiene Association: Fairfax, VA, 1998.

Mulhausen, J., Damiano, J. Comprehensive Exposure Assessment. In *The Occupational Environment—Its Evaluation and Control*; DiNardi, S. R., Ed. American Industrial Hygiene Association: Fairfax, VA, 1997.

Perkins, J. L. *Modern Industrial Hygiene—Volume 1, Recognition and Evaluation of Chemical Agents*; Van Nostrand Reinhold: New York, 1997.

Roach, S. *Health Risks from Hazardous Substances at Work—Assessment, Evaluation, and Control*; Pergamon Press: New York, 1992.

Exposure Sampling Strategies and Data Analysis

Ayer, H. E. Occupational Air Sampling Strategies. In *Air Sampling Instruments for Evaluation of Atmospheric Contaminants*, 7th ed.; Hering, S. V., Ed.; American Conference of Governmental Industrial Hygienists: Cincinnati, OH, 1989.

Corn, M. Strategies in Air Sampling. *Scand. J. Work, Environ. Health* **1985**, *11*, 173–180.

Corn, M.; Lees, P. S. J. The Industrial Hygiene Audit: Purposes and Implementation. *Am. Ind. Hyg. Assoc. J.* **1983**, *445*, 135–141.

DiNardi, S. R., Ed. *The Occupational Environment—Its Evaluation and Control*; American Industrial Hygiene Association: Fairfax, VA, 1997.

Guest, I. G.; Cherrie, J. W.; Gardner, R. J.; Money, C. D. *Sampling Strategies for Airborne Contaminants in the Workplace*; British Occupational Hygiene Society Technical Guide No. 11 (ISBN 0 948237 14 7), H and H Scientific Consultants Ltd: Leeds, United Kingdom, 1993.

Harris, R. L. Guideline for Collection of Industrial Hygiene Exposure Assessment Data for Epidemiologic Use. *Appl. Occup. Environ. Hyg.* **1995**, *10*, 311–316.

Leidel, N. A.; Busch, K. A. Statistical Design and Data Analysis Requirements. In *Patty's Industrial Hygiene and Toxicology*, 3rd ed., Vol. 3, Part A; Harris, R. L., Cralley, L. J., Cralley, L. V., Eds.; Wiley & Sons, Inc.: New York, 1994.

Leidel, N. A.; Busch, K. A.; Lynch, J. R. *Occupational Exposure Sampling Strategy Manual*; National Institute for Occupational Safety and Health (NIOSH) Publication No. 77-173 (available from the National Technical Information Service (NTIS), Publication No. PB274792, or the NIOSH Web site as a .pdf file), 1977.

Lippmann, M. Exposure Assessment Strategies for Crystalline Silica Health Effects. *Appl. Occup. Environ. Hyg.* **1995**, *10*, 981–990.

Lynch, J. R. Measurement of Worker Exposure. In *Patty's Industrial Hygiene and Toxicology*, 3rd ed., Vol. 3, Part A; Harris, R. L., Cralley, L. J., Cralley, L. V., Eds.; Wiley & Sons, Inc.: New York, 1994.

OSHA (Occupational Safety and Health Administration): *Code of Federal Regulations* 29; Part 1910.1048 Formaldehyde (Appendix B), 1994.

Perkins, J. L., Rose, V. E., Eds. *Case Studies in Industrial Hygiene*; Wiley-Interscience: New York, 1987.

Rappaport, S. M. Interpreting Levels of Exposures to Chemical Agents. In *Patty's Industrial Hygiene and Toxicology*, 3rd ed., Vol. 3, Part A; Harris, R. L., Cralley, L. J., Cralley, L. V., Eds.; Wiley & Sons, Inc.: New York, 1994.

Rock, J. C. Occupational Air Sampling Strategies. In *Air Sampling Instruments for Evaluation of Atmospheric Contaminants*, 8th ed.; Cohen, B. S., Hering, S. V., Eds.; American Conference of Governmental Industrial Hygienists: Cincinnati, OH, 1995.

Still, K. R.; Wells, B. Quantitative Industrial Hygiene Programs: Workplace Monitoring. (Industrial Hygiene Program Management series, Part VIII) *Appl. Ind. Hyg.* **1989**, *4*, F14-F17.

Statistical Data Analysis

Gilbert, R. O. *Statistical Methods for Environmental Pollution Monitoring*; Van Nostrand Reinhold: New York, 1987.

Exposure Measurement Issues

Cohen, B. S., Hering, S. V., Eds. *Air Sampling Instruments for Evaluation of Atmospheric Contaminants*, 8th ed.; American Conference of Governmental Industrial Hygienists: Cincinnati, OH, 1995.

Eller, P. M., Cassinelli, M. E., Eds. *NIOSH Manual of Analytical Methods*; National Institute for Occupational Safety and Health DHHS (NIOSH) Publication No. 94–113, 1994.

OSHA (Occupational Safety and Health Administration). *OSHA Technical Manual*; 1995.

Soule, R. D. Industrial Hygiene Sampling and Analysis; In *Patty's Industrial Hygiene and Toxicology*, 3rd ed., Vol. 1; Clayton, G. D., Clayton, F. E., Eds.; Wiley-Interscience: New York, 1978.

References

1. Hewett, P. Industrial Hygiene Exposure Assessment—Data Analysis and Interpretation. In *Handbook of Chemical Health and Safety*; American Chemical Society: Washington, DC, 2001.

2. Mulhausen, J., Damiano, J., Eds.; *A Strategy for Assessing and Managing Occupational Exposures*, 2nd ed.; American Industrial Hygiene Association: Fairfax, VA, 1998.

3. Mulhausen, J.; Damiano, J. Comprehensive Exposure Assessment. In *The Occupational Environment—Its Evaluation and Control*; DiNardi, S. R. Ed.; American Industrial Hygiene Association (1997): Fairfax, VA.

4. OSHA (Occupational Safety and Health Administration). Occupational Exposure to Asbestos; Final Rule. *Fed. Regist.* **1994**, *59*, 40964–41158.

5. Hawkins, N. C., Norwood, S. K., Rock, J. C., Eds. *A Strategy for Occupational Exposure Assessment*; American Industrial Hygiene Association: Fairview, VA, 1991.

6. CEN (Comité Européen de Normalisation). *Workplace atmospheres—Guidance for the assessment of exposure by inhalation of chemical agents for comparison with limit values and measurement strategy*; European Standard EN 689, effective no later than Aug. 1995 (English version) (Feb. 1995).

7. Leidel, N. A.; Busch, K. A.; Lynch, J. R. *Occupational Exposure Sampling Strategy Manual*; National Institute for Occupational Safety and Health (NIOSH) Publication No. 77-173 (available from the National Technical Information Service (NTIS), Publication No. PB274792, or the NIOSH Web site as a .pdf file), 1977.

8. Corn, M.; Esmen, N. A. Workplace Exposure Zones for Classification of Employee Exposures to Physical and Chemical Agents. *Am. Ind. Hyg. Assoc. J.* **1979**, *40*, 47–57.

9. Rappaport, S. M. Assessment of Long-term Exposures to Toxic Substances in Air. *Ann. Occup. Hyg.* **1991**, *35*, 61–121.

10. Rappaport, S. M. Interpreting Levels of Exposures to Chemical Agents. In *Patty's Industrial Hygiene and Toxicology*, 3rd ed., Vol. 3, Part A; Harris, R. L., Cralley, L. J., Cralley, L. V., Eds.; Wiley & Sons: New York, 1994.

11. Damiano, J. Quantitative Exposure Assessment Strategies and Data in the Aluminum Company of America. *Appl. Occup. Environ. Hyg.* **1995**, *10*, 289–298.

12. OSHA (Occupational Safety and Health Administration). Occupational Exposure to Benzene; Final Rule. *Fed. Regist.* **1987**, *52*(176), 34460–34578.

13. OSHA (Occupational Safety and Health Administration). Occupational Exposure to Cadmium; Final Rules. *Fed. Regist.* **1992**, *57*, 42336.

14. Lippmann, M.; Gomez, M. R.; Rawls, G. M. Data Elements for Occupational Exposure Databases: Guidelines and Recommendations for Airborne Hazards and Noise. *Appl. Occup. Environ. Hyg.* **1996**, *11*, 1294–1311.

15. Rajan, B., et al. European Proposal for Core Information for the Storage and Exchange of Workplace Exposure Measurement on Chemical Agents. *Appl. Occup. Environ. Hyg.* **1997**, *12*, 31–39.

16. Burton, D. J. General Methods for the Control of Airborne Hazards. In *The Occupational Environment—Its Evaluation and Control*; DiNardi, S. R., Ed.; American Industrial Hygiene Association: Fairfax, VA, 1997.

17. AIHA (American Industrial Hygiene Association). American Industrial Hygiene Association white paper: a generic exposure assessment standard. *Am. Ind. Hyg. Assoc. J.* **1994**, *55*, 1009–1012.

18. Dyjack D. T.; Levine, S. P. Critical Features of an ISO 9001/14001 Harmonized Health and Safety Assessment Instrument. *Am. Ind. Hyg. Assoc. J.* **1996**, *57*, 929–935.

19. Hewett, P. Interpretation and Use of Occupational Exposure Limits for Chronic Disease Agents. In *Occupational Medicine: State of the Art Reviews* **1996**, *11*(3).

20. ACGIH (American Conference of Governmental Industrial Hygienists). *1997 TLVs and BEIs—Threshold Limit Values for Chemical Substances and Physical Agents, Biological Exposure Indices*; ACGIH: Cincinnati, OH, 1997.

21. Fiserova-Bergerova (Thomas), V. Development of Biological Exposure Indices (BEIs) and Their Implementation. *Appl. Ind. Hyg.* **1987**, *2*, 87–92.

22. Roach, S. A.; Baier, E. J.; Ayer, H. E.; Harris, R. L. Testing Compliance with Threshold Limit Values for Respirable Dusts. *Am. Ind. Hyg. Assoc. J.* **1967**, *28*, 543–553.

23. Paustenbach, D. J. Occupational Exposure Limits, Pharmacokinetics, and Unusual Work Schedules. In *Patty's Industrial Hygiene and Toxicology*, 3rd ed., Vol. 3, Part A; Harris, R. L., Cralley, L. J., Cralley, L. V., Eds.; Wiley & Sons: New York, 1994.

24. Hewett, P. Mean Testing: I. Advantages and Disadvantages. *Appl. Occup. Environ. Hyg.* **1997**, *12*, 339–346.

25. OSHA (Occupational Safety and Health Administration). Occupational Exposure to Lead; Final Standard. *Fed. Regist.* **1978**, *43*(220), 52952–53014.

26. Rappaport, S. M.; Selvin, S.; Roach, S. A. A Strategy for Assessing Exposures with Reference to Multiple Limits. *Appl. Ind. Hyg.* **1988**, *3*, 310–315.

27. Roach, S. A.; Rappaport, S. M. But They Are Not Thresholds: A Critical Analysis of the Documentation of Threshold Limit Values. *Am. J. Ind. Med.* **1990**, *17*, 727–753.

28. Adkins, C. E., et al. Letter to the Editor. *Appl. Occup. Environ. Hyg.* **1990**, *5*, 748–750.

29. NIOSH (National Institute for Occupational Safety and Health). *Preventing Silicosis and Deaths in Rock Drillers*; DHHS (NIOSH) Publication No. 92-107, 1992.

30. Still, K. R.; Wells, B. Quantitative Industrial Hygiene Programs: Workplace Monitoring. (Industrial Hygiene Program Management Series, Part VIII). *Appl. Ind. Hyg.* **1989**, *4*, F14–F17.

31. Rock, J. C. Occupational Air Sampling Strategies. In *Air Sampling Instruments for Evaluation of Atmospheric Contaminants*, 8th ed.; Cohen, B. S., Hering, S. V., Eds.; American Conference of Governmental Industrial Hygienists: Cincinnati, OH, 1995.

32. Tuggle, R. M. The NIOSH decision scheme. *Am. Ind. Hyg. Assoc. J.* **1981**, *42*, 493–498.

33. Symanski, E.; Rappaport, S. M. An Investigation of the Dependence of Exposure Variability on the Interval between Measurements. *Ann. Occup. Hyg.* **1994**, *38*, 361–372.

34. OSHA (Occupational Safety and Health Administration). *Code of Federal Regulations*; 29, Part 1910.1048, Formaldehyde, 1994.

35. Occupational Safety and Health Act of 1970 (OSHAct). Public Law 91-596, 1970.

36. Federal Mine Safety and Health Act of 1977 (FMSHAct). Public Law 91-173, 1977.

37. Montgomery, D. C. *Introduction to Statistical Quality Control*, 3rd ed.; Wiley & Sons: New York, 1996.

38. Gressel, M. G., Heitbrink, W. A., Eds. *Analyzing Workplace Exposures Using Direct Reading Instruments and Video Exposure Monitoring Techniques*; National Institute for Occupational Safety and Health (NIOSH) Publication No. 92-104, 1992.

39. Nicas, M.; Spear, R. C. A Task-based Statistical Model of a Worker's Exposure Distribution: Part II—Application to Sampling Strategy. *Am. Ind. Hyg. Assoc. J.* **1993**, *54*, 221–227.

40. Andersson, I.; Rosén, G. Detailed Work Analysis for Control of Exposure to Airborne Contaminants in the Workplace. *Appl. Occup. Environ. Hyg.* **1995**, *10*, 537–544.

41. Smith, R. W.; Sahl, J. D.; Kelsh, M. A.; Zalinski, J. Task-Based Exposure Assessment: Analytical Strategies for Summarizing Data by Occupational Exposure Groups; *Am. Ind. Hyg. Assoc. J.* **1995**, *58*, 402–412.

42. Woskie, S. R., et al. The Real-Time Dust Exposures of Sodium Borate Workers: Examination of Exposure Variability. *Am. Ind. Hyg. Assoc. J.* **1994**, *55*, 207–217.

43. Kromhout, H.; Symanski, E.; Rappaport, S. M. A Comprehensive Evaluation of Within- and Between-Worker Components of Occupational Exposure to Chemical Agents. *Ann. Occup. Hyg.* **1993**, *37*, 253–270.

44. Rappaport, S. M.; Kromhout, H.; Symanski, E. Variation of Exposure Between Workers in Homogeneous Exposure Groups. *Am. Ind. Hyg. Assoc. J.* **1993**, *54*, 654–662.

45. Harris, R. L. Guideline for Collection of Industrial Hygiene Exposure Assessment Data for Epidemiologic Use. *Appl. Occup. Environ. Hyg.* **1995**, *10*, 311–316.

46. Hosey, A. D. General Principles in Evaluating the Occupational Environment. In *The Industrial Environment—Its Evaluation and Control*; National Institute for Occupational Safety and Health: Washington, DC, 1973.

47. BOHS (British Occupational Hygiene Society), *Sampling Strategies for Airborne Contaminants in the Workplace*. BOHS Technical Guide No. 11; H and H Scientific Consultants Ltd: Leeds, UK, 1993.

48. Hewett, P. Sample Size Formulae for Determining the True Arithmetic or Geometric Mean Exposure of Lognormal Exposure Distributions. *Am. Ind. Hyg. Assoc. J.* **1995**, *56*, 219–225.

49. Ayer, H. E. Occupational Air Sampling Strategies. In *Air Sampling Instruments for Evaluation of Atmospheric Contaminants*, 7th ed.; Hering, S. V., Ed.; American Conference of Governmental Industrial Hygienists: Cincinnati, OH, 1989.

50. Roach, S. *Health Risks from Hazardous Substances at Work—Assessment, Evaluation, and Control*; Pergamon Press: New York, 1992.

51. Hewett, P.; Ganser, G. H. Simple Procedures for Calculating Confidence Intervals Around the Sample Mean and Exceedance Fraction Derived from Lognormally Distributed Data. *Appl. Occup. Environ. Hyg.* **1997**, *12*, 132–142.

52. Corn, M. Strategies in Air Sampling. *Scand. J. Work, Environ., Health* **1985**, *11*, 173–180.

53. OSHA (Occupational Safety and Health Administration). *Code of Federal Regulations 29*; Part 1910.1048, Formaldehyde (Appendix B), 1994.

54. Gomez, M. R.; Rawls, G. Conference on Occupational Exposure Databases: A Report and Look at the Future. *Appl. Occup. Environ. Hyg.* **1995**, *10*, 238–243.

55. Rajan, B., et al. European Proposal for Core Information for the Storage and Exchange of Workplace Exposure Measurement on Chemical Agents. *Appl. Occup. Environ. Hyg.* **1997**, *12*, 31–39.

56. Lippmann, M.; Gomez, M. R.; Rawls, G. M. Data Elements for Occupational Exposure Databases: Guidelines and Recommendations for Airborne Hazards and Noise. *Appl. Occup. Environ. Hyg.* **1996**, *11*, 1294–1311.

57. Brief, R. S.; Scala, R. A. Occupational Exposure Limits for Novel Work Schedules. *Am. Ind. Hyg. Assoc. J.* **1975**, *36*, 467–469.

58. ACGIH (American Conference of Governmental Industrial Hygienists). *Documentation of the Threshold Limit Values and Biological Exposure Indices*, 6th ed.; ACGIH: Cincinnati, OH, 1991.

59. ACGIH (American Conference of Governmental Industrial Hygienists). *1996 TLVs and BEIs—Threshold Limit Values for Chemical Substances and Physical Agents, Biological Exposure Indices*; ACGIH: Cincinnati, OH, 1996.

60. OSHA (Occupational Safety and Health Administration). Occupational Exposure to Lead; Final Standard. *Fed. Regist.* **1978**, *43(220)*, 52952–53014.

61. OSHA (Occupational Safety and Health Administration). Occupational Exposure to Cadmium; Final Rules. *Fed. Regist.* **1992**, *57*, 42336.

62. Gunderson, E. C.; Anderson, C. C.; Smith, R. H.; Doemeny, L. J. *Development and Validation of Methods for Sampling and Analysis of Workplace Toxic Substances*; National Institute for Occupational Safety and Health (NIOSH) Publication No. 80-133, 1980.

63. Nicas, M.; Simmons, B. P.; Spear, R. C. Environmental versus Analytical Variability in Exposure Measurements. *Am. Ind. Hyg. Assoc. J.* **1991**, *52*, 553–557.

16

Industrial Hygiene Exposure Assessment—
Data Analysis and Interpretation

PAUL HEWETT

As discussed in Hewett[1] (see Chapter 15), industrial hygiene exposure assessments come in three varieties: qualitative, semiquantitative, and quantitative. Quantitative surveys, which involve the measurement of current worker exposure using personal sampling equipment or direct reading instruments, are often necessary for initial or baseline evaluations. Furthermore, periodic sampling and occasional audits are necessary for validating earlier assessments and for detecting upward trends in exposure. Consequently, an industrial hygienist is often faced with questions regarding the collection, analysis, interpretation, and management of occupational exposure data. Hewett[1] described the rationale behind exposure monitoring and covered data collection and data management. The purpose of this chapter is to suggest appropriate procedures for analyzing and criteria for interpreting exposure data.

As discussed in Hewett,[1] exposure-monitoring programs must be designed and tailored for a wide variety of work environments. But first, as noted by Roach[2] in 1967, "[it] is important that hygienic standards should not be given widely different interpretations." We should also agree that the goal of an effective exposure-monitoring program is to routinely and accurately characterize the exposure profile* of each worker. It is less critical that we adopt identical or similar data analysis and interpretation procedures. There are numerous data analysis techniques—parametric and nonparametric—that will yield similar *decisions* regarding the acceptability of the work environment. Regardless of the number of measurements collected or the sophistication of the analysis technique, there is always a role for professional judgment and common sense.[1]

The procedures presented here emphasize the calculation of the point estimate of a relevant exposure parameter, and in addition the 95% lower confidence limit (LCL) and the 95% upper confidence limit (UCL). Taken together, these confidence limits comprise a 90% confidence interval for the true parameter. The advantage of this interval estimate is that it can readily be used to gauge (a) the accuracy of our point estimate of the true parameter, and (b) the acceptability of the work environment by comparing the LCL or UCL to an exposure limit or other relevant criterion. These procedures are consistent with those recommended by the Exposure Assessment Strategies Committee (EASC)[3] of the American Industrial Hygiene Association (AIHA) and the Comité Européen de Normalisation (CEN).[4]

Each worker should expect a work environment devoid of unreasonable risks. Our goal is to protect *each* individual worker, but limited resources usually compel industrial hygienists to (a) aggregate workers into exposure groups; (b) determine which exposure groups warrant priority attention; and (c) evaluate the "exposure profile" of each exposure group in order of priority. Consequently, our data collection strategies (see Hewett[1]), and data analysis and interpretation procedures must be designed so that our *conclusions* regarding the exposure group are reasonably *predictive* of the exposures experienced by each member of the group.

Data Analysis—Goodness-of-Fit and Other Issues

We need to assure ourselves that our exposure measurements have *predictive value*. If we can demonstrate that (a) the exposure distribution is reasonably stable; (b) the data are reasonably independent and uncorrelated; and (c) the data are described reasonably well by a lognormal distribution, then it is logical to assume that they have predictive value for the exposure group. Furthermore, we often require that the exposure group be reasonably homogeneous with regard to the source and

---

*If we measured the full-shift TWA exposure of a single worker for each of the approximately 250 working days per year and plotted these measurements in a histogram, then the shape of the histogram would be an estimate of the true exposure distribution for that worker during that year. The term "exposure profile" is used to refer to this distribution. One can also conceptualize within-shift exposure profiles (for short-term exposures) or exposure profiles for periods shorter than or longer than a year.

conditions of exposure for each group member. Otherwise, the exposure profile for the group may not be predictive of the exposure profiles for one or more of the individuals within the group.

Is the Exposure Distribution Stationary?

When evaluating exposure measurements one always assumes that the underlying processes and work practices that generate and influence exposures will remain *stationary*, or relatively constant, until the next survey or evaluation. If exposures are currently acceptable, but trending upward, then a decision based on current measurements will have little or no predictive value. Symanski et al.[5] found that most exposure distributions will remain reasonably stationary for periods up to a year, after which systematic changes are more likely.

The "x-bar, r-charts" used in quality control are sometimes adapted for use in industrial hygiene.[6,7] Such charts are useful for demonstrating that an industrial process is stationary. Generally speaking, however, the number of measurements necessary to first establish statistical control limits and the number collected thereafter during each survey are usually beyond the means of most exposure-monitoring programs.[8] However, there are control charting techniques[9] designed for small sample sizes that could be considered.*

More useful, perhaps, are time series graphs such as recommended by the CEN,[4] Roach et al.,[2] and Roach.[10] The CEN recommended plotting moving or weighted averages for visually detecting trends. George et al.[11] discussed the use of formal statistical tests of stationarity, but cautioned that "visual investigation" always be part of any analysis.

Production line processes are often fairly stable. In such work environments conclusions based on current or recent exposure measurements may be reasonably predictive for periods up to a year or longer. However, the stability of a particular work environment may not be known until an exposure history of several years has been developed. Other work environments, such as min-

ing, are inherently unstable and subject to considerable change within a short time span. In such environments regular monitoring is necessary to detect trends toward higher exposures.

Are the Exposure Data Independent?

In principle, exposure data should be independent; that is, there is no linear relationship between measurements collected on successive days. However, exposure data collected during unusually elevated or depressed production rates may be correlated and unrepresentative of exposures in general. In such a situation, measurements collected randomly throughout the observation period would be more appropriate than the usual campaign type survey.

Analysis of autocorrelation requires large datasets of consecutive measurements. However, since many researchers[11,12] have found little evidence of significant autocorrelation in measurements collected on successive days, it is reasonable to assume, in the absence of compelling information or data to conclude otherwise, that full-shift exposure measurements are reasonably independent. However, because there are often patterns of exposures within a shift, it is logical to expect that short-term measurements collected sequentially within a shift will be somewhat correlated.

Is the Lognormal Distribution Assumption Valid?

Goodness-of-fit testing involves both graphical and analytical evaluations.[13] The data should be plotted so that (a) inconsistent data points can be identified and (b) the goodness-of-fit can be subjectively evaluated. This should be followed by an objective analytical test of the hypothesis of lognormality. The lognormal distribution model is often used when zero is the physical lower limit for possible values, large values occasionally occur, and the processes that generate or control exposures tend to interact in a multiplicative manner.[14] Experience has shown that multiple exposure measurements collected either from a single individual or from multiple individuals within an exposure group tend toward a lognormal distribution.[15] Therefore, it is reasonable to assume that the underlying distribution for workplace exposure data is the lognormal distribution unless there is a compelling reason to conclude otherwise. There are, however, instances where the normal distribution may be more appropriate, for example, when multiple measurements are simultaneously collected at the same location.

---

*Traditional x bar and r control charts can be adapted for use with the lognormal distribution. Such charts are not used to determine compliance with federal or authoritative OELs, but can be used to assess whether or not the process remains in "statistical control" (e.g., stable, stationary). The control limits are calculated using the log-transformed exposure measurements and then exponentiated. This results in asymmetric upper and lower control limits. Trends can then be evaluated using the resulting "geometric mean" control chart. Changes in variability can be detected using the resulting "ratio" control chart (referring to the ratio of the largest to smallest measurement in each dataset).

Log-Probability Plotting Techniques

Log-probability plotting is the traditional method for qualitatively assessing the adequacy of the lognormal distribution model assumption. Odd patterns in the data and inconsistent data points can be readily identified.[6] However, the procedure is subjective and, when done by hand, can be tedious, particularly for large datasets.

Procedure:

(1) Sort or rank order the data: $\mathbf{x} = \{x_1, \ldots, x_n\}$ where x_1 is the smallest value and x_n is the largest.

(2) Assign a rank (r) to each sorted value where r ranges from 1 to n, starting with x_1.

(3) Calculate a plotting position, p, (basically a pseudo-cumulative frequency) using Blom's formula:*

$$p_i = \frac{i - \dfrac{3}{8}}{n + \dfrac{1}{4}} = \frac{i - 0.375}{n + 0.25}, \quad 1 \le i \le n \tag{1}$$

(4) *Log-probability plot:* Using log-probability paper, plot p versus x. (Normal-probability paper can be used if the normal distribution assumption is of interest.)

Log-probit plot: Using regular graph paper or a spreadsheet program, plot m_i versus y_i, where $y = \ln(x)$ and m represents the probit (probability unit; also called "normal order statistic" or z-value) corresponding to p:†

$$m_i = \phi^{-1}[p_i]$$

For both the log-probability and log-probit plots, a straight line can be drawn emphasizing the influence of the central 80% of the data. That is, measurements in the tails should be given less weight when fitting the straight line. If *most* of the data fall along or near the straight line, then one can qualitatively state that the data appears to be lognormal.

Personal computer statistics or spreadsheet programs can be used to produce both log-probability and log-probit plot. It is also a relatively simple matter to have the program display a linear regression line along with the data to assist in the visual evaluation.‡ An alternative plot is the cumulative distribution function (CDF) plot. The plotting position, p, is graphed versus $\ln(x)$. Ideally, the data should appear to fall along a sigmoidal, or S-shaped curve. If *most* of the data fall along or near a sigmoidal curve, then one can qualitatively state that the data appears to be lognormal.

Formal Goodness-of-Fit Tests

There are numerous statistical tests for determining whether or not a particular set of data departs significantly from the normal distribution assumption. The one recommended here, Filliben's test[16] (as modified by Looney and Gulledge[17]), is easily implemented using a computer spreadsheet program or programmable calculator. This procedure is complementary to the graphical technique in that it incorporates the concept of the probability plot. The purpose of this test is to determine if a particular set of data departs significantly from the lognormal distribution assumption by evaluating whether or not the log-transformed values depart significantly from normal.

Filliben's test can be applied to sample sizes ranging from 3 to 100. Although not presented here, a similar procedure by Royston[13] can be applied to any sample size between 5 and 5000.§

Filliben's Test Filliben[16] developed a goodness-of-fit test based on normal order statistics. Looney and Gulledge[17] recalculated Filliben's critical values and recommended substituting Blom's formulae (see Step 3 below) for the plotting position formulae developed by Filliben. The CEN,[4] in their general guidance regarding exposure assessment, listed Filliben's test as one method for formally assessing the lognormal distribution assumption.

Procedure

(1) Sort or rank order the data: $\mathbf{x} = \{x_1, \ldots, x_n\}$ where x_1 is the smallest value and x_n is the largest.

---

*There are several formulae for estimating the plotting position, for example, $p_i = (i + 0.5)/n$ or $p_i = i/(n + 1)$. Blom's formula is preferred.

†For example, for a p_i of 0.95, $m_i = 1.645$; for a p_i of 0.50, $m_i = 0.000$. These values can be obtained from the cumulative normal distribution table found in texts on statistics or calculated using the statistical functions in a spreadsheet program.

‡Simple linear regression using $\ln(x)$ versus m can be used to produce a straight line through the datapoints.

§It can be shown that Filliben's and Royston's tests are essentially identical and yield consistent results for sample sizes ranging from 5 to 100, the range of overlap for the two procedures.

(2) Assign a rank (r) to each sorted value where r ranges from 1 to n, starting with x_1.

(3) For each x_i, calculate a plotting position, p_i, using Blom's formula (Eq. 1).

(4) Determine the corresponding normal order statistic, m, for each plotting position.

(5) Using simple linear regression, regress y versus m, where $y = \ln(x)$.

(6) Calculate the correlation coefficient r.

(7) Evaluate the following hypotheses by comparing r to a table of critical values (Table A1):

H_o: y is from a normal distribution.
H_a: y is not from a normal distribution.

(8) If r is less than or equal to the critical value, reject H_o with 95% confidence. Otherwise, accept H_o and conclude that y is normally distributed. Therefore, x is lognormally distributed.

How It Works If y is truly normally distributed, the calculated correlation coefficient r will tend to be near unity, or 1.0. Under the normal distribution assumption, Filliben determined the lower 5th percentile r value for sample sizes ranging from 3 to 100. If the calculated r value is less than or equal to this critical r value, then one can state, with 95% confidence, that the distribution from which these data were drawn is not normal. Consequently, if r is less than the critical value, reject H_o with a $(1 - \alpha)$ 100% confidence level. Otherwise, there is not enough evidence to reject H_o. If H_o cannot be rejected, then the conclusion is that y is normally distributed or at least approximately normally distributed. This being the case, one can infer that x is lognormally distributed.

Filliben's test can be applied to both the actual values and the log-transformed values. One could then use the magnitude of the correlation coefficient to select between the normal or lognormal distribution assumptions. Sometimes both the lognormal and normal distribution assumptions can be rejected. It also should be noted that an observed correlation coefficient, or r value, that is less than the critical value does not necessarily mean that the underlying data do not stem from lognormal distributions. It may be that the data reflect two or more underlying lognormal distributions, reflecting perhaps the inadvertent combination of two or more distinctly different exposure groups.

Is the Exposure Group Reasonably Homogeneous?

Ideally, each member of an exposure group should have an identical exposure profile, although on any single day the exposures will vary. In practice, identical exposure profiles are unlikely to be observed. If an exposure group is reasonably homogeneous with respect to the conditions of exposure (agent/jobs/task/controls), then we expect that exposures primarily reflect the influence of process and work environment and can be reduced or controlled through direct control of the process and/or by general ventilation. If the exposure group is decidedly heterogeneous, then we expect that exposures tend to reflect the effectiveness of individual ventilation controls; differences in the number and duration of assigned tasks; and/or the influence of individual work practices, in which case exposures are reduced by focusing on individual work environments and individual work practices.

The process of grouping workers using observational skills has been described as the "observational approach" and criticized for being subjective and prone to classification errors.[15,18] How then does one objectively determine if a particular exposure group is sufficiently homogeneous so that decisions based on an analysis of group exposures are relevant to each member of the group, whether measured or not? There are no recognized criteria for objectively grouping workers or for distinguishing between a reasonably homogeneous exposure group and a clearly heterogeneous exposure group, although at least one has been proposed.[15] The CEN[4] offered the following "rule of thumb":

> [I]f an individual exposure is less than half or greater than twice the arithmetic mean [of the n measurements collected from the homogeneous group], the relevant work factors should be closely re-examined to determine whether the assumption of homogeneity was correct.

As a practical measure, many industrial hygienists use a process of continuous improvement to increase worker similarity within an exposure group. Basically, the observational approach is used to devise initial, logical exposure groups that are homogeneous with respect to process, agent, job/task, and type of controls. If, after the baseline survey, the exposure profile for the group appears acceptable, then periodic follow-up surveys are planned to evaluate the individual work practices of randomly selected workers or workers who are suspected, based upon previous measurements or professional judgment, to experience generally higher exposures. This continual, cyclic evaluation and modification of individual work practices, habits, and controls is expected to result in exposure groups that become more homogeneous over time.

Within-group homogeneity is less of an issue in those situations where the group exposures can be rated highly controlled or well-controlled (see the Hewett[1] scheme for rating exposures as minimal, well-controlled, controlled, poorly controlled, and uncontrolled). Considerable heterogeneity within an exposure group may not matter if it is highly likely that all individual exposure profiles are appropriately controlled.[3] A similar reasoning applies to the exposure group where exposures are rated poorly controlled or uncontrolled: remedial action is necessary regardless.

For those in-between situations, where the group exposure profile appears to be controlled or borderline poorly controlled, differences in individual exposure profiles may be profound for some workers. It may be that only a fraction of the workers experience the majority of the overexposures. Identification of maximum-risk employees within the group, through personal sampling or the use of direct reading instruments, may lead to splitting the group into two or more groups, each containing workers with similar exposure profiles. ANOVA-based procedures, such as those used by Woskie et al.[19] and recommended by Rappaport,[15] should be considered when experiencing difficulty establishing reasonably similar exposure groups (see subsection titled Analysis of Repeat Measurements below).

Data Analysis—Descriptive and Compliance Statistics

When characterizing workplace exposures industrial hygienists are interested in accurately estimating the population parameters for the exposure profile associated with a particular work environment. These exposures may be specific to an individual employee or an exposure group.

Population parameters are almost always unknown and must be estimated from a sample of n measurements. Estimates of the population parameters are called "sample statistics" or "point estimates." Table 16.1 contains those population parameters and associated sample statistics that usually are of interest to industrial hygienists. Most of the statistics in Table 16.1 are considered "descriptive statistics," useful for characterizing the location and shape of the underlying distribution of exposures. Several statistics identified as "compliance" statistics are useful for determining whether a particular work environment is acceptable, unacceptable, or in need of further evaluation. (Here the term "compliance" is used to apply to the determination that an *exposure profile* complies with or conforms to some accepted definition of "acceptable." This use is different from the concept that compliance implies that each and every exposure must be less than the applicable federal OEL.)

Statistics can either be parametric or nonparametric. Parametric statistics are based on the assumption that the underlying distribution of exposures can be reasonably described by a known probability distribution function, such as the lognormal or normal distribution. Nonparametric statistics are not based on any distributional assumption and for this reason are sometimes called distribution-free statistics.

Parametric Statistics

Descriptive Statistics

Arithmetic Mean and Standard Deviation The sample mean ($\bar{x}$) is the most commonly used measure of central tendency. It is an unbiased estimate of the true population mean, regardless of the underlying distribution.

Geometric Mean and Geometric Standard Deviation Authorities are in general agreement that exposure profiles are often best described by the lognormal distribution.[3,4,6,14,15] The sample geometric mean represents an estimate of the median, or 50th percentile, of the expo-

---

TABLE 16.1 Exposure Profile Parameters and Sample Statistics of Interest to Industrial Hygienists

	Population Parameter	Sample Statistic
Descriptive	mean (μ)	sample mean ($\bar{x}$)
	standard deviation (σ)	sample standard deviation (s)
	geometric mean (GM)	sample geometric mean (gm)
	geometric standard deviation (GSD)	sample geometric standard deviation (gsd)
	median ($\tilde{\mu}$)	sample median ($\tilde{x}$)
Compliance	95th percentile ($X_{0.95}$)	sample 95th percentile ($x_{0.95}$)
	exceedance fraction (θ)	sample exceedance fraction (f)
	mean (μ)*	sample mean ($\bar{x}$) or MVUE*

*The 95% upper or lower confidence limit for the true mean of the exposure profile is compared to a long-term average OEL (LTA OEL), not the TWA OEL.

sure distribution. The sample geometric mean will always be less than the sample arithmetic mean. The sample geometric standard deviation is a measure of the spread or degree of dispersion in the data. It can be interpreted as the ratio of the 84th percentile to the 50th percentile (geometric mean) exposure, or the 50th percentile to the 16th percentile:

$$gm = \exp\left[\frac{1}{n}\sum_{i=1}^{n}\ln x_i\right] = \exp\left[\frac{1}{n}\sum_{i=1}^{n}y_i\right]$$

$$gsd = \exp \quad (s_y)$$

where

$$s_y = \sqrt{\frac{\sum_{i=1}^{n}(\ln x_i - \ln gm)^2}{n-1}} = \sqrt{\frac{\sum_{i=1}^{n}(y_i - \bar{y})^2}{n-1}}$$

The lowest GSD is a theoretical 1.0, indicating absolutely no variability in the log-transformed values (the exponential of zero is one). Exposure variability can be classified according to the following rules-of-thumb:

low-exposure variability	GSD $\leq$ 1.5
moderate-exposure variability	1.5 < GSD $\leq$ 2.5
high-exposure variability	GSD > 2.5.

An exposure profile for an exposure group may have a large GSD because it contains a number of dissimilar workers. True GSDs greater than 4 are unusual, particularly for individual workers. As a rule-of-thumb, sample GSDs of 3 or more should be checked to see if dissimilar workers or activities have been combined (e.g., indoor and outdoor activities), if there is seasonal variation, or simply too few data.

Minimum Variance Unbiased Estimator If the underlying distribution for the data is approximately lognormal, which is assumed to be the case with most exposure data, then the minimum variance unbiased estimator (MVUE) is the *preferred* point estimate of the true mean, particularly when the sample size is small and/or the sample geometric standard deviation is large.[20,21] The MVUE is calculated using the following formula:

$$\bar{x}_m = \exp(\bar{y})\,\psi\left(\frac{s_y^2}{2}\right) = gm \cdot \psi\left(\frac{s_y^2}{2}\right)$$

The ψ function is defined for any argument g as:

$$\psi(g) = \left[1 + \frac{(n-1)g}{n} + \frac{(n-1)^3 g^2}{n^2(n+1)2!} + \frac{(n-1)^5 g^3}{n^3(n+1)(n+3)3!}\right.$$
$$\left. + \frac{(n-1)^7 g^4}{n^4(n+1)(n+3)(n+5)4!} + \ldots\right]$$

The above equation is easily calculated using a programmable calculator or personal computer. Calculation to at least five terms is accurate to the third decimal place.[21] (There is often little difference between the simple arithmetic mean and the MVUE. Because of its familiarity the arithmetic mean may be preferred for presentation and reporting purposes.)

Compliance Statistics

The exceedance fraction and 95th percentile statistics are useful for evaluating whether or not an exposure profile is acceptable, relative to some evaluation criterion, or OEL.

Exceedance Fraction A point estimate of the exceedance fraction (f), or fraction of exposures greater than the OEL, for lognormal distributed data can be calculated using the following equation:

$$f = P(c > OEL) = P(Z > z)$$

where

$$z = \frac{\ln OEL - \bar{y}}{s_y} \tag{2}$$

and $\bar{y}$ and s_y are the sample mean and sample standard deviation of the log-transformed data calculated from a sample of n measurements. Equation 2 can be read as "the (sample) exceedance fraction equals the probability that a future concentration exceeds the OEL." This probability can be determined in the usual fashion by consulting a Z-value table found in any statistics textbook or by using the inverse z-function found in most computer statistics or spreadsheet programs.

95th Percentile Exposure The point estimate of the ith percentile of the underlying distribution for a sample of n measurements is estimated by

$$x_{1-\alpha} = \exp[\ln gm + Z_{1-\alpha} \cdot \ln gsd]$$
$$= \exp[\bar{y} + Z_{1-\alpha} \cdot s_y]$$

where α is the area under the distribution curve to the right of the ith percentile. When estimating the 95th percentile $\alpha = 0.05$ and the Z-value is replaced by $Z_{1-0.05}$, or 1.645.

Confidence Intervals

Sample statistics are rarely identical to the true population parameters. One can gain insight into the precision of the statistic or point estimate by calculating the 90% confidence interval around the sample statistic. The 90%

confidence interval can be thought of as the interval in which we will, 90% of the time, find the true value or population parameter. If the sample size is large and/or the variability in the exposures is low, then the confidence interval can be narrow, providing assurance that the point estimate is not far from the true value. However, if the sample size is small and/or the variability is large, then the confidence interval can be quite broad, suggesting that the true value may be considerably different from our point estimate.

The bounds of the 90% interval are the 95%LCL and 95%UCL. The 95%LCL and 95%UCL are useful in that they can be directly compared to target acceptable values. For example, if the 95%UCL is less than a target value, then one can be at least 95% confident that the true value of the statistic (the population parameter) is less than the target value. Conversely, if the 95%LCL is greater than a target value, then one can be at least 95% confident that the true value of the statistic is greater than the target value.

Arithmetic Mean Two procedures are presented. The first procedure can be used for sample data where it is assumed that the underlying distribution is normal. The second is preferred when it is assumed that the underlying distribution is lognormal.

Normal Distribution Assumption This procedure is fairly robust, that is, it works well for many non-normal distributions, especially as the sample size increases, producing reasonably accurate confidence limits.

Procedure:

(1) Calculate the sample mean ($\bar{y}$) and sample standard deviation (s).
(2) Calculate the 95% upper or lower confidence limit:

$$CL = \bar{x} + t \cdot \frac{s}{\sqrt{n}}$$

where $t = t_{0.95,n-1}$ for the 95%UCL and $t = t_{0.05,n-1}$ for the 95%LCL.

Taken together, the 95%LCL and 95%UCL form a 90% confidence interval for the true arithmetic mean when the data are normally distributed and an approximate 90% confidence interval when the distribution departs from normality.

Lognormal Distribution Assumption The following procedure was adapted from Land[22] and is preferred when the underlying distribution is assumed to be reasonably lognormal. (See Hewett[23] for a review of alternative procedures.)

Procedure:

(1) Calculate an estimate of the mean of the lognormal distribution:*

$$\bar{x}_m = \exp\left(\bar{y} + \frac{1}{2}s_y^2\right)$$

(2) Obtain the appropriate C-factor. Table A2a lists the C-factors necessary to estimate the 95%UCL: $C(s_y; n, 0.95)$. Table A2b lists the C-factors used to estimate the 95%LCL: $C(s_y; n, 0.05)$. Linear or Lagrange interpolation within and between sample sizes may be required.
(3) Calculate the 95% lower or upper confidence limit:

$$CL = \exp\left[\ln(\bar{x}_m) + C\frac{s_y}{\sqrt{n-1}}\right]$$

where

CL = the upper or lower confidence limit, depending upon the choice of C

$C = C(s_y; n, 1-\alpha)$ for the UCL where $\alpha = 0.05$

$\quad = C(s_y; n, \alpha)$ for the LCL where $\alpha = 0.05$

Taken together, the 95%LCL and 95%UCL form a 90% confidence interval for the true arithmetic mean of lognormal distribution data. See Hewett and Ganser[24] for approximation formulae for calculating the appropriate C-factor. Also, interpolation can be avoided by using a table s_y value that is slightly greater than the calculated value. This will result in a confidence interval that is slightly wider than 90%.†

Exceedance Fraction The following procedure for calculating the 90% confidence interval around the exceedance fraction (for lognormally distributed data) was adapted from Odeh and Owen.[25]

---

*There are several formulae for estimating the mean, or average, of a lognormal distribution. See Reference 23 for a related discussion.

†Occasionally when the sample GSD is large and/or the sample size is small the 95%UCL for μ will be quite large, sometimes greater than the 95%UCL for the 95th percentile. While seemingly illogical, the 95% upper confidence limit for the true mean is still valid: 95% of the time the true mean will be less than the calculated 95%UCL. The best advice here is to reexamine the data and the exposure group definition. Perhaps high- and low-exposure jobs or tasks are represented in the dataset, resulting in an unusually large sample GSD. Otherwise, collect more data and recalculate.

Procedure:

(1) Calculate z using Equation 2
(2) Using z and the sample size, n, the 95% LCL for f can be read from Table A3. However, interpolation (linear or Lagrange) will usually be necessary to obtain reasonable accuracy. The 95%UCL can also be determined from Table A3. Obtain the table value using n and the negative of z. The 95%UCL for f is the complement of this value (complement = 1 − value).

Interpolation can be avoided by (a) using a simplified procedure[24] or (b) using a table z-value that is slightly greater than the calculated z-value for the LCL and UCL calculations, respectively. The latter option will result in a confidence interval that is slightly wider than 90%, but simplifies the calculations.

95th Percentile Exposure The 95%LCL and 95%UCL for the $x_{0.95}$ (assuming the exposure data are lognormally distributed) are easily estimated using K-factors developed by Odeh and Owen.[25] Unlike the exceedance fraction, the 95th percentile is not specific for any particular OEL. Consequently, it is useful when there are multiple exposure limits that can be applied (e.g., a substance may have an OSHA PEL, a NIOSH REL, and an ACGIH TLV).

Procedure:

(1) Calculate the sample mean ($\bar{y}$) and sample standard deviation (s_y) of the log-transformed data where $y = \ln(x)$.
(2) Calculate the 95% lower or upper confidence limit:

$$95\%\text{LCL} = \exp\lfloor \bar{y} + K_{0.05,0.95,n} \cdot s_y\rfloor$$

$$95\%\text{UCL} = \exp\lfloor \bar{y} + K_{0.95,0.95,n} \cdot s_y\rfloor$$

See Table A4 for the appropriate 95%LCL and 95%UCL K values. (The K values for the 95%LCL are nearly identical to the K' (K prime) values described by Tuggle.[26]) Taken together, the 95%LCL and 95%UCL form a 90% confidence interval for the true 95th percentile ($X_{0.95}$), assuming a lognormal distribution. Note that the 95% UCL for the sample 95th percentile exposure is identical to the 95% upper tolerance limit recommended by several authorities.[3,6,27,28]

Nonparametric Statistics

There may be situations where there is compelling evidence that the lognormal distribution assumption does not apply, for example, when the log-probability plot is far from linear and the data fails a formal goodness-of-fit test such as Filliben's test presented above. In this situation the preferred approach would be to evaluate the data and supporting documentation for indications that the exposure group definition needs to be reevaluated. Perhaps several distinctly different exposure groups are represented in one larger group. Another approach would be to apply nonparametric procedures for estimating the median, 95th percentile, and exceedance fraction. One disadvantage to using nonparametric, or distribution-free, statistics is that the confidence intervals are wider than those estimated assuming a particular distribution, such as the lognormal or normal distribution. Perhaps this is why nonparametric statistics are not often reported in the industrial hygiene literature. Readers interested in nonparametric statistics should review the recommendations by Rock,[29] Esmen,[30] and the EASC.[3,6] A good general reference is that of Conover.[31]

Descriptive Statistics

The first step in a nonparametric analysis is to sort the data from low to high values. The ith ordered observation will be referred to as x_i.

Mean The simple arithmetic mean is an unbiased estimate of the center of mass of a distribution, regardless of the shape of the distribution.

Median The 50th percentile of any distribution is known as the median. If the sample size is odd, then the sample median is simply the middle value. If the sample size is even, the sample median is the average of the two middle values. Specifically:

for $n = 2k + 1$ (odd n): $\tilde{x} = x_{k+1}$

for $n = 2k$ (even n): $\tilde{x} = \dfrac{1}{2}(x_k + x_{k+1})$

The median exposure is one measure of central tendency of the exposure profile. It has no current use in determining compliance.

Compliance Statistics

Exceedance Fraction The observed, or nonparametric, exceedance fraction ($\tilde{f}$) is simply the ratio of the number of overexposures (i.e., measurements greater than the OEL) to the sample size: $\tilde{f} = m/n$, where m = number of overexposures and n = sample size. The observed exceedance fraction can often be zero for small sample sizes, even when the actual probability of an overexposure is quite large.*

95th Percentile Exposure The nonparametric 95th percentile can be estimated for $n \geq 20$. Linear interpolation between rank-ordered data is usually necessary to estimate the 95th percentile exposure.[32,33]

$$\tilde{x}_{0.95} = x_i + (0.95n - i)(x_{i+1} - x_i)$$

where i = integer portion of $0.95 \cdot n$

Confidence Intervals

Median For $n \leq 30$ confidence intervals for the median (50th percentile) exposure can be calculated using the following procedure.[31] Let q equal the quantile (percentile) of interest, 0.5 in this case. The ordered rank of the 95%LCL, l, is determined by finding l such that

$$\sum_{x=0}^{l-1} b\,(x;\, n,\, q) = 0.05 \tag{3}$$

where $b\,(x;\, n,\, q)$ is the binomial function: $b\,(x;\, n,\, q) = \binom{n}{x} q^x(1-q)^{n-x}$, $x = 0, 1, 2, \ldots, n$. The ordered rank of the 95%UCL, u, is determined by setting q to 0.5 and finding u such that

$$\sum_{x=0}^{u-1} b\,(x;\, n,\, q) = 0.95$$

Table A5 contains the rank of the 95%LCL and 95%UCL for the sample median for sample sizes ranging from 5 to 30. If $n > 30$, the following formulae can be used to estimate the rank of the 95%LCL or 95%UCL value, where q equals 0.5, and $Z = 1.645$[32,34]:

$$l = q(n + 1) - Z\sqrt{nq(1 - q)} \tag{4}$$
$$u = q(n + 1) + Z\sqrt{nq(1 - q)} \tag{5}$$

The resulting rank—l or u—should be rounded to the next smaller or larger integer, respectively. Alternatively, linear interpolation between the ordered values can be used to determine the 95%LCL or 95%UCL. Equations 4 and 5 are conservative, thus resulting in a confidence interval slightly wider than 90%.

Exceedance Fraction Confidence limits for the nonparametric exceedance fraction can be determined from Table A6 for $n \leq 30$. Confidence intervals for larger sample

sizes are determined by finding θ (pronounced "theta"; $0 < \theta < 1$) such that the following equations are true[35]:

$$\sum_{x=m}^{n} b\,(x;\, n,\, \theta_{\mathrm{LCL}}) = 0.05$$

$$\sum_{x=0}^{m} b\,(x;\, n,\, \theta_{\mathrm{UCL}}) = 0.05$$

For $n > 30$ the following equations can be used to calculate reasonably accurate confidence intervals.[32,36] The 95%LCL and 95%UCL is estimated by substituting 1.645 for Z in the following equations:

$$\theta_{\mathrm{LCL}} = \frac{1}{n + Z^2} \cdot$$
$$\left\{ (m - 0.5) + \frac{Z^2}{2} - Z \cdot \sqrt{(m - 0.5) - \frac{(m - 0.5)^2}{n} + \frac{Z^2}{4}} \right\}$$

$$\theta_{\mathrm{UCL}} = \frac{1}{n + Z^2} \cdot$$
$$\left\{ (m + 0.5) + \frac{Z^2}{2} + Z \cdot \sqrt{(m + 0.5) - \frac{(m + 0.5)^2}{2} + \frac{Z^2}{4}} \right\}$$

except if
$\quad m = 0$, then $\theta_{\mathrm{LCL}} = 0$
$\quad m = 1$, then $\theta_{\mathrm{LCL}} = 1 - (1 - \alpha)^{1/n}$ where $\alpha = 0.05$ for the
$\quad\quad$ 95%LCL(θ)
$\quad m = n$, then $\theta_{\mathrm{UCL}} = 1$.

95th Percentile The 95%LCL can be calculated for sample sizes greater than 4 (although 19 measurements are necessary before the point estimate can be calculated). In contrast, the sample size has to exceed 58 before the more useful 95%UCL can be calculated. The 95%LCL can be determined using Equation 3: let $q = 0.95$ and determine l such that the equation is true. This procedure was used to determine the values in Table A5. This table contains the rank (r) of the 95%LCL for the sample 95th percentile for sample sizes ranging from 5 to 30. If $n > 30$, Equation 4 is adequate for estimating the rank of the 95%LCL[32]: let $q = 0.95$ and $Z = 1.645$ and calculate l. If $n > 58$, Equation 5 may be used in a similar manner to estimate the rank of the 95%UCL.

Comments Confidence intervals for the nonparametric 95th percentile require large sample sizes, particularly for the 95%UCL, thus reducing their usefulness for risk-management decision making. In contrast, 90% confidence intervals for the exceedance fraction can be calculated for virtually any sample size. Note that these confidence intervals will often be considerably broader

---

*For large sample sizes, say $n > 30$, the calculated and observed exceedance fraction should be similar if the data are approximately lognormal.

than those estimated using the parametric lognormal assumption. Consequently, nonparametric statistics should only be used when one is confident in rejecting the lognormal (or normal) distribution assumption. Reevaluating the exposure group definitions and the range of tasks or activities within the exposure group may lead to new exposure groups where the lognormal assumption is reasonable.

Data Analysis Issues

Censored Data

Measurements less than the limit of detection (LOD) (or limit-of-quantification) often occur, particularly for well-controlled work environments and when the exposure limit is close to the LOD. Simply ignoring the LOD values will bias the sample GM upward and the sample GSD downward. There are various substitution techniques for estimating distribution parameters from datasets containing censored data: each censored datum is simply replaced with the (1) LOD, (2) half of the LOD, or (3) LOD divided by $\sqrt{2}$. The CEN[4] recommended using half of the LOD. Simple substitution works well when the percentage of LOD values is small. More sophisticated techniques generally produce more accurate sample estimates, but require tables and intermediate calculations. The reader is advised to consult the references before adopting any particular scheme.[32,37,38]

Inconsistent Data Points

Inconsistent data points, or what some might call outliers, should be carefully considered. A true outlier represents a gross error in either sampling or analysis, or an aberrant condition in the work environment that is clearly unlikely to be repeated. Unless strongly justified, inconsistent data should be not be eliminated from a dataset. Such values most likely are part of the actual exposure profile and reflect exposure conditions that should be investigated.[3,7]

Analysis of Repeat Measurements

Exposure variability within an exposure group can be divided into within-worker and between-worker components.[3,15] Within-worker factors that influence exposure variability include differences in assigned tasks or time at task, work practices, and individual exposure controls. Between-worker factors include production level variation and general ventilation. There is growing interest in the use of analysis-of-variance (ANOVA) or components-of-variance techniques to estimate these within-

and between-worker components of overall variability. Rappaport[15] illustrates the calculation techniques and advocates their use in assessing the degree of heterogeneity within exposure groups. Woskie et al.[19] described the application of similar techniques to short-term (within-shift) and full-shift measurements. Such analyses can guide decisions to focus on modifying general engineering/ventilation controls, or to focus on modifying individual engineering/ventilation controls or work practices. However, these techniques require the collection of repeat measurements for each of n randomly selected workers in each exposure group. Others have described ANOVA-based techniques for assessing compliance.[39,40] These proposals merit consideration, but apply only to true long-term average OELs. In summary, when repeat measurements are available for workers, the techniques described in these and similar papers may be useful for gaining insight into the relative contribution of the various sources of workplace variability and help direct intervention efforts.

Data Interpretation

The underlying goal of any data interpretation scheme is to *reliably* determine, given the available data, that the exposure profile of *each* worker in *each* exposure group is either acceptable or unacceptable for the current observation period.[1] Often an industrial hygienist may defer a final decision until additional information or data can be collected. However, unless action is taken to reduce exposures, a decision to defer is by default a decision that the work environment is acceptable. It is generally accepted that decisions in the presence of uncertainty should always be made in favor of the workers.[3]

For those exposure profiles judged acceptable, the decision should be made with a high degree of confidence. How much confidence is needed depends on the numerous factors involved. For example, if resampling intervals are close, then less confidence can be tolerated during each survey, such as when using a control chart approach or a simple decision scheme. If resampling occurs infrequently, then a high degree of confidence is needed such as provided by a formal statistical test using the 95%UCL. If the OEL is uncertain, such as when a working or provisional[1] OEL is used or the legal OEL is dated and no longer considered protective, then a high degree of confidence is needed. The same can be said for situations where the substance is particularly toxic or the toxicological information is uncertain.

Data interpretation schemes generally fall into one of three categories: nonparametric decision logics, parametric decision logics, and control chart techniques.

Nonparametric Decision Logics

Nonparametric decision logics are based on simple decision rules or the use of non-parametric statistics. Rule-based decision logics require few, if any, statistical calculations, and no underlying distribution is assumed.

Rule-Based Decision Logics

Historically, a common procedure for determining the acceptability of a "work environment" is to collect one or several measurements from one or more maximum-risk employees and apply simple decision rules. As the quantity of exposure data has increased, industrial hygienists have adopted ever more sophisticated and statistically defensible data analysis procedures. However, simple decision rules have a legitimate place in federal regulations[41] and when designing exposure surveillance programs (where one or a few exposures over the OEL or an action limit triggers a more comprehensive evaluation).

The 1977 NIOSH[42] sampling strategy and decision logic is summarized in Table 16.2. Notice that a decision can be rapidly reached with one or a few measurements. NIOSH introduced the concept of an "Action Level," or half of the OEL. Measurements above the Action Level, but below the OEL, triggered additional sampling. This scheme permitted rapid and efficient decision making. However, as noted by Tuggle,[27] the decision may not be the right decision, even when the true exceedance fraction is 0.25 or greater.

More recently the CEN[4] described the use of simple decision rules. For datasets containing less than six measurements, collected from a specific "homogeneous expo-

sure group," *each* measurement should be compared to the OEL. A single overexposure should trigger an investigation and remedial action, if necessary. The CEN provided several examples for determining when to re-sample and how many measurements to collect. One example is summarized in Table 16.3.

Nonparametric Statistical Analysis

Nonparametric statistical analysis has been recommended for those instances where one cannot assume that the data are derived from a lognormal distribution.[3,6,29,30] In general, larger sample sizes are needed before statistical significance is reached. Because of this limitation, judgment is often needed when balancing the expense of further sampling against the expense of installing additional controls.[3,30]

If the nonparametric 95th percentile exposure is less than the OEL, then one has evidence that the true 95th percentile is less than the OEL, but the confidence level is unknown. At least 20 measurements are needed just to estimate the 95th percentile, while 58 measurements are necessary to estimate the corresponding 95%UCL. Is this a reasonable use of resources? Perhaps attention should be given to defining exposure groups where parametric statistics can be used.

If the nonparametric exceedance fraction ($\tilde{f} = m/n$) is less than 0.05, then again one has evidence that the true exceedance fraction is less than 0.05. However, nonparametric exceedance fractions can be extremely misleading when the sample size is small. For example, consider a dataset where $n = 2$ and both measurements are below the OEL. The point estimate of the exceedance

---

TABLE 16.2 Example Decision Logic for Small Sample Sizes

Sampling Strategy:

Collect one or more measurements (C) from one or more maximum risk employees (MRE). Compare each measurement to the Permissible Exposure Limit (PEL) and the Action Level (AL = 0.5 · PEL).

Decision Rules	Decison/Action
1. If C < AL	then conclude the work environment is *acceptable* for the exposure group presented by the MRE.
2. If C > PEL	then conclude that the work environment is *unacceptable*, take corrective action, and re-sample at least monthly.
3. If AL ≤ C ≤ PEL	then collect an additional measurement at least every two months until either two consecutive measurements are less than half the PEL (and conclude that the exposures are acceptable) or any single measurement is above the PEL (and conclude that exposures are *unacceptable* and take appropriate actions to reduce exposures).

Note. Adapted from NIOSH.[42]

TABLE 16.3 Example Decision Logic for Small Sizes

Sampling strategy:

Collect a single measurement every two months from each exposure group. Compare *each* measurement to four action limits:

$N1 = 0.40 \cdot OEL$
$N2 = 0.70 \cdot OEL$
$N3 = 1.00 \cdot OEL$
$N4 = 1.50 \cdot OEL$

Decision Rules	Decision/Action
1. If $C \leq N1$ twice consecutively	then collect one measurement every 6 months.
2. If $C \leq N2$	then continue collecting one measurement every two months.
3. If $N2 < C \leq N4$	then collect two measurements every two months.*
4. If $N2 < C \leq N4$ twice consecutively	then collect two measurements every two months for eight months.†
5. If $N3 \leq C \leq N4$ twice consecutively	then immediate action is warranted to reduce exposures.
6. If $C > N4$	then immediate action is warranted to reduce exposures.

Note. Adapted from CEN.[4]

*Every overexposure ($C > N3$) should be investigated, the reasons for the overexposure identified, and appropriate measures taken.

†If the two measurements were collected on the same survey, then immediate action is warranted to reduce exposures.

fraction is 0.0, but because the 95%UCL is 0.777 (see Table A6) we have little confidence that the true exceedance fraction is less than 0.05.

Parametric Decision Logics

In 1977 NIOSH[42] stated that the goal of an effective exposure-monitoring program is to "attain 95% confidence that no more than 5% of employee days are over the standard." NIOSH was not implying that it is permissible to overexpose each employee once every twenty shifts, but was merely providing a "statistical" goal for designing an exposure monitoring program. A similar goal has been suggested by numerous organizations,[3,4,6] authoritative individuals,[10,29,43,44,45] and corporations.[46,47]

According to the AIHA[3,6] the exposure profile for a similar exposure group is usually deemed acceptable if it is highly likely that only a small percentage of the measurements exceed the OEL. This is the case if the parametric or nonparametric upper confidence limit on the 95th percentile is less than the OEL, or the upper confidence limit on the exceedance fraction is less than 0.05.

If six or measurements are randomly collected from an exposure group, the CEN[4] recommends that one use statistics to estimate the probability of overexposure for individuals within the HEG. If this probability exceeds 5%, then corrective action should take place. Otherwise, periodic monitoring should be used to confirm that the point estimate of the probability of overexposure remains less than 5%.

The above recommendations are pertinent to TWA OELs. When dealing with long-term average OELs (LTA OEL; see Chapter 15) some researchers advocate the use of ANOVA techniques when analyzing exposure measurements.[39,40] Such techniques require the characterization of the distribution of individual long-term mean exposures *within* an exposure group. This necessitates collecting multiple measurements from each of n randomly selected workers per exposure group. ANOVA techniques are then used for determining the probability that the long-term mean of any single worker in the exposure group exceeds LTA OEL.[15] Rappaport et al.[39] suggest that the exposure profile of the exposure group may be judged acceptable if this probability is 0.10 or less. Tests for comparing the average exposure for an exposure group or individual worker have also been described.[48]

Consistent with these authoritative recommendations, Table 16.4 contains generic criteria for determining whether a dataset suggests that the exposure profile for a particular exposure group or work environment is currently acceptable or unacceptable relative to a TWA OEL or LTA OEL. The evaluation statistic utilized can be either the point estimate of the exposure parameter of interest or its associated upper or lower confidence limit. For example, if the point estimate of the 95th percentile exposure is less than the TWA OEL, then one has evidence that the work environment is acceptable, but not necessarily compelling evidence. Compelling evidence exists when the 95% upper confidence limit for the 95th percentile is less than the TWA OEL. A similar logic ap-

TABLE 16.4 Evaluation Criteria for Testing Whether a Dataset Represents "Acceptable" or "Unacceptable" Exposure Conditions

Evaluation Criteria			
TWA OEL or STEL OEL	**LTA OEL**	**Exposure Profile**	**Appropriate Action**
$95\%\mathrm{UCL}(x_{0.95}) \leq \mathrm{OEL}$ or $95\%\mathrm{UCL}(f) \leq 0.05$	$95\%\mathrm{UCL}(\bar{x}) \leq \mathrm{LTA\ OEL}$	(clearly) acceptable	Periodically re-sample.
$x_{0.95} \leq \mathrm{OEL}$ or $f \leq 0.05$	$\bar{x} \leq \mathrm{LTA\ OEL}$	acceptable	Periodically re-sample.
$x_{0.95} > \mathrm{OEL}$ or $f > 0.05$	$\bar{x} > \mathrm{LTA\ OEL}$	unacceptable	Take steps to reduce exposures. Re-sample.
$95\%\mathrm{LCL}(x_{0.95}) > \mathrm{OEL}$ or $95\%\mathrm{LCL}(f) > 0.05$	$95\%\mathrm{LCL}(\bar{x}) > \mathrm{LTA\ OEL}$	(clearly) unacceptable	Take *immediate* steps to reduce exposures. Re-sample.

plies to the exceedance fraction and to the long-term mean (when a LTA OEL applies).

Note that concluding that an exposure profile is currently "acceptable" relative to an OEL permits one to *state* that the exposure profile "appears to be controlled." If the 95%UCL is compared to the OEL one could state that the exposure profile "is controlled, with at least 95% confidence." The OEL referred to in Table 16.4 should be replaced by either half or a tenth of the actual OEL in order to evaluate whether or not the true exposure profile can be rated "well-controlled" or "highly controlled" (see Table 15.3, Chapter 15[1]).

Control Chart Techniques

Control charts or time series plots can be used for recognizing trends[4,32] and cycles, to assess the stability of the work environment, and for visually comparing measurements to OELs and Action Limits. Roach et al.[2] and Roach[10] recommended the use of time series plots coupled with simple decision rules. For example, a warning line could be established at or below the OEL. Measurements above the warning line should elicit a repeat visit. Two measurements in a row above the warning line should result in immediate action. The CEN[4] also recommended a time-series plot for routinely collected measurements. Both the individual measurements and the moving average is plotted against time. In this manner the acceptability of individual values can be assessed, as well as any trends toward higher exposures.

The combination of time-series plots and simple decision rules should be considered when dealing with dynamic work environments where significant change, as part of the normal work process, is expected. Such techniques are also suited to work environments that are rated "controlled" (see Table 15.3, Chapter 15[1]). This is because continual surveillance and continuous improve-

ment are necessary until such time that statistical tests permit the exposure profile to be rated "well-controlled" or "minimal."

The application of statistical process control techniques to the control of exposures in the work environment is encouraged. The TWA OEL could serve as an upper specification limit that, in principle, should not be exceeded. For practical purposes, the upper specification limit could be defined as the 95th percentile of a "controlled" exposure profile. As measurements accumulate the annual (or less) long-term mean should be one-third or less of the TWA OEL, and not be permitted to exceed half the single-shift limit. It is even conceivable for stable processes to calculate traditional control limits for gauging drift and trends. As exposures are controlled, the upper specification limit could be progressively reduced, encouraging further reductions in exposure, thereby minimizing risk to the workers and the future liability risk of the employer.

Data Interpretation Issues

Interpretation of a Single
Exposure Measurement

Authoritative sources are unanimous in recommending that *each* overexposure be *investigated*, regardless of the level of sophistication of the exposure-monitoring program or the past exposure history[4,6] (see related discussion in Ref. 41). If there is compelling or convincing past exposure data to suggest that the overexposure is most likely a random occurrence in an otherwise controlled exposure profile, then it is reasonable to take no action beyond merely documenting the investigation. However, if no rational explanation can be found for the overexposures, one is compelled to conclude that a systematic change of some sort *may* have occurred; after all, in a "controlled" work environment overexposures should be infrequent to rare. Follow-up actions may consist of fine tuning existing controls, installation or modification of controls, or an evaluation of individual work practices. Regardless, additional measurements are usually warranted in order to verify the need for additional controls or to evaluate the effectiveness of any intervention.

As a rule-of-thumb, one should always be suspicious when one or more measurements exceeds 50% of the OEL in a small dataset, particularly when $n < 6$. This suggests that the exceedance fraction of the current exposure profile may be unacceptable (i.e., greater than 0.05).

Dual Limits

Many substances have dual limits; that is, both a TWA OEL and an STEL OEL. OSHA[49] noted that if the full-shift, TWA exposure is less than the PEL, then the short-term exposures are also likely controlled. Along similar lines, Spear et al.[50] concluded, using mathematical analysis, that the exceedance fraction for short-term, 15-minute average exposures will be limited to 5% or less if the exceedance fraction for full-shift, TWA exposure is limited to 5% or less. However, one cannot always rely upon a satisfactory TWA measurement to indicate that within-shift excursions are also acceptable. Short-term exposures should always be evaluated when there are predictable or recognizable within-shift cycles and episodes of high exposure.

In Great Britain vinyl chloride is subject to a 7 ppm single shift limit and a 3 ppm annual average limit.[51] In the United States there are no examples of explicit dual limits that consist of a TWA OEL and a LTA OEL. However, in a "controlled" work environment the long-term average exposure for each individual worker should be much less than the OEL, particularly for chronic disease agents, and the single shift limit should be infrequently exceeded, if at all. As discussed in Hewett[1] (Chapter 15), it is reasonable to devise a provisional LTA OEL using the AIHA recommendation of one third the single shift limit (or TWA OEL). The long-term average should be less than the LTA OEL and at no time exceed one half of the single shift limit.

Examples

Example 1

Table 16.5 contains total welding fume data collected in three welding departments at an automobile frame manufacturing plant. The data represent measurements collected on a single day in 1987 from randomly selected welders. Each measurement represents the full-shift, 8-hour TWA exposure to medium-steel welding fumes for each of n welders in the three departments. The applicable OEL is the 1987 ACGIH TLV of 5 mg/m^3 for welding fumes (total particulate,* not otherwise classified). Measurements were also collected on other days during this week as part of a research study,[52] but the data from a single day will be used to illustrate the calculations presented in this chapter. Assume that these data were collected as part of an initial baseline survey of these welding operations. It was not possible to monitor all welders, so a reasonable number were selected from each department. Since all the welders in each department were engaged in similar tasks, used similar welding consumables and equipment, and were subject to similar ventilation controls, it was logical to initially consider each "department" as an exposure group. What then can be said from an analysis of these data?

Goodness-of-Fit

Figure 16.1 shows the log-probit curves for these data. This graph was created using a standard computer spreadsheet program. The x-axis can be displayed in "normal" or "log" terms. The "log" x-axis is usually selected under the usual assumption that the data are lognormal distributed. The y-axis is in terms of the z-value equivalent of the plotting position (see Goodness-of-Fit section). In all three cases, the data appear "reasonably" linear with no obvious outliers or data points that need individual investigation. All three datasets pass the revised Filliben's goodness-of-fit test: the calculated correlation coefficient, r, exceeded the appropriate critical value found in Table A1. Since neither the subjective graphical evaluation nor the objective goodness-of-fit test suggest that the data do not come from a lognormal distribution, we may calculate descriptive and compliance statistics with reasonable assurance that our point estimates will be valid. Of course, with limited data, say n less than 10, it is difficult to reject any distributional assumption.

Descriptive Statistics

All three departments exhibited moderate exposure variability as indicated by the sample GSDs, which ranged from 1.56 to 1.67 (see Table 16.6). The sample GMs

---

*Since welding fumes are typically smaller than 1 μm in diameter, the "total particulate" specified in 1987 is essentially equivalent to the "inhalable particulate" specified in the 1997 TLV booklet.

TABLE 16.5 Data Used in the Examples (Sorted in Ascending Order)

Automobile Frame Manufacturing (welding fumes, mg/m³)[52]			Chemical Plant (inorganic lead, μg/m³)[53]		
Department B	Department C	Department E	Worker A		
0.21	1.63	6.39	3.9	12.4	21.5
0.42	2.02	6.89	7.9	12.9	21.9
0.49	2.04	9.59	8.6	13.0	22.2
0.58	2.32	10.89	9.0	14.4	24.6
	4.28	19.97	9.0	15.0	25.4
	6.04		9.5	15.9	25.6
			10.0	17.1	25.7
			10.0	18.6	28.9
			10.2	19.1	30.4
			10.4	19.5	34.0
			11.3	19.6	46.9
			11.4	20.2	56.4

are useful for estimating the median, or 50th percentile exposure, of a lognormal exposure distribution. The sample arithmetic mean is useful for indicating whether or not the long-term average exposure will be controlled to the ideal range of less than one-third of the OEL. In this example only Department B had a sample mean near this range. The confidence intervals about the sample mean were quite broad, reflecting the small sample size. Consequently, we have to interpret the estimates of the sample mean with some caution, as the true values may be considerably different from the sample estimates. It will require periodic monitoring over a period of a year or more to determine conclusively whether or not the long-term mean for Department B is routinely maintained in this range.

If the underlying distributions are reasonably lognor-

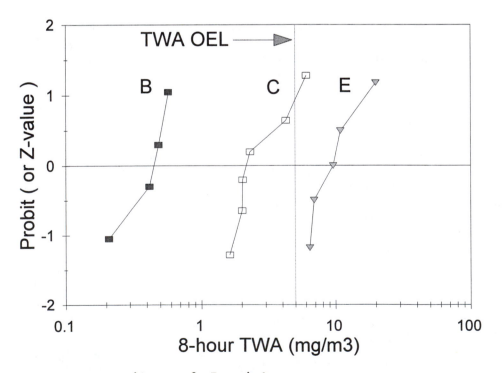

FIGURE 16.1 Log-probit curves for Example 1.

TABLE 16.6 Parametric Descriptive and Compliance Statistics for Examples 1 and 2 (the 90% confidence intervals are given in parentheses)

	Auto Frame Manufacturing Plant[52] (welding fumes, mg/m^3)			Chemical Plant[53] (inorganic lead, µg/m^3)
	Department B	Department C	Department E	Worker A
n	4	6	5	36
min	0.21	1.63	6.39	3.9
max	0.58	6.04	19.97	56.4
Descriptive statistics				
$\bar{x}$	0.43	3.06	10.75	18.7
	(0.29–1.06)	(2.15–5.66)	(7.58–20.60)	(16.1–22.4)
gm	0.40	2.72	9.83	16.2
gsd	1.56	1.67	1.57	1.73
mvue	0.43	3.02	10.66	18.7
Compliance statistics				
f_{OEL}	<0.01	0.12	0.93	0.02
	(<<0.01–0.03)	(0.02–0.40)	(0.63–0.99)	(<0.01–0.06)
$x_{0.95}$	0.83	6.29	20.75	39.7
	(0.55–3.95)	(4.25–18.01)	(14.25–66.33)	(32.5–52.5)
Goodness-of-fit				
log-probit graph	reasonably lognormal	reasonably lognormal	reasonably lognormal	lognormal
Filliben's test	$0.933 \geq 0.868$*	$0.937 \geq 0.889$*	$0.956 \geq 0.880$*	$0.987 \geq 0.969$*
r (lognormal)	∴ lognormal	∴ lognormal	∴ lognormal	∴ lognormal

*Critical value. If r is less than the critical value, then there is evidence that the underlying distribution is not lognormal.

mal, then the MVUE is the best point estimate of the true mean. In most cases, it will be similar to the simple arithmetic sample mean, but may vary substantially when the sample size is small and the sample GSD is large.

Compliance Statistics

The sampled welders working in Department B experienced exposures all of which were less than 12% of the OEL. In contrast, the sampled welders in Department E all experienced exposures greater than the OEL. In Department C all of the measurements were greater than 10% of the OEL, with one out of the six exceeding the OEL.

For Department B the compliance statistics suggest the point estimate of the exceedance fraction is virtually zero, with a 95%UCL of 0.03. Even with this small sample size we can be more than 95% confident that the true exceedance fraction is less than 0.05. Similarly, both the point estimate of the 95th percentile and its 95%UCL were less the OEL. For Department C the point estimate of the exceedance fraction is 0.12 while the point estimate of the 95th percentile exceeds the OEL. For De-

partment E the 95%LCLs for the 95th percentile and exceedance fraction exceeded the TWA OEL and 0.05, respectively.

Conclusions and Recommendations

With small sample sizes such as these all of the measurements should be less than the OEL for one to declare that the exposure profile appears "controlled" and that the work environment is currently "acceptable" (see Table 16.4). Even so, the confidence intervals will often be broad with 95%UCL values well into the unacceptable range. This necessitates occasional remonitoring to verify or validate the initial assessment and to determine if higher risk employees were missed in the baseline survey.

Exposures in Department B *appear* to be "well-controlled" (see Hewett,[1] Chapter 15, Table 15.3, for the statistical interpretations of these terms). Because the 95%UCL for the 95th percentile is less than the TWA OEL, we can *conclude*, with at least 95% confidence, that the exposure profile for this exposure group is "controlled." Occasional remonitoring and reassessment should take place in order to confirm this assessment, preferably with a mix of repeat sampling and sampling of other welders.

Regarding Department C, exposures routinely were substantial (i.e., greater than 10% of the OEL), with a point estimate of 12% overexposures. Steps should be taken to reduce exposures and to evaluate others in this exposure group. It is possible that adjusting existing controls and modifying individual work practices will be sufficient to permit, after follow-up sampling, a rating of "controlled" or better. However, periodic remonitoring will probably be necessary until the stability of the process can be ascertained and an exposure history is developed over a period of several years.

Exposures in Department E are obviously unacceptable without any calculations. We can expect approximately 93% of the exposures in this department to exceed the OEL. Immediate action is warranted, to include the use of interim respiratory protection. No further monitoring is necessary until improvements have been made in either the ventilation system and/or work practices.

Observations

As a general rule, it is recommended that one have at least 6–10 measurements before using statistical techniques for characterizing an exposure profile.[4,6] While fewer than six measurements often lead to highly variable *estimates* (i.e., statistics) of the true exposure profile parameters, it is often possible to reach a highly accurate *decision* regarding the question "Is the exposure profile acceptable?" Here we see that for Departments B and E sound conclusions can be reached, despite the fact that less than six measurements were used. If exposure profiles can be controlled to the point that the exposures can be rated "well-controlled" or "highly controlled" and the process is relatively stable, then the future sampling burden will be minimal.

The nonparametric descriptive and compliance statistics in Table 16.7 provide several interesting comparisons to Table 16.6. For Department B we calculated a parametric 95%UCL for the exceedance fraction of 0.03. If we used nonparametric statistics, we would be compelled to accept the proposition that the true exceedance fraction might be as great as 0.53, despite the fact that the maximum value was only 12% of the TWA OEL. Furthermore, the nonparametric 95th percentile cannot be estimated with such small sample sizes.

Example 2

Table 16.5 also contains inorganic lead personal exposure data collected at an alkyl lead manufacturing plant.[53] The measurements were collected from a single

worker over the course of 6 weeks. These data were collected as part of a research project, but are useful as an example of the analysis of a larger dataset collected over a longer span of time. We will use the 1978 OSHA PEL of 50 $\mu g/m^3$ as the relevant OEL.

Goodness-of-Fit

Figure 16.2 shows the log-probit plot of the data. The regression line was determined by simple linear regression of the log-transformed concentration values and the corresponding z-value. The data appear to follow the regression line suggesting that the lognormal distribution assumption is reasonable. This assessment is corroborated by Filliben's test where the correlation coefficient for the lognormal assumption was greater than the critical value. Consequently, we can conclude that there is no reason to reject the lognormal distribution and proceed to calculate parametric descriptive and compliance statistics.

Descriptive Statistics

Here we have a substantial number of measurements, permitting us to calculate reasonably accurate estimates of distribution parameters (Table 16.6). If the process is reasonably stable, these estimates should have considerable predictive value. The sample GM suggests that the true median will be approximately 16 $\mu g/m^3$. The sample GSD of 1.73 suggests a moderately variable work environment. The average exposure is 37% of the OEL, close to our goal of one-third of the TWA OEL or less.

Compliance Statistics

The point estimate of the exceedance fraction is 0.02, but the 95%UCL is slightly above 0.05. Similarly, the point estimate of the 95th percentile is less than the OEL, but the 95%UCL slightly exceeds the OEL. In comparison, the observed, or nonparametric, exceedance fraction (see Table 16.7) was 0.03 (one of 36 measurements exceeded the OEL). If the underlying distribution is truly lognormal, the observed and calculated exceedance fractions should, with increasing sample size, converge to the same value.

Conclusions and Recommendations

This exposure profile *appears* to be "controlled"; that is, the point estimates of the parametric 95th percentile and exceedance fraction are less than the target values. However, the 95%UCLs are slightly greater than the target values. Consequently, regular monitoring is necessary to confirm the designation of "controlled" exposure

TABLE 16.7 Nonparametric Description and Compliance Statistics for Examples 1 and 2

	Auto Frame Manufacturing Plant[52] (welding fumes, mg/m³)			Chemical Plant[53] (inorganic lead, μg/m³)
	Department B	Department C	Department E	Worker A
Descriptive statistics				
$\tilde{x}$ (median)	0.455	2.18	9.59	16.5
	(NA–NA)	(1.63–6.04)	(6.39–19.97)	(12.4–20.2)
Compliance statistics				
$\tilde{f}_{OEL} = m/n$	0.00	0.17	1.00	0.03
	(0.00–0.53)	(0.01–0.58)	(0.54–1.00)	(<0.01–0.13)
$\tilde{x}_{0.95}$	NA	NA	NA	34.1
(95th percentile)	$n < 20$	$n < 20$	$n < 20$	(28.9–NA)*

The 90% Confidence Intervals are given in parentheses. In addition, the sample size, minimum and maximum value, and arithmetic mean should also be reported.
*Because $n < 58$, only the 95%LCL can be estimated.

profile. Notice that even with 36 measurements the 95%UCL for the observed (i.e., nonparametric) exceedance fraction (see Table 16.7) is much larger than the parametric 95%UCL.

Analysis of work practices may reduce peak exposures within each shift, thus reducing the full-shift TWA exposures. In this manner, the long-term mean could be reduced to below the target value of one-third the TWA OEL. Furthermore, the 95%UCL for the either the 95th percentile or exceedance fraction may then fall below the target values—the TWA OEL and 0.05, respectively—permitting one to *conclude*, with at least 95% confidence, that the exposure profile is "controlled," or even "well-controlled."

Observations

These measurements were collected sequentially from a single worker; therefore, without additional information, our conclusions apply only to this worker. Let's assume

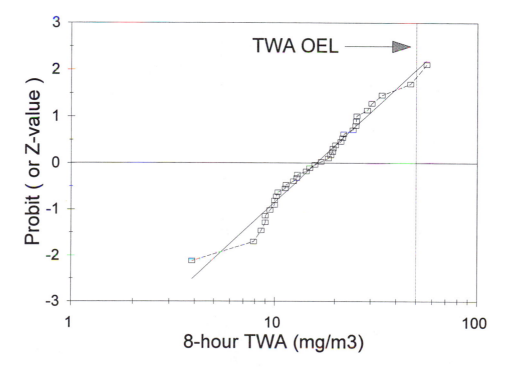

FIGURE 16.2 Log-probit curve for Example 2.

that the data represent measurements collected from one or more presumably maximum-risk employees. In this scenario we could state that the exposure profiles of these members of the exposure group, and by extension, all members of the exposure group, appear to be "controlled." Our selection of maximum-exposed workers should be verified by collecting measurements from randomly selected members of the exposure group during future surveys. Otherwise, our conclusion may not apply to all of the unmeasured members of the exposure group.

Alternatively, let's assume that the data represent measurements collected from randomly selected workers drawn from an exposure group where it is difficult to select maximum risk employees. Somewhat more caution is warranted. We could conclude from our statistical analysis that the exposure profile for the group appears to be "controlled." However, we are ultimately interested in each individual exposure profile. Whether or not this conclusion applies to all exposure profiles within the group depends on how homogeneous the exposure group is with respect to exposure controls and work practices. Ideally, we should reserve judgment until an exposure history has been developed. In the meantime, we can state that the exposure profiles appear to be "controlled," but that periodic surveillance is warranted to confirm our tentative rating.

In either case—selection of presumed maximum exposure employees or random selection—task analysis and assessment of individual work practices should increase the degree of homogeneity within each exposure group, thus increasing the predictive value of our measurements. If more than one measurement is available per employee, then an analysis of within-worker and between-worker variability may be useful for determining whether exposures are primarily affected by factors common to all workers in the exposure group or by factors specific to each worker.[15,19]

APPENDIX

TABLE A1 Filliben's Test Critical r Values ($\alpha = 0.05$) for Determining Goodness-of-Fit for the Normal or Lognormal Distribution Assumption (Ref. 17)

n	r	n	r	n	r
3	0.879	23	0.956	42	0.973
4	0.868	24	0.957	43	0.974
5	0.880	25	0.959	44	0.974
6	0.888	26	0.960	45	0.974
7	0.898	27	0.961	46	0.975
8	0.906	28	0.962	47	0.976
9	0.912	29	0.963	48	0.976
10	0.918	30	0.964	49	0.976
11	0.923	31	0.965	50	0.977
12	0.928	32	0.966	55	0.979
13	0.932	33	0.967	60	0.980
14	0.935	34	0.968	65	0.981
15	0.939	35	0.969	70	0.983
16	0.941	36	0.969	75	0.984
17	0.944	37	0.970	80	0.985
18	0.946	38	0.971	85	0.985
19	0.949	39	0.971	90	0.986
20	0.951	40	0.972	95	0.987
21	0.952	41	0.973	100	0.987
22	0.954				

TABLE A2a C Factors for Estimating the 95%UCL ($\alpha = 0.05$) for the Mean of Lognormally Distributed Data (Ref. 22)

s_y*	Sample Size (n)								
	3	4	5	6	7	8	9	10	15
0.01	2.415	2.054	1.918	1.849	1.807	1.779	1.759	1.745	1.706
0.1	2.750	2.222	2.035	1.942	1.886	1.849	1.822	1.802	1.749
0.2	3.295	2.463	2.198	2.069	1.992	1.943	1.908	1.881	1.809
0.3	4.109	2.777	2.402	2.226	2.125	2.058	2.011	1.977	1.882
0.4	5.220	3.175	2.651	2.415	2.282	2.195	2.134	2.089	1.968
0.5	6.495	3.658	2.947	2.638	2.465	2.354	2.277	2.220	2.068
0.6	7.807	4.209	3.287	2.892	2.673	2.534	2.439	2.368	2.181
0.7	9.120	4.801	3.662	3.173	2.904	2.735	2.618	2.532	2.306
0.8	10.43	5.414	4.062	3.477	3.155	2.952	2.813	2.710	2.443
0.9	11.74	6.038	4.478	3.796	3.420	3.184	3.021	2.902	2.589
1.0	13.05	6.669	4.905	4.127	3.698	3.426	3.239	3.103	2.744
1.25	16.33	8.265	6.001	4.990	4.426	4.068	3.820	3.639	3.163
1.5	19.60	9.874	7.120	5.880	5.184	4.741	4.433	4.207	3.612
1.75	22.87	11.49	8.250	6.786	5.960	5.432	5.065	4.795	4.081
2.0	26.14	13.11	9.387	7.701	6.747	6.135	5.710	5.396	4.564
2.5	32.69	16.35	11.67	9.546	8.339	7.563	7.021	6.621	5.557
3.0	39.23	19.60	13.97	11.40	9.945	9.006	8.350	7.864	6.570
3.5	45.77	22.85	16.27	13.27	11.56	10.46	9.688	9.118	7.596
4.0	52.31	26.11	18.58	15.14	13.18	11.92	11.03	10.38	8.630

s_y*	Sample Size (n)								
	20	30	40	60	101	201	401	601	1001
0.01	1.689	1.673	1.666	1.659	1.653	1.649	1.647	1.647	1.646
0.1	1.725	1.702	1.6911	1.680	1.670	1.662	1.658	1.656	1.654
0.2	1.776	1.744	1.728	1.712	1.697	1.685	1.677	1.674	1.671
0.3	1.838	1.796	1.775	1.753	1.733	1.716	1.705	1.700	1.090
0.4	1.977	1.850	1.832	1.803	1.777	1.755	1.740	1.734	1.728
0.5	1.999	1.932	1.898	1.862	1.830	1.802	1.784	1.776	1.769
0.6	2.097	2.015	1.974	1.930	1.891	1.857	1.835	1.825	1.816
0.7	2.205	2.108	2.058	2.007	1.960	1.919	1.892	1.881	1.870
0.8	2.324	2.209	2.151	2.090	2.035	1.988	1.957	1.944	1.931
0.9	2.451	2.318	2.251	2.181	2.117	2.062	2.027	2.012	1.997
1.0	2.586	2.434	2.357	2.277	2.205	2.143	2.102	2.085	2.068
1.25	2.952	2.750	2.648	2.542	2.447	2.364	2.310	2.288	2.266
1.5	3.347	3.094	2.966	2.832	2.713	2.609	2.542	2.514	2.486
1.75	3.763	3.457	3.303	3.142	2.997	2.872	2.791	2.757	2.723
2.0	4.193	3.835	3.654	3.465	3.295	3.148	3.053	3.013	2.974
2.5	5.079	4.617	4.384	4.139	3.920	3.729	3.605	3.553	3.503
3.0	5.988	5.424	5.138	4.838	4.569	4.334	4.183	4.119	4.057
3.5	6.910	6.244	5.907	5.552	5.233	4.956	4.776	4.700	4.627
4.0	7.841	7.074	6.685	6.276	5.908	5.588	5.380	5.293	5.208

*$s_y = \ln(gsd)$

TABLE A2b C Factors for Estimating the 95% LCL ($\alpha = 0.95$) for the Mean of Lognormally Distributed Data (Ref. 22)

s_y*	Sample Size (n)								
	3	4	5	6	7	8	9	10	15
0.01	−2.355	−2.022	−1.896	−1.831	−1.791	−1.766	−1.747	−1.734	−1.697
0.1	−2.130	−1.898	−1.806	−1.759	−1.731	−1.712	−1.699	−1.690	−1.666
0.2	−1.949	−1.791	−1.729	−1.697	−1.678	−1.667	−1.658	−1.653	−1.640
0.3	−1.816	−1.710	−1.669	−1.650	−1.639	−1.633	−1.629	−1.627	−1.625
0.4	−1.717	−1.650	−1.625	−1.615	−1.611	−1.610	−1.610	−1.611	−1.617
0.5	−1.644	−1.605	−1.594	−1.592	−1.594	−1.596	−1.599	−1.603	−1.618
0.6	−1.589	−1.572	−1.573	−1.578	−1.584	−1.591	−1.597	−1.602	−1.625
0.7	−1.549	−1.550	−1.560	−1.572	−1.582	−1.592	−1.600	−1.608	−1.638
0.8	−1.521	−1.537	−1.555	−1.572	−1.586	−1.599	−1.610	−1.620	−1.656
0.9	−1.502	−1.530	−1.556	−1.577	−1.595	−1.611	−1.625	−1.637	−1.680
1.0	−1.490	−1.530	−1.562	−1.588	−1.610	−1.628	−1.644	−1.658	−1.707
1.25	−1.486	−1.549	−1.596	−1.632	−1.662	−1.687	−1.708	−1.727	−1.793
1.5	−1.508	−1.590	−1.650	−1.696	−1.733	−1.764	−1.791	−1.814	−1.896
1.75	−1.547	−1.647	−1.719	−1.774	−1.819	−1.857	−1.889	−1.916	−2.015
2.0	−1.598	−1.714	−1.799	−1.864	−1.917	−1.960	−1.998	−2.029	−2.144
2.5	−1.727	−1.877	−1.986	−2.070	−2.138	−2.193	−2.241	−2.283	−2.430
3.0	−1.880	−2.065	−2.199	−2.301	−2.384	−2.452	−2.510	−2.560	−2.740
3.5	−2.051	−2.272	−2.429	−2.550	−2.647	−2.727	−2.795	−2.855	−3.067
4.0	−2.237	−2.491	−2.672	−2.810	−2.922	−3.015	−3.093	−3.161	−3.406

s_y*	Sample Size (n)								
	20	30	40	60	101	201	401	601	1001
0.01	−1.682	−1.668	−1.661	−1.655	−1.651	−1.647	−1.646	−1.646	−1.645
0.1	−1.656	−1.648	−1.646	−1.643	−1.642	−1.643	−1.644	−1.645	−1.645
0.2	−1.637	−1.636	−1.636	−1.638	−1.641	−1.646	−1.649	−1.651	−1.653
0.3	−1.626	−1.631	−1.635	−1.641	−1.648	−1.656	−1.663	−1.666	−1.669
0.4	−1.624	−1.634	−1.641	−1.651	−1.662	−1.674	−1.684	−1.688	−1.693
0.5	−1.629	−1.644	−1.655	−1.668	−1.683	−1.699	−1.711	−1.717	−1.723
0.6	−1.640	−1.661	−1.675	−1.692	−1.711	−1.731	−1.746	−1.753	−1.760
0.7	−1.658	−1.684	−1.700	−1.721	−1.744	−1.768	−1.786	−1.795	−1.804
0.8	−1.680	−1.711	−1.731	−1.756	−1.783	−1.811	−1.832	−1.842	−1.853
0.9	−1.708	−1.744	−1.767	−1.795	−1.826	−1.859	−1.884	−1.895	−1.907
1.0	−1.740	−1.781	−1.807	−1.839	−1.874	−1.912	−1.940	−1.953	−1.966
1.25	−1.835	−1.889	−1.923	−1.965	−2.012	−2.060	−2.097	−2.114	−2.131
1.5	−1.949	−2.016	−2.058	−2.111	−2.169	−2.229	−2.275	−2.296	−2.318
1.75	−2.078	−2.158	−2.208	−2.272	−2.341	−2.414	−2.469	−2.494	−2.514
2.0	−2.218	−2.311	−2.370	−2.445	−2.526	−2.611	−2.675	−2.705	−2.736
2.5	−2.525	−2.645	−2.721	−2.817	−2.921	−3.032	−3.115	−3.154	−3.194
3.0	−2.856	−3.003	−3.096	−3.214	−3.342	−3.478	−3.581	−3.628	−3.677
3.5	−3.204	−3.377	−3.488	−3.627	−3.780	−3.940	−4.062	−4.119	−4.177
4.0	−3.563	−3.764	−3.892	−4.052	−4.228	−4.414	−4.555	−4.620	−4.688

*$s_y = \ln(gsd)$

TABLE A3 Confidence Limit Values for Estimating the 95%LCL and 95%UCL for the Proportion in the Tail of a Normal (or Lognormal) Distribution (Ref. 25; Table 7.4)

	Sample Size (n)								
z	2	3	4	5	6	7	8	9	10
3.0	0.00000	0.00000	0.00000	0.00000	0.00000	0.00000	0.00001	0.00001	0.00001
2.8	0.00000	0.00000	0.00000	0.00000	0.00001	0.00001	0.00002	0.00003	0.00004
2.6	0.00000	0.00000	0.00001	0.00002	0.00003	0.00005	0.00007	0.00009	0.00011
2.4	0.00000	0.00001	0.00003	0.00006	0.00010	0.00014	0.00020	0.00026	0.00031
2.2	0.00000	0.00003	0.00009	0.00018	0.00029	0.00041	0.00054	0.00067	0.00080
2.0	0.00002	0.00011	0.00029	0.00052	0.00079	0.00107	0.00135	0.00163	0.00191
1.8	0.00008	0.00038	0.00085	0.00140	0.00199	0.00257	0.00315	0.00371	0.00424
1.6	0.00030	0.00115	0.00226	0.00344	0.00461	0.00574	0.00682	0.00783	0.00879
1.4	0.00106	0.00311	0.00543	0.00772	0.00987	0.01187	0.01372	0.01543	0.01701
1.2	0.00317	0.00753	0.01188	0.01589	0.01951	0.02277	0.02571	0.02837	0.03079
1.0	0.00820	0.01629	0.02364	0.03006	0.03565	0.04055	0.04488	0.04874	0.05220
0.8	0.01831	0.03164	0.04292	0.05236	0.06034	0.06719	0.07315	0.07839	0.08305
0.6	0.03544	0.05534	0.07131	0.08424	0.09493	0.10394	0.11168	0.11840	0.12433
0.4	0.05997	0.08771	0.10903	0.12581	0.13941	0.15071	0.16028	0.16853	0.17574
0.2	0.09006	0.12725	0.15462	0.17559	0.19228	0.20594	0.21740	0.22718	0.23566
0.0	0.12240	0.17114	0.20542	0.23099	0.25095	0.26707	0.28044	0.29175	0.30148
−0.2	0.15403	0.21648	0.25856	0.28913	0.31255	0.33120	0.34649	0.35932	0.37027
−0.4	0.18342	0.26118	0.31177	0.34762	0.37460	0.39580	0.41299	0.42728	0.43938
−0.6	0.21024	0.30413	0.36360	0.40483	0.43534	0.45898	0.47795	0.49357	0.50671
−0.8	0.23469	0.34491	0.41329	0.45976	0.49359	0.51947	0.54001	0.55677	0.57076
−1.0	0.25712	0.38345	0.46048	0.51188	0.54869	0.57648	0.59829	0.61592	0.63052
−1.2	0.27784	0.41982	0.50506	0.56090	0.60025	0.62954	0.65226	0.67045	0.68537
−1.4	0.29713	0.45416	0.54703	0.60671	0.64806	0.67839	0.70162	0.72002	0.73498
−1.6	0.31518	0.48661	0.58642	0.64929	0.69205	0.72292	0.74625	0.76451	0.77921
−1.8	0.33219	0.51732	0.62332	0.68867	0.73223	0.76315	0.78616	0.80395	0.81812
−2.0	0.34828	0.54640	0.65781	0.72489	0.76866	0.79913	0.82146	0.83849	0.85187
−2.2	0.36356	0.57398	0.68997	0.75806	0.80144	0.83104	0.85234	0.86834	0.88077
−2.4	0.37814	0.60013	0.71989	0.78826	0.83072	0.85905	0.87906	0.89384	0.90516
−2.6	0.39207	0.62495	0.74766	0.81563	0.85668	0.88341	0.90192	0.91535	0.92547
−2.8	0.40543	0.64851	0.77335	0.84028	0.87950	0.90440	0.92125	0.93325	0.94215
−3.0	0.41828	0.67087	0.79706	0.86237	0.89942	0.92229	0.93741	0.94796	0.95565
−3.2	0.43064	0.69210	0.81886	0.88204	0.91666	0.93740	0.95077	0.95990	0.96643
−3.4	0.44258	0.71224	0.83886	0.89946	0.93146	0.95003	0.96168	0.96946	0.97491
−3.6	0.45411	0.73134	0.85713	0.91478	0.94405	0.96048	0.97049	0.97701	0.98149
−3.8	0.46527	0.74945	0.87376	0.92818	0.95468	0.96903	0.97752	0.98291	0.98652
−4.0	0.47609	0.76661	0.88885	0.93983	0.96357	0.97596	0.98306	0.98744	0.99031
−4.2	0.48658	0.78285	0.90249	0.94988	0.97095	0.98151	0.98737	0.99088	0.99313
−4.4	0.49677	0.79821	0.91476	0.95849	0.97701	0.98592	0.99069	0.99346	0.99519
−4.6	0.50668	0.81272	0.92577	0.96584	0.98196	0.98938	0.99321	0.99537	0.99668
−4.8	0.51633	0.82641	0.93560	0.97205	0.98595	0.99207	0.99510	0.99676	0.99774
−5.0	0.52572	0.83932	0.94434	0.97727	0.98915	0.99413	0.99651	0.99776	0.99848
−5.2	0.53488	0.85147	0.95208	0.98164	0.99169	0.99570	0.99754	0.99847	0.99899
−5.4	0.54381	0.86290	0.95890	0.98525	0.99368	0.99688	0.99829	0.99897	0.99934
−5.6	0.55252	0.87363	0.96489	0.98823	0.99524	0.99776	0.99882	0.99932	0.99958
−5.8	0.56103	0.88369	0.97013	0.99067	0.99644	0.99841	0.99920	0.99955	0.99973
−6.0	0.56935	0.89311	0.97468	0.99265	0.99736	0.99888	0.99946	0.99971	0.99983

(continued)

TABLE A3 *Continued*

	Sample Size (*n*)								
n	11	15	21	30	40	60	80	100	120
3.0	0.00002	0.00003	0.00006	0.00011	0.00016	0.00024	0.00030	0.00036	0.00040
2.8	0.00005	0.00009	0.00017	0.00027	0.00037	0.00054	0.00068	0.00078	0.00087
2.6	0.00014	0.00025	0.00041	0.00064	0.00085	0.00118	0.00143	0.00163	0.00180
2.4	0.00038	0.00062	0.00097	0.00142	0.00182	0.00244	0.00290	0.00325	0.00353
2.2	0.00094	0.00145	0.00214	0.00299	0.00373	0.00481	0.00559	0.00618	0.00665
2.0	0.00218	0.00320	0.00447	0.00597	0.00723	0.00904	0.01029	0.01123	0.01197
1.8	0.00476	0.00660	0.00881	0.01130	0.01334	0.01617	0.01809	0.01951	0.02062
1.6	0.00969	0.01281	0.01641	0.02031	0.02341	0.02761	0.03040	0.03243	0.03401
1.4	0.01848	0.02341	0.02889	0.03463	0.03909	0.04499	0.04884	0.05162	0.05375
1.2	0.03301	0.04030	0.04814	0.05612	0.06219	0.07005	0.07511	0.07872	0.08147
1.0	0.05534	0.06545	0.07602	0.08653	0.09436	0.10434	0.11066	0.11513	0.11852
0.8	0.08723	0.10046	0.11396	0.12710	0.13674	0.14884	0.15640	0.16172	0.16572
0.6	0.12959	0.14604	0.16250	0.17822	0.18958	0.20367	0.21238	0.21846	0.22302
0.4	0.18211	0.20173	0.22101	0.23913	0.25206	0.26790	0.27760	0.28433	0.28935
0.2	0.24310	0.26577	0.28768	0.30795	0.32224	0.33956	0.35007	0.35732	0.36270
0.0	0.30997	0.33553	0.35982	0.38197	0.39740	0.41592	0.42705	0.43467	0.44032
—0.2	0.37977	0.40802	0.43444	0.45816	0.47450	0.49389	0.50544	0.51331	0.51911
—0.4	0.44981	0.48048	0.50867	0.53361	0.55056	0.57048	0.58223	0.59020	0.59604
—0.6	0.51795	0.55063	0.58014	0.60583	0.62308	0.64311	0.65482	0.66270	0.66846
—0.8	0.58264	0.61677	0.64703	0.67294	0.69010	0.70978	0.72116	0.72877	0.73431
—1.0	0.64284	0.67773	0.70809	0.73361	0.75027	0.76913	0.77990	0.78706	0.79224
—1.2	0.69787	0.73280	0.76255	0.78708	0.80283	0.82041	0.83033	0.83686	0.84156
—1.4	0.74741	0.78161	0.81008	0.83305	0.84755	0.86347	0.87233	0.87811	0.88225
—1.6	0.79132	0.82407	0.85068	0.87164	0.88461	0.89861	0.90628	0.91123	0.91475
—1.8	0.82967	0.86037	0.88463	0.90325	0.91454	0.92648	0.93291	0.93702	0.93991
—2.0	0.86268	0.89084	0.91243	0.92855	0.93809	0.94797	0.95319	0.95649	0.95879
—2.2	0.89067	0.91596	0.93473	0.94831	0.95614	0.96407	0.96818	0.97074	0.97251
—2.4	0.91407	0.93630	0.95224	0.96338	0.96963	0.97581	0.97893	0.98085	0.98216
—2.6	0.93334	0.95248	0.96569	0.97460	0.97946	0.98412	0.98642	0.98781	0.98875
—2.8	0.94896	0.96512	0.97582	0.98276	0.98642	0.98984	0.99148	0.99246	0.99311
—3.0	0.96145	0.97481	0.98328	0.98855	0.99123	0.99366	0.99480	0.99546	0.99590
—3.2	0.97127	0.98210	0.98866	0.99256	0.99447	0.99615	0.99691	0.99735	0.99763
—3.4	0.97888	0.98750	0.99246	0.99527	0.99660	0.99772	0.99821	0.99849	0.99867
—3.6	0.98469	0.99141	0.99508	0.99706	0.99796	0.99869	0.99900	0.99917	0.99927
—3.8	0.98906	0.99420	0.99685	0.99821	0.99880	0.99926	0.99945	0.99955	0.99962
—4.0	0.99229	0.99615	0.99803	0.99894	0.99932	0.99960	0.99971	0.99977	0.99980
—4.2	0.99464	0.99748	0.99879	0.99938	0.99962	0.99979	0.99985	0.99988	0.99990
—4.4	0.99633	0.99839	0.99927	0.99965	0.99979	0.99989	0.99992	0.99994	0.99995
—4.6	0.99752	0.99898	0.99957	0.99981	0.99989	0.99994	0.99996	0.99997	0.99998
—4.8	0.99835	0.99937	0.99975	0.99989	0.99994	0.99997	0.99998	0.99999	0.99999
—5.0	0.99892	0.99962	0.99986	0.99994	0.99997	0.99999	0.99999	0.99999	1.00000
—5.2	0.99930	0.99977	0.99992	0.99997	0.99999	0.99999	1.00000	1.00000	1.00000
—5.4	0.99955	0.99987	0.99996	0.99999	0.99999	1.00000	1.00000	1.00000	1.00000
—5.6	0.99972	0.99992	0.99998	0.99999	1.00000	1.00000	1.00000	1.00000	1.00000
—5.8	0.99983	0.99996	0.99999	1.00000	1.00000	1.00000	1.00000	1.00000	1.00000
—6.0	0.99989	0.99998	0.99999	1.00000	1.00000	1.00000	1.00000	1.00000	1.00000

(continued)

TABLE A3 *Continued*

z	Sample Size (n)		
	240	600	1000
3.0	0.00058	0.00080	0.00090
2.8	0.00121	0.00160	0.00178
2.6	0.00239	0.00307	0.00338
2.4	0.00455	0.00567	0.00617
2.2	0.00831	0.01007	0.01084
2.0	0.01453	0.01718	0.01832
1.8	0.02437	0.02818	0.02979
1.6	0.03926	0.04447	0.04665
1.4	0.06075	0.06757	0.07038
1.2	0.09039	0.09892	0.10240
1.0	0.12939	0.13964	0.14378
0.8	0.17844	0.19028	0.19501
0.6	0.23737	0.25056	0.25579
0.4	0.30504	0.31931	0.32493
0.2	0.37941	0.39445	0.40033
0.0	0.45772	0.47323	0.47926
−0.2	0.53687	0.55255	0.55860
−0.4	0.61380	0.62932	0.63527
−0.6	0.68583	0.70084	0.70656
−0.8	0.75088	0.76502	0.77037
−1.0	0.80759	0.82053	0.82539
−1.2	0.85536	0.86683	0.87109
−1.4	0.89425	0.90406	0.90767
−1.6	0.92484	0.93295	0.93589
−1.8	0.94811	0.95456	0.95687
−2.0	0.96521	0.97016	0.97191
−2.2	0.97736	0.98102	0.98229
−2.4	0.98571	0.98831	0.98920
−2.6	0.99125	0.99303	0.99363
−2.8	0.99481	0.99598	0.99636
−3.0	0.99701	0.99776	0.99800
−3.2	0.99833	0.99879	0.99893
−3.4	0.99910	0.99937	0.99945
−3.6	0.99953	0.99968	0.99973
−3.8	0.99976	0.99984	0.99987
−4.0	0.99988	0.99993	0.99994
−4.2	0.99994	0.99997	0.99997
−4.4	0.99997	0.99999	0.99999
−4.6	0.99999	0.99999	1.00000
−4.8	1.00000	1.00000	1.00000
−5.0	1.00000	1.00000	1.00000
−5.2	1.00000	1.00000	1.00000
−5.4	1.00000	1.00000	1.00000
−5.6	1.00000	1.00000	1.00000
−5.8	1.00000	1.00000	1.00000
−6.0	1.00000	1.00000	1.00000

TABLE A4 K Factors for Calculating the 95%LCL and 95%UCL for the 95th Percentile of Normally or Lognormally Distributed Data (Ref. 25; Tables 1.10.1–1.10.4 and 1.4.1–1.4.4)

$K_{0.05, 0.95, n}$				$K_{0.95, 0.95, n}$			
n	K	n	K	n	K	n	K
2	0.475	40	1.297	2	26.260	40	2.125
3	0.639	41	1.300	3	7.656	41	2.118
4	0.743	42	1.304	4	5.144	42	2.111
5	0.818	43	1.308	5	4.203	43	2.105
6	0.875	44	1.311	6	3.708	44	2.098
7	0.920	45	1.314	7	3.399	45	2.092
8	0.958	46	1.317	8	3.187	46	2.086
9	0.990	47	1.321	9	3.031	47	2.081
10	1.017	48	1.324	10	2.911	48	2.075
11	1.041	49	1.327	11	2.815	49	2.070
12	1.062	50	1.329	12	2.736	50	2.065
13	1.081	55	1.343	13	2.671	55	2.042
14	1.098	60	1.354	14	2.614	60	2.022
15	1.114	65	1.364	15	2.566	65	2.005
16	1.128	70	1.374	16	2.524	70	1.990
17	1.141	75	1.382	17	2.486	75	1.976
18	1.153	80	1.390	18	2.453	80	1.964
19	1.164	85	1.397	19	2.423	85	1.954
20	1.175	90	1.403	20	2.396	90	1.944
21	1.184	95	1.409	21	2.371	95	1.935
22	1.193	100	1.414	22	2.349	100	1.927
23	1.202	120	1.433	23	2.328	120	1.899
24	1.210	140	1.447	24	2.309	140	1.879
25	1.217	160	1.459	25	2.292	160	1.862
26	1.225	180	1.469	26	2.275	180	1.849
27	1.231	200	1.478	27	2.260	200	1.837
28	1.238	300	1.507	28	2.246	300	1.800
29	1.244	400	1.525	29	2.232	400	1.778
30	1.250	500	1.537	30	2.220	500	1.763
31	1.255	600	1.546	31	2.208	600	1.752
32	1.261	700	1.553	32	2.197	700	1.744
33	1.266	800	1.559	33	2.186	800	1.737
34	1.271	900	1.563	34	2.176	900	1.732
35	1.276	1000	1.567	35	2.167	1000	1.727
36	1.280	1500	1.581	36	2.158	1500	1.712
37	1.284	2000	1.590	37	2.149	2000	1.703
38	1.289	3000	1.600	38	2.141	3000	1.692
39	1.293	5000	1.610	39	2.133	5000	1.681
		10000	1.620			10000	1.670

TABLE A5 Ranks (r) for the Nonparametric 95%LCL and 95%UCL for the 50th Percentile and the 95%LCL for the 95th Percentile (taken together, the LCL and UCL form a 90% confidence interval for the true median)

	Median (50th Percentile)		95th Percentile		Median (50th Percentile)		95th Percentile
n	r LCL	r UCL	r LCL	n	r LCL	r UCL	r LCL
5	1	5	4	18	6	13	15
6	1	6	5	19	6	14	16
7	1	7	6	20	6	15	17
8	2	7	6	21	7	15	18
9	2	8	7	22	7	16	19
10	2	9	8	23	8	16	20
11	3	9	9	24	8	17	21
12	3	10	10	25	8	18	22
13	4	10	11	26	9	18	23
14	4	11	12	27	9	19	24
15	4	12	13	28	10	19	25
16	5	12	14	29	10	20	25
17	5	13	14	30	11	20	26

Note. The sample size has to exceed 58 before a 95%UCL rank can be calculated for the 95th percentile.

TABLE A6 Nonparametric 95% Lower and Upper Confidence Limit on the Fraction Exceeding an OEL [The left and right values represent the 95%LCL and 95%UCL, respectively. Taken together, they form the nonparametric 90% confidence interval for the true exceedance fraction (m refers to the number of overexposures out of n measurements).]

m	$n = 1$		$n = 2$		$n = 3$		$n = 4$		$n = 5$	
0	0.000	0.950	0.000	0.777	0.000	0.632	0.000	0.527	0.000	0.451
1	—	—	0.025	0.975	0.016	0.865	0.012	0.752	0.010	0.658
2	—	—	—	—	0.135	0.984	0.097	0.903	0.076	0.811
3	—	—	—	—	—	—	—	—	0.189	0.924

m	$n = 6$		$n = 7$		$n = 8$		$n = 9$		$n = 10$	
0	0.000	0.393	0.000	0.348	0.000	0.313	0.000	0.283	0.000	0.259
1	0.008	0.582	0.007	0.521	0.006	0.471	0.005	0.429	0.005	0.395
2	0.062	0.729	0.053	0.659	0.046	0.600	0.041	0.550	0.036	0.507
3	0.153	0.847	0.128	0.775	0.111	0.711	0.097	0.655	0.087	0.607
4	—	—	0.225	0.872	0.193	0.807	0.168	0.749	0.150	0.697
5	—	—	—	—	—	—	0.251	0.832	0.222	0.778

m	$n = 11$		$n = 12$		$n = 13$		$n = 14$		$n = 15$	
0	0.000	0.239	0.000	0.221	0.000	0.206	0.000	0.193	0.000	0.181
1	0.004	0.365	0.004	0.339	0.003	0.317	0.003	0.297	0.003	0.280
2	0.033	0.470	0.030	0.438	0.028	0.410	0.026	0.386	0.024	0.364
3	0.078	0.565	0.071	0.528	0.066	0.495	0.061	0.466	0.056	0.440
4	0.135	0.651	0.122	0.610	0.112	0.573	0.104	0.540	0.096	0.511
5	0.199	0.729	0.181	0.685	0.165	0.646	0.152	0.610	0.141	0.578
6	0.271	0.801	0.245	0.755	0.224	0.713	0.206	0.675	0.190	0.641
7	—	—	—	—	0.287	0.776	0.263	0.737	0.243	0.700
8	—	—	—	—	—	—	—	—	0.300	0.757

(continued)

TABLE A6 *Continued*

m	n = 16		n = 17		n = 18		n = 19		n = 20	
0	0.000	0.171	0.000	0.162	0.000	0.154	0.000	0.146	0.000	0.140
1	0.003	0.264	0.003	0.251	0.002	0.238	0.003	0.227	0.002	0.217
2	0.022	0.344	0.021	0.327	0.020	0.311	0.019	0.296	0.018	0.283
3	0.053	0.417	0.049	0.396	0.047	0.377	0.044	0.360	0.042	0.344
4	0.090	0.485	0.084	0.461	0.079	0.439	0.075	0.420	0.071	0.401
5	0.132	0.549	0.123	0.522	0.116	0.498	0.109	0.476	0.104	0.456
6	0.177	0.609	0.166	0.581	0.156	0.554	0.147	0.530	0.139	0.508
7	0.226	0.667	0.211	0.636	0.199	0.608	0.187	0.582	0.177	0.558
8	0.278	0.722	0.260	0.690	0.244	0.660	0.229	0.632	0.217	0.607
9	—	—	0.310	0.740	0.291	0.709	0.274	0.680	0.258	0.653
10	—	—	—	—	—	—	0.320	0.726	0.302	0.698

m	n = 21		n = 22		n = 23		n = 24		n = 25	
0	0.000	0.133	0.000	0.128	0.000	0.123	0.000	0.118	0.000	0.113
1	0.002	0.207	0.002	0.198	0.002	0.191	0.002	0.183	0.002	0.177
2	0.017	0.271	0.016	0.260	0.015	0.250	0.015	0.240	0.014	0.231
3	0.040	0.330	0.038	0.316	0.036	0.304	0.034	0.293	0.033	0.282
4	0.067	0.385	0.064	0.369	0.061	0.355	0.059	0.342	0.056	0.330
5	0.098	0.437	0.094	0.420	0.089	0.404	0.085	0.390	0.082	0.376
6	0.132	0.488	0.126	0.469	0.120	0.451	0.114	0.435	0.110	0.420
7	0.168	0.536	0.160	0.516	0.152	0.497	0.145	0.479	0.139	0.463
8	0.205	0.583	0.195	0.561	0.186	0.541	0.178	0.522	0.170	0.504
9	0.245	0.628	0.232	0.605	0.221	0.584	0.211	0.563	0.202	0.544
10	0.285	0.672	0.271	0.648	0.258	0.625	0.246	0.604	0.235	0.584
11	0.328	0.715	0.311	0.689	0.296	0.665	0.282	0.643	0.269	0.622
12	—	—	—	—	0.335	0.704	0.319	0.681	0.305	0.659
13	—	—	—	—	—	—	—	—	0.341	0.695

m	n = 26		n = 27		n = 28		n = 29		n = 30	
0	0.000	0.109	0.000	0.105	0.000	0.102	0.000	0.099	0.000	0.095
1	0.001	0.170	0.001	0.164	0.001	0.159	0.001	0.154	0.001	0.149
2	0.013	0.223	0.013	0.216	0.012	0.209	0.012	0.202	0.011	0.196
3	0.032	0.272	0.031	0.263	0.029	0.255	0.028	0.247	0.027	0.239
4	0.054	0.319	0.052	0.308	0.050	0.298	0.048	0.289	0.046	0.280
5	0.079	0.363	0.075	0.351	0.073	0.340	0.070	0.329	0.068	0.319
6	0.105	0.406	0.101	0.393	0.097	0.380	0.094	0.368	0.090	0.357
7	0.133	0.447	0.128	0.433	0.123	0.419	0.119	0.406	0.115	0.394
8	0.163	0.487	0.156	0.472	0.150	0.457	0.145	0.443	0.140	0.430
9	0.194	0.527	0.186	0.510	0.179	0.494	0.172	0.479	0.166	0.465
10	0.225	0.565	0.216	0.547	0.208	0.530	0.200	0.515	0.193	0.500
11	0.258	0.602	0.248	0.583	0.238	0.566	0.229	0.549	0.221	0.533
12	0.292	0.638	0.280	0.619	0.269	0.601	0.259	0.583	0.249	0.566
13	0.326	0.674	0.313	0.653	0.300	0.634	0.289	0.616	0.278	0.599
14	—	—	—	—	0.333	0.667	0.320	0.648	0.308	0.630
15	—	—	—	—	—	—	0.352	0.680	0.338	0.662

References

1. Hewett, P. Industrial Hygiene Exposure Assessment—Data Collection and Management. In *Handbook of Chemical Health and Safety*; American Chemical Society: Washington, DC, 2001; Chapter 15.

2. Roach, S. A.; Baier, E. J.; Ayer, H. E.; Harris, R. L. Testing Compliance with Threshold Limit Values for Respirable Dusts. *Am. Ind. Hyg. Assoc. J.* **1967**, *28*, 543–553

3. Mulhausen, J., Damiano, J., Eds. *A Strategy for Assessing and Managing Occupational Exposures*, 2nd ed.; American Industrial Hygiene Association: Fairfax, VA, 1998.

4. CEN (Comité Européen de Normalisation). Workplace atmospheres—Guidance for the assessment of exposure by inhalation of chemical agents for comparison with limit values and measurement strategy. *European Standard EN 689*, effective no later than Aug. 1995 (English version) (Feb. 1995).

5. Symanski, E.; Kupper, L. L.; Kromhout, H.; Rappaport, S. M. An Investigation of Systematic Changes in Occupational Exposure. *Am. Ind. Hyg. Assoc. J.* **1996**, *57*, 724–735.

6. Hawkins, N. C., Norwood, S. K., Rock, J. C., Eds. *A Strategy for Occupational Exposure Assessment*; American Industrial Hygiene Association: Fairview, VA, 1991.

7. Oudyk, J. D. Review of an Extensive Ferrous Foundry Silica Sampling Program. *Appl. Occup. Environ. Hyg.* **1995**, *10*, 331–340.

8. Hawkins, N. C.; Landenberger, B. D. Statistical Control Charts: A Technique for Analyzing Industrial Hygiene Data. *Appl. Occup. Environ. Hyg.* **1991**, *6*, 689–695.

9. Montgomery, D. C. *Introduction to Statistical Quality Control*, 3rd ed.; Wiley & Sons: New York, 1996.

10. Roach, S. *Health Risks from Hazardous Substances at Work—Assessment, Evaluation, and Control*; Pergamon Press: New York, 1992.

11. George, D. K.; Flynn, M. R.; Harris, R. L. Autocorrelation of Interday Exposures at an Automobile Assembly Plant. *Am. Ind. Hyg. Assoc. J.* **1995**, *56*, 1187–1194.

12. Francis, M.; Selvin, S.; Spear, R.; Rappaport, S. The Effect of Autocorrelation on the Estimation of Workers' Daily Exposures. *Am. Ind. Hyg. Assoc. J.* **1989**, *50*, 37–43.

13. Royston, P. A Pocket-Calculator Algorithm for the Shapiro-Francia Test for Non-normality: An Application to Medicine. *Stat. Med.* **1993**, *12*, 181–184.

14. Esmen, N. A.; Hammad, Y. Y. Log-normality of Environmental Sampling Data. *J. Environ. Sci. Health* **1977**, *A12*, 29–41.

15. Rappaport, S. M. Interpreting Levels of Exposures to Chemical Agents. In *Patty's Industrial Hygiene and Toxicology*, 3rd ed., Vol. 3, Part A; Harris, R. L., Cralley, L. J., Cralley, L. V., Eds.; Wiley & Sons, Inc.: New York, 1994.

16. Filliben, J. J. The Probability Plot Correlation Coefficient Test for Normality. *Technometrics* **1975**, *17*, 111–117.

17. Looney, S. W.; Gulledge, T. R. Use of the Correlation Coefficient with Normal Probability Plots. *The American Statistician* **1985**, *39*, 75–79.

18. Rappaport, S. M. Assessment of Long-term Exposures to Toxic Substances in Air. *Ann. Occup. Hyg.* **1991**, *35*, 61–121.

19. Woskie, S. R. et al. The Real-Time Dust Exposures of Sodium Borate Workers: Examination of Exposure Variability. *Am. Ind. Hyg. Assoc. J.* **1994**, *55*, 207–217.

20. Aitchison, J.; Brown, J. A. C. *The Lognormal Distribution with Special Reference to Its Uses in Economics*; Cambridge University Press: New York, 1957.

21. Attfield, M. D.; Hewett, P. Exact Expressions for the Bias and Variance of Estimators of the Mean of a Lognormal Distribution. *Am. Ind. Hyg. Assoc. J.* **1992**, *53*, 432–435.

22. Land, C. E. Tables of Confidence Limits for Linear Functions of the Normal Mean and Variance. In *Selected Tables in Mathematical Statistics*, Vol. III; Harter, H., Owen, D., Eds.; 1975; pp. 385–419.

23. Hewett, P. Mean Testing: II. Comparison of Several Alternative Procedures. *Appl. Occup. Environ. Hyg.* **1997**, *12*, 347–355.

24. Hewett, P.; Ganser, G. H. Simple Procedures for Calculating Confidence Intervals Around the Sample Mean and Exceedance Fraction Derived from Lognormally Distributed Data. *Appl. Occup. Environ. Hyg.* **1997**, *12*, 132–142.

25. Odeh, R. E.; Owen, D. B. *Statistics: Textbooks and Monographs Series*; Vol. 32—Tables for Normal Tolerance Limits, Sampling Plans, and Screening, 1980.

26. Tuggle, R. M. Assessment of Occupational Exposure Using One-Sided Tolerance Limits. *Am. Ind. Hyg. Assoc. J.* **1982**, *43*, 338–346.

27. Tuggle, R. M. The NIOSH Decision Scheme. *Am. Ind. Hyg. Assoc. J.* **1981**, *42*, 493–498.

28. Leidel, N. A.; Busch, K. A. Statistical Design and Data Analysis requirements. In *Patty's Industrial Hygiene and Toxicology*, 3rd ed., Vol. 3, Part A; Harris, R. L., Cralley, L. J., Cralley, L. V., Eds.; Wiley & Sons, Inc.: New York, 1994.

29. Rock, J. C. Occupational Air Sampling Strategies. In *Air Sampling Instruments for Evaluation of Atmospheric Contaminants*, 8th ed.; Cohen, B. S., Hering, S. V., Eds. American Conference of Governmental Industrial Hygienists: Cincinnati, OH, 1995.

30. Esmen, N. A. A Distribution-free Double-sampling Method for Exposure Assessment. *Appl. Occup. Environ. Hyg.* **1992**, *7*, 613–621.

31. Conover, W. J. *Practical Nonparametric Statistics, Second Edition*; Wiley & Sons: New York, 1980; pp. 111–112.

32. Gilbert, R. O. *Statistical Methods for Environmental Pollution Monitoring*; Van Nostrand Reinhold: New York, 1987.

33. Mood, A. M.; Graybill, F. A. *Introduction to the Theory of Statistics*, 2nd ed.; McGraw-Hill Book Company, Inc.: New York, 1963.

34. Snedecor, G. W.; Cochran, W. G. *Statistical Methods*, 7th ed.; Iowa State University Press: Ames, IA, 1980.

35. Beyer, W. H. *Handbook of Tables for Probability and Statistics*. Chemical Rubber Company (CRC): 1968.

36. Blyth, C. R.; Still, H. A. Binomial Confidence Intervals. *J. Am. Stat. Assoc.* **1983**, *78*, 108–116.

37. Hornung, R. W.; Reed, L. D. Estimation of Average Concentration in the Presence of Nondetectable Values. *Appl. Occup. Environ. Hyg.* **1990**, *5*, 46–51.

38. Perkins, J. L.; Cutter, G. N.; Cleveland, M. S. Estimating the Mean, Variance, and Confidence Limits from Censored (<Limit of Detection), Lognormally-distributed Exposure Data. *Am. Ind. Hyg. Assoc. J.* **1990**, *51*, 416–419.

39. Rappaport, S. M.; Lyles, R. H.; Kupper, L. L. An Exposure-assessment Strategy Accounting for Within- and Between-worker Sources of Variability. *Ann. Occup. Hyg.* **1995**, *39*, 469–495.

40. Lyles, R. H.; Kupper, L. L.; Rappaport, S. M. A Lognormal Distribution-based Exposure Assessment Method for Unbalanced Data. *Ann. Occup. Hyg.* **1997**, *41*, 63–76.

41. Hewett, P. Interpretation and Use of Occupational Exposure Limits for Chronic Disease Agents. *Occupational Medicine: State of the Art Reviews* **1996**, *11*(3).

42. Leidel, N. A.; Busch, K. A.; Lynch, J. R. *Occupational Exposure Sampling Strategy Manual*; National Institute for Occupational Safety and Health (NIOSH) Publication No. 77-173 (available from the National Technical Information Service (NTIS), Publication No. PB274792, or the NIOSH Web site as a .pdf file), 1977.

43. Corn, M.; Esmen, N. A. Workplace Exposure Zones for Classification of Employee Exposures to Physical and Chemical Agents. *Am. Ind. Hyg. Assoc. J.* **1979**, *40*, 47–57.

44. Still, K. R.; Wells, B. Quantitative Industrial Hygiene Programs: Workplace Monitoring. (Industrial Hygiene Program Management series, Part VIII). *Appl. Ind. Hyg.* **1989**, *4*, F14–F17.

45. Ayer, H. E. Occupational Air Sampling Strategies. In *Air Sampling Instruments for Evaluation of Atmospheric Contaminants*, 7th ed.; Hering, S. V., Ed.; American Conference of Governmental Industrial Hygienists: Cincinnati, OH, 1989.

46. Damiano, J. Quantitative Exposure Assessment Strategies and Data in the Aluminum Company of America. *Appl. Occup. Environ. Hyg.* **1995**, *10*, 289–298.

47. McHattie, G. V.; Rackham, M.; Teasdale, E. L. The Derivation of Occupational Exposure Limits in the Pharmaceutical Industry. *J. Soc. Occup. Med.* **1988**, *38*, 105–108.

48. Hewett, P. Mean Testing: II. Comparison of Several Alternative Procedures. *Appl. Occup. Environ. Hyg.* **1997**, *12*, 347–355.

49. OSHA (Occupational Safety and Health Administration). *Code of Federal Regulations 29*; Part 1910.1048, Formaldehyde (Appendix B), 1994.

50. Spear, R. C.; Selvin, S.; Francis, M. The Influence of Averaging Time on the Distribution of Exposures. *Am. Ind. Hyg. Assoc. J.* **1986**, *47*, 365–368.

51. HSC (Health and Safety Executive). *EH40/97 Occupational Exposure Limits 1997*; HSE Books: Suffolk, Great Britain, 1997.

52. Hewett, P. Characterization of Exposures to Welding Fumes and Gases during Production Line Welding. Presented at the 1989 American Industrial Hygiene Conference, 1989.

53. Cope, R. F.; Pancamo, B. P.; Rinehart, W. E.; Haar, G. L. Personnel Monitoring for Tetraalkyl Lead in the Workplace. *Am. Ind. Hyg. Assoc. J.* **1979**, *40*, 372–379.

17

Reproductive Hazards in the Workplace

ELIZABETH ANNE JENNISON

Reproductive disorders rank among the 10 leading work-related illnesses and injuries in the United States, with an estimated 14 million workers having potential occupational exposure to known or suspected reproductive hazards.[1] Unfortunately, efforts to prevent these disorders face major gaps in knowledge. Disorders of reproduction represent an interaction between individual genetic makeup, environmental conditions, and the intensity, duration, and timing of exposure to those conditions. A single toxicant can produce a variety of adverse outcomes depending on the specific conditions of exposure. Conversely, each class of reproductive outcomes can result from a variety of different agents, acting through several biologic mechanisms. Workers are often exposed to more than one agent, so there may be interactive effects from complex mixtures in the workplace or environmental agents outside the workplace. Finally, disorders of reproduction arising from occupational factors may be difficult to distinguish from those with nonoccupational etiologies.[2]

The term *reproductive hazard* is properly restricted to hazards that interfere with or prevent conception. Reproductive hazards may have adverse effects on libido, sexual behavior, any aspect of spermatogenesis or oogenesis, hormonal activity or physiological response that would interfere with the capacity to fertilize, fertilization itself, or the development of the fertilized ovum up to and including implantation.[3] They are distinct from *developmental hazards*, which produce structural abnormalities, functional deficits, pathological alterations to growth, or death. Using this distinction, reproductive hazards are those that affect the worker and their effects may be reversible. The effects of developmental hazards will be confined to the fetus or offspring and are almost invariably permanent. A given agent, for example, ionizing radiation, may present both a reproductive and a developmental hazard.[4]

Despite the OSHA Hazard Communication Standard,[5] which requires employers to provide workers with Material Safety Data Sheets (MSDSs) and training pertaining to hazardous substances used on the job, worker education regarding potential reproductive effects appears to be inadequate.[6] This may be related, in part, to the observation that a significant percentage of MSDSs contain no information on reproductive risks.[7] In addition, potential reproductive toxicity has not been assessed for an estimated 95% of the chemical and physical agents included in the NIOSH Registry of Toxic Effects of Chemical Substances.[8]

The United States currently regulates only three chemicals on the basis of reproductive or developmental effects, specifically lead,[9] 1,2-dibromo-3-chloropropane (DBCP),[10] and ethylene oxide.[11] Developmental or reproductive effects in animals were noted in the bases for recommended ACGIH standards in only 15 of the 575 chemicals for which standards existed as of 1980. A 1993 publication of the Agency for Toxic Substances and Disease Registry included 15 environmental toxicants for which epidemiological studies, occupational exposure data, or animal data indicate an association between adverse human reproductive or developmental effects[12] (Table 17.1). Many additional agents are suspected or known reproductive developmental hazards; a list of 200 such agents was developed for the California Proposition 65 Advisory Panel.[4]

The Organization of Teratology Information Services (OTIS) provides medical consultation on prenatal exposures to individuals of reproductive age and health care providers. Services are provided through state and regional programs throughout the United States and Canada. Information about the appropriate regional Teratology Information Service can be obtained through the main OTIS phone number (801-328-2229) or on their Web site (http://orpheus.ucsd.edu/otis/index.html). There is a collaborative relationship between OTIS and the European Network of Teratology Information Services (ENTIS).

REPROTOX is a database that includes information about the effects of the chemical and physical environment on human fertility, pregnancy, and fetal development. It includes information about occupational and environmental exposures, as well as prescription, over-the-counter, and recreational drugs. REPROTOX is available by subscription either on-line or in diskette format with quarterly updates. For further information about REPROTOX contact the Reproductive Toxicology Center in Washington, DC (www.reprotox.org).

TABLE 17.1 Environmental Toxicants and Adverse Reproductive Outcomes

Exposure	Associated Outcome
Aldrin	Spontaneous abortion Premature labor
Arsenic	Spontaneous abortion, low birth weight, congenital and developmental defects
Benzene	Spontaneous abortion, low birth weight, menstrual disorders, ovarian atrophy
Cadmium	Low birth weight
Carbon disulfide	Menstrual disorders, spontaneous abortion, adverse effects on sperm
Chlorinated compounds (Chloroform, TCE, PCE)	Eye, ear, and oral cleft anomalies, perinatal death, childhood leukemia, CNS disorders
DBCP	Adverse effects on sperm, sterility
Dichloroethylene	Congenital heart defects
Dieldrin	Spontaneous abortion, premature labor
Hexachlorocyclohexane	Hormonal imbalances, premature labor, spontaneous abortion
Lead	Stillbirth, preterm delivery, low birth weight, spontaneous abortion, neurobehavioral deficits, mental retardation, adverse effects on sperm and testes
Mercury	Menstrual disturbances, spontaneous abortion, blindness, deafness, mental retardation, developmental delays, brain damage
Polycyclic aromatic hydrocarbons	Decreased fertility
Polychlorinated biphenyls	Preterm delivery, low birth weight, growth deficiencies, reduced head circumference, neurobehavioral effects
Trichloroethylene	Congenital heart disease

Reproduced with permission from Reference 12.

Evaluating Reproductive and Developmental Toxicity

Reproductive and developmental toxicology are difficult to assess because of the complexity of the physiologic processes involved and the significant background levels of infertility and congenital anomalies. Several recent review articles provide good background information on the male[13] and female[14] reproductive processes and their hormonal regulation,[15] preimplantation development,[16] and developmental biology.[17] Approximately one in 13 couples of reproductive age is infertile, that is, unable to produce a viable pregnancy during one year of unprotected intercourse. Of recognized pregnancies, 10–15% are lost to spontaneous abortion prior to the 20th week of gestation; probably at least that many are lost prior to recognition.

Major birth defects are noted in about 3% of live births, and account for more than 21% of infant deaths. Approximately 60% of birth defects are of unknown etiology, including neural tube defects, cardiovascular system malformations, and oral-facial clefts, defects that account for a large proportion of infant morbidity and mortality.[18] Further complicating matters, definitions of spontaneous abortion, congenital malformations, semen abnormalities, and developmental delay are not applied consistently in clinical practice. In contrast to the chronic nature of most occupational diseases, which require repetitive exposures and often have long latency periods, effects on reproduction and development are often the result of short-term exposures during critical periods of vulnerability, such as fetal organogenesis.

Male

The effects of reproductive hazards on males include adverse effects on libido, spermatogenesis, sexual behavior, and other hormonal or physiologic factors. By far the easiest of these to assess is adverse effects on spermatogenesis and sperm physiology. While semen evaluation provides information on sperm production by the testes, patency of the reproductive tract, activity of the accessory sex glands, and capability for ejaculation, interpretation of semen parameters may be complicated.[19] Traditional measurements of semen quality are poorly quantitated; there is surprisingly little information on the distribution of seminal characteristics in the general population, and laboratory methodologies are not well standardized. Although the World Health Organization (WHO) has given normal limits for semen characteristics, it recommends that each laboratory determine its own range of normals using WHO-defined standard laboratory procedures. Clinical assessment of men who have "abnormal" laboratory findings but who have not attempted to achieve a pregnancy is problematic.

Female

The assessment of female reproductive health is perhaps more difficult than that of the male. Many of the events of importance, such as ovulation, implantation, and luteal function, occur without the woman being aware of them. Assessment of these functions may require physical examination, biopsy, or repeated hormone assays of blood or urine. While useful in the evaluation of reproductive disorders in an individual patient, most of these studies are not appropriate, practical, or ethical for large-scale studies.[20] In addition, it is difficult to establish absolute limits for normal and abnormal urinary steroid metabolite concentrations that can be applied to all individuals. The occurrence of regular, episodic vaginal bleeding in the form of menses, while easily observable, is not sufficient to establish normal ovarian function. More easily observed reproductive outcomes include time to conception, standardized fertility ratios (ratio of observed to expected births), and pregnancy outcomes such as prematurity, low birth weight, size for gestational age, spontaneous abortion, and stillbirth.

OSHA-Regulated Reproductive Toxins

Lead

Adverse effects of lead on the male reproductive system include low sperm counts, poor sperm motility, malformed sperm, and possible chromosomal aberrations.[21] There is also evidence that suggests that paternal exposure to lead is associated with an increased risk of spontaneous abortion.[22] Women exposed to lead may experience menstrual disturbances, and a higher frequency of sterility, premature births, miscarriages, and stillbirths. Lead can cross the placenta and may cause neurobehavioral impairments in the offspring. While OSHA has established a regulatory level of 40 µg lead/100 g whole blood as sufficient to protect workers against adverse health effects from lead exposure, 30 µg/100 g is listed as the maximum level recommended in both males and females who wish to bear children. The appendices to the OSHA lead standard are written in layman's terms and can provide a useful means of informing workers of the adverse reproductive effects of lead exposure (see Box 1).

Box 1
Medical Requirements of OSHA Lead Standard, 1910.1025

Biological Monitoring
  Blood lead level
  Zinc protoporphyrin level

Medical Examinations and Consultations*
  Occupational and Medical History
    Attention to past lead exposure, personal habits (smoking,
      hygiene), past gastrointestinal, renal, cardiovascular,
      neurological, and reproductive problems
  Physical Exam
    Blood pressure
    Attention to teeth, gums, hematologic, gastrointestinal, renal,
      cardiovascular, and neurological systems
  Laboratory Tests
    Blood lead level
    Zinc protoporphyrin level
    Blood urea nitrogen, serum creatinine
    Urinalysis with microscopic evaluation
    Hemoglobin, hematocrit, red cell indices and peripheral blood
      smear morphology
    Pregnancy testing or laboratory evaluation of male fertility if
      requested by employee

*See text regarding requirements for medical examinations versus biological monitoring

---

Biological monitoring is required for all employees who are or may be exposed to lead above the action level for more than thirty days per year. A complete medical examination is required prior to assignment to an area with airborne lead levels above the action level, for employees with blood lead levels at or above 40 µg/100 g at any time during the preceding 12 months, to evaluate signs or symptoms of lead intoxication, and for employees requiring medical removal under the standard. Medical consultation is also required if an employee desires medical advice concerning the effects of current or past exposure to lead on their ability to procreate a healthy child. Upon employee request, medical examinations conducted pursuant to the lead standard should include pregnancy testing and/or laboratory evaluation of male fertility.

DBCP

As far back as 1961, DBCP was demonstrated to cause sterility in rats and testicular atrophy in rats, guinea pigs, and rabbits. Unfortunately, this knowledge was not extended to the occupational setting until 1977 when decreased sperm counts and resulting infertility were demonstrated in a group of workers at a pesticide formulation plant in California. In March, 1978 OSHA issued an Emergency Temporary Standard with a permissible exposure limit (PEL) of 1 ppb, a limit which was retained in the final standard (see Box 2). While the OSHA standard for DBCP is designed to protect workers exposed during the manufacture of DBCP, it explicitly does not apply to exposure which results solely from the application and use of DBCP as a pesticide. Agricultural use of DBCP was banned in the continental United States in 1979, with the ban being extended to Hawaii in 1985. Internationally, approximately 1500 Costa Rican workers were sterilized as a result of DBCP use on banana plantations operated by U.S.-based transnational fruit companies.

DBCP acts via direct toxicity to the testes and causes an exposure duration-dependent effect on sperm counts. Testicular biopsy of men exposed to DBCP reveals significant atrophy of the seminiferous epithelium. The precise mechanism of action of DBCP is not completely understood, but a correlation between DNA damage and testicular necrosis suggests that DNA damage is the initiating event.[23] Unfortunately, the ban on DBCP led to its replacement by ethylene dibromide, a known testicular toxicant that also has carcinogenic potential.[24]

Ethylene Oxide

Ethylene oxide is an alkylating agent that binds to DNA and may cause cellular mutations. It has been shown to cause chromosomal aberrations in both animals and humans after inhalation exposure. Women exposed to ethylene oxide have an increased risk of spontaneous abortions. Labels on containers of ethylene oxide must comply with the Hazard Communication Standard and include the following legend: DANGER. CONTAINS ETHYLENE OXIDE. CANCER HAZARD AND REPRODUCTIVE HAZARD. (See Box 3.)

Box 3
Medical Requirements of OSHA
Ethylene Oxide Standard, 1910.1047

Medical and Occupational History
  Attention to pulmonary, hematologic, neurologic, and reproductive systems, eyes and skin
Physical Examination
  Attention to pulmonary, hematologic, neurologic and reproductive systems, eyes and skin
  Complete blood count with white cell differential, hematocrit, and hemoglobin
  Pregnancy testing or laboratory evaluation of fertility if requested by the employee and deemed appropriate by the physician

---

Fetal Protection Policies

In 1991 the United States Supreme Court, ruling in the case of *UAW* v. *Johnson Controls*, held that "fetal protection policies" under which employers excluded fertile or pregnant women from hazardous work sites constituted sex discrimination. Such policies were declared to be in violation of Title VII of the Civil Rights Act because the employer could not prove that a woman's fertility or pregnancy interfered with her ability to perform essential work assignments. While the Court's decision recognized discrimination in that women were being denied job opportunities, the legal issues of contention did not address whether men were receiving equal protection

from the toxicity of lead. Adverse reproductive effects in men were known, and OSHA had already recommended a maximum blood lead level of 30 μg/100 g for both males and females who wish to bear children. Essentially the Court said that employers could not avoid their obligation to maintain acceptable workplace conditions by discriminating against women workers.[25]

While the fetal protection policies in the Johnson Controls case centered around exposure to lead, such policies have been suggested for other reproductive toxicants. At the time the DBCP standard was being proposed, some industry representatives suggested that jobs producing or using DBCP "[m]ight be good for workers who did not desire more children as an alternative to planned surgery for vasectomy or tubal ligation, or as a means of getting around religious bans on birth control."[24] Company policies that require employees to sign waivers accepting fetal risk are not recommended as they appear to be in violation of the OSHA General Duty Clause.[25] In light of our increasing understanding of the susceptibility of both sexes to the effects of reproductive toxicants, in the words of labor leaders, it seems that "the best approach is to establish procedures that accommodate all health problems, risks, or other reasons for restriction and to fit the specific situation of exposure to reproductive hazards within that framework."[26]

Employer Guidelines

1. Consider potential reproductive and developmental hazards in reviewing new products or materials to be used in the workplace.
2. Obtain information about reproductive and devel-

Box 2
Medical Requirements of OSHA
DBCP Standard, 1910.1044

Medical history, including reproductive history
Occupational history
Physical examination
  General body habitus
  Examination of urogenital tract, including testicle size
Laboratory evaluations
  Serum follicle stimulating hormone (FSH)
  Serum luteinizing hormone (LH)
  Sperm count (males)
  Serum total estrogen (females)

opmental hazards already in use in the workplace. This may require additional information beyond that included on MSDSs.

3. Use appropriate controls for reproductive and developmental hazards following the industrial hygiene hierarchy:

    substitution with a less hazardous material
    engineering controls
    administrative controls including work practices
    personal protective equipment

4. Include discussions of reproductive and developmental toxicity in the hazard communication program. This should also include information about how personal choices and lifestyle factors can affect reproductive health.

5. Provide MSDSs and other toxicity information to concerned employees and their health care providers as requested.

6. Work with human resources personnel, worker representatives, and health care consultants to establish a policy regarding employee reassignment because of reproductive or developmental hazards. Avoid blanket policies which may be discriminatory. Risk levels and reassignment decisions should be made on a case by case basis.

7. Companies that conduct preplacement physical examinations or ongoing medical surveillance programs should include a reproductive history as part of the medical and occupational history and examination process.

The Future of Occupational Reproductive Toxicology

Efforts to prevent occupational disorders of reproduction are limited by significant gaps in our knowledge of the human reproductive process. Continued research is needed to improve our understanding of reproductive and developmental biology and to identify etiologic agents and populations at risk. Three components of the NIOSH "Proposed National Strategy for the Prevention of Disorders of Reproduction" are particularly applicable to the health and safety professional. First is the identification of appropriate control measures for known reproductive hazards and the dissemination of this information. Industry should take the lead in developing workplace-specific control systems based on the physiochemical properties of the hazard, conditions of use, and required exposure limits. Second is the development of appropriate public information and education programs.

Ideally, training materials should be specific for particular exposures found in a given workplace while heightening awareness of the influence of personal risk and lifestyle factors on an individual's reproductive health. Finally, enhanced professional information and education programs should place greater emphasis on the effects occupational exposures can have on reproductive health. This should include the education of occupational health and safety professionals, health professionals in general, and the business and engineering communities. "The disorders of reproduction that can be attributed to specific occupational hazards can be eliminated, or at least reduced, as the hazards are identified, as workers are protected from them, and as education succeeds in promoting healthy working environments and behaviors."[2]

References

1. CDC. Leading Work-related Diseases and Injuries—United States. *MMWR* **1985**, *34*, 219–22, 227.
2. NIOSH. *A Proposed National Strategy for the Prevention of Disorders of Reproduction*; DHHS (NIOSH) Publication 89–133, U.S. Department of Health and Human Services: Washington, DC, 1988.
3. Sullivan, F. M. The European Community Directive on the Classification and Labeling of Chemicals for Reproductive Toxicity. *J. Occup. Med.* **1995**, *37*(8), 966–969.
4. Johnston, J. D.; Jamieson, G. G.; Wright, S. Reproductive and Developmental Hazards and Employment Policies. *Br. J. Ind. Med.* **1992**, *49*, 85–94.
5. OSHA Hazard Communication Standard; 29 *CFR* 1910.1200.
6. Paul, M.; Daniels, C.; Rosofsky, R. Corporate Response to Reproductive Hazards in the Workplace: Results of the Family, Work, and Health Survey. *Am. J. Ind. Med.* **1989**, *16*, 267–280.
7. Paul, M.; Kurtz, S. Analysis of Reproductive Health Hazard Information on Material Safety Data Sheets for Lead and the Ethylene Glycol Ethers. *Am. J. Ind. Med.* **1994**, *25*, 403–415.
8. Gold, E. B.; Tomich, E. Occupational Hazards to Fertility and Pregnancy Outcome. *Occ. Med. State of Art Reviews* **1994**, *9*(3), 435–470.
9. OSHA Lead Standard; 29 *CFR* 1910.1025.
10. OSHA 1,2-dibromo-3-chloropropane Standard; 29 *CFR* 1910.1044.
11. OSHA Ethylene oxide Standard; 29 *CFR* 1910.1047.
12. ATSDR. *An Integrated Strategy to Evaluate the Relationship Between Illness and Exposure to Hazardous Substances*, Chapter 2: Birth Defects and Reproductive Disorders; U.S. Department of Health and Human Services: Atlanta, GA, July 1993.
13. Overstreet, J. W.; Blazak, W. F. The Biology of Human Male Reproduction: An Overview. *Am. J. Ind. Med.* **1983**, *4*, 5–15.
14. Takizawa, K.; Mattison, D. R. Female Reproduction. *Am. J. Ind. Med.* **1983**, *4*, 17–30.
15. Smith, C. G. Reproductive Toxicity: Hypothalamic-Pituitary Mechanisms. *Am. J. Ind. Med.* **1983**, *4*, 107–112.

16. Dean, J. Preimplantation Development: Biology, Genetics and Mutagenesis. *Am. J. Ind. Med.* **1983**, *4*, 31–49.

17. Swartz, W. J. Early Mammalian Embryonic Development. *Am. J. Ind. Med.* **1983**, *4*, 51–61.

18. Sever, L. E. Congenital Malformations Related to Occupational Reproductive Hazards. *Occ. Med. State of Art Reviews* **1994**, *9*(3), 471–494.

19. Overstreet, J. W. Clinical Approach to Male Reproductive Problems. *Occ. Med. State of Art Reviews* **1994**, *9*(3), 387–404.

20. Lasley, B. L.; Shideler, S. E. Methods for Evaluating Reproductive Health of Women. *Occ. Med. State of Art Reviews* **1994**, *9*(3), 423–434.

21. Thomas, J. A.; Brogan, W. C. Some Actions of Lead on the Sperm and on the Male Reproductive System. *Am. J. Ind. Med.* **1983**, *4*, 127–134.

22. Anttila, A.; Sallmen, M. Effects of Parental Occupational Exposure to Lead and Other Metals on Spontaneous Abortion. **1995**, *37*(8), 915–921.

23. Schrader, S. M.; Kanitz, M. H. Occupational Hazards to Male Reproduction. *Occ. Med. State of Art Reviews* **1994**, *9*(3), 405–414.

24. Infante, P. F. Occupational Reproductive Hazards: Necessary Steps to Prevention. *Am. J. Ind. Med.* **1983**, *4*, 383–390.

25. Clauss, C. A.; Berzon, M.; Bertin, J. Litigating Reproductive and Developmental Health in the Aftermath of *UAW versus Johnson Controls. Env. Health Persp. Supp.* **1993**, *101*(suppl. 2), 205–220.

26. Graham, T.; Lessin, N.; Mirer, F. A Labor Perspective on Workplace Reproductive Hazards: Past History, Current Concerns, and Positive Directions. *Env. Health Persp. Supp.* **1993**, *101*(suppl. 2), 199–204.

18

Epidemiology

NEELY KAZEROUNI

SHELIA HOAR ZAHM

ELLEN F. HEINEMAN

Epidemiology is the study of the occurrence of disease in human populations and the study of the factors that influence these disease patterns.[1] Epidemiologic observations are used to determine the cause of disease; to describe the natural history of disease; to provide data necessary for planning and evaluating public health care; to provide a foundation for developing and evaluating preventive programs; and to explain local disease occurrence. Epidemiology has its own methodology as well as utilizing methods from other disciplines, such as statistics, sociology, and biology. Epidemiologists try to identify causes of disease, whether chemical, physical, or biological (e.g., viral, bacterial, fungal) and relate them to host factors such as genetics, age, sex, and physiological status, examining individual effects and interactions.[1]

Methodology

As described in detail in Heineman and Zahm,[2] the epidemiologist studies associations between exposures and outcomes. Exposures may include age, genetic factors, diet, environmental chemicals—almost any characteristic of individuals or the environment. Any condition of

disordered health, for example, cardiovascular disease, cancer, or intermediate steps in a disease process such as chromosomal aberrations, is an outcome valid for study. Several study designs are available to evaluate possible associations between exposures and disease.

Study Designs

Experimental Studies

Experimental studies such as randomized clinical trials are experiments in which individuals are randomly assigned to two or more groups, known as the "treatment" and the "comparison" groups. The treatment group is given the treatment being tested such as a new drug or a new screening method and the comparison group is given the method in current use or a "placebo"—usually an inert substance such as a sugar pill.[1]

Observational Epidemiology

Observational epidemiology involves the study of the occurrence of disease in humans and making inferences about etiological factors that influence this occurrence.[1] Studies may be descriptive, correlational, or analytic.

Descriptive Studies enumerate diseases within populations by differing characteristics such as age, sex, race, time, and/or geography. They can stimulate further analytic investigations. Cancer mortality maps revealed so-called "hot spots" of lung cancer along the southeastern United States Atlantic coast, which stimulated further analytic research that established an association between lung cancer and asbestos exposures in shipbuilding, particularly during World War II. (Case reports are not epidemiologic studies, but may be informative and lead to in-depth investigations when clusters of disease occur in a limited time, the observed disease is relatively rare, and a history of common exposure is suspected.)

Correlational (or "Ecologic") Studies investigate correlations between disease rates and exposures among populations based on data available on each population as a group, not on an individual subject level. Because these studies are usually based on exposure and disease information obtained on a population level rather than from individuals, they typically cannot account for bias or the many other factors present in the community that might contribute to disease. They do, however, suggest clues to disease etiology, and provide leads for more detailed analytical studies.

Analytical Studies Analytical studies evaluate specific exposure-response relationships in the context of other factors, based on data available on an individual subject level. The two major approaches are the cohort and case-control study designs shown in Table 18.1.

The cohort study compares the rate of disease in an "exposed" group to that in a group without that exposure. In a retrospective cohort study, the exposure and the diseases to be observed in the population have occurred in the past. In a prospective cohort study, follow up of the population extends into the future. Studies may have both retrospective and prospective components.

Case-control studies identify study participants on the basis of their disease status and determine past exposures. The rate of exposure in the case group is compared to the rate of exposure in a similar population without the disease of interest.

Cross-sectional studies are the third approach, in which disease and exposure are both measured at one point in time. Because the time sequence of exposure and disease is not specified, this type of study is most useful for immediate biological responses, such as acute toxicity, or chronic illness in which the causes persist over time. It is also useful for generating hypotheses.

Criteria for Causality

Observation of an exposure-disease relationship does not automatically establish the cause of disease. The epidemiologist must seek out other explanations, such as chance, biased study methods, or other contributing factors. Several features strengthen a causal interpretation of the association (see Box 1).[3]

TABLE 18.1 Characteristics of Analytical Studies

	Cohort Studies	Case-Control Studies
Study group identification	Exposure status	Disease status
Outcome ascertained	Disease	Exposure
Health effects	Many	Few
Exposures	Few	Many
Advantages	Useful for rare exposures; Subject selection and measurement accuracy can be better controlled (prospective studies); Useful when focusing on a specific exposure common in a setting in which the exposed can be easily identified (e.g., vinyl chloride exposure in rubber workers)	Useful for rare diseases; Better opportunity for assessing a range of possible risk factors
Disadvantages	Often requires large sample size	Requires information on exposures that can be difficult to obtain; May not have adequate statistical power for evaluating rare exposure

```
┌─────────────────────────────────────────┐
│              Box 1                       │
│  Features Strengthening a Causal         │
│             Interpretation               │
│                                          │
│  • Strength of the association           │
│  • Exposure-response gradient            │
│  • Consistency of the observed           │
│    association across studies            │
│  • Temporality of the association        │
│  • Coherence                             │
│  • Specificity of the association        │
│  • Biological plausibility               │
└─────────────────────────────────────────┘
```

Specificity of association and biological plausibility need not be met for causality to be established. Specificity of a relationship between exposure and outcome strengthens confidence in a causal inference, but there are many examples of a single exposure with multiple results (e.g., smoking), and of a single result with multiple causes (e.g., cardiovascular disease) where causality is not ruled out. A plausible biological model enhances the credibility of the association, but biological plausibility depends on the biological knowledge of the day.

Strengths and Limitations of the Epidemiologic Approach

Several factors affect the utility of the epidemiological approach: study design and sample size, exposure assessment, time since first exposure, and potential for systematic bias.

Study Design, Magnitude of Effect, and Sample Size

We cannot randomize human subjects to receive possibly toxic or carcinogenic exposures. As a result, the observational nature of most epidemiological research restricts the data to preexisting types and levels of exposures, and limited subject numbers. These limitations can result in inadequate statistical power to evaluate an apparent association, and can be a barrier to studying certain exposures. In order to increase statistical power, researchers often increase the sample size of the study, which allows them to have greater confidence in study results. For some exposures, sample size cannot be increased, as when only a small number of workers experience a possibly hazardous occupational exposure. In addition, increasing the sample size increases costs and consequently may require compromise in the accuracy and detail of data collected on each subject.

Exposure Assessment

A major challenge in epidemiologic studies is obtaining accurate and precise exposure information, whether derived from interviews (with the study subjects, a next of kin, or a co-worker) or from records (e.g., medical or work history).[4,5] The quality of exposure information can be improved by a variety of means. Validation of questionnaire data improves the believability of study results. Historical employment records and monitoring data are often very useful. Nonetheless, they can be generally inaccurate, incomplete, or insufficient to determine the detailed exposure history of a worker for specific chemicals of interest. Limited exposure assessment can make risk assessment difficult, challenging the researcher's ability to define the health effects of exposure of individuals or populations to hazardous materials and situations. Industrial hygienists have attempted to minimize these limitations. Semiquantitative methods, such as ranking workers in broad exposure groupings of workers based on professional judgment or job-exposure matrices, allow epidemiologists to detect overall associations and exposure-response gradients. Quantitative exposure assessment, on an individual basis, based on periodic monitoring and accurate recordkeeping would be ideal.

Biomarkers of exposure can be used to increase the accuracy of an exposure assessment. For example, tobacco smoking has been linked with adverse reproductive effects based on questionnaire data. However, the association has been shown most convincingly when exposure has been estimated using measurements of urinary cotinine, a metabolite of tobacco smoke. For diseases with long latent periods, biologic exposure assessment should reflect long-term exposure, such as organohalides and metals accumulated in tissues. Current exposures may be irrelevant to the etiology of a disease with a long latency period. The use of biologic materials for exposure assessment can be employed in cohort studies, but collecting samples on all cohort members prior to disease development entails costly, long-term storage of a large number of samples. In case-control studies, biological materials collected at diagnosis may be affected by disease status. The difficulties in identifying appropriate biological markers are compounded when the mechanism of action of the agent under investigation is unclear.

Time Since First Exposure

Epidemiologic studies can assess the effects of newly introduced exposures only after the latent period for the disease of interest has elapsed. For diseases such as cancer this can require 20 to 40 years. Effects that appear a relatively short time after exposure, such as mutagenesis, teratogenesis, and other effects (e.g., alterations in immune function or neurotoxicity), can be studied in a more timely manner. The evaluation of chronic effects of chemical exposures could be fundamentally altered if we could identify with certainty the intermediate steps in the natural history of cancer or other chronic diseases. A biomarker of effect gives an assessment of an early effect of a chemical on a physiological or biochemical process and can shed insight into possible disease mechanisms. For example, assays on peripheral blood samples can reflect cytogenic damage due to a chemical exposure, which may be an intermediate step in carcinogenesis.

Potential for Systematic Bias

Bias is defined as a systematic error in the design, conduct, or analysis of a study resulting in a mistaken estimate of the relationship between exposure and disease. The most common forms of bias are described below.

Selection Bias

Selection bias occurs when there are systematic differences between the sample population chosen for study and the general population to the one that they represent. In case-control studies, it is important that the controls are selected from the same or a similar population to the one that gave rise to the cases. In cohort studies, a common form of selection bias, "healthy-worker effect," occurs when the working population selected for study is compared to a population of workers and nonworkers, such as the general population. Since the general population also contains very sick people (who are selected out of the workforce), the worker population appears to be at lower risk than the general population for certain types of disease, and has lower death rates, regardless of the hazards or benefits of their occupational setting.

Confounding

Confounding is "the distortion of a disease/exposure association brought about by the association of other factors with both disease and exposure."[6,7] For a confounding variable to distort the apparent magnitude of the effect of a study factor on risk, it must be a risk factor for the outcome and strongly associated with exposure. (Intermediate steps in the causal path between the exposure and the disease do not confound an association, as when *exposure* leads to *dysplasia* leads to *cancer*.)

Confounding may be controlled in either the design phase, the analysis phase, or both. A prior knowledge of probable confounders would allow them to be controlled in the study design by restriction, matching, or randomization. In the analysis of data, confounding is controlled by direct or indirect standardization of rates, stratification of data, and multivariate statistical techniques which simultaneously control for multiple factors.

Epidemiologists rely on prior epidemiologic and biologic studies and their own insights to identify factors that might be associated with both disease and exposure in their data. Controlling for confounding yields more accurate risk estimates. The fear of possible undetected confounding causing false positive associations may be overemphasized. In many situations, potential confounders may not make an important difference in the interpretation of the study. For example, smoking, although a strong risk factor for lung cancer, is rarely correlated with an occupational exposure of interest to such an extent that it could explain more than an extremely small elevation in risk.

Observation Bias

An over- or underestimate of the true risk can result from inaccurate classification of study subjects with respect either to disease or to exposure status. Sources include: (1) incomplete recall of prior exposures; (2) inaccurate reporting of exposure; (3) inaccurate recording or gathering of data by interviewers; and (4) inaccurate diagnosis of disease.

Epidemiology versus Toxicology

Toxicology, the study of the nature and effects of poisons, investigates the relationship between exposure and toxic effects through experimental research using animals or human tissue systems (Table 18.2). Because the research is experimental, exposures are determined by the investigator and usually well-characterized. Typically, the animals or other test systems are genetically homogeneous and the exposures other than the one under investigation are standardized. Test results, even long-term bioassays studying chronic effects such as cancer, are available in relatively short amounts of time due to the shorter life span of test animals than humans. The relevance of toxicologic research is limited, however, by the extremely high exposure levels used, the artifice of a pure high grade exposure that may not be comparable to the exposures experienced by humans, the sometimes incomparable route of administration of the exposure, and species differences in absorption, metabolism, target organ, and other factors.[2]

Epidemiology, while limited by its observational nature, often poorly characterized exposures, and heteroge-

TABLE 18.2 Epidemiology versus Toxicology

	Epidemiology	Toxicology
Design	Usually observational	Experimental
Dose	Exposure levels encountered naturally by humans	Many times higher than human exposure close to maximum threshold dose done to limit number of animals needed per test dose level
	No need to extrapolate dose	Must extrapolate dose to levels experienced by humans
Species	Humans Relevant target organ	Must extrapolate from animal species to humans Animal target organ may be different from human
Agent	Single or mixed Actual quality of exposures	Single Highly purified, standardized
Route	May involve unusual scenarios (e.g., worker take-home exposures) and single or mixed routes (e.g., inhalation alone or inhalation plus dermal)	Standard routes, typically only one per study
Interaction of exposure and host/environmental factors	Heterogeneity of exposed population (genetically, diet and lifestyle factors, other exposures, more and less susceptible individuals)	Homogeneous population
	Metabolic pathways in humans	Metabolic pathways in rodents, other nonhuman experimental species
Timetable for knowledge	Evidence of exposure-disease relationship and prevention activities can precede identification of specific agents	Specific agent must be identified and assessed before any preventive action may take place
	Evaluation can only take place after disease latency elapses, during which time many people may have been exposed	Information is available sooner because typically there is a shorter latent period in animals than humans
Exposure-disease relationship	Exposure well-quantified in some studies, but often poorly characterized	Exposure and response precisely quantified
Establish an upper limit on human risk	Yes—can identify agents, which, at usual doses in the human population, *do not* appreciably increase risk of human disease	No—an agent that does not appear to cause cancer in animals may nonetheless be carcinogenic in humans

neity of the population and their other life experiences and exposures, does provide data on the relevant species (humans), purity of exposure, dose level, relevant route of administration, and target organs.

Epidemiology can investigate the effects of exposures to which humans have already had exposure. It cannot help with the evaluation of the possible effects of agents newly introduced into the industrial or general environment. Toxicologic research on new compounds can indicate, sometimes with high reliability, what possible effects human populations might experience. The two disciplines are complementary in their contributions to protecting public health. In fact, over the past few decades, both epidemiology and toxicology have generated hypotheses that have been investigated further via the other discipline.

Conclusion

Epidemiology has made profound contributions to our understanding of disease and to risk assessment by identifying real-life human hazards, generating hypotheses to be tested in experimental animal studies, and by forging new awareness of the potential for gene-environment interaction in disease causation in human populations. Although epidemiology and toxicology are often complementary, epidemiology generates data that is immediately relevant to humans, without extrapolation over dose or species. As exposure assessment improves, epidemiologists' ability to study low-level effects will increase. The identification of genetically susceptible subgroups of the population will help epidemiologists identify new hazards and explain the variation across populations in their response to the same exposure. The future of epidemiology lies in increasing use of biochemical, genetic, or molecular data on exposure, effect, and susceptibility, which can provide improved insights into disease mechanisms and help disentangle gene-environment interactions and mixed exposures.

Acknowledgment

The authors acknowledge the helpful review of Dr. Patricia Hartge.

References

1. Lilienfeld, D. E.; Stolley, P. D. *Foundations of Epidemiology*, 3rd ed.; Oxford University Press: New York, 1994.

2. Heineman, E. F.; Zahm, S. H. *Toxic Substances J.* **1989**, 9, 255–277.

3. Hill, A. B. *Proc. R. Soc. Med.* **1965**, 58, 295–300.

4. Rothman, K. J. *Modern Epidemiology*; Little, Brown: Boston/Toronto, 1986; pp. 106–8.

5. Rappaport, S. M., Smith, T. J., Eds. *Exposure Assessment for Epidemiology and Hazard Control*; Lewis Publishers: Chelsea, MI, 1991.

6. Breslow, N. E.; Day, N. E. *Statistical Methods in Cancer Research. Vol. I—The Analysis of Case-Control Studies*; IARC Sci Pub No.32; International Agency for Research on Cancer: Lyon, France, 1980.

7. Breslow, N. E.; Day, N. E. *Statistical Methods in Cancer Research. Vol. II—The Design and Analysis of Cohort Studies*; IARC Sci Pub No. 82; International Agency for Research on Cancer: Lyon, France, 1987.

19

Carcinogenesis

ELIZABETH K. WEISBURGER

Carcinogenesis refers to the process by which a neoplasm or tumor arises in an organism. A neoplasm or new growth can be either benign or malignant and malignant tumors are often called cancers. Benign tumors are generally well contained and do not invade neighboring tissues. However, malignant tumors physically encroach on adjacent tissues and thus invade directly. Furthermore, groups of cells from malignant tumors are carried via the circulation to other organs where new colonies of malignant cells are established by the process called metastasis.[1]

Historical Development

Carcinogenesis, as manifested by cancer in animals and humans, is an ancient process. Bone tumors have been recognized in skeletons from extinct animals. Mummies from Egypt and Peru have shown different types of tumors. Several thousand years later, the same types of neoplasms continue to arise in these populations. Currently, since people are less likely to die from infectious diseases, they live longer and thus increase their risk of developing cancer. Carcinogenesis begins with a cell which by some change is no longer subject to the normal cellular inhibitory processes and thus neither stops growing nor dies at its usual appointed time. Other organs are invaded, either directly or through metastasis, and the cancer takes for itself essential nutrients from the host organism and continues to live until the death of the host.[1-3]

Types of Carcinogens

Agents that lead to the formation of cancers are called carcinogens. Some viruses, genetic changes, sunlight (ultraviolet light), ionizing and nonionizing radiation can also lead to cancer, but people often associate carcinogens with chemical compounds.

Direct-Acting Carcinogens These are capable of acting as such with cellular macromolecules as proteins or nucleic acids, the genetic material of the cell. Examples of direct-acting carcinogens are aziridine (ethylenimine), beta-propiolactone, bis(2-chloroethyl)sulfide, ethylene oxide, diepoxybutane, methyl (and ethyl) methane sulfonate, nitrogen mustards, and propane sultone. These substances alkylate nucleic acids, initially at the 7-position of guanine and then elsewhere without further need for metabolic intervention. But metabolism may make the original molecule more accessible to important receptors. Also, once in the organism, direct-acting molecules may be deactivated by water or other abundant cell constituents and never reach important targets or receptors. Very reactive direct-acting carcinogens often are not so potent as those of lesser reactivity since they are less likely to reach a receptor. Most direct-acting carcinogens do not present a hazard to the general population.[3]

Indirect-Acting Carcinogens These generally are more stable in the environment and thus are more likely to come in contact with people. These agents, often called procarcinogens, need to be activated metabolically before they can interact with cellular macromolecules. Metabolism, and thus the effect, can be influenced by many factors. In turn, cells have a certain ability to repair the defects caused by addition of foreign molecules to their DNA. Indirect-acting carcinogens include polycyclic aromatic hydrocarbons from combustion of organic matter, as typified by benzo(a)pyrene; aromatic nitro compounds, formed in engine exhausts; nitrosamines; traces of aromatic amines from dyes, polymer intermediates, and tobacco smoke; some fungal toxins; and traces of haloaliphatics from biomass burning, marine environments, or chlorination of water. Many carcinogens occur naturally at very low levels in plants, perhaps formed as protective agents against natural pests. In some cases, when fed at high levels to animals, these substances led to tumors. An example is safrole, which occurs in many spices, but there is no evidence that under the usual conditions of exposure safrole has led to cancer in people.

Genotoxic and Nongenotoxic Carcinogens or their metabolites that react with the genetic material of the cell are called genotoxic. Other carcinogens are known that do not react thus and are nongenotoxic. Examples are limonene, diethylstilbestrol, and ethylene thiourea. Different fibrous materials (asbestos) have been implicated as carcinogens. Tumors have been caused in animals from subcutaneous implantation of bits of solid material. In all these cases, various mechanisms that did not involve cellular DNA were responsible.

Initiation-Promotion-Progression

Initiation-promotion was originally noted from application of polycyclic aromatic hydrocarbons (PAHs) to mouse skin. The initiator was such a low dose of PAH that no tumors resulted. Following the initiator with regular application of a solution of the resin from croton plants (now identified as 12-0-tetradecanoyl phorbol-13-acetate) (TPA) led to numerous tumors. Initiation produces an irreversible rapid change in the cell genome

while the promoter causes expansion of the initiated cells.[1,4] Promoters are noncarcinogenic themselves, do not bind to DNA, and generally are not mutagens. Many promoting agents besides TPA have been identified, including phenol, dodecane, and various complex natural products. Initiation-promotion has been noted with other organ systems besides the skin, including colon, liver, mammary gland, pancreas, respiratory tract, and thyroid.[4]

Continuance of the initiated-promoted cells or tissue to a full-blown neoplastic state comprises the progression part of the process. Several stages have been implicated, but they are not clearly defined.[5]

Cocarcinogens

Besides promoters, compounds known as cocarcinogens enhance the action of a carcinogen when given concurrently with the carcinogen. They are not active alone. Ethanol appears to be a cocarcinogen in both animals and humans.[3,4]

Factors Influencing Carcinogenesis

Response to indirect-acting chemical carcinogens is governed largely by the degree of metabolic activation of the procarcinogen. This physiologic response is influenced by many factors.[2,3,6,7]

Age

Often newborn animals are more sensitive to chemical carcinogens than adult animals because the enzyme systems to detoxify xenobiotics are not well developed.

Species

Not all species are equally susceptible to a given carcinogen. Guinea pigs and steppe lemmings lack enzyme systems to make appreciable levels of the N-hydroxy derivatives of aromatic amines or amides and thus did not develop tumors after being fed 2-acetylaminofluorene. N-Hydroxylation is a first step in activation of such compounds, but feeding synthetic N-hydroxy-2-acetylaminofluorene to these species did yield tumors.[6]

Mice are usually susceptible to skin tumors from painting of benzo(a)pyrene but not monkeys, even after many years of skin painting.

Strain

There are appreciable differences among strains (within a species) in response to carcinogens. Female Sprague-

Dawley rats, given a single oral dose of 7,12-dimethyl-benz(a)anthracene (DMBA) at 50 days of age, develop a 100% incidence of mammary tumors in approximately 9 months. Under similar conditions, female Marshall rats developed no tumors.

Sex

Male and female animals may respond quite differently to carcinogens. In some cases males had higher levels of certain important enzymes, but this was not always so. With 2-diacetylaminofluorene, where male rats had high liver cancer rates versus low rates in females, hormonal manipulation reversed the effects.[7]

Diet

Deficient diets generally enhanced the action of carcinogens, while adequate diets had an inhibitory effect for some carcinogens. Animals on a restricted diet that provided about 80% of the usual calories, but all the essential vitamins and minerals, had a lower spontaneous tumor incidence. They also showed a lessened response to most chemical carcinogens. Of the usual dietary constituents, fat is most likely to influence a carcinogenicity study. High dietary fat appears to enhance skin, mammary, and colon cancers in experimental animals.[8,9] Epidemiologic studies have pointed toward an association between dietary fat and various types of cancer, but other studies were less definitive. Genetic factors are also involved, and in people the influence of fat is less definitive.[10]

Enzyme Inducers

If the enyme induced enhances detoxication, the response to a carcinogen is less, and the reverse situation also holds. Enzyme inducers occur in many plants as well as other environmental materials. Enzyme induction by natural constituents of plants is often the basis for the chemopreventive effects of many vegetables and fruits against cancer.[11,12]

Other

Other factors influencing carcinogenic response are interindividual variability and the activity of the immune system. The presence of an endogenous oncogene or of some viruses in animals predisposes them to tumor development.

Genetic Influences

In humans, carcinogenesis generally is a long process, requiring 10 to 30 years after exposure to some agent.

Genetic research has explained why some cancers arise without obvious exposure to a carcinogen. At least 10 or more genes associated with susceptibility to various cancers have been identified. Normally these genes act as tumor suppressors, but if there is an alteration in the gene, these genes allow the cancerous process to proceed or may facilitate it.[2] A listing includes: RB1 for retinoblastoma or eye tumors[13]; p53, involved in carcinomas of the breast, adrenal, brain, and other tumor types[14]; APC with colon, stomach or thyroid cancer[15]; WT1 with Wilms tumor of the kidney[16]; NF1 and NF2 with tumors of the nervous system; pl6 with familial melanoma; KA11 with prostate cancer[17]; and BRCA1 and BRCA2 with breast and ovarian cancer.[18] Only about 5% of the population carries defective genes, but their presence explains clusters of cancer cases in certain families.

From alterations in the cellular DNA of tumors, it can be inferred which carcinogen may have been involved.[19] Comparing the molecular changes in the tumor with those induced by specific carcinogens allows such a judgment. Such information cannot be obtained from histopathologic examination of processed tissue.

Structure

Since chemists synthesize new chemicals and study their properties before they can be tested for possible carcinogenicity, some clues on structures associated with this property may be useful.[20,21]

Alkylating Agents

These agents should be considered suspect: epoxy compounds; aziridines, strained lactones or sultones; alphahaloethers; nitrogen or sulfur mustards; esters with good leaving groups. Such compounds alkylate or cross-link DNA.

Aromatic Amines, Precursor Nitro Compounds, and Amino Azo Dyes

Aromatic amines (or precursors) where the amino or nitro group is on the most reactive position of the molecule are generally active carcinogens. Substituents, such as $-OH$, $-COOH$, $-SO_3H$, usually decrease the effect, but $-CH_3$, $-OCH_3$, $-F$, $-Br$ in certain positions may enhance carcinogenicity. Heterocyclic ring (N,O,S) analogs of aromatic amines often are carcinogens. Amino azo dyes undergo the same metabolic reactions as aromatic amines and are thus suspect. Azo dyes derived from benzidine can be reductively split by intestinal bacteria affording the original benzidine.

N-Nitroso Compounds

A few N-nitroso compounds such as nitrosoproline and N-methyl-N-nitroso-p-toluenesulfonamide are not carcinogens, but practically all others are suspect or active carcinogens. N-Nitrosoureas and amides are often active and have caused many unusual types of tumors in animals. Many interesting and unusual organ-specific tumors can be produced with nitrosamines.

Aliphatic or Aryl Hydrazines, Azo and Azoxy, and Related Compounds

Most of the aliphatic hydrazines have shown carcinogenic effects, especially in mice. Dialkylhydrazines are metabolized in steps to an active alkylating moiety, considered the ultimate carcinogen. All the compounds along this path are active, but their metabolic formation can be diminished by enzyme inhibitors.

Aromatic Hydrocarbons

Aromatic hydrocarbons and their heterocyclic analogs are carcinogenic when they consist of four or five aromatic rings, arranged in a nonlinear fashion with an area of 100–135 nm^2. Addition of substituents can influence the effect.[20,21]

Halogenated Aliphatics

These compounds are metabolized to active intermediates such as an epoxide with vinyl chloride, or a glutathione conjugate that acts like a sulfur mustard for 1,2-dibromoethane.

Other Compounds

Compounds of varied structure have shown carcinogenic activity. Aflatoxin, ethyl carbamate, limonene, and thiourea represent some of the diverse structures that, by one means or the other, have caused cancer in animals.

Physicochemical factors may modify or moderate the potential of a substance. Solubility, physical state, molecular weight, and chemical reactivity are involved.[20] Predictions from chemical structure alone provide guidelines but are not absolute.

Testing for Carcinogenicity

Various types of tests are available, both short term and long term or chronic. Short-term tests cannot be considered definitive but they often provide a clue as to which compounds are more likely to be effective in chronic tests.

Short Term

Short-term tests are conducted in various systems of which mutagenicity in bacteria are most often employed.

Mutagenicity

Mutagens are compounds that cause sudden heritable genetic changes. Some genetic changes are beneficial, but others are not. Interest in mutagenicity tests stems from the concept that mutation is involved in carcinogenesis, leading to the theory that mutagens are carcinogens and carcinogens are mutagens. The correlation is not absolute, for only about 50–60% of mutagens are carcinogens. Further, the relevance of tests in bacteria to humans is not always clear.

Bacterial Tests Most tests for mutagenicity are done in bacteria that have been specially modified to be sensitive to outside influences. The *Salmonella typhimurium* test is most widely used. The bacteria are grown in a culture medium deficient in histidine; if the test compound mutates the bacteria, they grow in the absence of histidine. The count of the bacterial colonies in the culture dish represents the mutagenic activity of the test material. Metabolic activation is usually supplied by adding an aliquot of a liver fraction from rats, mice, or hamsters. These tests are run in parallel with those where no activating system is added. Compounds are positive (mutagenic) if there is a dose-response relationship reaching some multiple of the background rate.[22,23] Besides *Salmonella*, some modified strains of *Escherichia coli* are often employed.[24] There have been efforts to produce even more sensitive strains of both *Salmonella* and *E. coli*. Unless such strains are readily available, they are not likely to be used until many validation studies have been done.

Other mutagenicity tests used on occasion include the use of mammalian cells in culture, especially the L5178Y mouse lymphoma and Chinese hamster ovary cells. However, a survey of parallel results from *Salmonella*, mouse lymphoma, chromosomal aberration, and sister chromatid exchange in Chinese hamster ovary cells showed that the *Salmonella* test was the best overall predictor of carcinogenicity. Combining the results with the other tests did not improve accuracy.[25,26] Since the *Salmonella* tests are easier and cheaper to perform, it appears that this system is the first choice for a quick predictive test.

Other Tests Additional short-term tests in vitro include induction of unscheduled DNA synthesis (DNA repair) in mammalian cells, induction of chromosomal aberrations, induction of liver cell proliferation, inhibition of intracellular communication, and transformation of mammalian cells in culture.

Short-term tests conducted in vivo include mutagenesis in fruit flies, induction of dominant lethal mutations in rodents, induction of sister chromatid exchange, chromosomal aberrations, and micronuclei, all in bone marrow of mice, unscheduled DNA synthesis or enzyme-altered foci in rodent liver, detection of DNA adducts, initiation/promotion studies, and gene mutation and/or activation in transgenic mice.[27-30] Still other short-term tests in vivo take advantage of an enhanced sensitivity of a particular species or strain of animal to produce a tumor in less time than usual. Lung tumors in strain A mice,[31] skin tumors in SENCAR mice,[32] or mammary tumors in 50-day-old Sprague-Dawley female rats[33] are examples.

Long Term

Long-term studies are generally conducted in rats, mice, or sometimes hamsters because these animals are small, easily cared for, and there is considerable information on their spontaneous tumor rates, longevity, and nutritional requirements. Larger animals such as dogs are used only under special circumstances. Essentially after range-finding studies on the acute and subacute effects of the test substance, animals are exposed to the material for a period of 2 years, which represents the greater portion of their lifespan. Ideally, the exposure route should be relevant to that for people. Inhalation or skin application are good for most industrial exposures or cosmetic ingredients. Feeding in the diet is appropriate for food additives. The usual guidelines specify at least 50 male and 50 female rats and equal numbers of mice, at zero dose levels (controls) plus three other dose levels. The top dose should be near the maximum the animals can tolerate. More complicated protocols start with more animals so that a sample can be taken for examination at intervals to follow the progression of any lesions. There are many reference works providing protocols.[34-38] The F344 rat (Fischer) and B6C3F1 mouse (hybrid of C57/BL female and C3H male) are the animal models currently used by the NTP for long-term studies.[39]

After the exposure period the animals are examined carefully for any adverse effects, including tumors. Comparison of tumor rates between controls and exposed animals, on a statistical basis, allows a judgment as to whether the test substance is a carcinogen in animals.

The laboratory or pilot plant worker should use proper protective equipment when working with untested materials to avoid possible exposure. Guidelines for handling putative hazardous materials are given in this book and others.[40]

Currently there is discussion on various means to simplify or decrease the cost of a chronic bioassay. One proposal is to use only male rats and female mice as test animals. Other suggestions involve testing in mice that carry or lack a crucial gene (transgenic mice) and that are now commercially available. In an exploratory study, C57BL/6 p53 deficient mice and TG·AC transgenic mice responded more rapidly to carcinogens than did the usual mice. However, they did not respond to noncarcinogens.[41]

Many more validation studies of these systems would be needed before they would be accepted by official bodies.

Use of Bioassay Data

Data from both short- and long-term studies, along with any epidemiologic data, are employed by various groups and regulatory agencies to estimate the possible risk from exposure to a particular substance. Certain caveats should be remembered: a compound that is a carcinogen in animals may not necessarily be one in humans; humans are more efficient in repairing damage to DNA than animals, route and levels of administration to animals may not be relevant to humans; there may be differences in pharmacokinetics, receptors, or gene targets; natural presence of an oncogene in some test animals.

The International Agency for Research on Cancer (IARC) has classified many industrial processes and separate compounds into the following classes: Carcinogenic in Humans; Probably Carcinogenic in Humans; Possibly Carcinogenic in Humans; and Cannot be Classified.[42] IARC often uses results from tests that are not relevant to the usual human exposure. Currently, over 60 compounds and processes are listed as carcinogenic to humans and over 50 as probables.[43] The IARC lists include substances that are essential micronutrients and others (estrogens) that are necessary for the cycle of reproduction. The NTP also publishes listings of substances considered to pose a carcinogenic hazard.[44] The EPA is the most prominent of the Federal agencies in performing risk estimates with the data from animal bioassay data. EPA extrapolates from the high animal doses to the low doses typical of human exposure using several models, namely, probit, Weibull, time-to-tumor, log-normal, and multistage. Somewhat different estimates can thus be obtained.[22] The EPA also does not recognize a threshold for a carcinogen, contrary to most toxicologic principles. Although the models are mathematical, carcinogenesis is a biological process. More attention to the biology of the process is needed.[45] Industrial hygiene groups such as AIHA and ACGIH also use bioassay data to estimate the levels of compounds that may be allowed in the workplace, but with the simpler safety factor approach.

The best approach for all is to avoid exposure. Thus, laboratory or pilot plant workers should use adequate protective gear and containment facilities. By following these precautions, the chemist developing new compounds can avoid overt exposure and reduce the chances of becoming a cancer statistic 30 years later. Following a healthy lifestyle also aids in lowering cancer risk.

APPENDIX. Discrete Compounds Evaluated by IARC and/or the NTP[a]

A. Carcinogenic in Humans	B. Probably Carcinogenic in Humans
Aflatoxin	Acrylamide
4-Aminobiphenyl	Acrylonitrile
Arsenic and compounds	Adriamycin
Asbestos	2-Amino-3-methylimidazo[4,5f]quinoline
Azathioprine	Azacytidine
Benzene	Benz(a)anthracene
Benzidine	Benzo(a)pyrene
Beryllium and compounds	Bischloroethylnitrosourea
Bis(chloromethyl) ether	1,3-Butadiene
1,4-Butanediol dimethane sulfonate	Captofol
Cadmium and compounds	Chloramphenicol
Chlorambucil	CCNU
Chlornaphazine	p-Chloro-o-toluidine
Chromium compounds	Chlorozotocin
Cyclophosphamide	Cisplatin
Cyclosporin	Dibenz(a,h)anthracene
Diethylstilbestrol	Diethyl sulfate
Erionite	Dimethylcarbamoyl chloride
Ethylene oxide	Dimethyl sulfate
Melphalan	Epichlorohydrin
8-Methoxypsoralen plus UV	Ethylene dibromide
Methyl-CCNU	N-Ethyl-N-nitrosourea
Mustard gas	Formaldehyde
2-Naphthylamine	5-Methoxypsoralen
Nickel and compounds	4,4'-Methylene bis(2-chloroaniline)
Radon	N-Methyl-N-nitrosourea
Thiotepa	N-Methyl-N'-nitro-N-nitrosoguanidine
Thorium dioxide	Nitrogen mustard
Treosulphan	N-Nitrosodiethylamine
Vinyl chloride	N-Nitrosodimethylamine
	Phenacetin
	Procarbazine hydrochloride
	Propylene oxide
	Silica, crystalline
	Styrene oxide
	Tetrachloroethylene
	1,2,3-Trichloropropane
	Tris(2,3-dibromopropyl) phosphate
	Vinyl bromide
	Vinyl fluoride

[a]Updated through Vol. 65 of the IARC Monographs.

References

1. Sirica, A. E., Ed. *The Pathobiology of Neoplasia*; Plenum Press: New York, 1989.

2. Pitot, H. C. III; Dragan, Y. P. In *Casarett & Doull's Toxicology. The Basic Science of Poisons*; Klaassen, C. D., Ed.; McGraw Hill: New York, 1996; pp. 201–267.

3. Weisburger, E. K. In *Carcinogenesis*; Waalkes, M. P., Ward, J. M., Eds.; Raven Press: New York, 1994; pp. 1–23.

4. Peraino, C.; Jones, C. A. In *The Pathobiology of Neoplasia*; Sirica, A. E., Ed.; Plenum Press: New York, 1989; pp. 131–148.

5. Fearon, E. R.; Vogelstein, B. *Cell* **1990**, *61*, 759–767.

6. Weisburger, J. H.; Weisburger, E. K. *Pharmacol. Rev.* **1973**, *25*, 1–66.

7. Weisburger, E. K.; Weisburger, J. H. *Adv. Cancer Res.* **1958**, *5*, 331–431.

8. Ip, C., Birt, D. F., Rogers, A. E., Mettlin, C., Eds. *Dietary Fat and Cancer*; Alan R. Liss: New York, 1986.

9. Freedman, L. S.; Clifford, C.; Messina, M. *Cancer Res.* **1990**, *50*, 5710–5719.

10. Sellers, T. A.; Gapstur, S. M.; Potter, J. D.; Kushi, L. H.; Bostick, R. M.; Folsom, A. R. *Am. J. Epidemiol.* **1993**, *138*, 799–803.

11. Hecht, S. S. In *Dietary Phytochemicals in Cancer Prevention and Treatment*; Am. Inst. Cancer Res., Ed.; Plenum Press: New York, 1996; pp. 1–11.

12. Wattenberg, L. W. *Cancer Res.* **1985**, *45*, 1–8.

13. Wang, J. Y. J.; Knudsen, E. S.; Welch, P. J. *Adv. Cancer Res.* **1994**, *64*, 25–85.

14. Ozbun, M. A.; Butel, J. S. *Adv. Cancer Res.* **1995**, *65*, 71–141.

15. Nakamura, Y. *Adv. Cancer Res.* **1993**, *62*, 65–87.

16. Kovacs, G. *Adv. Cancer Res.* **1993**, *62*, 89–124.

17. Dong, J.-T.; Lamb, P. W.; Rinker-Schaeffer, C. W.; Vukanovic, J.; Ichikawa, T.; Isaacs, J. T.; Barrett, J. C. *Science* **1995**, *268*, 884–886.

18. Takahashi, H.; Bahbakht, K.; McGovern, P. E. et al. *Cancer Res.* **1995**, *55*, 2998–3002.

19. Aguilar, F.; Hussain, S. P.; Cerutti, P. *Proc. Natl. Acad. Sci. USA* **1993**, *90*, 8586–8590.

20. Woo, Y.-T.; Arcos, J. C.; Lai, D. Y. In *Handbook of Carcinogen Testing*, 2nd ed.; Milman, H. A., Weisburger, E. K., Eds.; Noyes Publications: Park Ridge, NJ, 1994; pp. 2–25.

21. Weisburger, E. K. *Chem-Tech* **1987**, *17*, 422–424.

22. Beliles, R. P.; Schulz, C. O. In *Patty's Industrial Hygiene and Toxicology*, 4th ed., Vol. 2, Part A; Clayton, G. D., Clayton, F. E., Eds.; Wiley: New York, 1993; pp. 25–75.

23. Zeiger, E. In *Handbook of Carcinogen Testing*, 2nd ed.; Milman, H. A., Weisburger, E. K., Eds.; Noyes Publications: Park Ridge, NJ, 1994; pp. 83–99.

24. Marwood, T. M.; Meyer, D.; Josephy, P. D. *Carcinogenesis* **1995**, *16*, 2037–2043.

25. Tennant, R. W.; Margolin, B. H.; Shelby, M. D.; Zeiger, E.; Haseman, J. K.; Spalding, J. W.; Caspary, W.; Resnick, M.; Stasiewicz, S.; Anderson, B.; Minor, R. *Science* **1987**, *236*, 933–941.

26. Zeiger, E.; Haseman, J. K.; Shelby, M. D.; Margolin, B. H.; Tennant, R. W. *Environ. Mol. Mutagen.* **1990**, *16*, 1–14.

27. Auletta, A. In *Handbook of Carcinogen Testing*, 2nd ed.; Milman, H. A., Weisburger, E. K., Eds.; Noyes Publications: Park Ridge, NJ, 1994; pp. 58–82.

28. Morris, S. M.; Casciano, D. A.; Casto, B. C., in Ref. 20, pp. 100–115.

29. McQueen, C. A.; Williams, G. M., in Ref. 20, pp. 116–129.

30. Sivak, A.; Tu, A. S. In *Handbook of Carcinogen Testing*, 2nd ed.; Milman, H. A., Weisburger, E. K., Eds.; Noyes Publications: Park Ridge, NJ, 1994; pp. 130–149.

31. Stoner, G. D.; Shimkin, M. B. In *Handbook of Carcinogen Testing*, 2nd ed.; Milman, H. A., Weisburger, E. K., Eds.; Noyes Publications: Park Ridge, NJ, 1994; pp. 197–232.

32. Slaga, T. J.; Nesnow, S. In *Handbook of Carcinogen Testing*, 2nd ed.; Milman, H. A., Weisburger, E. K., Eds.; Noyes Publications: Park Ridge, NJ, 1994; pp. 248–268.

33. McCormick, D. L.; Moon, R. C. In *Handbook of Carcinogen Testing*, 2nd ed.; Milman, H. A., Weisburger, E. K., Eds.; Noyes Publications: Park Ridge, NJ, 1994; pp. 233–247.

34. Prejean, J. D. In *Handbook of Carcinogen Testing*, 2nd ed.; Milman, H. A., Weisburger, E. K., Eds.; Noyes Publications: Park Ridge, NJ, 1994; pp. 333–346.

35. Hamm, T. E., Jr. In *Handbook of Carcinogen Testing*, 2nd ed.; Milman, H. A., Weisburger, E. K., Eds.; Noyes Publications: Park Ridge, NJ, 1994; pp. 270–285.

36. *CFR* 21, part 58.

37. *IARC Monographs on the Evaluation of the Carcinogenic Risk of Chemicals to Humans*; Long-Term and Short-Term Screening Assays for Carcinogens: Critical Appraisal. Suppl. 2, IARC: Lyon, France, 1980.

38. Grice, H. C., Ciminera, J. L., Eds. *Carcinogenicity*; Springer: New York, 1988.

39. Goodman, D. G.; Boorman, G. A.; Strandberg, J. D. In *Handbook of Carcinogen Testing*, 2nd ed.; Milman, H. A., Weisburger, E. K., Eds.; Noyes Publications: Park Ridge, NJ, 1994; pp. 347–390.

40. Walters, D. B. In *Handbook of Carcinogen Testing*, 2nd ed.; Milman, H. A., Weisburger, E. K., Eds.; Noyes Publications: Park Ridge, NJ, 1994; pp. 298–332.

41. Tennant, R. W.; French, J. E.; Spalding, J. W. *Environ. Health Perspect.* **1995**, *103*, 942–950.

42. *IARC Monographs on the Evaluation of Carcinogenic Risks to Humans*; Overall Evaluations of Carcinogenicity: An Updating of IARC Monographs Volumes 1 to 42. Suppl. 7, IARC: Lyon, France, 1987.

43. *IARC Monographs on the Evaluation of Carcinogenic Risks to Humans*. Vol. 65; Printing Processes and Printing Inks, Carbon Black and Some Nitro Compounds. IARC: Lyon, France; 1996.

44. USDHHS, NTP. *Seventh Annual Report on Carcinogens*. Summary, 1994.

45. Clayson, D. B.; Iverson, F. *Regul. Toxicol. Pharmacol.* **1996**, *24*, 45–59.

20

Toxicology

ELIZABETH K. WEISBURGER

Toxicology is the branch of science that studies and deals with the adverse effects of substances, either natural or synthetic, on some target organism.

Background

Toxicology has a long history; in prehistoric times humans discovered by trial and error that certain plants could be eaten safely while others caused illness or even death. Likewise, the bites of certain insects and reptiles led to harmful results, while others may have inflicted pain or inconvenience but could be sustained by most humans. The experience thus gained became part of primitive toxicology. In the ancient world toxicology became associated with the practice of deliberate poisoning of humans. An example was the execution of Socrates by an extract of poison hemlock.[1-3]

The word "toxic" stems from the early Greek use of poisoned arrows. The word for the bow was toxon, and the Romans took this word as toxicum (for poison). Eventually toxicum became "toxin" and the knowledge or science of toxins became "toxicology."

The ancients thus knew of the toxic effects of many substances, but about 1530 Paracelsus emphasized that the dose of a substance was an important factor in the distinction between a therapeutic and a harmful effect. In 1700 Ramazzini summarized the harmful effects of various occupations, many of which involved exposure to chemical substances. However, Orfila (1815) is considered the founder of modern toxicology since he used quantitative methods in investigations of the harmful effects of chemicals.[4] An early publication by Kobert[5] recommended limits on the amounts of substances to which industrial workers should be exposed, a prelude to the concept of Threshold Limit Values® (TLVs) as promulgated by the ACGIH, the AIHA, and the Permissible Exposure Limits (PELs) of OSHA in this country or by equivalent groups in other countries.[6]

Types of Toxicity

Acute

Acute toxicity relates to the ability of some material, either a single substance or a mixture, to cause systemic damage to an organism after a single exposure, often of short duration.[2,7] Acute toxicity is often expressed in terms of the LD_{50}, or the dose that kills 50% of the test animals, expressed in mg/kg body weight. The LD_{50} is a statistically derived dose, due to the many factors that influence toxicity, and it represents the response of a group of animals under a specific set of conditions. LD_{50} values are used for substances given orally or dermally, while for inhalation exposure the LC_{50} (lethal concentration) or mg/cubic meter of air is more appropriate. LC_{50} can also refer to mg/liter of water, used to express toxicity to fish. LD_{50}s for humans cannot be measured, but accidental poisonings or suicide attempts often furnish data on what levels are likely to be harmful.

A small LD_{50} means that the chemical in question is quite toxic; a high LD_{50} shows that the material has lesser acute toxicity. Occupationally, compounds with an oral LD_{50} less than 1 mg/kg are considered as extremely toxic; from 1–50 mg/kg are highly toxic; 50–500 mg/kg is listed as moderately toxic; 0.5–5 g/kg slightly toxic; 5–15 g/kg practically nontoxic; over 15 g/kg is considered relatively harmless.[8] Some compounds that have an appreciable impact after one dose or exposure and are thought of as acutely toxic are chlorine, hydrogen cyanide, hydrogen sulfide, methyl isocyanate, hydrogen fluoride, and phosgene, for example. However, at certain levels even "relatively harmless" substances such as salt, sugar, and water can have adverse effects.

As with other types of toxicity, the LD_{50} depends to some extent on many factors, including exposure route, species, strain, sex, age, nutritional status, and interindividual variation. For practical reasons, most acute toxicity tests are done in small animals—mice, rats, or sometimes hamsters. Larger animals such as dogs or rabbits are used in some cases. It is important to use sufficient animals so that the results will be meaningful on a statistical basis and will meet the requirements of the FDA Good Laboratory Practice (GLP) precepts.[9] Proper animal care and husbandry are part of the GLPs.[10,11] At least 10 animals per group has been the accepted number previously. Pressure to lessen the number of animals used in research has led to an "up and down" or pyramid method by which the LD_{50} can be estimated with a smaller number of animals. This is also advisable when larger animals are the test subjects.*

---

*This method also described in Reference 27.

Subchronic

Subchronic studies are meant to determine any adverse effects from repeated exposure. Often they are a prelude to chronic or lifetime tests in order to furnish information on cumulative toxicity, which organs are affected, what dose leads to no apparent effects, and other parameters. Generally a 14-day study is the first part of a stepwise approach, followed by a 90-day trial. In a subchronic oral study the test substance can be given by gavage (stomach tube), mixed in the diet, by capsule (suitable for larger animals), or in the drinking water. Several dose levels are employed; careful observation is necessary and histopathologic evaluation of tissues is required at the end of the study.[12,13]

Chronic

Chronic toxicity refers to a harmful effect noted after long-term exposure to a substance at a low level.[14] Most such studies have been done in rats and mice, usually by mixing the test material in the feed. In some cases the test substance has been given as a solution in place of drinking water or by gavage. In a few cases with larger animals, the test compound has been given as a bolus in capsules. The interest in the oral route stems from the need to test food additives or contaminants such as pesticides. Furthermore, testing by the oral route is often easier and cheaper than by other routes. It may also simulate to some extent occupational exposures in a dusty situation where some of the material is swallowed. For most occupational exposures, inhalation or skin application are more relevant.

Because the effects of long-term exposure are relatively gradual, they often may not be distinguished from various other changes, such as those due to age. Thus control groups of animals are especially important.

In these toxicity tests, suitable groups of animals are given the test material for a period of 18 months to 2 years or more. The response is plotted against any effect; often no response is noted until a certain level that is a threshold or lowest observed adverse effect level (LOAEL). Any dose below the LOAEL comprises a no observed adverse effect level (NOAEL). LOAELs and NOAELs are used in extrapolating the possible effects to humans when exposure limits are set.[2,6,15] The dose-response curve usually rises to a level where no further response is noted and the curve levels off because there is saturation of the receptors or death of the test animals.

At the end of the test period, careful necropsy of the animals and histopathologic examination of tissues are necessary. Data should be treated statistically to ascertain whether any differences between test and control animals are actually significant.[16] Comparison with historical control data is also helpful in noting the variations (ranges) in disease incidence in control animals.

Exposure Routes

Oral

The oral route has been the standard for tests of proposed food additives or for pesticides where traces may remain in food crops.[12,17–19] The equipment required is relatively simple, food mixers, which for safety are used so that contamination of the surroundings or exposure to personnel do not occur. The test compound should be of known purity. The diet as mixed should be analyzed to ascertain that the test compound has been uniformly distributed and that the test compound and diet constituents have not reacted. Thus, only relatively stable, nonvolatile compounds can be given in the diet. The amount of diet eaten should be checked at intervals. Compounds which are less stable should be freshly dissolved in a suitable vehicle and be given by gavage. Administration of a test material in the drinking water is appropriate if it is stable in water. The amount of drinking fluid consumed should be measured at intervals to provide a better check on the amount of test substance actually administered.

Inhalation

Inhalation is the method of choice for testing gases and relatively volatile substances, and it is thus most relevant for many industrial exposures.[20–22] Any material which enters the lungs has access to the blood supply and thus the cardiovascular complex by means of the exchange system which is the function of the lungs. In the environment there is exposure to gases, vapors from volatile materials, aerosols, and dusts. Dusts and particles are likely to deposit in the alveolar spaces of the lungs and may eventually be absorbed and distributed. Substances delivered in the other forms are likely to distribute throughout the body more rapidly.

Inhalation testing requires more complicated equipment than tests by other routes. In chamber operations, cages of the test animals are placed in enclosures where the level of the test material can be controlled and measured at various points. Chambers can be of different sizes, ranging from 10–50 liters to very large ones with volumes of several cubic meters and capable of holding several hundred animals. Usually large chambers are made of nonreactive material such as stainless steel, with glass viewing ports. Small chambers for preliminary tests can be simple glass bell jars. In nose- or head-only exposure, the head or nose is exposed by placing the head in an exposure tube, or by passing the test material into a hood over the animal's head. Fewer animals can be treated thus, but the methods are useful for preliminary runs, if there is a low supply of test compound, or if large animals are the test subjects.

Skin Exposure

The skin acts as a barrier to many substances, but there are others which cause irritation or penetrate the skin at levels sufficient to cause systemic toxicity in the exposed subject.[23] Rabbits and guinea pigs are often used as test animals; they often are more sensitive than humans, affording an additional measure of safety for people. Most of the skin tests involve placing fairly concentrated solutions of the test material on both intact and abraded animal skin to ascertain whether abrasion increases the effect. The test site is covered with an impervious material and at 24 or 72 hours the degree of skin irritation is noted. There are variants of the test, depending on the situation.[24,25]

Similarly, testing of possible eye irritant effects constitutes the Draize test in which the substances are applied to rabbit eyes. The degree of irritancy or corrosion is read after specific time periods.[26] Adverse opinion on the true value of these tests has led to efforts to substitute bovine or porcine corneas or artificial protein membranes for both the skin and Draize tests.[27] However, many of the results with these membranes have not been fully validated, and for regulatory purposes, the standard tests still need to be performed.

Determinants of Toxicity

In a toxicity study there are two major components—the test compound and the test organism. The substances under study should be representative materials and of known purity; sometimes contaminants can lead to spurious results. The other component, the test organism, is more likely to afford unexpected results due to the many factors that influence toxicity.

Species

The species used as test subjects have all evolved somewhat differently, and what is poison to one species may not be so to another species. Methanol is quite toxic for primates, causing blindness, while nonprimates do not show such an action. Tri-o-cresyl phosphate affects nerve fibers in chickens and humans but not in rats and dogs. Many other examples are known.[28]

Strain

Within any species, especially in laboratory animals, there are strain differences. Strain effects are more prominent in long-term studies but should be considered in acute studies also. For example, Sprague-Dawley rats are often less affected in acute toxicity tests than are Fischer rats.[cf 27]

Age

Age of the test organism influences toxicity because in very young animals the enzyme systems that detoxify xenobiotics are often not developed. Old animals may also be deficient in enzyme activity; thus for most tests young adult animals are best employed.

Sex

Sex may not appear relevant in most acute and range-finding studies, but both males and females should be used for subchronic and chronic protocols, due to the differences in activities of the metabolic enzymes. For example, the dyestuff intermediate 2-aminoanthraquinone was toxic for female rats, but not for males, an effect which could be reversed by hormonal manipulation.[29] Conversely, d-limonene was toxic for male but not female rats, an outcome linked to a specific protein in the kidneys of male rats.[30]

Diet

Diet plays a greater role in chronic than in shorter tests since both the quality and quantity of diet may influence the outcome. It is important that the test animals receive a nutritionally adequate diet with the necessary levels of major dietary components plus the essential vitamins and trace minerals. Diets must be free of molds, filth, and the like. The quantity of diet also influences toxicity studies, for animals fed *ad libitum* are more likely to become obese, die earlier, and develop more chronic diseases than animals on a somewhat restricted diet.[31] The effect is related to levels of activating and DNA repair enzymes, expression of these enzymes, and other biochemical mechanisms which are influenced by dietary components.

Other

Other factors relating to the test organism are the status of the immune system, the general health of the animals, and interindividual variation, which occurs even in inbred strains of animals which are more uniform genetically than outbred strains.[7]

Metabolism

Metabolism is an important factor in the toxicity of many substances. Another point is that many xenobiotics are somewhat lipophilic and are converted in two phases to more water-soluble substances.[32-35]

Phase I

In this phase, a polar reactive group is added to the parent molecule by processes involving several enzyme systems. Oxidation often is basic for phase I reactions. Oxidation of aliphatic hydrocarbons or alkyl chains (as on aromatic hydrocarbons) may yield alcohols, ketones, and carboxylic acids. These steps may afford less toxic derivatives; toluene, for example, forms less toxic benzoic acid. Or more toxic metabolites may result, as is the case with hexane. The neurotoxic action of hexane has been ascribed to its 2,5-hexanedione metabolite; methyl butyl ketone has a similar neurotoxic effect, thus reinforcing the concept.

Oxidation of aromatic hydrocarbons apparently affords an initial epoxide or arene oxide. Rearrangement can yield phenols, while a reaction mediated through epoxide hydrolase leads to dihydrodiols; further oxidation of the dihydrodiols may yield another epoxide. These steps occur during metabolism of benzo(a)pyrene to an activated carcinogen.

Dealkylation involving O-, S-, and N-alkyl groups is also an oxidative-type reaction that occurs commonly in the metabolism of many pesticides and drugs. N-Oxidation of aromatic amines or amides affords the corresponding N-hydroxy derivatives. Depending on the aromatic structure and position of the amino or amido group, carcinogenic or toxic intermediates may form; 2-naphthylhydroxylamine is a carcinogenic intermediate, while phenylhydroxylamine is toxic. Additional oxidative metabolic pathways are deamination and S-oxidation. Some oxidative reactions are mediated by flavine-dependent enzymes and various peroxidases.

The most important enzyme system in the totality of metabolic reactions is the carbon monoxide binding protein called the cytochrome P450 system, in reality a heme-thiolate protein. Originally this was considered a single protein, but improved protein separation methods, molecular studies, and gene mapping have led to the discovery of almost 500 P450 genes (see Table 20.1), in

different types of organisms. The subgroups of the P450 family are named according to their gene locus.[36,37]

Phase II

In this second metabolic phase, the substrate that has been oxidized, dealkylated, or otherwise modified is conjugated with some endogenous agent, affording a relatively water-soluble product that is then excreted.

Glucoside and Glucuronide Conjugation In these steps, uridine diphosphate glucose or uridine diphosphate glucuronic acid react with the substrate, forming O-glucosides or glucuronides from alcohols and phenols, some acids and hydroxylamines. N- and S-glucuronides have also been formed from some aromatic amines and thiophenols. This enzyme is glucuronyl transferase; among the common domestic or research animals, the cat is deficient in this enzyme and thus excretes none or extremely low levels of these conjugates.

Sulfate Conjugation Formation of a sulfate ester occurs with alcohols, phenols, N-hydroxyarylamines, and some arylamines. The actual sulfating entity is adenosine 3′-phosphate-5′-phosphosulfate and the resulting sulfate esters are usually fairly water soluble.

Glutathione Conjugation The small peptide, reduced glutathione, reacts with electrophilic centers in xenobiotics, mediated by glutathione-S-transferase, to yield mercapturic acids. Generally this is a detoxication reaction, but for some halogenated compounds such as 1,2-dibromoethane, glutathione conjugation leads to an activated intermediate akin to a sulfur mustard; this intermediate then attacks the DNA to form adducts.

Amino Acid Conjugation Foreign carboxylic acids form amides, largely with glycine or glutamine, a mammalian-type reaction. Other amino acids act as the conjugating entities in other organisms. Formation of hippuric acid from benzoic acid and glycine is an example of this reaction.

Acylation Foreign amines, hydroxy and sulfur compounds can be acylated through acetyl-coenzyme A, yielding the corresponding acyl derivatives. Genetic and species differences exist in the levels of the acetyl transferase enzyme. Among humans, rapid acetylators are thought to be less susceptible to bladder cancer from arylamines.

Other Enzymes Various other enzymes are involved in metabolism of xenobiotics, including reductases, which

TABLE 20.1 Some P450s

Gene Symbol	Function	Species	Substrate	Inducer	Inhibitor
CYP 1A1	Metabolism of PAHs, aromatic amines, halogenated hydrocarbons	Dog, guinea pig, hamster, human, rabbit, rat, macaque	PAHs	Dioxin	7,8-Benzoflavone, Ellipticine
CYP 1A2	Metabolism of drugs and environmental chemicals	Dog, guinea pig, hamster, human, rabbit	Acetanilide, Caffeine, Warfarin	PAHs, Cruciferous vegetables	7,8-Benzoflavone, Furafylline
CYP 1B1		Human, rat			
CYP 2A 1-3		Rat			
CYP 2A6		Human	Coumarin, Nicotine	Barbiturates	Diethyldithiocarbamate
CYP 2A8-9		Hamster			
CYP 2B1-3		Rat		Phenobarbital	
CYP 2B6	Metabolism of drugs	Human	Cyclophosphamide		Orphenadrine
CYP 2C1-6		Rabbit			
CYP 2C8	Steroid hydroxylation	Human	Taxol, Carbamazepine		Quercetin
CYP 2C9		Human	Phenytoin, Tolbutamide, Warfarin	Rifampin	Sulfaphenazole
CYP 2C18		Human			
CYP 2C19		Human	Phenytoin, Hexobarbital	Rifampin	Tranylcypromine
CYP 2D1-5		Rat			
CYP 2D6		Human	Codeine, Fluoxetin		Quinidine Ajmalicine Yohimbine
CYP 2E1	Metabolism of many toxins	Hamster, human, macaque, marmoset, rabbit, rat	Alcohols, Aniline, Caffeine, Dapsone, Nitrosamines, Styrene	Ethanol, Isoniazid	Aminoacetonitrile, Disulfiram, 4-Methylpyrazole
CYP 3A4	Metabolism	Human	Aldrin, Acetaminophen, Lidocaine, Quinidine, Steroids, Taxol, Warfarin	Dexamethasone, Phenobarbital, Phenytoin	Ethinylestradiol, Gestodene, Naringenin, Troleandomycin
CYP 4	Fatty acid hydroxylation	Vertebrates			
CYP 5	Thromboxane synthase	Vertebrates			
CYP 7A	Cholesterol hydroxylation	Vertebrates			
CYP 8	Prostacyclin synthesis	Vertebrates			
CYP 11	Cholesterol metabolism	Vertebrates			
CYP 17	Steroid hydroxylation	Vertebrates			
CYP 19	Androgen aromatization	Vertebrates			
CYP 21	Steroid hydroxylation	Vertebrates			
CYP 24	Steroid hydroxylation	Vertebrates			
CYP 27	Steroid hydroxylation	Vertebrates			

Data from References 32 and 37 and Guengerich, F. P. in Reference 1, pp. 1259–1313.

Many P450s have been identified in insects; plants have an equivalent system.
PAH = polycyclic aromatic hydrocarbon.

reduce nitro or azo compounds; hydrolases that split esters; and many others that are active in more special cases. In the intestinal tract many xenobiotics or their metabolites become substrates for the enzymes found in intestinal bacteria. These organisms are often responsible for reductive splitting of azo dyes or reduction of some nitro compounds. The final product from administration of a xenobiotic to a mammal may represent a composite of mammalian and bacterial metabolism.

Enzyme Induction and Inhibition

The phenomenon of enzyme induction was discovered during studies involving pesticides and in others with chemical carcinogens.[38] Inadvertent exposure of rats to certain pesticides led to increases in the rate at which xenobiotics were metabolized. Treatment with the carcinogens benzo(a)pyrene or 3-methylcholanthrene decreased the effects of other carcinogens, contrary to expectations. Follow-up on these results opened a new area in toxicology and pharmacology. Many compounds were noted as capable of increasing the expression (induction) of oxidases and other enzymes. One of the most potent is the environmental contaminant 2,3,7,8-tetrachlorodibenzo(p)dioxin (TCDD), but other substances such as phenobarbital, polychlorinated biphenyls, some polycyclic aromatic hydrocarbons, many chlorinated pesticides (chlordane, mirex, DDT), and many plant constituents (terpenes, monoterpenes, various sulfur compounds) all have this property. Small amounts of enzyme inducers occur naturally in foods; of note are those in cruciferous vegetables, watercress, garlic and onions, and citrus fruits. Generally, these substances may have a differential effect on phase I enzymes and increase the activity of phase II enzymes, especially glucuronyl transferase and glutathione-S-transferase. The beneficial activity of these dietary constituents against experimentally induced cancer has led to many recommendations to increase dietary consumption of fruits and vegetables.[39]

Conversely, there are substances that inhibit the actions of drugs and other xenobiotics. Examples are diallyl sulfide, disulfiram, metyrapone, and piperonyl butoxide. This action is not always clearcut since a substance that induces one enzyme may inhibit another. Diallyl sulfide from garlic suppresses the level of rat P450 2EI, considered to be involved in metabolic activation of various carcinogens, but it increases P450 2Bl, involved in detoxication.[40]

The matter of enzyme induction/inhibition assumes special importance for patients whose response to their usual drugs is altered by another drug or by some environmental or dietary constituent such as cheese, red wine, or citrus fruit. Handbooks for pharmacists and physicians now include alerts on drug interactions.[41]

Subspecialties in Toxicology

As with other areas of science, toxicology has developed to the extent that various subspecialties have arisen within the field. Some are concerned with various organ systems or functions. For example, reproductive toxicology, a matter of great public interest at present, is covered elsewhere in this volume, as is carcinogenesis, a type of long-term toxicity. In order to demonstrate the wide range of toxic materials which affect organ/function systems, several are discussed briefly, with examples of the toxins. Moreover, some compounds are toxic for certain of these systems, but not for others. Other substances act as generalized toxins for most of the body systems.

Blood

The blood is responsible for transporting oxygen to tissues and removing waste products. Carbon monoxide, cyanides, or hydrogen sulfide are toxins for the blood since they bind to hemoglobin, the oxygen carrier, leading to serious and fatal consequences.[42]

Cardiovascular System

The several components of this system are affected differently. Thrombosis is produced by homocysteine and endotoxins; toxicity to platelets by serotonin, epinephrine, and testosterone; toxicity to clotting factors by epinephrine, tyramine, ACTH, mercuric chloride, and corticosteroids. The vascular system is altered adversely by allyl amine, heavy metals, carbon monoxide, carbon disulfide, 1,3-butadiene, oxygen, and other substances. Cardiotoxicity is exhibited by alcohols, catecholamines, various antibiotics, heavy metals, and reactive oxygen species.[43]

Dermal/Eye

Although the skin is considered a barrier, many xenobiotics of a lipophilic nature can penetrate the skin. Corrosives such as ammonia, calcium oxide, strong acids and alkalis, hydrogen fluoride, and phenol are toxic to the skin, and especially to the eye.[23-25]

Some drugs, such as chlorpromazine, sulfonamides, and tetracyclins, although taken internally, increase the

phototoxic effects of sunlight, as do the methoxypso-
ralens. Besides the sensitivity of the eye to irritating and
corrosive agents, steroids, chlorpromazine, thallium, bu-
sulfan, and 2,4-dinitrophenol are linked to formation of
cataracts. Methanol is a toxin for the eye in humans due
to metabolism to formate, causing blindness. Carbon disulfide, chloroquine, organomercurials, phenothiazine,
retinoids, and tamoxifen are also somewhat toxic for the
eye.[26]

Immune System

The immune system is less well defined, but numerous
substances can suppress its action. Heavy metals, organophosphates, organotins, polychlorinated biphenyls,
polycyclic aromatic hydrocarbons, pulmonary irritants,
and tobacco smoke are among these.[44,45]

Kidney

Specific nephrotoxicants include acetaminophen, cis-
platin, cyclosporin A, halogenated hydrocarbons (bro-
mobenzene, chloroform, tetrafluoroethylene), some my-
cotoxins, and nonsteroidal anti-inflammatory agents.[46,47]

Liver

Besides being the major metabolic organ, the liver is the
glycogen storage depot for the body. Toxins for the liver
are allyl alcohol, carbon tetrachloride, chlorpromazine,
dimethylformamide, diethyl- and dimethylnitrosamine,
methylenedianiline, the hormone estradiol, and the natural product aflatoxin B_1.[48,49]

Nervous System

The response of the neurons and axons of the peripheral
nervous system may differ only slightly. Methyl mercury, trimethyl tin, and doxorubicin affect the neurons.
Acrylamide, carbon disulfide, colchicine, diketones, as
from metabolism of hexane, disulfiram, heavy metals,
lead, especially, hydralazine, 3,3'-iminodipropionitrile,
and kepone plus others target the neurons.[50] Although
the brain is partly protected by the blood-brain barrier,
1-methyl-4-phenyl-1,2,3,6-tetrahydropyridine (MPTP),
a contaminant of an illicit drug, affects the brain. MPTP
is converted by an amine oxidase to 1-methyl-4-phenyl
pyridine (MPP$^+$), which leads relatively quickly to Par-
kinsonian-type symptoms in both humans and experimental animals.[51]

Respiratory System

The lungs serve as a gas exchange system with oxygen
entering and carbon dioxide leaving the organism. Injury to the lungs may lead to fibrosis, emphysema, and
lung cancer. Lung toxins include particulate materials
such as asbestos, coal and cotton dust, iron oxides, ka-
olin, silica, and talc plus volatiles such as chlorine, methyl
isocyanate, ammonia, nitrogen oxides, ozone, paraquat,
and phosgene.[52]

Applications of Toxicology

The applications of toxicology and the interactions with
other disciplines serve to emphasize the varied and broad
scope of toxicology. Some examples are presented here.

Analytical toxicology is best known through the application of the techniques as a necessary aspect of forensic
toxicology, which is important in the legal and law enforcement areas.[53,54] Degree programs in this special area
are provided at several universities.

Clinical toxicology often is a matter of life or death
for those dealing with management of poisoning, drug
overdoses, adverse drug interactions, bites from venom-
ous creatures, suicide attempts, and the like. In such instances, the correct rapid treatment is necessary. Often
the supporting clinical laboratory must act quickly to determine what the toxic material may be, which ties into
analytical toxicology. Cooperation between the laboratory and clinical personnel is essential.[41]

Industrial toxicology relates to application of the discipline for the protection of workers. As such it blends into
the practice of industrial hygiene, an area of interest to
workers, labor unions, ACGIH, AIHA, management,
NIOSH, and OSHA, a regulatory agency.[6]

Regulatory toxicology comprises the arena where toxicologists must interact with legal and regulatory personnel.[55-57] This area also has international ramifications as
countries and industries attempt to standardize various
guidelines and regulations.

Future Directions in Toxicology

The current emphasis on using risk estimates during regulatory decision making will require a knowledge of comparative toxicology and the greater use of physiologically
based pharmacokinetic models.[58-60] Well-conducted risk
estimates will lead to obtaining actual data rather than
employing estimates and default assumptions.[61,62]

The growth of molecular biology will also influence
toxicology, with molecular toxicology assuming greater
prominence. The exact determination of which cellular
entities are affected by a toxin (chromosomes, genes,
amino acid sequences) will become more commonplace.

More knowledgeable means to counteract against toxins would then be possible.

Resources in Toxicology

Many resources in toxicology are available, from printed to electronic media. The Registry of Toxic Effects of Chemical Substances (RTECS), published by NIOSH, has data on thousands of chemicals. It is now available on CD-ROM.[63] There are several extensive reference books on toxicology.[1,3,4,11] Some of these provide exact protocols for many of the laboratory tests.[1,64] One of these reference works lists contract and academic programs in toxicology and organizations within the field.[11] The documentation for the Threshold Limit Values® of ACGIH contains relevant toxicologic information[65] as do the individual chapters in *Patty's Industrial Hygiene and Toxicology*.[66] Several journals are devoted to various aspects of toxicology, including *Chemical Research in Toxicology, Fundamental and Applied Toxicology,* * *Journal of Applied Toxicology, Toxicology, Toxicology and Applied Pharmacology,* and others. The National Library of Medicine has available the TOXLINE data system, which is part of the electronic media. The CCRIS system at NLM has data on long-term effects of chemicals.

Thus, there are many resources available to anyone wishing to explore toxicology. This pursuit can be a fascinating journey in the area of the interaction of chemicals with living organisms.

References

1. Borzelleca, J. F. In *Principles and Methods of Toxicology,* 3rd ed.; Hayes, A. W., Ed.; Raven Press: New York, 1994; pp. 1–17.
2. Ottoboni, A. M. *The Dose Makes the Poison,* 2nd ed.; Van Nostrand Reinhold: New York, 1991.
3. Ballantyne, B., Marrs, T., Turner, P., Eds. *General & Applied Toxicology*; Macmillan/Stockton: New York, 1993.
4. Gallo, M. A. In *Casarett & Doull's Toxicology: The Basic Science of Poisons,* 5th ed.; Klaassen, C. D., Ed.; McGraw-Hill: New York, 1996; pp. 3–11.
5. Kobert, R. *Kompendium der praktischer Toxikologie zum Gebrauche für Artze, Studierende und Medizinalbeamte,* 5th ed.; F. Enke: Stuttgart, 1912.
6. Zapp, J. A.; Doull, J. In *Patty's Industrial Hygiene and Toxicology,* 4th ed., Vol. 2, Part A; Clayton, G. D., Clayton, F. E., Eds.; Wiley: New York, 1993; pp. 1–23.
7. Eaton, D. L.; Klaassen, C. D. In *Casarett & Doull's Toxicology: The Basic Science of Poisons,* 5th ed.; Klaassen, C. D., Ed.; McGraw-Hill: New York, 1996; pp. 13–33.
8. Sowinski, E. J.; Cavender, F. L. In *Patty's Industrial Hygiene and Toxicology,* 4th ed., Vol. 2, Part A; Clayton, G. D., Clayton, F. E., Eds.; Wiley: New York, 1993; pp. 77–105.
9. 21 *CFR,* Part 58.
10. Reynolds, S. A.; Burger, G. T. In *Principles and Methods of Toxicology,* 3rd ed.; Hayes, A. W., Ed.; Raven Press: New York, 1994; pp. 497–544.
11. Siglin, J. C.; Rutledge, G. M. In *CRC Handbook of Toxicology*; Derelanko, M. J., Hollinger, M. A., Eds.; CRC Press: Boca Raton, 1995; pp. 1–49.
12. Stevens, K. R.; Mylecraine, L. In *Principles and Methods of Toxicology,* 3rd ed.; Hayes, A. W., Ed.; Raven Press: New York, 1994; pp. 673–695.
13. Auletta, C. S. In *CRC Handbook of Toxicology*; Derelanko, M. J., Hollinger, M. A., Eds.; CRC Press: Boca Raton, 1995; pp. 51–104.
14. Wilson, N. H.; Hayes, J. R. In *Principles and Methods of Toxicology,* 3rd ed.; Hayes, A. W., Ed.; Raven Press: New York, 1994; pp. 649–672.
15. Beliles, R. P.; Schulz, C. O. In *Patty's Industrial Hygiene and Toxicology,* 4th ed., Vol. 2, Part A; Clayton, G. D., Clayton, F. E., Eds.; Wiley: New York, 1993; pp. 25–75.
16. Gad, S. C.; Weil, C. S. In *Principles and Methods of Toxicology,* 3rd ed.; Hayes, A. W., Ed.; Raven Press: New York, 1994; pp. 221–274.
17. Reddy, C. S.; Hayes, A. W. In *Principles and Methods of Toxicology,* 3rd ed.; Hayes, A. W., Ed.; Raven Press: New York, 1994; pp. 317–360.
18. Saunders, D. S.; Harper, C. In *Principles and Methods of Toxicology,* 3rd ed.; Hayes, A. W., Ed.; Raven Press: New York, 1994; pp. 389–415.
19. Kotsonis, F. N.; Burdock, G. A.; Flamm, W. G. In *Casarett & Doull's Toxicology: The Basic Science of Poisons,* 5th ed.; Klaassen, C. D., Ed.; McGraw-Hill: New York, 1996; pp. 909–949.
20. Kennedy, G. L., Jr.; Valentine, R. In *Principles and Methods of Toxicology,* 3rd ed.; Hayes, A. W., Ed.; Raven Press: New York, 1994; pp. 805–838.
21. Newton, P. E. In *CRC Handbook of Toxicology*; Derelanko, M. J., Hollinger, M. A., Eds.; CRC Press: Boca Raton, 1995; pp. 217–276.
22. McClellan, R. O., Henderson, R. F., Eds. *Concepts in Inhalation Toxicology,* 2nd ed.; Taylor & Francis: Washington, DC, 1995.
23. Rice, R. H.; Cohen, D. E. In *Casarett & Doull's Toxicology: The Basic Science of Poisons,* 5th ed.; Klaassen, C. D., Ed.; McGraw-Hill: New York, 1996; pp. 529–546.
24. Patrick, E.; Maibach, H. In *Principles and Methods of Toxicology,* 3rd ed.; Hayes, A. W., Ed.; Raven Press: New York, 1994; pp. 767–803.
25. Marzulli, F. N., Maibach, H. I., Eds. *Dermatotoxicology,* 5th ed.; Taylor & Francis: Washington, DC, 1996.
26. Potts, A. M. In *Casarett & Doull's Toxicology: The Basic Science of Poisons,* 5th ed.; Klaassen, C. D., Ed.; McGraw-Hill: New York, 1996; pp. 583–615.
27. Chan, P. K.; Hayes, A. W. In *Principles and Methods of Toxicology,* 3rd ed.; Hayes, A. W., Ed.; Raven Press: New York, 1994; pp. 579–647.
28. Albert, A. *Selective Toxicity,* 7th ed.; Chapman and Hall: London, 1985.
29. Gothoskar, S. V.; Weisburger, E. K. *Med. Biol.* **1980**, *58*, 281–284.

*Now *Toxicological Sciences.*

30. Lehman-McKeeman, L.; Rodriguez, P. A.; Takigiku, R.; Caudill, D.; Fey, M. L. *Toxicol. Appl. Pharmacol.* **1989**, *99*, 250–259.

31. Hart, R. W., Neumann, D. A., Robertson, R. T., Eds. *Dietary Restriction: Implications for the Design and Interpretation of Toxicity and Carcinogenicity Studies*; ILSI Press: Washington, DC, 1995.

32. Parkinson, A. In *Casarett & Doull's Toxicology: The Basic Science of Poisons*, 5th ed.; Klaassen, C. D., Ed.; McGraw-Hill: New York, 1996; pp. 113–186.

33. Abou-Donia, M. B. In *CRC Handbook of Toxicology*; Derelanko, M. J., Hollinger, M. A., Eds.; CRC Press: Boca Raton, 1995; pp. 539–589.

34. Rozman, K. K.; Klaassen, C. D. In *Casarett & Doull's Toxicology: The Basic Science of Poisons*, 5th ed.; Klaassen, C. D., Ed.; McGraw-Hill: New York, 1996; pp. 91–112.

35. deBethizy, J. D.; Hayes, J. R. In *Principles and Methods of Toxicology*, 3rd ed.; Hayes, A. W., Ed.; Raven Press: New York, 1994; pp. 59–100.

36. Ioannides, C., Ed. *Cytochrome P450. Metabolic and Toxicological Aspects*; CRC Press: Boca Raton, 1996.

37. Nelson, D. R.; Koymans, L.; Kamataki, T. et al. *Pharmacogenetics* **1996**, *6*, 1–42.

38. Conney, A. H. *Pharmacol. Rev.* **1967**, *19*, 317–366.

39. American Institute for Cancer Research. *Dietary Phytochemicals in Cancer Prevention and Treatment*; Plenum Press: New York, 1996.

40. Brady, J. F.; Wang, M. H.; Hong, J. Y. et al. *Toxicol. Appl. Pharmacol.* **1991**, *108*, 342–354.

41. Bryson, P. D. *Comprehensive Review in Toxicology for Emergency Clinicians*, 3rd ed.; Taylor & Francis: Washington, DC, 1996.

42. Smith, R. P. In *Casarett & Doull's Toxicology: The Basic Science of Poisons*, 5th ed.; Klaassen, C. D., Ed.; McGraw-Hill: New York, 1996; pp. 335–354.

43. Ramos, K. S.; Chacon, E.; Acosta, D., Jr. In *Casarett & Doull's Toxicology: The Basic Science of Poisons*, 5th ed.; Klaassen, C. D., Ed.; McGraw-Hill: New York, 1996; pp. 487–527.

44. Dean, J. H.; Cornacoff, J. B.; Rosenthal, G. J.; Luster, M. I. In *Principles and Methods of Toxicology*, 3rd ed.; Hayes, A. W., Ed.; Raven Press: New York, 1994; pp. 1065–1090.

45. Burns, L. A.; Meade, B. J.; Munson, A. F. In *Casarett & Doull's Toxicology: The Basic Science of Poisons*, 5th ed.; Klaassen, C. D., Ed.; McGraw-Hill: New York, 1996; pp. 355–402.

46. Davis, M. E.; Berndt, W. O. In *Principles and Methods of Toxicology*, 3rd ed.; Hayes, A. W., Ed.; Raven Press: New York, 1994; pp. 871–894.

47. Goldstein, R. S.; Schnellmann, R. G. In *Casarett & Doull's Toxicology: The Basic Science of Poisons*, 5th ed.; Klaassen, C. D., Ed.; McGraw-Hill: New York, 1996; pp. 417–442.

48. Moslen, M. M. In *Casarett & Doull's Toxicology: The Basic Science of Poisons*, 5th ed.; Klaassen, C. D., Ed.; McGraw-Hill: New York, 1996; pp. 403–416.

49. Plaa, G. L.; Charbonneau, M. C. In *Principles and Methods of Toxicology*, 3rd ed.; Hayes, A. W., Ed.; Raven Press: New York, 1994; pp. 839–870.

50. Anthony, D. C.; Montine, T. J.; Graham, D. G. In *Casarett & Doull's Toxicology: The Basic Science of Poisons*, 5th ed.; Klaassen, C. D., Ed.; McGraw-Hill: New York, 1996; pp. 463–486.

51. Sirinathsinghji, D. J. S.; Heavens, R. P.; McBride, C. S. *Brain Res.* **1988**, *443*, 101–116.

52. Witschi, H. R.; Last, J. A. In *Casarett & Doull's Toxicology: The Basic Science of Poisons*, 5th ed.; Klaassen, C. D., Ed.; McGraw-Hill: New York, 1996; pp. 443–462.

53. Poklis, A. In *Casarett & Doull's Toxicology: The Basic Science of Poisons*, 5th ed.; Klaassen, C. D., Ed.; McGraw-Hill: New York, 1996; pp. 951–967.

54. Kaye, B. H. *Science and the Detective*; VCH: Weinheim, 1995.

55. Powers, W. J., Jr. In *CRC Handbook of Toxicology*; Derelanko, M. J., Hollinger, M. A., Eds.; CRC Press: Boca Raton, 1995; pp. 751–769.

56. Merrill, R. A. In *Casarett & Doull's Toxicology: The Basic Science of Poisons*, 5th ed.; Klaassen, C. D., Ed.; McGraw-Hill: New York, 1996; pp. 1011–1023.

57. Munro, I. C.; Morrison, A. B. In *Progress in Predictive Toxicology*; Clayson, D. B., Munro, I. C., Shubik, P., Swenberg, J. A., Eds.; Elsevier: Amsterdam, 1990; pp. 373–387.

58. Krishnan, K.; Andersen, M. E. In *Principles and Methods of Toxicology*, 3rd ed.; Hayes, A. W., Ed.; Raven Press: New York, 1994; pp. 149–188.

59. Renwick, A. G. In *Principles and Methods of Toxicology*, 3rd ed.; Hayes, A. W., Ed.; Raven Press: New York, 1994; pp. 101–147.

60. Medinsky, M. A.; Klaassen, C. D. In *Casarett & Doull's Toxicology: The Basic Science of Poisons*, 5th ed.; Klaassen, C. D., Ed.; McGraw-Hill: New York, 1996; pp. 187–198.

61. Faustman, E. M.; Omenn, G. S. In *Casarett & Doull's Toxicology: The Basic Science of Poisons*, 5th ed.; Klaassen, C. D., Ed.; McGraw-Hill: New York, 1996; pp. 75–88.

62. Paustenbach, D. J., Ed. *The Risk Assessment of Environmental Hazards*; Wiley: New York, 1989.

63. Sweet, D. V., Ed. *RTECS Registry of Toxic Effects of Chemical Substances*; NIOSH, U.S. Government Printing Office: Washington, DC, 1985–86.

64. Hayes, A. W., Ed. *Principles and Methods of Toxicology*, Student Edition; Raven Press: New York, 1982.

65. *Documentation of the Threshold Limit Values and Biological Exposure Indices*, 6th ed., ACGIH: Cincinnati, 1991.

66. Clayton, G. D., Clayton, F. E., Eds. *Patty's Industrial Hygiene and Toxicology* Vol. II, Parts A–F. Wiley & Sons: New York, 1994.

21

Process Safety Reviews

DENNIS C. HENDERSHOT

A process safety review is an organized effort to understand the hazards of a chemical process, to evaluate the effectiveness of protective systems, and to identify the need for additional protection. The process safety review may also identify areas where existing knowledge is insufficient to understand process hazards. A process safety review does not focus on general workplace hazards—those not directly related to the chemical process. These hazards are also important and must be addressed in an overall safety program, but other review and management techniques are more effective in dealing with them.

Process safety reviews are important components of a Process Safety Management (PSM) program. In the United States, process safety reviews are required for processes covered by the OSHA Process Safety Management regulation,[1] the EPA Accidental Release Prevention Program,[2] and several state regulations. Similar regulations are in place or are being adopted in most countries. Industry standards such as API RP 750[3] and ISA-S84.01[4] also include requirements for process safety reviews.

Process Safety Review Procedure

Figure 21.1 shows the important steps in a process safety review. A number of process hazard analysis techniques are available[5] and are discussed below, but the general steps are the same regardless of the methodology. The preparation and follow-up steps are particularly important. A review can only be as good as the team, and the information available to the team. And, without complete follow up on all actions and recommendations, the process safety review becomes a paper exercise which results in no real improvements to the plant or process.

Objectives and Scope

Objectives

In general, a process safety review has the following major objectives:

- identify hazards and potential hazardous incident scenarios

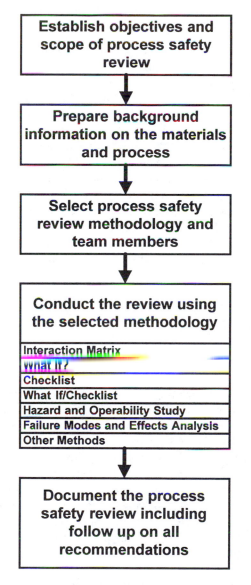

FIGURE 21.1 General procedure for a Process Safety Review.

- identify protective systems to prevent or mitigate the effects of potential incidents, and evaluate their effectiveness
- recommend process modifications or additional protective systems if judged to be needed

- recommend a more detailed study or additional research if the hazard or effectiveness of protective systems is not adequately understood (*Note*: The process safety review is not considered complete until the additional study recommended is also completed, and the results assessed)
- document the process safety design basis for ongoing process safety management and management of future process changes

Most process safety review methods are team-oriented brainstorming techniques. Teams are good for brainstorming and creative activities, but are not as effective for decision making or designing a solution to a specific problem. For this reason, the focus of a process safety review should be on identifying hazards, potential incidents, and areas where existing protective systems may not be adequate, rather than on designing solutions. Those responsible for follow up on the action items can devote appropriate time and resources to designing a specific solution outside the process safety review meeting. The solution should itself be the subject of a process safety review to ensure that it addresses the original concern, and does not introduce any new hazards.

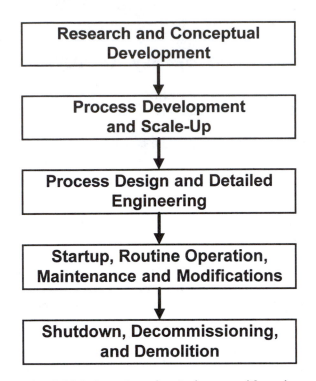

FIGURE 21.2 Steps in a chemical process life cycle.

When Should a Process Safety Review Be Done?

The life cycle of a chemical manufacturing process begins in the laboratory, and proceeds through development, plant construction and operation, and ultimately to plant shutdown (Fig. 21.2). Process safety reviews are appropriate at all stages of the life cycle. As the process moves through the life cycle, more information and knowledge will be available, and the process safety review will be able to consider the process in more detail. Most process hazard analysis techniques can be adapted for use at any stage in the process life cycle. However, some more formal techniques, such as Hazard and Operability Studies (HAZOP), Failure Modes and Effects Analysis (FMEA), and Fault Tree Analysis (FTA), are better suited to later stages in the life cycle, when more information is available.

The best opportunities for improving the safety of chemical processes occur early in the life cycle, when the product and basic chemistry are defined. At this stage there may be considerable freedom in the selection of chemical synthesis routes. The chemist and engineer should aggressively look for inherently safer chemical routes, using less hazardous raw materials and intermediates, less energetic reactions, smaller reactors, and ambient temperature and pressure reaction conditions.[6,7] As a process passes into later stages of the life cycle, oppor-

tunities for implementing inherently safer process chemistry become more difficult, although not impossible.

Scope Definition

To perform an effective process safety review, everybody involved must clearly understand the scope of the study. This allows the review team to focus on the specific process or process sections to be reviewed. Scope definition includes several areas:

- Process steps—The scope may cover an entire process from receipt of raw materials through the shipment of final product, or it may focus on smaller sections of a complex process or plant. Often, a large and complex process is divided into several sections. Different teams, each expert on a particular process step, may review the various sections. Scope definition for each team becomes especially important to ensure that no important process operations are missed in the review, and that interactions among the sections are considered.
- Physical scope—The physical boundaries of the process to be reviewed. This might include the geographical boundaries of the process, as well as the

boundaries of utility and support systems to be considered.

- Type of hazards—Most process safety reviews are intended to identify and evaluate all types of hazard, but there may be circumstances in which a review will be focused on a particular type of hazard. For example, following the Bhopal accident, most major chemical companies conducted some type of process safety review focused on identifying potential incidents which could result in the release of a large toxic gas cloud.

Background Information

Available process data will vary—and generally increase—through the process life cycle. A wide range of process and facility data, such as the information listed in Table 21.1, should be considered for a process safety review. At early stages in the process development, much of the plant specific data may not be available because the plant has not been designed or built.

Select Methodology and Team

Methodology

A number of process hazard analysis techniques are available. Checklist, What If, and What If/Checklist methods can be used at any stage of the life cycle. More formal techniques such as HAZOP, Failure Modes and Effects Analysis, and Fault Tree Analysis are usually applied at later stages in the process life cycle. CCPS[5] offers guidance on how to select an appropriate methodology based on the process life cycle and the perceived level of risk.

Form a Process Safety Review Team

The composition of the team is critical to a good review. An ideal team includes six to eight members—generally enough to cover the required areas of expertise, but not so large that the team works inefficiently. The majority of the team should consist of full-time members who participate in all meetings, thereby maintaining

TABLE 21.1 Examples of Data That Should Be Considered for a Process Safety Review[5]

Process Chemistry	Plant and Process
Raw materials	Process flow diagrams
Intermediates	Piping and instrumentation diagrams
Products	Material and energy balances
By-products and impurities	Design basis of process equipment, including
Catalysts	safety devices
Solvents	Safety alarms and interlocks
Desired reactions	Fire protection systems
Side reactions and undesired alternate reactions	Site utility data
Thermochemistry and kinetics	Plot plans and building layouts
Material Properties	Major material inventories and storage
Physical properties	conditions
Properties related to safety, health, and	Raw material and product transportation
environmental concerns	information
Flammability	Detailed equipment drawings and vendor data
Toxicity (acute and chronic)	(maintenance, operation)
Thermal stability	Process control information—control
Environmental impacts	philosophy, computer control system design
Corrosion	data, software functional descriptions
Plant Operation Data	Site Data
Operator training	On site and off site population
Operating instructions and procedures	Sensitive nearby populations
Emergency response plans and procedures	Environmental impact areas
Management of change procedures	Surrounding geography
Previous safety reviews and studies	Meteorological data
Incident reports and investigations	Hydrology and geological data
Maintenance procedures	Regulatory
Computer control software maintenance	Applicable codes and standards (internal to the
procedures	company and external)
	Applicable regulations
	Permit limits and restrictions
	Variances

continuity for the review. However, the team should be able to call upon other experts to participate part-time when reviewing sections of a process where a particular expertise is required. A single team member may represent several areas of expertise—for example, a process engineer may be knowledgeable in the unit operations of the process, operator training, and the process control system. Table 21.2 lists some areas of expertise which should be considered for participation in a Process Safety Review.

Every process safety review team must include the following critical roles:

- Facilitator—leads the meetings. The facilitator need not be knowledgeable about the manufacturing process, but must be trained in the use of the hazard analysis method, have a good general background and experience with chemical processing, and have good meeting and team leadership skills.
- Recorder—documents the process safety review using a format appropriate to the methodology. The recorder may be a team member, or a stenographer who documents the discussions in the proper format and is not expected to be a participant in the discussions. It is usually not suitable to ask a team member with critical expertise to serve as recorder—the recorder is busy trying to document the discussions and may not be able to participate effectively.
- Owner—responsible for the process at the current life cycle stage, and able to authorize follow-up on any actions. The owner may be a research chemist at the early stages, a process design team leader during plant design, or a production manager for

an operating plant. If the process owner participates in the review he/she will understand the basis for the actions.

- Process experts—knowledgeable in the chemistry and process technology.
- Process operator—an important process expert, especially for an existing plant. The operator knows how the plant actually works (or doesn't work!). Participation by an operator (employee representative) is required for a PHA to meet OSHA[1] or EPA[2] regulatory requirements.

For additional discussion on the composition of process safety review teams, see CCPS[5] and Frank et al.[8]

Conduct Review: Process Safety Review Methodologies

Interaction Matrix

The Interaction Matrix[5,7] is a simple tool for identifying interactions between specific parameters (materials, energy sources, environmental conditions, people). The parameters are listed on the axes of a matrix, as shown in Figure 21.3, and the team asks what the consequence of each interaction would be. As a practical matter, the interactions are generally limited to two parameters because the number of potential interactions increases rapidly as more simultaneous interactions are considered. However, higher order interactions could be considered, for example, by including stable mixtures of chemicals as parameters in the matrix. A computer program, CHEMPAT, is available from the American Institute of Chemical Engineers to assist in constructing interaction matrices, and in storing information on interactions in a database for future reference.[9]

What If?

What If analysis[5,10–12] is a brainstorming technique in which a team with expertise on the process asks "what if" questions about the process to identify potential hazards or incident scenarios. What If is a free brainstorming process, is very flexible, and can be applied at any stage in the process life cycle. The team meets and reviews the process, and then asks questions and raises concerns about various safety aspects. Table 21.3 is an example of the results of a What If analysis.

The unstructured nature of What If analysis can be both an advantage and a disadvantage. With an experienced and knowledgeable team, the technique can be powerful, with the discussion and interaction among

TABLE 21.2 Potential Areas of Expertise for Process Safety Review Team Members

Analytical chemist	Operator training specialist
Control computer programmer	Process chemist
Corrosion/materials specialist	Process control engineer
Electrical engineer	Process engineer
Environmental specialist	Process operator
Equipment manufacturer representative	Product chemist
Human factors expert	Product integrity specialist
Industrial hygienist	Production manager
Instrument engineer	Project engineer
Instrument technician	Raw material supplier experts
Maintenance engineer	Research process engineer
Mechanic or pipe fitter	Safety specialist
Mechanical engineer	Stenographer
Medical/toxicology expert	Unit foreman
	Unit operation experts

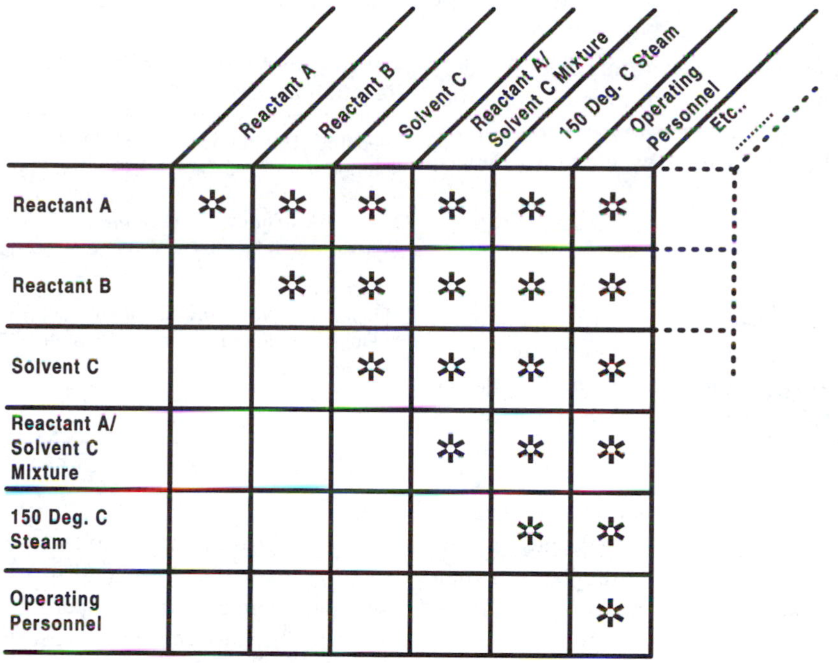

	Reactant A	Reactant B	Solvent C	Reactant A/ Solvent C Mixture	150 Deg. C Steam	Operating Personnel	Etc..
Reactant A	*	*	*	*	*	*	
Reactant B		*	*	*	*	*	
Solvent C			*	*	*	*	
Reactant A/ Solvent C Mixture				*	*	*	
150 Deg. C Steam					*	*	
Operating Personnel						*	

Note: Describe in detail the interaction of each pair of matrix elements (indicated by the asterisks in the matrix).

FIGURE 21.3 An example of an Interaction Matrix.

TABLE 21.3 Example of the Results of a What If Process Safety Review for a Batch Polymerization Process

What If . . . ?	Hazard or Consequence	Safeguards	Recommendations
1. Monomer feed rate is too high?	1. Heat generation rate exceeds reactor heat removal capability, leading to increased temperature, potential runaway reaction.	1. High monomer flow rate interlock shuts down monomer feed. High reactor temperature interlock shuts down monomer feed. Rupture disk sized adequately to protect reactor for maximum monomer feed rate with no cooling.	1. Establish testing program to ensure that high monomer flow rate and high reactor temperature interlocks are reliable.
2. Reactor temperature is too low?	2. Reaction may stall resulting in buildup of unreacted monomer. If temperature subsequently increases or reaction initiates, there is potential for a runaway reaction.	2. Low reactor temperature alarm warns operator of abnormally low temperature.	2. Evaluate rupture disk size—how much monomer buildup can the rupture disk provide adequate protection. Based on results, determine if additional safeguards are required.

team members in the meeting generating a lot of information about potential hazards. However, the unstructured nature also means that the analysis may be incomplete if the technique is used by an inexperienced team and/or facilitator.

Checklist

A Checklist[5,10] is a list of items used to verify that a plant or process is designed and operated consistently with a predetermined set of good practices embodied in the checklist. A checklist is often used to confirm that a plant complies with codes, standards, or regulations. Checklist analysis can be applied at any stage in the life cycle of a process. Checklists can vary from very general, dealing with broad process chemistry issues, to extremely detailed listings of specific requirements of complex codes and standards. Many checklists are simple, requiring only "yes/no" answers.

The use of Checklist analysis depends on the availability of suitable checklists for the process being reviewed. Good checklists are most likely to be available for common types of installations such as flammable solvent storage facilities, and are unlikely to be available for unique, one of a kind process operations. Completeness depends on the experience of the checklist authors.

The output of a Checklist analysis is a list of responses to the checklist questions, with areas of noncompliance highlighted. Recommendations for bringing the facility into compliance, or identification of reasons why the particular requirement is inappropriate for the facility should also be included.

What If/Checklist

The What If and Checklist techniques can be used together.[5,11,12] The checklist ensures that all hazards covered by the checklists are discussed. The What If analysis provides an opportunity for a creative process safety review team to identify hazards and potential incidents unique to the specific design of the facility being reviewed, and possibly not covered by checklists.

The What If/Checklist can be applied in two ways:

- The checklist is reviewed in detail, followed by a brainstorming "What If" session to identify hazards which were not covered by the checklists.
- A "What If" review is done first, and, when the team runs out of questions or concerns, checklists are used to ensure that the What If analysis has covered all of the critical safety areas.

A What If/Checklist review may also use a more creative type of checklist. The questions are open-ended,

TABLE 21.4 Example HAZOP Intention for a Continuous Reaction Process

Feed 150–160 pounds/hour of 42–44% aqueous Raw Material A solution from Feed Tank F-101 using Pump P-11 through line 10436 and flow controller FIC-310-01 to Reactor R-310. Raw Material A solution temperature is 20–35°C. While feeding, Reactor R-310 agitator is running at 50 rpm.

Reference:
  Operating procedure dated October 15, 1996, Steps 37–38
  Piping and Instrumentation Diagrams 36-10456,
    Revision B, dated August 3, 1990, and 36-10457,
    Revision C, dated January 30, 1989
  Instrument Diagram 35-301-01, dated May 12, 1988

without simple answers, and are intended to promote "What If" creative thinking. Some good checklists for common chemical process operations, well suited for a What If/Checklist review, have been published by the Center for Chemical Process Safety.[5,7,13]

Hazard and Operability Study (HAZOP)

A Hazard and Operability Study (HAZOP)[5,10,14,15] is a *guide-word-based hazard evaluation technique*. The technique begins with the premise that the process is safe if operated as intended (the team must agree that this is indeed the case!), and that incidents arise from deviations from the intended operation. *Guide words* are used in conjunction with the specified process operating parameters to identify potential deviations, and the process safety review team determines the consequences of those deviations.

To do a HAZOP, the process is first divided into sections, or nodes, which are analyzed individually. A node might be a transfer line from one vessel to another, a piece of process equipment such as a reactor or heat exchanger, or a step in a batch process. The team precisely states the intended operation of each process node, including specific values for all of the process parameters—the process "intention." Table 21.4 is an example of a process intention for a HAZOP study.

The team then applies guide words to the parameters in the process intention for the first node to identify potential deviations from intended operation. The basic guide words are listed in Table 21.5. As an example, the guide word "MORE" can be combined with the specified material flow rate in the intention to arrive at the deviation "MORE FLOW." As the HAZOP technique has been further developed, many extensions to the basic guide words have been developed to make the HAZOP technique easier to apply to specific kinds of chemical processes. As an example, modified guide words have been

TABLE 21.5 Basic HAZOP Guide Words

Guide Word	Meaning	Example Deviation (based on the example intention)
NO	No part of the intention is accomplished, but nothing else happens at the same time	No flow of Raw Material A solution from Feed Tank F-101 to Reactor R-310.
MORE	An increase in a parameter specified in the intention	Feed more than 160 pounds per hour of Raw Material A from Feed Tank F-101 to Reactor R-310.
LESS	A decrease in a parameter specified in the intention	Feed Raw Material A from Feed Tank F-101 to Reactor R-310 at 10° C.
AS WELL AS	Everything specified in the intention occurs, and something else also occurs at the same time	While feeding Raw Material A, Intermediate B is also fed to Reactor R-101.
REVERSE	The opposite of something specified in the intention occurs	Reactor R-310 contents flow back into Raw Material A Feed Tank F-101.
PART OF	Only a part of the intention is achieved; part of it is not	Raw Material A is fed to Reactor R-310 at the proper conditions, but the R-310 agitator is not running.
OTHER	Something completely different than the desired intention occurs	Raw Material C is fed to Reactor R-310 instead of Raw Material A.
	Many HAZOP leaders use the guide word OTHER for a What If brainstorming session to identify any potential deviations not already discussed	Consider impact of items such as external fire, utility failure, human error, maintenance, startup, shutdown, etc.

used to adapt HAZOP to computer-controlled processes.[16,17]

Once a deviation has been identified, the team determines as many potential causes of the deviation as possible. For example, the deviation "MORE FLOW" will have a number of possible causes—a flow control valve stuck open, incorrect set point, and others. The team lists the consequences of each deviation-cause combination, and any existing safeguards. The team qualitatively judges the effectiveness of those safeguards to determine if they are adequate, based on the potential incident likelihood and consequences. If the existing safeguards are judged to be inadequate, the team should recommend appropriate action to mitigate the potential hazard.

The team continues to apply the guide words to each node until no additional deviations can be identified. These steps are repeated for each process node, until the entire process has been reviewed. Table 21.6 shows a part of the output of a typical Hazard and Operability Study.

HAZOP is best applied when specific process and plant information is available—for example, a detailed plant design or an operating plant. However, the guide word approach can be a useful thought process for a process safety review at any stage in the process life cycle.

Failure Modes and Effects Analysis (FMEA)

Failure Modes and Effects Analysis[5,10,18] is based on identifying the ways in which each piece of process equipment can fail to perform as designed, and determining how that failure will impact the process. FMEA and

HAZOP are very similar methodologies. The main difference is the starting point for identifying potential hazardous incident scenarios. HAZOP starts by postulating a deviation in the value of a process parameter (e.g., more flow), and asking what kind of equipment failures or operating errors might have caused that deviation, and what the process impact will be. FMEA starts by postulating a known equipment failure mode (e.g., control valve stuck open), and asks what impact this failure will have on the operation of the process.

An FMEA starts with a detailed functional description of each piece of process equipment, and then asks the review team to identify ways in which that piece of equipment might fail to perform as designed. The team must have a good understanding of the equipment and all potential failure modes. The FMEA team determines how the process will respond to the potential equipment failure, determines if a potentially hazardous incident will result, identifies existing safeguards, evaluates their effectiveness, and develops recommendations for action where appropriate. These steps are very similar to the corresponding steps in a HAZOP study. Table 21.7 shows a part of the output from a typical FMEA study.

Other Techniques

Process safety review techniques related to What If methodologies include:

- Safety Reviews[5]
- Preliminary Hazards Analysis (PHA)[5,10,19]

Other techniques are based on logic models to aid in understanding how hazardous process incidents may oc-

TABLE 21.6 Example of Partial Results of a HAZOP Review

Deviation	Causes	Consequences	Safeguards	Recommendations
1. MORE than 160 lb/hour Raw material A flow	A. Wrong set point (too high) on FIC-301-01 B. FIC-301-01 control valve stuck open C. FIC-301-01 flow sensor miscali-brated—reads low	A. Reactor R-310 temperature increases, heat balance indicates there is no potential for runaway reaction, even for maximum possible flow rate; product will not meet specifications if R-310 temperature increases above 90°C B. Same as A. C. Same as A.	A. 1-High flow alarm on FIC-301-01 warns operator. 2-High temperature alarm TAH-310-05 on Reactor R-310 warns operator. B. Same as A. C. A-2 only applies. A-1 will not provide an alarm because the flow sensor is miscalibrated. Plant experience indicates this is not likely.	A. Confirm that operator training includes proper response to FIC-301-01 and TAH-310-05 high alarms.

Note. See Table 21.4.

TABLE 21.7 Example of Partial Results of a Failure Modes and Effects Analysis

Item and Description	Failure Mode	Effects	Safeguards	Recommendations
1. Control Valve FIC-301-01 Pneumatically operated valve, fails closed on air supply failure	A. Valve stuck in open position	A. Flow of Raw Material A may exceed desired flow rate. If this occurs, Reactor R-310 temperature increases, heat balance indicates there is no potential for runaway reaction, even for maximum possible flow rate; product will not meet specifications if R-310 temperature increases above 90°C.	A. High flow alarm on FIC-301-01 warns operator. High temperature alarm TAH-310-05 on Reactor R-310 warns operator.	A. Confirm that operator training includes proper response to FIC-301-01 and TAH-310-05 high alarms. Review maintenance procedures and make sure Valve FIC-301-01 is regularly inspected and tested.
	B. Air failure to FIC-301-01	B. Valve will close, stopping flow of Raw Material A to Reactor R-310. No hazard, loss of production.		

Note. See Table 21.4 for a general description of the system.

TABLE 21.8 Potential Contents of Process Safety Review Report

Study objectives
Scope of the study
Methodology used
Names of study team members
   Position
   Expertise
Meeting dates
Brief process description
List of reference material used for the Process Safety Review
   Procedures
   Drawings
   Material and safety data references
   All other documents used for the Process Safety Review,
     with dates and revision numbers
Summary of action items
   Prioritization
   Responsible party for follow up
   Action plan for confirming follow up
Detailed documentation of the Process Safety Review (e.g.,
   interaction matrices, response to checklists, What If questions
   and responses, HAZOP or FMEA tables, etc.)

cur. These methods are often used to aid in understanding a particular hazard in a well-developed plant design or an existing plant.

- Fault Tree Analysis (FTA)[5,10]
- Event Tree Analysis (ETA)[5,10]
- Cause Consequence Analysis[3]
- Human Reliability Analysis[5,10,20]

Documentation and Follow-Up

To obtain the full value from a process safety review, it must be fully documented.[21,22,23] All discussions, including those which did not lead to discovery of a hazard or an action item, should be documented. This will be valuable in the future when the process safety review is updated, or if new information on the process becomes available. Table 21.8 summarizes some of the information which should be considered for inclusion in the Process Safety Review report.

A process safety review will generate a list of recommendations and action items. If there is no follow up on these action items, the resources invested in doing the review have been wasted. All action items should be assigned to a responsible party, and tracked until they are resolved. The resolution should be described in detail, and may be any of the following:

- A specific design or process change was implemented as recommended.
- An alternative solution to the one suggested was implemented to address the concern identified by the review team.
- An issue was resolved by implementation of a process or design change identified subsequent to the process safety review.
- Further study indicated that the potential concern identified by the review team did not require process modifications or additional safeguards.

Any process modification or protective feature added to a process as a result of a process safety review should itself be the subject of a process safety review. All changes to a process, even those intended to improve safety, have the potential to introduce new hazards or potential incident scenarios. There are numerous examples of safety modifications causing incidents by introducing new, unanticipated failure modes into a system.

Process safety reviews should be updated whenever a process undergoes a change, as part of a management of change program. In addition, the process safety review should be updated periodically to allow consideration of the cumulative impacts of the "small" changes which inevitably occur in any operating plant, and to incorporate new knowledge and experience from the plant operation into the review. For example, OSHA[1] and EPA[2] regulations require a revalidation of the process safety review for covered processes every 5 years.

References and Additional Reading

References

1. OSHA. Process Safety Management of Highly Hazardous Chemicals; Explosives and Blasting Agents, 29 *CFR* 1910.119.
2. EPA. Risk Management Programs for Chemical Accidental Release Prevention Requirements, 40 *CFR* 68.
3. API RP 750. Management of Process Hazards; American Petroleum Institute: Washington, DC, January, 1990.
4. ISA-S84.01. Application of Safety Instrumented Systems for the Process Industries; Instrument Society of America: Research Triangle Park, NC, February 15, 1996.
5. Center for Chemical Process Safety (CCPS). *Guidelines for Hazard Evaluation Procedures, 2nd Edition With Worked Examples*; American Institute of Chemical Engineers: New York, 1992.
6. Kletz, T. A. *Plant Design for Safety*; Taylor & Francis: Bristol, PA, 1991.
7. Bollinger, R. E.; Clark, D. G.; Dowell, A. M.; Ewbank, R. M.; Hendershot, D. C.; Lutz, W. K.; Meszaros, S. I.; Park, D. E.; Wixom, E. D. In *Inherently Safer Chemical Processes: A Life Cycle Approach*; Crowl, D. A., Ed.; American Institute of Chemical Engineers: New York, 1996.

8. Frank, W. L.; Giffin, J. E.; Hendershot, D. C. *International Process Safety Management Conference and Workshop*; American Institute of Chemical Engineers: New York, 1993; pp. 129–146.

9. *CHEMPAT: A Program to Assist Hazard Evaluation and Management*; American Institute of Chemical Engineers: New York, 1995; Publication Z-1.

10. Greenberg, H. R., Cramer, J. J., Eds. *Risk Assessment and Risk Management for the Chemical Process Industry*; Van Nostrand Reinhold: New York, 1991.

11. Burk, A. F. *Chem. Eng. Prog.* **1992**, *88*(6), 90–94.

12. Goodman, L. *Chem. Eng. Prog.* **1996**, *92*(7), 75–79.

13. Center for Chemical Process Safety (CCPS). *Guidelines for Design Solutions to Process Equipment Failures*; American Institute of Chemical Engineers: New York, 1997.

14. *A Guide to Hazard and Operability Studies*; Chemical Industries Association: London, 1977.

15. Knowlton, R. E. *A Manual of Hazard and Operability Studies*; Chemetics International Company, Ltd.: Vancouver, BC, Canada, 1992.

16. Kletz, T.; Chung, P.; Broomfield, E.; Shen-Orr, C. *Computer Control and Human Error*; Gulf Publishing Company: Houston, TX, 1995.

17. Nimmo, I. *Chem. Eng. Prog.* **1994**, *90*(10), 32–44.

18. Moubray, J. *Reliability-Centred Maintenance*; Butterworth-Heineman: Oxford, UK, 1991.

19. MIL-STD-882B. Military Standard System Safety Requirements; Department of Defense: Washington, DC.

20. Lorenzo, D. K. *A Manager's Guide to Reducing Human Errors*; Chemical Manufacturers Association: Washington, DC, 1990.

21. Center for Chemical Process Safety (CCPS). *Guidelines for Process Safety Documentation*; American Institute of Chemical Engineers: New York, 1995; pp. 73–106.

22. Freeman, R. A. *Plant/Operations Progress* **1991**, *10*(3), 155–158.

23. Hendershot, D. C. *Plant/Operations Progress* **1992**, *11*(4), 256–263.

Additional Reading

Freeman, R. A.; Lee, R.; McNamara, T. P. *Chem. Eng. Prog.* **1992**, *88*(8), 28–32.

Gujar, A. M. *J. Loss Prev. Process Ind.* **1996**, *9*(6), 357–361.

Kletz, T. A. *HAZOP & HAZAN: Notes on the Identification and Assessment of Hazards*, 2nd ed.; Institution of Chemical Engineers: Rugby, Warwickshire, UK, 1986.

Lees, F. P. *Loss Prevention in the Process Industries*, 2nd ed.; Butterworth-Heinemann: Oxford, UK, 1996.

Leveson, N. G. *Safeware: System Safety and Computers*; Addison-Wesley: New York, 1995.

Process Safety Management; Chemical Manufacturers Association: Washington, DC, 1985.

Skelton, B. *Process Safety Analysis: An Introduction*; Gulf Publishing Company: Houston, TX, 1997.

Sutton, I. S. *Process Reliability and Risk Management*; Van Nostrand Reinhold: New York, 1992.

Wells, G. *Hazard Identification and Risk Assessment*; Institution of Chemical Engineers: Rugby, Warwickshire, UK, 1996.

22

Safe Entry into Confined Spaces

LAURA L. HODSON

Entry into an untested confined space can be fatal. One of the more tragic things about confined spaces is that they often take the lives of those people who are attempting to rescue someone who has been overcome by one of the many dangers in confined spaces. NIOSH has reported that more than 60% of confined space fatalities occur among would-be rescuers.[1] The most important thing a Safety and Health professional can do is label all of the confined spaces at their facility and properly train all employees who will be working either in or around confined spaces. Being able to recognize and plan appropriately for entry into a confined space will significantly lower the potential health and safety risks.

This chapter addresses the difference between confined spaces and *permit-required* confined spaces, what hazards may exist in confined spaces, how to measure and control the potential risks to provide for a safe entry, and how to maintain the proper paperwork to ensure compliance with the OSHA regulation 29 *CFR* 1910.146, Permit-required Confined Spaces.[2]

OSHA Definition

OSHA 29 *CFR* 1910.146 defines a confined space as having the following characteristics:

1. Is large enough for a person to bodily enter and perform work;
2. Has limited or restricted means of entry or exit;
3. Is not designed for human occupancy.

A *permit-required* confined space is a confined space that also has one or more of the following characteristics:

1. Contains or has the potential to contain a hazardous atmosphere
2. Contains a material that has the potential for engulfing an entrant
3. Has an internal configuration such that an entrant could be trapped or asphyxiated by inward converging walls, or by a floor which slopes downward and tapers to a similar cross section
4. Contains any other recognized serious safety or health hazard

The main difference between a *permit-required* and nonpermit confined space, is that nonpermitted spaces do not contain or, with respect to atmospheric hazards, do not have the potential to contain any hazard capable of causing death or serious physical harm. Examples of permit-required confined spaces include rail tank cars, truck transport cars, reaction vessels, grain silos, pits, and sewers. Nonpermit confined spaces may include crawl spaces, or air handlers that have been properly locked out/tagged out.[3]

A space classified as a permit-required confined space may be reclassified as a nonpermit confined space if the permit space poses no actual or potential atmospheric hazards and if all nonatmospheric hazards within the space are eliminated without entry into the space.

Written Plans

All employers must evaluate the workplace to determine if any spaces are permit-required confined spaces. The employer must inform exposed employees, by posting danger signs or by any other equally effective means, of the existence and location of and the danger posed by the permit spaces. An example of a properly labeled permit-required confined space is shown in Figure 22.1. Additionally, if the employer decides that employees will enter permit spaces, the employer must develop and implement a written permit-required space entry-program.

If the employer decides that employees will not enter permit spaces, then effective measures must be taken to prevent any inadvertent entry.

Dangers of Confined Spaces

Confined spaces are considered a significant health and safety risk due to chemical (atmospheric), physical/mechanical, and biological hazards (Table 22.1). Atmospheric hazards may cause suffocation or instantaneously explode. Physical hazards may trap an entrant in an unreachable location or expose him/her to unshielded electricity. Biological hazards may expose an entrant to infectious viruses.

Hazardous Atmospheres

A hazardous atmosphere means an atmosphere that may expose persons to the risk of death, incapacitation, injury, acute illness, or impair the ability to self-rescue.

Oxygen

Oxygen Deficiency Normal ambient air contains an oxygen concentration of 20.9%.[7] Oxygen deficiency (concentrations less than 19.5%) can be encountered in confined spaces or any poorly ventilated areas where the air may be displaced by gases, vapors of volatile materials, or where oxygen may be consumed by chemical or biological reactions. Oxygen concentrations below 16% can impair thinking and muscle coordination and eventually cause death. In addition to chemical process facilities where oxygen may be displaced, it is also important to think of common occurrences such as the decomposition of vegetables in a root cellar, which could present a potentially oxygen-deficient environment.

Oxygen Enrichment When the oxygen concentration rises above 23.5%, the atmosphere is considered enriched and is prone to become unstable. Enriched oxygen increases the likelihood of a flash fire or explosion.

Combustion

Combustion is traditionally defined by a fire triangle where three elements (fuel, oxygen, and a source of ignition) are necessary to support a fire. As the ratio of gas to air changes, the atmosphere passes through three ranges: lean, explosive, and rich. The explosive concentration is expressed as the range between the Lower Explosive Limit (LEL) and the Upper Explosive Limit (UEL). Care must be taken whenever a mixture is too rich and the UEL has been reached because dilution with fresh air could bring the atmosphere into the explosive range. Due to the potential of combustion in confined spaces, the confined space sampling equipment must be intrinsically safe so that it does not generate a spark and become the source of ignition.

Chemical Hazards

Atmospheric concentrations of any substance known to be in the confined space for which a permissible expo-

FIGURE 22.1 A properly labeled permit-required confined space.

sure limit is set should be measured. Entry into areas that have atmospheric concentrations in excess of the exposure limits should only be done if proper respiratory protection is provided.

While the chemical transported in a tanker car or mixed in a reaction vessel will be known, often confined spaces contain unknown toxic gases and vapors. Some of the more frequently encountered gases in confined spaces include carbon monoxide (CO), hydrogen sulfide (H_2S), methane (CH_4), sulfur dioxide (SO_2), and ammonia (NH_3).[8] Additionally, it is important to consider what chemicals may be introduced by work performed in the confined space; for example, painting could introduce

TABLE 22.1 Types of Hazards Encountered in Confined Spaces

Chemical hazards	Oxygen <19.5% or >23.5%[4]
	Flammable gases or vapors >10% of the LEL[5]
	Combustible dusts >10% of the LEL
	Atmospheric concentrations > exposure limits[6]
	Changing atmospheres from work practices
	Poor ventilation
	Any other atmospheric condition that is immediately dangerous to life and health
Physical hazards	Engulfment by dry or wet materials
	Energy and material releases
	Water (slippery surfaces)
	Temperature extremes
	Poor lighting
	Restrictive space
Biological hazards	Molds, mildews, and spores
	Bird and rat feces
	Poisonous snakes and spiders
	Bacteria and viruses

hydrocarbons, and welding could introduce ozone and metal fumes. Any gasoline-powered equipment has the potential to generate carbon monoxide. Keep vehicle exhaust from being inadvertently drawn into a confined space; do not allow any cars to idle near manhole/sewer entries.

Physical Hazards

Engulfment

Permit-required confined spaces, such as grain silos or bulk powder containers, can contain material which has the potential for engulfment of a person. Engulfment hazards either block the employee's airways or compress the upper body to the point where suffocation takes place. Engulfment hazards frequently exist where there is loose material such as grains, flour, sawdust, sand, or crushed stone. While these materials may appear to give a solid footing, there are often hidden air pockets that can collapse under the weight of a person.

Energy and Material Release

Unintentional energy or material releases in a confined space can cause death or serious physical harm. Confined spaces may contain gears, turning shafts, conveyer belts, or other moving parts. Whenever possible all energy sources should be locked out/tagged out,[3] mechanical devices should be blocked or disconnected, and pipes should be bled.

Other Hazards

Since confined spaces are not meant for continuous human occupancy, the areas are not built with comfort in mind. Confined spaces may have temperature extremes, slippery floors, poor lighting, and tight restrictive areas. It may be very awkward to complete a given task in the area.

Biological Hazards

Biological hazards such as molds, mildews, and spores can frequently be found in dark, damp confined spaces. Feces from birds and rodents may be in the area. If the confined space is a sewer there is also the potential to be exposed to human bacteria and viruses. Poisonous snakes or spiders may also occupy the space. There may be insect nests. The best protection is to use good lighting and always look where the hands and feet are to be placed.

Confined-Space Entry

Confined-Space Permit

The OSHA regulation requires employers to maintain a written permit-required confined space program, and use entry permits. The entry permits must be completed prior to entry and are provided by the employer to allow and control entry into permit spaces.

There are 15 specific items which must be on the entry permit; they are as follows:

1. The permit space to be entered
2. The purpose of entry
3. The date and authorized duration of the entry permit
4. The names of the authorized entrant(s)
5. The name of the authorized attendant(s)
6. The name and signature of the authorized supervisor(s)
7. The hazards of the permit space to be entered
8. The measures used to isolate the permit space, and to eliminate or control permit space hazards before entry
9. The acceptable entry conditions
10. The results of the initial and periodic atmospheric testing, and the name of the tester
11. The rescue and emergency services that can be summoned and the phone numbers or other means of contacting the rescue services
12. The communication procedures for use between the entrant and the attendant
13. A list of all equipment such as personal protective equipment (PPE), testing equipment, and rescue equipment

14. Any other information whose inclusion is necessary in order to ensure employee safety
15. Any additional permits, such as for hot work, that may have been issued to authorize work in the permit space

Confined-space permits are canceled when entry is completed and must be retained for at least one year to allow for an annual review. A confined space may be entered without the need for a written permit provided that the space is determined *not* to be a permit-required confined space. However, it would be a good idea to utilize a form of the entry permit to use as a notification system for entry into nonpermit spaces. For example, if an employee will be working in an isolated area of the facility (such as in a crawl space), and the employee does not return at quitting time, the confined-space notification form would inform the supervisor of where to look for the employee.

Atmospheric Testing

Before entry into a permit-required confined space can be made, the atmosphere must be tested with calibrated, direct reading instrument(s) for the following conditions in the order given:

1. Oxygen content
2. Flammable gases and vapors
3. Potential toxic air contaminants

Battery-powered, direct reading instruments are practical and easy to use. Many manufacturers are providing models which can do simultaneous monitoring of at least the oxygen content and LEL/UEL. An instrument with a long probe is most useful for monitoring manholes and remote areas. If a flammable mixture is known to be or previously had been stored in a confined space, use extreme caution in opening the entry portal. The atmosphere should be tested through a small entry port hole if at all possible. Use intrinsically safe (non-spark-producing) equipment.

Atmospheric testing should be conducted at all lengths and depths of a confined space due to the various gas densities. Some flammable gases may be located at the bottom of a confined space and some may be located at the top of the confined space. Results of the atmospheric testing is entered on the permit and are used to determine if respiratory protection or ventilation is necessary.

In order for the permit-required confined space to be entered, the oxygen levels must be between 19.5% and 23.5%, and the atmosphere must be less than 10% of the LEL. Supplied air respiratory protection will not provide adequate protection from an oxygen-enriched environment nor from an environment that is in excess of 10% of the LEL, due to the potential for fire. If entry is required into these atmospheric conditions, then the space must be ventilated prior to and continuously throughout the entry.

Atmospheric testing should be made continuously throughout the entry. An audible alarm for percent oxygen and flammable gas should be attached to the entrant, or measured in close proximity to the entrant.

Each authorized entrant or that employee's authorized representative must have the opportunity to observe the pre-entry and any subsequent testing or monitoring of the permit spaces.

Ventilation

If the permit-required confined space does not have an acceptable atmosphere, continuous forced air ventilation may be used to eliminate the atmospheric hazards. The forced air supply must be from a clean source and personnel may not enter the space until the forced air ventilation has eliminated any hazardous atmosphere. The forced air ventilation shall be directed so as to ventilate the immediate area where a person may be present, and must continue until all persons have left the space. The atmosphere must be continually monitored to ensure that the continuous forced air ventilation is preventing the accumulation of a hazardous atmosphere.

Personnel

Entry into a confined space always takes the coordination of three persons. There must always be an entrant, an attendant, and a supervisor. Each of these persons has specific responsibilities to ensure a safe entry.

Duties of Entrant(s)

All authorized entrant(s) must:

1. Know the hazards that may be faced during entry, including information on the mode, signs or symptoms, and consequences of the exposure.
2. Properly use personal protection and retrieval equipment.
3. Maintain communication with the attendant as necessary to enable the attendant to monitor the entrants' status.
4. Alert the attendant whenever any potential dangers are encountered.
5. Exit the permit space as quickly as possible when-

ever an order is given by the attendant or entry supervisor, or whenever the entrant recognizes any warning sign or symptom of exposure to a dangerous situation, or the entrant detects a prohibited situation, or an evacuation alarm is activated.

Duties of Attendant(s)

All authorized attendant(s) must:

1. Know the hazards that may be faced during entry, including information on the mode, signs or symptoms, and consequences of the exposure.
2. Be aware of possible behavioral effects of hazard exposure to authorized entrants.
3. Continuously maintain an accurate count of authorized entrants in the permit space and ensure that the entrants are all authorized.
4. Remain outside the permit space during entry operations until relieved by another attendant.
5. Communicate with the entrant(s) as necessary to monitor entrant status and to alert entrant of the need to evacuate if necessary.
6. Monitor the activities inside and outside the space to determine if it is safe for entrants to remain in the space.
7. Summon rescue and other emergency services as soon as the attendant determines that authorized entrants may need assistance to escape from permit space hazards.
8. Take actions to keep unauthorized persons away from the permit space;
9. Perform *nonentry rescues* as specified by the employer's rescue plan.
10. Perform no duties that might interfere with the attendant's primary duty to monitor and protect the authorized entrant(s).

Duties of Supervisor(s)

All authorized supervisor(s) must:

1. Know the hazards that may be faced during entry, including information on the mode, signs or symptoms, and consequences of the exposure.
2. Verify by checking that the appropriate entries have been made on the permit, that all tests have been conducted and that all procedures and equipment specified by the permit are in place before endorsing the permit and allowing entry to begin.
3. Terminate the entry and cancel the permit.

to a rescue tripod of some sort; this body harness provides both a means of rescue/retrieval if the entry is into an area with more than 5 ft. vertical drop, and a means of fall protection if the entrant is in a position to fall more than 6 ft.[10]

Head protection, protective gloves for electrical safety or other reason(s), steel toe safety shoes, safety glasses, protective clothing, hearing or respiratory protection, and communication equipment may all be necessary. Each confined space will need to be evaluated and the correct PPE selected by a qualified person.

Rescue

Employers may have either employees or persons other than the host employer's employees perform permit-space rescue. If the employees are providing the responsible rescue service, then the employers must make sure that the members of the rescue team:

1. Are trained and provided with personal protective equipment and rescue equipment
2. Practice making permit-required confined-space rescues at least once every 12 months
3. Are trained in basic first aid and cardiopulmonary resuscitation (CPR); at least one member of the rescue team must be holding current certification in first aid and CPR

FIGURE 22.2 An attendant overseeing an entrant entering a permit-required confined space.

If persons other than employees will be providing the rescue services (such as the local fire department), the host employer must:

1. Inform the rescue service of the hazards they may confront in the confined space.
2. Provide rescue services with access to all permit spaces from which rescue may be necessary, so that the rescue service can develop appropriate rescue plans.
3. Evaluate a prospective rescue service's ability, in terms of proficiency with rescue related tasks and equipment, to function appropriately while rescuing entrants from the particular permit space or types of permit spaces identified.

4. Verify that rescue services are available and that the means for summoning them are operable.
5. Remove unauthorized individuals who enter or attempt to enter a permit space during entry operations.
6. Determine whenever responsibility for a permit space entry operation is transferred and at intervals dictated by the hazards and operations performed, that entry operations remain consistent with terms of the entry permit and that acceptable entry conditions are maintained.

Personal Protection Equipment

The OSHA personal protection equipment (PPE) regulation[9] lists requirements for assessing the need for PPE. The PPE required for permit-required confined-space entry will vary depending on the space. Frequently, there will be the need to wear a full body harness connected

To facilitate nonentry rescue, retrieval systems or other methods must be used whenever an authorized entrant enters a permit-space, unless the retrieval equipment would increase the overall risk of entry or would not contribute to the rescue of the entrant. Retrieval equipment includes a chest or full body harness with a retrieval line usually attached at the center of an entrant's back, near shoulder level. Figure 22.2 demonstrates an entrant properly fitted for a permit-required confined-space entry. The retrieval line must be attached to a mechanical device or fixed point outside the permit space in such a manner that rescue can begin as soon as the rescuer becomes aware that rescue is necessary. A mechanical device must be available to retrieve personnel from vertical-type permit spaces more than 5 feet deep.

Conclusion

Just about every facility has a confined space (or several hundred confined spaces) and it is important to note that confined-space entry done incorrectly can be fatal. It is important to stress to your employees that over 60% of the confined-space fatalities have occurred among would be rescuers. Proper identification of confined spaces, training of employees, atmospheric testing, ventilation, personal protection equipment, retrieval lines, and communication devices can ensure safe entry into and out of confined spaces.

References

1. U.S. Department of Health and Human Services, National Institute for Occupational Safety and Health (DHHS/NIOSH). *Request for Assistance in Preventing Occupational Fatalities in Confined Spaces*, HS/PHS/CDC/NIOSH: Cincinnati, OH, 1986.
2. OSHA Permit Required Confined Space Standard, 29 *CFR* 1910.146.
3. OSHA The Control of Hazardous Energy (lockout/tagout) Standard, 29 *CFR* 1910.147.
4. Clayton, G., Clayton, F., Eds. *Patty's Industrial Hygiene and Toxicology*, 4th ed.; Wiley and Sons: New York, 1994.
5. McKinnon, G., Tower, E., Eds. *Fire Protection Handbook*, 18th ed.; National Fire Protection Association: Boston, MA, 1997.
6. *NIOSH Pocket Guide to Chemical Hazards*; U.S. Department of Health and Human Services: Washington, DC, 1994.
7. Weast, R. C., Ed. *CRC Handbook of Chemistry and Physics*, 76th ed.; CRC Press: Boca Raton, FL, 1995.
8. *Closing In On Confined Spaces: A Primer On Hazards and Equipment*; MSA: Pittsburgh, PA, 1993.
9. OSHA Personal Protective Equipment Standards, 29 *CFR* 1910, Sections 132–138.
10. OSHA Fall Protection Standards, 29 *CFR* 1926, Sections 500–503.

23

The Control of Hazardous Energy (Lockout/Tagout)

DAVID J. VAN HORN

Energy is necessary to perform work, but it must be controlled.

The term "Lockout and Tagging" became part of the OSHA Electrical Standard[1] in 1970. This requires "qualified" persons (those with specific training) to follow safety-related work practices. One practice is lockout and tagging to prevent electric shock or other injuries from either direct or indirect contact with electric circuits.

Over the years, many people have been injured or killed by contact with a large variety of uncontrolled sources of hazardous energy, not just electrical energy. In 1989, the OSHA General Environmental Controls Standard, "The Control of Hazardous Energy,"[2] became effective. This law is commonly referred to as the Lockout/Tagout Standard, or LOTO, and addresses the control of all sources of hazardous energy. It redefined what is required to use lockout or tagout to control sources of hazardous energy. Some organizations have nicknamed

the standard "LOTO/TEST" as a reminder that if you use lockout or tagout, it is required to test and verify that the procedure followed was effective in isolating the energy source.

There is a functional definition of an accident published in 1980[3] that is helpful in understanding the role of energy in accidents and provides ideas for controls. It says, in part, an accident is "An unwanted transfer of energy, Because of a lack of barriers and/or controls, Producing injury to persons, property, or process, Proceeded by sequences of planning and operational errors."

Energy Sources

Most laboratory and pilot plant employees handling chemicals generally associate the following energy sources with their work or the source of past injury and/or damage. Stored, or potential, energy must also be considered.

Chemical	Kinetic, linear, and rotational
Mechanical	Pneumatic and hydraulic
Electrical	Thermal

Gravity, biological, and radiation energy sources are also mentioned.

From my own analysis of injury data in research laboratories, the personal use of excessive force was a major cause of accidents. That is to say, the more energy a person puts into a task, the more likely that energy will be released in an unexpected way, causing an injury. "Excessive force," for this analysis, was defined as more energy than normally required to perform a task. These accidents are a result of personal behavior and cannot be prevented by LOTO. I mention this only because it is an example of a different source of energy causing accidents that is worth considering in training or coaching on expectations concerning personal behavior. Recognizing that excessive force is a high-risk activity allows the person to stop and figure out a safer way to do the task (a mental rather than physical disconnect that can prevent an accident).

Common Strategies Used to Control Hazardous Energy

The following strategies are generally considered during the hazard analysis conducted to, among other things, determine what hazardous energy exists in the planned work, and to establish controls to prevent injury or damage due to the uncontrolled release of that energy.

- Substitution (use lower-hazard rated materials, intrinsically safe equipment, etc.)
- Limit the amount used to that just needed for the work (quantity, volume, voltage, height, temperature, pressure, etc.)
- Prevent the release (increased safety factor of containment, *lockout, tagout*, etc.)
- Barriers (machine guards, electrical insulation, personal protective equipment, interlocks, disconnects, etc.)
- Separation (remote controls, distance, etc.)
- Control rate of release (safety value, bleed off, etc.)

A Process Hazard Review can lead to selecting an appropriate control, or combination of controls, for the planned work that can be incorporated into training, procedures, and inspections.

When Lockout/Tagout (LOTO) Does Not Apply

The LOTO standard does not apply to normal operations in the following instances:

- When normal intended adjustments are made that are routine, repetitive, and integral to the use of the equipment. Generally, this means there is machine guarding, SOP, training, or other practices in place resulting in a history of safe operations
- When the equipment has a sole source of energy supplied by a cord and plug that when unplugged remains under the exclusive control of the person authorized to work on the equipment
- When performing Hot Tap operations on pressurized gas, steam, water, or petroleum products distribution equipment where continuity of service is critical, shutdown is impractical, and effective procedures are followed

When to Use LOTO

> The LOTO standard applies only to the control of energy where "any employee performs any servicing or maintenance on a machine or equipment where the *unexpected* energizing, startup, or release of stored energy could occur and cause injury."

There are two triggers to alert one when LOTO is required:

1. When an employee is required to remove or bypass a guard or other safety device
2. When an employee is required to place any part of his/her body into an area on a machine or piece of equipment where work is actually performed on the material being processed, or where an associated danger zone exists during a machine operating cycle

When a machine "jams" or starts sounding "funny" is a particularly bad time if an employee does not have specific instructions to follow or training that will keep him/her from making a bad decision. Most employees are conscientious and may make a decision to take a shortcut (generally violating triggers 1 or 2) if they think that is what supervision wants to keep the process going versus doing the right thing and thinking of their safety.

Refer to Table 23.1 for examples of hazardous energy and LOTO methods for protecting employees from their accidental flow.

TABLE 23.1 Examples of Hazardous Energies and Methods for Protecting Personnel from Their Accidental Flow

1. Electrical
   a. Where provision is available for locking a shutoff device, lock/tag it.
   b. Where there is no such provision, fuses must be removed from the circuit and the fuse box locked/tagged or an electrician must disconnect and tag the wires at the operating equipment.
2. Gravitational
   a. Lower the elevated hazard to a stable surface.
   b. Block the elevated hazard in position to prevent accidental movement.
3. Hydraulic
   a. Divert the fluid downstream of the pump back to the reservoir and lock the valve open.
   b. Disconnect and tag the battery cables.
4. Internal Combustion Engine
   a. Turn off the engine and remove the ignition key.
   b. Disconnect and tag the battery cables.
5. Pneumatic
   a. Disconnect the air supply, vent the supply line, and lock the valve open.
   b. Close the shutoff-cocks to block the air source and vent the line leading to the actuator.
6. Spring Tension
   a. Release the spring tension
   b. Physically restrain to immobilize either the spring or the mechanical equipment attached to it.

LOTO Compliance Requirements

> The employer must establish an Energy Control Program consisting of documented training, written energy control procedures, and documented periodic inspections.

Since 1990, all new or modified machines or equipment must be equipped with energy isolating devices designed to accept a lockout device.

This sounds, and is, relatively straightforward, but there are four sections of the LOTO standard that describe a number of steps required for compliance.[4] The standard does not define how to identify processes utilizing hazardous energy that meet the requirements for LOTO, or how to keep your program current. It is expected that the organization has a Hazard Analysis Program for new or modified equipment and a Process Hazard Review for planned work, as well as some Change Management or Document Control procedures to keep the program current. The standard also contains an "Appendix A" that is a helpful example of a "Typical Minimal Lockout Procedure."

Once each energy source, or combination of energy sources, is identified for every piece of equipment and every process where you decide the LOTO standard applies, you are ready to put in place the Energy Control Program.

Training Requirements

The Employer must provide training on the purpose of the Energy Control Program. Those involved with LOTO specifics must receive the knowledge and skills to carry out LOTO for the hazardous energy sources and methods of energy isolation for the machines, equipment, and processes with which they are authorized to work. *Training and retraining must be documented* and include subject, employee name, and date.

Affected employees are those who work on or around machines or equipment on which servicing or maintenance is performed following LOTO energy control procedures. They need to understand the Program and how it affects their work.

Authorized employees are those who are trained and authorized by supervision to perform LOTO. Should an authorized employee be requested to perform LOTO on equipment or a process with which he/sher is not familiar, he/she cannot do that until trained on that particular hazardous energy and isolation method(s). An affected employee becomes an authorized employee when his/her duties change to include servicing and maintenance and he/she is trained.

Other employees are those who may occasionally be in an area where there is equipment covered by the LOTO

Standard. They need to know the basics of the program and that it is a serious safety violation to tamper with or remove any locks or tags.

Retraining must be provided to those whose job involving LOTO has changed, or if there is any change in equipment, machines, or process that presents a new hazardous energy or control, or any change in the Energy Control Procedures.

Energy Control Procedures

Written procedures need to be developed to control any potential hazardous energy release for the servicing or maintenance steps. Authorized employees should be involved in the writing of procedures. Refer to Table 23.2 for a summary of Energy Control Procedures.

Standard Operating Procedures (SOPs) SOPs are generally developed for the operation of equipment and process

TABLE 23.2 Machine-Equipment-Specific Energy Control Procedures

Machine/Equipment Type: _____   Manufacturer: _____

Building: _____   Location: _____   RL#: _____

Preparation Date: _____   Revision Date: _____

A. Notify all personnel who work on or around the equipment that it is being de-energized.
B. Locate and identify all hazardous energy sources and the associated energy isolating devices.

Electrical (Voltage): _____ volts   Gravity: _____
1. _____   1. _____
2. _____   2. _____
3. _____   3. _____

Pneumatic (Pressure): _____ psi   Chemical (identify) (Gas, Liquid, Solid): _____
1. _____   1. _____
2. _____   2. _____
3. _____   3. _____

Hydraulic (Pressure): _____ psi   Other (Identify): _____
1. _____   1. _____
2. _____   2. _____
3. _____   3. _____

C. If the equipment is in operation, shut it down by the normal operating procedure.
D. Operate the energy isolating devices and isolate the equipment from all hazardous energy sources.
E. Lockout/tagout the energy isolating devices. List the number of devices: _____
F. All potentially hazardous stored energy shall be relieved, restrained, or otherwise rendered safe.
G. Verify that isolation of hazardous energy sources was effective by testing with the appropriate operating controls or test equipment. After ensuring that no personnel are exposed, operate the normal operating controls to make certain that the equipment will not start-up or cycle. Return all controls to the neutral or off position after verification.
H. Perform work task.
I. Release from Lockout/Tagout
   1. Notify and clear all affected employees within the work area.
   2. Physically check the work area to ensure that all tools are removed, all shields are properly reinstalled, and that all interlocks have been restored.
   3. The employee who attached the lockout/tagout devices shall remove them.
   4. Operate the energy isolating devices to restore energy to the equipment.

LOTO Devices/Hardware

Devices must be *Durable* (for the environment and exposure time), *Standardized* (by color, shape, size, and be used for no other purpose), and *Substantial* (to prevent removal without excessive force).

Locks: Every authorized person must have his/her personal lock and one key (there can be a secure department set of keys). Lock must contain person's name or identifying number. It can also have a picture and be color coded by craft if that is helpful.

Tags: Must contain an appropriate warning, like DO NOT OPERATE, or what makes sense for operations. The attachment of the tag must be done by hand, be non-reusable, self-locking, and non-releasable with minimum unlocking strength of no less than 50 pounds.

Besides locks and tags, other hardware can be chains, wedges, key blocks, adapter pins, self-locking fasteners, or whatever the employer approves to isolate, secure, or block out energy sources.

---

steps where it is important that the procedure be followed the same way every time for safety or quality reasons, regardless of who follows them or when they are executed. SOPs are also used to train new employees or when retraining is required. Many organizations find it beneficial to incorporate the service/maintenance LOTO energy control procedures into the department operating SOPs. If there are no operating SOPs, LOTO SOPs still must be developed.

Safety Permits In many organizations special SOPs, or permits, are written for particularly hazardous operations like Hot Work, Confined Space Entry, Asbestos Removal, and LOTO. These are usually used when Maintenance or Service personnel or Contractors enter the department to perform a requested service requiring LOTO. The Safety Permit requires both maintenance and operating personnel to sign off that it is safe for both before the procedure is followed.

Application of LO or TO Device, SOP Format The SOP must describe specific steps in the following sequence.

> *Preparation for Shutdown.* Authorized employee makes sure he/she is trained for the required servicing and notifies affected employees.

Equipment Shut Down. Authorized or affected employee follows the normal shutdown operating procedures.

Machine Isolation. Authorized employee locates and operates all isolating devices so as to isolate equipment from energy source(s). Examples: circuit breaker, disconnect switch, valves, or anything that can block or isolate an energy source.

Affix Lockout or Tagout Device. Authorized employee affixes a LO device to each isolating device to hold it in a safe or "off" position. Look at any major safety supply catalog and you'll find a LO device for most every energy-isolating device made.

Stored energy must be relieved, disconnected, restrained, and otherwise rendered safe after the LO or TO devices are affixed.

Verification. Prior to starting work, authorized employee must test to insure LO or TO and deenergization of equipment was effective. Make sure the equipment is left in an "off" position or neutral after verification so it will not start up unexpectedly after servicing and energization.

Release from LOTO Following servicing or maintenance and before LOTO devices are removed, the following procedures must be followed by the authorized employee.

Inspect work area to insure machine components are intact and no nonessential items are left behind.

Check affected and other employees to ensure they are in a safe position and notify them that LO or TO devices are being removed.

Employee who applies device, removes device. If the employee who applied the device is not available to remove it, there must be a procedure to contact the employee (on or off site) and determine status and actions. If employee cannot be contacted, supervi-

---

Lockout versus Tagout

It is generally agreed the preferred and more sure method is to use locks rather than tags. However, where lockout devices are not possible or practical, tags may be used. If tags are used they must be affixed at the isolating device, or as close thereto as possible. Some organizations with a TO program have been allowed to continue it with OSHA approval where it can be shown to be as effective as following LO procedures. To demonstrate *full employee protection* this can mean added safety measures than those required for LO.

TABLE 23.3 Lockout/Tagout Inspection/Audit Checklist

Tool/ID #: _____ Equipment Name: _____

Date of Inspection: _____ Authorized Inspector's Name: _____

Names of Authorized and Affected Employees Inspected: _____

_____

_____

Removal of Equipment from Service	____Yes	____No	____N/A
1. Were all affected employees notified that the machine or equipment was going to be Locked/Tagged out?			
2. Were all the hazardous energy sources and the associated energy isolating devices correctly identified and located?	____Yes	____No	____N/A
3. Was the equipment shutdown performed correctly?	____Yes	____No	____N/A
4. Were the energy isolating devices operated so that the equipment was isolated from all hazardous energy sources?	____Yes	____No	____N/A
5. Were lockout/tagout devices placed on all the energy isolating devices?	____Yes	____No	____N/A
6. Were approved lockout/tagout devices used?	____Yes	____No	____N/A
7. Were all potentially hazardous stored energies relieved, restrained, or otherwise rendered safe?	____Yes	____No	____N/A
8. Was the isolation of hazardous energy sources verified to be effective by testing with appropriate instrumentation and by operating the normal equipment controls after ensuring that no personnel were exposed?	____Yes	____No	____N/A
9. Were the equipment controls returned to the neutral or off position after verifying that the equipment would not start up or cycle?	____Yes	____No	____N/A
10. Were additional procedures dealing with shift or personnel change, group lockout/tagout, and testing/positioning of equipment followed?	____Yes	____No	____N/A
11. Were unique lockout/tagout requirements for this equipment written in the procedure and followed?	____Yes	____No	____N/A
Release from Lockout/Tagout	____Yes	____No	____N/A
1. Was the work area checked to ensure that all tools were removed, all shields were properly reinstalled, all interlocks were restored, and that the area was clear of hazards and personnel before the equipment was re-energized?			
2. Were all affected employees notified that the equipment was going to be returned to service?	____Yes	____No	____N/A
3. Were the lockout/tagout devices removed by the authorized employee who attached them?	____Yes	____No	N/A
4. Were the energy isolating devices operated to restore energy to the equipment?	____Yes	____No	____N/A
General Requirements	____Yes	____No	____N/A
1. Does the lockout/tagout procedure provide adequate employee protection?			
2. Did the authorized employees correctly explain their responsibilities under the lockout/tagout procedure being inspected?	____Yes	____No	____N/A
3. Did the affected employees correctly explain their responsibilities under the lockout/tagout procedure being inspected?	____Yes	____No	____N/A

NOTES:

1. All no responses require corrective action with dates for completion. Use the space below to specify.
2. Department manager shall retain the completed inspection forms for audit purposes.

_____

_____

_____

_____

_____

_____

_____

_____

_____

_____

I certify the completion of the inspection:

Manager's (Signature) _____ Date _____ Dept # _____

sion must access hazard status and decide about removing the device. Notify employee if device is removed before he/she returns to work.

An authorized or affected employee can operate the energy-isolated devices to restore energy to the equipment when needed.

Special Situations

Group LOTO When servicing or maintenance is done by a crew or several people, it generally involves multiple energy-isolating devices and different tasks being performed. This work should be coordinated by the supervisor of the department where the equipment is and can be controlled by a specific procedure and a group lockout, like a department lock. Each authorized employee still has to apply, verify isolation, and remove his/her lock following procedures, but the job is not complete until the final lock (department lock) is removed.

Shift or Personnel Changes Under no circumstances should a lock be removed from an incomplete job without replacing it with another person's lock who will continue to follow the procedure. The department supervisor is responsible for an orderly transfer of locks and/or tags and work.

Contractor Lockout Procedures need to be worked out with contractors to follow the organization's LOTO program, including agreement on the lockout hardware to be used.

Periodic Inspection (the Audit of LOTO Procedures and Use)

What Gets Audited and How Often At least annually, every authorized employee must be observed by another authorized employee (other than one authorized on that specific procedure) to determine if the employee knows his/her LO or TO responsibilities and is following the procedure properly. It is not necessary to "audit" everything requiring LO or TO, or every piece of equipment or process for which the employee is authorized. But, every authorized employee must be audited at least once per year following at least one Energy Control Procedure. Table 23.3 provides a LOTO Inspection/Audit Checklist.

Inspection Documentation The employer is responsible for ensuring that the annual audit occurs and that the documentation includes the machine or equipment on which the Energy Control Procedure was being utilized, the date, the employee(s) inspected, and the person(s) performing the inspection.

References

1. OSHA Electrical Standard, 29 *CFR* 1910.333(b) (2).
2. OSHA General Environmental Standard, 29 *CFR* 19 10.147.
3. MORT: Safety Assurance Systems (Management Oversight and Risk Tree) 980 ISBN 0–8247-6897-3; Marcel Dekker, Inc.: New York; Published in cooperation with the National Safety Council.
4. OSHA General Environmental Standard, 20 *CFR* 1910.147(c)–(f).

SECTION III

EMERGENCY MANAGEMENT

RUSSELL PHIFER

SECTION EDITOR

24

Emergency Response Planning and Training

LOUIS N. MOLINO, SR.

Emergency response planning and training are often overlooked or undervalued and allowed to "slip" in many communities and businesses. All too often this becomes evident only after an emergency has occurred and lives are lost or property is destroyed needlessly. The good news in regard to this is that the value and importance of proper planning and training for emergency response is amplified all the more with each passing incident and the better news is that in many cases history has shown that lessons are being learned. Proper planning and training will not eliminate the possibility of an emergency occurring, but studies have shown that communities and businesses that have adequate planning and training in place will be able to recover from the incident much quicker than those without an equal level of preparedness.

Which First: Planning or Training?

Often the question comes up which should be done first, the planning or the training? The not so simple answer to this question is BOTH!

Effective Emergency Response Plans (ERPs) are based on expected conditions at the time of an emergency. In order to plan for emergency incidents the planner must therefore have a thorough understanding of response conditions that are realistic yet widely diverse due to the unpredictable nature of emergency incidents and the ever-present Murphy's Law: whatever can go wrong will go wrong.

The only way to obtain this information is to conduct a thorough hazard analysis for the area covered by the plan. After the hazard analysis is completed you will often find that there are potential risks that were not readily apparent without this in-depth hazard analysis. The hazard analysis may indicate that there are some

hazards present that involve training around issues that may be obviously lacking or you may identify areas where you will need to train personnel prior to performing the hazard analysis. This is particularly true in cases where the plan will involve a large and complex plant facility of an entire community.

Hence, you may need to do training in order to start the planning process or vice versa. The only constant is that this starting point will be different for each facility or community, and furthermore, it will depend to a large degree on what planning and training elements are currently in place. Since this is more than likely out of your control you will need to make several adjustments to your strategy before getting into the main thrust of the entire process, which is to improve overall response by the business or agency involved.

The Importance of ERPs

History of Emergency Response Planning

Several major incidents have occurred involving hazardous chemical releases in the past 25 years that have proved conclusively that there was a direct correlation between the magnitude of the human suffering and property losses and the amount of preparedness on the part of the responding agencies. An old quote is that "if you fail to plan, you plan to fail." This statement could not be any truer for any topic than for emergency response planning.

Legal Requirements

In the United States one of the major motivations for the creation of emergency response plans is the legal requirements stemming from several regulatory agencies, most notably OSHA and the EPA. Major ERP regulations are outlined below.

OSHA

OSHA General Industry Standard 1910.120(q) requires that all emergency response organizations prepare and utilize ERPs.[1] Paragraph (l) of the same standard outlines the planning requirements for hazardous waste sites. These plans must be written and must outline both general and specific procedures for use both for on- and off-site emergencies. In many ways this requirement is somewhat redundant to the Superfund Amendments and Reauthorization Act of 1986 (SARA). This redundancy was done to ensure that there were no gaps in legal requirements between old and new regulations and the jurisdictions of OSHA and the EPA. Over the years several compliance directives from OSHA have upheld that if your facility or agency meets the requirements of SARA you are also deemed to be compliant with the OSHA requirements.

The areas that must be addressed by the ERP include pre-emergency coordination and planning with other agencies that may be involved in a response, roles of response personnel, lines of authority and communications, training issues, recognition and identification of incidents involving hazardous materials, ERP implementation procedures, generic standard operating procedures for such things as access to Emergency Medical Services (EMS), decontamination, evacuation routes, and so forth.

Provisions also require governmental and nongovernmental cooperation in ensuring that their respective plans do not conflict and are otherwise compatible.

EPA

In addition to the requirements as described above, the EPA also regulates provisions of the Resource Conservation and Recovery Act of 1976 (RCRA) which outline ERP requirements for hazardous waste treatment, storage, and disposal facilities.

Case Studies

Community Plans: Miamisburg, Ohio

On July 8, 1986, a train derailed in Miamisburg, Ohio, a city of roughly 18,000 people. One of the cars that derailed was carrying white phosphorus. As the rail car left the tracks a fire began in this car, causing a thick and very toxic cloud to begin to spread over a wide area. The ERP that the city had developed in conjunction with the local industries and the other levels of government were instrumental in preventing the loss of life and reducing the loss of property. The Police Chief of Miamisburg was quoted as saying, "Here in the city of Miamisburg, we have worked hand in hand with Monsanto Research Corp. and we go through periodic training with them. . . . [I]t [the training] was instrumental in helping us set up command posts and EOC (Emergency Operations Center) operations because we'd gone through the mock disasters and we knew what to expect and what needed to be done" (Bob Goenner, Police Chief, City of Miamisburg).[2]

Emergency Response Plans versus Standard Operating Procedures

Some organizations are operating under the mistaken assumption that they have an ERP in place when in reality they only have Standard Operating Procedures (SOPs). The primary difference between an ERP and an SOP is the scope which each covers and the ability to "stand alone" in actual use.

ERPs are designed to provide for a systematic escalation of response capability to an emergency incident. The ERP can be implemented on a small-scale incident and only utilized to the extent needed for that incident while at the same time affording the responding agency with the capability of full-scale implementation should that level of response be needed. Also, in an Integrated Emergency Management System (IEMS) corporate, local, state, and federal plans are able to "mesh" together in major disaster scenarios such as a multi-facility incident, natural disasters, and even in time of war. This allows for a coordinated response at many levels from multiple response agencies ensuring that no areas of need go unattended.

SOPs, on the other hand, are more task-oriented and in many cases are written more for day-to-day operations and events. An example of an SOP would be a standing order that while operating a piece of equipment an operator must wear specific personal protective equipment. Most companies and governmental entities have a multitude of SOPs and some are operating under the incorrect assumption that this collection of SOPs serves as their ERP. This mistake could result in the entity receiving fines from enforcing agencies and, worse, it could result in property loss, injuries, and loss of life!

Emergency Response Plans

In order for an IEMS to function at peak effectiveness in all types of incidents each key component in the government and business sectors responsible for their own emergency planning must coordinate with all of the other components and ensure that their plans are compatible. Roles of the federal, state, and local governments and the private sector are outlined below.

Types of Emergency Response Plans

Several major differences exist in the four areas of emergency planning responsibility in the United States. Contrary to some popular beliefs held amongst the masses, the various levels of government are not the ultimate authority in terms of the responsibility to plan for emergency response to any emergency incident. Fortunately, most people in the business world are keenly aware that they need to be prepared to respond to their own emergencies as well as any natural disasters which may befall their facility. The responsibility of each entity in ascending order or responsibility is provided below.

Private Sector Planning As stated above, in the private sector most business people realize that they must take the primary responsibility to ensure that they will be able to return to normal operations after an emergency of any magnitude occurs. An adequately prepared and maintained ERP will go a long way to ensure that a normal return to daily activities is possible as soon as possible dependent on the type and magnitude of the incident and its direct impact on daily operations.

The person responsible for the creation and maintenance of the ERP and the coordination of it with other plans will vary depending on the size of the operation as well as the nature of the business. Some firms will have dedicated personnel whose primary responsibility will be emergency planning. Examples of businesses who use this type of system are larger chemical companies and nuclear power plants. Other smaller facilities and firms may assign a member of their safety or operations department to undertake the management aspects of the ERP.

In any system, clear responsibility for the ERP should be established to ensure that all levels of the corporate structure are aware of the importance of the ERPs creation and maintenance.

Local Government Planning On a daily basis local governments provide and are expected to provide various levels of emergency services. Police, fire, and EMS services are the primary example of services found in nearly all local jurisdictions. The system types and levels of service range from fully staffed and paid services to volunteer services with every conceivable combination in between. One area that many "outsiders" are unaware of is Emer-

gency Management. Many communities include the responsibilities of Emergency Management with the daily operations of the above-referenced services.

Local government planning efforts are based on the SARA Title III: Emergency Planning and Community Right-To-Know. Title III defines the tiered planning system for all levels of government as well as outlining the responsibility for facilities covered under the legislation to report on quantities and the nature of releases to the Local Emergency Planning Committee (LEPC).

The LEPC is comprised of members of the local emergency management community along with business and industry leaders as well as the general public. The LEPC is required to develop the local ERP and coordinate this plan with the other levels of government and the private sector. The plan must include the following[3]:

- identification of facilities that use Extremely Hazardous Substances (EHS) and potential transportation routes for EHSs and other hazardous materials
- emergency response procedures for response to facility of on-site incidents
- designation and identification of community and facility responsible for the creation
- implementation, and maintenance of the ERP
- emergency notification requirements and procedures
- descriptions of and availability of community, mutual aid, commercial and industrial emergency
- response equipment, supplies, and activation procedures
- evacuation plans and procedures
- descriptions of training and exercise schedules to ensure the viability of the ERP

The LEPC is the organization where industrial emergency managers are able to integrate their ERPs with the local ERP and coordinate through to the higher levels of government response and planning. The LEPC serves as the foundation and backbone for effective emergency response planning and response to emergency incidents of all types.

State Government Planning At the state level each state is required to maintain a State Emergency Response Commission (SERC). The SERC is similar to the LEPC in that its membership is broad based in nature and designed to bring members of the statewide emergency response team and chemical community together in their respective planning efforts.

The first responsibility of an SERC is to delineate the boundaries of local planning districts—these are the areas served by the LEPCs referenced above. Once the planning districts are established (originally done by July

17, 1987, but updated as needed) the SERC will be responsible for the following:

- supervision and coordination of the activities of the LEPCs
- establishing procedures for the dissemination of public information about state and local emergency planning efforts
- reviewing and coordinating local ERPs
- coordination with other states' SERCs and federal response planning authority on multijurisdiction issues

While not directly required under Title III, most SERCs also provide some type of funding and educational services as well as other support services for the improvement of the LEPCs' overall efforts.

Federal Government Planning Efforts Several federal agencies have jurisdiction and regulations concerning emergency planning. Examples are OSHA, the EPA, and the Department of Transportation. Each of these agencies publishes information to the business community to assist it in complying with their respective regulations.

The overall responsibility for federal planning falls to the Federal Emergency Management Agency (FEMA). FEMA coordinates the activities of SERCs and LEPCs directly and indirectly through its 10 regional offices and is responsible for providing both indirect and direct assistance during actual incidents. Funding is provided for training and exercising plans as well as to assist in the development of additional support materials. An example of this might be a state using federal money to develop and deliver a state guide for the creation of ERPs to the local jurisdiction and a similar guide for use by the private sector to ensure compatibility with governmental plans. FEMA also will coordinate any federal response to an actual incident up to and including services rendered in accordance with a Presidential Disaster Declaration.

Exercising ERPs

In order to ensure the effectiveness of an ERP, it will need to be exercised. Several types of exercises are available to planners to measure the effectiveness and viability of an ERP. These include:

- Orientation seminar—Basically a walk-through of the ERP to ensure that all participants understand the plan and their roles in the plan. Often done with new plans or when major revisions are undertaken to determine if additional work is needed on the ERP.

- Tabletop exercise—These exercises are excellent ways to test the response capability of the ERP without the cost and problems associated with functional and full-scale exercises. Participants are able to respond to realistic scenarios in a non-hostile, nonthreatening environment and make changes to plans and response procedures.
- Functional exercise—A functional exercise will allow practice of those components of an ERP that can not be tested using the Orientation seminar or Tabletop exercise. An example of this might be having the response team don protective equipment and stop a simulated leak in a process pipe. This can be done with limited disruption of normal operations.
- Full-scale exercise—A full-scale exercise will often entail actual simulated response of outside agencies such as local fire department and hazardous materials teams. While these are certainly effective in testing the depth and breadth of an ERP, they are difficult to coordinate and expensive as they often require a disruption of normal operations as well as capital expenditures in order to be effective.

Whichever exercise option is used, the goal is to evaluate either part of or the entire ERP and make adjustments so that when the real event occurs the plan functions as is intended.

Training for Emergency Response to Chemical Emergencies

Training is the single most important aspect of emergency preparedness. Only by training on every aspect of an ERP and practicing the skills needed to actually respond to an emergency such as spill control, fire fighting, and the like will an organization be able to effectively respond to and cope with an incident.

Legal Training Requirements

Federal, State, and Local Requirements

There are numerous federal, state, and in some cases even local training requirements for emergency response. Some of these are outlined below.

The single most significant requirements can be found in OSHA General Industry Standard 1910.120(q).[1] A tiered system of training has been developed to ensure that a wide range of employees can be trained to recognize and respond to emergency incidents involving hazardous materials.

Responders are divided into four levels of capability: Awareness, Operations, Technician, and Specialist. Each level builds on the next as do the training requirements. One of the major differences in the level of capability is what types of tactics each is permitted to employ under the law.

Generally two major types of incident control tactics are used to control incidents, Offensive and Defensive. Defensive techniques or tactics are those that do not require any direct product involvement. Offensive operations are those which place the responder in potential contact with the products(s) involved. Therefore offensive operations are inherently more dangerous than defensive operations.

An example of a defensive tactic would be control of a break in a pipeline by shutting off a valve above the break, allowing the material to drain, and having other responders clean the spill. An offensive tactic on the same incident might be to have a hazardous materials response team attach a control valve at the point of rupture when the flow in the pipe cannot be stopped by any other means. Detailed descriptions for each level of responder are found below.

Awareness is the most basic level of emergency responder. Awareness-level training is designed to allow a responder to recognize that an incident that potentially contains hazardous materials has occurred and then activate the ERP including evacuation of the immediate area of the incident. Responders at this level will not be directly involved with any defensive or offensive incident control.

Operations An operations-level responder is trained in basic defensive practices to control incidents using existing equipment such as control valves and other engineering controls as well as basic spill and incident countermeasures such as use of spill clean-up kits and dicing materials.

Training in personal protective equipment is based on the equipment that their employers commonly have available to be used in incidents; however, operational responders are only permitted to use defensive techniques.

Technician Technician-level responders are highly trained and are able to utilize a wide range of personal protective equipment along with incident control equipment and to employ a wide variety of response techniques in both defensive and offensive modes. This allows the technician to respond to a variety of incident scenarios ranging from small spills of highly toxic substances to major train derailment with multiple chemical involvement.

Specialist The fourth level is reserved for persons who have specific knowledge regarding specific products or other aspects of emergency response such as modes of transportation. Railroad employees who have been trained in emergency response and have extensive knowledge and training in the operation of a railroad as a system and of all of the associated components are an example of specialist employees. Specialists often serve as liaisons for actions taken with federal, state, local, and other government agencies.

Incident Commander The incident commander is the highest ranking official on the site of the incident and has overall command responsibility for the incident above the Awareness level. This individual assumes command of the Incident Command System, the use of which is mandated by OSHA at all incidents involving hazardous materials. The Incident Commander acts as the central point of contact for all response agencies and is the focal point for all incident-related communications.

OSHA General Industry Standard 1910.120 outlines the specific areas that must be included in each level of training, refresher requirements, and minimum training times. The times stated in the standard are considered to be absolute minimums and many organizations require additional training for specific operations.

Minimum Training Times[4]

Level	Initial Training Time	Annual Refresher
Awareness Level	Competency*	Demonstration of Competency
Operations Level	8 Hours	Demonstration of Competency
Technician Level	at least 24 Hours	Demonstration of Competency
Specialist Level	at least 24 Hours	Demonstration of Competency
Incident Commander	at least 24 Hours	Demonstration of Competency

* = "As needed to develop competencies" (OSHA Definition)

Selection of Emergency Response Training Providers

The selection of the training provider is critical to the success of the overall emergency response preparedness program. The training provider should be evaluated on several factors such as overall experience in emergency response training, individual experience of the instructors, quality of instructional materials used such as books and audio-visual aids, and the use of training props for simulation of incidents.

Instructors must have knowledge of adult education principles and must also have demonstrable experience and education in those areas in which they will be instructing. Many times the level of consideration given to the instructor's experience is overlooked during the selection process. Examine the level and type of Train-the-Trainer education and experience of the instructors who are under consideration.

One solution to this problem is to utilize trainers who have met the certification requirements of National Environmental Training Association (NETA) Certified Environmental Trainers (CET) program. NETA has developed the CET program, which is an examination-based credential to ensure that both technical competency and knowledge and understanding of instructional technology are mastered by the trainer. Other professional credentialing programs offer additional proof of technical competence, but the CET is the only certification that includes instructional technology.

Consideration to the question of whether the training should be done on an internal or external basis is important. Internal resources may be developed so that much or perhaps even all of the training can be done on an internal basis, but in situations where experience is not readily available then use of external training providers is necessary.

References

1. OSHA General Standard, 29 *CFR* 1910.120
2. *Miamisburg: Anatomy of a Response*; Chemical Manufacturers Association: Washington, DC, 1987
3. National Response Team, *Hazardous Materials Planning Guide*, NRT-1, 1987.
4. Lori P. Andrews, P. E. Ed., *Emergency Response Training Manual for the Hazardous Materials Technician*; Center for Labor Education and Research: Birmingham, AL, 1992.

25

Emergency Equipment

JOHN C. BRONAUGH

The general public believes that emergency equipment will never be used unless they work for a chemical manufacturer or use highly concentrated chemicals at their workplace daily. In some instances, employers have emergency equipment because they have been required to have it by a government inspector or agency or by an insurance risk-management group. Fire extinguishers and automatic ceiling sprinkler systems fall into a similar category; however, they are mandated by state law. Even though emergency equipment is not (yet) so mandated it is nevertheless necessary.

Introduction

Six Basic Units

The six types of emergency equipment are eyewashes, eye/face washes, drench showers, combination units, personal eyewashes, and hand-held drench hoses. This chapter will relate this equipment to the widely accepted American National Standard for Emergency Eyewash and Shower Equipment, ANSI Z358.1 of 1998.[1] The Standard deals with minimum performance requirements for this type of equipment for employees working where chemicals of any kind are being used, handled, or stored.

Emergency Safety Equipment Terminology

Note these two words in the title of the Standard: "Emergency" and "Equipment." We often hear and see the words emergency equipment and safety equipment used interchangeably. It may be a matter of semantics, but from a legal as well as an industry point of view, the word "safety" implies that by using such equipment you will not suffer any bodily harm. Unfortunately, this is not true. The damage begins instantly upon contact. You then literally have seconds to minimize the injury. This is why these products are referred to as emergency equipment. The ANSI Standard has always used this terminology.

Ten Factors

The seconds start ticking the instant a chemical comes into direct contact with a victim. The extent of the injury is related to one or more of these factors:

1. How accessible is the emergency equipment?
2. How quickly did the victim get to the unit?
3. What is the configuration of the equipment?
4. Is it adequate for this emergency?
5. Was it installed according to the manufacturer's instructions?
6. When was it last tested?
7. Was it properly maintained?
8. Was the victim properly trained in advance of its use?
9. Was someone aware of the emergency and did they assist?
10. Has professional medical aid been summoned?

For most contaminants, the best possible immediate first-aid treatment is adequate, copious flushing for a minimum for 15 minutes with potable water or a flushing fluid designed for this purpose. The ANSI Standard requires it. Chemical contaminants in the eyes are not only the most common but the most serious type of injury.[2] Alkaline chemicals, especially the hydroxides, are often more dangerous than acids.[3] Most medical experts agree that flushing within the first 10 seconds after the incident is the most critical in order to prevent severe and possible permanent damage to the body and/or eyes.

The Standard

Six Types

The ANSI Standard provides a listing and definition for six different types of emergency eyewash and drench shower equipment:

1. Hand-Held Drench Hose
2. Personal Eyewash
3. Plumbed Eyewash
4. Plumbed Eye/Face Wash
5. Emergency Shower
6. Combination Unit

The ANSI Standard defines these types as follows:

Hand-Held Drench Hose "A flexible hose connected to a flushing fluid supply and used to irrigate and flush face and body areas." This unit is often seen in laboratories and some industrial applications. A drench hose is handy for small area rinses when a full drenching is not required or where the victim is unconscious or cannot stand. However, per the ANSI Standard, "hand-held drench hoses provide support for emergency showers and eyewash units but shall not replace them."

Personal Eyewash "A supplementary eyewash that supports plumbed units, self contained units, or both, by delivering immediate flushing fluid." There are many types of personal eyewash devices such as various types of squeeze bottles and some 6 gallon self-contained eyewashes. They do not deliver water within the minimum ANSI Standard and therefore these types should never be relied upon or installed as your primary emergency equipment but only as a first aid device. Again, the ANSI Standard states, "personal eyewash equipment supports plumbed and self-contained units but shall not replace them."

Plumbed Eyewash "A device used to irrigate and flush the eyes." This usually consists of two water orifices at each side angled toward the center. A bowl is often included to catch the diluted chemical from splashing onto other parts of the body. The water flow is initiated by an instant-on, stay-open ball-type valve actuated by a push flag and/or foot treadle located close to the front or to one side of the unit. Per ANSI, the minimum flow rate must be 0.4 GPM (1.5 liters/minute) for 15 minutes. This piece of emergency equipment is probably the most common (Fig. 25.1).[1]

Plumbed Eye/Face Wash "A device used to irrigate and flush both the face and eyes." This fixture allows the flushing of the eyes and the face simultaneously using a

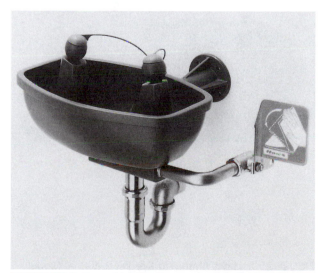

FIGURE 25.1 Haws Uni-Flo Eyewash—Model 7100BT.

face-spray ring in conjunction with the eyewash heads, large eye/face wash heads or multiple heads. A bowl or receptor is often included. Since splashed chemicals are not just limited to the eyes, this unit may be preferable to an eyewash-only unit. ANSI Standard requires a minimum flow rate of 3.0 GPM (11.4 liters/minute) for 15 minutes (Fig. 25.2).[1]

Emergency Shower "An assembly consisting of a showerhead controlled by a stay open valve and operated by an approved control valve actuator." It is known as a drench shower or deluge shower. The shower head must provide a minimum of 20 gallons of water per minute (75.7 liters/minute) for 15 minutes as long as the appropriate flow pattern is maintained over the victim using an instant-on, stay-open ball valve that is activated by either a pull rod, cord, or chain and ring. We often forget this excellent piece of equipment, which can be used in areas that have a potential for clothing fires. It is designed to quickly drench the entire body. Because of the vigorous downward flow of water produced, it should never be used as an eyewash (Fig. 25.3).[1]

Combination Unit "An interconnected assembly of emergency equipment supplied by a single source of flushing fluid." Often, this piece of emergency equipment is a multipurpose unit designed so that all components can be operated independently or simultaneously from a common fixture line. Make sure that the eyewash or eye/face wash delivers a full minimum flow rate as previously given and a proper pattern and flow rate when

FIGURE 25.3 Haws Drench Shower—Model 8123H.

the drench shower is concurrently on. The ANSI Standard requires this (Fig. 25.4).[1]

Tepid (Tempered) Water

Tepid (tempered) water is now recommended for all types of emergency equipment. This topic has generated a lot of discussion and questioning over the years. The wording on Delivered Flushing Fluid Temperature in the current Appendix B section of the ANSI Standard reads as follows: "Continuous and timely irrigation of affected tissues for the recommended irrigation period are the principal factors in providing first aid treatment. Providing flushing fluid at temperatures conducive to use for the recommended irrigation period is considered an integral part of providing suitable facilities. Medical recommendations suggest a flushing fluid at tepid temperatures be delivered to affected chemically injured tissue. Temperatures in excess of 38°C (100°F) have proven to be harmful to the eyes and can enhance chemical interaction with eyes and skin. While cold flushing fluid temperatures provide immediate cooling after chemical con-

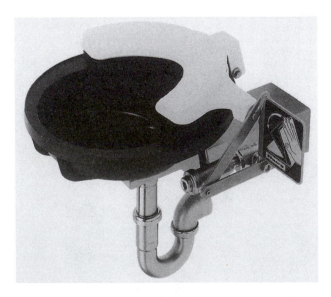

FIGURE 25.2 Haws Omni-Flo Eye/Face Wash—Model 7000BT.

tact, prolonged exposure to cold fluids affects the ability to maintain adequate body temperature and can result in the premature cessation of first aid treatment."

"Delivered flushing fluid temperature shall be tepid." This language is common for each piece of equipment. It should be noted that tepid water is defined by ANSI on page eight under definitions as "moderately warm; lukewarm." There is no temperature range given in the Standard. This is because there is no data available to substantiate any range.

The reason for the concern is obvious. Under extreme conditions, the shock of dumping lots of very cold water at 20 to 30 gallons per minute could be as serious as the splashing of the chemical and its effects. On the other extreme, in hot climates, water standing in long runs of piping exposed to the sun can become hot and dangerous enough to cause at least first-degree burns.

Hypothermia

Cold-water flushing compounds the emergency situation of the victim who is in a panic state and trying desperately to find immediate first aid. Cold air and wind in conjunction with a deluge shower can lead to excessive body heat loss and hypothermia. Symptomatically, there is numbness, confusion, disorientation, and uncontrolled shivering. This condition can quickly lead to a very serious situation especially in a remote job location. Even if the water temperatures are not that extreme, they can be uncomfortable enough to cause the injured person to terminate his or her first aid treatment before he or she has had adequate time to flush away the chemicals. Reflection upon such potential discomfort could cause a worker to preclude against his or her using equipment long before there is an emergency.

Thermostatic Mixing Valve

Tempered water and heated enclosures are the solutions. Both are engineered products. Thermostatic tempering or mixing valves with several redundant backup or protection by-pass systems must be used and are at the heart of any tempered water system (Fig. 25.5).[1] There are advantages and disadvantages to various types of systems. They often must be customized to a specific local hazardous situation. Further discussion is beyond the scope of this chapter. Consult an emergency equipment manufacturer for additional information.

Which Unit?

There are no hard and fast rules. Generally two main considerations must be evaluated. What are the characteristics of the specific potential hazard or hazards? In other words, what types of hazards are present and how dangerous are they? Thus, identify the worst case scenario. Will personnel be exposed to acids, alkalis, or other corrosives, flammable solvents, oils and greases, biotoxins or radioactive materials?[4]

To determine the number of units that should be installed, ask what is the maximum number of personnel that potentially could be working in the hazardous area at any given time when the unthinkable might occur.

Accessibility

According to the ANSI Standard, the location requirements of eye washes, eye/face washes, emergency show-

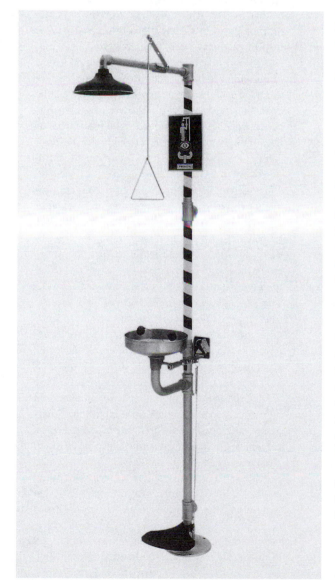

FIGURE 25.4 Haws Combination Unit—Model 8300.

FIGURE 25.5 Haws Tempered Water Blending System—Model TWBS.SH.

ers, and combination units are the same. All of these units "shall be in accessible locations that require no more then 10 seconds to reach." Now, there is no distance requirement. The effectiveness of any piece of emergency equipment is in direct relation to the time in which it is activated. Remember, the most important aspect in the treatment of a chemical exposure is obtaining first aid promptly, within the first 10 seconds. From a practical standpoint, if you can place the emergency equipment even closer to the hazard, do so. If possible, a permanently installed plumbed unit should be chosen over a gravity operated or self-contained portable pressurized unit. There are several reasons. The main reason is that it is easy to test. When portables are used, there is an ongoing maintenance cost of replenishing the units on a regular basis with a flushing fluid. Also, personnel become accustomed to the permanent location of the unit, thus making it easier to locate in the case of an emergency.

Another way to determine where you should place a piece of emergency equipment is to stand blindfolded at the center of the potentially dangerous activity. Can you

reach the unit within 10 seconds? Are there any obstacles precluding accessibility?

Eleven Practical Considerations[5]

1. There should be a safe distance between the emergency unit and anything electrical. (Although the reason for it is obvious, in this author's experience, this precaution is often violated and potentially fatal.) A worker could even decide in advance not to use the emergency equipment because of possible electrocution versus a chemical splashing.

2. The ANSI Standard requires that the units be clearly marked with signs and the area properly lighted. Use safety green paint (see Section IV, below) on the wall and floor area around the unit for 3 feet in all directions to mark the area. Inspect regularly with stiff reprimands for anyone who obstructs the marked area with clutter. In remote areas or in hazardous locations where there are few people, an alarm system should be installed so that when the shower or eyewash is activated, help is summoned. Audible alarms or flashing lights can be wired to sound or flash locally, but it is better if they are activated simultaneously in a manned control room or security office as well as locally so medical help can immediately be sent. A constant stay-on area light mounted on or near the unit is still another way to constantly call attention to the emergency unit's location. It guides the victim and those employees who will be assisting as well as those who will be rendering professional medical aid.

3. The emergency equipment should be on the same level or floor as the potentially hazardous area and the travel path must be free of obstructions per the ANSI Standard. It should never be separated by walls that could provide a maze effect or doors that could unknowingly become locked. If practical, emergency equipment should be near an exit in case of an explosion. Emergency aid personnel (firefighters, paramedics, medical technicians, etc.) must also have quick and easy access to the victim.

4. Provide for adequate potable water, as required by the ANSI Standard. Wherever possible, add strainers to the lines to remove potentially troublesome particles or debris or purchase emergency equipment that already has an integral in-line strainer or filter. Quite often, remodeling, repair work, or new construction is the direct cause for debris to be in the lines. It only takes

a particle the size of 20 microns to damage delicate and sensitive eye tissue.

5. Once these fixtures are installed, proper maintenance is mandatory.

   The ANSI Standard states that emergency equipment should be tested weekly. This does three things:

   a. Verifies that the equipment itself is operational.

   b. Verifies that a cutoff valve somewhere between the water supply and the unit has not been inadvertently closed.

   c. Flushes the lines to insure clean, debris-free water.

6. It is important to keep a record of the date of the weekly inspection and the inspector's initials. An annual inspection is now required to make sure each unit conforms to the ANSI Standard. A waterproof test record tag should be attached to each unit to show this information. Unfortunately units in a plant or even a lab are subject to vandalism. Check for this along with leaks due to loose connections. Repair or replace broken or worn parts immediately. Specification, maintenance and parts manuals should be filed in a known, easy to reach location for maintenance personnel to obtain in order to facilitate a rapid repair. Installation manuals are required per the ANSI Standard in order to prove that the owners have the literature to properly complete repairs. If a repair cannot be done immediately, install warning signs around the inoperative unit and state the location of the next closest unit. Or better yet, place a temporary self-contained, portable emergency unit at the site until the regular unit is repaired, tested and is again operable.

7. Provide adequately sized floor drains. While the victim is of prime importance, the minimum flow of 20 gallons of water per minute from a drench shower can cause a lot of water damage and also presents a slippery hazard not only to the victim but to those who are assisting. If it is an eyewash or eye/face wash, make sure the spray can be contained within the receptor or bowl for the same reason.

8. The ANSI Standard requires training for all employees who might be exposed to a chemical splash. Do not assume that because you know how to use a piece of emergency equipment others also know how. When flushing the eyes, continuously hold both upper and lower eyelids away from the eyeball with thumb and forefinger and, while flushing, roll the eyeballs up, down, and sideways without stopping so as to flush contaminants out of the eye socket and from under the eyelids. Train all personnel on how to get to the unit and use it properly. The unit looks simple enough to use, but under emergency conditions, seconds count. Panic often sets in even if training drills have been held. Be ready to assist the victim.

9. Maintain a relationship with emergency equipment sales personnel. They can be valuable in assisting you in product selection, placement, personnel training, and maintenance of each unit including the supply of parts, installation, and maintenance manuals.

10. Emergency equipment is not a substitute for proper primary protective devices such as eye and face protectors and protective clothing.

11. Last, but certainly not least, install and use only those products that meet or exceed the ANSI Standard. Look for a nationally or internationally recognized third-party certification that states that the piece of equipment meets or exceeds the ANSI Standard such as the Canadian Standards Association (CSA). You might want to ascertain that the manufacturer of your emergency equipment is also a ISO-9001 certified company. You can then be assured that they have met and continue to have a quality standard recognized worldwide. There are some emergency equipment units on the market today that are poorly designed and poorly made. Compare those products to manufacturers that have a reputation for quality.

Common Features

There are some common features for most of this type of emergency equipment that you should look for:

1. The stay-open valve is a must. The ANSI Standard requires it for three reasons:

   a. It allows the injured person to use both hands to open the eyelids for through flushing.

   b. It also allows the removal of clothing, if necessary.

   c. It will not shut off prematurely. The unit should continue to flush with a constant minimum pattern for at least 15 minutes.

2. You should look for a soft, gentle, and regulated flow of water in eyewash or eye/face wash units. The unit should deliver the flow equally from both eyewash heads, so that streams of water into the

eyes do not injure the soft eye tissue. A flow control is necessary and should be part of the unit in order to compensate for wide variations in the water supply pressure. Flow restrictors cannot adequately perform this function in varying pressures.

3. The ANSI Standard requires that the spray heads be protected from airborne particles. There are two types of dust covers. One covers each spray head while the other covers the entire unit. Both types of covers are designed to automatically remove dust particles when the stay-open valve is activated.

Major Guidelines on Selection

The Basic Principle

Optimum employee protection is using the best available equipment.[6] Initial purchase and installation costs are very minor items. What price will you place on pain, suffering, medical bills, time off, adverse publicity, hiring/training temporary replacement, or the ultimate—death of a fellow employee with possible future legal liability? Cost should never influence your buying decision. Buy the best equipment in quality, reputation, and what is adequate for the degree of hazard. Take the time now to compare different types, manufacturers, and their units including availability of repair parts. It will pay dividends later. The ideal time to plan, select, and install emergency equipment is when new facilities are in the design stage. An entire safety system can be evaluated and considered. Don't wait until the building is almost completed.

Background

Here is a little background into the ANSI Standard that we have referred to throughout this chapter. Approximately 20 plus years ago, the Emergency Eyewash and Shower Group of the Industrial Safety Equipment Association began to work on this standard. The group was composed of people from companies in the United States who manufactured emergency equipment at that time and were knowledgeable in its design, installation, and use. What prompted this was an urgent need for a minimum equipment performance standard. The purpose was to eliminate duplication as well as conflicting standards and form a single, nationally accredited standard. At that time and even today, federal OSHA, unfortunately, has over 25 vague general regulations relating to emergency equipment scattered over several manuals.

Five Categories

Five categories were addressed by the Emergency Eyewash and Shower Group for each of the major pieces of emergency equipment previously discussed:

1. Performance criteria
2. Testing procedures
3. Installation
4. Maintenance
5. Training

ANSI Standard Z358.1

Approved and published in June 1981 was the first American National Standard for Emergency Equipment Eyewash and Shower Equipment, a 24-page booklet. It was better known as ANSI Z358.1-1981. It was later revised and published again in 1990. ANSI requires that a standard be reviewed every 5 years. The latest revised edition of the standard was published as of April, 1998.[7] Remember, the Standard is still voluntary and is not a code or public law. It does represent a general agreement among most manufacturers, scientific, technical, professional individuals, and various governmental organizations. It is the only known reference on the subject . . . in the world. This is why federal OSHA, many state OSHAs, and other local regulatory bodies constantly and consistently are referring to this Standard.

State OSHAs

It should be noted that approximately half the states have their own OSHA agencies that are directly involved in the regulation of worker safety. Their requirements on emergency equipment vary. Many have either adopted the ANSI Standard or have modified it. Contact your state OSHA office to obtain a copy of their current regulations.

Standard Copies

For a fee, you can obtain a copy of the current ANSI Standard by contacting the Industrial Safety Equipment Association, 1901 N. Moore St., Suite 808, Arlington, VA 22209; Telephone: (703) 525-1695, Fax: (703) 528-2148.

Organization

A closer look at the revised ANSI Standard will reveal how it is organized. Two pages of definitions of the equipment and terminology are given. Next follows nine pages of the basic Standard for all six types of emergency equipment. Each page of the Standard is laid out in a single-column format. There are also nine pages of illustrations. The last three pages, the Appendix sections A and B, discuss the 11 items that follow. It should be noted that none of the following are considered part of the ANSI Standard but are included for information only.

A1. Personal Eyewash Unit
A2. Flushing Fluid Quality
A3. First Aid Practices
A4. Waste Disposal
A5. Personal Protection Equipment
B1. Supply Lines
B2. Water Capacity
B3. Valve Operation
B4. Alarm Devices
B5. Placement of Emergency Equipment
B6. Delivered Flushing Fluid Temperature

Again, these 11 items are important but are for information purposes only. They are *not* part of the Standard. Some of these portions may be incorporated into future revisions of the Standard. Tempering water (Delivered Water Temperature) in the previous 1990 Standard is a classic example.

ANSI Standard Z535.1 of 1991

There are a multitude of colors being used on emergency equipment including, green, yellow, orange, red, and blue. Yellow seems to be the predominant color used inside any industrial setting; therefore, it is not the color to use for emergency equipment.

Though not as well known, there is an ANSI Standard, ANSI Z535.1 of 1991,[8] which recommends:

"Safety Green shall be the color for the identification of Safety location of First Aid and Safety Equipment." Paragraph 6.4.1(5) specifically states Safety Deluge Showers as an example.

Americans with Disabilities Act

Public Law 101-336, better known as The Americans with Disabilities Act (ADA), became effective January 26, 1992. In Title III of the ADA, there is no language regarding eyewashes, drench showers, combination units, or related emergency equipment. However, since disabled workers are part of the workplace, it behooves specifiers and owners to consider drench showers and eyewashes designed with the disabled in mind. Some manufacturers have products in their catalogs designed especially for the disabled.

References

1. Haws Emergency Equipment Catalog 0003241583R.
2. Wohlen, J. P. Using Emergency Wash Stations to Reduce Injuries. In *Plant Engineering*; August 19, 1982.
3. Blais, B. R. Treating Chemical Eye Injuries. *Occup. Health Saf.* **1996**, 23–26.
4. Weaver, L. A., III. Eyewashes and Showers: Ensuring Effectiveness. *Occup. Health Saf.* **1983**, 13–19.
5. Kelly, P. T. Emergency Eyewashes and Showers—A Check List for Effective Use. In *Occupational Health and Safety—Canada, 1994 Buyers Guide*, p. 58.
6. Jonathan, R. A. Selection and Use of Eyewash Fountains and Emergency Showers. *Chemical Engineering*, September 15, 1975, pp. 147–150.
7. American National Standard for Emergency Eyewash Shower Equipment (ANSI Z358.1-1998).
8. American National Standard for Safety Color Code (ANSI Z535.1-1991).

26

Emergency Evacuation/Shelter-in-Place Plans

JOHN J. McNAMARA

Chemicals themselves and processes that use chemicals are inherently dangerous. A great deal of attention has been paid to process safety and mitigation. However, the responses of persons not involved in mitigation are rarely discussed in detail. It seems that evacuation and sheltering in place are considered simple actions to execute. In reality, they deserve as much consideration as prevention and mitigation.

Attitude: What If? Or, It Can't Happen

Before any other action is taken, a vulnerability analysis should be conducted. This analysis should do a great deal to quantify the risks involved. However, even if the level of risk is low, it would seem a prudent approach to accept the probability of accidents. Even at those facilities with excellent safety records, something will eventually go wrong. In fact, the resolve to plan for what could/will happen is often weakened by long periods of safe operation. It is not suggested that accidents should be scheduled but rather that there should be continual maintenance of the planning effort. Realistic exercises will contribute to these efforts, especially if they include off-site responders. When the motivation for emergency planning weakens, consider the liabilities that could result from a lack of comprehensive planning.

Impediments to Decision Making

One of the hardest decisions that may face a manager or incident commander is the decision to do something. Both evacuation and sheltering-in-place are options which have their own real and perceived risks, as well as benefits. A situation may resolve itself, but not always. A premature decision may be as detrimental as one made too late when protective actions are necessary.

Hierarchy

It is not uncommon for an employee to lack the authority to decide on a course of action or even to recommend one to responders from off-site. This is evident in many facility planning documents. Often, there are a myriad of telephone calls that must be made before the employees on the scene can even join together to consider a proper response to the incident itself. A decision to recommend evacuation or sheltering-in-place may be reserved to a management level higher than might be found on-site, especially at night or on a weekend. Also, alternates may not be listed in the event the primary manager is not available. All these issues must be weighed in writing a comprehensive plan.

Work Culture

Maintenance of a favorable work culture is extremely important. Are employees truly empowered? Will they be afraid to report a small problem in the hope that it can be mitigated without much notice? Are they aware of possible consequences if a process goes awry and safety barriers are broken down? Do they care? Are they aware of their own limitations? Will they remain in a danger area rather than moving to safety? Will they forget the training they have received and vacate the area rather than mitigating the accident or trying to prevent further escalation of the problem? Consider an example of this situation—a company that was processing chemicals in a reactor vessel when the reaction went out of control. There was a continual release to the atmosphere. When the fire company arrived on scene after being alerted by neighbors near the plant, the shift manager refused to tell the Incident Commander the type of chemicals being used and the nature of the process because he believed the process involved a trade secret. He was right, although in this case his employer had failed to inform him that this did not apply in an emergency. Trust, accurate information and procedures, and proper training are all components of preparedness.

Minimization

The attraction of least disruption or lower response level is always present. In fact, they may have more attraction the higher one goes in management. Public image, media attention, and interruption of normal business seem to be concerns that may prevent the recommendation or ordering of protective actions. These attitudes may be more prevalent when there have been a series of minor accidents which have escaped notice. If a protective action will attract particular attention to an accident preceded by a series of accidents, there may well be hesitation to initiate actions.

Emergency Evacuation

Based on the vulnerability and risk analysis that should be accomplished before any emergency planning is commenced, the threat posed by a particular chemical or process should be known. Yet chemical accidents are but one reason to evacuate employees. Others include fire and bomb threats. Usually, planning for other events is based on the fire plan—not a bad place to start. However, do we really practice even that threat model completely? Is there a good census of personnel after an evacuation? Do we report this to the designated person? Is that person available or already involved in direct response to the accident? Are employees given further direction, or do they just hang around without knowing what they are expected to do? Basically, the fire plan is probably a solid foundation for procedures that address other types of events. If, however, we permit erosion of this foundation, the entire structure of planning will crumble.

Some planning considerations for evacuation are listed below. Not all of these will apply to every locality, but those that do could be important to the plan.

- The impact or threat of various accident scenarios should be known. If would be helpful if they can be translated into time and distance factors as well.
- The need for a warning or alerting signal—public address system, alarm, or intercom. Should the same signal or message be used for fire and other incidents? Is there a need to cover areas outside of buildings?
- Isolation of an area as opposed to evacuation of the entire facility.
- Directions on what employees should take with them when they evacuate. There would probably be time to take the required personal items such as purses, coats, boots, or umbrellas. Obviously, these are important to employee comfort and health in inclement weather. If there is a quick resolution of the problem, the fact that the employees are close to the facility could make resumption of normal

operations easier. Purses are important in that they probably contain car keys. Recently, a building was evacuated when a fire alarm was sounded. After assembling at their designated locations, the employees were informed that it was a bomb threat. Subsequently, because the search of the building would be lengthy and since it was midafternoon, the employees were told to go home. Many had not brought their car keys with them. Since reentry was not possible, a lot of plans and arrangements suffered and management took the blame.

- Special considerations that may make a problem worse, such as air-handling equipment spreading contamination.
- Arrangements for an accurate and complete census report to management. It would be helpful if the person who should receive these results be readily identifiable. A bright jacket, colored hat, or pennant are possibilities.
- Information and direction from management to employees on a timely basis. Cellular telephones or portable radios would permit information from the accident scene to be relayed outside for the benefit of the evacuees.
- Cellular phones would also be useful in communicating with suppliers so that deliveries to the facility can be rescheduled. Conversely, customers may need to be notified that deliveries to them may be delayed.
- Entry to the facility grounds is likely to be curtailed. Someone should be designated to provide information to delivery trucks, salesmen, visitors, who may be expected.
- Will the evacuation interfere with a payday?
- Notification to those employees who may be sent home regarding reopening of the facility or extension of the evacuation, as applicable. Announcements may be made over a designated radio station, or phone trees might be effective.
- Refreshments, hot or cold, as appropriate, for on-scene personnel as well as evacuated personnel who are expected to remain in the vicinity for an extended period.
- Prior identification of key employees who may need to be instantly available.
- An evacuation, even if it is only one building on a site, will draw attention. There should be a designated spokesperson and controlled location for interface with media representatives.
- Even if the evacuated employees are to remain onsite, consideration should be given to the safety of

their vehicles. Will the release or other accident result in harmful deposits or other contamination? Is there time to permit the employees to move their vehicles? Is there an alternative parking area close by in the event employees are expected to remain in the vicinity?

- Many facilities are located in or near residential areas. Residents will probably be concerned if they see groups of employees standing outside of buildings or leaving the area at unusual times. Alarms, sirens, and other alerting devices will be heard at a distance, so the residents will be impatient for information.

- Even if not required, should emergency responders and the 911 Center or dispatcher be notified? This could enable them to answer any questions and would provide them with basic information should the incident escalate.

- It is also a good idea to participate in planning with the Fire Chief and other potential responders. They should be provided with the information necessary to preplan responses. They should know where to find MSDSs, blueprints, or other information they may need. Too often the fire trucks roll up to a scene and either the information is not available or there is no one to meet them with appropriate data.

Most of the preceding considerations are applicable whether or not evacuation of the public is required. As soon as off-site impact is evident, other considerations come into play. These include:

- Consideration of the need for sirens or other devices to alert neighbors.
- While police and fire personnel will assist in the evacuation of the public and obtain a shelter/mass care center, an evacuated facility still has an important role. Can employees be made available to accompany fire and police staff? If not, can they serve in a public information capacity?
- A proven technique is to have a company team go to the shelter/mass care center to brief the evacuees moved there. Updating briefings should be scheduled at that time so the public will know when additional information will be available. This will decrease anxiety and assure them that the company is concerned about their welfare. When reentry is possible, the company team should brief the evacuees on the extent of any contamination which may have affected their homes.
- Evacuation of the public will cause more intense media interest. The company media information ef-

fort should be reinforced. Coordination with the Incident Commander, other response agencies, and employees is critical. The team that is briefing evacuees at the shelter/mass care center should be aware of any releases that are made to the media. Reporters will contact everyone if they want to develop their stories. It will not be good press if a responder or member of the public claims that "the company hasn't told us anything" or gives perceptions contrary to the official pronouncements.

- SARA has mandated plans for certain facilities which use Extremely Hazardous Substances (EHS). These plans are good models for other facilities because they are written to facilitate public evacuation.

- Are there vehicles available to move that segment of the public who would not have automobiles available? These can be obtained through the Incident Commander with the support of municipal or county Emergency Operations Centers.

- Are area hospitals aware of the possible effects from releases? Prior coordination is important. Even more important is contact during an incident. Some of the public will go to an Emergency Room as soon as an evacuation is ordered even if there is no evidence of any possible physical effects. If a release is accompanied by odors or visible vapor clouds, even more people will go to hospitals. The community emergency response system will work to evenly distribute these walk-in patients. However, the hospitals need current information on possible symptoms.

- The municipal or County Emergency Operations Center can assist by relaying to Emergency Alert System (EAS) broadcast stations critical information for the public. They should be assisted by providing clear and accurate information at a level that can be understood by the average person.

Shelter-in-Place

There are many things done in evacuations that apply to sheltering-in-place. However, the sheltering-in-place option has several unique challenges.

- Time is a major consideration. If an event is moving so fast that evacuation is impractical, sheltering-in-place is the other protective action available.
- If risk analysis indicates that certain types of accidents, such as small- to medium-size spills, might

only have a limited chance of affecting another building or area, consideration could be given to sheltering in a building separate from the accident location instead of opting for an evacuation.

- The theory of sheltering-in-place is very simple. Close or shut down doors, windows, air handling equipment or other sources of outside air. In a business or industrial setting this may not be difficult given the availability of comprehensive procedures. Experience indicates the weak link is usually maintenance or physical plant personnel. One or two might be familiar with the necessary control devices. However, they may be on vacation or away from the plant property when an incident occurs. Backup personnel are necessary.

- Essentially, the public carries out the same tasks in their residences—shutting doors and windows and closing off vents or air conditioning systems. In much of the training material on this subject, the public is instructed to preselect a particular room which has limited sources of outside air. Usually this is a bedroom with, if possible, a connecting bathroom, and with a kit that includes a portable radio, flashlights, fresh batteries, snacks, water (if not connected to a bathroom). Sections of precut plastic sheets and masking tape should be available in this room to seal over the doors and windows. The radio should be available to receive information and guidance concerning the accident.

- While the availability of a sheltering-in-place kit is a good concept, its use demands a great deal of public education. Even then, the public will not take the time to assemble a kit unless their experience indicates the probability of an accident. The first block or group of residences might be convinced to have this kit available, but the farther from the possible sources, the harder it will be to convince them.

- A key goal is to keep people from transiting through or into an area where it has been necessary to shelter-in-place. Coordination with police so that access to the area can be controlled is essential.

- When evacuation is considered advisable or necessary, the establishment of alternate routes and evacuation centers must be considered, as primary routes or centers may be (or later become) inaccessible. In addition, if the emergency situation involves airborne chemical release(s), the primary route and center may be in the plume.

In summary, both protective actions have their own utility and should be considered. Detailed plans and procedures should be made and practiced. It is not as easy as just telling an employee or resident to leave or shelter. There is a great deal of coordination necessary prior to, during, and at the conclusion of an event. As a final warning, always make sure the event is really over before permitting reentry or lifting sheltering. People are very reluctant to follow instructions when they have been told there is no more danger.

References

1. Antokol, N.; Nudell, M. *The Handbook for Effective Emergency and Crisis Management*; Lexington Books: Lexington, KY, 1988.
2. Herman, R. E. *Disaster Planning for Local Government*; Universe Books: New York, 1982.
3. Christiansen, J. R.; Blake, R. H.; Garrett, R. L. *Disaster Preparedness*; Horizon Publishers: Bountiful, UT, 1984.
4. Auf Dere Heide, E. *Disaster Response*; C.V. Mosby Press: St. Louis, 1989.
5. Drabek, T. E., Hoetmer, G. J., Eds. *Emergency Management: Principles and Practice for Local Government*; International City Management Association: Washington, DC, 1991.

27

Flood Contingency Plans

RUTH A. HATHAWAY

If a facility is located in an area that is prone to flooding (either flash flooding or a slow-rising flood, due to a stream or a river), or is protected by a levee, the facility must have a written flood contingency plan in place. A flood contingency plan is relatively simple, but requires forethought and prior planning. Below are considerations that should be included.

Introduction

This section should explain why this plan has been developed. It should state that the management will abide by the provisions of the plan and participate in its development and administration. It should also note that the preservation of life is of prime importance, that all procedures will be carried out in a manner that minimizes risk to personnel, and that rescue and medical activities have priority over all other actions.

Within this section the concept of emergency operations should be defined. This includes the three elements: preparedness, response, and recovery.

Facility Background Information

This section provides information regarding the facility. This includes its location, the surrounding area, the potential flood threat, and its history. If the facility is protected by a levee system, this section should also include information regarding the history, construction, maintenance, and ownership of the levee. There should be a written agreement between the ownership of the levee and the facility. The contact person(s) and phone numbers should be listed. Some levees are built and/or equipped with warning devices that provide advance warning of an impending breach. This possible time frame should be noted and explanation for the length of time noted.

Planning

This section details individual responsibilities for the facility's emergency operations planning and implementation.

Plant Manager has final authority for the implementation of this plan. Responsibilities are as follows:

- ensures continued compliance
- approves the plan's provisions and all subsequent revisions
- reviews and approves all information released to the media or public
- approves all agreements with any governmental, community, or industrial group
- assures the availability of adequate resources
- monitors the effectiveness of response activities and ensures that all appropriate procedures are followed
- determines when a full site evacuation is necessary
- determines the end of the emergency and gives the "all clear"

Assistant Plant Manager assists the plant manager and assumes his duties if the plant manager is unable to perform his duties.

Production Manager oversees and directs all on-site response activities. Duties are as follows:

- appoints personnel to emergency duties.
- conducts postemergency investigations.

Safety Manager's responsibilities are as follows:

- develops the facility's emergency management plans and procedures
- formulates, reviews, and ensures the implementation of the plan
- trains all employees concerning the provisions of the plan
- tests the plan through the conduct of drills and exercises
- coordinates emergency planning and response activities with local governmental, community, or private organizations
- administers safety and health review programs
- prepares and submits emergency-related reports to management as required

This section should also outline the maintenance and distribution of the plan.

Preparedness

This section describes the purpose and content of training of the employees regarding the plan. It spells out the purpose, type of (i.e., tabletop exercises, functional drills, and/or full-scale exercises), and frequency of drills and exercises. A list of all written (there should be no verbal) agreements should be listed. This includes contracts for the removal of hazardous materials and finished product prior to flooding and for hazardous materials response (if needed), cleanup, and recovery after flooding.

When the potential for flooding occurs, the media and the public will want information. A public information policy should be in place and either included or referenced in this section.

Response

This section will take the longest amount of time to prepare. It requires a great deal of thought and input. It must consider both the immediate and potential threat, the workers on-site, the raw materials, and finished product at that location. This section should include the following.

Notification, Direction, and Control

This outlines the provisions for the direction, control of emergency operations, and the request for additional community-response assistance. It spells out where the control center will be, both on-site and off-site. It must note who activates (along with two alternates) the plan, and those individuals who are responsible for headcounts at the evacuation station(s).

Procedures

This section is crucial. It will provide a specific list of steps to be taken by each member of the emergency-response team. An example of such a list might be:

The Plant Manager will:

1. notify the engineering manager
2. notify the production manager
3. notify the office manager
4. notify the maintenance supervisor
5. notify the corporate office
6. notify the State Emergency Response Office
7. prepare a statement for the media

The list of action steps will differ if there is time before the flooding or if there is an immediate threat. Both types of lists must be prepared. The action steps must be prioritized such that the highest priority items are performed first in case time runs out before the full list can be accomplished.

When time is available (for example, a 10-hour window), all hazardous materials, especially those with any rating of 3 or 4 on the NFPA 904 scale, finished product, and all records that cannot be replaced (such as personnel files) should be removed. The goal is to remove as much as possible from the area to a safe location for storage or a temporary working site. There must be a written agreement with the owner of the temporary location, if it is not owned by the facility. The logic for hazardous materials is to remove them so as to not have a hazardous waste cleanup. The logic for the finished product is to provide a cash flow during the incident. Office records that are lost and cannot be replaced may cause a legal headache in years to come.

When a river is involved, it is recommended that when the river stage has a prediction equal to or greater than a certain level (such as the 100-year flood level, or a level that would place water within the facility), special contingencies be made. This would include stating a level, lower than that determined above, at which all finished product would be moved off-site. This also includes stating a level at which the receiving of all shipments would be halted.

Evacuation and Personnel Accountability

This must address the specific threat. If it is immediate, such as flash flooding or unanticipated levee breech, there will be no warning. It may simply state that once an evacuation alarm is sounded (if time permits), all employees will leave the premises and seek higher ground. Obviously, everyone is on their own. If this is a possible scenario, a phone number should be given to all employees to report to or a location for them to assemble at for a head count. The facility could establish with their insurance carrier an emergency 800 number for such a situation. If this is done, the company must ensure that the carrier has a current employee list. The employee should give the following information when calling:

- name and employee number
- his/her location and condition
- brief description of what happened and when
- his/her immediate plans and phone number where he/she can be reached

If the possible threat may be due to a levee breach, then the evacuation plan must consider a levee breach above, at (e.g., within a half-mile radius), and below the facility. The evacuation route detailed in the plan will be dependent on the location of the breach. Therefore, more than one evacuation route and/or evacuation sites may be needed.

If the facility has time prior to evacuating, emergency shut-down procedures should be detailed or referenced in this section. Nonessential personnel should be sent home at the earliest possible time. The facility must consider how to secure equipment, LP tanks, and storage tanks. Written contracts for the removal of all materials must be in place. It is recommended that if the facility has LP tanks, a contract be drawn up requesting the supplier to remove the product and secure the tank(s) if not removed.

Emergency Public Information

This section provides for the managed release of information to the public during and following the emergency situation. A press center off-site should be selected. The press release should state:

- the nature and extent of the emergency situation
- response actions underway
- impact on off-site areas
- coordination with other off-site officials

News stories should be monitored and any misinformation or rumors should be quickly corrected or dispelled.

Recovery

This section should include:

- establishing a recovery team (who, each member's responsibilities, and function)

- damage assessment team (who, each member's responsibilities, and function)
- cleanup and restoration operations (who, each member's responsibilities, and function)
- postemergency recovery reports (who prepares, what it includes, and to whom it is submitted)

As before, written contracts must be in place both for cleanup assistance and for supply replacement.

One final comment: the plan and all contracts must be reviewed and updated annually. The plan must be tested either by exercises or drills. It can only be corrected and improved via an exercise or in an actual event. Yes, drills and exercises cost money, but they save lives and the environment. No plan is ever perfect; it is a continually evolving document, and therefore, time and finances must be designated for its viability.

References

EPA. *Guide to Exercises in Chemical Emergency Preparedness Programs*; U.S. Environmental Protection Agency: Washington, DC, May 1988.

EPA. *The Day Before . . . Chemical Response Planning through Simulation: A Production Guide*; U.S. Environmental Protection Agency, Region VII: Kansas City, KS.

Guidelines for Technical Planning for On-Site Emergencies; American Institute of Chemical Engineers: New York, NY, 1995.

Hance, B. J.; Chess, C.; Sandman, P. M. *Industry Risk Communication Manual: Improving Dialogue with Communities*; Lewis Publishers: Boca Raton, FL, 1990.

Kelly, R. B. *Industrial Emergency Preparedness*; Van Nostrand Reinhold: New York, 1989.

O'Reilly, J. T. *Emergency Response to Chemical Accidents: Planning and Coordinating Solutions*; McGraw-Hill: New York, 1987.

28

Seismic Safety

STANLEY H. PINE

Glassware rattles, reagent bottles "walk" the shelf and clank together—or even fall, equipment slides across the workbench, the noise is deafening: it is an earthquake. For many laboratory workers in seismically active regions, this is an experience that they would hope to avoid. But alas, the realities of an earthquake are ever present. Over the last few decades, a moderate or major earthquake has occurred in California every two to three years. Luckily, not all have been centered in high-density population areas of this most populous state, yet their impact has been enormous. The 1994 "Northridge" earthquake, measuring 6.6 on the Richter scale but described by most native Californians as the strongest quake they had ever experienced, caused multibillion dollar damage. Its devastation has led to an increasing attention to building design and even to the nature of the earthquake insurance industry. Interestingly, this same area of Los Angeles's San Fernando Valley experienced a similar, though not as costly, earthquake in 1970, centered in Sylmar just north of Northridge.

Although many industrial and academic laboratories experienced minor to major damage, the chemistry facilities at California State University Northridge probably sustained the most severe total losses. Here, a combination of fire, structural failure, and just severe shaking, resulted in damage that closed the university for several weeks and then required about two years of repairs to return to some degree of normality.

This Northridge earthquake was another reminder to engineers and seismologists that earthquake motion is quite variable and thus structural safeguards must be multifaceted. Of the various motions of this earthquake, a strong vertical (up-down) movement cracked heavy steel support anchors on one campus building, while horizontal (back and forth) movement may have been responsible for the collapse of a parking structure. By contrast, the 1987 "Whittier" earthquake, centered just east of downtown Los Angeles, had more of a horizontal motion. The "floating" Physical Sciences building at California State University Los Angeles served its designed purpose by moving with the quake rather than failing structurally. However, the movement, particularly to-

ward the upper part of this eight-story structure, threw unsecured equipment and glassware off of benches and shelves (Fig. 28.1) and caused considerable superficial damage to plaster walls.

Though we may leave it to the architects and structural engineers to design "earthquake-safe" buildings, it

FIGURE 28.1 Equipment and supplies that fell from benches. Note that the chemical containers on the background shelves are intact, thanks to the shelf restraints.

is the laboratory directors and workers who must ensure that quake damage and personal injury are minimized within their facility. Securing essentially everything that can move is usually within the capabilities and resources of the laboratory management. Let's look at what kinds of things have proven effective.

The stability of heavy compressed gas cylinders is of concern in all laboratories, even those not located in seismically active regions. If these cylinders fall for any reason, they can injure people, damage equipment by hitting it, or even drag things down that are connected to the cylinder. In some instances, a fallen cylinder can rupture and become a moving missile. Most laboratories secure their compressed gas cylinders with a bench-mounted clamp containing a cloth strap that wraps around the cylinder. One such support, located about one-third the way down from the top of the cylinder, is typical. But the common commercially available clamps that use a compression screw to hold to a benchtop are not adequate for the sideways (horizontal) motion of a quake. Even those that are tightly secured can slide off of the bench. And this can also happen in nonearthquake situations where the typically finger-tightened clamp screw is not sufficiently secure to hold a heavy cylinder that might be knocked sideways by a worker or possibly be hit by a laboratory cart.

Bench clamps should be bolted to the benchtop. Wall-mounted clamps should be secured to a structural support, not simply toggled into the plasterboard, or into flimsy partition walls. Another concern is the composition of the strap of the cylinder clamp. Most commercial systems have a cloth strap, which is strong enough to hold the cylinder but is flammable. A laboratory fire can burn that kind of strap and leave the heavy cylinder vulnerable to falling. A metal strap or chain is the recommended kind of restraint for all cylinder supports.

A still unresolved question is whether one or two cylinder supports are necessary. Experience has demonstrated that cylinders almost never fall when one properly installed support is used. However there are some who advocate a second support about one-third the way up from the floor. This is believed to keep the lower portion of the cylinder from sliding to the side as the total cylinder moves downward within the upper support. In most cases the priority for installing such a lower second support will not be high on the scale of the many different safety efforts most laboratories must consider.

Another concern about compressed gas cylinder safety involves situations where many cylinders are stored together such as in a storage area or even within a laboratory. Groups of cylinders must be individually secured, not as a group, for example with one chain around that group. Cylinder racks with the ability to individually secure each cylinder of a group are recommended. Because many of the compressed gas cylinders used in laboratories contain flammable materials, there may be addi-

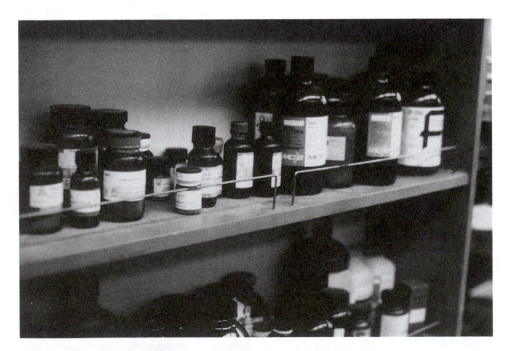

FIGURE 28.2 Wire shelf restraints effectively held chemical containers.

tional restraints imposed by local building codes or fire departments.

Equipment such as balances, centrifuges, spectrometers, computers, and so forth, are commonly located on laboratory benches. The motion of an earthquake can move them along the benchtop and potentially off the edge, resulting in serious damage or even personal injury. Preventing such movement has often been accomplished by bolting the equipment to the bench. This approach is often a required part of the installation of equipment such as centrifuges. But such bolting limits the ready movement of equipment which may be needed at multiple locations in the facility. And even when such movement can be effectively accomplished by unbolting and rebolting, unsightly holes are left in the bench. A common alternative is to "tether" the equipment to a secure location by a metal strap or chain.

Since the mid-1980s a new industry has developed to provide a wide variety of earthquake holddown devices. These include various kinds of portable materials such as Velcro™ straps and puttylike material commonly known as museum wax that is used in households to secure on shelves vases and similar breakable items.

Clear silicone sealant provides a very effective and inexpensive restraining material. A small portion of sealant is placed on the feet of the equipment which is then sealed to the bench. The advantage of this and the other simple holddown devices is that they are easily removed (the silicone sealant remains elastic and is easily cut away) so that mobility is not seriously restricted.

For theft security reasons, laboratory balances and computers are now commonly anchored to their bench location in many academic institutions. This also provides earthquake stability. Interestingly, the silicone sealant and other relatively invisible holddown materials can serve to discourage theft by a less than determined person.

Breakage of containers of chemicals may be one of the most serious potential results of an earthquake and can lead to serious damage, danger, or injury to personnel, contamination of the environment, and expensive cleanup. Shelving should include some type of restraint to prevent containers of chemicals from moving off of the shelf. The restraint should not, however, so inhibit the handling of containers that they are spilled by catching on the restraint during routine laboratory use.

Often a front lip of about 1 inch is sufficient to stop containers in an earthquake. Wires or rods strung across the front of a shelf 1 to 2 inches above its surface are also effective. When shelving is moveable, use a front restraint that moves with the shelf. A long, low rod mounted on the front of the shelf accomplishes all of these requirements and does not complicate cleaning of the shelf surface (Fig. 28.2).

Shelving within closed cabinets should also be equipped with front edge restraints. If movement of the containers takes place when the doors are closed, the containers may move against the doors. Then when the cabinet is opened, those containers can fall out. Since it is common to keep the more hazardous kinds of chemicals in closed cabinets, such an event can lead to very serious spills. It is also advisable to ensure that cabinet door latches are effective to keep the doors closed in an earthquake in order to further minimize opening and potential spills (Fig. 28.3).

Many kinds of cabinet door restraints are available for use in the home, particularly to prevent dishes or glassware from falling in an earthquake. Interestingly, some of the best cabinet door restraints are designed to keep small children from opening cabinets.

FIGURE 28.3 Unsecured cabinet doors opened and allowed containers to slide out and break.

Don't forget the common sense practice of not storing large containers, usually greater than 1 liter, on high shelves. Even if you do not experience the effects of an earthquake, the weight of these containers placed high up will often destabilize the shelving. This practice also minimizes ergonomic problems associated with routine handling of these heavier objects.

Chemical storage in refrigerators has similar hazards to storage in cabinets. Doors, often held closed only by a magnetic latch, can open in an earthquake and spill contents. A further potential for spills is due to unstable refrigerator shelf racks. Since most refrigerators used in laboratories, even those properly modified to minimize explosion hazard, have wire shelves, small containers easily fall over. Spills within a refrigerator can contaminate other materials in the area and be difficult and costly to clean up. Plastic food containers, readily and inexpensively available, provide excellent containment devices. And they encourage organization within the refrigerator.

Larger plastic containers and stainless steel trays can serve the same function on storeroom shelves. As budgets permit such localized containment and organization, this can be a beneficial goal of the chemicals-storage process. When choosing such containers it is important to consider their composition. Be certain that the container will hold a spill and not be dissolved by the chemicals it holds. In this regard, note that flammable plastic containers may not be capable of containing a large flammable solvent spill. If a fire develops, the plastic container may melt and further increase the danger.

Laboratory shelving, cabinets, and even office bookshelves should be anchored securely to the wall. The stability of these typically tall, narrow, furniture units are very vulnerable to the movement caused by an earthquake and the spilling and breakage of the contents of an entire cabinet can have serious consequences.

Laboratory workers must not forget the importance of securing glassware and other apparatus that is in use for a reaction or chemical process. It often seems reasonable just to set a vessel containing reactants on the bench or on a stirring or heating apparatus. But these can easily be knocked over in an earthquake or even through other activity on the bench. Secure such apparatus with appropriate clamps and stands. Labeling of any reactions set up in the laboratory should also be a required procedure.

Fire can be a major cause of damage resulting from an earthquake. Historians will recall that most of the damage to San Francisco in the famous 1906 earthquake resulted from the subsequent fires. In the 1987 Whittier earthquake and the 1994 Northridge earthquake, university laboratory fires were believed to have been started by active metals, sodium or potassium, be-

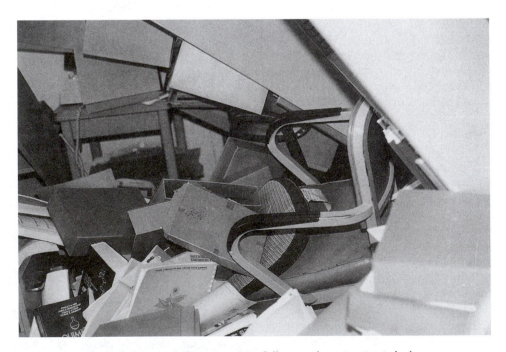

FIGURE 28.4 Bookshelves and their contents fell onto the scientists' desk areas.

ing used to dry hydrocarbon solvents. This common laboratory practice requires particular attention to the stability and security of the containers.

Fire safety cabinets proved very valuable at the Northridge laboratories in 1994. They protected the contents inside the cabinets. Their stability resisted tipping and the possibility of spills into the surrounding area.

An earthquake can lead to building damage that is quite serious within a laboratory facility. Utilities such as water, gas, and electricity may be damaged. Ideally, each laboratory area should be equipped with readily accessible utility shutoffs so as to isolate problems. It is often advisable to shut off utilities to the entire building, then carefully reopen them to one controlled area at a time as damage is assessed. The movement of various instruments and apparatus can easily damage electrical connections or gas piping that will lead to danger when utilities are turned on. Shutting off utilities to assess damage and hazards, however, must be balanced against the need for these services in case additional emergencies develop. For example, a laboratory building without water can lead to further disaster if a fire occurs during this damage-assessment process.

We must not forget that rooms other than laboratories can be affected by the movement of an earthquake. Most scientists will have office space somewhere within their working facility and those office spaces typically contain bookshelves and other kinds of cabinets. Although there may not be a danger of falling chemical containers, personal injury can be sustained when books, storage boxes, or even the shelving fall (Fig. 28.4). Bookshelves and cabinets should be secured to the walls.

The major emergency that may exist after an earthquake can affect a large geographical area. The availability of sufficient numbers of site-knowledgeable response personnel may be limited. This is complicated by the demands for needs of one's employment facility and of home and family. Mutual aid procedures can bring help of persons from long distances who are not familiar with the specific site situations. It is imperative that information be readily available that will enable these helpers to effectively participate in the emergency-response operation. Although appropriate training for sufficient numbers of local people to handle such major emergencies is desirable, it would be advisable, within the pertinent regulatory framework, to include in the emergency operations procedures the provision to make use of any personnel with useful knowledge about the damage or cleanup.

Safety precautions designed for seismic activity can have application to those in nonseismically active areas. However, we must keep in mind that even commonly "quiet regions" have experienced earthquakes. The most severe earthquakes reported in modern times in the contiguous 48 states occurred in the period of 1811–1812 in the New Madrid region of Missouri!

References

Pine, S. H. Earthquake Safety. *J. Chem. Health Saf.* **1994**, *1*, 10.

Pine, S. H. Laboratory Safety and Emergency Preparedness, The Lessons of an Earthquake. *J. Chem. Ed.* **1988**, *65*, A98.

Prudent Practices in The Laboratory—Handling and Disposal of Chemicals; National Academy of Sciences: Washington, DC, 1995; pp. 51, 73, 74, 122.

29

Accident/Incident Investigation

STEPHEN SICHAK

An *accident* can be described as an unplanned event that interrupts the completion of an activity and results in, or suggests the possibility of, personal injury, property damage, production interruption, diminished health, or environmental damage.

"Every accident has to be investigated." It matters not whether it's a major incident leading to property damage and/or lost time injuries or a minor incident—a so called "near miss"—that could have been an accident or a minor injury event; every accident should be investigated! Minor injuries and near-accidents must be reported as well as lost time accidents, because nothing is learned from unreported accidents.

Near accidents are often predictors of future accidents.

For example, for every 331 times a safety rule or procedure is violated, 300 times nothing will happen, 30 times a close call or minor incident will occur, and 1 time an accident/injury will happen. Rarely does a serious accident/injury happen for the first time—generally there have been a number of "close calls" to warn of an impending serious accident. "Near misses" should serve as opportunities to examine and correct problems before a serious accident occurs.

The investigation of accidents is a critical part of the accident-prevention program. It identifies accident causes so that similar accidents can be prevented through mechanical improvements, better supervision, or improved employee training. It determines the causes of human errors. Unsafe conditions and/or human errors, alone or in sequence, cause accidents. Generally, unsafe conditions account for approximately 10% of accidents and human errors account for 90%.

Not all accidents are treated with equal time, effort, and personnel. The amount of resources used should be related to the actual or potential severity of the accident. An investigation of all OSHA lost time ANSI Z16 disabling injuries, or serious property-damage accidents should be conducted (American National Standards Institute Z16.2 Method of Recording Basic Facts Relating to the Nature of Occupational Work Injuries).

The front line supervisor is always the key person in accident investigations:

1. Supervisors are closest to, and most knowledgeable of, jobs, working conditions, and workers.
2. A supervisor's sense of responsibility will be reinforced by accident investigation. After each investigation, supervisors should become more aware of their safety procedures and accident-prevention responsibility.
3. Supervisors learn more about accident causes by investigating accidents.
4. Supervisors are the primary instrument of accident prevention.

Responsibility for accident reporting and investigation should remain as close to the scene as possible; that is, responsibility should rest with the supervisor(s) of an employee injured in an accident or with the location manager at the site of an accident. However, as stated previously, the frontline supervisor is always the key person in accident investigation.

In large-scale investigations, the frontline supervisor is usually one of a team of experts. The team of experts may include a trained accident investigator, an engineer, a maintenance supervisor, an upper-level manager, and a safety professional. A team approach lends credibility to the investigation and is more likely to yield feasible action recommendations. Accident investigations should be used to find real, underlying causes, identify trends, and point out possible corrective actions.

The frontline supervisor should make an immediate report of every injury requiring medical treatment and any other accidents he or she may be directed to investigate. The supervisor is on the scene and knows more about the accident than anyone else and, in most cases, must put into effect whatever measures need to be adopted to prevent similar accidents.

The safety professional, a representative of the safety department, should verify the team's findings by making sure that the four forms of evidence are collected—the so-called "four Ps": People (interviews with injured or affected workers, eyewitnesses, and others who may be familiar with the accident area), Positions (measurements and photographs of the scene), Parts (chemicals in the area, broken pieces from the equipment), and Paper (forms on file such as inspection records, equipment specifications, and documentation of training).

The accident investigator's specialized training and analytical experience enables him/her to search for all the facts, apparent and hidden, and to submit a report free from bias or prejudice. The accident investigator has no interest in the investigation other than to get information that can be used to prevent a similar accident. To accomplish this, the accident investigator must:

1. Find relevant facts.
2. Evaluate the factors for their impact on the accident.
3. Reconstruct the accident sequence.
4. Analyze information.
5. Develop conclusions.
6. Report the findings.
7. Develop a list of remedial needs.
8. Recommend corrective actions.

The purpose of collecting accident data is not to fix blame, but to find accident causes and prevent their recurrence. Investigators must emphasize that their investigation is a fact-finding mission. Otherwise, the workers and supervisors who have the information may conceal it to protect themselves and their fellow workers.

Thoroughness in an investigation and attention to underlying causes are all important. When an accident or incident occurs, the supervisor and other members of the investigating team should be on the scene "immediately" or "as soon as possible." The first responsibility is to secure the area in three ways:

1. The first concern is for the injured. They should have immediate access to first aid and medical facilities.

Unless the injured are well enough to be questioned at the scene, they should not be further upset with questions.

2. Guard against the possibility of a more dangerous secondary event by removing or securing the hazardous energy or substance and/or evacuating other personnel from the area.

3. Restrict access to the accident scene so nothing or no one else will be harmed and so that the scene will not be disturbed before it can be examined.

The actual evidence-gathering process should begin immediately thereafter—not at some later time. Fast response investigations may yield easy remedies that can be implemented overnight, but more important, the fragility of evidence is very real. Witnesses forget much of the detail of an incident within 24 hours because conditions at the time of the incident are only a "snapshot in time."

- Assess the extent of the damage and personal injury.
- Evaluate operating conditions just prior to the accident.
- Make a preliminary evaluation of the area to identify any remaining potential hazards.
- Eliminate or guard against any hazards that are identified.

Be sure to gather enough information at the scene such as taking photographs from all angles. Photographs will prove useful both as a record and a basis for analysis. Each photograph should be numbered and the position of the camera and its direction of vision should be indicated on a map or sketch.

The investigation should be conducted so as to not destroy the evidence. The wreckage contains valuable evidence, which, if correctly identified and assessed, will provide factual evidence necessary for the determination of the cause of the accident.

The active cooperation of all survivors, eyewitnesses, and persons familiar with the material or equipment involved can provide valuable information for the investigation, particularly in reconstructing the sequence of accident events. Employees should be told that the investigation is being conducted to learn the facts so that future accidents can be prevented. They should be reassured that an investigation is not an attempt to pinpoint blame. Fairness and impartiality are absolutely essential. Witnesses are an important source of information, as in the case of a fatal injury they may be the only source of information. Likewise, a witness may be the only source if the accident victim is too badly injured to give a clear account.

An emphasis should be placed on the value of the witnesses' cooperation. Standard introductory questions at the beginning can help get the witness to talk. After gaining the confidence of the witness, the interviewer should ask precisely what happened before, during, and after the accident. It is usually desirable to keep the flow of conversation going, with the interviewer interrupting only for such things as to pinpoint times in a sequence or to introduce a new direction in questioning.

Witnesses may be brought to a central interview room, or may have to be questioned where they can be reached. However, witnesses should be interviewed at the scene of the accident, whenever possible, and as soon as possible after the accident while memories are fresh and before they have been affected by discussions with others. For that reason, witnesses should be interviewed separately. Group interviews, if planned, might be preferable after individual interviews have been completed.

The witness must be reassured that his/her statements are taken in confidence and that ascertaining the basic facts is important in preventing future accidents. If the witness requests the presence of an advisor, the advisor should be informed that the interview is part of an official investigation.

Obtain the witnesses' versions with minimal interruption. Ask open-ended questions whenever possible to find out not only what happened but also why it happened. The interviewer should continue to ask witnesses (and himself or herself) "why" until underlying causes surface. Again, witnesses should be interviewed at the scene of the accident, whenever possible, and should be asked to recall the incident in their own words. Interrupt only if there is a question of clarity. Direct specific questioning to clarify or amplify the witness' account.

Summarize the statements from witnesses and clarify any questionable areas so that the witness' account and the investigator's understanding of it are in agreement.

Once the on-scene interviews have been completed, the incident can be documented in a report or series of reports. Contents usually include a narrative description of what happened, the facts about the immediate causes (unsafe acts, unsafe conditions, or human errors), and the measures that must be taken to deal with the root causes to prevent a recurrence.

Accident reports should be written for the reader. The general rule of thumb for any report is to include as many facts as possible in a clear, readable format. The report should summarize the findings, then go into detail and, also, should be supplemented with maps, schematic diagrams, and photographs. Remember, accident investigation and reporting are part of your safety program. You must be able to use those reports.

The final facts to be established are the identity of those directly involved and the time the accident occurred (date and exact time). These facts permit identification of possible witnesses, the production period, weather conditions, supervisors in charge, and other factors that may be important. If more than one person is injured in the same accident, a report must be completed for each one involved.

The report should identify the location or scene of the accident because it can be used to identify high-risk areas. Usually, a concise descriptive phrase is all that is needed.

The injured person's occupation, position, or job title should be included in the report so that risk hazard by job or occupation can eventually be evaluated. (Include physical capability; did he/she have sufficient experience to perform the work? Was the job full-time, part-time, or seasonal?)

Describe the accident fully. What tasks were being performed when the accident occurred? A narrative description should be prepared: it should include what the person was doing; what objects or substances were involved; and actions or movements that led to the injury. This is elaborated into a detailed accident sequence that starts within the injuring event and works backwards in time through all the preceding events that directly contributed to the accident. The data also include a description of any product or equipment that was directly involved with the accident sequence and the task being performed. The task involved can be used to identify specific hazards of jobs or equipment and job procedures, as well as workers' capability and need for additional training. Any other conditions such as temperature, light, noise, and weather that pertains to the accident should also be noted.

What other elements may have contributed to the accident or incident?

What personal protection equipment was being used and did the employee's apparel affect the accident sequence? What kind of training did the injured person have for the task he or she was performing? Did standards or procedures exist for the task? Were they written? Were they followed? Were all guards in place and in use? What was the nature of supervision at the time of the accident? What immediate remedial actions were taken to prevent recurrence?

The nature of the injury or injuries and the portion of the body affected must be recorded as well as the OSHA severity class. If the accident resulted in some permanent impairment it, should be noted.

How can future accidents of this type be prevented? An accident investigation is not complete until corrective action has been determined. Action to prevent recurrences should be identified on the basis of direct and indirect causes of the accident under study. Recommendations for remedial action should be generally described with an explanation as to their purposes.

30

Chemical First Aid

JOHN R. McINERNEY

Industry, agriculture, and military defense produce over 25,000 chemicals that are capable of producing harmful effects in humans. Chemical exposure can occur by four routes:

- direct contact to skin or eyes
- inhalation
- ingestions
- injection, or puncture wounds

In some instances, the exposure could be a combination of routes. For example, when a person swallows a chemical, he could also inhale some as well as have some come in contact with the face or lips.

Overview

First aid can be thought of as four processes:

- anticipate
- evacuate
- ventilate/irrigate
- evaluate

In most cases of chemical injury, these processes will not occur in a stepwise fashion. Instead, the processes will run concurrently.

Anticipate

Prevention is extremely important! It is much easier to minimize chemical injury than it is to treat it. Identify all the chemicals in the workplace or home. Obtain all the information on the chemicals that you can, using the Material Safety Data Sheet (MSDS) or the label. Keep the chemicals in the original containers, clearly identified. If you don't know what is in a container, do not open the container. *If you don't know it, don't touch it!*

Make sure the appropriate administrative and engineering controls are in place prior to the use of any chemical. Make sure to use the correct Personal Protective Equipment (PPE) when using the chemical.

Identify the closest exit. Identify the closest source of water.

Do you have all of the MSDSs? On the package label and MSDSs, there will be a general instruction: consult a physician. The best source of information for chemical exposures may be a 24-hour emergency room or the regional Poison Control Center. Have the names and number of these resources attached to your phone. Think through how you would access medical care. Is there a physician on-site? Should you drive yourself to the doctor? In most cases, it will be best to activate the EMS (Emergency Medical Services) system. What number do you call for an ambulance? Although 911 is quite common, in some areas of the country a regular phone number is used. What do you do if you are away from home?

Safety walkthroughs are helpful. Be quizzed on what the chemicals are in your area. What steps would you take if you splashed a particular chemical on your arms?, in your face? Where is the closest safety shower, eyewash fountain, or other source of water? Where is the closest exit? Who do you notify? How do you access the local EMS?

Are there special types of chemicals that you work with that will affect how treatment of the injury is handled? Examples of special situations are:

- cyanide
- hydrofluoric acid
- organophosphates
- mercury
- sodium and related metals
- phenols
- white phosphorus
- radioactive materials

If there are chemicals that require specific treatment, have a plan of treatment in effect before exposure.

Evacuate

The first aid treatment of chemical exposure should begin *immediately* by the victim himself. He should remove himself from the source. All clothing that has been exposed to the chemical and all solid material should be removed by the victim himself. If risk to others exists, the general area should be evacuated.

Rescuers beware! Do not expose yourself to harm in an attempt to be a hero! Do not expose yourself to the chemical. Then there are two victims. Needless deaths have occurred when the first worker is overcome by gas in a closed, confined space. The second worker also perished when he went in to rescue the first. Medical personnel have sustained significant organophosphate exposure when they undressed a patient who had been drenched by insecticide. They were not using gloves. They became ill.

Ventilate

If there has been a chemical inhalation exposure, get the victim to fresh air. Ideally, have him walk himself. If the victim is unresponsive, check for breathing and a pulse. If either is absent, activate the EMS system. A rescuer trained in CPR should take over and initiate CPR.

If there was a chemical ingestion, *do not give the victim anything to eat or drink. Do not try to make the victim vomit.* Call a knowledgeable physician, the Emergency Department, or the Poison Control Center before you do anything.

Irrigate with Lots of Water

As soon as possible after chemical exposure to the skin or eyes, all areas that have been exposed should be flooded with large amounts of water. Again, all clothing that has been exposed to the chemical should be removed, ideally by the victim himself. Water must be used immediately to be successful. Delay of more than 3 minutes has been shown to cause significantly more damage than if water was used within 1 minute.

Regardless of the chemical or degree of damage, the treatment goal is the same: minimize the area of permanent damage and maximize the salvage of borderline damage.

Ideally the water should be low pressure to reduce the risk of splash injury. Ideally the water should be body temperature and protected with an antiscald limiting control. Think about special situations. If there is chemical contamination only on the top of the head, it would be better to rinse the hair over a sink than to step into a shower and rinse the chemical down over the entire body. If goggles or a respirator are being worn, keep them on until all the chemical has been flushed off the face and the top of the head. If a safety shower or eyewash fountain are not readily available, it may be necessary to improvise: buckets of water or a garden hose. Continue the flushing with water for 15 minutes.

Eyes

Chemical burns to the eyes are common and considered as a true medical emergency. Immediate water irrigation should begin and be continued during transport to the closest emergency facility. When using an eyewash fountain, use the thumb and index finger to hold the eyelids open if necessary. Roll the eyes continually up and down and from side to side. This will remove as much of the chemical as possible from the surface of the eyeball. If you begin using the eyewash, use it for a minimum of 15 minutes.

Special Situations

Dry chemical particles such as lime should be brushed away before irrigation. Lime can react with water, releasing considerable heat that could cause more damage.

Some elemental metals can either ignite spontaneously or react with water to give off tremendous heat that can cause more damage (sodium, lithium, potassium, phosphorus, and calcium). These types of metals should be covered with mineral oil before flushing with water.

Burning metal embedded in the skin can be extinguished with a class D fire extinguisher or smothered with sand.

Water irrigation should be continued for a minimum of 15 minutes.

Evaluation

While first aid is occurring, the chemical should be identified and medical consultation obtained. Identify the substance and its route of entry into the body. Many substances may require no treatment other than first aid.

Sometimes it can be very difficult to identify the substance or its toxicity. The container may be unlabeled. There may have been a number of chemicals around. The MSDS may be unavailable or inaccurate. The chemical may smell different to different individuals. Remember to treat the patient, not the label or MSDS.

In first aid, the most important piece of equipment is a calm human brain. Emergency situations generate a tremendous amount of anxiety. Anxiety is highly contagious. Anxiety feeds on itself, easily leading to a criticality reaction, or panic. It is not easy to remain calm when you see your co-workers running to the exits, screaming. It is quite intimidating to be in your street clothes talking to a firefighter in his fully enclosed hazmat moon suit with his speech interrupted with the hiss of supplied air.

Summary of First Aid Treatment for a Chemical Exposure

Direct Contact to Skin or Eyes (Chemical Burns)

Chemical exposure can occur with direct contact to the skin or eyes. This is the most common kind of chemical exposure. Chemicals can produce burns, skin rashes, allergic reactions, thermal injury, or generalized poisoning (systemic toxicity). There are over 60,000 emergency room visits per year for chemical burns.

Burns produced by most chemicals are similar. Different chemicals damage the skin at different rates and with varying degrees of pain. In some cases, the initial damage done may be deceptively mild due to lack of pain (alkali), only to be followed by extensive skin damage and even generalized poisoning. Chemical burns can cause redness, blisters, or discoloration of the skin. Eyes can be red, watery, or turn milky white. With chemical burns, flush with water for 15 minutes, and consult with a knowledgeable physician, Emergency Department, or Poison Control Center.

Inhalation

Chemical inhalation occurs when a chemical is inhaled into the lungs. Inhalation may cause such symptoms as dizziness, nausea, slurred speech, stumbling gait, unconsciousness, cough, sore throat, trouble breathing, or change in mental status. Although some chemical inhalations produce symptoms shortly after exposure, many chemicals produced delayed symptoms. With chemical inhalation, get the victim to fresh air and consult with a knowledgeable physician, emergency department, or poison control center.

Ingestion

Ingestion of a chemical can occur intentionally, such as a suicide attempt. A child can ingest a chemical by mistake. With those exceptions, the ingestion of a significant quantity of a hazardous chemical is rare. With a chemical ingestion, do not let the victim eat or drink anything. Do not induce vomiting. Do not give syrup of ipecac. Consult with a knowledgeable physician, Emergency Department, or Poison Control Center immediately.

Injection

Injection exposures occur when a chemical is injected through the skin with a sharp object such as a needle or a shard of glass or metal. If needed, apply direct pressure to control bleeding. Consult with a knowledgeable physician, Emergency Department, or Poison Control Center.

If there is only one phone number you want to have for chemical exposure, let it be the number of the Poison Control Center.

References

Emergency Medicine, A Comprehensive Study Guide; 1996, pp. 899–905.

Emergency Medicine: Concepts and Clinical Practice; 1992, pp. 965–968.

The Merck Manual of Diagnosis and Therapy; 1992, pp. 2681–2682.

Wilderness Medicine, Auerbach, 1995; pp. 246, 250.

SECTION IV

LABORATORY EQUIPMENT

DOUGLAS B. WALTERS*

SECTION EDITOR

*This work (Section IV) was done by Douglas B. Walters acting in his private capacity. No official support or endorsement by the National Institute of Environmental Health Sciences is implied or should be inferred.

31

Specialized Instrumentation and Monitors

DAN AGNE

SAL AGNELLO

Safe practices are exercised best in regulated environments such as laboratories, clean rooms, and isolation rooms. These functional environments contain entire control systems dedicated to the detection and removal of the specific hostile agents—by design.

The bigger challenge is the indoor environment that is controlled primarily by temperature. Contaminant measurement has not been the usual practice unless specific contaminant risks have been painstakingly identified. Virtually any residence or common office building falls into this category.

There are many factors contributing to a safe, quality building environment that make indoor air quality (IAQ) a multifaceted issue. A common *misconception* presumes that using contaminant-specific monitoring will ensure occupant safety. However, the *single biggest step* in gaining proper and safe control of the overall IAQ is to pay particular attention to a very specialized, yet common, building system: the heating, ventilation, and air conditioning (HVAC) system. Once this primary consideration has been properly managed, specific contaminant issues are fewer, easier to isolate, and less complicated to correct.

Poor Indoor Air Quality: A Developing Area of Risk

Overview

The fact that a building's internal environment can impact the *health, comfort, and productivity of occupants* has been an increasing concern of the public in general, government, health experts, and our customers. After all, these are the reasons *why buildings are built in the first place.*

Results of poor environmental quality have ranged from occupant dissatisfaction (odors, thermal discomfort, irritation) to life-threatening illness (*Legionella*, toxic vapors). Expert estimates suggest that at any given time, 20–30% of commercial buildings have indoor environmental problems. In addition, 10–20% of buildings without known problems are estimated to be vulnerable for such conditions.

The most publicized problems have been reported in government and commercial office buildings. Major problems have also turned up in schools and hospitals. Poor IAQ has become a risk in modern buildings, whose insulation and sealed windows have made occupants dependent on proper design and performance of HVAC and other building systems. The general public's awareness of this problem has also been increasing. Cases of Building-Related Illness (BRI) among employees have received increasing attention. Medical and scientific experts have classified problems associated with poor indoor environments into 3 major categories:

Building-Related Illness (BRI) is the most serious health and safety problem related to IAQ. This refers to diagnosed diseases or disabling conditions that result from harmful agents in the building and persist after leaving the building. These could require immediate medical attention. Examples include:

- Legionnaire's Disease—*Legionella* bacteria–based pneumonia (water-borne bacteria)
- Pontiac Fever—Legionnella-based flulike disease
- Hypersensitivity Pneumonitis—mold-induced lung inflammation (HVAC component contamination)
- Carbon Monoxide Poisoning—combustion-exhaust-based nervous system illness

Sick Building Syndrome (SBS) refers to various symptoms experienced while an occupant is in the building, but that usually disappear when the individual leaves. SBS is suspected when a "substantial portion" of occupants experience similar on-site discomfort symptoms such as those listed below—but some conditions could in fact be BRI that has not yet been diagnosed. Such symptoms include:

- eye, nasal, or skin irritation
- nausea
- dizziness
- headache
- sleepiness/fatigue
- thought/emotional difficulties
- nonspecific complaints—general discomfort

Multiple Chemical Sensitivity (MCS) is a condition in which chemically sensitive individuals experience debilitating physical, and possibly mental, symptoms in various low-level chemical environments that do not affect others. It is medically controversial since there are many symptoms and differences among affected individuals (estimated at 15% of the U.S. population), and it is not associated with specific diagnoses. Recent studies have shown uncertain cause and effect between physical and emotional symptoms (i.e., the same part of the human brain affects both taste/smell and cognitive/affective functions).

Generally, a combination of internal pollutants, "airtight" buildings, and HVAC deficiencies have been shown to cause such conditions. These problems can occur when a building's HVAC system actually generates contaminants and fails to eliminate pollutants associated with interior materials, equipment, or occupant activity.

Examples of these types of pollutants are highlighted in Table 31.1.

Even though other environmental conditions (lighting, noise, layout, psychosocial factors, etc.) can cause discomfort and stress, the overwhelming majority of cases result from indoor air problems, and HVAC deficiencies have been the major factor. Table 31.2 lists some of the most common problems, causes, and risks.

Financial Consequences of IAQ Problems

Litigation Our legal system establishes obligations on building owners, facility managers, contractors, and HVAC and energy management consultants to provide safe and healthy environments. Indoor air quality problems have recently led to a number of now well-known cases that have resulted in significant dollar awards to plaintiffs. Individuals, workers, spouses, and visitors of buildings have initiated litigation. Indoor air quality awards to plaintiffs have ranged from over $500,000 to tens of millions. In addition, legal fees are usually incurred for the defense, even if the owner wins the suit.

Remedial Repair and Retrofit Table 31.3 shows data based on a 1993 EPA survey of IAQ diagnostic and remediation firms. The summary shows the *average* costs related

TABLE 31.1 Health Impact from Various Building Materials-Related Sources

Sample Sources	Stressor	Health Impact Examples
Mold, mildew, humidifiers with stagnant water, water-damaged surfaces, HVAC system coils and drip pans	Bacteria, molds, other microbes	Allergic reactions Respiratory problems Odors Sinus problems Legionnaire's Disease Pontiac Fever Acute Humidifier Fever Hypersensitivity Pneumonitis
Particleboard, plywood, furniture, cleaners, glues/resins	Formaldehyde	Eye and skin irritation Suspected carcinogen
Solvents, paints, building materials, waxes/polishers, binders, combustion products	Volatile organic compounds (VOCs)	Odor Headache Eye and respiratory irritation
Dust, frayed fiberglass insulation	Particulates	Eye and skin irritation Respiratory problems
Poor heating, cooling, and humidity control	Thermal and humidity	Discomfort Fatigue

TABLE 31.2 Health Risks from Heating, Ventilating, and Air Conditioning Problems

HVAC Factor	Typical Causes	Health Risks
Insufficient supply of outdoor air	Broken damper linkages Fan casing corrosion Broken or maladjusted dampers, diffusers Thermal loads too high for current system capacity Pressure imbalances	CO_2 accumulation Humidity Chemical vapor accumulation Spread of airborne infections Microbial accumulation
Insufficient distribution of outdoor air	Malfunctioning or blocked diffusers and registers Improper or nonexistent air testing and balancing Improper pressurization	Same as for insufficient supply Need to get the ventilation to individual occupants
HVAC bringing in outdoor contaminants	Location of outdoor air intakes Polluted or untreated outdoor air (e.g., no filtration) Cooling-tower-contaminated mist	Carbon monoxide Odors Microbial Excess moisture
HVAC as source of pollutants	Generally poor design, installation, maintenance Filthy drain pans, fan coils, ductwork interiors Debris in filters and filter media Corroded equipment parts Dirty mechanical rooms	Microorganisms Particulates Odors Thermal discomfort (e.g., clogged coils)
HVAC distributing pollutants	Unfiltered air or filter bypass Recirculation of air from pollutant source areas Re-entrainment of exhaust air into outdoor air intake	Particulates/debris Chemical vapors Odors ETS
Failure of HVAC to control other pollutants	Lack of local exhaust Improper air testing and balancing "Thermal bridges" (cold areas combined with high humidity causing moisture) Lack of make up air Improper boiler combustion	Fumes and vapors from machinery and equipment Mold and mildew from condensation Specialized use areas odors and vapors Combustion gases (e.g., CO)

TABLE 31.3 Sample Costs of Correcting Air Quality Problems Relating to HVAC Problems

	Median High Cost[a]	Mean % Cases Performed[b]	Average Cost per Building
Ventilation system modification	$10,000	35	$3,500
Air balancing	$3,000	32	$960
Air cleaning/filtration equipment	$3,000	36	$1,080
Ventilation system decontamination	$5,000	32	$1,600
Building maintenance	$5,000	38	$1,900
Construction/renovation	$20,000	30	$6,000
Source removal/modification	$7,000	40	$2,800
Duct cleaning	$8,000	48	$3,840
Radon mitigation	$3,000	32	$960
Total			$22,640

[a]U.S. EPA. *Survey of Indoor Air Quality Diagnostic and Mitigation Firms* (1993).
[b]Reflects actual mitigation performed by respondent firms; excludes asbestos removal.

to correcting IAQ problems once they occur in commercial and public buildings. Individual cases show that in some cases it can cost in excess of $30.00 per square foot of building to correct the IAQ problems.

Regulatory Risks include OSHA's proposed IAQ rule, building codes' incorporation of ASHRAE 62–1989 ventilation rates, and various state regulations. OSHA visits in response to employee complaints can result in fines ($7,000/day per employee) if contaminants exceed current OSHA contaminant safety limits or reveal other unsafe conditions. Impending IAQ regulations may call for additional penalties for a single violation in addition to state regulations and codes.

Workers' Compensation claims have arisen from illness and injury attributable to indoor air pollution in the workplace. Nationally, occupational *illness* claims have grown by 12% and represent every 10th reported injury.[1]

Illness and Absenteeism About half of sickness-based absenteeism, or 4.75 days/person per year, is attributable to upper respiratory tract infections (URI) such as colds, flu, and other conditions that can be spread by air recirculation.[2] Research has shown that proper ventilation can decrease the incidence of such conditions by 50%.[3] Based on this figure, a building with appropriate ventilation and microbial aerosol dilution/filtration could cut absenteeism down by about 2.4 days per employee annually. Less absenteeism contributes to greater business efficiency and larger, incremental costs for substitute labor.

Health of occupants has become an increasingly important concern among facilities' managers. A recent survey by IFMA indicated that 57% of facilities managers are monitoring for health problems related to indoor air pollution, with 70% indicating health concerns as the primary reason for their assessing ventilation sufficiency.[4]

Productivity and Performance The most significant cost factor in a commercial building is people. The significant cost in an owner-occupied office building is the salaries of the people working in it. Employee productivity studies have shown associations between indoor environments and human performance for years. Specific financial benefits will depend on types of jobs, salary levels, and other factors. One example that has been internationally cited is the Rensselear Polytechnic Institute study of the Johnson Controls Personal Environments Module (PEM) installation at West Bend Mutual Insurance Company. Using its own definitions of productivity, the customer attributed at least a 2.8% increase in its worker productivity attributable to the improved environment.[5] With a payroll at the time of $10 million, the PEMs added $280,000 to the customer's bottom line annually.

Occupant Complaints and Investigation One Canadian study estimated an annual cost of $0.25–0.50/square foot related to work time lost in handling or processing occupant complaints.[6]

Other costs can be significant to an organization even though they cannot readily be associated with dollar figures. Managers seek to avoid many of the above risks

TABLE 31.4 Typical Air Handling Equipment and Accessories Servicing Tasks to Reduce Safety Risk

Service Task	Objective	Risk
Routine replacement or cleaning of filters	Prevent or correct: Filter sag due to overload with dirt Unfiltered air bypass Filter blow out	Dirty ducts and air Microbial growth Insufficient overall air flow Thermal discomfort due to clogged coils
Inspection and repair of filter tracks for gaps	Prevent bypass of dirty air	
Control system monitoring of air pressure across filters	Promote timely filter servicing	
Clean clogged coils	Unrestricted airflow and heat transfer	Insufficient ventilation Thermal discomfort
Inspection and replacement of duct lining in air handlers	Replace moist or contaminated duct lining	Microbial contamination and BRI
Component cleaning such as drain pans	Prevent/clean mold	
Installation of noise attenuators or electronic noise control	Prevent need for acoustical duct lining which can become contaminated with microbes when moist or dirty	Excessive mechanical noise Microbial contamination and BRI

TABLE 31.5 Typical Humidification and Dehumidification Equipment Tasks to Reduce Safety Risk

Service Task	Objective	Risk
Check and maintain appropriate humidity levels during heating and cooling seasons	Provide appropriate humidity for season and occupancy loads for heating and cooling, thus preventing condensation	Thermal discomfort Microbial growth Static electricity
Retrofit system to provide appropriate humidification and dehumidification capacity as required		
Retrofit system for providing clean steam rather than treated boiler water	Prevent air distribution of water contaminants	Toxic aerosols in air Odors

because they are undesirable regardless of their dollar cost. Poor publicity, charges of fraud, negligence, and so forth, are risks to be avoided by both facilities management and general management of reputable organizations.

Actions to Move Toward Safer Indoor Environments

Preventative Strategies

Effective preventive measures reduce the risk of poor indoor air quality. These general objectives include:

- proper ventilation in accordance with ASHRAE standards
- clean and properly functioning HVAC equipment and systems
- proper temperature and humidity control

- integrating related expertise (e.g., architecture, industrial hygiene) as needed for non-HVAC building problems

Regular inspection and maintenance of HVAC equipment help keep the IAQ factors in check. The inspections should verify proper operation, replace worn components, and clean air-path exposed surfaces. Some of the IAQ-critical mechanical system components to have inspected regularly in a preventive maintenance program are:

- outdoor air intakes
- mixed air plenum and outdoor air controls
- air handlers and related equipment
- humidification and dehumidification equipment
- supply fans
- ducts
- air terminal devices
- return air systems (nonducted plenums)

TABLE 31.6 Typical Building Controls and Instrumentation Tasks to Reduce Safety Risk

Service Task	Objective	Risk
Regular maintenance and calibration	Proper temperature, humidity, building pressurization, and system modulation control	Thermal discomfort
Proper control location and zoning	Proper heating and cooling distribution	Unintended pressurization and air flow: Positive pressure—drafts and contaminants from pollutant areas Negative pressure—infiltration of outdoor contaminants
Institute special control applications: Ventilation Humidity Filter monitoring	Proper levels of: Humidification Outdoor air ventilation Filtration functionality	Contaminant accumulation resulting in: Odors Sick building Syndrome discomfort Negative pressurization-related infiltration of outside contaminants Particulate and microbial contamination Airflow reduction

TABLE 31.7 Typical Boiler Servicing Tasks to Reduce Safety Risk

Service Task	Objective	Risk
Adjust for proper combustion	Minimize carbon monoxide production	CO poisoning
Provide sufficient outdoor air for combustion	Proper pressurization of boiler room and functioning of flue gas and exhaust systems	Backflow of flue gasses and exhaust gases
Adjust temperature		CO poisoning
Maintain gaskets and breaching	Prevent leakage of combustion products	
Maintain flue lines	Prevent fuel line leaks	Contaminant fumes and odors Fire
Retrofit or relocate exhaust	Prevent reentrance of exhaust gases into the building	Contaminant fumes and odors
Retrofit or relocate outdoor air intake		CO poisoning

- exhaust/relief fans and air outlets
- self-contained heating and cooling units
- environmental control systems (thermostats, controllers, etc.)
- boilers
- cooling towers
- chillers

Inspections and Maintenance

Inspections and maintenance tasks need to be customized for the specific building equipment manufacturer, application, and environment. Listed below are some sample task listings of major building systems that are tuned to specifically prevent adverse IAQ factors from occurring. The impact of ignoring these service tasks can be deadly.

- Air Handling Equipment and Accessories Servicing (Table 31.4)
- Humidification and Dehumidification Equipment (Table 31.5)
- Building Controls and Instrumentation (Table 31.6)
- Boilers (Table 31.7)

Conclusion

It can be hard to think of the common building's "air condition equipment" as a specialized environmental instrumentation and monitoring system, but it is. The quality of air that it delivers determines the health of the occupants, efficiency of the business, and liability of the business stakeholders.

More Information

Published standards for design, operation, and maintenance are available from the American Society of Heating, Refrigeration, and Air-conditioning Engineers (http://www.ashrae.com):

- 62-1989, Ventilation for Acceptable Indoor Air Quality
- 55-1992, Thermal Environmental Conditions for Human Occupancy
- 52-1992, Method of Testing Air Cleaning Devices Used in General Ventilation for Removing Particulate Matter
- Guideline 1-1989, Guideline for Commissioning of HVAC Systems
- EPA/NIOSH Guide, "Building Air Quality"

References

1. International Federation Of Employee Benefit Plans. *Legislative Trends In Workplace Stress Claims*, 1996.
2. Holcomb, L. F.; Pedelty, J. F. *Comparison of Employee Upper Respiratory Absenteeism Costs with Costs Associated with Improved Ventilation*. Report from Holcomb Environmental Services: Olivet, MA, 1992.
3. Brundage, J. F. et al.; Building-Associated Risk of Febrile Acute Respiratory Diseases in Army Trainees. *JAMA*, 1988.
4. IFMA. *Environmental Issues in the Workplace II. 1994*. Quoted in *Air Conditioning, Heating and Refrigeration News*, April 24, 1995.
5. Kroner, W. et al. *Using Advanced Office Technology to Increase Productivity*. Center for Architectural Research, Rensselear Polytechnic Institute: Troy, New York, 1992.
6. Levin, H. Comments On OSHA Federal Register-Indoor Air Quality Proposal, August 12, 1994.

32

Inert Atmospheres Work

GERALD W. BOICOURT

Inerting

Combustion is a process of oxidation that occurs at a rate fast enough to produce heat, and, usually, light, in the form of either a glow or flames. The combustion process also develops pressure if confined, or turbulence is generated in unconfined vapor/gas deflagrations. For combustion to occur the three elements of fuel, oxidant, and an ignition source must be present simultaneously. The fuel can consist of suspended solids (dusts), gas, flammable vapor, or any combination thereof. The oxidant is usually oxygen present in air. Some fuels can combust even when molecular oxygen, such as found in air, is not present (i.e., ethylene oxide). Ignition can come from a number of different sources such as open flames, static electricity, hot surfaces, and so on. Removal of any one of the three elements of this "combustion triangle" (see Fig. 32.1) prevents the combustion process from occurring.

Inerting is the process of adding an inert gas to a combustible mixture to reduce the concentration of oxygen below the point (called the limiting oxidant concentration, or LOC) where a flame can propagate beyond the point of ignition. In effect, this is the removal of the oxidant in the "combustion triangle." Typical inert gases used are nitrogen, helium, carbon dioxide, flue gas, and water vapor (steam). The LOC is a function of the type of inert gas used as illustrated in Tables 32.1 and 32.2 and is typically 10–12 vol% for vapors, gases, and dusts.

Where the concentration of oxygen is continuously monitored, NFPA 69[1] recommends a maximum oxygen concentration 2 vol% below the LOC unless the LOC is less than 5 vol%. In that situation the equipment should be operated at no more than 60 vol% of the LOC. Where the oxygen is not continuously monitored, the oxygen concentration should be no more than 60 vol% of the LOC, or 40 vol% of the LOC if the LOC is below 5 vol%.

The LOC is illustrated by means of a flammability dia-

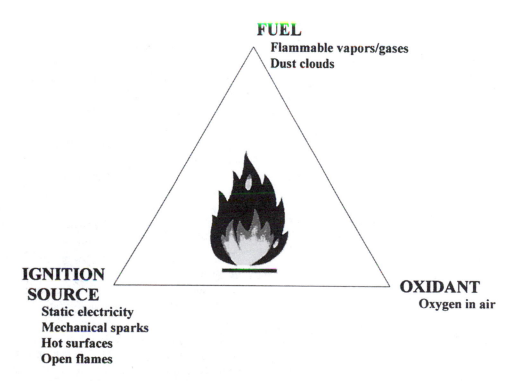

FIGURE 32.1 Combustion triangle.

TABLE 32.1 Flammability Characteristics of Liquids and Gases[1,17-21]

Gas or Vapor	Flashpoint (°F)	LFL (% in air)	UFL (% in air)	LOC (N_2/air)	LOC (CO_2/air)
Acetone	0	2.5	13	11.5	14
Ethane	−211	3	12.5	11	13.5
Propane	gas	2.1	9.5	11.5	14.5
Butane	−76	1.6	8.4	12	14.5
Pentane	−40	1.51	7.8	12	14.5
Hexane	−15	1.1	7.5	12	14.5
Heptane	24.8	1.1	6.7	11.5	14.5
Ethylene	gas	2.7	36.0	10.0	11.5
Propylene	−162	2.0	11.1	11.5	14
Benzene	12.0	1.3	7.9	11.4	14
Toluene	40	1.2	7.1	9.5	—
Styrene	87	1.1	7.0	9.0	—
Carbon monoxide	gas	12.5	74	5.5	5.5
Ethanol	55	3.3	19	10.5	13
Hydrogen	gas	4.0	75	5.0	5.2
Methanol	54	6	36	10	12
Methyl acetate	15	3.1	16	11	13.5
Propylene oxide	−35	2.3	36	7.8	—
Methyl ethyl ketone	24	1.4	11.4	11	13.5

gram shown in Figure 32.2, which is that for mixtures of methane, oxygen, and nitrogen at atmospheric pressure.[5,20,25] Any mixture of composition within the envelope bounded by points ABCDE is flammable. The locus of concentrations represented by line ABC is defined as the lower flammable limit (LFL). The locus of points rep-

TABLE 32.2 Flammability Characteristics of Combustible Dusts[1]

Dust	LOC (N_2/air)	LOC (CO_2/air)
Cornstarch	15.8	17
Sucrose	11.9	14
Phenothiazine	15.8	17
Charcoal	15.8	17
Bituminous coal	15.8	17
Aluminum	0	2
Iron	6.7	10
Silicon	9.3	12
Zinc	6.7	10
Cellulose	10.6	13
Sulfur	9.3	12
Isopthalic acid	11.9	14
Pthalic anhydride	11.9	14
Hard rubber	13.2	15
Sodium resinate plastic	11.9	14
Acrylonitrile plastic	10.6	13
Methyl methacrylate plastic	8	11
Polyethylene plastic	9.3	12
Polystyrene plastic	11.9	14

resented by line CDE is defined as the upper flammable limit (UFL). Air is represented by point F consisting of 79 vol% N_2 and 21 vol% O_2. If the fuel methane is added to air, the composition of the resultant mixture must lie on the line represented by FBDM. As the first amount of methane is added, the mixture is below the LFL and is too fuel-lean to support combustion. As more methane is added, the LFL is reached and combustion is possible for those mixture compositions represented by line BD. With the addition of still more methane, the UFL is exceeded and the mixture becomes too fuel-rich to support combustion. The LOC is the minimum concentration of oxygen in the inert gas and fuel that allows combustion to occur, regardless of the fuel concentration. This is represented by point C on the flammability diagram in Figure 32.2. In this case, if the vol% of O_2 is kept below 12%, combustion is not possible. If methane is added to an atmosphere containing 13 vol% O_2 and 87 vol% N_2, the locus of resultant mixtures must lie to the right of line HCM and combustion is not possible.

Flammability information can be represented in a variety of other ways, such as that illustrated in Figure 32.3 where the flammability of ethane in nitrogen-diluted air mixtures is presented. The LFL is represented by line ABC. The UFL is represented by line CD. The envelope of flammability includes all mixtures within the region represented by ABCD. The LOC at point C is found by difference as being approximately 52 vol% air, or approximately 11 vol% O_2.

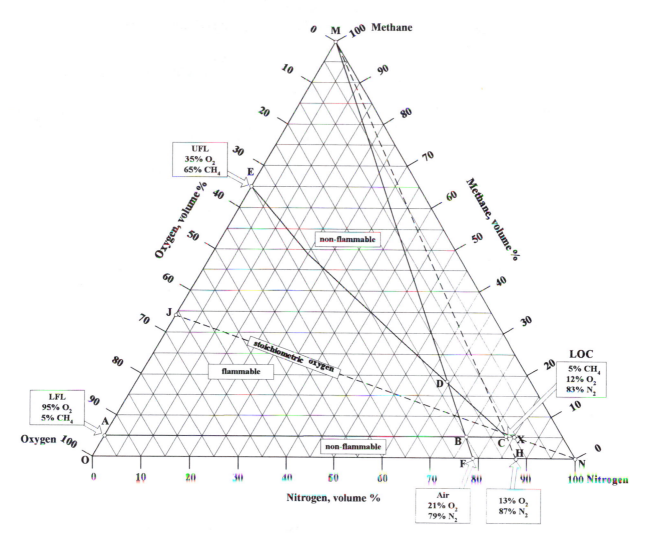

FIGURE 32.2 Flammability of oxygen, nitrogen, and methane mixtures.[25]

Flammability limits (UFL and LFL) and LOCs can be measured experimentally or estimated. Both are discussed.

Estimating Flammability Limits of Vapors and Gases in Air

Lacking experimental data the UFL and LFL can be estimated[6] from the stoichiometric oxygen required in the combustion process. In calculating the stoichiometric oxygen the following assumptions are made:

1. Nitrogen and sulphur in the fuel mixture react to form NO_2 and SO_2.
2. Chlorine reacts to form HCl.
3. Excess chlorine forms Cl_2.
4. Hydrogen reacts to form water.

Writing a generalized combustion correlation as

$$C_cH_hO_oCl_dN_nS_s + \left[c + n + s + \frac{(h-a)}{4} - \frac{o}{2}\right]O_2$$

$$\rightarrow cCO_2 + aHCl + nNO_2$$

$$+ sSO_2 + \frac{(d-a)}{2}Cl_2 + \frac{(h-a)}{2}H_2O$$

The stoichiometric number of oxygen moles required per mole of combustible gas is

$$O_s = c + n + s + \frac{(h-a)}{4} - \frac{o}{2}$$

Defining C_{st} as the volume % of fuel in the mixture of fuel plus air, C_{st} is:

$$C_{st} = \frac{\text{Moles Fuel}}{\text{Moles Fuel} + \text{Moles Air}} \times 100$$

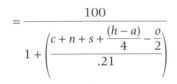

$$= \cfrac{100}{1 + \left(\cfrac{c + n + s + \cfrac{(h-a)}{4} - \cfrac{o}{2}}{.21} \right)}$$

For many aliphatic hydrocarbons in air the LFL and UFL expressed in vol% are proportional to the stoichiometric concentration C_{st} of fuel and can be estimated[20] as:

$$LFL = 0.55\, C_{st}$$
$$UFL = 4.8\, \sqrt{C_{st}}$$

Estimating the LOC for Vapors and Gases

If experimental data are not available, the LOC can be estimated[4] using the stoichiometry of the combustion reaction and the LFL as

$$LOC = LFL\, O_s$$

where the LFL is expressed as vol% fuel and O_s is the stoichiometric number of moles of O_2 per mole of fuel. If the LFL is correct, the LOC so calculated typically yields conservative values since it represents the intersection of

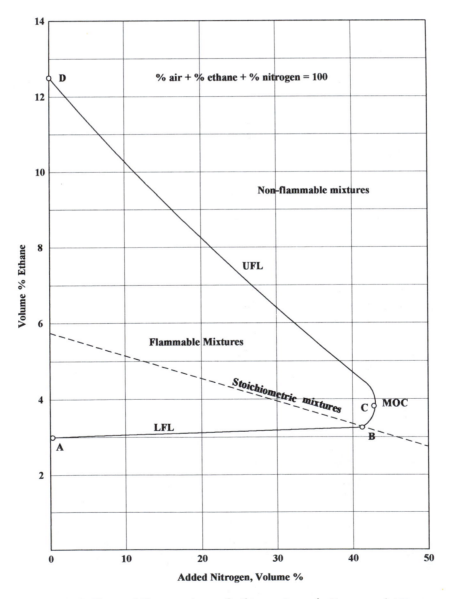

FIGURE 32.3 Flammability envelope of ethane, air, and nitrogen mixtures at ambient conditions.[2]

the LFL in air (line ABX—Fig. 32.2) with the locus of the stoichiometric oxygen concentrations (line JCX—Fig. 32.2) at the point X. The maximum errors in estimating the LOC in this manner are approximately ±3 vol% oxygen.[5]

Example 1

The LFL of methane in air at atmospheric pressure is 5.0 vol%. Estimate the LOC for methane. The stoichiometry for this reaction is:

$$CH_4 + 2O_2 \rightarrow CO_2 + 2H_2O$$

So that $O_s = 2.0$ and the LOC is 10%. The measured LOC for methane is 12%.

Another method for estimating the LOC of combustible gases has been suggested by Subramanian and Cangelosi.[6] Their method involves the calculation of the total number of moles of nitrogen necessary to reduce the stoichiometric moles of oxygen to the LOC. A group contribution correlation of 40 components for which the LOCs are known was developed and summarized here. The group contributions are summed and defined as

$$N_T = \sum (n_i G_i)$$

where n is the number of a particular bond group and G is the group contribution factor associated with bond group i, both identified in Table 32.3. For combustible gases containing chlorine, an adjustment factor, related to the number of chlorine molecules (d) in the fuel, must be made to N_i as

$$N_T = \sum (n_i G_i) - 19.181 \left(\frac{d}{O_s}\right)^{0.0715}$$

The LOC is calculated as:

$$LOC = \frac{O_s}{1 + O_s + N_T} 100$$

Example 2

Estimate the LOC of methane using the methodology of Subramanian and Cangelosi.[6] From Example 1, $O_s = 2$ moles of O_2/mole methane.

$$N_T = \sum (n_i G_i) = (1)(13.67) = 13.67$$

$$LOC = \frac{O_s}{1 + O_s + N_T} 100 = \frac{2}{1 + 2 + 13.67} 100 = 12\%$$

LFL and UFL of Mixtures of Vapors and Gases

The LFL and UFL for mixtures of combustible gases and vapors can be estimated using Le Chatelier's equation[7,8] as:

$$LFL_{mix} = \frac{1}{\sum \frac{y_i}{LFL_i}}$$

$$UFL_{mix} = \frac{1}{\sum \frac{y_i}{UFL_i}}$$

where y_i is the mole fraction of component i. Le Chatelier's equation is empirically derived and subject to limitations.[9]

Effects of Temperature and Pressure on the Flammability Limits of Vapors and Gases

The limits of flammability generally widen (the LFL decreases and the UFL increases) with increasing temperature[10] and pressure.[11] The effect on the lower flammable limit is slight but the effect on the upper flammable limit is more pronounced. The following empirical expressions may be used to estimate the effects.[20]

$$LFL_T = LFL_{25} - \frac{0.75 (T - 25)}{\Delta H_c}$$

$$UFL_T = UFL_{25} + \frac{0.75 (T - 25)}{\Delta H_c}$$

where

T	= temperature, °C
ΔH_c	= net heat of combustion, kcal/mole
P	= pressure, mega pascals absolute
UFL_{25}	= upper flammable limit at 25°C, vol%
LFL_{25}	= lower flammable limit at 25°C, vol%

Experimental Determination of Flammable Limits of Vapors and Gases

The UFL, LFL, and LOC can be determined experimentally using techniques and equipment given in two ASTM Standards. Standard E 681[14] is applicable to the determination of limits at atmospheric pressure and temperatures to 150°C. A diagram of the apparatus used is shown in Figure 32.4. Standard E 918[15] is applicable to the determination of limits at pressures to 200 psia and temperatures up to 200°C. A diagram of the apparatus used is shown in Figure 32.5. The limits obtained experimentally are dependent on a great number of factors, some of which are (1) strength of ignition source; (2) type of ignition source; (3) size of test chamber; and (4) turbulence levels at time of ignition. Some gases such as halogenated hydrocarbons present particular experimental difficulties due to their large quenching distances

TABLE 32.3 Derived Values of Group Contributions for Determining the Total Nitrogen Required per Mole of Combustible Gas[6]

Group	G	Group	G	Group	G
H–C–H (with H above and H below)	13.67	–O–	–1.02714	–N with H and H	45.3875
H–C– (with H above and H below)	13.1169	–S–	–34.5144	–N with H	–66.5034
–C– (with H above and H below)	11.1453	–Cl	6.1377 (per chloride group) –19.181 (number of chlorine atoms/stoichiometric moles of O_2)$^{0.0715}$	–N	7.09294
–C– (with H above)	6.31407	=C with H and H	13.000	=N–	76.572
–C–	–0.499077	=C with H	9.11694	–C≡N	16.9444
O=C with H	4.51884	=C	3.76617	H_2	8.5
O=C	4.36966	H–C≡	17.7812	CO	7.59
O=C with O–	1.08435	H–C (benzene ring)	9.60497	H_2S	17.5
–OH	–0.426518	–C (benzene ring)	27.444	CS_2	56

and difficulty of ignition. A complete discussion of experimental techniques and interferences is beyond the scope of this summary.

A gas deflagration index K_G is defined[13] as

$$K_G = \left(\frac{dP}{dt}\right)_{max} V^{\frac{1}{3}}$$

where

P = pressure (bar)
t = time (seconds)
V = volume (m^3)
K_G = index (bar m/sec)

In the case of quiescent gases, K_G is not a constant but increases with the volume of the test chamber.[16]

Experimental Determination of Flammable Limits of Combustible Solids (Dusts)

The deflagration characteristics of combustible dusts can be determined experimentally using techniques and equipment given in ASTM Standard E 1226.[12] A diagram of the apparatus used is shown in Figure 32.6. As with gases and vapors, the deflagration characteristics of a dust cloud are dependent on a great number of factors such as

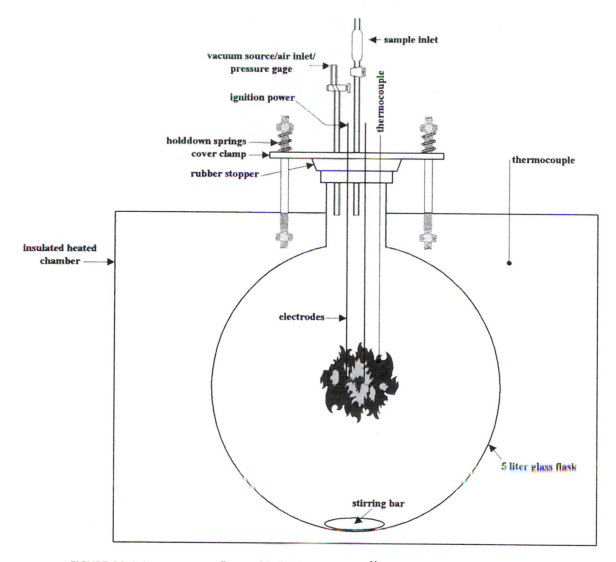

FIGURE 32.4 Low-pressure flammable limits apparatus.[14]

(1) energy of the ignition source; (2) turbulence levels; (3) size of test chamber; (4) particle size and distribution of the dust; (5) moisture content of dust; and (6) presence of flammable gas or vapor (called hybrid systems).

The pressure-time characteristics of a typical dust deflagration are shown in Figure 32.7. A deflagration index K_{St} is defined as:

$$K_{St} = \left(\frac{dP}{dt}\right)_{max} V^{\frac{1}{3}}$$

where
P = pressure (bar)
t = time (seconds)
V = volume (m^3)
K_{St} = index (bar m/sec)

K_{St} is approximately constant provided that the volume of the test chamber is greater than 20 liters and other test variables are controlled within prescribed limits. The utility of the K_{St} is that it can be used to size deflagration vents using well established procedures.[13]

Dusts are given an St classification that is based on the K_{St} as shown in Table 32.4. The St class is a measure of the explosive violence of the dust. A device that gives a semiquantitative measure of a dust's St Class is the 1.2 liter Hartmann Tube shown in Figure 32.8. Much early dust explosivity information was generated on the Hartmann tube. No direct measurement is made of the rate of pressure rise, but rather the force of the deflagration lifts a flapper on the top of the tube, which is attached to a motion indicator. The Hartmann Tube may be used

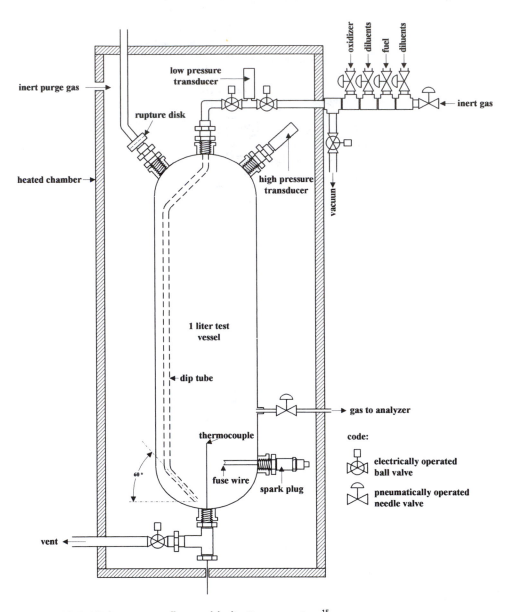

FIGURE 32.5 High-pressure flammable limits apparatus.[15]

as a screening device to give an indication of a dust's explosiveness, but direct application to vent design is not allowed.[13]

Inerting Procedures

Many techniques may be employed to ensure that the atmosphere in a vessel or enclosure is noncombustible. Two general methods are *batch purging* which includes syphon, pressure, vacuum, and sweep-through purging where the atmosphere is rendered noncombustible by one or multiple staged operations, at the conclusion of which no further inerting is conducted throughout whatever processing/manufacturing operations are being conducted, and *continuous purging*, which includes fixed-rate and demand-rate applications where the inerting gas is introduced throughout the processing/manufacturing operations.

Batch Purging

Vacuum Purging

Vacuum purging is the most common inerting procedure for vessels. It is applicable only for those vessels

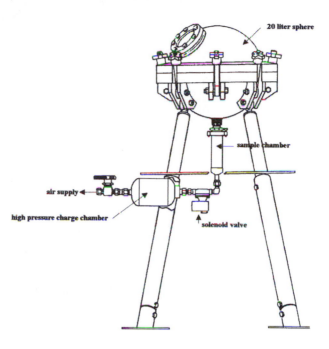

FIGURE 32.6 20-liter dust sphere.

that are rated for full vacuum, thus excluding most atmospheric storage vessels from this method. As the name implies, vacuum purging consists of a series of staged operations where vacuum is first drawn on the vessel, followed by vacuum relief with the inerting gas. The process is repeated until the oxygen level is below the LOC. The following expression can be used to estimate the number of purge cycles that are necessary:

$$y_j = y_o \left(\frac{P_L}{P_{atm}} \right)^j$$

where

y_j = mole fraction of oxygen after j purge cycles, dimensionless

y_o = initial mole fraction oxygen before purge cycles, dimensionless

P_L = low pressure limit, psia

P_{atm} = atmospheric pressure or high pressure limit, psia

j = number of purge cycles

This expression assumes (1) the pressure limits P_{atm} and P_L are identical for each cycle; (2) the inert gas is free of oxidant; (3) perfect mixing in the vessel; and (4) the vacuum is relieved to P_{atm} each cycle.

Pressure Purging

Vessels may be pressure purged by adding inert gas under pressure. Pressure purging consists of a series of staged operations where the vessel is pressurized to a given level with inert gas, followed by relief to atmospheric pressure. The process is repeated until the oxygen level is below the LOC. The following expression can be used to estimate the number of purge cycles that are necessary:

$$y_i = y_o \left(\frac{P_{atm}}{P_H} \right)^{j+1}$$

where

y_j = mole fraction of oxygen after j purge cycles, dimensionless

y_o = initial mole fraction oxygen before purge cycles, dimensionless

P_H = high pressure limit, psia

P_{atm} = atmospheric pressure or low pressure limit, psia

j = number of purge cycles

This expression assumes (1) the pressure limits P_H and P_{atm} are identical for each cycle; (2) the inert gas is free of oxidant; (3) there is perfect mixing in the vessel; and (4) the pressure is relieved to P_{atm} each cycle.

Siphon Purging

In siphon purging the vessel is first completely filled with liquid followed by the introduction of the inert gas to replace the liquid as it is drained from the vessel.

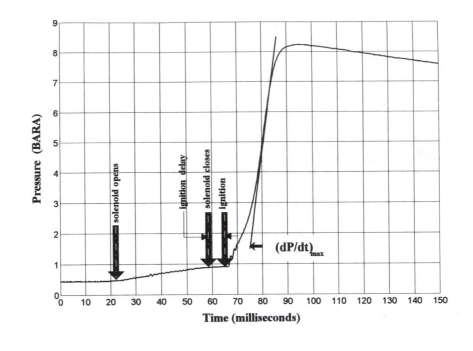

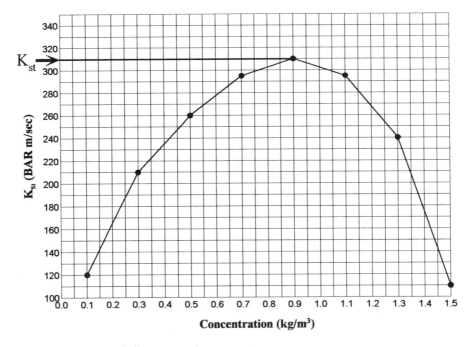

FIGURE 32.7 Dust deflagration characteristics.

TABLE 32.4 St Class and Deflagration Index K_{st}

K_{st} (bar m/sec)	St Class
0	0
1 → 200	1
201 → 300	2
>301	3

Sweep-Through Purging

Sweep-through purging involves adding inert gas at one location in the vessel and withdrawing it at a different location. The time required to reduce the oxidant to the required level can be estimated by the following expression.

$$t = \frac{V \ln\left(\frac{C_1 - C_o}{C_2 - C_o}\right)}{Q}$$

where

C_1 = initial concentration of oxygen in vessel, vol%
C_o = concentration of oxygen in incoming purge gas, vol%

C_2 = final concentration of oxygen in vessel, vol%
V = vessel volume, ft^3
Q = volumetric flow rate of purge gas, ft^3/min
t = time, min

This expression assumes that the incoming gas mixes instantaneously in the volume (V) to be inerted. This process is most commonly used for vessels that cannot tolerate either vacuum or pressure. This inerting operation is most easily conducted at atmospheric pressure. Particular attention must be paid to the inert gas intake/exhaust configuration to ensure that channeling (the passage of the inert gas directly from the intake to the exhaust without mixing in the vapor space) does not occur.

Continuous Purging

Fixed Rate Applications

This method involves the introduction of the purge gas into the vessel at a constant rate throughout the processing/manufacturing operations. Measures must be taken to ensure that the LOC is reduced to a level below the combustible limit before operations are initiated. The

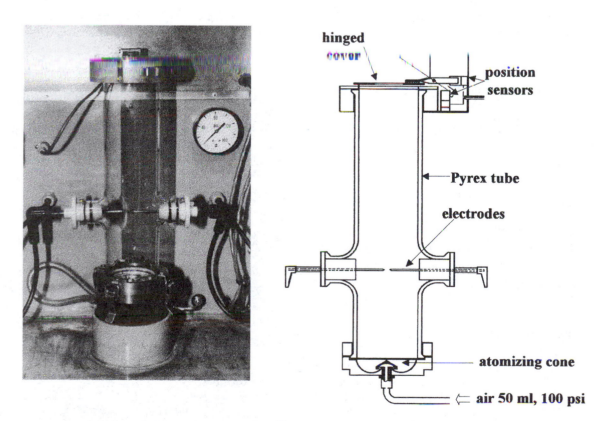

FIGURE 32.8 Modified Hartmann tube.[24]

rate at which the purge gas is introduced must be sufficient to at least equal the maximum rate of efflux from the vessel in product and/or gas/vapor.

Demand Rate Applications

This method involves the introduction of the purge gas into the vessel at a volumetric rate that matches the volumetric efflux from the vessel in product and/or gas/vapor. Typically the vessel is placed under a slight positive pressure with the supply of inerting gas being regulated by a PIC (pressure indicating control) device. Such a system has the advantage that the quantity of inert gas required is minimized; however, proper operation and maintenance of the PIC controller at low pressure differentials is difficult.

Measurement of Oxygen Concentrations

Process oxygen analyzers use four basic methods for the measurement of oxygen:

- measurement of the paramagnetic property of oxygen
- catalytic combination of oxygen and hydrogen
- oxidation of certain characteristic chemical compounds by oxygen
- effect of oxygen on the operation of a galvanic cell

As a result, there is a wide range of commercial analyzers available. In matching the analyzer to the process consider:

- operating temperature and pressure
- response times
- accuracy
- environmental interferences (i.e., dust clogging and interfering with the diffusion of oxygen through polorographic cell membranes)
- maintenance requirements

The location of the analyzer in the process vessel needs to be addressed with care. Placement in "dead" zones where little mixing occurs must be avoided.

Ignition Sources

A tabulation of ignition sources and their frequency in initiating dust cloud deflagrations is shown in Table 32.5. This can be used as a guide for assessing source and frequency of ignition sources for other combustible fuel-oxidant mixtures. Whether the elimination of ignition sources can realistically be effected needs to be evaluated.

TABLE 32.5 Frequency Distribution of Ignition Sources for 357 Dust Explosions[22]

Ignition source	%
Mechanical sparks	29
Smoldering particles	10
Mechanical heating	9
Static electricity	9
Hot surfaces	5
Fires	8
Autoignition	5
Miscellaneous or unknown	25

Inert Gas Sources

Inerting gas can be obtained from a number of sources which include, but are not limited to:

- High pressure tanks or cylinders of commercially available inert gases such as nitrogen, CO_2, argon, or helium.
- Flue gases made by burning fuel gas or fuel oil or combustion products from process or boiler furnaces. The possibility of contaminants needs to be considered and appropriate measures taken if that possibility exists.
- Steam. Consideration needs to be given to the interaction of water vapor or liquid with the product being inerted. Temperatures must be maintained sufficiently high to prevent condensation of the steam which can move the oxygen concentration back into the flammable region.
- Inert gas supplied by the removal of oxygen from air by absorption, adsorption, chemical reaction, or membrane permeation technologies.

Personnel Safety

Any inerting system needs to be properly maintained. The potential for leaks into operating areas where personnel can be present needs to be considered. Personnel need to be trained in vessel-entry procedures and those procedures need to be strictly adhered to.

When to Inert

In the final analysis the decision of whether or not to inert is a judgment call based on an assessment of the relative risks involved. Despite the fact that inerting systems can be costly, require maintainence and a significant infrastructure to be in place (such as vessel entry procedures, training and preventative maintenance (PM) programs), these are usually minor issues in the decision process. Inerting is done to protect personnel from injury and death, and secondarily to prevent property loss by the prevention of a deflagration. Inerting is not without its dangers also. Deaths have resulted from faulty inerting systems or from failing to properly follow vessel (containing inert gas)-entry procedures. When considering whether of not to inert, a judgment must be made as to the relative risks. In making that assessment one needs to:

Consider	How to Address
1. Whether a flammable atmosphere can possibly exit within the vessel at any time during the process	Appropriate Process Hazards Analysis
2. All possible ignition sources and their potential for occurring when a flammable atmosphere is present	May require consultation of trained safety personnel
3. The impact of a deflagration should it occur	Pressures developed during deflagrations can be determined experimentally; appropriate Material and Safety Engineers can assess the impact on equipment

If the possibility of a catastrophic deflagration cannot be eliminated, consideration should be given to (1) venting the deflagration to prevent pressures developing in the vessel to dangerous levels; (2) designing the vessel to withstand the pressures that develop during the deflagration (typically gas and dust deflagrations develop maximum pressures on the order of 150 psia); (3) suppression systems that inject an extinguishing agent into the developing fireball that quench the combustion process.

Corporate safety policies and standards should be reviewed if available. Removal of the ignition source may in *some* situations be considered a sufficient safety measure. If ignition sources are not present, a deflagration is not possible.

The decision to inert should be made only after careful consideration of possible chemical reaction consequences. For example, some inhibitors used for monomer stabilization require the presence of oxygen. Complete removal of oxygen when inerting such systems may result in catastrophic runaways.

References

1. NFPA-69. *Explosion Prevention Systems*; National Fire Protection Association: Quincy, MA.
2. Hansel, J. G.; Mitchell, J. W.; Klotz, H. C. Predicting and Controlling Flammability of Multiple Fuel and Multiple Inert Mixtures. *Plant/Operations Progress* **1992**, *11*(4) 213–217.
3. Jones, G. W. Inflammation Limits and Their Practical Application in Hazardous Industrial Operations. *Chem. Rev.* **1938**, *22* (1), 1–26.
4. Bodurtha, F. T. *Industrial Explosion Prevention and Protection*; McGraw Hill Book Co.: New York, 1980.
5. Britton, L. G. Operating Atmospheric Vent Collection Headers Using Methane Gas Enrichment. *Process Safety Progress* **1966**, *15*(4).
6 Subramaniam, T. K.; Cangelosi, J. V. Predict Safe Oxygen In Combustible Gases. *Chem. Eng.* **1989**, 108.
7. Le Chatelier, H. Estimation of Firedamp by Flammability Limits. *Ann. Mines* **1891** *19*, ser. 8, 388–395.
8. LeChatelier, H.; Boudouard, O. Limits of Flammability of Combustible Vapors. *Compt. Rend.* **1898**, *126*.
9. Coward, H. F.; Jones, G. W. Limits of Flammability of Gases and Vapors. *U.S. Bureau of Mines Bulletin* **1952**, *508*, 6.
10. Zabetakis, M. G.; Lambiris, S.; Scott, G. S. Flame Temperatures of Limit Mixtures. *Seventh Symposium on Combustion*; Butterworths: London, 1959; p. 484.
11. Zabetakis, M. G. Fire and Explosion Hazards at Temperature and Pressure Extremes. *AIChE-Inst. Chem. Engr. Symp., Ser. 2, Chem. Engr. Extreme Cond. Proc. Symp.* **1965**, 99–104.
12. ASTM Standard E 1226-94. Standard Test Method for Pressure and Rate of Pressure Rise for Combustible Dusts. *Annual Book of ASTM Standards*; Volume 14.02, 1996.
13. NFPA-68. *Venting of Deflagrations*. National Fire Protection Association: Quincy, MA.
14. ASTM Standard E 681-94. Standard Test Method for Concentration Limits of Flammability of Chemicals. *Annual Book of ASTM Standards*; Volume 14.02, Section 14, 1995.
15. ASTM Standard E 918-83. Determining Limits of Flammability of Chemicals at Elevated Temperature and Pressure. *Annual Book of ASTM Standards*; Volume 14.02, Section 14, 1995.
16. Britton, L. G.; Chippett, S. Practical Aspects of Dust Deflagration Testing. Paper 58d, 17th Annual Loss Prevention Symposium: American Institute of Chemical Engineers: Houston, March 24–28, 1985.
17. Windholtz, M. W., Ed. *The Merck Index: An Encyclopedia of Chemicals, Drugs, and Biologicals*, 12th ed.; Merck and Company: Rahway, NJ, 1996.
18. Hawley, G. G., Ed. *The Condensed Chemical Dictionary*, 13th ed.; Van Nostrand Reinhold: New York, 1997.
19. Wadden, R. A.; Scheff, P. A. *Engineering Design for the Control of Workplace Hazards*; McGraw-Hill Book Company: New York, 1987.
20. Zabetakis, M. G. Flammability Characteristics of Combustible Gases and Vapors. *U.S. Bureau of Mines Bulletin* **1965**, *627*,

21. U.S. Department of Commerce, National Technical Information Service PB-294 250. *Matrix of Combustion-Relevant Properties and Classifications of Gases, Vapors, and Selected Solids*; 1979.

22. Bruderer, R. E. Ignition Properties of Mechanical Sparks and Hot Surfaces in Dust/Air Mixtures. *Plant/Operations Progress* 19 8(3).

23. Crowl, D. A.; Louvar, J. F. *Chemical Process Safety-Fundamentals with Applications*; PTR Prentice-Hall, Inc.: Englewood Cliffs, NJ, 1990; ISBN 0-13-129701-5.

24. ASTM Standard E 789-95. Standard Test Method for Dust Explosions in a 1.2 Liter Closed Cylindrical Vessel. *Annual Book of ASTM Standards*; Volume 14.02, 1996.

25. Mashuga, C. V.; Crowl, D. A. Application of the Flammability Diagram for Evaluation of Fire and Explosion Hazards of Flammable Vapors. Paper presented at the American Institute of Chemical Engineers 32nd Loss Prevention Symposium: New Orleans, LA, March 1998.

33

Preventive Maintenance

JERRY R. HINES

The efficient and safe operation of a simple or a complex laboratory and pilot plant facility is a product of a combination of facility design, management of change, the detection and correction of facility and equipment problems, and the performance of a planned preventive maintenance process. The scope of maintenance activity can include the basic physical structure of buildings, utility systems, process and laboratory equipment, hazard detection equipment, fire and emergency protective systems, and other life safety needs.

It is important to note that maintenance is often focused only on the upkeep and repair of property and equipment (often, what is broken) and not on proactive maintenance to prevent unnecessary and unwanted failure. The scope of a "preventive" maintenance program should be to anticipate the potential for failure and to take action before operations are interrupted unnecessarily and at often greater expense. The NSC *Accident Prevention* (Administration and Control—Ninth Edition) *Manual* states that "preventive maintenance is the orderly, uniform, continuous, and scheduled action to prevent breakdown and prolong the useful life of equipment and buildings."

Planning for Preventive Maintenance

Planning for preventive maintenance should include a process that determines needs, sets priorities, defines management responsibilities, assigns who will be performing the maintenance, defines how often preventive maintenance is to be provided, provides for maintenance records, and inspects and audits the process. The process should also ensure that the maintenance is provided by trained personnel and within constraints of regulatory, local jurisdictions and site-specific requirements (see Fig. 33.1). The determination of preventive maintenance priorities and maintenance cycles can be assisted by being aware of and using available resources.

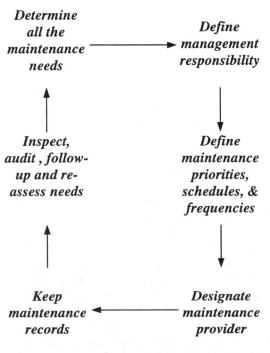

Determine all the maintenance needs → *Define management responsibility*

Inspect, audit, follow-up and re-assess needs

Define maintenance priorities, schedules, & frequencies

Keep maintenance records ← *Designate maintenance provider*

FIGURE 33.1 Preventive maintenance process.

Examples include:

- manufacturers specifications and recommendations for equipment and products
- regulatory standards and industry consensus standards
- internal company requirements, policies, and procedures
- facility maintenance inspection and audit programs to surface maintenance concerns
- maintenance records and service histories
- an information system to collect employee reporting of problems

Consideration should also be given to the advantages of having spare parts and other consumables on hand for immediate use and/or information of where parts and needed supplies can be obtained.

Safety Considerations

As with all work activities, maintenance work should be conducted within the boundaries of safe prudent practices, company policies and procedures, and regulatory health, safety, and environmental standards. The safety, health, and well-being of maintenance workers and craftsmen, the protection of the physical facilities and resources, the protection of the environment, and the well-being of the communities surrounding a work location are integral parts of the preventative maintenance process.

Some of the basic considerations to ensure that preventative maintenance work is performed safely include:

- the communication of chemical and physical hazards to those performing the maintenance
- the control of all types of energy during the performance of maintenance
- the training and qualifications of personnel performing maintenance activities
- assessment of personal protective equipment needs
- a work permit system, and
- a safety, health, and environment audit and inspection program

Assessing Needs and Assigning Preventive Maintenance

Assessment

Why is maintenance a part of the work process? I suspect that we all know that all things will wear out or at least not perform as well as expected at some time. The saying, "if it isn't broke, don't fix it" may have short-term validity, but in the long run, attention should be paid to correct problems and provide repair before situations begin to affect the efficiency and cost-effectiveness of work. Traditionally, scheduled shutdowns and turn-arounds have been used in industry to provide time for necessary and needed service and repairs.

Failures can be expected sooner or later because of the effects of wear and tear, corrosion, erosion, degradation, heat, cold, power fluctuations, contamination, and chemical reactions. Also, there is a need to inspect, test, and recalibrate instrumentation. Maintenance prevents and/or delays failure and replaces and upgrades equipment and facilities.

Providing Preventive Maintenance

The scope of maintenance work in a laboratory and pilot plant facility will most likely be shared by several groups. Examples of the occupations that exist to counter and control the common causes for failure include painters, electricians, pipefitters, welders, sheet metal workers, roofers, insulators, plumbers, pressure equipment technicians, electronic technicians, radiographic technicians, instrument technicians, ventilation technicians, computer technicians, and engineers.

The traditional maintenance group which usually exists within a facility's engineering or services group serves to provide the maintenance for the building and utilities which support the general operation of the facility. Test shops and instrument shops can often provide the technical knowledge and "know-how" to maintain more complex detectors, instrumentation, computers, control systems, relief devices, pressure vessel testing, and so forth. Vendor or supplier expertise may be needed to provide maintenance for fire detection and protection systems, and for the more specialized equipment and devices used in lab and pilot plant work. In some cases individual laboratory and pilot plant employees could be assigned responsibility to provide the preventive maintenance of specialized equipment and systems because of special knowledge or qualifications.

Maintenance for Laboratories and Pilot Plants

Providing maintenance for laboratory and pilot plant facilities will most likely include most or all of the maintenance job titles and will include all areas of the facility. The physical plant and buildings (including streets, walking surfaces, offices, storage areas, warehouse, shops, meeting rooms, etc.), utilities and services areas, laboratories, and pilot plant process areas are examples of areas where preventive maintenance will be needed because they all provide the service and support for the laboratory and pilot plant work.

Buildings and Facilities

It is generally accepted in industry and it is stated in OSHA Subpart D-1910.22 General Requirements (walking working surfaces, streets, doors, windows) that work places should be functional and maintained in a clean, orderly, and sanitary condition. The daily use of facilities by employees, equipment, and vehicles and the effects of corrosion, weather, and everyday wear and tear are examples of activity that can affect the appearance, safety, and the efficient and functional operation of a facility. It is also stated in Subpart E-OSHA 1910.37 Means of Egress, General, that "the means of egress shall be continuously maintained free of all obstructions or impediments to full instant use in the case of fire or other emergency."

With time, all of the above-mentioned elements of a facility will most likely require repair, possible replacement, and should be in a preventive maintenance process. The maintenance tasks can include cleaning, painting, lubricating, welding, torch use, grinding, brushing, and the repair or replacement of building materials, roofing materials, piping, glass, and electrical equipment. In addition, the tools and equipment used to provide maintenance work must also be maintained.

Some of the common items that are included in a preventive maintenance program for buildings and facilities are provided in the following list. Safety concerns have also been noted for some of the items.

Maintenance Items in Buildings and Facilities

- Walkways and walking surfaces should be clean, free of dirt and debris, free of holes, uneven surfaces, and slippery surfaces.
- All doors should open and close freely and latch.
- Fire door closures and fusible links should be in working order at all times.
- Masonry and mortar joints need periodic inspection and repair.
- Glass in windows and doors should be replaced.
- Lighting in exits, exit paths, hallways, offices, meetings areas, laboratories, and pilot areas should be inspected and maintained to meet life safety standards.
- Drainage paths from roofs and other work areas should be clear and open to prevent excessive loading.

- Building surfaces should be kept clean and painted to prevent corrosion.
- Electrical grounding paths, circuit breakers, and electrical switch gear should be maintained in working order.
- Stairs, stair step treads, railings, and landings should be clean, connected, and mounted securely.
- Painting a contrasting color stripe just before the first step on stair step treads can help to overcome a depth perception problem created by walkways and stairs being constructed of similar materials and painted the same color.
- Indoor walking surfaces should be clean and free of slipping hazards.
- Light fixtures should be clean, and burned-out bulbs replaced as needed.

Safety Notes

- Any penetration through fire walls by electrical conduit, piping, tubing, and so forth must be re-sealed to meet original fire wall design specifications.
- Maintenance work that requires the removal of insulation should take into account the possibility of asbestos insulation.
- Fusible links must not be painted. The temperature at which the link will melt when exposed to heat can be affected.
- Broken glass should be replaced with the correct type of glass. This is especially important in fire door applications.
- The existence of lead- or cadmium-based paints should be considered when removing paint. Testing of paint samples should be done to detect lead or cadmium. If found, safety precautions to prevent exposure to maintenance workers and other workers should be used.
- Care should be taken to prevent the use of waxes with low friction factors.
- Exit lights should be maintained on a regular basis.

Utility Service

Reliable utility services are essential for the operation of laboratories, pilot plants, and supporting operations. Electrical power (whether purchased or generated), lighting systems, heating, ventilation and air conditioning (HVAC) systems, water systems (potable, industrial, and fire), steam boilers, compressed air, compressed gases, cryogenic gases, and waste treatment plants are examples of utilities that should be included in routine preventative maintenance programs. All these services are essential for the operations of lab and pilot plant and process areas.

These areas will need periodic and scheduled preventive maintenance. All of the utilities systems noted are made of component parts that require maintenance. A partial list of individual equipment that make up these systems includes pressure vessels, motors, electrical breakers, electrical generators, electrical panes, pumps, piping, conduit, relief devices, backflow preventors, chemical mixing systems, filters, treatment systems, and sampling systems.

Laboratory and Pilot Plant Facilities

Laboratories, especially laboratories using chemicals, depend on reliable equipment and utility services. Continuing and uninterrupted performance of laboratory hood and local exhaust systems depend on the fan motors, fans, flow sensors, exhaust ducts, dampers, and controls to operate efficiently and as designed. The negative air pressure balance of laboratories as compared to surrounding work areas is of particular importance to prevent the transfer of airborne laboratory contaminants or materials to adjoining work areas. Pilot plant and related process work areas usually have maintenance similar to laboratory areas, but because of the scale of operations their maintenance operations may require increased maintenance efforts and attention.

Pilot plant areas usually require the use of larger volumes of flammable and/or potentially toxic materials, increased needs for electrical power, heating or cooling mediums (water, steam, oil), pressure relief systems, venting systems (including flare systems), leak and/or fire detection systems, and larger exhaust ventilation systems. Because of the scale of the pilot plant/process area, the need for maintenance of relief devices, pressure vessels, piping and distribution systems, electrical power distribution (substations and breakers), industrial sewers systems, and so forth can be significant. In many newer operations and facilities, the maintenance of computer control systems can also be important in maintenance operations.

The NFPA 45, Standard on Fire Protection for Laboratories Using Chemicals, indicates that maintenance is needed for utilities (gas, water, steam), detectors and alarms, electrically operated equipment, laboratory hoods, air supply and exhaust systems, compressed and liquefied gases, storage and piping systems, laboratory apparatus, cryogenic fluids systems, refrigeration equipment, cooling equipment, ovens, and pressure equipment. In most case the codes and standards calls for at least annual inspection, testing, and maintenance. In addition to annual maintenance requirements, deficiencies should be corrected when observed or the hood use must be restricted if not needed. Water systems for laboratory use should be clean and protected from corrosion. Maintenance procedures should prevent, detect, and control and/or eliminate cross connections between potable (drinking) water systems and lab, industrial, and fire water systems. Other systems which may require maintenance in laboratories include steam generators, water and oil baths, industrial and laboratory sewer systems, and lighting systems.

Safety and Emergency Systems

The life safety and protection of personnel in laboratories, pilot plant process areas, as well as service and administrative support personnel, is essential. In the event that an emergency situation should occur, the systems that have been designed and established for use by employee and response personnel must operate as needed. These safety and emergency systems include manual and automatic alarm reporting systems, public address systems, emergency lighting systems, emergency power generation systems, fire detection and reporting systems, combustible and toxic chemical detection systems, and fixed fire protection systems.

Maintenance Frequency

See Table 33.1, Examples of Maintenance Needs, Priorities, Codes & Standards, for an example of how needs and priorities could be summarized. The frequency in which maintenance is to be performed is subject to many parameters. It is common to find that annual (and sometimes more frequent) maintenance is required by codes, standards, or regulations. Meeting these minimum requirements may satisfy regulatory requirements, but doing so may not provide the necessary maintenance that ensures that facilities or equipment will operate efficiently and safely. It is especially important that systems and equipment that directly contribute to life safety be maintained to be functional at all times. Maintenance schedules can be based on regulations and manufacturer recommendations, but better and more importantly by the evaluation of site-specific information gained through inspections, audits, and experience.

Safety and Training Considerations

Required Training

Manufacturers' specifications, internal requirements, policies and procedures, consensus standards, and regula-

TABLE 33.1 Examples of Maintenance Needs, Priorities, Codes, and Standards

Maintenance Needs	Maintenance Priorities, Frequency	Maintenance Work Assignments	Reference, Standard, or Code
Sprinkler systems, standpipe and hose systems, private fire service (water) mains, fire pumps, water storage pumps, water spray fixed systems	High priority Annual inspection or testing, as needed, or as required by local jurisdictions	Plant maintenance, outside service contractor	NFPA 25
Substations switch gear, assemblies, switchboards, panel boards, motor control centers and disconnect switches, fuses and circuit breakers, rotating equipment, batteries and battery rooms, premises wiring, hazardous (classified) locations, portable electric tools and equipment	High priority Annual or as needed and set by maintenance schedules	Plant maintenance or as needed by outside contractors	NFPA 70 E
Exit doors, emergency lighting, alarm systems, PA systems, etc.	High (critical) priority Permanently maintained—testing, inspection, immediate repair of problems, failures	Plant maintenance, outside contractors, health, safety and environment groups	NFPA 101 OSHA 1910.37
Boilers, pressure vessels	Routine priority Inspection, testing, and repair of problems	Plant maintenance, contractors	ASME Boiler and Pressure Vessel Code
Combustible vapor detectors	Routine priority Routine testing, recalibration and repair	Instrument test shops, identified, skilled laboratory, pilot plant personnel, outside contractors	ISA-RP12.13, Part II
Continue the process until all maintenance items are identified, priorities are set, responsibilities are determined			

tory standards are the sources needed to help determine training needs for maintenance workers. Electrical safety, the control (lockout and tagout) of hazardous energy, permit-required confined space entries, hazard communication, MSDS use, fall protection, PPE, work-permitting, hand and power tools, excavations, welding and cutting, and basic fire fighting/first aid are examples of specific training needs. Of these requirements, lockout and tagout of all types of hazardous energy are especially important. These requirements are necessary for in-house maintenance providers as well as contract maintenance providers.

The Occupational Safety and Health Standards as contained in 29 *CFR* Part 1910 for General Industry and Part 1926 for the Construction Industry provide specific safety requirements for maintenance activities.

Management of Change

As maintenance is performed, the "management of change" has significant importance. Whenever equipment is replaced, special attention should be given to changes that are not "changes in kind." For example, if a 25 GPM pump is replaced with a 50 GPM pump, the "effect" of the change should be evaluated for possible performance and, most importantly, the safety of operations. The larger capacity pump could exceed the designed pressure or flow limitations. The same would hold true for electrical devices, pumps, or other equipment. Although normally applied on larger industrial plants and processes, the OSHA 1910.119 Process Safety Management of Highly Hazardous Materials standard is an excellent source of information on the important issue of "Management of Change."

General References

Accident Prevention Manual for Industrial Operations, Administration and Programs, Ninth Edition; National Safety Council.

Environmental Management Checklist, Business & Legal Reports, Inc., 39 Academy Street, Madison, CT 06443-1513.

Industrial Ventilation Manual, 21st ed.; ACGIH: 6500 Glenway Avenue, Bldg. D-7, Cincinnati, OH 45211, 1992.

Inspection of Piping, Tubing, Valves, and Fittings, June 1990, API: 1220 L Street, NW, Washington, DC 20005.

Installation, Operation, and Maintenance of Combustible Gas Detection Instruments, ISA-RPI12.13, Part II-1987.

NFPA 25—*Inspection, Testing, and Maintenance of Water Based Fire Protection Systems*, 1995 edition.

NFPA 45—*Fire Protection for Laboratories Using Chemicals*, 1996 edition.

NFPA 70—*National Electric Code*, 1999.

NFPA 70B—*Electrical Equipment Maintenance*, 1998.

NFPA 221—*Fire Walls & Fire Barrier Walls*, 2000.

NFPA 90A—*Air Conditioning & Ventilating Systems, Fire Dampers, Smoke Dampers, Ceiling Dampers, Filters, Ducts, Plenums, Fans, and Fan Motors*, 1996 edition.

NFPA 101—*Code for Safety to Life from Fire in Buildings and Structures*, 1994 edition.

NFPA 221—*Fire Walls & Fire Barrier Walls*, 2000.

Occupational Safety and Health Standards for the Construction Industry (29 *CFR* 1926), 1995; CCH Incorporated: 4025 W. Peterson Ave., Chicago, IL 60646-6085.

Occupational Safety and Health Standards for General Industry (29 *CFR* 1910), 1995; CCH Incorporated: 4025 W. Peterson Ave., Chicago, IL 60646-6085.

Pressure Vessel Inspection Code (Maintenance, Inspection, Rating, repair, and Alteration), June 1990; API: 1220 L Street, NW, Washington, DC 20005.

34

Health and Safety in the Microscale Chemistry Laboratory

ZVI SZAFRAN

MONO M. SINGH

RONALD M. PIKE

Microscale chemistry has become an increasingly common way of promoting laboratory health and safety. It is possible to carry out all standard laboratory procedures at the microscale level, including physical measurements (mp, bp, density), analytical measurements (titrations, gas collection, spectroscopy), and chemical synthesis (including workup, purification, and characterization).

By working with microquantities of chemicals, several health and safety advantages quickly become apparent:

- Exposure to potentially toxic materials is minimized.
- Air quality is markedly improved.
- Ventilation requirements are reduced.
- Chances of laboratory explosions, fires, and major spills are essentially eliminated.
- Microscale glassware is more durable, reducing chances of breakage and consequent injury.
- The possibility of accidents occurring is minimized.

Laboratory workers trained in microscale techniques become more psychologically aware of health and safety issues, further promoting laboratory safety.

What Is Microscale Chemistry?

Performing an experiment at the microscale level involves using sharply reduced quantities of both solid reagents and solvents. In a typical microscale procedure, 50–100 mg of solid and 100–500 µL of solvent will be used. While it is true that even smaller quantities could have been selected, this scale was chosen for several reasons:

- The reagents can be easily measured using standard equipment (mg balance, automatic delivery pipets). Expensive or exotic equipment is not necessary.
- Sufficient quantities of reagents are used to easily see the reactions taking place. Observations can be made and results obtained without necessarily using electronic equipment to monitor the course of the reaction.
- Sufficient quantities of material are used to provide the same level of analytical rigor as with standard laboratory procedures (i.e., four significant figures in a titration).
- Sufficient quantities of materials are used to provide sufficient product for multistep syntheses or for standard characterizations of products (IR, NMR, TGA).

Working at the microscale level requires the use of special small-sized glassware, as the wetting characteristics of most solvents render most standard-sized equipment unusable. Typically, glassware of 10-mL volume or less is used. This glassware is commercially available from several firms (Ace Glass, Kimble/Kontes to name two), and microscale equivalents for most standard glassware are available. Common examples include beakers, Erlenmeyer flasks, volumetric flasks, pipets, condensers, round-bottom flasks, conicals vials, distillation units, burets, and so on. Other smaller-sized equipment is also necessary, such as magnetic stirring hot plates, microclamps, microspatulas, microcrucibles, and so forth.

It is tempting to think that all one needs to do is reduce the quantity of chemicals and shrink the glassware. In general, this is not true. Experimental techniques at the microscale level are considerably different from their conventional-scale equivalents, and have been developed with an eye toward laboratory safety. Consider, for example, the simple process of mixing two compounds and heating them to reflux at the microscale level. The microscale procedure differs from the standard scale in several ways.

Measurement of Chemicals

First, the appropriate quantities of the chemicals to be used are measured using balances (solids) or pipets (liquids). Triple-beam and single-pan balances are replaced at the microscale level by commercially available electronic digital balances, capable of weighing to ±0.01, 0.001, or 0.0001 g. These balances are usually automatically taring, and are simple to use, at costs ranging from \$500–\$2000 depending on accuracy. Volumetric pipets are replaced by automatic delivery pipets at the microscale level. These adjustable pipets are available in two standard ranges: 10–100 µL and 100–1000 µL. External exposure to chemicals while using the automatic delivery pipet is minimized due to the presence of an ejectable disposable tip. Semiautomatic pipets are available at a cost of \$150–\$250, with fully automatic pipets being considerably more expensive.

Glassware Selection

Standard-scale reaction vessels are usually round-bottom flasks fitted with 24/40, 19/22, or 14/20 standard taper joints. These round-bottom flasks would be replaced with conical vials (shown in Fig. 34.1) at the microscale level, although 14/10 standard-taper 5- and 10-mL round-bottom flasks could also be used. The flat bottom of the conical vial gives it more stability (the vial will not roll over) than a round-bottom would, crucial when the equipment is that small. If two phases are present in the vial, a second advantage becomes apparent: the interface line is much easier to see, due to the

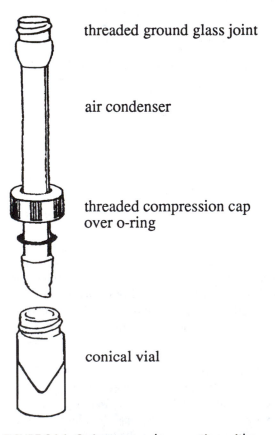

threaded ground glass joint

air condenser

threaded compression cap over o-ring

conical vial

FIGURE 34.1 O-ring cap seal connection with conical vial and air condenser.

tapered cone within the vial giving a greater height to the same quantity of liquid relative to a nontapered vial. This is crucial at the microscale level, given that the quantity of a given phase may be quite small.

Connecting Glassware

Standard-size glassware are usually connected to each other using lubricating grease. While a tiny amount of grease as a containment in 15 g of product would not be particularly problematic, it could spell disaster to a microscale synthesis. Thus, all connections at the microscale level must be greaseless.

The microscale glassware used in our process connects in a very different manner, shown in Figure 34.1 (see Ref. 6 for further illustrations). Most pieces of microscale glassware are threaded on the top of the outside, and have a 14/10 standard female taper joint on the inside. Condensers and other equipment which might be attached to the conical vials or round-bottom flasks carry a vacuum-tight 14/10 male standard taper joint at the bottom. The male joint fits within the female joint without needing any grease as a lubricant. A threaded plastic cap supported by an O-ring sits on the shoulder of the male joint, and screws onto the outer threads mentioned above. Collectively, this is called an O-Ring Cap Seal Connector. A great advantage of this type of connector is that the entire glassware assemblage can be held by a single three-prong clamp. Further, the connection is vapor-tight, eliminating losses of solvent or generated gases. This improves product yield and prevents contamination of the laboratory air quality.

In the case of the process being discussed, we would connect a condenser to a round-bottom flask (or conical vial) containing the reaction mixture. A drying tube could also be attached to the top of the condenser, if the mixture was moisture-sensitive. The assembly would then be clamped in a sand bath set atop a magnetic-stirring hot plate (see Heating and Stirring, below).

Transferring Chemicals

At the standard laboratory scale, a pipet is sufficient to transfer liquids from one container to another. At the microscale level, dust, dirt, and other filterable impurities can cause serious contamination problems. This difficulty is removed by use of a Pasteur filter pipet, prepared by inserting a small piece of balled-up cotton into the long tip of the pipet using a wire for tamping. The plug acts as a filter, and also guards against back-pressure. Transfers can be made efficiently in this manner, without the normal leakage that occurs when volatile liquids are transferred.

Solids are usually transferred directly into the reaction vessel, which can be placed on the electronic balance and tared out. The transfer is effected using a micro-spatula. This process minimizes the amount of dust in the laboratory air.

Heating and Stirring

It is now necessary to heat and stir the reaction mixture. The use of a conventional mechanical stirrer is problematic at the microscale level. Instead, solutions are stirred magnetically using spin vanes or bars. These come in several configurations, appropriate for use in either conical vials (where the vane must fit into the cone) or in round-bottom flasks (where a spin bar is used). A magnetic-stirring hot plate is used to spin the magnetic bar as well as to heat the solution.

Since the surface area in contact with the hot plate is small, it is desirable to increase the contact area with the heating surface. This is usually done by using a heating bath (held in a small crystallizing dish) or a metal block (made from any good conductor, such as Al or Cu). At the microscale level, sand is used in place of oil in the heating baths. This improves safety in several ways, as the sand cannot be oxidized or catch fire (a problem with oil baths), and does not spread in a dangerous manner. If a spill does occur, sand is cleaned up much more easily than oil. Gas microburners are also available, but their use is minimized in the microscale laboratory to prevent risks from fires.

Changed Reaction Conditions

Chemical reactions take place at the surface of the chemicals, making an analysis of mass transfer imperative at large operating scales. When the quantity of solid reagent is reduced to the microscale level, however, the ratio of surface area to bulk quantity rises dramatically. This results in a proportionally larger reaction interface, which in turn results in faster reaction times. Coupled with time savings generated by easier material measurements, faster glassware assembly, and shortened workups, a cumulative time savings of approximately 50% is readily achieved.[1]

The faster reaction rates have several potential drawbacks, however. Undesirable reactions (oxidation, hydrolysis, etc.) will also occur more readily. The problem of spot heating is essentially eliminated at the microscale level, if efficient stirring is provided. This problem is much more difficult to solve at the standard or industrial scale.

Health and Safety Benefits from Microscale Chemistry

Legally acceptable exposure levels have consistently dropped over the past decades, while the number of chemicals that must be monitored has sharply increased. Maintenance of a healthful and safe laboratory environment is a top priority in all chemical operations. Traditionally, the laboratory environment has been improved through the application of large scrubbing, air handling, and heating capacities. Microscale laboratory techniques provide an alternative to this way of doing things.

Improvement of Air Quality

In a typical standard laboratory procedure, 100–250 mL of solvent would be used. At the microscale level, the quantity of solvent generally ranges from 200–1000 μL, a 100-fold to 1000-fold decrease. This has important consequences on laboratory air quality.[2,3] Consider the use of acetone as a solvent in an example.[4]

The TLV level for acetone is 1780 mg/m^3. Suppose the laboratory in question has a volume of 75 m^3, and that the air is exchanged twice each hour. Since the air in the laboratory is not ideally mixed, it would be a safe assumption to conclude that about 30% of the air is available for dilution of chemical fumes (the rest being stagnant). The amount of air available for dilution is therefore:

$$(75 \text{ m}^3)(2 \text{ cycles/hour})(0.3) = 45 \text{ m}^3/\text{hr}$$

Suppose further that the laboratory contains 10 chemists, each using 250 g of acetone solvent, and that each experiences a 10-g loss due to evaporation over the course of 1 hour. The cumulative evaporation loss of acetone is therefore 100 g, and the exposure level would be:

$$(100,000 \text{ mg}/1 \text{ hr})(1 \text{ hr}/45 \text{ m}^3) = 2,200 \text{ mg/m}^3$$

This exceeds the TLV value by more than 25%, necessitating the use of hoods or other additional air circulation devices.

At the microscale level, the quantity of solvent would be reduced by a factor of at least 100, with a consequent reduction of evaporation loss. Assuming this also dropped by a factor of 100, cumulative evaporation loss of acetone is 1 g, and the exposure level would now become:

$$(1,000 \text{ mg}/1 \text{ hr})(1 \text{ hr}/45 \text{ m}^3) = 22 \text{ mg/m}^3$$

This is, of course, well below the TLV value, and hoods would no longer be necessary.

Many organic solvents have TLV values much lower than acetone. Microscale chemistry makes their use less problematic in maintaining a healthy laboratory environment. The faster reaction times associated with microscale experiments further reduce exposure levels.

Reduction of Internal Exposure to Chemicals

By reducing exposures from organic solvents to small fractions of their TLV values at the microscale level, long-term health problems resulting from prolonged breathing of chemical vapors can be all but eliminated. Problems associated with the inadvertent ingestion of solid chemicals are similarly reduced.

Another way of looking at this is to consider the range of chemicals that can be safely handled. Chemicals with intermediate LD-50 values can now be handled without extreme measures being taken, due to the decreased exposure afforded by the use of micro quantities.

Reduction of External Exposure to Chemicals

In most laboratories, some risk exists of exposure to chemicals that have splashed onto the skin, or that have inadvertently been left on the benchtop or equipment. The use of microquantities of chemicals clearly minimizes this possibility.

Reduction of Injury Due to Physical Hazards

Many organic solvents are flammable or explosive, making their use a potential safety concern. The reduction to microscale levels essentially eliminates these risks. Even if a fire or explosion were to occur, given the small quantities of chemical involved, the associated danger would be relatively small. In over 10 years of microscale technique use at Merrimack College and NMC[2], there has not been a single case of injury from this source.

It has also been observed that microscale glassware is more durable than standard glassware, with breakage costs decreasing by 30–35%. Pieces of equipment are connected using O-Ring Cap Seals. Thus, the chances of greased glassware slipping and breaking are reduced, minimizing chances of cuts or laceration from broken glass.

Reduction of Risks in Stockroom

The use of microscale chemistry allows much smaller quantities of chemicals to be ordered and stored. This, coupled with the smaller sized glassware and equipment, lessens stockroom space requirements, and improves stockroom air quality. The safety improvements described above also, of course, apply to the stockroom workers.

Elimination of Chemical Wastes

The health and safety aspects of handling chemical wastes and the regulation of the chemical environment are subjects of increasing concern. The problems associated with using chemicals are increasingly in the public eye. They have led to a form of chemophobia among nonscientists in general, and the mass media in particular. The most visible problems have been associated with the disposal of chemical wastes. Chemical wastes historically have been disposed of by incineration (generating air pollution), disposal through sewage systems (generating water pollution), or disposal to landfills (generating groundwater and soil pollution). All these methods have health hazards associated with them. The Times Beach, Missouri, dioxin disposal and the chemical waste dump at Love Canal, New York, are familiar examples.

As the population has become more aware of chemical-waste-disposal problems, the traditional ways of dealing with wastes are no longer politically viable. Siting of new incinerators or waste disposal dumps are increasingly problematic. The emphasis of the regulatory agencies has shifted from regulation of the disposal site to elimination of the waste at the generation site, called *source reduction*. Microscale chemistry is a method of source reduction, applicable to the academic laboratory, industrial research and development. and quality control applications.

How Much Product Is Needed?

In all chemical processes, the question of how much product is needed for analysis or for subsequent reaction must be considered. Table 34.1 lists the amounts of product that were produced in representative standard-scale inorganic laboratory experiments.[5] The amount of product produced is compared to the amount that was actually needed for the analysis described in the same experiment.

In each case, the amount of product actually needed for analysis or subsequent reaction is quite small—100 mg or less. The amounts of product generated in the experiments exceeded the necessary amounts by factors of 20–400. The rest became waste, triggering the problems associated with waste disposal. The obvious lesson to be learned is "Never make more product than is required for subsequent work or characterization."

The Three Rs and Green Chemistry

Microscale chemistry upholds the principles of the three Rs: reduce, recover, and recycle. The "reduce" aspect is covered by the 100- to 1000-fold reduction in scale of operation. Recovery becomes a more viable option, given the smaller amounts of material involved. Consider, as an example, a laboratory operation involving 10 chemists, each performing repetitive experiments three times per day using 250 mL of ether as a solvent. The amount of solvent waste generated is 7.5 L per day. Over the course of one week's time, 37.5 L of waste, or nearly a 10-gallon drum, would be generated. It would be a major undertaking to recover this quantity of laboratory waste on a weekly basis. At the microscale level, the amount of waste generated would be less than 0.5 L. This small amount of waste facilitates recovery and offers one several options as to how to accomplish it. For example, the waste could quickly and easily be recovered on a weekly basis, or alternatively, accumulated for several months and recovered in one shot. Once the solvents have been recovered, they can be purified and reused with little effort. More expensive solid reagents can similarly be recovered.

TABLE 34.1 Product Generation Relative to Product Need

Product	Amount Generated	Amount Needed for Analysis	% Disposed
NH_4BF_4	5.0 g	0.1 g for ^{19}F-NMR	98%
SnI_4	2.7 g	5 mg for melting point	99+%
$(NH_4)_2PbCl_6$	5.0 g	2 g for stability tests*	60%
$CpFe(CO)_2CH_3$	2.0 g	5 mg for melting point, 0.1 g for ^{1}H-NMR 10 mg for IR	94%
$(C_2H_5)_4Sn$	10 g+	0.1 g for ^{1}H-NMR†	99%
$Cu(NH_3)_4SO_4$	15 g	50 mg for solubility test	99+%
BF_3	40 g	12 g to make adduct*	70%

*Amount used in published experiment. Much less could have been used to do the same analysis or reaction at the microscale level.
†Some product used in second reaction, also at large-scale.

Microscale Laboratory Techniques

In order for microscale-level laboratory work to be carried out, with consequent enhancement to laboratory health and safety, microscale equivalents to standard techniques have been developed. These techniques can be considerably different from their standard equivalents. In many cases, the microscale glassware will also have important modifications relative to their larger counterparts. Reaction conditions may also be significantly altered. A representative sampling of microscale techniques are described here. A fuller description and additional methods are described in References 6–9.

Measurement of Boiling Points

Using standard-scale methods, several milliliters of a liquid would be required in order to determine a boiling point. At the microscale level, boiling points can be determined using only a fraction of one drop. There are two common methods of obtaining a boiling point.

In the *micro-capillary bell method*, a closed-end capillary tube is cut 2–5 mm from the bottom, forming a small "bell." The bell is inserted open-end down in a small (6×55 mm) test tube. A few drops of the analyte liquid are added to the test tube, which is then attached (using a twist tie or rubber band) to a calibrated thermometer. The assembly is placed in a heating bath. As the boiling point is approached, small bubbles will form at the bottom edge of the bell. The bubbles will form slowly at first, and then become a rapid stream. Heating is discontinued at this point. The bubbling rate decreases, and when the last bubble "hangs" on the edge of the bell and is then drawn inside the bell, the thermometer temperature is read. This is the boiling point of the liquid. Boiling points within $0.5°$ of literature values are readily obtained.

In the *ultra-micro method*, the mid-section of an open-ended capillary tube is heated using a microburner, and is then drawn out to form a thin filament. This section is broken in the middle, and the resulting two open ends are sealed. The appropriate length for a bell (3–5 mm) is broken off from each of the two sections, thereby obtaining two ultramicro bells. The bell is then inserted, open-end down, inside a closed-end capillary tube. Using a capillary delivery pipet, a microdrop of the analyte is added to the capillary tube. It may be necessary to centrifuge the assembly to bring the bell and liquid to the bottom of the capillary. The capillary is then attached to a thermometer, and the boiling point obtained as in the method above. Both procedures are shown in Figure 34.2.

Measurement of Density

Densities are obtained at the Microscale level using micropycnometers,[10] which require only about $100 \mu L$ of liquid. Extremely accurate results are obtained in this manner. The micropycnometer is constructed by heating a 5″ Pasteur pipet in the oxidation zone of a microburner flame, 1 cm from the constricted part of the pipet stem as shown in Figure 34.3. The pipet is rotated in the flame until it becomes *very soft* (so that it has almost liquefied). It is then removed from the flame, and the two ends are pulled apart to form a 10- to 25-cm-long thin capillary. The capillary is strongly heated about 0.5 cm from the rear of the bulb to detach it from the rest of the pipet. The pipet (which now has a capillary stem) is saved to transfer the solutions into the micropycnometer. This detached part is called a *microdelivery pipet*. The sealed part of the pycnometer is heated strongly to form a smooth, rounded end. Finally, the opposite end of the pycnometer is cut to widen its mouth, and the opening is fire-polished. The mouth of the pycnometer must be large enough to allow the pycnometer to be filled easily by the microdelivery pipet.

The dry pycnometer is weighed, and some of the analyte is transferred into it using the microdelivery pipet. As the pycnometer is filled to the top, the microdelivery pipet should be slowly removed to prevent an air bubble from forming. The outside of the pycnometer is cleaned and dried, and the pycnometer is then reweighed.

Using the same microdelivery pipet, the analyte is removed from the pycnometer. The pycnometer is then rinsed once with acetone and twice with distilled water, and then refilled with distilled water. The water-filled pycnometer is then weighed.

Using the density (from a table) and the mass of water, the volume of water, and, hence, of the micropycnometer is calculated. Since the mass and volume of the analyte are both known, the density can be calculated.

Titration

At the microscale level, titration is carried out using a microburet, which is assembled in the following way. A 5- or 10-mL syringe body is attached to the top of a 2-mL calibrated (± 0.01 mL) pipet using a 1″ length of Tygon tubing. A yellow (10- to 100-μL size) automatic-delivery pipet tip is pushed onto the bottom of the pipet tip (to allow for smaller drop formation). The buret is easily filled from the bottom, by inserting the tip into the appropriate liquid, and pulling up on the syringe plunger. With a little practice, single microdrops of liquid are easily delivered.

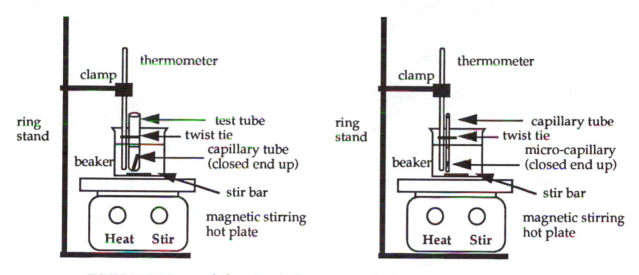

FIGURE 34.2 Micro and ultramicro boiling point methods.

Titrations carried out using microburets of this type give results that have the same precision (four significant figures) as those done with a 50-mL buret. Since the time required for a single titration run is much shorter with the microburet, additional replicate titrations can be run in the same amount of time, resulting in greater accuracy in the results.[11] Acid-base, oxidation-reduction, and complexometric titrations can be successfully carried out in this manner.

The microscale titration method is somewhat different from that at the standard scale in that indicators (which are themselves acids or bases) must be added to the 10 mL of stock solution of titrant, rather than to the

analyte. This is done to avoid titration errors, given the small amount of analyte.

Distillation

There are two types of distillation carried out at the microscale level. The simple distillation apparatus consists of a conical vial or round-bottom flask attached to a Hickman still. The still head can be further connected to a condenser and drying tube, if desired. The apparatus is shown in Figure 34.4a. As the mixture is heated, the more volatile fraction collects in the column collar, capable of holding up to 500 μL of liquid. The distillate may

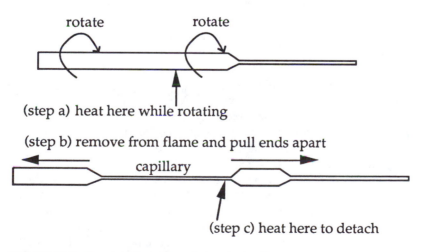

FIGURE 34.3 Construction of a micropycnometer.

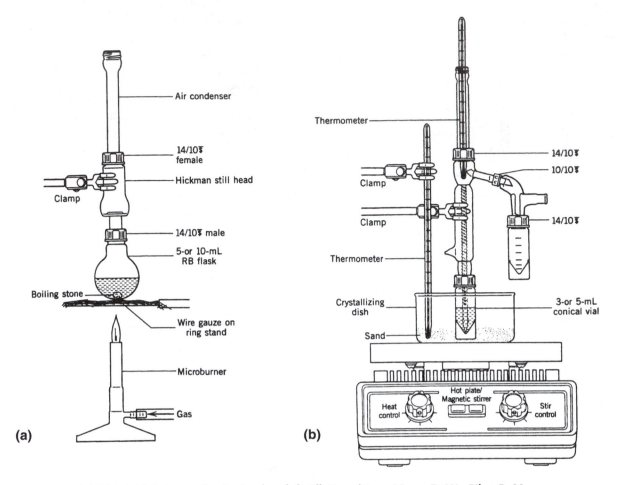

FIGURE 34.4 Hickman and spinning-band distillation. (From Mayo, D. W.; Pike, R. M.; Butcher, S. S.; Trumper, P. K. *Microscale Techniques for the Organic Laboratory*; Wiley: New York, 1991. Reprinted by permission of John Wiley & Sons, New York.)

be removed from the collar using a Pasteur pipet. Separations of liquids with boiling point differences of greater than 30°C can be accomplished in this manner. Later versions of the Hickman still have a side arm, which greatly aids in removal of the distillate from the collar (see Ref. 6).

More difficult separations require that spinning-band distillation be used. The system consists of a conical vial containing a spinning-band, attached to a one-piece combination air-jacketed condenser with a Claisen head and thermometer adapter. A side-arm-vacuum adapter and collection vial completes the apparatus, which is shown in Figure 34.4b. The spinning band is made of Teflon, and has a magnet embedded in it at the tip, which in turn is cut to fit in the cone of a conical vial. The band is spun magnetically using a magnetic-stirring hot plate, which also serves as the heat source for the distillation. As the material is heated, the spinning band forces the more volatile fraction to partially condense,

and to mix intimately with the rising vapor fraction. This process is known as rectification, and allows even as small a column as this to develop nearly 12 theoretical plates, the highest resolution of any standard distillation apparatus (including at the standard scale).

Other Types of Separation

Even smaller quantities (down to 5–10 μL) can be separated and purified using preparative gas chromatography. The analyte is injected into a gas chromatograph. A special stainless steel heat sink fitting (available from Ace Glass, Inc.) is attached to the outlet port of the chromatograph. The fitting is threaded on the inside to screw onto the outlet port, while at the other end, the fitting emulates a female ground glass joint on the inside, and is threaded on the outside to accept an O-Ring Cap Seal of a collector tube. The collector tube is 10-cm long, with two blown-out bulbs to hold the eluent. The chromatogram trace (or other output device) indicates when collection begins and ends, although it is possible to see the collection with the naked eye. Once separation is complete, the collector tube is removed from the fitting, attached to a 100-μL conical vial (using the same Cap Seal) and centrifuged. In this way, the separated fraction is collected in the conical vial for further work or analysis (see Ref. 6).

Purification by Crystallization

Craig tubes are used to carry out crystallizations of 10–100 mg of solids at the microscale level. The Craig tube consists of two parts: a small glass tube with an imperfect ground glass interior at the end, and a Teflon barrel. The interior is imperfectly ground so that the barrel does not completely seal the tube, which allows the mother liquid to pass through while the solid is retained.

The solid to be purified is dissolved in a minimum of hot solvent inside the test tube portion of the Craig tube.

A magnetic stirring hot plate is used as the heat source. The mixture is rapidly stirred and agitated using a microspatula to speed the dissolution process. The Teflon barrel is inserted (to keep out dust), the Craig tube is placed into a small beaker, and allowed to cool slowly to room temperature (or, if desired in an ice bath). A piece of string is tied around the stem of the Teflon barrel and a centrifuge tube is placed over the assemblage, which is then inverted. Centrifuging removes the mother liquor from the recrystallized product (see Ref. 6).

A Brief History of Microscale Chemistry

Carrying out chemistry experiments at microscale levels was attempted in the mid-1800s, with the work of Emish and Pregl in Germany. Pregl received the Nobel Prize for his microscale work in 1917. Microscale chemistry gained a foothold in the United States shortly after World War II. Cheronis and Ma at Brooklyn College,

---

Box 1
National Microscale Centers

National Microscale Chemistry Center

Merrimack College	Director: Mono M. Singh
315 Turnpike Street	Phone: (508) 837-5137
North Andover, MA 01845	Fax: (508) 837-5017
	Web: www.silvertech.com/microscale
	email: msingh@merrimack.edu

Nordic Microscale Chemistry Center (Finland)
Director: Touko Virkkala
web: sun1.kokpoly.fi/ketek/kemia/nmc2.html
email: touko.virkkala@kokpoly.fi

Swedish National Microscale Center
Director: Christer Gruvberg
email: christer.gruvberg@kask.se

Australasian Microscale Chemistry Center
Director: Enrico Mocellin
web: apamac.ch.adfa.oz.au/AMCC
email: erasec@deakin.edu.au

Mexican Microscale Chemistry Center
Director: Arturo Fregoso
email: afregoso@uibero.uia.mx

Netherlands Microscale Chemistry Center
Director: Martin Goedhart
email: goedhart@chem.uva.nl

Benedetti-Pichler and Schneider at Queens College, and Stock at the University of Connecticut taught microscale techniques and published several papers and texts.

These microscale programs, however, were not widely adopted in either industry or academia. This was due to several factors:

- Enviromental concerns were not in vogue before 1970.
- Laboratory air quality and the risks associated with chemical exposure were not too well understood until recent times.
- Chemical costs were low, and few regulations governed waste disposal.
- Microscale quantities of materials were very difficult to measure accurately, since electronic balances and automatic delivery pipets were not available.

Microanalysis was therefore a delicate, tedious, and specialized process, practiced to a small degree only in research and graduate areas. It was not until the early 1980s, when environmental concerns had risen to the forefront and the electronic milligram balance became available, that using microscale experiments became practically possible.

The initial development of the modern form of microscale chemistry was carried out by Dana W. Mayo and Samuel S. Butcher (from Bowdoin College) and Ronald M. Pike (from Merrimack College) in an effort to solve a laboratory air-quality problem in the organic chemistry laboratory. The Bowdoin Chemistry Department first considered retrofitting the laboratory with improved air-handling capacity, but the cost was about $300,000. This led to the inevitable question, "Isn't there a better way?" resulting in the rebirth of microscale chemistry.

The necessary techniques, materials, and experiments were then developed for the organic chemistry laboratory in 1982–3. The first microscale chemistry textbook, *Microscale Organic Laboratory*, was published in 1986.[6] Microscale chemistry was expanded to the area of Inorganic Chemistry by Zvi Szafran, Ronald M. Pike, and Mono M. Singh (at Merrimack College) in order to expand the range of experimental coverage to include such important areas as organometallic chemistry of the heavy metals, catalysis, and bioinorganic chemistry. The textbook *Microscale Inorganic Chemistry: A Comprehensive Laboratory Experience* appeared in 1991.[7]

Microscale conversion in organic and inorganic chemistry is currently widespread, with about 2000 colleges and universities across the United States doing laboratory work at the microscale level. Most recently, a National Microscale Center has been established at Merrimack College, to further the worldwide dissemination of microscale chemistry technology, and to encourage the incorporation of these techniques into academia and industry. National centers are also in place in Mexico, Sweden, Finland, the Netherlands, and Australia (see Box 1).

References

1. Pickering, M.; LePrade, J. E. Macro- vs. Micro-Lab: A Controlled Study of Time Efficiency. *J. Chem. Educ.* **1986**, *63*, 535.
2. Butcher, S. S.; Mayo, D. W.; Pike, R. M.; Foote, C. M.; Hotham, J. R.; Page, D. S. Microscale Organic Laboratory I: An Approach to Improving Air Quality in Instructional Laboratories. *J. Chem. Educ.* **1985**, *62*, 147.
3. Mayo, D. W.; Butcher, S. S.; Pike, R. M.; Foote, C. M.; Hotham, J. R.; Page, D. S. Microscale Organic Laboratory II: The Benefits Derived from Conversion to a Microscale Organic Laboratory Program and Representative Experiments. *J. Chem. Educ.* **1985**, *62*, A238.
4. Example drawn from: Mayo, D. W.; Pike, R. M.; Butcher, S. S.; Trumper, P. K. *Microscale Techniques for the Organic Laboratory*; Wiley & Sons: New York, 1991.
5. Szafran, Z.; Singh, M. M.; Pike, R. M. The Inorganic Microscale Laboratory: Safety, Economy and Versatility. *J. Chem. Educ.* **1989**, *66*, A263.
6. Mayo, D. W.; Pike, R. M.; Trumper, P. K. *Microscale Organic Chemistry*, 4th ed.; Wiley & Sons: New York, 2000.
7. Szafran, Z.; Pike, R. M.; Singh, M. M. *Microscale Inorganic Chemistry: A Comprehensive Laboratory Experience*; Wiley & Sons: New York, 1991.
8. Szafran, Z.; Pike, R. M.; Foster, J. C. *Microscale General Organic Chemistry Laboratory with Selected Macroscale Experiments*. Wiley & Sons: New York, 1993.
9. Singh, M. M.; Pike, R. M.; Szafran, Z. *Microscale and Selected Macroscale Experiments for General and Advanced General Chemistry: An Innovative Approach*. Wiley & Sons: New York, 1995.
10. Szafran, A.; Pike, R. M.; Singh, M. M. *Microscale Chemistry for High School*, Kendall/Hunt Publishing Co.: Dubuque, IA, 1998; Vol. 2.
11. Singh, M. M.; McGowan, C.; Szafran, Z.; Pike, R. M. Comparative Study of Microscale and Standard Burets. *J. Chem. Educ.* **2000**, *77*, 625.

35

Laboratory Scale-Up and Pilot Plant Operations

AMY L. ROMANOWSKI

Scale-up is generally a sign of a successful research or development project because at this stage a new product or process shows enough promise to warrant work on a larger scale. Sometimes the purpose is to produce larger batches of a prototype material. More often, scale-up units are built to provide data for commercial design. Decisions made during the scale-up process can have profound effects on the health, safety, and environmental (HS&E) aspects of a commercial plant.

Managing the safety of scale-up activities is challenging. By their nature, research and development operations are constantly changing, and consequently, the properties of the materials and reactions are not always completely understood. Scale-up operations fall between the better-defined territories of good laboratory practices and commercial process design.

Although there can be a tendency to slight the safety review process for these operations, other factors dictate that the review process be taken seriously and that it focus on the unique aspects of these operations. Compared to commercial plants, scale-up units are operator-intensive with their frequent start-ups, shutdowns, material handling, and sampling. While scale-up operations generally lack the potential for a major catastrophe, handling even a small amount of a hazardous material under the wrong conditions is enough to cause serious injury. This chapter provides an overview of health, safety, and environmental issues for the scale-up of laboratory and pilot plant operations.

Hazard Identification and Assessment

The safe and effective scale-up of laboratory operations begins with a thorough understanding of the hazards involved. The best design, procedures, and training cannot effectively address unrecognized hazards. A good method of hazard identification is a systematic review meeting conducted by a variety of people with different backgrounds. The group should include people familiar with the chemistry, design, control systems, equipment, safety and regulatory standards, and day-to-day opera-

tions and maintenance. This hazard assessment review is best done when the design basis is defined but before equipment-specification and construction stages are complete. A checklist is a good way to ensure that no areas are overlooked. Figure 35.1 contains items to consider for inclusion in a hazard identification checklist.

The timing is important. If the review is performed before the operating conditions, chemistry, and flows are defined, hazards can be missed. If it is performed after equipment specification and purchase, changes can be expensive and can delay the project. Although the purpose of the meeting is to identify safety issues, the results usually include ideas for improved operability and more effective process design. These improvements contribute to the safety of the unit by making it easier to operate. The quality of work done at this stage not only contributes to the safe operation of the development plant, but also can have significant benefits at the commercialization stages.

Understanding the Chemistry

In novel reaction systems, conditions that cause hazardous reactions may not be well understood. Start-up difficulties, such as contamination or an overlong cooldown period, can trigger temperature runaways or other hazardous reactions in operations that have been run successfully in the lab many times.

The first line of defense is a thorough understanding of the chemistry, including reactions at normal operating conditions and for reasonably anticipated deviations. Various calorimetry methods should be used to characterize systems before scale-up when the potential exists for hazardous exothermic reactions and the mechanisms and triggering events are not well understood.

Impurities

Laboratory operations often use research-grade materials of high purity. At some point in the scale-up, the continued use of such materials becomes impractical or prohibitively expensive. When switching to lower-grade

1. **Physical Hazards.** Check all that apply:

Confined areas	Heat stress	Radiation, ionizing
Cryogenics	Hot surfaces	Radiation, non-ionizing
Dust hazard	Moving equipment	Rotating equipment
Electrical hazard	Noise	Sharp objects
Falling objects	Overhead hazards	Slippery surfaces
Flying objects	Pinch points	Tripping hazard

2. **Chemical Exposure Hazards.** List all chemicals and the following properties for this operation:
   a. Name: Include CAS number and concentration if present in a mixture
   b. NFPA rankings for health, flammability and reactivity
   c. Volume and rate used or generated
   d. Operating temperature, pressure, and physical state
   e. Hazard type if applicable (corrosive, poison, irritant, carcinogen)
   h. Exposure routes (inhalation, skin, ingestion) and exposure limits (OSHA, ACGIH)
   i. Activity that may result in exposure (spill, leak, transferring between containers, sampling, slopping waste)

3. **Chemical Reactivity Hazards**
   a. Describe the chemistry of the operation including the energy of reaction. List specific chemical reactions when applicable
   b. Can the chemistry of the operation produce hazardous exotherms, side reactions, by-products, or other hazardous conditions. If yes, what are the conditions of temperature, contamination, and so forth that produce the hazardous condition? Is calorimetry needed to further define this reaction system?

4. **Regulatory and Compliance Issues**
   a. What electrical classification does the operation require?
   b. Are additional chemical or flammable storage facilities required?
   c. Is any special ventilation or containment required?
   d. Is an experimental chemical not on the TSCA inventory being used, imported, generated, or shipped?
   e. Does the work involve the use of contractors?
   f. Does the operation require a steam boiler or a state-licensed pressure vessel?
   g. Does the operation use a fired heater?
   h. Does the operation contain a confined space?
   i. Does this operation qualify for coverage under the OSHA PSM or EPA RMP standards?
   j. What specific training is required for this operation?
   k. Do any special material-handling requirements need to be met?
   l. What medical surveillance is needed?
   m. What personal protective equipment is needed?

5. **Other Hazards**: List any other hazards associated with the operation, including items from previous lab, pilot plant, or commercial experience with this operation.

FIGURE 35.1 Items to consider for inclusion on a hazard assessment form.

materials, researchers must determine the type and quantity of impurities and consider their potential effects on the safety of the operation. Impurities often reflect the source of the chemical, and the vendor can usually supply a list of the expected contaminants and their range of concentration. Other impurities, such as water and oxidation products, can be introduced by poor handling and storage practices.

Cross-contamination is a potential problem with scale-up operations because equipment is often switched from process to process as the need arises. Simple flushing cannot always remove material accumulated in crevices and dead spots, and some disassembly may be required. Cleaning solvents should be included in the hazard review; they are easily overlooked and may have toxic or corrosive properties.

Another type of impurities to consider are those that may be introduced by incompatible materials of construction in the storage or processing equipment. Although most often a product-purity problem, corrosion by-products may have an unexpected catalytic potential and cause a runaway reaction.

Flammables

Flammable materials fall into several categories: gases, liquefied gases, and liquids. Table 35.1 shows the further classification of flammable and combustible liquids as defined by the NFPA.[1,2] In the process of scale-up, the quantity of flammable materials can quickly overtake the safe limits of an area and previous practices. For example, the transfer of flammable liquids in quantities greater than about 5 gal requires bonding and grounding to protect against fire from a static electric spark.[3] Grounding is a connection through conductive materials to an earth ground; it permits the dissipation of electrical charges safely to the ground. Bonding is a conductive connection between two vessels to allow the equilibration of electrical charges during the flammable liquid transfer. Fatal fires have occurred when the user scaled up without understanding the importance of preventing static charge buildup by proper electrical grounding and bonding.

Laboratories are generally considered to be unclassified electrically,[4] based in part on the scale of the operations. The NFPA 45 Standard on Fire Protection for Laboratories Using Chemicals recommends specific limits on the quantities of flammable and combustible liquids in laboratories. At some point in the scale-up process, operations involving flammables will require the electrical classification of areas. The result of this classification is the requirement for special electrical equipment and installation. Guidelines are available in the National Electric Code and from the American Petroleum Institute. At any scale of operation, electrical classification and minimization of ignition sources must be carefully considered before working with flammables.

Design and Equipment Safety Issues

Design Review Process

The design review process in scale-up work can be considered in three parts: the paper review of specifications, drawings, and procedures; the physical review of the as-built unit; and the ongoing review of change.

Paper Review

The paper review is sometimes called a *P&ID (piping and instrumentation diagram) review* or *engineering design review*. It takes place after the initial hazard assessment when the design specifications and drawings are complete. The purpose of the review is to verify that good engineering practices and appropriate standards have been used in the design so that the unit meets its process objectives safely. This review typically consists of a line-by-line examination of the process flow diagram or P&ID conducted by a group of reviewers familiar with the operations, equipment, control systems, and applicable codes and regulations. A list of standard questions is a good aid to ensure that all aspects of safe design and regulatory compliance are included in the review.

At this stage of the scale-up process, a summary document of design safety information should be compiled. This document generally includes a list of pressure vessels with pressure and temperature rating, relief valve sizing, limits, alarms, interlocks, hazardous gas monitors, guards and shields, and any other safety-specific design information. In addition to its use during the initial review, this summary is invaluable for later reviews of modifications.

During the engineering design review, all aspects of the process must be considered, including start-up, shutdown, and maintenance. Equipment must be fully compatible, not only with the process materials but also with any other materials used for support operations such as cleaning. Chemical compatibility of parts will vary across a temperature range. Particular attention should

TABLE 35.1 Flammable and Combustible Liquid Properties[1,2]

Property	Flammable Liquid			Combustible Liquid		
Liquid class	IA	IB	IC	II	IIIA	IIIB
Flashpoint, °F (°C)	<73 (<22.8)	<73 (<22.8)	≥73 (≥22.8) and <100 (<37.8)	≥100 (≥37.8) and <140 (<60)	≥140 (≥60) and <200 (<93.3)	≥200 (≥93.3)
Boiling point, °F (°C)	<100 (<37.8)	≥100 (≥37.8)	—	—	—	—
NFPA Flammability Ranking	4	3	3	2	2	1

be paid to o-rings, valve seats, and gaskets to ensure chemical compatibility with all materials at all operating temperatures.

Corrosion in development plants is a critical consideration. Small equipment, particularly thin-walled tubing, can corrode to failure very quickly, resulting in a release of corrosive material. In addition, the by-products of corrosion can contaminate the operation and cause plugging of lines, valves, and pressure relief devices.

Sometimes the detailed methods of hazard analysis and risk assessment, such as a HAZOP (Hazards and Operability Analysis), What If?, or FMEA (Failure Modes and Effects Analysis), are applied to hazardous scale-up operations. These methods require a trained facilitator, and a detailed description is beyond the scope of this chapter.

Physical Review

The physical review can be called the *precommissioning* or *pre–start-up safety review*. The main purpose of this review is to confirm that the unit was built according to the specifications approved in the previous reviews. Because Research and Development (R&D) is a dynamic process, changes may be introduced during the construction phase. All deviations from the approved design must be identified during this phase of the project and reviewed by a competent authority to be sure the changes do not compromise the safety of the operation. These changes and their endorsement must be documented.

A second objective of the pre–start-up review is to be sure the equipment works as planned and that the unit is safe for operation. This review is the time to verify that the alarm set points match the specifications and that the lights and bells go off when they are supposed to. During this review items such as sharp edges and uninsulated lines are caught before they cause injury. This review is also the time to make sure the aisles are clear, the fire extinguisher is the right type, and the safety shower and eyewash are plumbed and working. Once again, a checklist is the best way to make sure all areas are covered.

Ongoing Review

Ongoing safety reviews of scale-up operations are the most difficult part of the design review process because change is constant in research and development work. Once assembled, a laboratory unit or pilot plant is modified based on results until the process goals are met or determined to be unmeetable. Then everyone moves on to the next project.

Common oversights in making modifications that can lead to hazardous conditions include underestimating the effect of temperature, failure to consider backflow, and using incompatible materials of construction. This is where the design summary document prepared during the design review process should be used to confirm that the proposed changes fall within the safe design limits of the equipment. The changes need to be reviewed and approved by a competent authority and they must be documented. The design summary document and the P&ID must be updated.

An effective ongoing review program starts with management commitment and a culture that values safety. It also helps to keep the paperwork simple and remind the users that changes that can cause safety problems can also cause process problems.

Specific Design Issues in Scale-up Operations

Space Issues

Few facilities have the luxury of excess space. Operations that scale up gradually can easily overwhelm their site and create a crowded and hazardous condition. Equipment in hoods can become too dense to allow adequate air flow. A release near the front of an overcrowded hood can escape into the room and the breathing space of an unsuspecting operator. Interactions of operations that are in close proximity can affect safety. Without proper planning, a leak of flammable material on one unit can be ignited by an adjacent operation.

Building codes specify aisle widths for the main exit hallways in a building, but few hard rules for aisle space in laboratories and pilot plants exist. Recommendations for aisle width vary from 18 in. to 6 ft., depending on the type of access needed and the amount of traffic. The tasks performed on all sides of the unit by operations and maintenance personnel need to be considered.

When allocating space for scale-up operations, planners often overlook the room needed for support functions. Larger equipment may require ladders where none were needed before. The risk of personnel injury increases as tools, equipment, and chemical containers get larger and heavier. The weight of the items being handled and their positioning need to be considered. The risk increases if workers are maneuvering in close quarters or reaching at awkward angles because of space constraints. Checklists used in the hazard assessment and design review phases of the project can identify these concerns and ensure that they are appropriately managed in the final design.

Ventilation

Laboratory experiments are often performed in fume hoods to minimize personnel exposure to toxins and the accumulation of flammable vapors and associated fire hazards. Commercial operations located outdoors have natural ventilation. However, many scale-up operations are located indoors in laboratories or in dedicated pilot plant space. In many chemical laboratories, most of the exhaust ventilation comes from the hoods. This air-flow pattern is a potential problem for process units located in a general lab area because a gaseous release is drawn to the hoods through the area operators are likely to be working in. To minimize the chance of operator exposure to toxic chemicals, a distributed ventilation system that includes movable, flexible trunks is the best option for these laboratories and pilot plant buildings.

Mixing

The scale-up of mixing processes must be carefully designed to be effective and avoid potential safety problems. Mixer design is complex. To achieve an equally well-mixed system at 10 gallons compared to 1 gallon does not mean increasing everything by ten. Some scale-up ratios do go up linearly; but others increase in proportion to the square or the cube of the linear dimension. Equipment manufacturers are good sources of assistance in how to design a mixing system for a larger tank. Although poor mixing more often causes problems with product quality than with safety, the recognition of potential problems is important.

Stratification can occur when materials with very different densities are inadequately mixed. In exothermic reactions, one component is usually added at a controlled rate to a well-mixed vessel to allow the heat of reaction to evolve slowly enough to be controlled by the reactor's cooling system. With poor mixing, stratification can occur and the reaction may not be gradual, but may occur suddenly when mixing is improved. In this case the rapid heat generation can overwhelm the capacity of the cooling system and cause pressure build-up. This situation can happen when the bottom dense layer becomes high enough to reach the impeller or, worse, when the material is drained from the system by an unsuspecting operator.

Stagnant spots in specific equipment can also cause problems. For example, glass-lined vessels usually have a hollow at the bottom of the vessel leading to the outlet valve because the glass-lining process needs gradual curves in the vessel geometry.

When scale-up operations make use of existing equipment, the batch should be sized to match the mixer's capacity. If the operation includes the gradual addition of a component to a well-mixed system, the volume at the *start* of mixing must be high enough in the vessel to cover the impeller.

Heat Transfer

Another case of geometry affecting safety is in heat transfer. In scaling-up process equipment, the ratio of surface area to volume usually decreases for standard geometries. In other words, less external surface area is available for heat transfer per pound of material. The heat-up and cooldown rates for a larger reactor of the same shape are longer. In the worst case, the rate of heat generation exceeds the rate of heat loss, and a hazardous runaway reaction occurs. This phenomenon also occurs outside of reaction vessels and can be a problem in large-scale storage. Fires and explosions have been reported from the bulk storage of ammonium nitrate and even wood chips.[5]

Control Systems and Safety Interlocks

The quantity and complexity of instrumentation, control systems, and safety interlocks generally increase as operations scale up in size. At each step, the decisions made about the numbers and types of control points, alarms, and interlocks must be based on a careful analysis of the hazards.[6]

Distinguishing between control and safety systems is important. Although related, they have different functions. Control systems are designed to maintain operations within specifications for optimum data generation or product quality. Safety interlocks are designed to intervene to keep a process out of a hazardous range of operation. Critical safety interlock systems must be kept separate from control systems because they are most likely to be needed when the control system is not functioning properly. The proliferation of microprocessor-based control systems has brought changes to safety interlock system design. In many cases physical relays are being replaced by lines of programming code that need to be carefully protected from inadvertent changes.

Control and interlock systems must be designed to fail in a safe manner if power or other utilities are lost. Battery backups are not reliable enough to be counted on for safety-critical systems. The best use of backup systems is to provide power to key sensors and screens so that the operator has information on unit status during upset conditions. Determining the appropriate fail-safe mode requires some analysis and a thorough understanding of the hazards.

The failure mode of individual components must also be considered, particularly with a microprocessor-based system. For example, temperature sensors are generally set to fail to the high end of the temperature scale so that heaters will stop supplying power if the sensor fails or is disconnected.

As a final note, what happens in the restart mode when power returns needs to be considered. Safety interlocks should not automatically restart but should require the operator to verify that the unit is safe for restart. When a battery backup supplies power to the microprocessor if the plant goes down, the program may continue to advance even though the plant equipment is not responding. Consequently, potential problems can occur when power returns.

Regulatory Issues

Many regulations for chemical processes come into effect when operations exceed a given threshold size or quantity. Requirements from other sources, such as internal policy or insurance carriers may also be based on size. A good way to recognize when a scale-up operation reaches these trigger points is to incorporate pertinent questions into the review checklists. A few examples of codes and regulations that come into effect above a defined threshold quantity include:

- OSHA Process Safety Management (PSM), 29 *CFR* 1910.119
- EPA Risk Management Plan (RMP), 40 *CFR* 68
- ASME Pressure Vessel Code—Insurance carriers and state laws usually require pressure vessels above a certain size to be fabricated in an ASME-certified shop and equipped with ASME-certified pressure-relief valves.

This is not meant to be comprehensive. Users need to make a careful survey of regulations and policies to determine what is applicable to their operation.

Inherent Safety

When scaling up a process, fundamental decisions are made about the design of the commercial plant. This stage is the best time to incorporate inherent safety features into the design. Inherent safety is based on the idea that avoiding hazards entirely is better than protecting against them. The 10 features of inherent safety summarized below come from Kletz's book.[7]

Intensification: The practice of reducing the inventory of hazardous materials, most effectively done in storage and reactor design

Substitution: The use of a safer material in place of a more-hazardous one, such as replacing a flammable solvent with a nonflammable material

Attenuation: The use of a hazardous material in a less-hazardous form. Solids that form explosive dusts are more safely handled as a slurry.

Limitation of effects: Equipment or design that results in a better "worst case scenario." Steam heating is limited in effect compared to electrical heaters.

Simplification: Fewer opportunities for mistakes or equipment failure

Avoid add-on effects: Equipment siting should be based in part on minimizing the consequences of likely failures on the surrounding equipment and environment.

Make incorrect assembly impossible: To avoid the installation of two lines in the wrong ports in close proximity, use different styles or sizes of fittings.

Making status clear: The use of equipment, such as ball valves, that gives the operator quick and obvious information

Tolerance: A measure of how well a design or piece of equipment forgives poor installation or operation. Metal vessels are more forgiving than glass

Ease of control: Processes designed with a flat and slow response to change

Human Factors

The human-equipment interface is an integral part of process safety, and factors that affect human reliability must be considered in hazard analysis and design. OSHA has recognized the importance of human factors and included a requirement for considering them in the hazard analysis done for processes that are covered by the Process Safety Management standard, 29 *CFR* 1910.119. These factors range from excess noise and heat in the workplace to the value placed on safety in the corporate culture. The *Guidelines for Preventing Human Error in Process Safety* provides a thorough account of human error considerations for process safety.[8] Human reliability factors that should be given special consideration in the design and operation of laboratory scale-up experiments and pilot plants include:

- *Insufficient information transfer*, usually between the project scientist and the operations staff, either because the information is not known or the knowledge gap is not recognized
- *Frequent changes*, so that keeping process information, procedures, and labels up to date is difficult

Window of Opportunity for an Accident

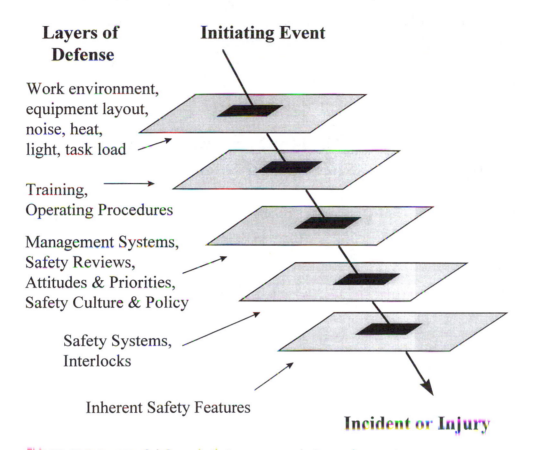

FIGURE 35.2 Layers of defense built into process design and operation to prevent accidents or injury.

- *Task complexity and unfamiliarity* because of frequent start-ups, shutdowns, and working with processes that are not yet well-defined
- *Working conditions* such as crowding and poor lighting that arise over time as equipment setups evolve and change

Good process design builds many layers of protection to minimize the chance of a single failure causing a major incident. Figure 35.2 is a representation of the layers of defense that usually stand between an initiating event and a major incident or injury. Starting at the bottom, the fundamental layer of defense is the use of inherent safety features. The next layer is equipment-related such as pressure-relief valves and safety interlocks. The remaining layers of defense are human factors: management systems and the priority placed on safety reviews, operator training and procedures, and finally the work

environment including factors such as heat, noise, and the individual's taskload.

The best defense against accidents and injuries in the workplace is an environment where management clearly demonstrates safety as a core value and where operators are well trained in the process and equipment and are involved in the design and safety analysis of new and modified operations.

References

1. NFPA 30. Flammable and Combustible Liquids Code; National Fire Protection Association: Quincy, MA, 1993.
2. NFPA 704. Standard System for the Identification of the Fire Hazards of Materials; National Fire Protection Association: Quincy, MA, 1990.
3. NFPA 77. Recommended Practice on Static Electricity; National Fire Protection Association: Quincy, MA, 1993.
4. NFPA 45. Standard on Fire Protection for Laboratories Using

Chemicals; National Fire Protection Association: Quincy, MA, 1996.

5. Bisio, A.; Kabel, R. L. *Scaleup of Chemical Processes*; Wiley & Sons: New York, 1985.

6. Palluzi, R. P. *Pilot Plant and Laboratory Safety*; McGraw-Hill: New York, 1994.

7. Kletz, T. *Plant Design for Safety: A User-Friendly Approach*; Hemisphere: New York, 1991.

8. *Guidelines for Preventing Human Error in Process Safety*; Center for Chemical Process Safety of the American Institute for Chemical Engineers: New York, 1994

36

Reducing Electrostatic Hazards Associated with Chemical Processing Operations

VAHID EBADAT

JAMES C. MULLIGAN

Introduction to Static Electricity

Static Electricity

Static electricity is the charge that is developed when two contacting surfaces move relative to each other. Perhaps the most familiar manifestation of static electricity is that which is developed as one walks across a carpeted floor, particularly on dry days. In this instance, the two contacting surfaces are the carpet and the soles of your shoes. You become aware of the static electricity developed when you reach out to touch a doorknob and receive a shock. In fact, rather than receiving the shock, you are really originating it by discharging the static charge accumulated in your body to the doorknob—a point at a lower electrical potential.

The need for this chapter is evidenced by the fact that static electricity is developed during many common chemical processing operations, such as when liquids and powders move relative to other surfaces. The mechanisms by which electrostatic charge is developed and accumulated during chemical processing operations are described in the following sections.

The primary concern posed by static electricity in an industrial setting is the risk of fire and explosion due to the ignition of flammable atmospheres by electrostatic discharges. Flammable gases, liquids, and powders are used in many chemical processes and precautions are required to prevent their ignition. One area of precautions involves the exclusion or elimination of potential electrostatic ignition sources from locations where flammable atmospheres may be expected to exist. It is this area that is the focus of this chapter.

Electrostatic Charge Development

There are several mechanisms by which electrostatic charge is developed: (1) contact electrification; (2) double-layer charging; and (3) induction charging.

Contact Electrification

Contact electrification—also referred to as triboelectrification—occurs when two solid materials come into contact and then separate. There is a rearrangement of electrons at the contacting surfaces. One surface gains electrons and is said to be negatively charged, while the other surface is left electron-deficient and therefore is said to be positively charged. In cases involving two conductive surfaces, depending on the speed and nature of the separation, most of the charge may recombine as the surfaces are separated, and therefore the magnitude of the resulting net charge may be small. However, when one or both of the surfaces are insulating, the charge recombination is limited.

Contact electrification occurs when, for example, powder particles collide or slide with other surfaces during handling and when the soles of footwear make contact with insulating flooring. The speed and pressure of the contact, nature of the contacting surfaces, and the relative humidity, among other factors, influence the level of charge developed by contact electrification.

Double-Layer Charging

Double-layer charging can occur at liquid/solid and liquid/liquid interfaces. Ions of uniform (like) polarity are adsorbed at the interface. The adsorbed ions attract ions of opposite polarity within the liquid, thus forming a layer near the interface. This layer is thin for conductive liquids and more diffuse in relatively less conductive liquids due to reduced ion mobility. As the liquid flows relative to the interface, the oppositely charged ions repel and move apart, thus increasing the potential within the liquid. This effect—combined with the competing effect of ion recombination at the vessel wall—results in an unbalancing of charge within the liquid. This is the type of static charging that occurs as a liquid flows through piping.

The principal difference between low- and high-conductivity liquids is the rate at which ion recombination occurs. However, both low- and high-conductivity liquids can develop static charge during processing. The amount of charge developed is dependent primarily on the distribution and mobility of the ions in the liquid and on the liquid-flow characteristics. Relevant flow characteristics include turbulence and flow velocity. For example, turbulent flow typically causes higher static charging than laminar flow.

The principal indicator of ion mobility in liquids is conductivity. Liquids having conductivity less than 50 picosiemens per meter (pS/m) are generally considered insulating. However, liquids having a conductivity much greater than 50 pS/m are capable of developing high levels of static charge under certain conditions. For example, charge development in liquids can be aggravated by:

1. immiscible solids and liquids
2. filters
3. settling of immiscible materials
4. splashing and spraying
5. stirring and mixing

Under certain conditions, it is possible to control charge development in liquids by limiting flow velocity.

Induction Charging

Induction charging occurs when a conductive object, which is electrically isolated from ground, enters an electric field, such as that emanating from an adjacent charged object. The charged object causes a polarization of the otherwise balanced charge in the conductor. If an isolated conductor becomes charged by induction and is subsequently exposed to ground, an electrostatic spark discharge may result.

Electrostatic Charge Accumulation

Electrostatic charge can accumulate in liquids, on insulating solids, and on conductive objects and personnel that are electrically isolated from ground.

Charge Accumulation in Liquids

When the rate of electrostatic charging exceeds the rate at which ions recombine at the interface, charge accumulates in a liquid. Charge relaxation can be quite rapid for conductive liquids in suitably grounded vessels and piping. Low-conductivity liquids can retain charge for somewhat longer periods (microseconds to a few minutes), even when conveyed and contained in grounded vessels and piping.

Charge Accumulation on Insulating Solids

The propensity of a solid material to accumulate static charge is reflected in its volume and surface resistivity. Materials having a volume resistivity on the order of 10^9 ohm-meters or greater or a surface resistivity on the order of 10^{10} ohms per square or greater, measured at <30% RH, are generally considered insulating (high-resistivity) in electrostatic terms. Typical insulating materials include polyvinylchloride (PVC), polytetrafluoroethylene (PTFE), polyethylene (PE), polypropylene (PP), glass, and many bulk powders and solids.

While insulating materials are considered hazardous primarily due to their ability to accumulate charge, they can also create a hazard by isolating conductors from ground. For example, PTFE gaskets can cause metal piping to become isolated from ground if proper precautions are not taken. Metal piping isolated from ground can accumulate electrostatic charge upon the flow of liquids or solids.

Charge Accumulation on Isolated Conductors

Typical conductive materials that can become isolated from ground and accumulate static charge include metal piping, flange couplings, vessels, tools, other metal plant items, and personnel.

Electrostatic Discharges

Electrostatic charge accumulation may give rise to hazardous electrostatic discharges. Electrostatic discharges occur between objects or surfaces at different potentials (voltages). As an object or surface accumulates charge, the electrostatic field strength above it intensifies. If the field strength exceeds the breakdown strength of the atmosphere, an electrostatic discharge will occur.

There are principally six types of electrostatic discharges:

1. Spark discharges
2. Brush discharges
3. Propagating brush discharges
4. Corona discharges
5. Cone discharges
6. Discharges from charged mists or dust clouds (in relatively large vessels)

Spark Discharges

Spark discharges occur between two conductors at different potentials. Typically, one of the conductors is charged while the other is grounded. Examples of spark discharges include discharges from metal piping, flanges, vessels, tools, and other metal plant items. Since the human body is considered conductive in electrostatic terms, discharges from charged personnel are also a form of spark discharge.

Spark discharges are characterized by a distinct energy channel between the conductors. Charge flows rapidly through the channel until almost all of the previously stored energy is dissipated or until the potential difference between the conductors is equalized.

The energy of a spark discharge is dependent on the capacity of the charged conductor to store charge and the potential to which the conductor is charged. This relationship is described by the following equation:

$$E = \frac{1}{2} CV^2 \tag{1}$$

where

E = stored energy (joules)
C = capacitance (farads)
V = voltage (volts)

The energy of a spark discharge from a person under normal charging conditions is less than 100 millijoules (mJ). The energy of spark discharges from charged metal plant items can be greater. In contrast, the minimum ignition energy (MIE) of common organic solvent vapors is typically less than 1 mJ. Consequently, spark discharges can pose a significant ignition threat in industrial flammable atmospheres.

Brush Discharges

Brush discharges are produced between a charged insulating surface and a grounded conductor having a relatively large radius of curvature (>1 mm), that is, a rounded or blunt electrode. Since charge mobility on insulating surfaces is impeded by the resistivity of the material, only charge from a limited surface area participates in a brush discharge. Also, charge cannot flow to form one distinct discharge channel and instead a number of smaller channels is formed. The term brush discharge refers to the appearance of these discharge channels, which resemble a brush. Brush discharges are known empirically to exhibit equivalent energies which are unlikely to exceed 4 mJ. Examples of brush discharges include discharges from plastic drums, drum liners, and implements.

Propagating Brush Discharges

When the opposite side of a highly charged surface of a thin insulating material is in close proximity to a grounded object, a double layer charge is created. If a grounded conductor approaches the highly charged insulating surface, a discharge may occur. This type of discharge—known as a propagating brush discharge—can exhibit an effective energy of the order of several Joules—significantly higher than an ordinary brush discharge. Consequently, propagating brush discharges can pose a significant ignition risk in flammable atmospheres. Propagating brush discharges typically occur where thin insulating materials are used in applications where high levels of electrostatic charge is developed, such as liners in metal vessels. Normally, only materials possessing a certain dielectric strength, that is, breakdown voltage of 4 kilovolts (kV) or greater, and a thickness less than 8 mm are capable of supporting propagating brush discharges.

Corona Discharges

Corona discharges arise when a pointed conductor, that is, a conductor having a sufficiently small radius of curvature (<1 mm), is raised to a high potential or is inserted into a strong electric field when grounded. In either case, the field strength at the tip of the pointed conductor is very strong and the atmosphere within the vicinity of the tip is ionized. Charged species are then accelerated either toward or away from the pointed electrode depending on their polarity. The term corona refers to the faint glow of the electron plasma which forms around the tip of the pointed conductor.

The effective energy of corona discharges has not been adequately described. However, these discharges are believed to pose an ignition hazard only to very sensitive flammable atmospheres, for example, hydrogen, acetylene, carbon disulfide, and other flammable gases and vapors in oxygen-enriched atmospheres.

Cone Discharges

Cone discharges, also known as bulking discharges, are discharges that occur on the surface of highly charged accumulations of high-resistivity bulk solids and powders during vessel-filling operations. As a bulk solid or powder accumulates in a vessel, the electric field on the surface of the material intensifies—a phenomenon known as charge compaction. The strong electric field causes the ionization of air just above the material surface, giving rise to cone discharges. The name derives from the typical cone shape in which powders and bulk solids accumulate in vessels. The effective energy of cone discharges is dependent on the mass charge density, particle size, and volume resistivity of the accumulating powder and the diameter of the vessel.

Lightning-like Discharges

In nature, lightning is produced by highly charged liquid droplets forming clouds or highly charged ash erupting from a volcano with volumes of the order of thousands of cubic meters. However, there is no evidence that lightning-like discharges can occur in industrial processes. Consequently, the probability that lightning-like discharges will occur in industrial equipment such as silos, tanks, reactors, and dryers is considered to be very low.

Flammable Atmospheres

As previously suggested, perhaps the greatest hazard posed by electrostatic discharges is as a potential ignition source for flammable atmospheres. Flammable atmospheres are atmospheres that are capable of being ignited and supporting combustion. Fuels commonly associated with industrial fire and explosion incidents include hydrocarbon gases, organic solvent vapors, mists, and dust clouds. The principal oxidant is the oxygen in air. Other oxidants include halogen gases, such as chlorine and bromine, and nitrogen and sulfur oxides (NO_x and SO_x).

The MIE of most hydrocarbon gases and organic vapors ranges from 0.1–1.0 mJ, while the MIE of most dispersed dusts ranges from <1–5000+ mJ. While some literature data for MIE of gases, vapors, and dusts are available, it is generally suggested that such data be developed on a process specific basis since variations in composition or physical and chemical characteristics can have significant effects. For example, the MIEs of powders and dusts decrease as particle size decreases. The type of electrical circuit used for determining the MIE of a flammable atmosphere can also influence the measurement results.

Inert gases, such as nitrogen and carbon dioxide, can be introduced to a vessel in order to decrease the relative oxygen concentration to a level below which a fire or explosion cannot occur. This oxygen concentration is referred to as the limiting oxygen concentration or LOC.

Each substance has its own particular LOC. It is prudent to apply a safety factor to the LOC so as to establish a maximum allowable oxygen concentration for a vessel. If properly implemented, inerting can be equally effective for all types of flammable atmospheres, including gases, vapors, and dispersed powders.

Controlling Electrostatic Hazards Associated with Specific Equipment and Operations

Metal Reactors, Tanks, Vessels

As described in earlier sections of this chapter, electrostatic charge can be developed during processing of liquids and bulk solids. Such charge may be retained when liquids and solids are introduced to reactors, tanks, and other vessels. Charge retained on liquids and solids can pose an electrostatic hazard when flammable vapors or dust clouds are created within the vessel. Potential electrostatic ignition sources include:

1. Electrostatic discharges from the surface of charged liquids and solids
2. Spark discharges from isolated conductors
3. Spark discharges from isolated personnel

Suggestions for reducing electrostatic hazards associated with metal reactors, tanks, and other vessels include the following:

General

1. All metal vessels and ancillary piping and equipment should be suitably grounded such that their electrical resistance to ground is less than 10 ohms to prevent the accumulation of the electrostatic charge. The suitability of grounding should be checked regularly by measurement.

2. Cleaning of vessels should be performed in a manner which minimizes the creation of flammable sprays and mists.

3. Other than controlled vent lines, all manways and other openings of vessels should be closed to prevent the egress of flammable vapor or dust and the influx of air.

4. Personnel working around vessels in which flammable gases or liquids or explosible solids are processed should be suitably grounded.

Liquids

5. The interior of vessels should be visually inspected and debris should be removed. Such items may float and

become isolated from ground and charged as liquid is introduced to the vessel.

6. Efforts should be made to avoid splashing when flammable liquids are introduced to a vessel in order to avoid the formation of charged sprays and mists. This is typically accomplished through the use of a grounded metal fill pipe or by introducing liquids through a valve located at or close to the bottom of the vessel. A fill pipe should extend sufficiently close to the bottom of the vessel such that it is immediately submerged beneath the liquid upon filling.

7. In general, the flow velocity of low conductivity liquids through pipes should not exceed 7 m/s at any time during vessel filling. For pure or single-phase liquids having a conductivity of 100 pS/m or less, and liquids having a conductivity of 10,000 pS/m or less and containing solids, immiscibles, or multiple phases, the flow velocity should be limited to 1 m/s or less. If higher flow velocities are required, the use of an antistatic additive to raise the conductivity of the liquid above 100 pS/m for a pure or single-phase liquid, or above 10,000 pS/m for liquids containing solids, immiscibles, or multiple phases should be considered.

Bulk Solids and Powders

8. Although it depends on individual cases, as a general precaution against electrostatic ignition, inerting or explosion protection should be considered during vessel filling when the bulk solid or powder being processed is insulating and the MIE of the material is 10–25 mJ or less.

Metal Vessels with Internal Linings or Coatings

Many operations involving liquids necessitate the use of vessels and piping having chemical resistant linings or coatings such as glass or polytetrafluoroethylene (PTFE). Such materials are highly insulating and may present an electrostatic hazard, particularly when present in vessels receiving and containing flammable liquids or explosible solids.

High-resistivity linings and coatings may become electrostatically charged when liquids or solids are introduced to the vessel. Such linings and coatings may also inhibit the relaxation of the electrostatic charge from liquids and solids. Linings and coatings may also become charged during mixing or agitation of the vessel contents or other operations. Potential electrostatic ignition sources include:

1. Electrostatic discharges from the surfaces of charged liquids or solids
2. Spark discharges from electrically isolated conductors

3. Propagating brush discharges through high resistivity linings and coatings
4. Brush discharges from the liner or coating

Flammable atmospheres may include flammable vapors, dust clouds, and mists developed during vessel filling or cleaning.

Many of the suggestions outlined above for unlined metal vessels remain applicable for metal vessels having internal linings or coatings. Additional specific precautions for reducing electrostatic hazards associated with such vessels include the following:

1. For metal vessels having a high-resistivity lining or coating, provision should be made for a suitable route to ground to promote relaxation of electrostatic charge from the vessel contents. This is often accomplished by installing a conductive bottom discharge valve or a small grounded tantalum patch.

2. Cleaning of vessels having high-resistivity internal linings should be performed in a manner which minimizes static charging of the internal lining and development of flammable sprays and mists. In particular, the use of flammable solvents for cleaning should be avoided and personnel should not scrub or rub the lining during cleaning.

Small Metal and Plastic Containers

Small containers can pose an electrostatic hazard when used in flammable atmospheres. Electrostatic charge can be developed when a low-conductivity liquid or a high-resistivity bulk solid or powder is dispensed into a container. Transfer of charge from charged liquids or solids or handling by personnel may cause electrically isolated metal and plastic containers to become charged.

Potential electrostatic ignition sources include:

1. Spark discharges from isolated metal containers
2. Discharges from charged liquids and solids in the container
3. Brush discharges from plastic containers
4. Spark discharges from charged personnel and implements

Flammable atmospheres may include flammable vapors or dust clouds evolved during dispensing of liquids and solids or other nearby flammable gases, vapors, or dust clouds.

Suggestions for reducing electrostatic hazards associated with small containers include the following:

Small Plastic Containers

1. The use of containers made of plastic or other high-resistivity materials in flammable atmospheres hav-

ing an MIE of 4 mJ or less is generally not recommended. The use of such containers may be allowable, however, when it can be demonstrated that the level of electrostatic charge will be insufficient to produce discharges capable of igniting the flammable atmosphere in question.

For example, because of their chemical resistance, it is sometimes desirable to use plastic containers for small quantities of flammable solvents. In these instances, measures may be necessary to limit the development of electrostatic charge. Such measures may include: (1) limiting the rate at which the container is filled; and (2) a grounded metal fill pipe that extends sufficiently close to the bottom of the container. Also, care should be taken to avoid rubbing or handling insulating containers in a way that causes development of static charge.

Small Metal Containers

2. When flammable atmospheres are present, metal containers should be suitably grounded during filling and emptying such that the resistance-to-ground is less than 10 ohms.

3. For pure and single-phase liquids having a conductivity of 100 pS/m or less and liquids having a conductivity of 10,000 pS/m or less and containing solids, immiscibles, or multiple phases, the flow velocity of liquid entering a grounded metal container should be limited to one 1m/s or less.

4. If it is not possible or it is undesirable to limit the flow velocity of a low-conductivity liquid during dispensing, the use of an antistatic additive to raise the conductivity of the liquid should be considered.

Plastic Items and Implements

Operators sometimes use various plastic items and implements such as tools and scoops during materials-handling operations. Items made from plastics and other high resistivity materials can pose a hazard when used in or around certain flammable atmospheres. Plastic items and implements can become charged by contact electrification during use and by charge transfer when in contact with other charged materials. Charged plastic items may give rise to brush discharges when exposed to grounded plant, personnel, or equipment.

Suggestions for reducing electrostatic hazards associated with plastic items and implements include the following:

1. The use of implements and items made from plastics and other high-resistivity materials is generally not suggested in or around flammable atmospheres having an MIE of 4 mJ or less. Such items may be used, how-

ever, when it can be demonstrated that the level of electrostatic charge will be insufficient to produce discharges capable of igniting the flammable atmosphere in question.

2. Alternatively, grounded metal or antistatic plastic items and implements should be used in or around flammable gases, vapors, or dust clouds having MIEs of 4 mJ or less. The resistance-to-ground of grounded metal items and implements should be less than 10 ohms. Suitable antistatic items and implements should be made from materials having a volume resistivity on the order of 10^9 ohm-meters or less, or a surface resistivity on the order of 10^{10} ohms per square or less, measured at <30% RH.

Piping and Hoses

As liquids and solids flow through piping and hoses, electrostatic charge can be developed. Liquids typically become charged due to double-layer charging, while solids become charged by contact electrification. Piping and hoses made from high-resistivity (insulating) materials can also become charged due to external charging, such as when personnel rub or brush against such piping. Static charge can accumulate on low conductivity liquids, high-resistivity solids, metal piping and components isolated from ground, high-resistivity lining materials used in piping, and nearby personnel if electrically isolated from ground.

Potential electrostatic ignition sources include:

1. Spark discharges from isolated conductors, such as metal piping, flanges, connectors, valves, and clamps
2. Brush discharges from charged insulating materials, such as high-resistivity piping, and hoses
3. Propagating brush discharges from insulating piping and hoses and through high-resistivity linings

Flammable atmospheres may include flammable gases, vapors, and dusts external to piping and hoses, leaks from piping and hoses conveying flammable liquids and powders, and residual vapors from flammable liquids inside piping and hoses upon draining.

Suggestions for reducing electrostatic hazards associated with piping and hoses include:

1. All metal piping and hoses should be suitably grounded. The resistance of such piping should be checked periodically. Metal piping and hoses should have a resistance to ground of less than 10 ohms.

2. Some piping and hoses constructed from high-resistivity materials may have some associated metal elements, such as flanges, connectors, fasteners, clamps, sheaths, and reinforcing spirals and meshes. Such con-

ductive elements should be grounded or bonded such that their resistance to ground is less than 10 ohms.

3. The use of piping and hoses constructed from high-resistivity materials is generally not suggested for conveying a low-conductivity liquid due primarily to the risk of ignition from brush and propagating brush discharges.

4. Piping and hoses constructed from high-resistivity materials may be used to convey liquids having a conductivity greater than 10,000 pS/m as long as the following conditions are met:

a. All metal sections of such piping and hoses are grounded in accordance with Suggestion (2) above.

b. External mechanisms by which such piping and hoses may become charged can be excluded.

5. For pure or single-phase liquids having a conductivity of 100 pS/m or less, and liquids having a conductivity of 10,000 pS/m or less and containing solids, immiscibles, and multiple phases, the potential electrostatic hazard posed by such liquids flowing through metal piping having a high-sensitivity lining or coating will be low, as long as a continuous pathway to electrical ground has been provided for the liquid.

6. When a pure or single-phase liquid having a conductivity of 100 pS/m or less or a liquid having a conductivity of 10,000 pS/m or less and containing solids, immiscibles, or multiple phases is conveyed in metal piping having a high-resistivity lining, charge will likely be retained by the liquid as well as the lining. If a sufficient potential is developed, a discharge may occur through the lining to the grounded outer metal piping forming a pinhole in the lining. Such a discharge may not pose a fire or explosion hazard when the pipe is full of liquid due to the absence of vapor space in which to form a flammable atmosphere. However, such discharges may pose a fire and explosion hazard if they arise during draining of the pipeline.

Centrifuges

A centrifuge is one means for separating a solid product from a liquid. The liquid-solid slurry is typically transferred into the main chamber of the centrifuge. During operation, the main chamber spins about its center axis. The slurry is accelerated to the sides of the chamber by centrifugal force. The sides of the chamber are perforated and covered with a filter media which allows the liquid filtrate to pass through, leaving behind the solid filter cake. After the filtrate passes through the filter media, it is discharged from the centrifuge.

In a batch centrifuge, the filter cake is typically removed manually. The centrifuge is opened and an oper-

ator reaches into the main chamber and dislodges the filter cake from the side walls using a scoop and/or removing the filter media. The filter cake is collected in a container or transferred directly to the next operation.

Electrostatic hazards may arise during centrifuge operation, principally due to static charging of the filtrate as it is accelerated in the chamber and as it passes through the filter media. Static charging will be especially pronounced for low-conductivity filtrates. Potential electrostatic ignition sources include:

1. Spark discharges from isolated conductors, such as metal centrifuge components, tools used to remove filter cake, and the operator

2. Brush discharges from insulating filter media which may become charged during operation or upon removal of filter cake

3. Discharges from the filter cake

Flammable atmospheres may include flammable vapors from filtrates and explosible dust evolved from dry filter cake. Fire- and explosion-prevention during operation is typically based on inert gas blanketing. However, during removal of the filter cake at the end of a batch, the centrifuge must be opened, and thus safety generally cannot be based on inerting. Consequently, safety may be based on exclusion of ignition sources. In this regard, suggestions for controlling electrostatic ignition sources associated with centrifuge operation and otherwise improving operation safety include the following:

1. All metal components of the centrifuge should be suitably grounded. Grounding should be checked routinely to verify a resistance to ground of less than 10 ohms.

2. The use of insulating (high-resistivity) filter media with flammable liquids and wash solutions is generally not suggested. Insulating filter media may become charged due to contact with charged liquid or filter cake by scraping during the removal of solids or during separation from the centrifuge. Instead, antistatic or conductive media are suggested. Acceptable filter media should have a surface resistivity on the order of 10^{10} ohms per square or less.

3. The centrifuge should be opened only after a suitable waiting period to allow for relaxation of electrostatic charge. The waiting period should be based on the relaxation time for the filter cake in place on the filter media. Effective charge relaxation will occur during the waiting period only if all metal components of the centrifuge are grounded and if the filter media is antistatic, conductive, or—if insulating—wet with a conductive filtrate.

4. Filter cake should be removed from filter media using only a grounded metal scoop or tool and only after

the prescribed waiting period. Filter cake should be removed gently, using slow, steady motions to minimize the development of static charge.

5. Once removed from the filter media, filter cake wet with a flammable liquid should be collected only in a grounded metal or antistatic vessel or container. Metal containers should be grounded such that the resistance-to-ground is less than 10 ohms. If filter cake is collected in a fiber drum, the metal chimes of the drum should be suitably grounded. Only antistatic plastic liners—having a surface resistivity on the order of 10^{10} ohms per square or less—should be used in such drums.

6. All personnel working around the open centrifuge should be suitably grounded such that their resistance-to-ground is less than 1×10^8 ohms. This is typically accomplished by the specification of antistatic footwear for personnel and antistatic or conductive flooring for the area around the centrifuge. Grounding of personnel may also be accomplished through the use of grounding straps. For more ignition-sensitive flammable filtrates or wash solutions having an MIE of 0.2 mJ or less, personnel should be grounded such that their resistance to ground is less than 1×10^6 ohms.

Manual Addition of Solids to Vessels Containing Flammable Liquids

Bulk solids and powders are sometimes introduced through open manways located on the top of vessels. A variety of methods are used to transfer solids to vessels, including sacks, drums, scroll conveyors, and flexible intermediate bulk containers (FIBCs). The contents of sacks and smaller drums are often poured manually into a vessel by operators. Larger drums are poured using a drum hoist or lift. FIBCs are sometimes suspended above a vessel using a hoist or forklift.

The introduction of bulk solids and powders to vessels containing flammable liquids can pose a significant electrostatic hazard unless suitable precautions are taken. The pouring of bulk solids and powders can generate hazardous levels of static charge. Charge may accumulate on sacks, drums, FIBCs, chutes, funnels, receiving vessels, and electrically isolated personnel. Potential electrostatic ignition sources include:

1. Spark discharges from isolated conductors and personnel
2. Brush discharges from insulating materials such as plastic liners, containers, sacks, and FIBCs

Flammable atmospheres include vapors evolved from liquid in the receiving vessel, dust clouds created in the area of the manway, and hybrid mixtures of vapors and dusts.

Suggestions for reducing electrostatic hazards associated with the manual addition of bulk solids and powders to vessels containing flammable liquids include the following:

1. In general, the loading of solids directly into a vessel containing a flammable liquid is not suggested. Instead, suitable intermediate equipment should be used to remove the solids-handling operation—a potential electrostatic ignition source—from the flammable atmosphere. A wide variety of intermediate transfer equipment is available, including rotary valves or bag dump hopper/scroll feeder systems.

2. All metal and conductive items and components should be suitably grounded. Conductive items of concern include metal drums, metal rims of the fiberboard drums, buckets, scoops, chutes, funnels, receiving vessels, and intermediate transfer equipment.

3. Only antistatic plastic or paper sacks should be used in or around flammable gases, vapors, or dusts having minimum ignition energies (MIEs) of 4 mJ or less. Antistatic plastic and paper sacks should be grounded. Suitable antistatic sacks will be made from materials having a surface resistivity on the order of 10^{10} ohms per square or less. If antistatic sacks cannot be grounded directly, it may be possible to accomplish suitable grounding through the operator if he or she is suitably grounded and handles the sacks with bare hands or antistatic gloves.

4. The paper used in paper sacks is sometimes coated or treated to improve its durability or moisture resistance. These coatings and treatments can sometimes increase the surface resistivity of the paper to the point where it may be capable of accumulating a potentially hazardous level of charge. Only paper sacks having a surface resistivity on the order of 10^{10} ohms per square or less qualify as antistatic.

5. The use of containers and other items made from insulating materials in or around flammable gases, vapors, or dust clouds having MIEs of 4 mJ or less is generally not suggested. Items of concern include plastic scoops, tools, chutes, and funnels. Instead only grounded metal or antistatic plastic items should be used.

Blending and Mixing

Many chemical processes involve the blending or mixing of liquids, solids, or liquid-solid mixtures. Liquids mixing equipment include agitated tanks, in-line mixers, and jet mixers. Solids mixing equipment includes vee, cone, ribbon, and orbital blenders.

Electrostatic hazards may arise during blending and mixing as a consequence of static charging of materials due to continuous contact at liquid-liquid, liquid-solid, and solid-solid interfaces. Static charging will be pronounced for electrically insulating bulk solids or powders and low-conductivity liquids, especially those containing solids, immiscibles, or multiple phases. Static charge can accumulate on insulating liquids and solids and on isolated conductors, such as metal agitators, baffles, flanges, and valves. Potential electrostatic ignition sources include:

1. Spark discharges from isolated conductors
2. Discharges from charged liquid surfaces
3. Discharges from bulk solids and powders after blending

Suggestions for reducing electrostatic hazards associated with blending and mixing operations include the following:

1. All metal components of blenders and mixers should be suitably grounded. Items of concern include agitators, baffles, flanges, valves, and the vessel itself. Grounding should be checked routinely to verify a resistance-to-ground of less than 10 ohms.

2. When flammable liquids or explosible bulk solids and powders are blended or mixed, all personnel who may be exposed to the flammable atmosphere should be suitably grounded such that their resistance-to-ground is less than 1×10^8 ohms.

3. Virtually all types of blending and mixing operations—particularly those involving low-conductivity liquids or high-resistivity bulk solids—are characterized by the development of high levels of electrostatic charge. Consequently, the exclusion of electrostatic discharges may prove difficult due to the risk of discharges from the surface of charged liquids or solids. Therefore, for low conductivity liquids and electrically insulating powders having an MIE of less than 25 mJ, the preferred basis of safety for blending and mixing operations is inerting or explosive protection in these cases.

4. When pure or single-phase liquids having a conductivity greater than 100 pS/m, or liquids containing immiscibles, contaminants, or multiple phases having a conductivity greater than 10,000 pS/m, are blended or mixed in a metal vessel, grounding of the vessel and all metal appurtenances can generally control the accumulation of electrostatic charge.

5. For blending and mixing vessels having an internal insulating lining such as glass or PTFE, a path to ground should be provided to allow for the relaxation of electrostatic charge from the vessel contents. Suitable approaches may include a grounded tantalum patch or a conductive discharge outlet and valve below the liquid level.

Drying

After filtration or centrifugation, it is often necessary to dry the wet cake. There are many different types of industrial dryers, including:

1. spray dryers
2. fluidized-bed dryers
3. rotary dryers
4. band dryers
5. tray ovens
6. vacuum dryers

Electrostatic hazards may arise during drying operations due to static charging of solids during processing, particularly in dryers where the wet product is fluidized or agitated. Potential electrostatic ignition sources in dryers include:

1. Spark discharges from isolated conductors, such as metal spray heads, baffles, flights, and flanges
2. Brush discharges from insulating items and coatings

Flammable atmospheres may include dust clouds or flammable vapors evolved from the drying solids.

Suggestions for reducing electrostatic hazards associated with drying operations include:

1. All metal components of drying equipment should be suitably grounded.

2. As a precaution against electrostatic ignition, inerting and/or explosion protection measures such as venting or containment should be considered when the bulk solids or powders being processed have an MIE of 10–25 mJ or less or are wet with a flammable liquid.

3. After drying, material should be transferred only to a grounded metal or antistatic container. If the material is collected in a fiber drum, the chimes of the drum should be suitably grounded. Only antistatic plastic liners should be used in such drums. Coatings used on the interior or exterior surfaces of such drums should also meet this criterion.

4. While not necessarily relevant to electrostatic hazards, the thermal stability of bulk solids and powders should be assessed by laboratory testing before full-scale drying operations are conducted.

Milling and Grinding

The purpose of milling and grinding operations is to reduce the size of solid materials. Solids are transferred into the main chamber of the unit where size reduction is typically accomplished by repeated direct or frictional contact with steel balls, blades, hammers, and other surfaces. Solids are transferred from the main chamber pneumatically, centrifugally, or by gravity and are then collected in a container or conveyed directly to the next operation.

Electrostatic hazards may arise during milling and grinding due to static charge developed by continuous contact of processed solids. Static charging will be especially pronounced for high-resistivity solids. Potential electrostatic ignition sources include:

1. Spark discharges from isolated metal components of mills, grinders, or hoppers
2. Discharges from powders accumulating in downstream vessels
3. Spark discharges from operators isolated from ground

Flammable atmospheres may include dust clouds within or external to the mill or grinder or dust clouds in downstream vessels.

Suggestions for reducing electrostatic hazards associated with milling and grinding operations include the following:

1. All metal components of mills and grinders should be suitably grounded. Grounding should be checked routinely to verify a resistance to ground of less than 10 ohms.

2. Material discharged from a mill or grinder will likely be highly charged. Therefore, solids should be accumulated only in a grounded metal or antistatic vessel or container. If the material is collected in a fiber drum, the chimes of the drum should be grounded. Only antistatic plastic liners—having a surface resistivity on the order of 10^{10} ohms per square or less—should be used in such drums.

3. While it is not an electrostatic discharge hazard, bearing failure and foreign objects are common sources of ignition for dusts in mills and grinders. Therefore, mills and grinders should be routinely maintained to prevent mechanical and other failure, and efforts should be made to exclude foreign objects from feed hoppers.

4. Since charge development and frictional heating and sparks are intrinsic to milling and grinding operations, inerting and/or explosion protection such as containment should be considered.

Acknowledgments

The authors acknowledge the contributions of Dr. Paul Cartwright to this chapter.

General References

Code of practice for Control of undesirable static electricity, Part 1. General considerations, British Standard BS 5958: Part 1: 1991.

Code of practice for Control of undesirable static electricity, Part 2. Recommendations for particular industrial situations, British Standard BS 5958: Part 2: 1991.

Cross, J. *Electrostatics: Principles, Problems and Applications*; Adam Hilger: Bristol, 1987.

Lüttgens G · Glor M. *Understanding and Controlling Static Electricity*; Ehningen bei Böblingen: Expert Verlag, 1989.

NFPA 69 Standard on Explosion Prevention Systems; 1997 Edition.

NFPA 77 Recommended Practice on Static Electricity; 2000 Edition.

37

Centrifuge Safety

MARTHA A. McRAE

Centrifugation employs centrifugal force to accelerate the separation of nonsoluble suspended particles from a liquid. Liquids of varying density (specific gravity) may also be separated through centrifugation.

In general, there are two centrifuge configurations, floor model and benchtop, and three classifications (based on the speed of the rotor): low-speed (2,000–6,000 rpm), high speed (18,000–30,000 rpm), and ultra-speed (35,000–150,000 rpm). Specialized centrifuges such as microfuges (used extensively in microbiological and radioimmunoassay applications for pelleting of small samples) and continuous-flow batch centrifuges may also be found in the lab.

All centrifuges have three basic components: a rotor,

a drive shaft, and a motor. A cabinet usually surrounds and supports these parts. The cabinets provide varying degrees of protection to the operator from tube breakage or any metal parts failure during operation.[1] Construction of centrifuges in many countries is governed by regulations intended to assure user safety. However, a good design cannot ensure that a misused product will perform without failure. Accidents involving centrifuges do happen. An estimated 90% of all centrifuge failures are caused by user error. The most common user errors are improper seating of the rotor on the drive shaft, centrifuging without the rotor lid in place, and improper securing of the rotor lid.[2] Prior to using the centrifuge, all lab personnel should read (and/or be trained on) the operating requirements specific to the instrument and the accessories in use, paying particular attention to the safety information. Strict observance of the operating procedures and precautions will help avoid actions by lab personnel that could result in damage or adversely affect instrument performance. Precautions for the use and handling of the centrifuge's associated rotors are also imperative (see rotor safety, under mechanical hazards, and the rotor maintenance sections below).

Centrifuge Safety Considerations

Centrifugation safety procedures can be broadly separated into the following categories: installation and maintenance; electrical, mechanical, and fire hazards; and the hazards of the materials being centrifuged. Safety information can be found in the user's manual as well as on labels on the packaging or on the centrifuge itself.

Centrifuge Installation and Maintenance

Ultracentrifuge and other floor model centrifuges are typically installed by the manufacturer's personnel. Prior to installation, the manufacturer will advise you of any required facility modifications, such as dedicated electrical circuits, remote emergency switches, and so forth. Benchtop models and microfuges are usually installed by the user. Carefully following the installation instructions included with the instrument will help assure safe installation of these centrifuges.

If an anchoring system is provided, be sure to use it to secure the centrifuge in place. The anchoring system is designed to reduce the possibility of injury or damage that could result from centrifuge movement in event of a major rotor mishap.

Minor preventive maintenance servicing (such as replacing worn belts or vacuum pump oil) is often performed by in-house maintenance personnel. The preventive maintenance and servicing information section of the user's manual should be reviewed before any servicing is attempted or the manufacturer should be contacted for assistance. Ensure that the instrument is turned off, unplugged, and locked and tagged out before servicing. Dispose of waste materials produced during servicing (such as used pump oil) in a manner consistent with federal, state, or local regulations. Generally, the manufacturer must always be contacted for assistance in moving and reinstalling any floor model centrifuge *and* for any major servicing of any centrifuge.

Electrical Hazards

As with all electrically powered equipment, improper use of the centrifuge may result in electrical shock. Centrifuges should use a three-wire electrical cord and plug to connect the equipment to earth-ground. It is important to ensure that the matching wall outlet receptacle is properly wired and earth-grounded.

- Prior to installation, check that the line voltage agrees with the voltage listed on the name rating plate affixed to the centrifuge or provided with the centrifuge's preinstallation instructions.
- Never use a three-to-two plug adapter for benchtop models. Never use a two-wire extension cord or a two-wire grounding type of multiple outlet receptacle strip.[3]

Mechanical Hazards

Mechanical failure during centrifugation can be serious. Mechanical failure is primarily due to problems associated with the rotor or its contents, although drive breakdown may also be the cause. Even the most careful design will not protect a rotor against misuse or abuse. Rotors may fail due to stress, metal fatigue, or corrosion. However, many of these can be eliminated by proper rotor use and maintenance.

Stress

Centrifugal force generates a load on the rotor, which causes it to stretch and change in size. Rotor materials can handle some amount of stress (called elastic strain) and will recover to their original dimensions after the load is removed. At a certain level of stress (called plastic strain), the elastic limit of the material is exceeded and the rotor will not return to its original configuration. Operating rotors under this kind of load can result in microstructure cracks. In order to minimize adverse stress effects, all rotors are designated with a maximum speed and sample density rating. It is imperative that these parameters not be exceeded to prevent the chance of rotor failure.

Metal Fatigue

All metal under stress will eventually suffer fatigue. Repeated accelerations and decelerations cause changes in the metal's microstructure due to the cyclic stretching and relaxing of the metal. Depending on how close the operating speed is to the elastic limit of the metal, and how many alternating cycles are performed, these microstructural changes will eventually result in microscopic cracks. With continued use, the cracks will enlarge and rotor failure will ultimately occur.[4] To prevent the possibility of a rotor mishap resulting from metal fatigue, it is a good practice to retire rotors and/or accessories at the expiration of the safe service life established by the manufacturer.

Corrosion

Metal rotor surfaces can be attacked by moisture, chemicals, or alkaline solutions. Salts are especially corrosive to aluminum rotors and components. Bleach solutions (used for biological decontamination) and high-alkaline detergents (used for radioactive decontamination) are also very corrosive. The resultant pitting leaves the rotor with less metal available to bear operational stress. When corrosion occurs in a high-stress area, such as the bottom of a swinging bucket rotor, the increased load on the remaining metal results in a "stress concentrator." This combination of stress and corrosion is called "stress corrosion."[4] Stress corrosion results in rotor failure sooner and at much lower stress levels than for an uncorroded rotor. In order to eliminate or reduce corrosion, proper rotor selection and maintenance is imperative.

Other means of reducing the chance for mechanical failure are:

- Ensure that the rotor and centrifuge are compatible (i.e., use only rotors that have been tested for use in the centrifuge to prevent inadvertent rotor failure). A list of the compatible rotors for each centrifuge should be in the manufacturer's operating manuals. In addition, use only components and accessories designed for use in the rotor being centrifuged. (These should be listed in the rotor manual.) Contact the manufacturer if you have questions.
- Ensure that the buckets (or carriers) are properly balanced (Fig. 37.1) and attached to the yoke of

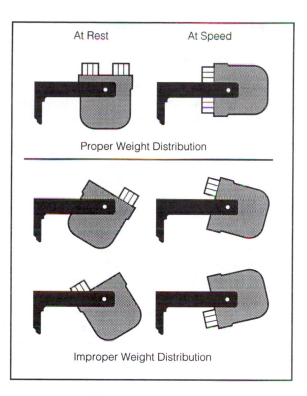

At Rest   At Speed

Proper Weight Distribution

Improper Weight Distribution

FIGURE 37.1 Proper balancing of rotors is critical to assuring safe centrifugation. This figure shows proper and improper weight distribution for a swinging bucket rotor. Refer to the rotor user manuals for proper balancing information for the rotors to be used. (Reproduced with permission from Reference 5. Copyright 1996, Beckman Instruments, Inc.)

swinging bucket rotors. An improperly hooked bucket cannot swing freely to a horizontal position during centrifugation and could cause a rotor failure. Attach all buckets, loaded or empty, following the instruction in the rotor manual.

- Before loading the rotor into the centrifuge, be sure the chamber bowl is dry and the drive shaft is clean. Once installed, ensure that the rotor is properly seated on the drive shaft. If the centrifuge requires a rotor tie-down, ensure that the device is securely fastened before start-up.
- All rotors are designed to carry a maximum load at a specific maximum speed. Do not exceed the maximum rated speed or load of the rotor. A well-maintained rotor log is essential for continued safe operation of an ultracentrifuge rotor (although newer models of ultracentrifuges are designed to eliminate the need for a user-maintained log, by automating the process through the data entry, the use of magnetic readers, etc.). Be sure to verify

any record-keeping requirements prior to using the equipment.

- Never attempt to override the door interlock system when the rotor is spinning or to stop or slow a rotor by hand.
- Do not place materials in locations where they could fall into the centrifuge, and never run a centrifuge with the door open unless it is a model designed for zonal separations (i.e., separation of sample particles through a solution of less density than the particles themselves which results in the particles being separated into distinct sedimentation zones due to their different rates of sedimentation), and it is in the zonal mode.
- Do not lift or move the centrifuge while the drive motor is spinning.[3] If the centrifuge is on wheels, ensure that brakes, chocks, or load levelers are on or in place prior to start-up.
- In the event of a power failure, do not attempt to retrieve the sample from the centrifuge for at least 1 hour.[3] Some centrifuges will automatically lock to prevent access after a power failure. Do not attempt to override the lock. The centrifuge user's manual should include specific instructions for recovery, or contact the manufacturer for assistance.
- The strength of glass and plastic containers can vary between lots and will depend on handling and usage.[5] Pretesting labware with buffer or gradient of equivalent density to the intended sample solutions is highly recommended. Examine glass tubes for scratches or nicks as even microscopic scratches can significantly weaken glass. Properly discard any tubes with these flaws.
- If the glass tubes break in the chamber bowl, using the appropriate personal protective equipment, remove the glass very carefully from adapters, buckets, or rotor cavities. Examine and clean the centrifuge gasket and/or chamber bowl with care because glass fragments may have become embedded.
- Do not use sharp tools on the rotor as this could cause scratches in the rotor surface. Corrosion begins in scratches and may open fissures in the rotor with continued use.[5]
- Inspect rotors regularly and lubricate rotor components, if required, as specified in the user's manual. If rough spots, pitting, or any other abnormality is noted during the rotor inspection, check with the manufacturer before using the rotor (Fig. 37.2). If rotors (especially those made of composite materials) are dropped, thoroughly inspect for chips, cracks, or other areas that may have been weakened and could subsequently fail under stress.

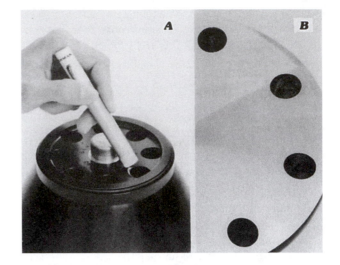

FIGURE 37.2 Before use, inspect rotors for rough spots, pitting, or other abnormalities. A. A penlight can be used to illuminate tube cavities. B. A cutaway of an aluminum rotor shows the pitting which has occurred at the bottom of tube cavities. (Reprinted with permission from Reference 4. Copyright 1983, Beckman Instruments, Inc.)

- If the centrifuge is located in an area away from the lab, it is a good practice to stay nearby until operating speed is reached to ensure no problems occur during start-up. If a rotor mishap does occur, contact the manufacturer before proceeding.

Fire Safety

Typically, centrifuges, rotors, and accessories are not designed for use with materials capable of developing flammable or explosive vapors. Do not centrifuge such materials in or handle or store them within 30 cm (1 ft) of the clearance envelope surrounding the centrifuge.[3] If centrifugation of such materials is necessary, contact the centrifuge manufacturer about developing a custom application.

Chemical, Radioactive, and/or Biological Hazards

Centrifugation may result in aerosol generation.[6–10] Therefore, observe proper safety precautions for aerosol containment. Handle all infectious substances according to the appropriate Biosafety Level recommended by the U.S. Centers for Disease Control and Prevention and the National Institutes of Health.[11] Ask your laboratory safety officer to advise you about the Biosafety Level required for your application and about the proper decontamination or sterilization procedures to follow if fluids escape from containers. As a general rule, the following guidelines should be used:

- Observe all caution information printed on the original containers of the materials being centrifuged.
- Aspirate rather than pouring the supernatant from centrifuge tubes. If pouring is unavoidable, wipe the rim afterward so subsequent centrifugation steps do not result in aerosolization of remaining droplets. (Use a disinfectant wipe for biological application.)
- Screw-caps or caps that fit over the rim of the centrifuge tubes are safer than plug-type or flip-top types of closures. Heat-sealed tubes may be used in some applications. Proper sealing should be verified by gently squeezing these tubes before loading the rotor.
- Loading and unloading biohazardous samples should take place in a biosafety cabinet.
- Centrifuges that are used with biological hazards and which draw a vacuum in the chamber should

be fitted with an in-line HEPA filter for the vacuum pump. (Note: This may require a retrofit. Contact the centrifuge manufacturer for assistance.)

- Centrifuge O-rings or gaskets are not designed as containment seals for aerosol or liquid containment. In situations where hazardous samples are being centrifuged, contain the samples in safety trunnion cups, safety buckets (safety cups), or sealed rotors (Fig. 37.3). Allow at least 10 minutes after the centrifuge has stopped for any aerosols present in the chamber to settle before proceeding.
- In the case where safety containers are not available, placing the centrifuge in a ventilated enclosure (biosafety cabinet, hood, or specially designed cabinet for floor models) can decrease the potential of resultant aerosol contaminating the lab. When appropriate secondary containment is not available, evacuating the chamber before the centrifuge is opened is recommended. This can be accomplished using specially ventilated centrifuges or attaching a vacuum pump via an appropriate tubing connection to one of the capped ports available on some models. (If a port is not available, contact the

FIGURE 37.3 In situations where hazardous samples are centrifuged, select sealed tubes, safety buckets (safety cups), safety trunnion cups or sealed rotors, such as the types shown here. (Reprinted with permission from Reference 12. Copyright 1994, Beckman Instruments, Inc.)

centrifuge manufacturer to see if a custom application can be developed.) A disinfectant trap or in-line HEPA filter should be used to protect the pump from contamination.[10]

- Large bulk or zonal rotors and continuous-flow centrifuges are more difficult to seal. Due to the large volumes of materials processed, if hazardous components are present, the entire centrifuge system should be placed in a ventilated enclosure.[10] Additional safety precautions may be required due to the nature of materials generally processed through these systems.[8] Contact your lab biosafety officer to determine if your use of these kinds of centrifuges warrants additional facility design, engineering controls, or use of personal protective equipment beyond that normally in use in the lab.
- Clean and decontaminate, as appropriate, all safety containers prior to reuse. In addition, implement a good preventative maintenance program on these devices to assure that they continue to operate op-

timally. (For example, regularly inspect O-rings or gaskets and replace those that have become cracked or dry.)

- Centrifuges, rotors, and accessories can be damaged by decontamination chemicals. Prior to using the centrifuge, procedures should be developed, in conjunction with your lab safety officer, for rotor and centrifuge decontamination and for any resultant waste disposal. In general, titanium and composite rotors are more resistant to chemical degradation than aluminum rotors, and stainless steel tubes and caps are more resistant than most plastics.
- Many currently produced rotors can be safely autoclaved, which avoids the need to use biohazard decontaminating agents. Autoclaving composite rotors or canisters may result in an increase in the number of surface cracks, an increased visibility of existing surface cracks, a change in canister color due to oxidation, or the formation of a white pow-

---

Box 1
General Rotor Maintenance Information

- Periodically inspect the rotor body, especially the inside cavities, for rough spots, pitting, or heavy discoloration. If any of these are present, do not use the rotor. Contact the manufacturer for information about rotor inspection or rotor repair programs.
- Do not clean rotor cavities and buckets with an ordinary bottle brush which has sharp wire ends.
- Avoid using alkaline detergents or cleaners on aluminum rotors, as the anodized coating can be removed by high alkalinity. (Note that most radioactive decontamination solutions are highly alkaline.[4]) Unless recommended by the manufacturer, do not wash rotor components and accessories in a dishwasher or soak in detergent solutions for long periods of time. Do not use organic solvents (such as acetone or methyl ethyl ketone) on composite materials at any time as solvents may damage the surface material.[13] Contact the manufacturer for assistance in choosing a cleaner compatible with your rotor.
- Particular care must be taken with aluminum swinging-bucket or fixed-angle rotors that have been used with uncapped tubes of CsCl or other salt.[4] Any salt crystals must be washed away as soon as possible to prevent corrosion.
- After the rotor has been cleaned and thoroughly rinsed with distilled water, it should be air dried with the buckets or cavities upside down.[4,13]
- Store rotors in a dry environment, not in the centrifuge. Fixed-angle and vertical-tube rotors should be stored upside down with the lids or plugs removed. Swinging-bucket rotors should be stored with the bucket caps removed.[4]

dery residue on the outside wall of the canister.[13] These conditions are usually cosmetic in nature. Contact the manufacturer if you have questions about the autoclavability of your rotor and/or accessories.

Development of Safe Use Practices

It is imperative to review all pertinent safety information and to develop standard operating procedures (SOPs) for each laboratory use of the centrifuge. Using a process safety analysis (PSA) or job hazard analysis (JHA) approach to evaluate experimental design will assist the lab worker in identifying and eliminating (or minimizing) the hazards associated with centrifuge use. PSAs or JHAs are also valuable tools in the development of SOPs and can be used along with an SOP as a resource for training, process review, auditing, and accident investigation.

Rotor Maintenance

The safety of a rotor depends largely on the user. Proper rotor maintenance (Box 1) is necessary to ensure the rotor's life expectancy is met. Improper care may result in corroding of some parts, resulting in weaknesses in the rotor, potentially leading to a rotor failure.

Summary

By establishing SOPs for using centrifuges, rotors, and their accessories and by following the manufacturer's centrifuge and rotor operating manuals, centrifugation can be safely conducted in the lab. Work with your institute's lab safety officer to develop SOPs prior to using the centrifuge. If you have questions or require additional assistance, contact the centrifuge and rotor manufacturer.

References

1. *A Centrifugation Primer*; Spinco Division, Beckman Instruments, Inc.: Palo Alto, CA, 1980.
2. *Centrifugation Hazards* (video); Howard Hughes Medical Institute: Chevy Chase, MD, 1995.
3. *Centrifuge Safety*; Spinco Business Center of Beckman Instruments, Inc.: Palo Alto, CA, 1994.
4. *Rotor Safety Guide*; Spinco Division, Beckman Instruments, Inc.: Palo Alto, CA, 1983.
5. *Rotors and Tubes for Beckman J2, J6 and Avanti J Series Centrifuges, A User's Manual*; The Spinco Business Center of Beckman Instruments, Inc.: Palo Alto, CA, 1996.
6. McGarrity, G. J.; Hoerner, C. L. In *Laboratory Safety: Principals and Practices*, 2nd ed.; Fleming, D. O., et al, Eds.; American Society for Microbiology: Washington, DC, 1995; Chapter 9, p. 122.
7. *Lab Safety Monographs, A Supplement to the NIH Guidelines for Recombinant DNA Research*; U.S. Dept. Health, Education, and Welfare, PHS, NIH: Washington, DC, 1979.
8. *Centrifuge Biohazards: Proceedings of a Cancer Research Safety Symposium*; National Cancer Institute: Bethesda, MD, 1973.
9. Collins, C. H. *Laboratory-Acquired Infections*, 3rd. ed.; Butterworth Heinemann: Oxford, England, 1993; pp. 70–76.
10. Kuehne, R. W., et al. In *Laboratory Safety: Principals and Practices*, 2nd ed.; Fleming, D. O., et al., Eds.; American Society for Microbiology: Washington, DC, 1995; Chapter 11, p. 146.
11. U.S. Department of Health and Human Services, Public Health Services, Centers for Disease Control and Prevention and the National Institutes for Health. *Biosafety in Microbiological and Biomedical Laboratories*. 4th ed.; U.S. Government Printing Office: Washington, DC, 1999.
12. *New J-Lite Rotor JLA-10.500 Data Sheet*; Beckman Instruments, Inc.: Fullerton, CA, 1994.
13. *J-Lite JLA-10.500 Fixed Angle Rotor Assembly*; The Spinco Business Center of Beckman Instruments, Inc.: Palo Alto, CA, 1996.

38

Safe Use of Laboratory Glassware

ROBERT J. ALAIMO

The safe use of laboratory glassware is an important component of any laboratory safety program[1]. Laboratory glassware is generally designed for specific applications. It possesses the qualities of toughness, safety, corrosion, and heat resistance. It is important to remember that most laboratory glassware is designed for a particular use and you should be sure you have the correct piece of glassware for the job. The use of a laboratory glass product for an application other than the one for which it was designed can be dangerous. Using normal household glass products or packaging materials for chemical reactions can result in serious consequences. The following section provides guidance on the safe handling, use, cleaning, and storage of laboratory glassware.

Chemicals

When working with volatile materials, remember that heat causes expansion, and confinement of expansion results in explosion. Remember also that danger exists even though external heat is not applied.

Do not mix sulfuric acid with water inside a cylinder. The heat from the reaction can break the base of the vessel because of the thickness of the base and the seal.

Perchloric acid is especially dangerous because it explodes on contact with organic materials. Do not use perchloric acid around wooden benches or tables. Keep perchloric acid bottles on glass or ceramic trays having enough volume to hold all the acid in case the bottle breaks. Always wear protective clothing when working with perchloric acid.

Glass will be chemically attacked by hydrofluoric acid, hot phosphoric acid, and strong hot alkalis, so it should never be used to contain or to process these materials.

Always flush the outside of acid bottles with water before opening. Do not put the stopper on the counter top where someone else may come in contact with acid residue.

Never fill a receptacle with material other than that specified by the label. Label all containers before filling. Dispose of the contents of unlabeled containers properly.

Handling Glassware

Hold beakers, bottles, and flasks by the sides and bottoms rather than by the tops. The rims of beakers or necks of bottles and flasks may break if used as lifting points. Be especially careful with multiple-neck flasks.

Use a rubber policeman on glass, or use Teflon rods to prevent scratching the inside of a vessel.

To avoid breakage while clamping glassware, use coated clamps to prevent glass-to-metal contact, and do not use excessive force to tighten clamps.

Examine all glassware carefully before using and do not heat glassware that is etched, cracked, nicked, or scratched—it is more prone to break. Discard damaged glassware in a proper disposal container.

When pieces are not to be used for an extended period of time, take apart stopcocks, ground joints, flask stoppers, and joints to prevent sticking. Remove the grease from the joints. Teflon stoppers and stopcocks should be loosened slightly. For easy storage and reuse, put a strip of thin paper between ground joint surfaces.

If a ground joint sticks, this procedure will generally free it. Immerse the joint in a glass container of freshly poured, carbonated liquid. You will be able to see the liquid penetrate between the ground surfaces. When the surfaces are wet (allow 5 to 10 minutes submersion), remove the joint and rinse with tap water. Wipe away excess water. Then gently warm the wall of the outer joint by rotating it for 15 to 20 seconds over a low Bunsen burner flame or with a heat gun. Wear heat-resistant gloves to avoid burns. Be sure that 50% of the inner surface is wet before inserting the joint in the flame or heat source. Remove from the heat and gently twist the two members apart. If they do not come apart, repeat the procedure. Never use force when separating joints by this method. If the joints do not separate easily it is advisable to discard the glassware rather then risk injury.

Glass stopcocks on burets and separatory funnels should be lubricated frequently to prevent sticking. If one does stick, a stopcock plug remover, available from laboratory supply houses, should be used.

Wet both tubing and stopper with glycerin or water

when trying to insert glass tubing into a rubber stopper. Wear a protective glove and wrap glass in a towel to prevent injury.

Fire polish rough ends of glass tubing before inserting into flexible tubing or through a stopper.

If it becomes impossible to remove a thermometer from a rubber stopper, it is best to cut away the stopper rather than to risk breaking the thermometer or injury.

In using lubricants, it is advisable to apply a light coat of grease completely around the upper part of the joint. Use only a small amount and avoid greasing that part of the joint which contacts the inner part of the apparatus.

Three types of lubricant are commonly used on standard taper joints. A hydrocarbon grease is the most widely used. It can be easily removed by most laboratory solvents, including acetone. Because hydrocarbon grease is so easily removable, silicone grease is often preferred for higher temperature or high-vacuum applications. It can be removed readily with chlorinated solvents. For long-term reflux or extraction reactions, a water-soluble, organic-insoluble grease, such as glycerin is suitable. Water will clean glycerin.

Heating

Bunsen Burners

Adjust Bunsen burner to get a large soft flame. It will heat slowly but also more uniformly. Uniform heat is a critical factor for some chemical reactions.

Adjust the ring or clamp holding the glassware so that the flame touches the glass below the liquid level. Heating above the liquid level does nothing to promote even heating of the solution and could cause thermal shock and breakage of the vessel. A ceramic-centered wire gauze on the ring will diffuse the burner flame to provide more even heat.

Rotate test tubes to avoid overheating one particular area. Uniform heating may be critical to your experiment.

Heat all liquids slowly. Fast heating may cause bumping, which in turn may cause the solution to splatter.

Heat Guns

Heat guns have become commonplace in the laboratory. They are very useful for drying chromatography plates and glassware or heating surfaces of distillation equipment. They should never be used near flammable liquids, and the heating element should never come in contact with the glassware. Heat guns recommended for laboratory applications should be the only type used.

Hot Plates

There are several types of hot plates. Some are electrical, some are water heated. They may be ceramic or metal topped. You should consult your instruction manual before using a hot plate for the first time.

Always use a hot plate larger than the vessel being heated.

Thick-walled items such as jars, bottles, cylinders, and filter flasks should never be heated on hot plates.

Evaporation work should be observed carefully. Be careful when handling a vessel that has been heated after evaporation has occurred. It may crack unexpectedly.

Do not heat glassware directly on electrical heating elements. Excessive stress will be induced in the glass and this can result in breakage.

Heating Mantles

Heating mantles were developed for the heating of round-bottomed flasks. The heating elements are covered by layers of fiberglass fabric and provide uniform heating of the surface. Heating mantles are to be used with variable voltage control transformers. Care should be taken when using heating mantles that the manufacturer's recommendations are followed. Additional information regarding heating mantles can be found from the manufacturer.[2]

Heating Thick-Walled Vessels

Glassware with thick walls such as bottles and jars should not be heated over a direct flame or comparable heat source. The use of an immersion heater to heat the liquids within these vessels is recommended.

Temperature and Temperature Extremes

Do not put hot glassware on cold or wet surfaces, or cold glassware on hot surfaces. It may break with the temperature change. Cool all labware slowly to prevent breakage, unless you are using Vycor which can undergo extreme temperature changes without damage to the glass.

Use tongs or gloves to remove all glassware from heat. Hot glass can cause severe burns. Protective gloves, safety shoes, aprons, and goggles should be worn in case of chemical accidents, spilling, or splattering.

Suggestions for Cleaning and Storage

Good laboratory technique demands clean glassware, because the most carefully executed piece of work may give an erroneous result if dirty glassware is used. In all instances, glassware must be physically clean; it must be chemically clean; and in many cases, it must be bacteriologically clean or sterile.

All glassware must be absolutely grease-free. The safest criteria of cleanliness is uniform wetting of the surface by distilled water. This is especially important in glassware used for measuring the volume of liquids. Grease and other contaminating materials will prevent the glass from becoming uniformly wetted. This in turn will alter the volume of residue adhering to the walls of the glass container and thus affect the volume of liquid delivered. Furthermore, in pipets and burets, the meniscus will be distorted and the correct adjustments cannot be made. The presence of small amounts of impurities may also alter the meniscus.

Wash labware as quickly as possible after use. If a thorough cleaning is not possible immediately, put glassware to soak in water. If labware is not cleaned immediately, it may become impossible to remove the residue.

Most new glassware is slightly alkaline in reaction. For precision chemical tests, new glassware should be soaked several hours in acid water (a 1% solution of hydrochloric or nitric acid) before washing.

When washing, soap, detergent, or cleaning powder (with or without an abrasive) may be used. Cleaners for glassware include Alconox, Dural, M&H, Lux, Tide, and Fab. The water should be hot. For glassware that is exceptionally dirty, a cleaning powder with a mild abrasive action will give more satisfactory results. The abrasive should not scratch the glass. During the washing, all parts of the glassware should be thoroughly scrubbed with a brush. Brushes with wooden or plastic handles are recommended as they will not scratch or abrade the glass surface. This means that a full set of brushes must be at hand—brushes to fit large and small test tubes, burets, funnels, graduates, and various sizes of flasks and bottles. Motor-driven revolving brushes are valuable when a large number of tubes or bottles are processed. Do not use cleaning brushes that are so worn that the spine hits the glass. Serious scratches may result. Scratched glass is more prone to break during experiments. Any mark in the uniform surface of glassware is a potential breaking point, especially when the piece is heated. Do not allow acid to come into contact with a piece of glassware before the detergent (or soap) is thoroughly removed. If this happens, a film of grease may be formed.

Grease is best removed by boiling in a weak solution of sodium carbonate. Acetone or any other fat solvent may be used. Strong alkalis should not be used. Silicone grease is most easily removed by soaking the stopcock plug or barrel for 2 hours in warm decahydronaphthalene. Drain and rinse with acetone or use fuming sulfuric acid for 30 minutes. Be sure to rinse off all of the cleaning agents.

It is imperative that all soap, detergents, and other cleaning fluids be removed from glassware before use. This is especially important with the detergents, slight traces of which will interfere with serologic and cultural reactions.

After cleaning, rinse the glassware with running tap water. When test tubes, graduates, flasks, and similar containers are rinsed with tap water, allow the water to run into and over them for a short time, then partly fill each piece with water, thoroughly shake, and empty at least six times. Pipets and burets are best rinsed by attaching a piece of rubber tubing to the faucet and then attaching the delivery end of the pipets or burets to a hose, allowing the water to run through them. If the tap water is very hard, it is best to run it through a deionizer before using. Rinse the glassware in a large bath of distilled water. To conserve distilled water, use a 5-gallon bottle as a reservoir. Store it on a shelf near your clean-up area. Attach a siphon to it and use it for replenishing the reservoir with used distilled water. For sensitive microbiologic assays, meticulous cleaning must be followed by rinsing 12 times in distilled water.

Glassware which is contaminated with blood clots, such as serology tubes, culture media, petri dishes, and so forth, must be sterilized before cleaning. It can best be processed in the laboratory by placing it in a large bucket or boiler filled with water, to which 1–2% soft soap or detergent has been added, and boiled for 30 minutes. The glassware can then be rinsed in tap water, scrubbed with detergent, and rinsed again. You may autoclave glassware or sterilize it in large steam ovens or similar apparatus. If viruses or spore-bearing bacteria are present, autoclaving is absolutely necessary.

Dry test tubes, culture tubes, flasks, and other labware by hanging them on wooden pegs or by placing them in baskets with their mouths downward and allowing them to dry in the air; or place them in baskets to dry in an oven. Drying temperatures should not exceed 140°C. Line the drying basket with a clean cloth to keep the vessel mouths clean. Do not apply heat directly to empty glassware which is used in volumetric measurements. Such glassware should be dried at temperatures of no more than 80 to 90°C.

Dry burets, pipets, and cylinders by standing them on a folded towel. Protect clean glassware from dust. This is done best by plugging with cotton, corking, taping a heavy piece of paper over the mouth, or placing the glassware in a dust-free cabinet.

Store glassware in specially designed racks. Avoid breakage by keeping pieces separated.

Do not store alkaline liquids in volumetric flasks or burets. Stoppers or stopcocks may stick.

Cleaning Specific Types of Glass Labware

Burets

Remove the stopcock or rubber tip and wash the buret with detergent and water. Rinse with tap water until all the dirt is removed. Then rinse with distilled water and dry. Wash the stopcock or rubber tip separately. Before a glass stopcock is placed in the buret, lubricate the joint with stopcock lubricant. Use only a small amount of lubricant. Burets should always be covered when not in use.

Culture Tubes

Culture tubes which have been used previously must be sterilized before cleaning. The best method for sterilizing culture tubes is by autoclaving for 30 minutes at 121°C (15 p.s.i. pressure). Media that solidifies on cooling should be poured out while the tubes are hot. After the tubes are emptied, brush with detergent and water, rinse thoroughly with tap water, rinse with distilled water, place in a basket, and dry.

If tubes are to be filled with a media which is sterilized by autoclaving, do not plug until the media is added. Both media and tubes are thus sterilized with one autoclaving.

If the tubes are to be filled with sterile media, plug and sterilize the tubes in the autoclave or dry air sterilizer before adding the media.

Dishes and Culture Bottles

Sterilize and clean as detailed under Culture Tubes. Wrap in heavy paper or place in a petri dish can. Sterilize in the autoclave or dry air sterilizer.

Pipets

Place pipets, tips down, in a cylinder or tall jar of water immediately after use. Do not drop them into the jar. This may break or chip the tips and render the pipets useless for accurate measurements. A pad of cotton or glass wool at the bottom of the jar will help to prevent breaking of the tips. Be certain that the water level is high enough to immerse the greater portion or all of each pipet. The pipets may then be drained and placed in a cylinder or jar of dissolved detergent or, if exceptionally dirty, in a jar of chromic acid cleaning solution. After soaking for several hours, or overnight, drain the pipets and run tap water over and through them until all traces of dirt are removed. Soak the pipets in distilled water for at least 1 hour. Remove from the distilled water, rinse, dry the outside with a cloth, shake the water out, and dry. To prevent breakage when rinsing or washing pipets, cylinders, or burets, be careful not to let tips hit the sink or the water tap.

Blood-Cell-Count Diluting Pipets

After use, rinse thoroughly with cool tap water, distilled water, alcohol, or acetone, and then ether. Dry by suction. Do not blow into the pipets as this will cause moisture to condense on the inside of the pipet. To remove particles of coagulated blood or dirt, a cleaning solution should be used. One type of solution will suffice in one case, whereas a stronger solution may be required in another. It is best to fill the pipet with the cleaning solution and allow to stand overnight. Sodium hypochlorite (laundry bleach) or a detergent may be used. Hydrogen peroxide is also useful. In difficult cases, use concentrated nitric acid. Some particles may require loosening with a horse hair or piece of fine wire. Take care not to scratch the inside of the pipet.

Automatic Pipet Washers

Where a large number of pipets are used daily, it is convenient to use an automatic pipet washer. Some of these, made of metal, can be connected directly by permanent fixtures to the hot and cold water supplies. Others, such as those made with polyethylene, can be attached to the water supplies by rubber hose. Polyethylene baskets and jars may be used for soaking and rinsing pipets in chromic acid cleaning solution. Electrically heated metallic pipet dryers are also available.

After drying, place pipets in a dust-free drawer. Wrap serologic and bacteriologic pipets in paper or place in pipet cans and sterilize in the dry air sterilizer. Pipets used for transferring infectious material should have a cotton plug placed in the top end of the pipet before sterilizing. The plug will prevent the material being measured from being drawn accidentally into the pipetting device.

Serological Tubes

Serological tubes should be chemically clean, but need not be sterile. However, specimens of blood which are to be kept for some time at room temperature should be collected in a sterile container. It may be expedient to sterilize all tubes. To clean and sterilize tubes containing blood, discard the clots in a waste container and place the tubes in a large basket. Put the basket, with others, in a large bucket or boiler. Cover with water, add a fair quantity of soft soap or detergent, and boil for 30 minutes. Rinse the tubes, clean with a brush, rinse and dry with the usual precautions. It is imperative when washing serological glassware that all acids, alkalis, and detergents be completely removed. Acids, alkalis, and detergents in small amounts interfere with serologic reactions. Serologic tubes and glassware should be kept separate from all other glassware and used only for serologic procedures.

Slides and Cover Glass

It is especially important that microscope slides and cover glass used for the preparation of blood films or bacteriologic smears be perfectly clean and free from scratches. Slides should be washed, placed in glacial acetic acid for 10 minutes, rinsed with distilled water, and wiped dry with clean paper towels or cloth. Once the slides have been washed, place them in a wide jar of alcohol. As needed, remove from the jar and wipe dry. If the slides are dry stored, wash them with alcohol before use.

Appendix

Glass Terminology

Anneal: To prevent or remove objectionable stresses in glassware by controlled cooling.

Binder (Fibrous Glass): Substances employed to bond or hold the fibers together.

Blister: An imperfection, a relatively large bubble or gaseous inclusion.

Check: An imperfection, a surface crack in a glass article.

Chill Mark: A wrinkled surface condition on glassware, resulting from uneven contact in the mold prior to forming.

Chip: An imperfection due to breakage of a small fragment from an otherwise regular surface.

Cord: An unattenuated glass inclusion, possessing optical and other properties differing from those of the surrounding glass.

Cullet: Waste or broken glass, usually suitable as an addition to raw batch.

Devitrification: Crystallization in glass.

Dice: The more or less cubical fracture of tempered glass.

Fiber: An individual filament made by attenuating molten glass. A continuous filament is a glass fiber of great or indefinite length. A staple fiber is a glass fiber of relatively short length (generally less than 44 cm).

Fusion: Joining by heating.

Glass Ceramic: A material melted and formed as a glass, then converted largely to a crystalline form by processes of controlled devitrification.

I.D.: Inside diameter.

Lampworking: Forming glass articles from tubing and rod by heating in gas flame.

Lap: (1) An imperfection, a fold in the surface of a glass article caused by incorrect flow during forming. (2) A process used for mating ground surfaces.

Liquidus Temperature: The maximum temperature at which an equilibrium exists between the molten glass and its primary crystalline phase.

Mat (Fibrous Glass): A layer of intertwined fibers bonded with some resinous material or other adhesive.

O.D.: Outside diameter.

Out-of-Round: Asymmetry in round glass articles.

Sealing: See Fusion.

Seed: An extremely small gaseous inclusion in glass.

Softening Point: The temperature at which a uniform fiber 0.5 to 1.0 mm in diameter and 22.9 cm in length elongates under its own weight at a rate of 1 mm per minute when the upper 10 cm of its length is heated in a prescribed furnace at the rate of approximately 5°C per minute. For a glass of density near 2.5, this temperature corresponds to viscosity of 107.6 poises.

Standard Taper:

1 is the symbol used to designate interchangeable glass joints, stoppers, and stopcocks complying with the requirements of ASTM E-676, and requirements of ASTM E-675. All mating parts are finished to a 1:10 taper.

3 is the designation for spherical (semi-ball) joints complying with ASTM E-677.

2 is the designation for tapered stopcocks using a fluorocarbon plug complying with ASTM E-911. All mating parts are finished to a 1:15 taper. The size of a particular piece appears after the appropriate symbol. Due primarily to the greater variety of apparatus equipped with fittings, a number of different types of identifications are used as follows:

Joints: A two-part number, *1* 24/40, with 24 being the approximate diameter in mm at the large end of the taper and 40 the axial length of taper, also in mm.

Stopcocks: A single number, *1* 2, with 2 being the approximate diameter in mm of the hole or holes through the plug.

Bottles: A single number, *119*, with *19* being the approximate diameter in mm of the opening at the top of neck.

Flasks: (Other than most boiling flasks) a single number, *119*, with *19* again being the approximate diameter in mm at top of neck. For dimensional details of the various stoppers, see the individual listings in Corning's general catalog of laboratory products.

> The complete designation of a spherical joint also consists of a two-part number, *312/2*, with *12* being the approximate diameter in mm of the ball and *2* the bore in mm of the ball and the socket.

> Finally, for the fluorocarbon plug, a single number is used, as with *1* stopcocks. Thus *22* means a stopcock with a hole of approximately 2 mm in the plug.

Stone: An imperfection; crystalline contaminations in glass.

Stria: A cord of low intensity, generally of interest only in optical glass.

Tempered Glass: Glass that has been rapidly cooled from near the softening point, under rigorous control, to increase its mechanical and thermal strength.

Thermal Endurance: The relative ability of glassware to withstand thermal shock.

Weathering: Attack of a glass surface by atmospheric elements.

Working Range: The range of surface temperature in which glass is formed into ware in a specific process. The "upper end" refers to the temperature at which the glass is ready for working generally corresponding to a viscosity of 10^3 to 10^4 poises. The "lower end" refers to the temperature at which it is sufficiently viscous to hold its formed shape, generally corresponding to a viscosity greater than 10^6 poises. For comparative purposes and when no specific process is considered, the working range of glass is assumed to correspond to a viscosity range from 10^4 to $10^{7.6}$ poises.

References

1. Adapted from the Technical Information Section of the Corning Labware & Equipment Catalog, PYREX®, PYREXPlus®, VYCOR®, and COREX®. Reprinted with permission of Corning Incorporated 1997, Science Products Division, Corning, NY 14831. More technical information is available from Corning Incorporated and at the Corning Internet Website www.corninglabware.com.

2. Glass-Col Laboratory Products, www.glascol.com.

39

Pressure/Vacuum-Containing Systems and Equipment

KENNETH K. MILES

The pressure discipline, or the use of equipment and systems which contain pressure or vacuum, is entirely related to hardware which is either commercially available or designed and fabricated for a specific need. Uses range from research (often small, one-of-a-kind laboratory-scale experiments) to large or multiple systems for industrial applications. Most of the National Consensus Codes and Standards are either hardware- or application-specific, that is, one Code may apply to one application and not apply to others. As such, the discipline is fraught with issues of applicability and interpretation. Similarly, industry practices apply to both hardware and application. Practices recommended for use in general industry (the use of industrial grade gases and threaded fittings) would not be used in the high-technology applications such as the microelectronics industry (where ultra-high purity electronic grade gases and welded fittings are used). Pressure systems are often thought of as operating at either high or low pressures. Once again, the pressure rating is tied to both process application and the hardware. Low-pressure systems are not always inherently safe, and many high-pressure system failures may be rather innocuous. The definition of high or low pressure varies according to the application, and the distinction between the two is not uniformly agreed upon. System design, fabrication, type of tubing connection, rules for use, and level of scrutiny is a function of the energy contained in the system, other hazards involved with the various processes, the potential impact on the safety and health of personnel and equipment, and the potential damage to the environment.

Codes and Standards

The National Consensus Codes have evolved over the years to provide minimum, acceptable, and conservative construction techniques and guidance. Manufacturers have responded to the needs of industry, and their products reflect their association and interaction with those particular customers. A generalization of the pressure Codes and Regulations associated with these industries may be broken up into the following general categories (recognizing that there is some crossover):

Manufacturing

The American Society of Mechanical Engineers (ASME) Boiler and Pressure Vessel Code[1] pertains to the design, materials of construction, fabrication methods, and sometimes erection (boilers, piping) of components and/or systems and includes recommended rules for the care and operation of power and heating boilers. Some components or systems may not fit the basic design parameters, and Code interpretations and special cases are available through the various Code Committees. Piping and fittings are built to pressure ratings calculated in accordance with the ANSI Power Piping design. New facilities invoke the use of the Uniform Building Code for the installation and the structural aspects of heating, ventilating and air conditioning (HVAC), or plant systems. Other vessels or systems which are not specifically approved under the Code, but which are not prohibited, must be qualified for use within the general guidelines of the Code. These vessels are built to the intent of the Code with proper materials and fabrication, adequate safety margin, quality-process controls, and post-test inspection and testing. Many state or local jurisdictions require that an engineering package approved by a professional engineer be prepared prior to the operation or sale of such vessels/systems.

Inspection and Repair

The National Board of Boiler and Pressure Vessel Inspectors is an organization which promotes uniformity in construction, installation, inspection, and repair of boil-

ers and other pressure vessels, and is responsible for enforcing and administering the ASME Boiler and Pressure Vessel Code. They also address the in-use issues such as deterioration due to corrosion, fatigue, structural weaknesses, and the functionality of safety devices, and so on, for plant systems (boilers, air receivers). The DOT regulations on transportation, 49 *CFR*, address requirements for over-the-road containers, including fabrication, stamping, and re-qualification procedures for compressed gas cylinders, tube trailers, and some vessels.

Process Controls

The *Code of Federal Regulations* for OSHA, 29 *CFR* 1910, contains limited guidelines on the use of air receiver tanks (inspection and testing of relief device), compressed gas cylinders, and a few of the more common gaseous systems (hydrogen, acetylene, oxygen, LPG, etc.). Included by reference are the other codes (ASME, ANSI, DOT, NFPA, etc.)[2,3], and industry practices based on technical information compiled by organizations such as the Compressed Gas Association,[4] which publishes guides, pamphlets, technical bulletins, and offers contacts within the industry. The National Fire Protection Association (NFPA) Standards provide guidelines on the use of flammable liquids and gases for both operations and storage, including restrictions on quantities. Other groups have been formed to self-regulate facilities and processes which were not generally recognized by the standing codes. The Semi-Conductor Safety Association (SSA) is one such group that works directly with local and national jurisdictions to incorporate adequate design and safety features into the semiconductor production facilities. National and local conferences, meetings, and seminars are held, often including the participation of major exhibitors of the latest technologies available. The Association magazines, technical standards, and other reference material provide timely information on key technical and safety issues. A major benefit gained from participating in these organizations is the development of personnel contacts on pertinent technical and safety issues. The Compressed Gas Association (CGA) and the American Vacuum Society (AVS) perform similar functions for their respective disciplines. The American Petroleum Institute (API)[5] addresses similar issues associated with the petroleum industry. Major suppliers of industrial and specialty gases and cryogenic liquids provide booklets, pamphlets, guides, technical bulletins, and Material Safety Data Sheets (MSDS) on their products.[6,7] Depending on the industry or process involved, the appropriate code or organization should be referenced and/or contacted to incorporate their requirements and experience into the mix.

Hardware

Pressure and vacuum components can be placed into two general categories, those commercially available and manufactured to national consensus standards (components and mass-produced vessels), and those which must be specifically designed and built for special applications. Usually the latter case does not directly meet the Code because of shape, size, weight, operational characteristics, or unique applications. Very high-pressure systems such as closure-type experimental volumes, reactor assemblies, intensifiers, and so on, can be treated as "off the shelf" components because of the manufacturers' experience in designing and producing these components, and extensive in-service user experience. Similarly, small vacuum systems made up of components such as chambers, flanges, fittings, and valves, are considered "off the shelf." The use of these components is acceptable within the following general guidelines:

1. Work closely with the manufacturers and vendors. They know their hardware and can recommend the proper application. Most have technical publications and engineering data on their hardware which is useful in determining the proper operational and safety characteristics of systems design. It is important to get as much information from the vendor at the time of purchase, including recommendations on inspection criteria and intervals, corrosion allowances, replacement intervals, and so forth.

2. Keep track of the pressure ratings, compatibility with working fluids, system design, protection schemes, and so forth, so that the history of the hardware is not lost. This is important for modifications, reevaluations, training of operators, and validation of proper component application and system design.

3. Users/operators should be the most knowledgeable about their systems. They should participate in the design, selection of hardware, assembly, and documentation. Additional effort is required for preassembled tools and apparatus purchased directly from the vendor. These items are normally supplied with operator manuals which often do not detail the specifics of hardware ratings and protection schemes, and the user must work with the supplier to validate and document the basics.

4. When assembling hardware, owners must know how to make up fittings. Suppliers often provide local training seminars for the particular fittings they sell. Once again, this is based on customer demand, and if not available, ask for it.

The second category of components includes unique, one-of-a-kind design or special-use items which are not mass produced and do not have an in-service experience history. The following guidelines should be used:

1. Use the reputable manufacturer/vendor concept. Fabricators with experience in similar designs or applications probably meet the majority of the requirements, for example, a Code shop provides assurances of quality controls even for non-Code-stamped items.

2. Specify and define the operating conditions:

- Maximum Allowable Working Pressure (MAWP)
- Normal operating temperatures and temperature extremes
- Cyclic nature of pressure or structural loads
- Presence of a hostile environment, internal or external

3. Select the design procedure:

- Following existing design standards as much as possible
- The design safety factor must be adequate for the manufacturing techniques and intended use. Aviation and space applications usually have a low safety factor because of weight considerations. Other controls, such as advanced manufacturing and welding techniques, compensate for the more conservative Code design used for mass-produced vessels.

4. Select or specify the materials and processes:

- Material selection is critical based on the operating limitations such as temperature, pressure, cyclic loads, and materials-compatibility with the working substance and the environment.
- Manufacturing processes or techniques will be based on the design and the capabilities of the manufacturer. Since Code oversight is probably not invoked for most of these manufacturers (non-Code-stamp manufacturers), validation of these parameters falls to the purchaser:
  - materials specifications
  - manufacturing techniques
  - weld and welder qualification
  - postmanufacturing testing

The concept of a reputable, experienced manufacturer is often satisfactory for validating the quality and safety of product, but it remains the purchaser's responsibility to make this judgment.

5. Specify postmanufacturing testing:

- Overpressure test
- Nondestructive evaluation—often a good partner with an overpressure test
- Quality control process

6. Paperwork: Generally, paperwork associated with the item is only received with ASME Code-Stamped vessels and relief devices (such as ASME Code-stamped ves-

sels, which are registered with the NBIC on a U-1 form). Any documentation prepared by the manufacturer must be specifically contracted for.

Safe Practices

Most pressure/vacuum systems which are properly designed and assembled can be operated safely in the intended environment without additional safeguards. The following practices should be considered and applied as necessary:

1. Identify all hazards and the consequences of principal failure modes.
2. Minimize energy (pressure and/or volume).
3. Use proven hardware compatible with the internal and external environment.
4. All pressure systems should be protected by a pressure relief device (PRD) unless it is impossible to overpressurize the system.

- PRD should relieve at or below MAWP of the lowest rated component in the system.
- Correct sizing and placement is imperative.
- Regulators do no provide overpressure protection.
- Perform periodic functional checks.

A typical Safety Manifold protection scheme for a high-purity gas system is shown in Figure 39.1.

5. Operate within the design intent of the process (pressure, temperature, compatibility of materials).
6. Isolate hazardous substances, components, and operations through the use of barricades, shielding, gas cabinets, or remote operation if the integrity of the system is in doubt or the consequences of component failure or a process gas leak are unacceptable.
7. Provide warning alarms to alert personnel of a potential hazardous situation (toxic gas sources and supply controls are placed in ventilated, alarmed gas cabinets).

This manifold design provides the standard functions of regulating high pressure, providing pressure relief, and includes vent & fill valves,
- also provides a "cycle purge" {alternate vacuum / nitrogen pressure cycles} technique to:
1) remove atmospheric gases from the piping system
2) remove adsorbed water from gas wetted surfaces
3) remove hazardous process gases before opening system

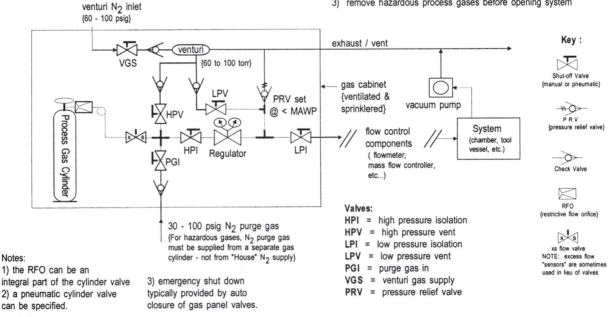

FIGURE 39.1 Typical safety manifold for a high-purity gas system.

8. Know the properties and characteristics of the process medium. Recommended practices, codes or regulations, and the hardware available are directly related. The major suppliers of industrial and specialty gases and cryogenic liquids provide booklets, pamphlets, guides, technical bulletins, and Material Safety Data Sheets which contain pertinent information on recommended practices and precautions.

9. Develop procedures and training (available commercially for common activities such as welding, cryogenics, compressed gas cylinder handling, etc.).

The level of effort applied to these practices (rigor) is dependent on an assessment of the consequences of the potential systems failures. There is a wealth of material available to assist in making these determinations. Depending on the industry and process involved, responsible personnel must collect and assimilate the information necessary from the sources mentioned until an acceptable level of hazard mitigation is reached.

Training/Qualification

Even the simplest system requires a minimum level of qualification for operators. This can be done with on-the-job training by experienced operators, in conjunction with formal corporate or vendor orientation or knowledge-based training. The more sophisticated systems, or those with potentially critical process or exposure implications, should definitely have a method of qualifying operators. The use of qualification cards or forms is usually adequate.

Documentation

Many of the requirements of the preceding paragraphs will be lost or forgotten if not kept as a permanent record with the pressure/vacuum system, in particular special-use components and vessels for which additional effort and expense was made to validate their integrity and level of safety. Off-the-shelf components and apparatus can often be adequately documented with a copy of the page describing the component taken from the vendor's catalog. Include any tables associated with the component such as the working pressures, temperature range, or materials compatibility. Highlight the selected component and information. Sometimes additional technical bulletins are also needed. Technical information on the gases, or system fluid being used, should also be included. A brief summary page at the beginning of the documentation package should justify the basic design and hardware issues. Any system documentation remains pertinent for reference and training.

Emergencies

Any sophisticated piece of equipment should have an easy way to be shut down in a safe manner. Emergency stop buttons or process controls such as excess flow valves will shut down the process if operating conditions are exceeded. Highly toxic gas systems must give adequate warning to personnel. These systems should have a dual-level alarm. The low-level alarm alerts operator personnel to a condition which they may be able to correct. The high-level alarm alerts personnel to evacuate the area in the event of a system leak or failure. Flammables are controlled by safe location and minimizing volume. Potential releases should be confined to areas that include automatic fire system sprinklers or to a remote area. Pressure-relief devices should discharge to a safe area or be vented to a scrubber or a recovery system if the system fluid is of high value, corrosive, or environmentally unfriendly. Emergency-response personnel should be trained in the potential release scenarios and have the capability to restore systems to a safe condition. Users of toxic or corrosive gases must be able to remove leaking cylinders through the use of a hazardous materials handling team and a cylinder containment coffin.

References

1. Chuse, R.; Carson, B. E., Sr. *The ASME Code Simplified, Pressure Vessels*, 7th ed.; McGraw-Hill: New York, 1993.
2. DOT Standards for Transportation, 49 *CFR* 100-199; Superintendent of Documents, U.S. Government Printing Office: Washington, DC, 20402.
3. *NFPA Codes*; National Fire Protection Association: Quincy, MA 02269.
4. Compressed Gas Association, Inc. *Handbook of Compressed Gases*, 3rd ed.; Van Nostrand Reinhold Company: New York, 1990.
5. API 1104, Standard for Welding Pipe Lines and Related Facilities; American Petroleum Institute: 1220 L Street, N.W., Washington, DC, 20005.
6. *Guide to Safe Handling of Compressed Gases*, Matheson: Lyndhurst, NJ, 1983.
7. Braker, W.; Mossman, A. L. *Matheson Gas Data Book*, 6th ed., Matheson: Lyndhurst, NJ, 1980.

40

Laboratory Ovens and Furnaces

ROBERT J. ALAIMO
STEPHEN A. SZABO

Ovens and furnaces are used to produce an elevated, regulated environment for certain laboratory operations. Some may use the terms synonymously, but typically ovens operate at temperatures up to 600°C, and furnaces operate as high as 2000°C. Laboratory ovens and furnaces have the appearance of being very simple systems, but that does not relieve the users and operators from taking all necessary precautions to prevent damage and injury.

Laboratory Ovens

Simple ovens with operating temperatures below 600°C usually transfer the heat mainly by convection. Simple ovens have heaters mounted near the bottom of the chamber to heat the cooler air near the bottom that naturally sinks due to convection. More sophisticated ovens are designed with a means to circulate the atmosphere inside the box in order to achieve more precise internal temperature control and more uniform internal temperature. Some specialty ovens have been built with an external, remote unit that heats a gas which is then pumped into the oven to provide the heating.

"Laboratory ovens should be constructed such that their heating elements and their temperature controls are physically separated from their interior atmospheres. Small household ovens and similar heating devices usually do not meet these requirements and, consequently, should not be used in laboratories."[1]

Some larger ovens, usually used for drying materials, are gas-fired. These have additional concerns around adequate air for combustion; proper exhaust of products of combustion; ensured burner ignition; and fuel safety shutdown. Gas-fired ovens often do not provide as precise control of temperature as achieved with electric resistance heaters.

Oven doors may sometimes be equipped with manual or automatic locks to prevent opening under unsafe circumstances. Some larger industrial ovens have door-operated switches that turn off circulating fans when opened to protect the operator from the hot air release. Typically this is not an issue or hazard with small laboratory units.

Laboratory Furnaces

Typical laboratory furnaces have higher operating and maximum temperatures than do ovens. They are designed to transfer heat by radiant heating. Generally the cost of the furnace is related to the temperature rating. Furnaces come in a variety of sizes and shapes. They include ashing and burn-off box furnaces, general box furnaces, high-temperature box furnaces, tube furnaces, and high-temperature inert gas and vacuum furnaces. One of the basic, and possibly the original, laboratory furnaces is the tubular type.

Tubular Furnaces

Tubular furnaces are typically manufactured by winding a base metal resistance wire onto a helically grooved ceramic tube. The wire is tied down at each end of the coil and usually cemented in place. This heating element is then wrapped in alumina wool and the assembly encased in a sheet-steel shell. The ends of the wire coil and shunt connections, if required, are brought out through ceramic bushings in the metal shell and terminated in suitable connections. Generally, tubular furnaces are engineered to run directly on standard line voltages, that is, 120 V or 240 V. Because of these line voltages the terminal connection area should be completely enclosed during furnace operation. Note that shunt resistors when employed are both electrically and thermally hot. For this reason when profiling the heating element, the furnace must be disconnected from the power source before opening the terminal enclosure to adjust shunt resistors. Modern units have either double-skin construction or an outer mesh guard to promote natural air cooling and protect the operator from hot surfaces. Figure 40.1 shows an example of a laboratory tubular furnace.

Box Furnaces

The typical box furnace can come in various sizes and in top-, front-, or bottom-loading versions depending on the need. The temperature range needed dictates the

FIGURE 40.1 Laboratory tubular furnace showing cutaway view of terminal connection box with shunt resistors installed.

heating element selection. The lower operating temperature units have heating elements of FeCrAl resistance wire and can be used up to 1300°C. Silicon Carbide (SiC) elements are used up to 1600°C. Temperature ranges from 1600 to 1800°C require elements of molybdenum-disilicide ($MoSi_2$), lanthanumchromite ($LaCrO_3$), or platinum wire. When operating above 1800°C, graphite, molybdenum, or zirconia (ZrO_2) are used. Recent years have seen the proliferation of furnaces using $MoSi_2$ heating elements. These elements are mostly used in the form of hanging loops that are exposed to the interior of the furnace. Since these elements usually operate at power line voltages, the connections must be in enclosed boxes to prevent accidental contact. For convenience in laboratory work, these air-operating furnaces are com-

monly loaded and unloaded with the hot zone at high temperature. Caution must be taken to use suitable tongs and protective gloves for this operation as the elements are obviously hot and electrically conductive. $MoSi_2$ elements are also very brittle by nature, and accidental contact will almost certainly break them. Figure 40.2 shows an example of a typical laboratory box furnace.

Before connecting any laboratory furnace to the power line, carefully check that no short circuits exist. Use a "voltmeter" check between the element connectors and the furnace vessel to ensure that there is a high resistance, and no shorts. Also, verify that all electrical contacts are tight and properly secured. Electrical installation must be done in accordance with codes, such as the

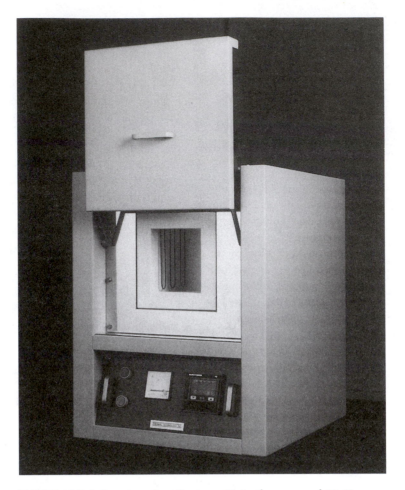

FIGURE 40.2 Laboratory box furnace. Note the exposed MoSi$_2$ heating elements.

National Electrical Code (NFPA 70), and as required by any other applicable regulations that are in effect at the installation site.

Laboratory furnace operation is generally relatively simple, but should not be approached in a casual manner. The operator must be trained according to the manufacturer's instructions in the operation of the functional controls as well as the user's safety regulations. Typically, laboratory furnaces operate in an ambient-type atmosphere and therefore can be loaded and unloaded while at temperature. The operator must take precautions to ensure protection against burns from the furnace itself or the work parts being handled. Furnaces present a higher thermal hazard than ovens, but (anecdotally) the incident rate appears not to be greater than the incident rate for ovens. A possible reason for this is that the worker perceives a much higher risk with the associated higher temperatures and is therefore more diligent in applying risk-reducing techniques. These techniques include the use of furnace gauntlets, furnace tongs, face shields, and body protection. At high temperatures the radiation of light can be damaging to the eyes. For this reason, if observation of the hot zone is to be made, the observer should use appropriate welding glasses to protect the eyes. A face shield should be used to protect against hot particle emission, and appropriate protective clothing, gloves, and goggles should be worn and work parts handled with tongs or other suitable equipment.

Furnace gauntlets are typically constructed of a flame- and heat-resistant outer layer (treated leather or KEVLAR has replaced asbestos in modern furnace gloves) over one or more layers of insulating material (such as wool felt). These gauntlets sometimes have a removable inner liner (cotton) to help with glove sanitation. In very hot environments, the back and sides of the hand and fingers and the entire cuff are covered with a layer of reflective (aluminized) material to reduce the impact of the radiant heat. Furnace gauntlets usually extend ⅓ to ⅔ of the way from the wrist to the elbow.

Furnace tongs are longer and made of heavier-gage material than standard laboratory (oven) tongs. Furnace tongs should not be plated (galvanized, cadmium-coated, etc.), as the plating may volatilize in the furnace heat and add to contamination problems. Furnace tongs should be long enough to reach the back wall of the furnace without the laboratorian's hand breaking the plane of the front (entrance) of the furnace.

When the worker will need to spend an extended time (anything more than a very few minutes) working in front of an open furnace, appropriate face and body protection may be needed to avoid radiant heat burns. Face protection can vary from a simple heat-reflective face shield to an air-supplied aluminized hood that covers the entire head, neck, and shoulders. Safety or industrial hygiene professionals should be consulted to evaluate the hazard and recommend appropriate face protection.

Body protection can vary from a simple, fire-retardant lab coat to a full insulation-lined aluminized long coat. Again, professional evaluation and recommendation should be sought. Figure 40.3 shows an example of a high-temperature laboratory furnace.

High-Temperature Inert Gas and Vacuum Furnaces

To achieve high temperatures under controlled atmospheres, furnaces with graphite and refractory metal hot zones have been developed. These hot zones are enclosed in vacuum-tight, water-cooled vessels manufactured from wrought aluminum or stainless steel. Since the heating elements have a very low resistance, they operate at relatively low voltage, typically less than 40 volts, and high current, 600 to 1000+ amperes. This requires the use of step-down power transformers and large cross-section, often water-cooled, power cables between

FIGURE 40.3 Laboratory high-temperature furnace, typically configured.

the transformer and the element connections. The power connections to the element are electrically insulated from the metal vessel by ceramic or other techniques. With the current flow being high, care must be taken to avoid shorting across the power connection with a conductive material. A metal tool creating a short across the power terminals will become white hot within seconds and cause severe burns. Power connections that can be accidentally shorted with typical tools must be covered with an insulating material. The enclosing vessel being designed to operate with a vacuum atmosphere will therefore safely withstand internal atmospheric pressure of 15 psi gauge and more.

Hazards

Hazards associated with the use of laboratory ovens and furnaces include:

- thermal burns
- ignitable atmospheres and explosions
- pressure relief and explosion venting
- fires
- toxic atmospheres
- other considerations

Thermal Burns

Although hot objects are an obvious hazard, every year there are several instances of burns or dropped equipment or ruined samples stemming from someone not perceiving that "It's HOT" and attempting to lift something out of an oven with his/her bare hands. Ovens operating in the drying range of less than 100°C seem to be especially prominent in "setting people up" for this.

Another situation is when a person is using lab tongs to remove items from an oven or furnace, reaches into the unit for a piece at the back of the chamber, and contacts the hot surface of the interior with his/her hand (often the back of the hand against the rack above or against the roof of the unit).

Although usually not disfiguring, these incidents often result in second-degree burns to the worker as well as damage to the sample or equipment. These incidents can be avoided by the routine use of appropriate gloves. "Oven gloves" should provide:

- protection against contacting the hot surface
- protection extending past the wrist
- dexterity for handling tongs and equipment
- sufficient "tackiness" for holding equipment without dropping it

Oven gloves need not be as protective as the furnace gauntlets described previously, but should give protection at 150% of the maximum temperature achievable by any oven in the facility. Where oven gloves may be used by more than one worker, each person should have, and be required to wear, their own personal glove liners (i.e., light-cotton weighing gloves) inside the oven gloves.

In addition, oven tongs, suitable for the materials being manipulated and sized for the depth of the oven in use, should be kept close to the oven where the tongs are easily accessible by the laboratory worker.

An offshoot of thermal burns is the increased risk of heat-related illnesses. This is especially true in installations with several ovens and/or furnaces in the same work area. This risk of heat-related illness also increases as the level of protective clothing increases.

Ignitable Atmospheres and Explosions

Ovens and Furnaces

The possibility of forming an ignitable atmosphere (a confined vapor cloud) exists whenever a furnace or oven is used to "dry" equipment that has been in contact with an organic material and whenever a furnace is used to "ash" a sample containing organic material. "To avoid explosion, glassware that has been rinsed with an organic solvent should not be dried in an oven until it has been rinsed again with distilled water. Potentially explosive mixtures can be formed from volatile substances and the air inside an oven."[1] A typical scenario would be that the organic material is vaporized and mixes with the air in the oven—the mixture passes the LEL—a slight spark occurs as the oven controller closes the heater contacts—the oven explodes!!! or there is no sparking contact until the lab worker opens the oven door, then—!!! or there is no sparking of contacts, but the surface on the resistance heaters gets hot enough to ignite the vapor-air mixture. There are several techniques that may be used to avoid this problem. The most obvious is to not use an oven for the volatilization or heating of an organic material. When this is not feasible, other techniques may be applied. Some of these are:

1. Install a continuous air sweep through the oven to a controlled exhaust. The air flow must be great enough to ensure that volatilized materials are exhausted out of the oven before they can reach 50% LEL (some say 10% LEL). This technique can lead to some problems in maintaining a consistent and constant temperature in the oven if the ventilation system is not carefully engineered.

2. Establish an inert atmosphere in the oven before beginning the heating cycle. Maintain a slight positive

pressure on the oven chamber with the inert gas during the heating cycle. Before opening the oven, exhaust the oven chamber to a suitable exhaust (i.e., fume hood exhaust system), flushing the oven chamber with inert gas. After the inert flushing, the oven chamber is refilled with room air. Caution must be taken to select an appropriate exhaust and to perform the initial flush and inerting and the final flushing and refill in a way that does no damage to any samples that might be in the oven. This technique is often combined with that in item 3 to provide maximum safety.

3. Isolate all ignition sources in the oven from the possibility of contact with the vapor-air mixture by isolating them in purged enclosures ("z-purging"). The contacts and heating elements are obvious items for enclosure, but each oven should be evaluated individually to ensure that no ignition sources have been overlooked. The efficacy of the isolation should be proved with a nonignitable tracer gas before the oven is used for heating ignitables. It is recommended that this technique be combined with either item 1 or item 2 to ensure greater safety.

4. Construct a system that remotely heats a gas (nitrogen is first choice) and then pumps the nitrogen into the oven chamber to heat it. The exhaust nitrogen is removed through an appropriate exhaust, such as the fume hood exhaust system. In plants with excess steam available, this can be developed into a very inexpensive means for providing oven heat, although it is best used for lower oven temperatures.

5. Use RF (microwaves or induction) for heating. The oven box can be fairly easily isolated from arc-producing oven components to avoid ignition.

The use of furnaces creates much of the same concerns as found with ovens, except that the interior surfaces of the furnace must be considered as possible ignition sources. Therefore, the most usable alternative is to inert the furnace interior and to provide a controlled inert flush during the heating. Flush system design is important: the design must allow control of the flow rate of the flush gas; must minimize "dead" pockets in the *loaded* furnace; must provide for the temperature equilibration (cooling) of exhaust flush gases before discharge into an appropriate exhaust system (i.e., lab hood); must not interfere with the operation of (use of, access to) the furnace; and must not compromise any pressure-relieving features of the furnace.

Pressure Relief/Explosion Venting

Where pressure buildup inside an oven or furnace is a possibility, appropriate means for venting this unwanted pressure *before opening the oven or furnace* must be provided. Also, anything that might be blown out of the furnace by rapid release of the pressure (i.e., the thermometer in the roof of some ovens or the thermocouple probe in some furnaces) must be attached in a way that will prevent it from becoming a missile. Alternately, a substantial means to "catch and hold" the missile may be provided.

If you use "magnetic" catches on doors as part of the venting device, you must make sure that the door is restrained from flying completely open by some means that is at least as substantial as the hinges on the door. It is questionable whether the "suitcase" latches provided as original equipment on some ovens would stand up to this duty. The design and adjustment of this restraint takes some care; the restraint must allow the door to open enough to relieve the internal pressure safely, yet must hold the door closed enough that the explosive opening does not throw the door in the way of the worker and enough that the door deflects any internal components of the oven from blowing into the lab.

A good alternative for ovens is to cut a "relief panel" in the rear of the oven. This requires some sheet metal and mechanical work. The panel is often cut to be approximately 90% of the internal surface of the oven. This (insulated) panel is held in place with fasteners, the strength of which (summed up) is less than 80% of the failure strength of the next weakest oven component that might see excess pressure. A substantial cage is constructed around the exterior back of the oven. The cage is sized to allow the panel to relieve yet not leave the vicinity of the oven (cage depth is approximately 2 × the thickness of the relief panel). In some instances the cage is designed so that the panel falls down and away from the opening, leaving the cage to contain any missiles from inside the oven. This alternative moves the venting action to the rear of the oven, away from the lab worker. Furnaces will usually have to be allowed to vent at the door. Similar door constraints as with ovens need to be observed.

Fires

Fires must always be recognized as a possibility whenever heaters are in use. Electric resistance heaters may be more liable to have/cause fires than other types of heating, but all types of heating can cause fire. Redundant controls, over-temperature shutdowns, pilot sensing devices (on gas ovens), and nonflammable construction all help to reduce the probability of fires. However, appropriate fire extinguishing equipment (appropriate both to the oven/furnace and to its contents) should be readily at hand. Individuals working in the oven/furnace work area should be trained in the correct use of this fire suppression equipment.

Toxic Atmospheres

Some laboratory processes require the use of inert- or reducing (flammable)-atmosphere gases. These gases can cause toxic exhaust hazards and, in the case of reducing atmospheres (hydrogen), a combustible hazard. All exhaust fumes should be properly vented from the building in accordance with local regulations. When flammable gases are to be used, a suitable safety system with a burn-off provision must be utilized. Consult with the furnace manufacturer for availability of suitable safety operation systems. In any case, ventilation of the surrounding area is mandatory for any furnace operation. Certain product-processing in the furnace may cause combustion or the evolution of toxic fumes. The user should give consideration to any such chemical reactions and take all appropriate measures for the protection of personnel.

With the exception of vacuum drying ovens, laboratory ovens rarely have a provision for preventing the discharge of the substances volatilized in them into the laboratory atmosphere. Thus, it should be assumed that these substances will escape into the laboratory atmosphere and may also be present in concentrations sufficient to form explosive mixtures with the air inside the oven. This hazard can be reduced by connecting the oven vent directly to an exhaust system. In addition to the more obvious concerns about volatilization of hazardous materials, one must consider the possibility of toxic thermal degradation products, especially where complete oxidation cannot be assured. Consideration also needs to be given to whether or not the "explosion gases" need to be controlled and exhausted from the work area.

The best protection against these is the inert-gas purged heating chamber that is vented to an appropriate exhaust (see "Ignitable Atmospheres and Explosions"). As discussed earlier, this can be developed either for an oven or for a furnace.

Note that improper use of an inert-gas purging system in a closed, poorly ventilated room can present its own potential hazard, namely, the depletion of the oxygen concentration in the room. When using inert-gas purging, always ensure that the vent gas is exhausted away from the work space and that the work space is well ventilated with FRESH AIR (recirculated air does not protect against oxygen depletion).

Other Considerations

The use of laboratory equipment for the consumption, storage, or preparation of food should be strictly prohib-

ited. Laboratory ovens or furnaces are designed for specific purposes and should never be used for food preparation.

The use of glass thermometers inserted through the vent of an oven to confirm the oven temperature is a risky practice. The bulb of the thermometer is very subject to being broken as items are moved into and out of the oven. This situation is made even worse if the glass thermometer is mercury-filled, requiring special precautions should cleanup become necessary. Should a mercury thermometer be broken in an oven of any type, the oven should be closed and turned off immediately, and it should remain closed until cool. All mercury should be removed from the cold oven with the use of appropriate cleaning equipment and procedures (see Chapter 5, section 5.C.l I .8)[1] in order to avoid mercury exposure. After removal of all visible mercury, the heated oven should be monitored in a fume hood until the mercury vapor concentration drops below the threshold limit value (TLV).

The user should be aware that most furnaces are made from oxides of aluminum (Al_2O_3) and silicon (SiO_2) and can be attacked by some materials. The most common agents are the low-melting-point oxides of lead, sodium, and potassium; fluxes used in melting such as lithium and borax and some other salts; sulfur and its compounds; and the halogens chlorine, fluorine, and iodine.

The system must be operated within limits recommended by the manufacturer and periodically calibrated.

The oven or furnace user should use caution and check on maximum temperatures to ensure that a specific product can be operated without causing damage to the furnace or a hazard to the operator.

In cases of operation as well as maintenance, deposits of materials of unknown substance found on components should be treated as hazardous until otherwise determined.

It is recommended that the user become familiar with certain standards, and one recommendation is to obtain a copy of NFPA 86.[2] This is the national standard governing ovens and furnaces which in turn references other standards of importance to safe operation.

References

1. *Prudent Practices in the Laboratory*; National Research Council, National Academy Press: Washington, DC, 1995.
2. NFPA-86. *Standard for Ovens and Furnaces*; National Fire Protection Association: Quincy, MA, 1999.

41

Refrigerator, Freezer, and Cold Room Use in Chemical Laboratories

ROBERT J. ALAIMO

The presence of refrigerators, freezers, and walk-in cold rooms in the laboratory environment has become commonplace. This equipment is generally used for the storage of reagent chemicals and other materials requiring lower temperature conditions. They are also frequently used for cooling reactions and in some cases to allow crystallization to take place. The selection of the proper equipment must be the first order of business prior to the introduction of freezers or refrigerators into the laboratory. The next step is to develop appropriate safe storage practices for refrigerated chemicals in the laboratory. These practices are as important as any other work practice. The consequences of poor equipment selection and storage practices can be devastating. Refrigerator temperatures are most often higher than the flashpoints of the flammable liquids being stored. Vapor accumulation, spills, and leaking containers can present the fuel for an explosion. The domestic refrigerator can supply the ignition source with its lights and switches, thermostats, heater strips, and compressor circuitry.

Selection Criteria

In order to assure that the correct refrigerator or freezer is purchased for the laboratory the decision should be subject to a hazard review. The researcher and site safety specialist, not just a purchasing agent or design engineer, should first evaluate the requirements of the equipment. Questions to be addressed include what temperature range is needed, what classes of chemicals will be stored, what size is needed, and where the unit will be located in the laboratory. In many cases common domestic refrigerators or freezers are not appropriate for laboratory use. These units were designed specifically for the storage of food products. They were not designed to store flammable, corrosive, or radioactive chemicals. The use of the household refrigerator for the storage of typical laboratory solvents presents a significant hazard to the laboratory work area. Explosion-proof or explosion-safe units are specifically engineered for various applications. Underwriters Laboratories–approved units should be used whenever refrigerated flammable stor-

age is needed. Laboratory equipment manufacturers can provide this certification. *Explosion proof* units are refrigerators/freezers designed to prevent against ignition of flammable vapors both inside and outside the storage compartment. These units are specially designed and wired to be used in locations, such as pilot plants and laboratories where all electrical equipment needs to meet Article 501 of NFPA 70, National Electrical Code and NFPA 45 Standard on Fire Protection for Laboratories Using Chemicals, Section A-9-2.2.2.[1] *Explosion safe* or *flammable material storage* refrigerators/freezers are suitable for the typical laboratory environment. The principle feature of these units is to eliminate the potential for ignition of flammable vapors within the storage area by electrical sources within the same compartment. These units may also possess features such as spill-protection thresholds, self-closing doors, magnetic gaskets, and special interior materials. The compressor and thermostatic controls are located on the top of the unit to prevent the ignition of floor level vapors.[1]

Household, domestic refrigerators do have a utility in the laboratory, for the storage of nonflammable liquids, solids requiring low-temperature storage, or media and films. Unless the storage of flammable chemicals is an impossibility, the presence of these household units in the laboratory is an accident waiting to happen. Examples of serious explosions and fires resulting from the storage of flammable materials in household refrigerators have been documented.[2]

The refrigerator or freezer should not be a self-defrosting unit, since these units are usually not explosion safe. Manual defrost units may be able to be modified to become suitable for storage of flammable materials, but these modifications must be done by qualified personnel. Self-defrosting models cannot be successfully modified to provide even minimum safeguards against vapor ignition.[1] In order to modify a manual defrost unit for use for storage of flammable liquids, all potential sources of electrical discharge inside the refrigerator must be eliminated. The modification must include:

- removal of the interior light and switch unit located on the door and sealing of the interior

- relocation of all temperature controls, thermostats, and fans to outside the storage compartment and sealing of all points where capillary tubing or wiring previously entered the storage compartment
- replacement of the mechanical door latches with nonsparking magnetic door gaskets

The selection of the right interior surface material is also important if corrosives or radioactive materials might be stored in the refrigerator. Surfaces should be nonporous, corrosion-resistant, and easily decontaminated.

Walk-in cold rooms are usually not the norm in most laboratories, but the need for these units in special situations warrants a brief discussion. Generally these units are used for the storage of serums and biological samples and not flammable chemicals. If used for storage of flammables, the same selection criteria would hold true. Prefabricated rooms are far superior in quality to built-in rooms and are far more flexible in use because they can be moved.[3] The ideal units should have audible and visual alarm systems to report temperature changes outside the ranges required. The door should be self-closing with magnetic gaskets for a tight seal and a heating wire to prevent freezing. The door hardware must include provisions to allow for easy opening from the inside and have available other emergency options for quick escape. These might include breakaway handles or emergency alarm systems. Light fixtures should be vapor-tight to keep condensation out and include cold-temperature-rated lamps.[3] Depending on the application, explosion-proof light fixtures might be necessary. Emergency lighting should be included to allow for easy escape in case of power failure. Since these rooms are designed and built to be closed air circulation systems, any release of toxic substances can pose a serious danger to anyone working inside. If the potential for low oxygen levels or a flammable atmosphere is possible, flammable gas detectors and oxygen monitors are recommended.[4] Allowing samples received in dry ice to be stored in a cold room before removing the dry ice could elevate CO_2 levels to unacceptable levels.

Safe Practices

Prior to putting a laboratory refrigerator or freezer into service it should be checked to assure that it is properly grounded. This can be done by using a voltage detector such as the Tic Tracer or a similar sensor.[5] This will protect workers from potential electrical shocks from improperly grounded equipment and static charge buildup. A fatality was reported involving a 25-year-old North Carolina man. The man slipped on a wet floor and reached out and grabbed the handle of an ungrounded refrigerator and was electrocuted.[6] The case of the unit was obviously charged. It is recommended that laboratory refrigerators be placed against fire-resistant walls, have heavy-duty cords, and preferably be protected by their own circuit breaker or ground-fault-interrupter circuit breaker.[4]

Regardless of its use every refrigerator, freezer, or cold room located in the laboratory should be labeled prominently with hazard warning or use suitability labels. These labels should indicate whether the unit is suitable for the storage of hazardous chemicals. All laboratory refrigerators should carry a sign indicating:

> NO FOOD OR BEVERAGES SHOULD BE STORED IN THIS UNIT or FOR CHEMICAL STORAGE ONLY. NO FOOD OR BEVERAGES ALLOWED.

Refrigerators used for the storage of radioactive materials must be labeled with the caution signs indicating:

> CAUTION, RADIOACTIVE MATERIAL. NO FOOD OR BEVERAGES MAY BE STORED IN THIS UNIT.

If possible it is advisable to provide refrigerators in the break areas and lunchrooms for food storage purposes. Making food storage units readily available will help compliance. These units should labeled as well with signage indicating:

> FOOD AND BEVERAGE STORAGE ONLY.

If the refrigerator is to be used for the storage of flammable materials it should be labeled:

> EXPLOSION PROOF or EXPLOSION SAFE, FLAMMABLE STORAGE PERMITTED.

Domestic refrigerators and freezers that have not been modified should be prominently labeled with wording such as:

> DO NOT STORE FLAMMABLE MATERIALS IN THIS REFRIGERATOR.

Numerous examples of explosions and fires caused by improper storage of flammable solvents in ordinary refrigerators have been reported.[2,4] A technician placed a beaker containing methyl alcohol and ether in a domestic refrigerator to cool the mixture. The unit was believed to have been equipped with a bimetallic thermostat which created a spark that set off the vapors inside the refrigerator compartment. The door was blown off the unit and the ensuing fire engulfed the entire laboratory and caused over $100,000 worth of damage. Fortunately, no one was injured and the quick response by fire services prevented more severe damage according to the insurance report. Another example involved an explosion and fire that caused extensive damage to an analytical laboratory when flammable vapors were ignited by the temperature-control device in the freezer compartment.[2] For similar reasons, volatile flammable solvents should not be used in cold rooms. The exposed motors of the circulation fans can serve as an ignition source and initiate an explosion. The use of volatile acids should also be avoided in cold rooms because the acids can corrode the cooling coils and cause leaks of refrigerant.[4] Special precautions and practices must be taken to assure rapid escape from cold rooms. Regular maintenance should include inspection and lubrication of the emergency release button. Condensation around the button could freeze and cause difficulty in opening of the door. One option would be to install an emergency alarm system which would be available to alert others of any problem. Another option could involve the use of a door or lock blocking pin that prevents the door from closing and locking completely when the cold room is occupied.

All materials stored in laboratory refrigerators, freezers, or cold rooms must be properly labeled with contents, appropriate hazard warnings, storage and expiration dates, and a contact person. Labels should be a permanent type and not subject to smudging. All containers placed in the units should be completely sealed or capped. Aluminum foil and other foils and corks or glass stoppers should be avoided. Foils have a tendency to decompose over time and the glass stoppers can freeze to the walls of the flask.

An inventory of the contents should be readily available for easy reference, and a periodic inspection of the contents should be performed. The unit should be cleaned at regular intervals and out-of-date and forgotten materials properly discarded. Refrigerators can easily become overcrowded and pose serious breakage and spillage possibilities. One option is to place containers in spill trays that can be removed from the unit in order to examine the contents. This eliminates the need for the researcher to reach into the refrigerator or freezer to remove materials, thus minimizing the potential of knocking containers off the shelves. Even worse may be the need to place one's head into the unit to find something, thus exposing the individual to chemical vapors. The spill trays also serve as containment for any spills or leaking containers. This protects the interior of the unit from possible damage from the spilled chemicals and prevents contamination of other materials.

Another common practice that should be avoided is use of a mercury thermometer to monitor interior temperatures. These thermometers are easily broken and the toxic mercury can spread everywhere. If the interior temperature is critical, non-mercury thermometers or temperature monitors should be utilized. Temperature alarms are also recommended in many situations to

FIGURE 41.1 Typical laboratory refrigerators and freezers.

alert the laboratory personnel of equipment failures. Emergency power should be available for those units that must be kept operating. Particular care must be taken when opening refrigerators or entering cold rooms following long power outages. Vapor levels can rise dramatically in these situations.

Periodic review of the storage practices and updating of written safe practices related to refrigerators, freezers, and cold rooms should be an integral part of any comprehensive safety program. Without due diligence these useful laboratory appliances can contribute to deadly consequences.

Examples of laboratory refrigerators and freezers are shown in Figure 41.1.

References

1. NFPA 70 National Electrical Code Article 501, NFPA Standard Code No. 45, *Fire Protection for Laboratories Using Chemicals*, Section A-9-2.2.2; National Fire Protection Association: Quincy, MA, 1996.
2. *Case Histories of Accidents in the Chemical Industry*; Manufacturing Chemists Association Case History No. 1794, American Chemistry Council: Washington, DC, April 1975.
3. *Handbook of Facilities Planning*, Vol. I; Ruys, T., Ed.; Van Nostrand Reinhold: New York, 1990; p. 361.
4. *Prudent Practices in the Laboratory*; National Academy Press: Washington, DC, 1995; pp. 113, 194.
5. Tic Tracer TIF Instruments, Inc., Miami, Florida 33150.
6. Division of Safety Research, NIOSH, Report No. FACE-86-43, 1986.

42

Specialized Laboratory Containment-Control Hoods

LOU DiBERARDINIS

Purpose of Ventilation

Ventilation is needed to provide an environment that is safe and comfortable. This is accomplished by providing controlled amounts of supply and exhaust air plus provisions for temperature, humidity, and velocity control. *Comfort ventilation* is a means of supplying measured amounts of outdoor air for breathing and maintaining design temperature and humidity.[1] This is sometimes referred to as general exhaust. It can be used for control of nonhazardous odors or low levels of toxic materials by diluting them to unobjectionable levels. *Local exhaust* or *contaminant control ventilation* is designed specifically for health and safety protection. The laboratory chemical hood (sometimes erroneously call a "fume" hood because it contains gases, vapors, and some aerosols as well as just "fumes") is an example of local exhaust ventilation. A complete discussion of laboratory chemical hoods is provided in Chapter 43. Good laboratory local exhaust ventilation contains or captures toxic contaminants at the source before they escape into the laboratory. It transports them out of the building by means of ductwork and a fan that will not contaminate other areas of the building by recirculation from discharge points to clean air inlets or by creating significant negative pressure in the building to subject inactive hoods to

downdrafts. A laboratory ventilation *system* will frequently be a combination of comfort and local exhaust ventilation. The components of a ventilation system include fans, ducts, air cleaning filters, inlet and outlet grilles, sensors, and controllers.

Types of Local Exhaust Ventilation

The laboratory chemical hood is a specialized form of local exhaust that totally encloses the emission source. It is an enclosing hood because it primarily contains materials generated within the enclosures. See Chapter 43 for further discussion of these hoods. Other examples of enclosures include glove boxes, toxic gas cylinder storage cabinets, biological safety cabinets, and equipment enclosures. This type of hood will only provide control of toxic materials generated within the enclosure. Enclosures are the preferable type of hood.

Often, total enclosure of the source is not possible, or is not necessary. In these cases, another form of local exhaust, a capture hood (sometimes referred to as an exterior hood) can be used. A capture hood prevents the release of toxic materials into the laboratory by capturing or entraining them at or close to the source of generation, usually a work station or laboratory operation.

Considerably less air volume is required than for the standard chemical hood. However, since the contaminant is generated outside the hood, it must be designed and used properly in order to be as effective as an enclosure.

Enclosure Hoods

While the laboratory chemical hood is the most common type of enclosure hood found in laboratories using chemicals, there are other types that have specific applications. Their selection and design is based on the particular materials being used and activities being performed.

Glove Box

Use of a totally enclosed and ventilated glove box is recommended for handling very hazardous materials (Figs. 42.1a, b). The use of glove boxes minimizes air volume and simplifies air treatment for environmental protection. Its design requirements are specified by the ACGIH Industrial Ventilation Manual.[2]

Biological Safety Cabinets

The biological safety cabinet is a special form of containment equipment used for work with cell and tissue cultures and parenteral drugs when the materials being handled must be maintained in a sterile environment and the operator must be protected from toxic chemicals and infective biological agents. The dual functions of protecting the worker and maintaining sterility are achieved by two separate cabinet flows: a turbulence-free downward flow of sterile HEPA-filtered air inside the cabinet for work protection, and an inward flow of laboratory air through the work opening to provide worker protection. In addition, all air exhausted from biological safety cabinets is filtered through HEPA filters to provide environmental protection. The inward airflow and the downward airflow are delicately balanced in the biological safety cabinet and great care must be exercised to maintain the design flow rate of each, as well as the ratio between the two.

Biological safety cabinets are designated as Class I, II, and III based on the level of containment needed.[3] Class II cabinets are the most commonly used and they are further subdivided into types A, B1, B2, and B3. The differences in each type is related to the quantity of airflow and the construction of the cabinet. Figures 42.2a and b show a Class II, Type A cabinet. For more information on types of cabinets and when to use them see the NIH Guidelines.[3]

Specialty Enclosures

Other enclosures can be custom designed for the particular equipment and operations. The general design guidelines are as follows:

1. Provide the most enclosure possible while allowing the user access to perform the necessary functions.
2. Provide enough exhaust airflow to create an inflow air velocity of between 50 and 200 feet per minute (fpm) through all openings. The high end of this range is needed for more toxic materials or if the material is released with a high velocity.

Capture Hoods

Capture hoods are used when it is not feasible to enclose the particular source of generation. The basic principle of operation of this type of hood is that the airflow through the hoods creates a "capture velocity" of air at the source of generation and "captures" the contaminant and draws it into the hood.

To work effectively, the air inlet of a capture hood must be of proper geometrical design and placed near the point of chemical release. The distance away will depend on the size and shape of the hood and the velocity of air needed at the generation source to "capture" the contaminant and bring it into the hood. Generally, the closer to the generation source the better. Design face or slot velocities are typically in the range of 50 to 200 and 1000 to 2000 fpm, respectively. Many design guidelines exist for this class of exhaust hoods in Chapter 3 of the Industrial Ventilation Manual[2] or in Burgess et al., "Ventilation for Control of the Work Environment."[4] Two types of capture hoods that find frequent application in the laboratory are "canopy" hoods and "slot" hoods.

Canopy hoods (Fig. 42.3) are used primarily for capture of gases, vapors, and aerosols released in one direction with a velocity that can be used to aid capture. These are sometimes called "receiving" hoods. This type of hood is generally used when the process to be controlled is at elevated temperatures to make use of the thermal updraft or when the emissions are directed upward by the process. Examples of operations that may be controlled include ovens, gas chromatographs, autoclaves (Fig. 42.3), and atomic absorption spectrophotometers. Many equipment manufacturers recommend specific capture hood configurations that are suitable for their units. They should be consulted for advice.

Slot hoods are used for control of laboratory bench operations that cannot be performed inside a contain-

FIGURE 42.1a Glove box. (*Source:* Labconco Corporation, Kansas City, Missouri.)

ment hood (chemical hood) or under a canopy hood. Typical operations include slide preparation, microscopy, biological specimen preparations, and chemical mixing. Examples are shown in Figures 42.4 and 42.5. The required flow can be calculated from a series of equations determined empirically depending on the size and shape of the hood, the distance of the hood from the source, and the physical properties of the source.

Another laboratory operation that needs special consideration is the weighing of highly toxic materials. Because of the nature of the materials used, it is advantageous to do this in a ventilated enclosure. A chemical laboratory hood is not the best location because its relatively high airflow may disrupt the weighing process and it is not an efficient use of ventilated space. A ventilated weighing station (Fig. 42.6) has been developed as part of the National Toxicology Program and is commercially available. Since this is almost a full enclosure, relatively little airflow (100 to 400 CFM) is needed compared to

the laboratory chemical hood (approximately 1000 to 1500 CFM).

Regulations

Those facilities covered by OSHA must comply with the Laboratory Standard.[5] This is a performance-based standard so it does not have specific requirements on laboratory ventilation. It simply requires that adequate engineering controls be in place to prevent overexposure to toxic chemicals and that they be operating effectively.

Guidelines

A standard on laboratory ventilation was adopted by the American National Standards Institute.[6] The purpose of this standard is to establish minimum requirements and

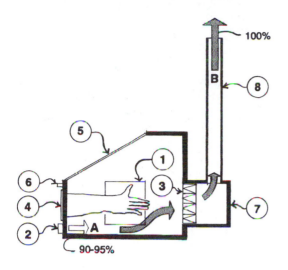

100%

B   8

5

6

4

2   A

90-95%

KEY

1   Air Lock Pass-Through
2   Constant Air In Leakage
     HEPA Filter
3   Roughing Filter
4   Glove Ports
5   Glass Window (Sealed)
6   Controls
7   Exhaust Blower
8   Exhaust Duct To Final
     Air Cleaning Filter.
□   Clean Air
■   Contaminated Air

AIR FORMULAS
A + Leakage = B

FIGURE 42.1b Glove box: section.

procedures for laboratory ventilation systems which are used to prevent overexposure of personnel to chemical contaminants generated within the laboratory.

The standard requires that each laboratory facility establish a Laboratory Ventilation Management Program and designate a "cognizant person" to coordinate and monitor the program. It emphasizes that management participation in the selection, design, and operation of laboratory ventilation systems is important.

There are no specific design guidelines or operational guidelines for local exhaust systems other than those for laboratory chemical hoods and glove boxes.

Measures of Hood Performance

Since the main purpose of a hood is to prevent the release of toxic materials into the laboratory environment, the ultimate measure of hood performance is to determine the concentration outside the hood of all the chemicals used inside the hood. If these concentrations are below relevant health standards with an adequate safety factor applied, the hood performance would be considered adequate. This is not practical to do because of the wide variety of chemicals and operations performed inside hoods. As a result other surrogate measures of hood performance are used.

Face or Slot Velocity

The common measure used is the hood face velocity for enclosures or slot velocities for capture hoods. This is the speed of airflow into the hood through the opening measured at the plane of the opening. There are no specific standards for what these velocities should be because they depend on the design and use requirements. The ACGIH Ventilation Manual can be used for guidelines. Unfortunately, there are other factors that effect hood performance that may not be uncovered by this method alone.

Smoke Visualization

This method is qualitative and involves observing that airflow is always into the hood by generating smoke or using dry ice. This can be valuable only if used in conjunction with other methods such as face velocity. One can also try to release the smoke at the source of generation and observe if it is contained (if enclosure) or captured and brought into the hood (capture hood).

Containment Testing

This method involves generating a known amount of a tracer gas (such as sulfur hexafluoride) inside the hood for an enclosure or at the point of generation for a capture hood and measuring the concentration outside the hood at the breathing zone of a hood user. The generation rate is selected based on the type of activity anticipated for the hood ranging from 1 lpm (very little manipulation) to 8 lpm (boiling an organic solvent). Essentially this determines the leakage. This is a cumbersome and time-consuming process. It is most valuable as a prepurchasing specification for hoods or as part of the installation acceptance criteria. Existing methods have been published for use on laboratory chemical hoods and could be modified to evaluate other forms of contaminant control hoods.[7,8]

For biological safety cabinets a standard performance test procedure has been adopted and is published by the National Sanitation Foundation.[9]

FIGURE 42.2a Class II Type A biological safety cabinet. (*Source:* The Baker Co., Sanford, Maine.)

Frequency of Testing

The ultimate desire is to allow the user to have continuous feedback of hood performance. Ideally, a continuous flow monitor should be on each hood. In the absence of continuous flow monitors, hoods should be tested at least annually. For operations involving highly toxic materials, the frequency may need to be increased to quarterly.

Communication to Hood Users

There should be a clear indication on the hood as to its measure of performance. If continuous monitors are used, acceptable versus unacceptable conditions should be clearly evident to the user. Where continuous monitors are not used, a label should be clearly visible indicating the last time the hood was checked, its status at that time, and who to call for up-to-date information.

Work Practices

Recent studies have indicated that the work practices followed by hood users have as much effect if not more

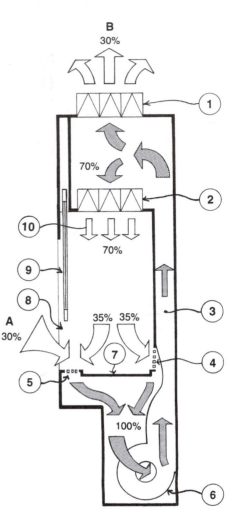

KEY
1   Exhaust Air HEPA Filter
2   Supply Air HEPA Filter
3   Exhaust Plenum
4   Rear Exhaust Grille
5   Front Exhaust Grille
6   Fan (1 Per Unit)
7   Work Surface
8   Front Opening
9   Vertical Sliding Sash
10 Filtered Recycled Air
11 Lights
☐   Clean Air
▨   Contaminated Air

AIR FORMULAS
A = B

FIGURE 42.2b Class II Type A biological safety cabinet. (Reprinted with permission from John Wiley & Sons.)

FIGURE 42.3 Canopy hood.

that the hood and laboratory design. Improper placement of equipment or your body can adversely effect the ability of the hood to contain the toxic materials within.

Conversely, careful arrangement of equipment and procedures performed can enhance containment capability. Below are examples of good practice that should be followed whenever possible.

1. Work as close to a capture hood as possible. Don't work farther away from the hood than it was designed for.
2. Never put your head inside an operating enclosure hood to check an experiment. The plane of the sash is the barrier between contaminated and uncontaminated air.
3. Work with the enclosure's sash in the lowest position possible. The sash will then act as a physical barrier in the event of an accident in the hood.
4. Do not clutter the enclosure with unnecessary bottles or equipment. Do not use a hood for storage. Only materials actively in use should be in the hood. This provides optimal containment and reduces the risk of extraneous chemicals being involved in a fire or explosion if it occurs.
5. Do not obstruct the slots of a capture hood. Clean the grille along the bottom slot of the hood regularly so it does not become clogged with objects or paper.
6. Note the potential impact of lab design (location of doors and windows) on hood containment. Containment can be affected by people walking

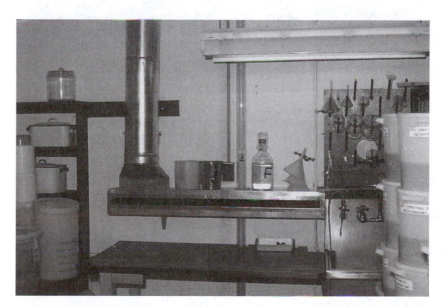

FIGURE 42.4 Slot hood.

FIGURE 42.5 Slot hood.

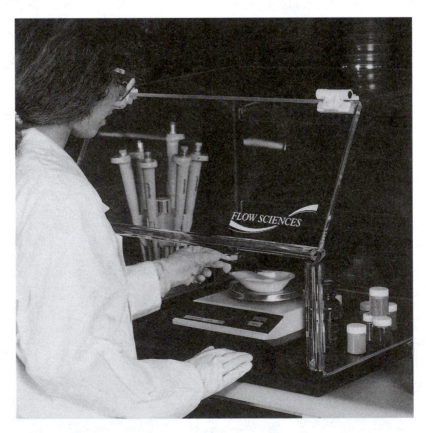

FIGURE 42.6 Vented balance safety enclosure. (*Source:* Flow Sciences Inc., Wilmington, NC.)

behind the work area and opening and closing doors or windows.

7. Do not dismantle or modify the physical structure of your hood or exhaust system in any way without first consulting the appropriate people in your organization.

8. Report any suspected hood malfunctions to the appropriate person immediately.

9. Don't use a hood for any function it was not designed for.

10. Use a hood only when you have determined that it is performing adequately.

11. Wear protective clothing at all times.

References

1. ANSI/ASHRAE Standard 62-1999. *Ventilation for Acceptable Indoor Air Quality*; American Society of Heating, Refrigerating Air-Conditioning Engineers: Atlanta, GA, 1999.

2. American Conference of Governmental Industrial Hygienists, 1998. *Industrial Ventilation, A Manual of Recommended Practices*, 23rd ed.; Committee on Industrial Ventilation: Cincinnati, OH, 1998.

3. *Biosafety in Microbiological and Biomedical Laboratories*; U.S. Department of Health and Human Services, Centers for Disease Control and Prevention, National Institutes of Health, 3rd ed., May, 1993, U.S. Government Printing Office: Washington, DC, 1993.

4. Burgess, W. A.; Ellenbecker, M. J.; Treitman, R. D. *Ventilation for Control of the Work Environment*; John Wiley and Sons: New York, 1989.

5. U.S. Department of Labor, Occupational Safety and Health Administration. Health and Safety Standard; Occupational Exposure to Hazardous Substances in Laboratories.

6. ANSI/AIHA Standard Z9.5. *Laboratory Ventilation*; American Industrial Hygiene Association: Fairfax, VA, 1993.

7. ASHRAE 110-1995. *Method of Testing Performance of Laboratory Fume Hoods*; American Society of Heating, Refrigerating and Air-Conditioning Engineers: Atlanta, GA, 1995.

8. Ivany, R.; First, M. W.; DiBerardinis, L. J. A New Method for Quantitative, In-Use Testing of Laboratory Fume Hoods. *Am. Ind. Hyg. J.* **1989**, *50*(5), 275–280.

9. National Sanitation Foundation Standard No. 49 for Class II, (Laminar Flow) Biohazard Cabinetry; NSF: Ann Arbor, MI, 1992.

43

Laboratory Chemical Hoods

LOU DiBERARDINIS

Purpose of Ventilation

Laboratory ventilation systems are crucial for the protection of the health and safety of the laboratory occupants and the condition of the facility. This is accomplished by providing controlled amounts of supply and exhaust air plus provisions for temperature, humidity, and velocity control. *Comfort ventilation* is a means of supplying measured amounts of outdoor air for breathing and maintaining design temperature and humidity.[1] The main purpose is to provide a work environment within a specific temperature, air exchange, and humidity range. It can be used for control of nonhazardous odors or low levels of toxic materials by diluting them to unobjectionable levels. This is frequently referred to as *general exhaust*. *Local exhaust* or *contaminant control ventilation* is designed specifically for health and safety protection. The laboratory chemical hood (sometimes erroneously call a "fume" hood because it contains gases, vapors, and some aerosols as well as just "fumes") is an example of local exhaust ventilation. Good laboratory local exhaust ventilation contains or captures toxic contaminants at the source before they escape into the laboratory. It transports them out of the building by means of ductwork and a fan that will not contaminate other areas of the building by recirculation. The ventilation system may serve to supply makeup air for exhaust systems and provide for a comfortable and safe work environment or there could be a separate system for each function.

Types of Local Exhaust Ventilation

The laboratory chemical hood is a specialized form of local exhaust that totally encloses the emission source. It is an enclosing hood because it primarily contains materials generated within the enclosures. Enclosures are the preferable type of hood.

Often, total enclosure of the source is not possible, and

in these cases, another form of local exhaust, a capture hood, can be used. A capture hood prevents the release of toxic materials into the laboratory by capturing or entraining them at or close to the source of generation, usually a work station or laboratory operation. Considerably less air volume is required than for the standard chemical hood. However, since the contaminant is generated outside the hood, it must be designed and used properly in order to be as effective as an enclosure. See Chapter 42 for a further discussion of these hoods.

Laboratory Chemical Hoods

The purpose of a laboratory hood is to prevent or minimize the escape of contaminants from the hood to the laboratory air. Successful performance depends on an adequate and uniform velocity of air moving through the hood face, commonly referred to as the face velocity. Hood performance is adversely affected by high-velocity drafts across the face, large thermal eddy currents at the opening, and poor operating procedures on the part of personnel using the hood. With the sash closed, the hood can minimize the effects of small explosions, fires, and similar events that may occur within, but it should not be depended upon to contain fires or explosions other than trivial ones. To function correctly, a chemical hood must be designed, installed, and operated according to well-established criteria.

It is essential for the laboratory hood users to understand thoroughly the functions that characterize a satisfactory laboratory hood and the several designs that are on the market. Not all are equally effective or efficient in the utilization of airflow and the way they function.

Well-designed fume hoods (Figs. 43.1a, b) have several important characteristics in common. First, air velocity will be uniform, that is, the velocity at any point will be plus or minus 20% of the average velocity over the entire work access opening. Second, all the hood surfaces surrounding the work access opening will be smooth, rounded, and tapered in the direction of airflow to minimize air turbulence at the hood face. Third, average face velocity will range from 80 to 120 fpm. Under the OSHA Laboratory Standard 1910.1450, promulgated in 1990,[2] OSHA did not specify a hood face velocity recognizing that face velocity alone may not be a good indicator of protection. Hood face velocities in excess of 120 fpm are not recommended because they cause disruptive air turbulence at the perimeter of the hood opening and in the wake of objects placed inside the work area of the hood.[3,4] Fourth, the face velocity of hoods with adjustable front sash will be maintained constant (within reasonable limits) by an inflow air bypass that proportions the air volume rate entering the

open face to the open area, or by some other method that produces a similar result.

There are a number of distinctive types of laboratory hoods in widespread use. Each is identified in the following sections.

Types of Laboratory Chemical Hoods

Bypass Type

The basic chemical hood incorporates the four principles enumerated above that characterize a well-designed hood. Most laboratory furniture suppliers have one or more models that will meet the listed criteria. Important characteristics of a bypass hood as shown in Figure 43.1b are:

- Bottom and side airfoils around the open face to produce turbulence-free airflow into the hood.
- A mechanism to minimize excessive velocities (300 fpm) when the total opening is 6 in. or less. This may be accomplished by an air bypass, by switching the fan to a lower speed, or by use of a variable air volume system.
- The sash will be constructed of shatterproof material.
- The material of construction of the hood will be resistant to damage by the materials to be used in the hood.
- There will be an inward airflow across the entire opening. Reverse flow can be detected by passing a smoke stick or equivalent smoke generator across the entire parameter of the face opening and looking for flow out of the hood.
- The airflow through the hood will provide an average face velocity of 80 to 120 fpm at the maximum face opening.
- No individual velocity measurement across the hood face with the sashes positioned to provide the maximum opening will be less than 20% or greater than 20% of the design's average face velocity.
- The airflow through the system should be continuously monitored by an in-line measurement device (Pitot tube, orifice meter) or by static pressure measurements in the duct just downstream of the hood or similar device and calibrated to the specific hood system. The output of the measurement instrument should be visible to the hood user.
- During renovations, when existing hoods are to remain, an attempt to modify them should be made, if necessary. This would include providing a bottom airfoil (if missing) and providing some means

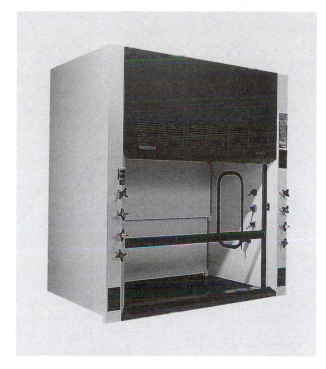

FIGURE 43.1a Bypass hood. (*Source:* Fisher Hamilton Inc., Two Rivers, WI 54241.)

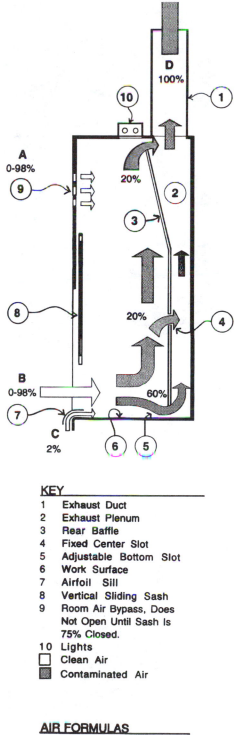

of controlling face velocity at all sash positions by means of a bypass or air volume control and providing an airflow monitoring system

Auxiliary Air Type

Auxiliary air hoods (Figs. 43.2a, b) are a basic type of hood that is acceptable for laboratory use. The major difference between the standard chemical hood and the auxiliary air hood is the method employed to provide makeup air to the hood. For standard hoods, all of the makeup air is provided by the room HVAC system, whereas, for the second type, sometimes called "supply air hoods," part of the air is introduced from a supply air grille just above and exterior to the hood face. It is recommended that the percentage of air delivered in this manner be kept below 60% with the remainder coming from the laboratory HVAC system. The auxiliary supply air should not be introduced behind the sash because the hood chamber is likely to become pressurized and blow toxic contaminants out the open front.

Auxiliary air hoods have two advantages over bypass hoods. First, the air supplied to the hood does not have to be cooled in warm climates, a significant energy saving. Second, auxiliary air hoods can be used in laboratories that contain so many hoods that the volume

KEY

1	Exhaust Duct
2	Exhaust Plenum
3	Rear Baffle
4	Fixed Center Slot
5	Adjustable Bottom Slot
6	Work Surface
7	Airfoil Sill
8	Vertical Sliding Sash
9	Room Air Bypass, Does Not Open Until Sash Is 75% Closed.
10	Lights
□	Clean Air
▓	Contaminated Air

AIR FORMULAS

A + B + C = D
B + C = D (Sash Fully Open)
A + C = D (Sash Fully Closed)

FIGURE 43.1b Bypass hood: section. (Reprinted with permission from John Wiley & Sons.)

FIGURE 43.2a Auxiliary air chemical hood. (*Source:* Fisher Hamilton Inc., Two Rivers, WI 54241.)

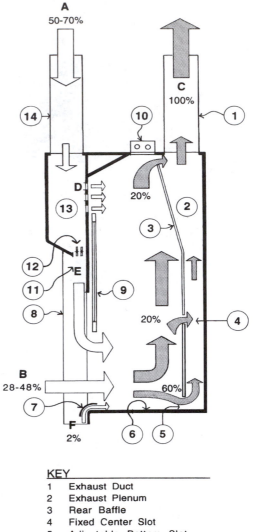

KEY

1	Exhaust Duct
2	Exhaust Plenum
3	Rear Baffle
4	Fixed Center Slot
5	Adjustable Bottom Slot
6	Work Surface
7	Airfoil Sill
8	Side Baffles, 6" Minimum
9	Vertical Sliding Sash, Closes Top Air Supply when Open.
10	Lights
11	Supply Air Slot, Velocity 250-300 FPM.
12	Air Turning Vanes In Plenum
13	Supply Air Plenum
14	Outside Supply Air Duct
☐	Clean Air
▨	Contaminated Air

AIR FORMULAS

$A + B + F = C$
$D + E = A$
$D + F = C$ (Sash Closed)
$B + E + F = C$ (Sash Open)
$A = E$ (Sash Open)

FIGURE 43.2b Auxiliary air chemical hood: section. (Reprinted with permission from John Wiley & Sons.)

of supply air, were they all standard chemical hoods, would result in many more room air changes per hour than are desirable and create excessive drafts. Providing auxiliary supply air, even to only some of the hoods, reduces room air velocities and supply air quantities to more acceptable levels.

There are disadvantages to the use of the auxiliary supply air hoods. First, they are more complex in design than the bypass hood and their correct installation is more critical to safe and efficient operation. Often only a small imbalance between room air and auxiliary air supplies can result in unsafe operating conditions. Second, unsafe conditions occur when the velocity of external auxiliary air supplied to the face of the hood is excessive. High-velocity air sweeping down across an open hood face can produce a vacuum effect and draw toxic contaminants out of the hood. Third, this type of hood must have two mechanical systems (separate exhaust and supply systems) for each hood. Inasmuch as some supply air will need to be added to the laboratory anyway, the auxiliary air hoods require two supply air systems instead of the one supply system that standard chemical fume hoods require.

Generally, the use of auxiliary air hoods is discouraged by safety and health professionals.[3] Use of auxiliary air hoods is particularly attractive for renovations when one or more hoods must be added to a laboratory with limited amounts of supply air capacity. Because of the complexity of this type of hood, the performance specifications are more stringent than for bypass hoods. It is extremely important to choose the particular auxiliary air hood that will be purchased before the supply air system is designed because not all auxiliary air hoods have the same requirements for room and auxiliary air volumes and some must use less than 60% auxiliary air to meet containment requirements.

Horizontal Sliding Sash Hoods

Economy in the utilization of conditioned air for laboratory hoods can be achieved most satisfactorily by maintaining the required face velocity but restricting the open area of the hood face. A transparent horizontal sliding sash (Fig. 43.3) arrangement can cut overall air requirements by 50% if two half-width panels are used. Similarly, if three panels are used, the maximum open area reduction is 67% for a two-track setup and 33% for a three-track arrangement. This design also has an advantage over the conventional vertical sash because the full height of the hood opening is available. If panels 14- to 16-in wide are used, they can also serve as safety shields.

Sometimes, turbulence occurs at sharp panel edges when inflow velocities exceed design values. Visual smoke trails may be used to test for inward airflow across the entire face opening and an absence of turbulence at sash edges. Note that removing the panels results in decreased face velocities and seriously compromises the safety of hood performance.

Perchloric Acid Hoods

Perchloric acid hoods require special construction, construction materials, and internal water-wash capability. Problems reported with hoods heavily used for perchloric acid digestions are associated with the formation of explosive organic perchlorate vapors that condense while passing through the hood exhaust system. They can detonate upon contact during cleaning, modification, or repair. Therefore, use of specially designed hoods is required for use with perchloric acid at elevated temperatures or in a manner that will produce an aerosol. Perchloric acid hoods should meet the same contaminant retention capabilities as the hoods described in the preceding sections. In addition, they should be constructed of stainless steel and have welded seams throughout. No taped seams or joints and no putties or sealers can be used in the fabrication of the entire hood and duct system. The perchloric acid hood also requires an internal water-wash system to eliminate the buildup of perchlorates. The water-wash system should consist of a water spray head located in the rear discharge plenum of the hood plus as many more as are needed to ensure a complete washdown of all surfaces of the duct work from the hood work surface to the discharge stack on the roof. To drain all of the water to the sewer in a satisfactory way during normal washdown operations and to facilitate

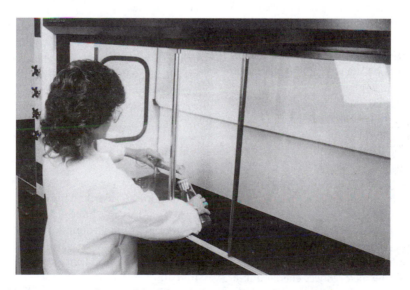

FIGURE 43.3 Chemical hood with horizontal sliding sash. (*Source:* Fisher Hamilton, Inc., Two Rivers, WI 54241.)

periodic examination for maintenance purposes, it is important to have the exhaust duct go straight up through the building with no horizontal runs.

For those cases where small amounts of perchloric acid are used, it is possible to construct an air scrubber in the hood. This can be done if the point of generation of perchloric acid fumes can be identified and a small capture hood designed. The National Safety Council recommends such a system.[5]

Variable-Air-Volume Hoods

Variable-air-volume hoods can be any one of the standard hoods discussed above. The major difference is that the exhaust quantity is not constant. Generally the exhaust quantity decreases as the hood face opening decreases. This can be accomplished by a variety of techniques.

The major advantage of this type of system is a reduced operating cost in the form of energy savings. The disadvantages relate to the relative sophistication of the control equipment, needed maintenance, and uncertainty about the long-term repeatability of the control systems. Variable-air-volume systems can be either on a single hood or part of a manifold system.

Some of the advantages of the variable-air-volume-hood exhaust system are as follows:

1. It reduces the number of exhaust fans in a building. Compared to one or two hoods to an exhaust fan concept, by its nature results in a multiple number of fans. Combining the hoods to a central system can reduce the need to one or two large fans.
2. One or more central fans can be installed as a backup to provide continuous ventilation if the other fans fail. In this manner, some ventilation is guaranteed to all hoods at all times. If one installs a single fan to a hood, there is no ventilation in that one hood if the fan fails.
3. As the number of hood stacks are minimized, they can be inexpensively increased in height, thus reducing the possibility of reentry of exhaust contaminants into the building.
4. If a bypass damper is provided at the inlet side of exhaust fans on the roof, the air exhausted is further diluted thereby minimizing the chemical concentration of air exhausted.
5. As the number of hoods are connected, concentration of a particular chemical is diluted.
6. A variable volume exhaust and makeup air system can be installed which reduces energy utilization by varying the air volume with respect to the demand.

The major disadvantages are:

1. In order to balance the system, there will be either manual or automatic dampers in the exhaust stream. These dampers can fail, although they are usually designed to fail in the open condition. If so, one or more hoods can be without proper exhaust for some time.
2. Mixing various or different types of chemicals and products in the exhaust stream may result in an unsafe situation. This is highly unlikely in most laboratory applications, but needs to be evaluated.
3. The control equipment used may consist of a hotwire element in a direct exhaust stream which may be hazardous in certain chemicals being exhausted. Additionally, the element may become corroded and nonfunctional.

Airflow Monitors

For safety, especially when working with hazardous materials that give no sensory warning by odor, visibility, or prompt mucous membrane irritation of their escape into the workroom from the laboratory hood, it is advisable to provide each hood with an airflow monitor capable of giving an easily observed visual display of functional status. This is required by ANSI[3] and is inferred by OSHA.[1] Installation of a hood airflow gauge at the site of use has the special advantage of placing responsibility for day-by-day monitoring of hood function with the primary users. Devices that may be used to monitor hood function include liquid-filled draft gauges or Magnehelic gauges, which measure hood static pressure. Other devices measure airflow velocity in the exhaust duct where velocity is high or at some point at the side of the hood opening as a surrogate measurement of total exhaust. In all cases, the type and location of monitors should be evaluated carefully because incorrect positioning of static pressure taps can give false readings and inline airflow devices may become clogged or corroded.

Regulations

Those facilities covered by OSHA must comply with the Laboratory Standard.[2] This is a performance-based standard so it does not have specific requirements on laboratory ventilation. It simply requires that adequate engineering controls be in place to prevent overexposure to toxic chemicals and that there is a mechanism to assure that the controls are effective. There are no specific requirements on number or type of hoods, face velocities, or air changes per hour.

Guidelines

A standard on laboratory ventilation was adopted by the American National Standards Institute.[3] The purpose of this standard is to establish minimum requirements and procedures for laboratory ventilation systems that are used to prevent overexposure of personnel to chemical contaminants generated within the laboratory.

The standard requires that each laboratory facility establish a Laboratory Ventilation Management Program and designate a "cognizant person" to coordinate and monitor the program. It emphasizes that management participation in the selection, design, and operation of laboratory ventilation systems is important.

There are no specific recommendations for an air exchange rate since the Committee feels that this is an inappropriate criterion to use for contamination control. Much emphasis is placed on the laboratory chemical hood and other means of local exhaust ventilation.

The recirculation of hood exhaust air is discouraged. The recirculation of general laboratory room air is allowed only if specific conditions are met. These conditions relate to the type of materials in use, the concentration of air contaminants generated by the "maximum credible accident," the type of air cleaning performed on the exhaust stream, the presence of a continuous air monitor and alarm system, and the ability to bypass or divert the recirculated air to atmosphere.

The general requirement for directional airflow is that air should flow from areas of low hazard to higher hazard unless the laboratory is used as a Clean Room. Where airflow from one area to another is critical to exposure control, airflow monitoring devices shall be installed to signal or alarm a malfunction.

Testing and monitoring of airflows and control systems is required. Records should be maintained for all inspections and maintenance activities and should be part of the Laboratory Ventilation Management Plan. Routine testing is recommended for all air cleaning devices, contaminant control ventilation systems, continuous monitoring equipment, and directional airflow if appropriate.

Another guideline that provides ventilation criteria is NFPA 45, *Fire Protection for Laboratories Using Chemicals*, 2000 edition.[6] Although this guideline primarily deals with fire protection, section 6 discusses ventilation issues.

Measures of Hood Performance

Since the main purpose of a hood is to prevent the release of toxic materials into the laboratory environment, the ultimate measure of hood performance is to determine the concentration outside the hood of all the chemicals used inside the hood. If these concentrations are below relevant health standards with an adequate safety factor applied, the hood performance would be considered adequate. This is not practical to do because of the wide variety of chemicals and operations performed inside hoods. As a result, other surrogate measures of hood performance are used.

Face Velocity

The common measure used is the hood face velocity. This is the speed of airflow into the hood through the open sash measured at the plane of the sash opening. Although there are no strict standards, several published guidelines[3,7] suggest this should be an average face velocity of 80–120 feet per minute (fpm) with no one measurement more or less than 20% of the average. Unfortunately, there are other factors that affect hood performance that may not be uncovered by this method alone.

Smoke Visualization

This method is qualitative and involves observing that airflow is always into the hood by generating smoke or using dry ice. This can be valuable only if used in conjunction with other methods such as face velocity.

Containment Testing

This method involves generating a known amount of a tracer gas (such as sulfur hexafluoride) inside the hood and measuring the concentration outside the hood at the breathing zone of a hood user. The generation rate is selected based on the type of activity anticipated for the hood ranging from 1 lpm (very little manipulation) to 8 lpm (boiling an organic solvent). Essentially this determines the leakage. This is a cumbersome and time-consuming process. It is most valuable as a prepurchasing specification for hoods or as part of the installation acceptance criteria. Existing methods have been published.[4,8]

Frequency of Testing

The ultimate desire is to allow the user to have continuous feedback of hood performance. See the section on continuous flow monitors for more detail. In the absence of continuous flow monitors, hoods should be tested at least annually. For operations involving highly toxic materials the frequency may need to be increased to quarterly.

Communication to Hood Users

There should be a clear indication on the hood as to its measure of performance. If continuous monitors are used, acceptable versus unacceptable conditions should be clearly evident to the user. Where continuous monitors are not used, a label should be clearly visible indicating the last time the hood was checked, its status at that time, and who to call for up-to-date information.

Work Practices

Recent studies have indicated that the work practices followed by hood users have as much effect if not more than the hood and laboratory design. Improper placement of equipment or your body can adversely effect the ability of the hood to contain the toxic materials within.

Conversely, careful arrangement of equipment and procedures performed inside the hood can enhance containment capability. Below are examples of good practice that should be followed whenever possible.

1. Set up work at least 6 inches behind the plane of the sash to avoid turbulence at sash edge (could use smoke to show the turbulence).
2. Never put your head inside an operating hood to check an experiment. The plane of the sash is the barrier between contaminated and uncontaminated air.
3. Work with the sash in the lowest position possible. The sash will then act as a physical barrier in the event of an accident in the hood.
4. Do not clutter the hood with unnecessary bottles or equipment. Do not use a hood for storage. Only materials actively in use should be in the hood. This provides optimal containment and reduces the risk of extraneous chemicals being involved in a fire or explosion if one occurs.
5. Do not obstruct the back slots of the hood. Clean the grille along the bottom slot of the hood regularly so it does not become clogged with objects or paper.
6. Note the potential impact of lab design (location of doors and windows) on hood containment.

Containment can be affected by people walking behind the work area and opening and closing doors or windows.

7. Do not dismantle or modify the physical structure of the hood or exhaust system in any way without first consulting the appropriate people in your organization.
8. Report any suspected hood malfunctions to the appropriate person immediately.
9. Do not use a hood for any function it was not designed for, such as perchloric acid digestion, radioisotope work, explosives work, or biological agents.
10. Use a hood only when you have determined that it is performing adequately.
11. If large equipment must be used in the hood raise it at least 2″ off the countertop on "legs" to allow air movement below.
12. Keep sash in fully closed position whenever possible.
13. Keep window sash clean and clear.
14. Wear protective clothing at all times.

References

1. ANSI/ASHRAE Standard 62-1999, *Ventilation for Acceptable Indoor Air Quality*; American Soceity of Heating, Refrigerating Air-Conditioning Engineers: Atlanta, GA, 1999.
2. U.S. Department of Labor, Occupational Safety and Health Administration. *Health and Safety Standard; Occupational Exposure to Hazardous Substances in Laboratories.*
3. ANSI/AIHA Standard Z9.5. *Laboratory Ventilation*; American Industrial Hygiene Association: Fairfax, VA, 1993.
4. Ivany, R.; First, M. W.; DiBerardinis, L. J. A New Method for Quantitative, In-Use Testing of Laboratory Fume Hoods. *Am. Ind. Hyg. J.* **1989**, *50*(5), 275–280.
5. *Controlling Perchloric Acid Fumes*, Research Development Section Fact Sheet; National Safety Council: Chicago, IL, May 1985.
6. NFPA Standard 45. *Fire Protection for Laboratories Using Chemicals*, 2000 ed.; National Fire Protection Association: Boston, MA, 2000.
7. American Conference of Governmental Industrial Hygienists, 1998. *Industrial Ventilation, A Manual of Recommended Practices*, 23rd ed.; Committee on Industrial Ventilation: Cincinnati, OH, 1998.
8. ASHRAE 110-1995. *Method of Testing Performance of Laboratory Fume Hoods*; American Society of Heating, Refrigerating and Air-Conditioning Engineers: Atlanta, GA, 1995.

44

Biological Safety Cabinets

RAYMOND W. HACKNEY, JR.

Laboratory techniques may produce aerosols, which can contain hazardous research materials, such as infectious agents or chemical carcinogens, that can be inhaled by laboratory workers. Biological safety cabinets (BSCs) are used as primary barriers to contain hazardous research materials in order to prevent exposure of laboratory personnel and contamination of the general environment.

Principles of Containment

Containment of hazardous aerosols in biological safety cabinets is achieved by the use of air barriers, physical barriers, and HEPA filtration.[1] Air barriers provide containment by use of directional airflow from the laboratory past the researcher into the cabinet via the work opening. Hazardous aerosols generated during experimental procedures inside the cabinet are captured and carried by the flow of air and then trapped in HEPA filters. Some biological safety cabinets provide protection of experimental procedures using uniform, unidirectional HEPA-filtered air, referred to as laminar airflow, that continuously flows over the work area. Turbulence inside the cabinet is minimized by the laminar airflow allowing for immediate removal of contaminants generated by the procedures. The integrity of the containment provided by air barriers can be compromised by the disruption of the airflow patterns in the cabinet.[8] Air barriers are therefore considered to provide only partial containment and should not be used with highly toxic or infectious materials requiring Biosafety Level 4 containment.

Physical barriers are impervious surfaces such as metal sides, glass panels, rubber gloves, and gaskets, which physically separate the experimental procedures from the researcher.[1] Biological safety cabinets incorporating all of these physical barriers (Class III BSC), and not relying on air barriers, can be used for higher risk agents since compromising containment is less likely than with air barriers.

High-efficiency particulate air (HEPA) filters are defined as filters with a filtration efficiency of 99.97% for thermally generated monodisperse dioctylphthalate (DOP) 0.3 μm diameter particles.[2,6,7] Because of their high efficiency, HEPA filters are used in biological safety cabinets to remove virtually all particulates, including hazardous microbiological and chemical aerosols, in the air stream passing through the filter. All biological safety cabinets have exhaust filters that remove contaminants as air is discharged from the cabinet. Some types (discussed below) of biological safety cabinets also have supply HEPA filters that provide clean air to the work area. HEPA filters are not effective in capturing chemical vapors.

Classification of Biological Safety Cabinets

There are three classes of biological safety cabinets, designated as Class I, Class II, and Class III.[1,2,3,5,6,7] Class I and II cabinets have a protective air barrier across the work opening that separates the laboratory researcher from the work area. Class II cabinets have an additional feature of providing a HEPA filtered, clean work area to protect the experiment from room contamination. There are several variations of Class II cabinets, which are described below. Class III biological safety cabinets have a physical barrier between the operator and the work area. Arm-length rubber gloves are sealed to glove ports on the cabinet to provide the operator with access to the work area. The distinctive features of the three classes of cabinets are discussed in more detail below.

Class I

- The Class I (see Fig. 44.1) cabinet is ventilated for personnel and environmental protection with an inward airflow away from the operator. It is similar in air movement to a chemical laboratory hood.
- The minimum airflow through the work opening is 75 feet per minute (fpm).
- The cabinet exhaust air is HEPA-filtered to protect the environment before it is discharged to the outside atmosphere.
- This cabinet is suitable for work with low- and moderate-risk biological agents, where no product protection is required.

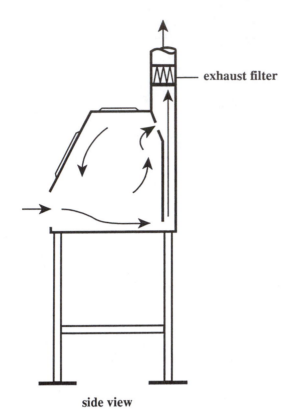

side view

FIGURE 44.1 Class I.

- Because of the popularity of Class II cabinets and the product protection they provide, use of Class I cabinets has declined.

Class II

- The Class II cabinet is ventilated for personnel, product, and environmental protection having an open front and inward airflow for personnel protection.
- Product protection is provided by HEPA-filtered laminar airflow from a diffuser located above the work area. The downflow air splits at the work surface, and exits the work area through grilles located at both the rear or front of the work surface, respectively.
- The cabinet has HEPA-filtered exhausted air for environmental protection.
- Types of Class II biological safety cabinets are designated: types A, B1, B2, and B3.

Class II, Type A

- The work opening is 8 to 10 inches in height.
- The type A cabinet (see Fig. 44.2) may have a fixed work opening, a sliding sash, or a hinged window.

- A minimum average inflow velocity of 75 fpm is maintained through the work area access opening.
- Approximately 70% of the cabinet air is recirculated through a HEPA filter into the work area from a common plenum, while approximately 30% of the air enters through the front opening and an amount equal to the inflow is exhausted from the cabinet through a HEPA filter.
- A fan located within the unit provides the intake, recirculated supply air, and the exhaust air.
- The cabinet may exhaust HEPA-filtered air back into the laboratory.
- The cabinets may have positive pressure contaminated plenums. Contaminated plenums under positive pressure must be gas tight.
- Type A cabinets are suitable for work with low- to moderate-risk biological agents in the absence of volatile toxic chemicals and volatile radionuclides.

Class II, Type B1

- The work opening is 8 inches in height, with a sliding sash that can be raised for introduction of equipment into the cabinet.
- Type B1 cabinets (see Fig. 44.3) have a minimum average inflow velocity of 100 fpm through the work area access opening.

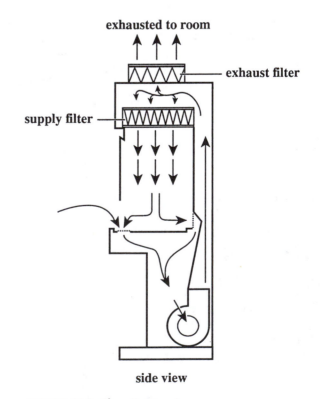

side view

FIGURE 44.2 Class II, type A.

- The HEPA-filtered downflow air is composed largely of recirculated inflow air.
- Supply fans located in the base of the cabinet below the work surface draw air through a grille at the front of the work surface and supply HEPA filters located directly below the work surface. The fans then force the filtered air through plenums in the sides of the cabinet and recirculate the air through a diffuser above the work surface. Some cabinets now have a secondary supply filter located above the work surface.
- Approximately 70% of the contaminated downflow air is exhausted through a HEPA filter and a dedicated duct and then is discharged outside the building.
- The remote exhaust fan is usually located on the roof of the building.
- All biologically contaminated ducts and plenums are under negative pressure.
- The type B1 cabinet is suitable for work with low- to moderate-risk biological agents. They may also be used with biological materials treated with minute quantities of toxic chemicals and trace amounts of radionuclides.

Class II, Type B2 (sometimes referred to as "total exhaust")

- The type B2 cabinet (see Fig. 44.4) has a sliding sash with an 8-inch opening.
- The type B2 cabinet maintains a minimum average inflow velocity of 100 fpm through the work area access opening.

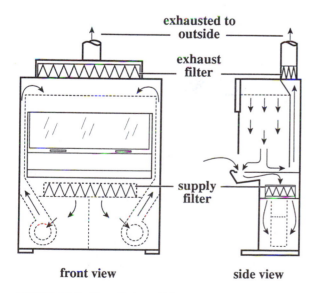

front view   **side view**

FIGURE 44.3 Class II, type B1.

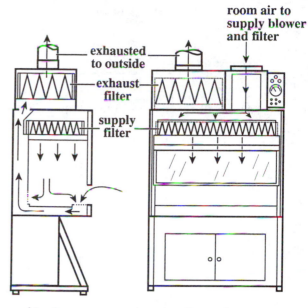

side view   **front view**

FIGURE 44.4 Class II, type B2.

- No air is recirculated within the cabinet.
- A supply fan draws air from the laboratory and forces it through a supply HEPA filter located over the work area.
- A remote exhaust fan, usually located on the roof, pulls all of the inflow air and all of the supply air through a HEPA filter and then discharges it outside the building. As much as 1200 cubic feet per minute may be exhausted from the cabinet.
- The cabinet has all contaminated ducts and plenums under negative pressure.
- Type B2 cabinets are suitable for work with low- to moderate-risk biological agents. They may also be used with biological materials treated with toxic chemicals and radionuclides.

Class II, Type B3

- Type B3 cabinets (see Fig. 44.5) are type A cabinets with the following conditions:
  - Type B3 cabinets have a minimum of 100 fpm average inflow velocity.
  - No contaminated plenums are under positive pressure.
  - Air exhausted from the cabinet is exhausted outside the building. A remote fan is required and must be sized precisely to avoid altering the balance of the cabinet airflow. The fan and ducting of type B3 cabinets are discussed under

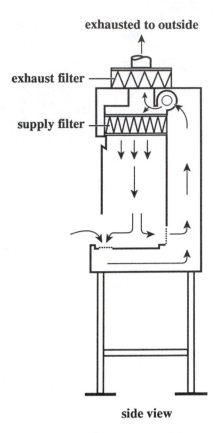

FIGURE 44.5 Class II, type A/B3.

the "Installation and Certification of Biological Safety Cabinets" section.
- Type B3 cabinets are suitable for work with low- to moderate-risk biological agents. They may also be used with biological materials treated with low concentrations of toxic chemicals and radionuclides.

Class III

- The Class III cabinet (see Fig. 44.6) is totally enclosed, ventilated, and of gas-tight construction.
- Operations in the cabinet are conducted through attached arm-length rubber gloves, which serve as physical barriers.
- The cabinet is maintained under negative air pressure of at least 0.5 inches water gauge.
- Supply air is drawn into the cabinet through HEPA filters.
- The exhaust air is treated by double HEPA filtration, or by HEPA filtration and incineration.
- Class III cabinets are used in maximum containment laboratories and may be used with agents of low, moderate, and high risk.

Laminar Flow Clean Benches

Horizontal and vertical laminar flow clean benches have sometimes been mistaken for biological safety cabinets. Clean benches provide product protection but no personnel protection. The horizontal flow clean bench discharges HEPA-filtered air across the work surface onto the operator. The less common vertical flow clean bench discharges air downward from a HEPA filter above the work surface. The airflow leaves the work area through the front opening where the operator is located. With both versions of the clean bench, the operator could be exposed to contaminants generated by the work performed on the work surface. Clean benches can be used for assembly of sterile apparatus. They should not be used for handling cell cultures, drug formulations, radioactive nuclides, or biological hazards. Exposure to some of these materials from manipulation on a clean bench can cause hypersensitivity.[2]

Selection of Biological Safety Cabinets

The selection of the class or type of biological safety cabinet is based on the degree or nature of the hazard associ-

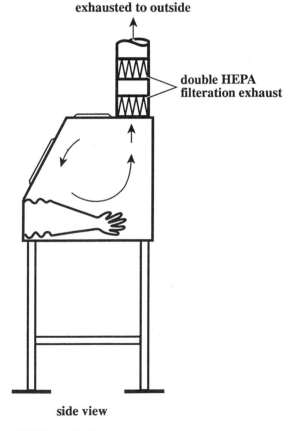

FIGURE 44.6 Class III.

ated with the experiment, the aerosol-producing potential of the laboratory procedures, and the requirement to protect the experiment from airborne contamination.[5] All purchases and installations of biological safety cabinets should be reviewed and approved by the institution's safety office. This is to ensure that the proper cabinet is selected for the intended use and that the cabinet will be installed to allow for proper functioning, maintenance, and certification.

Table 44.1 provides a summary and comparison of biological safety cabinet characteristics and applications. In general, the Class II, type A biological safety cabinet should be selected when the work involves infectious agents and cell culture work in the absence of volatile toxic chemicals and radionuclides. When small amounts of volatile chemicals or radionuclides are involved, the Class II, type B (types B1, B2, or B3) should be selected. The Class III cabinet should be selected when the work involves the handling of high-risk microbiological agents, concentrated amounts of carcinogens, or highly toxic chemicals.

The National Sanitation Foundation (NSF) lists the models of the various cabinet manufacturers that meet its standard criteria described in NSF Standard 49, *Class II (Laminar Flow) Biohazard Cabinetry*.[6] Under this standard there are design and construction requirements as well as performance specifications such as vibration; temperature rise; noise level; and personnel, product, and cross-contamination protection determined by a spore aerosol challenge. Selection of an NSF-approved model assures the purchaser that minimum construction and performance standards have been met.

Most manufacturers offer user options such as whether or not the cabinet has UV or germicidal lights. UV lights are not required, but if installed must be cleaned weekly to remove dust, which can reduce their germicidal effectiveness. The intensity of the lights must also be measured regularly to ensure that a germicidal

TABLE 44.1 Comparison of Biological Safety Cabinet Characteristics and Applications

Class, Type	Work Opening	Inflow Velocity (fpm)	Percentage Recirculated Air	Percentage Exhausted Air	Exhaust Volume (cfm) (approximate)	Exhaust Requirement	Application
Class I	fixed	75	0%	100%	4 ft—200 6 ft—300	Exhausted to the outside (remote fan) or to the room through a HEPA filter (integral fan)	Biosafety Level 1–3; small amounts of toxic chemicals or radionuclides (if exhausted to outside)
Class II, type A	fixed, sliding, or hinged	75–100	70%	30%	4 ft—300 6 ft—400	Exhausted to room through HEPA filter	Biosafety Level 1–3
Class II, type B1	sliding	100	30%	70%	4 ft—250 6 ft—400	Exhausted to outside, with remote fan; duct is hard-connected	Biosafety Level 1–3; small amounts of toxic chemicals or radionuclides
Class II, type B2	sliding, hinged	100	0%	100%	4 ft—600 6 ft—1000	Exhausted to outside, with remote fan; duct is hard-connected	Biosafety Level 1–3; small amounts of toxic chemicals or radionuclides
Class II, type B3	sliding, hinged	100	70%	30%	4 ft—300 6 ft—400	Exhausted to outside, with remote fan, utilizing thimble, or hard-connected duct	Biosafety Level 1–3; small amounts of toxic chemicals or radionuclides
Class III	glove ports	N/A	0%	100%	—[a]	Exhausted to outside, through 2 HEPA filters, with remote fan; duct is hard-connected	Biosafety Level 1–4; small amounts of toxic chemicals or radionuclides

[a]Class III cabinets should have approximately 20 air changes per hour or enough ventilation to accommodate the heat load. A negative pressure of 0.5 in w.g. must be maintained and 100 fpm should be maintained through a glove port, if a glove is accidentally removed.[1]
Adapted from References 1 and 2.

intensity is being maintained. Precautions must be taken to prevent exposure of the eyes and skin to damaging effects of UV light. UV lamps must be turned off when the room is occupied.[2]

The design of biological safety cabinets can be modified by the manufacturer to accommodate special applications. Examples include alteration of the front sash to accommodate a microscope eye piece, or the work surface can be adapted to include a centrifuge or animal waste handling capabilities.[2]

Installation and Certification of Biological Safety Cabinets

Installation

The cabinet should be located in a space that is free from drafts and traffic. Air conditioning vents, the opening of doors, and personnel traffic can produce air currents that may penetrate the air barriers at the front opening of the cabinet.[8] Ideally, cabinets should be located in a "dead-end" area of the laboratory.[1,2,6,7] Nearby HVAC vents should be directed away from the cabinet. A clearance of 12 to 14 in. above Class II, type A cabinets should be provided to allow access for accurate exhaust flow measurements and for filter replacement.[2,7]

The Class II, type A biological safety cabinet is designed to exhaust back into the laboratory and therefore it is generally best not to connect a duct to the cabinet. If a duct is connected, the remote exhaust fan must be sized to match precisely the exhaust air volume from the cabinet. An exhaust fan that removes too little or too much air from the cabinet will interfere with the supply/exhaust air balance in the cabinet. The exhaust duct may be either hard-connected to the cabinet or connected with a thimble utilizing a 1-in. air gap that reduces the likelihood of interference with the air balance of cabinet.[6] Also, the duct must not interfere with requirements for certification tests and filter replacement. Provisions must be made for access to the exhaust filter. The exhaust fan should also be interlocked with the cabinet fan switch so that both fans are either "on" or "off" at the same time. If the cabinet fan is "off" on a Class II, type A or B3 cabinet but the exhaust fan is "on," the exhaust fan will pull room air contaminants through the cabinet and the supply filter in the opposite direction of normal cabinet operation. The clean side of the supply filter is therefore contaminated. When the cabinet is eventually turned "on," the airflow through the supply filter is then reversed. Contaminants can be dislodged from the filter media into the clean work area of the cabinet.

Class II, types B1 and B2 cabinets by design must be ducted to the outside using a remote exhaust fan, usually located on the roof. To correctly size the exhaust fan, the cabinet manufacturer should be consulted to obtain the pressure drop through the cabinet with fully loaded filters.[1] For some models, the exhaust filter is also located remotely from the cabinet, often just upstream from the fan. A bag-in/bag-out filter assembly is used so that the maintenance technician does not have to handle directly a contaminated filter.[2] Upstream and downstream certification test ports and isolation dampers for formaldehyde gas decontamination should also be provided.

Certification

Each cabinet must be tested and its performance evaluated after it is installed in the laboratory, whenever it is moved, prior to use, and periodically thereafter.[2,4,5,6,7] The following field tests are usually performed: downflow velocity profile for the supply air, work access opening airflow, HEPA filter leak test, cabinet integrity test, and airflow smoke patterns. Certification field tests are described in NSF Standard 49.[6] NSF also accredits biological safety cabinet certifiers. Recertification should be performed when the HEPA filters are changed, maintenance repairs are required, or when a cabinet is relocated. The cabinet must be decontaminated with formaldehyde gas before maintenance work, before filter changes, after gross spills of biohazardous materials, and before moving the cabinet.[2,6]

Procedures for the Proper Use of a Class II Biological Safety Cabinet

Any laminar flow biological safety cabinet is only a supplement to good microbiological techniques, not a replacement for it. If the cabinet is not properly understood and operated, it will not maintain an adequate protective barrier between the operator and the experiment. Listed below are procedures for proper use of safety cabinets.[3,4,7]

1. Turn the cabinet fan and the fluorescent light on. Turn the UV light off. Confirm that the drain valve is closed.
2. Wipe the work surface with 70% ethanol or other appropriate disinfectant. Let the unit run for 5–10 minutes to clean itself.
3. Plan the work operation in advance. Place everything needed for the complete procedure in the cabinet before starting. Nothing should pass

through the air barrier, either in or out, until the procedure is complete. Arrange materials in a logical manner such that clean and contaminated materials are segregated. Materials or equipment not necessary for the particular procedure should be removed from the cabinet.

4. Avoid placing materials on the air intake grille, which disrupts the protective air barrier.

5. Keep equipment at least 4 in. inside the cabinet work area. Perform manipulations of hazardous materials as far back in the work area as possible.

6. After the procedure is completed, decontaminate all equipment in direct contact with the research agent with an appropriate disinfectant. Run the cabinet for at least 3 minutes with no activity to purge airborne contaminants from the work area.

7. Surface decontaminate the interior surfaces with 70% ethanol, or other appropriate disinfectant, after removal of all materials and equipment. Clean any spilled culture media which may support fungal growth and cause contamination in subsequent experiments.

8. Turn off the cabinet fan. The UV light may be turned on. Some researchers prefer to let the cabinet run continuously.

Summary

The use of biological safety cabinets (BSCs) is an important component of a laboratory safety program. BSCs provide effective primary barriers to protect both laboratory personnel and the general environment from exposure to infectious or chemical aerosols. However, to be effective, biological safety cabinets must be tested and certified that they are functioning according to the manufacturer's design, and BSCs must be understood and used properly by laboratory personnel.

References

1. AIHA. *Biosafety Reference Manual*, 2nd ed.; American Industrial Hygiene Association: Fairfax, VA, 1995.

2. CDC/NIH. *Primary Containment for Biohazards: Selection, Installation and Use of Biological Safety Cabinets*; DHHS (CDC/NIH): Washington, DC, September 1995.

3. CDC/NIH. *Biosafety in Microbiological and Biomedical Laboratories* (HHS Publication No. (CDC) 93-8395; 3rd ed.); DHHS (CDC/NIH): Washington, DC, May 1993.

4. DHEW. *Effective Use of the Laminar Flow Biological Safety Cabinet* (slide-tape presentation); Department of Health, Education, and Welfare, NIH: Washington, DC, (no date given).

5. NIH. *Laboratory Safety Monograph, A Supplement to the NIH Guidelines for Recombinant DNA Research*; NIH: Bethesda, MD, 1978.

6. NSF. *Class II (Laminar Flow) Biohazard Cabinetry*; National Sanitation Foundation International: Ann Arbor, MI, 1992, NSF 49.

7. *Personnel & Product Protection: A Guide to Biosafety Enclosures*; Labconco Corporation: Kansas City, MO, 1993.

8. Rake, B. W. Influence of Cross-Drafts on the Performance of a Biological Safety Cabinet. *Appl. Environ. Microbiol.* **1978**, *36*, 278–283.

SECTION V

CHEMICAL MANAGEMENT

JAY A. YOUNG

SECTION EDITOR

45

Material Safety Data Sheets

ROBERT J. ALAIMO

LYNNE A. WALTON

Current regulations require that chemical suppliers provide a copy of a material safety data sheet (MSDS) for each hazardous material they sell or distribute. The MSDS is to be provided with the initial shipment of the material and whenever a change in the order is made. The effective date for the U.S. regulation, the Hazard Communication Standard, which required the distribution of the MSDS to customers, was 11/25/85.[1] But this was not really the beginning of the story.

Samuel Kaplan presented a paper at the 191st American Chemical Society National Meeting (April 1986)—and later posted it on the Internet—in which he traced the origin of the MSDS back into the dawn of time when information was exchanged verbally on the materials used as medicines and for dyes.[2,3] With the random testing of the natural materials available to people at that time, largely through trial and error, humans gradually built up a large body of knowledge concerning the preparation of simple drugs and dyes, their storage parameters, application, and hazards of use. This information forms the basis of a chemical data sheet. Some highlights of Kaplan's paper are included in the following paragraphs.

Early written material has been found in the tombs of the Egyptians, either on the walls of their tombs or on papyrus records. These date back over 4000 years and include the prescriptions of Imhotep, the first great Egyptian physician. These data were basically a description of the materials used in the treatment of the various diseases prevalent at the time. They also included the sources, names, preparation, storage, and application procedures, as well as warnings against improper use and application. A few hundred years or so later, the early Sumarians developed a system of writing that has been preserved on clay tablets. They extended the wisdom of the Egyptians and added many more materials to this body of knowledge, especially in the area of dyes.

The Greeks, a thousand or so years later, began to record not only their own observations, but also some of their early experimental work (the original laboratory notebook). These records added to the growing knowledge of natural chemicals and their daily use. During the great era of medical inquiry, in the fourth and third centuries B.C. at Alexandria, actual experimentation into new drugs, dyes, bleaches, and other organic and inorganic materials occurred. During the Roman period, there was an increase in the supply of pharmaceuticals available to the physician because of the Roman Legions, which were needed to maintain the Roman Empire. Much of this work was recorded by Galen in his works on medicine and the human body.

During the Dark Ages, the fifth through the fourteenth centuries, much of the work of the previous centuries was maintained in the monasteries of Europe. The bulk of the knowledge of the Near East, Greece, and Rome was preserved and expanded by the Islamic nations, especially during their great renaissance of the ninth, tenth, and eleventh centuries. Much of this work was done and recorded in Alexandria and Baghdad. This material included for the first time some of the formulations of China, India, and the Far East.

By the end of the fourteenth century, much of this knowledge had been transferred to the southern parts of Italy and France and led to the European Renaissance, which brought about a resurgence of inquiry into the very nature of the materials used today. Up to this time every idea, formula, and the like had to be copied by hand, restricting the widespread use of "chemical data sheets." The development of movable type toward the end of the fifteenth century set the groundwork for the emergence of the modern chemical data sheet. But the safety world had to wait another couple of hundred years until standard units of measurement that were acceptable to the growing scientific community were developed. It is assumed that some of the more enlightened manufacturers of pharmaceuticals and dyes did pass on some safe handling information. Most likely some of the chemists of that era passed along handling and storage precautions in various written correspondence. Kaplan assumed that by the middle of the nineteenth century, manufacturers were supplying their customers with some form of data sheet. The earliest example of an MSDS, as reported by Kaplan, was dated 1906 and was supplied by Valentine and Company. In the following years more information related to flammability and fire fighting was added to the data sheet. Two world wars provided an impetus to add more health hazard and exposure-protection information to the data sheet. After World War II, the Department of Labor began to publish a series of documents under the title, "Controlling Chemical Hazards." These provided an information source for workers in the chemical industry. The Manufacturers Chemical Association (now Chemical Manufacturers Association, or CMA) began to publish their "Chemical Safety Data Sheets" in 1946. In 1968, the first governmental MSDS appeared. This was called a "Material Safety Data Sheet" and was developed to meet the needs of maritime workers. This led to the OSHA Form-20, which was issued in 1972. Subsequently, the publication of the OSHA Hazard Communication Standard defined the requirements for the MSDS.[1]

The modern data sheet is intended to communicate to the chemical user hazard information and protective measures for hazardous substances in commerce. OSHA requires that for every hazardous chemical on the premises an MSDS also be on the premises and available to any employee who requests it. The supplier of the hazardous chemical has the obligation of supplying an MSDS. The federal definition for hazardous substances includes any chemical that would present a hazard under normal use or foreseeable misuse conditions. Because there is some level of hazard associated with every chemical, OSHA has generally interpreted all chemicals as hazardous. OSHA imposes no restriction on the source of the MSDS, although a few state laws specify that only an MSDS obtained from the manufacturer of the chemical is acceptable.

Format

Although OSHA has made available a "suggested" standard format for material safety data sheets, any format is acceptable provided that it presents all of the information required to be in an MSDS as called for in 29 *CFR* 1910.1200.[1] This includes information stored in computer databases, provided that the data can be readily accessed. On the other hand, the voluntary MSDS standard, ANSI Z400.1,[4] mandates a specific, detailed format for an MSDS, but with a few exceptions, the multipage format that is prescribed has not been adopted by chemical suppliers. Although the Z400.1 format would provide information in an understandable style that seems to satisfy both the OSHA specifications and that of several other countries for their regulations on providing chemical hazard information, almost all currently published MSDSs use either the recondite OSHA-suggested format or a variation adopted by the individual chemical supplier. Although a specification of the United States OSHA standards 1910.1200 and 1910.1450[5] requires MSDSs to be in English, many international companies must also support other language and regulatory needs for non–U.S. work sites. There are some international harmonization efforts to bring standardization to several factors for businesses, including chemical information formats.

Reliability

The MSDS is a brief text or reference source that should, but does not always, delineate all of the hazardous properties, and some of the nonhazardous properties, of a chemical or mixture of chemicals, and in addition recommends appropriate precautions for the handling, use, storage, and disposal of the chemical or mixture. One way to evaluate the accuracy and completeness of the information in a MSDS is to examine it critically for internal consistencies or inconsistencies. For example, in the section dealing with ventilation, an MSDS may recommend that the ventilation be sufficient to keep the level of the vapor below the PEL or TLV. But the section dealing with these limits advises that neither a PEL nor a TLV has been established for the vapor! See Young for additional examples.[6]

When an evaluation indicates that an MSDS is defective, users should, if possible, obtain an MSDS for the same substance from another supplier. Or, in a more laborious procedure, users should refer to the several reference works that collectively supply the desired information. These references include the ACGIH Documentation of the Threshold Limit Values and Biological Exposure Indices,[7] Bretherick,[8] Gosselin,[9] Patty,[10] RTECS,[11] and online databases such as the Chemical Abstracts Service, Medline, and Toxline.

The International Agency for Research on Cancer (IARC) evaluates the carcinogenic risk to humans from exposure to chemicals, chemical mixtures, and other agents through critical review of existing research data. IARC publishes these reviews periodically in the form of monographs.[12] Although IARC is not a regulatory agency, OSHA makes the chemicals IARC considers to be carcinogens a part of its floor list under the Hazard Communication Standard.

Employee Training

The OSHA standards for Hazard Communication (29 *CFR* 1910.1200) and Laboratory Safety (29 *CFR* 1910.

1450) require that employees covered by these standards be trained on the physical and health hazards of the chemicals in the workplace. Many employers use MSDSs as the basis of the information provided to employees for this training. The information provided by the MSDSs, supplemented and summarized by the container labels, must be understood by persons who will use those chemicals. Therefore, employees must be given an explanation of the labeling system and the material safety data sheet, and how to obtain and use the appropriate hazard information. Training approaches should include the general format and terms of the material safety data sheet, as well as the other required training elements required by the OHSA standards.

Terms Used in an MSDS

The following notes and definitions for terms typically used in MSDSs will assist in the comprehension of the information provided in a MSDS:

CAS Registry Number This number is not required by OSHA but most state right-to-know laws require it. This number is assigned to each chemical by Chemical Abstracts Service. There are a few instances where a chemical has several different numbers and a few chemicals have no assigned number. Most mixtures do not have assigned numbers.

Chemical Name Usually the IUPAC or CAS chemical name is given, but it also may be a common name for the chemical (e.g., ethylene glycol is acceptable instead of 1,2-ethanediol). Trade names may be supplied but the chemical name is also required unless it is considered to be a trade secret.

Composition of Mixtures All hazardous materials present at concentrations greater than 1% and all carcinogens greater than 0.1% must be identified. Trade names can be used, but chemical names must also be included unless this information is considered a trade secret.

Date Prepared OSHA requires that the date of preparation or most recent revision be stated on the MSDS.

Fire and Explosion Hazard Data These data are often included:

- *Flash point*—The flash point of the chemical is the minimum temperature at which its vapor can be ignited under controlled conditions, including whether an open or closed cup, which should be specified, was used.

- *Flammability limits*—The lower and upper concentrations of the vapor in air below and above which the substance, if flammable, cannot be ignited under controlled conditions.
- *Auto ignition temperature*—The temperature at which a flammable chemical ignites spontaneously in the air under controlled conditions.
- *Recommended extinguishing media*—The type of fire suppression media or extinguisher to be used in fire situations.
- *Unusual fire and explosion hazards*—Any remarkable consequences resulting from exposure to fire or excessive heat.

Health Effects Identification of target organs or organ systems adversely affected by overexposure. Chronic effects and any existing condition that would be aggravated by exposure are also included under this terminology.

Health Hazard Data For toxic substances, this topic usually includes information concerning:

- *LD50 (lethal dose 50)*—This is the lethal single dose (usually oral) in mg/kg (milligrams of chemical per kilogram of animal body weight) of the chemical that is expected to kill 50% of a specified test animal population.
- *LC50 (lethal concentration 50)*—This is a concentration dose expressed as ppm for gases and vapors or, for dusts and mists, as milligrams of material per cubic meter of air expected to kill 50% of a specified test animal population by the inhalation route in one exposure of stated duration.
- *Primary route(s) of entry*—The means by which a toxic chemical is most likely to enter the body, for example, by inhalation, by ingestion, by absorption through the intact skin, through a body orifice other than the nose or mouth such as the socket of the eye or the ear canal, and/or by injection.
- *PEL (permissible exposure limit)*—This is an OSHA-established, mandatory time-weighted average limit for an 8-hour day and, for some chemicals, also includes a maximum concentration exposure limit. The figures may be in parts per million (ppm) or mg per cubic meter (mg/m^3).

Physical/Chemical Characteristics This typically includes some of the following properties:

- *Boiling point*—The value stated may be at atmospheric or reduced pressure and either in degrees Celsius or Fahrenheit.

TABLE 45.1 MSDS Phrases and Recommended Precautions

Phrase	Definition	Precaution
Allergic reaction	Some individuals may develop severe allergic reactions to all but undetectable quantities of certain types of chemicals.	For the susceptible individual, virtual total isolation from the chemical is necessary. Allergic evaluation by a qualified physician may be necessary.
Avoid contact	Take the necessary steps to assure there is no contact with the chemical.	Do not breathe vapors and avoid contact with skin, eyes, and clothing. Use a hood or respirator; wear gloves or other protective clothing.
Carcinogen	Substances that may cause, are suspected to cause, or are known to cause cancer.	Exercise extreme care when handling. Make sure that vapor, mist, or dust is not inhaled and that there is no contact with the skin, eyes, or clothing. Wear suitable protective equipment and use appropriate confining apparatus. OSHA has specified rules for handling "known" carcinogens. Refer to 29 *CFR* 1910 subpart Z.
Corrosive	Living tissue (skin, eyes, mucous membranes) will be destroyed by contact. Some corrosives act rapidly, within seconds; some corrosives do not cause pain as the tissue is being destroyed.	Do not breathe vapors and avoid contact with skin, eye, and clothing. Wear suitable protective equipment.
Danger	Applies to substances that have known harmful effects. Also applies to substances whose properties are unknown, but may have harmful effects.	Usually, the word "DANGER" is accompanied by one or more warning word or phrase in which case follow the precaustions that apply to those words or phrases. If there is no additional warning information, treat the chemical as if it were the most dangerous chemical; there may or may not be hazards associated with the chemical.
Explosive	Substances known to explode under some conditions (e.g., a rapid temperature change, or when unscrewing the cap on a container).	Avoid mechanical shock, friction, sparks, and heat. Do not drop. Isolate explosives from other chemicals that are hazardous if spilled.
Flammable	Substances that give off vapors that readily ignite under usual working conditions.	Keep away from heat, sparks, or flame. The vapors of most flammable substances will creep along the floor or ground for several feet. If the vapors reach an ignition source they can ignite and "flash back." Do not rely on the flammable limits stated in a MSDS to prevent a fire or explosion of vapor-air mixtures. The flammable limits of all flammable vapors vary with temperature and near an ignition source the temperature of the vapor-air mixture may be high enough to render the mixture, at that higher temperature, well within the explosive range.
Irritant	Substances that have an irritant or inflammatory effect on the skin, eyes, or respiratory tract.	Do not breathe the vapors and avoid contact with the skin and eyes. The irritant effect is often evident at low chemical concentrations.
Lachrymator	Substances that have a strong irritant effect or produce a "burning" sensation on the eyes and cause tearing. Opening the cap on a container often causes an immediate effect on the eyes.	Only open lachrymators in an operating and fully tested fume hood. Do not breathe the vapors. Avoid heating. Avoid contact with skin, eyes, and clothing.
Mutagen	Chemical or physical agents that cause genetic alterations.	Handle with extreme care. Do not breathe vapors, mists, or dusts and avoid contact with skin, eyes, and clothing. Wear suitable protective equipment.
Oxidizer	A substance capable of oxidizing a reducing agent. The ensuing reaction is often vigorous and exothermic.	Do not open the container, handle, use, or store the material without thoroughly understanding the potential reactions that the substance can undergo. The information is provided in the reactivity section of the MSDS.
Peroxide former	Substances that form peroxides or hydroperoxides upon standing or when in contact with air.	Many peroxides are explosives. See Chapters 52 and 59.

TABLE 45.1 Continued

Phrase	Definition	Precaution
Poison	Substances that have serious and often irreversible health effects including fatal exposures. Hazardous when breathed, swallowed or in contact with the skin.	Avoid all contact. When handling use suitable protective equipment.
Reactive	Substances that react vigorously with water, air, or other substances with which they come in contact.	Use care in opening the container and in handling the substance.
Teratogen	Substances that cause the production of physical defects in a developing fetus or embryo.	Handle with extreme care. Do not breathe the vapors, mists, or dusts and avoid contact with skin, eyes, and clothing. Wear suitable protective equipment.
Toxic	Substances that are hazardous to the health when breathed, swallowed, or in contact with the skin.	Avoid contact. When handling use suitable protective equipment.

- *Reactivity hazard data*—Information in this section of an MSDS describes whether the material is stable or unstable and, if the latter, the conditions under which instability exists. Incompatibilities with other chemicals, self-reactivity, explosive propensity, and hazardous decomposition products are also described, if applicable.

TLV (threshold limit value) A voluntary exposure limit established by the ACGIH similar to the permissible exposure limit. The measuring units are the same as those used for the PEL. The TLV list is updated each year by the ACGIH.

MSDS Phrases and Recommended Precautions

Material safety data sheets use words or phrases such as "avoid contact," "flammable," and others. Generalized descriptions of many of these phrases and precautions to be practiced are listed in Table 45.1.

References

1. Hazard Communication. *Code of Federal Regulations*, Part 1910.1200, Title 29, 1997.

2. Kaplan, S. A. Development of Material Safety Data Sheets. Presented at the 191st American Chemical Society National Meeting, 14 April 1986.

3. History of the MSDS, http://www.phys.ksu.edu/~tipping/msdshist.html.

4. ANSI Z400.1-1993. *Hazardous Industrial Chemicals Material Safety Data Sheets—Preparation*; American National Standards Institute, Inc.: New York, 1993.

5. Occupational to Hazardous Chemicals in Laboratories. *Code of Federal Regulations*, Part 1910.1450, Title 29, 1997.

6. *Improving Safety in the Chemical Laboratory—A Practical Guide*, 2nd ed.; Young, J. A., Ed.; Wiley-Interscience: New York, 1991; pp. 21–23.

7. *Documentation of the Threshold Limit Values and Biological Exposure Indices*, Vols. I, II, and III, 6th ed.; American Conference of Governmental Industrial Hygienists, Inc.: Cincinnati, OH, 1991, with annual up-dates.

8. *Bretherick's Handbook of Reactive Chemical Hazards*, Vols. 1 and 2, 6th ed.; Urben, P. G., Ed.; Butterworth-Heinemann: Boston, MA, 1998.

9. *Clinical Toxicology of Commercial Products*, 5th ed.; Gosselin, R. E., et al., Eds.; Williams and Wilkins: Baltimore, MD, 1984.

10. *Patty's Industrial Hygiene and Toxicology*, Vol. 2, Parts A through F, 4th ed.; Clayton, G. D., Clayton, F. E., Eds.; Wiley-Interscience: New York, 1993–1994.

11. *Registry of Toxic Effects of Chemical Substances* (RTECS); National Institute for Occupational Safety and Health: Washington, DC. Available through the National Library of Medicine Database.

12. International Agency for Research on Cancer (IARC), World Health Organization, Geneva, Switzerland.

46

Personal Protective Equipment

S. Z. MANSDORF

Personal protective equipment (PPE) is what a worker uses for protection from a hostile external environment. It is called "personal" because each item is intended to protect only the user who wears it. Examples include hearing protection (which attenuates noise transmission into the ear), air purifying respirators (which filter out or adsorb air contaminants), and chemical protective clothing (which provides a barrier through which the hazardous material cannot pass). PPE can range from simple and inexpensive hair nets to relatively expensive, fully encapsulating body suits with a self-contained breathing apparatus like those worn by emergency responders to chemical spills. Common PPE includes head protection (e.g., helmets), eye and face protection (e.g., face shield), respiratory protection (e.g., dust masks), protective clothing (e.g., chemically resistant suits), and protective footwear (e.g., steel-toed boots). PPE for protection from excessive noise is not covered in this chapter.

PPE can be used as an effective control measure when occupational hazards cannot be eliminated or otherwise controlled. PPE should not be relied upon unless other control measures are not feasible, since it does not eliminate the hazard. PPE relies on proper fit, use, and maintenance by the wearer. This means a mistake due to a lack of knowledge on proper use and limitations could render the PPE ineffective. It is also important to recognize that failure of PPE because of product deficiencies or improper selection, use, or maintenance is likely to result in an injury or illness. Other control measures that should be considered before relying on personal protective equipment include elimination of the hazard (e.g., substitution of a less or nonhazardous chemical), automation of the process, engineering controls (e.g., ventilation), and administrative controls (e.g., outdoor work in the early morning or early evening for heat stress control).

Laboratory and pilot plant operations commonly require the use of PPE. This chapter highlights examples of those applications while also covering more common industrial applications.

The remainder of this chapter covers deciding when to use PPE, followed by sections on protection for the head, eyes and face, respiratory system, and skin.

Deciding When to Use Personal Protective Equipment

A decision to rely on PPE should be based on a thorough understanding of the hazards and worker risks involved. OSHA calls this investigative process for deciding on PPE the "hazard assessment." Hazard assessments are a fundamental part of the practice of industrial hygiene encompassing the four basic tenets of anticipation, recognition, evaluation, and control of hazards.

OSHA Requirements

Overall requirements for personal protective equipment are described in the General Industry Standards, section 1910.132. Other requirements may be listed within other sections of the OSHA General Industry Standards or within the sector standards such as the Construction Standards and Maritime Standards. The exact location of the applicable standard will depend on the nature of the hazard (e.g., specifically regulated substance such as lead or physical hazard such as ionizing radiation) and the industry sector. Requirements for laboratories are also covered under the OSHA standard for occupational exposure to hazardous chemicals in laboratories.

General requirements for PPE include:

- maintenance of the equipment in a reliable and sanitary manner
- equipment of a safe design and construction for the work to be performed
- that a hazard assessment be completed before selection of the equipment
- that defective or damaged equipment not be used
- that the employee has received training and understands when the equipment is needed; the proper use and maintenance of the equipment; how properly to put on, wear, and take off the equipment; its useful life; and how properly to dispose of the equipment

In conducting the hazard assessment for PPE, OSHA suggests that the assessment start with a walkthrough survey to identify sources of hazards to workers. These hazards might include impact, penetration, compression, chemicals, heat, harmful dusts, electricity, and other hazards. They suggest that the data obtained from the walkthrough be followed by analysis of the information collected. Analysis would include type of hazard, level of risk, and seriousness of potential injury (or illness). This would be followed by evaluation of selection guidelines and references from expert sources and equipment suppliers. From this information, the selection would be made and the devices fitted to the user.

Another element not specifically mentioned but necessary is an assessment of the effectiveness of the device after selection (i.e., during use). OSHA does require that hazards be reassessed for changes as necessary. OSHA does not specifically state that the hazard assessment be written. They do require that the employer "certify" that the assessment has been done. A written assessment would be the best practice to meet this requirement. The OSHA standard for occupational exposure to hazardous chemicals in laboratories also requires that the employer identify "criteria . . . to determine and implement control measures to reduce employee exposures to hazardous chemicals. . . . " The appendix (not mandatory) to the OSHA standard also contains a recommendation that the laboratory supervisor "determine the required levels of protective apparel and equipment" as part of their responsibilities for chemical hygiene. These recommendations are from the National Research Council's *Recommendations Concerning Chemical Hygiene in Laboratories*.

As previously stated, many OSHA requirements are specific to the type of personal protective equipment used and the work environment. Major requirements or suggested guidelines are covered in the following sections on categories of PPE starting with head protection.

Head Protection

Head protection may be needed to protect workers from physical hazards such as falling or flying objects, bumping into objects, electrical shock, and getting long hair caught in moving machinery. Other specialized forms of head protection can be obtained for other types of hazards, such as for use in abrasive blasting, but are not detailed in this section. There are laboratory and pilot situations that may require head protection to include work below overhead hazards, when working with rotating machinery, and in similar situations. The specific type of protection necessary will depend on the nature of the inherent hazards and the tasks to be accomplished. The American National Standards Institute has published a voluntary consensus standard for head protection in their standard Z89.1, while OSHA regulates head protection under their General Industry Standards

(1910.135), Construction Industry Standards (1926. 100), and Maritime Standards. Common forms of head protection include helmets, bump caps, hairnets, and caps.

Helmets

Helmets, or as they are commonly called, "hard hats" or "safety hats," are required when there is a risk of injury from falling or moving objects that may come in contact with the worker's head (see Fig. 46.1). A special class of head protection is that used to insulate the head from electric shock. Hard hats can also serve to prevent injury from moving into low hanging objects (same function as the bump cap). Hard hats will not prevent injuries to the neck or lower body parts.

Hard hats are available in a variety of shapes, sizes, and colors. Differences in color are commonly used for designating the role of the worker (e.g., white hard hats for supervisors). Helmets consist of three basic components, which are the shell, a suspension system, and a means to secure the helmet to the head. The shell can be constructed of molded thermoplastic, fiberglass, or metal. Most hard hats today are made of impact-resistant plastic. The suspension is typically a multiple point system to keep the helmet from direct contact with the head. Since the purpose of the suspension within the helmet is to distribute the force of any blow to the head, the headbands and suspension webs inside the helmet should have at least 1 inch of clearance to the helmet shell. Configurations that are available vary but can include those with face shields and hearing protection integrated into the helmet. Additionally, ANSI and OSHA

FIGURE 46.1 Example of a hard hat. (Courtesy of MSA.)

require the helmets to be labeled as to the helmet class and manufacturer name. OSHA regulates the general characteristics and performance of hard hats by requiring they meet the specifications of the ANSI standard Z89.1. This ANSI standard specifies three classes of hard hats. These are:

- Class A: helmets intended to protect the head from the force of impact of falling objects and from electric shock during contact with exposed low-voltage conductors
- Class B: helmets intended to protect the head from the force of impact of falling objects and from shock during contact with exposed high-voltage conductors
- Class C: helmets intended to protect the head from the force of impact of falling objects

All helmets should be routinely inspected for cracks, excessive wear, broken suspensions, chemical damage, and other signs of potential defects. Damaged helmets should be replaced.

Bump Caps

Bump caps are not helmets since most do not meet the ANSI standards for protection. They are principally used to prevent injury from bumping into stationary objects rather than falling objects or are used to identify an affiliation such as being a team member. They are usually lighter, may not have full suspension systems, and may not have brims.

Hairnets and Caps

Hairnets and caps are used to prevent contamination of the product (food or other products such as electronics) and from getting hair caught in moving machinery. Most food-processing industries require the use of hairnets or caps for sanitation. Some laboratories and health care industries are also common work environments that require caps or hairnets. They should not be used for protection from electrical hazards, falling objects, or chemicals.

Eye and Face Protection

Eye and face protection are required if there is a risk of injury to the worker from flying objects or particles, infrared and ultraviolet radiation, chemical vapors or splashes, or other hazards to the eye. ANSI has a voluntary standard, Z87.1—*Practice for Occupational and Educational Eye and Face Protection*, which provides guidance on protective eyewear. OSHA regulates eye and face protection under the General Industry Standards (1910.133), Construction Industry Standards (1926.102), or Maritime Standards. General classes of eye protection include:

- eyeglasses
- shaded eyeglasses
- goggles
- hand and face shields
- welding helmets

Full facepiece respirators can also provide eye protection. They are discussed in the section of this chapter on respiratory protection.

Eyeglasses

Eyeglasses or spectacles (commonly called "safety glasses") are used to protect the wearer against frontal impact of flying objects (see Fig. 46.2). Operations that commonly require use of protective eyewear include machining, grinding, hammering, sawing, or other work activities in areas that might have the potential for flying objects that could damage the eyes. Lenses can be used to correct vision (prescription lenses) or simply for protection (known as plano or flat lenses). Lenses may be an integral part of the protective eyewear (e.g., molded plastic eyewear) or lenses of tempered glass or shatter-resistant plastic in an eyeglass frame. Again, OSHA uses an ANSI standard to establish the specifications and performance requirements for this protective device. The ANSI standard Z87.1 establishes the impact-resistance requirements for the lenses and the frames. Safety glasses should be used with side shields to prevent entry of flying objects from the side of the eyeglass frame. Side shields may be made of plastic, wire mesh, or other ma-

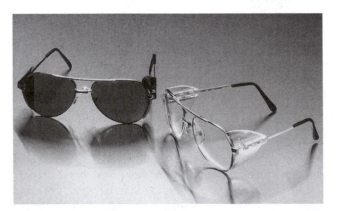

FIGURE 46.2 Examples of safety eyewear. (Courtesy of MSA.)

terials. They should completely cover the opening to the eye from approximately the temple of the head to the eyeglass lenses. For employees who require prescription lenses, the glasses must incorporate that prescription unless plano glasses can be worn over the regular prescription lenses. It is also important to note that while "safety glasses" are commonly required in most laboratory situations, they do not provide adequate protection of workers for splash or vapor hazards to the eyes.

Shaded Eyeglasses

Shaded lenses are commonly used to protect against natural sunlight and can be used in applications for protection from other forms of intense radiation. They are commonly used for protection from direct eye exposure to lasers, for example. They should not be used as a protective measure when damage to the face from radiation may also exist. To be effective, the configuration of the glasses and attachments should protect the eye from stray radiation at all angles. The selection of shading and color of the lenses is determined by the spectrum of the radiation hazard. Eyecup lenses are also used in some special applications for protection against non-ionizing radiation.

Goggles

Goggles provide frontal impact protection and can provide protection against other hazards such as radiation, mists, and splashes (not for chemicals that could damage the skin) depending on the type of goggle (see Fig. 46.3). Goggles are commonly used for mixing chemicals, pipetting, washing glassware, and work environments where there are high ambient dust levels since glasses could

FIGURE 46.3 Example of vented goggles and a hard hat. (Courtesy of MSA.)

FIGURE 46.4 Example of a face shield and hard hat. (Courtesy of MSA.)

allow entry. Goggles come in a variety of styles, lens shadings, and configurations ranging from tight-fitting eyecups to chemical goggles. Ventilation for the goggles is usually provided to prevent fogging. However, goggles used for protection against chemical splash should have indirect ventilation that will not allow the penetration of liquids.

Hand and Face Shields

Neither glasses nor goggles provide protection for the face. When the entire face must be protected, a hand or face shield is the appropriate choice. Hand shields are less dependable as a protective device than face shields since the worker must use them at the appropriate time, although stationary shields can be effectively used (e.g., in laboratories as a barrier for chemical reactions). Accessories for face shields include neck and chin protectors for some applications. Face shields can also be obtained to fit onto helmets (see Fig. 46.4). They are most commonly used for protection against potential chemical splashes of corrosive materials and other situations where the facial skin is also at risk. Since face shields typically do not meet the ANSI Z87 requirements for impact resistance, safety glasses may also be required under the face shield. Face shields may be constructed of clear thermoplastic or wire mesh (molten metals industries) that is curved to protect the face and, in some configurations, the sides of the head.

FIGURE 46.5 Examples of welding helmets. (Courtesy of MSA.)

Welding Helmets

Welding is not a common hazard for laboratories but is more common to pilot plant maintenance. Welding and cutting operations (from plasma arch, etc.) can present hazards from infrared and ultraviolet radiation as well as from hot flying objects. Welding and cutting requires the protection of the entire face with materials that are fire and heat resistant (see Fig. 46.5). The degree of protection needed from radiation (shading of the helmet lens) is dependent on the type of welding or other operation being performed. It should also be remembered that welding can generate respiratory hazards and high levels of noise requiring that other protective devices be used in conjunction with the use of the welding helmet.

Foot Protection

Foot protection is not usually necessary for laboratories but may be required for some special operations and for pilot plants. Foot protection is necessary whenever there is a potential for injury to the feet from falling or rolling objects, objects piercing the sole of the foot, electrical hazards, and other such hazards. This may include such jobs as the handling of heavy drums and cylinders. Specialized applications include conductive shoes for work in explosive atmospheres, slip-resistant footwear for wet work, and thermally protective footwear. Most of the heavy industries as well as the construction industry require the use of protective footwear. Laboratories that potentially generate explosive atmospheres may require conductive footwear. ANSI has a consensus standard on protective footwear, Z41.1, that provides guidance to users and is incorporated into the OSHA standards. OSHA regulates foot protection under their General Industry Standards (1910.136), Construction Industry Standards (1926.96), and Maritime Standards. Protective footwear includes:

- shoes
- boots
- toe caps and metatarsal guards

Shoes and Boots

Shoes can be obtained with varying designs, materials of construction, and features. The most common types are those with protected toe areas (usually by including a metal cover over the toe area inside the shoe or by the toecaps) and are commonly called "safety shoes." A modern development is the evolution of the safety sneaker or tennis shoe. These are commonly used for protection where the hazard is primarily to the toes but not from crushing injuries or to other parts of the foot. Shoes can also include a steel or metal plate incorporated in the sole to prevent penetration injuries as well as slip-resistant and chemically resistant soles. Leather and other natural fiber shoes should not be worn for work around hazardous chemicals since they may absorb and retain them. The molten metals industries also have some special requirements for shoes and boots including the requirement that no openings be present (no tongues or requirements for covers).

In general, boots may be obtained in a wide variety of sizes and shapes with all the features mentioned above. In some industries, such as the steel industry, metatarsal guards may also be required. Water-resistant or waterproof boots or overboots may be necessary for wet work, while chemically resistant boots should be used for work where contact with chemicals is likely. Again, natural fiber and leather boots should not be used for chemical protection.

Toe Caps and Metatarsal Guards

Toecaps can be worn over shoes or boots to provide protection from physical hazards. Metatarsal guards can be worn over the boots to protect the rest of the foot from crushing injuries. They are somewhat similar to the guards worn over skates by hockey players. Spats, gaiters, or leggings can be used to protect the other portions of the leg not protected by shoes or boots.

Respiratory Protection

The most common route of entry for hazardous materials in the industrial environment is by the respiratory route. There are many processes that can generate airborne particulate, mists, vapors, and gases, which might present a respiratory hazard. These can be found in most private and many public sector work activities including laboratory and pilot plant operations. In fact, NIOSH has estimated that as many as seven million workers use respirators at some time each year as protective devices against airborne contaminants and lack of adequate breathing air. OSHA has estimated their new regulatory standards will affect five million users in 1.3 million workplaces. The statistics from NIOSH and OSHA demonstrate the wide use and importance of respiratory protection in the workplace.

Types of Respiratory Hazards

Airborne particulate contamination can include dusts, fibers, fumes, and mists. Gaseous contaminants include the categories of vapors and gases. Dusts are generated by the mechanical breakup of solids through actions such as grinding, mixing, and cutting. Air contamination can also be created during the application of particles such as spray painting and blasting to clean parts or through the mechanical processing of powders. Airborne fibers can be generated by some specialized laboratory tasks, industrial processes, construction work, and so forth. Fumes are common contaminants of welding and cutting of metals through the condensation of heated solids. Mists are common to heated processes such as plating, metals descaling, and other processes where vapors are condensed or through the mechanical action of applicators. Vapors, produced when liquids (or solids with low to moderate vapor pressures) evaporate, are common to processes involving use of volatile liquids. Gaseous contaminants are commonly generated in many laboratories and industrial processes such as distilling, chemical mixing, refining, specialty gas production, brewing, and others.

Oxygen-deficient atmospheres can be created naturally through oxygen displacement by decomposition, the depletion of oxygen through oxidation, the addition of other gases, or through other means. This situation is most common to confined spaces that do not allow normal diffusion and gas mixing or in high-altitude situations.

Standards for Respiratory Protection

ANSI Z88.2 is the voluntary consensus standard that provides guidance on respiratory protection. OSHA has just recently revised their requirements and regulates respiratory protection in the General Industry Standards under a new section 1910.139 (from 1910.134). This now incorporates the requirements under the Construction Industry Standards, Shipyard, Longshoring and Maritime Standards. The key general requirements include:

- identifying and evaluating respiratory hazards in the workplace
- use of control methods other than respirators unless not feasible
- providing employees with suitable respirators for the purpose intended
- establishment and maintenance of a written respiratory protection program
- medical evaluation of employees required to use respirators
- providing for fit testing of tight-fitting respirators
- providing training for respirator users
- designating a qualified program administrator
- use of respirators that have NIOSH approval
- providing for maintenance and care of respirators
- providing Type I, Grade D breathing air for atmosphere-supplying respirators
- conducting inspections of emergency use respirators
- meeting special requirements for work in IDLH atmospheres

Types of Respirators

Respirators can be classified into two general categories: air purifying and air supplied. The air-purifying respirators filter or otherwise "purify" the ambient air to remove contaminants; however, they do not provide breathing air. Air-supplied systems actually provide breathing air from either a remote source or from compressed air sources. A special class of respirators can actually produce breathing air (oxygen) from a chemical reaction and are used in special escape applications. These respirators are also known as self-rescuers.

The simplest forms of respirators are those that are air purifying. These fall into three categories: particulate (and aerosol) respirators, gas and vapor removing respirators, and respirators that incorporate a combination of the particulate and gas and vapor features. Particulate respirators may be powered or not powered. Powered air-purifying respirators (PAPRs) utilize a blower to pump air to the user while nonpowered air-purifying respirators rely on the negative and positive atmospheric pressures generated by the user's respiration to move air through the filters or chemical cartridges and exhalation valves. Full facepiece respirators that use large canisters instead of the smaller cartridges worn on the respirators are commonly called gas masks. Respirators can be obtained for a single use or for reuse.

Respirator facepieces (also called masks) may partially or fully cover the face. These are commonly referred to by the amount of the face they actually cover, typically ranging from one-quarter to one-half to a full facepiece.

Atmosphere-supplying respirators consist of supplied-air respirators (hose mask or airline) that provide breathing air to a loose fitting facepiece (such as a sand blasting hood) or a tight-fitting facepiece, and the self-contained breathing apparatus (SCBA). The SCBA incorporates all parts of the respiratory protection in a configuration that is worn by the user (hence the term self-contained). They use either compressed air sources (tanks), which are called open circuit, or are rebreathers (provide oxygen and scrub carbon dioxide), which are called closed circuit.

Air-Purifying Respirators

There has been an enormous growth of disposable particulate respirator use from the earlier use of quarter facepiece or half facepiece rubber respirators that used cartridges or filter pads. NIOSH has estimated that employers annually purchase over 110 million disposable respirators. These new generation disposable respirators are generally easy to wear, relatively inexpensive, and lightweight. They are typically manufactured as one piece of filter media molded to the face, although they are available in other configurations (see Fig. 46.6). Because of this, NIOSH has just recently changed their certification process for particulate respirators based, in part, on the efficiency of the filters.

Particulate respirators are now classified by three levels of efficiency based on testing against 0.3 μm sodium chloride or dioctylphthalate (DOP) aerosols. The new

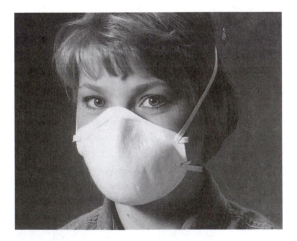

FIGURE 46.6 Example of a disposable particulate respirator. (Courtesy of MSA.)

TABLE 46.1 NIOSH Certification Classes and Uses for Particulate Respirators

Filter Class	Use	Efficiency
N-95	Aerosols not containing oil	95%
N-99		99%
N-100		99.97%
R-95	Aerosols that may contain oil (with restrictions)	95%
R-99		99%
R-100		99.97%
P-95	Any aerosol type (P-100 for radionuclides)	95%
P-99		99%
P-100	(HEPA color designation of magenta)	99.97%

classifications have three letter designations and three levels of filter efficiency within each letter designation. The letter designations, called a series by NIOSH, are N, R, and P. The letter designation N represents "not resistant to oil," the letter designation R represents "resistant to oil," while P represents "oil proof." The efficiency designations are 95%, 99%, and 99.97% (labeled as 100), as shown in Table 46.1.

N series filters are tested against sodium chloride and are effective for solid and water-based particles but not oil-based aerosols that could degrade the filter media. The R and P series are tested against DOP and thus can be used against any particle or fiber contamination as appropriate for the level of efficiency selected (i.e., 95, 99, or 99.97). The P100 filter meets the HEPA (high-efficiency particulate air) criteria and will be the only filter of those described that will have the magenta color-coding used for absolute (HEPA) filters.

Air-purifying respirators that use chemical cartridges can be obtained in half-facepiece or full-facepiece configurations with single or twin cartridges that are attached to the facepiece or can be attached directly or by a remote connection, such as when wearing a welder's helmet (see Figs. 46.7 and 46.8). Gas masks that utilize larger chemical sorbent beds are full-facepiece respirators with a hose attached to a large chemical canister. Gas masks are most commonly used for escape purposes. OSHA has a color-coding requirement for chemical cartridges and canisters. This color coding can be used for determining that the right cartridge or canister is worn for the chemical contaminant present. Table 46.2 presents a summary of these color codes. It should be noted that the magenta color indicates a HEPA filter for use with highly hazardous particulate or aerosols and is not a chemical sorbent.

Atmosphere-supplying respirators provide breathing air from either a remote source or from a compressed

FIGURE 46.7 Examples of half-facepiece respirators with different cartridges. (Courtesy of MSA.)

FIGURE 46.8 Example of a full-facepiece, air-purifying respirator. (Courtesy of MSA.)

gas source. This includes respirator combinations such as those that have both the features of air-purifying cartridges and atmosphere-supplying ones. Air that is free from contaminants, oil, and odors (per the requirements of ANSI/Compressed Gas Association specification G-7) is required for atmosphere-supplying respirators. Air line respirators are the simplest of the atmosphere-supplying respirators. Airline respirators have a hose that supplies air from a stationary source (air pump, compressor, or compressed air source) through a regulator to a hood, helmet, or facepiece on the wearer (see Fig. 46.9). There are three types comprising demand, pressure demand, or continuous flow. Demand-air-supplied systems provide air with the negative pressure developed by inhalation from the user. Pressure-demand systems work in a similar fashion except the pressure in the facepiece is kept positive relative to the ambient atmosphere outside of the facepiece and is increased on inhalation. Continuous-flow systems provide a continuous flow of air to the user at all times, rather than only on demand. Some of

these systems may include an air-purifying cartridge for use for escape should the supply of air falter and providing the cartridges are appropriate for the ambient air concentration of contaminant. For those situations where the atmosphere is or could be immediately dangerous to life and health (IDLH), OSHA requires the use of a supplemental air bottle to be worn by the user for escape purposes. There are a number of other requirements for these systems including limits to pressures, hose lengths, air quality, and type of compressor that may be used. The OSHA regulations should be reviewed for these requirements (Part 1910.139).

The self-contained breathing apparatus (SCBA) with tight fitting full-facepiece represents the highest level of protection for the wearer (see Fig. 46.10). There are two types. These are open circuit and closed circuit. Open-circuit devices exhaust the exhaled air to the atmo-

TABLE 46.2 Summary of Color Codes for Chemical Cartridges and Canisters by Contaminant[1]

Color	Contaminant
White	Acid gases
White with a ½ in. green stripe	Hydrocyanic acid gas
White with a ½ in. yellow stripe	Chlorine gas
Black	Organic vapors
Green	Ammonia gas
Blue	Carbon monoxide gas
Yellow	Acid gases and organic vapors
Brown	Acid gases, ammonia, and organic vapors
Red with a ½ in. gray strip	Acid gases, ammonia, chlorine, carbon monoxide, and organic vapors
Magenta (purple)	Radioactive materials (except tritium and noble gases) and asbestos

1. See 29 *CFR* 1910.134(g)(6) for the complete table.

FIGURE 46.9 Example of an atmosphere-supplying (air line) respirator with air-purifying cartridges. (Courtesy of MSA.)

FIGURE 46.10 Example of an SCBA. (Courtesy of MSA.)

sphere, while the closed-circuit units scrub the carbon dioxide and add oxygen to replace that which has been used. The open-circuit units usually consist of a tank of compressed air of 30 to 60 minutes-use duration that is fed through a pressure regulator to the respirator facepiece. Some of the closed-circuit devices have smaller containers of compressed or liquid oxygen with a scrubbing system allowing up to 4 hours of use. Some of these units may be used in atmospheres that are IDLH. Escape SCBAs are also certified in combination with supplied-air, airline respirators.

Selection and Fit Testing of Respirators

Respirators should be selected based on the hazard assessment discussed previously. Once the conditions of exposure are known, the required protection factor and respirator equipment can be determined. Routine respirator use in laboratories and pilot plants is not common. Nevertheless, there are many times when respiratory protection must be used. Examples of categories of work requiring respiratory protection include maintenance work, for certain tasks that must be done outside of glove boxes and laboratory hoods or where other methods for control of emissions are not feasible, work in confined spaces, and for spills and emergency responses.

The protection provided by respirators can be determined experimentally by measuring the ambient concentration of contaminant outside the respirator and the concentration inside the respirator. Dividing the outside concentration by the inside concentration yields the protection factor (outside concentration/inside concentration = PF). Minimum expected PFs have been estimated for each class of respirator even though higher values

may be found with quantitative fit testing. These are experimental values that will vary based on the actual fit of the respirator to the user and the efficiency of the respirator actually selected. Nevertheless, it is good practice and a requirement that only the minimum values be used when determining the correct respirator to assign. The current OSHA standard does not establish "official" protection factors; however, they may be included in the next revision of their standard on respiratory protection. OSHA does require fit testing for tight-fitting respirator facepieces.

Table 46.3 developed from the ANSI Z88.2 standard data shows typical protection factors for common classes of respirators. The protection factors are used to calculate the maximum expected exposure to the worker. For example, if the air concentration of a dust were 15 mg/M^3 and a half-mask particulate respirator with a PF

TABLE 46.3 Protection Factors for Common Respirator Classes

Respirator	Protection Factor
Air purifying half-facepiece respirators	10
Powered air-purifying half-facepiece respirators	50
Air-supplied, demand-flow half-facepiece respirators	50
Air-supplied, continuous-flow half-facepiece respirators	50
Full-facepiece, demand-air-purifying respirators	100
Full-facepiece powered-air-purifying respirators for dusts and mists	100
Air-supplied, continuous-flow full-facepiece respirators	1000
Self-contained breathing apparatus full-facepiece-pressure demand/rebreathers	10000

of 10 were used, the theoretical exposure to the worker would be 1.5 mg/M^3 (outside concentration 15/10 PF = 1.5). An exposure limit for the substance above 1.5 mg/M^3 would make this an acceptable choice. If the exposure limit were below 1.5 mg/M^3, another more protective selection such as a full-facepiece respirator would have to be used.

For air purifying respirators, there are also other limitations on their selection. Since the chemical cartridges have a finite capacity to absorb or otherwise remove the contaminant and breakthrough is possible, an odor threshold or other warning property, such as irritation, must exist that is below the exposure limit. For some substances, such as the isocyanates, this eliminates the use of chemical cartridges since there is no odor threshold at the PEL. Even when there are adequate odor-warning properties, OSHA now requires a schedule for changing the chemical cartridges or canisters based on objective data on expected air concentration and the capacity of the units being used unless there is an end-of-service-life indicator. Particulate filters usually do not have this limitation since the dust loading will create additional breathing resistance, signaling the user to the need to change the respirator, and generally increase the efficiency of the filter. However, the increased breathing resistance can result in a greater likelihood of leakage around the facepiece.

Fit Testing of Respirators

OSHA regulations now require fit testing of respirators. This is because of the differences in the geometry and sizes of faces among users, fit patterns and respirator sizes among manufacturers. Additionally, anything that would prevent the sealing of the respirator to the face would prevent a proper fit. This includes eyeglasses for full-facepiece respirators, special facial features, and other factors affecting the sealing areas. OSHA now prohibits the wearing of tight-fitting respirators when the user has a beard or other factors that could prevent an effective facial seal.

There are two general types of fit testing: qualitative and quantitative fit testing. There is also a simple seal test for the user of tight-fitting respirators. This user seal check should be performed every time the respirator is put on (contained in Appendix B of the OSHA standard 1910.134). It consists of a positive and negative pressure check. The positive pressure check is done by closing off the exhalation valve and exhaling gently into the facepiece. The face fit is considered satisfactory if a slight positive pressure can be built up inside the facepiece without any evidence of outward leakage of air at the seal. The negative pressure check is done by closing off

the inlet opening of the canister or cartridge(s) by covering with the palm of the hand(s) or by replacing the filter seal(s) and inhaling gently so that the facepiece collapses slightly, and holding the breath for 10 seconds. The manufacturer's recommendations may be used instead of the OSHA procedure if they are demonstrated to be equally effective.

Qualitative fit testing involves the use of a simulant to actually test the fit of the respirator under use conditions. There are several simulants that OSHA will accept. These are irritating cold smoke (stannic chloride), banana oil (isoamyl acetate), saccharin aerosol, and denatonium benzoate (trade name Bitrex). In these test procedures, the subject is first tested for sensitivity to the simulant. If sensitivity is demonstrated, the test is performed. First, the user is given a choice of respirators appropriate for his/her work. After the selected respirator has been properly fitted with the applicable cartridge and the negative and positive pressure user check performed, the subject is placed in an environment of the simulant with the respirator on (usually using a hood or similar device to hold the simulant). The subject then is asked to perform a number of exercises to simulate a work environment. These may be found in the OSHA standard.

Once the test is completed, the respirator user is usually asked to remove the respirator to determine if they are still sensitive to the simulant. At any point in the test procedure during which the user detects the simulant, the test should be stopped and another respirator (different size or manufacturer) tried.

Quantitative fit testing is a more accurate measure of the level of protection provided the user since it actually provides a protection factor by measuring the simulant or ambient concentration difference inside and outside of the respirator or by measuring inward leakage under negative pressure. Simulants include the use of several different aerosols such as sodium chloride, di-2-ethyl hexyl sebacate (DEHS), corn oil, and polyethylene glycol 400 (PEG 400). Ambient aerosols are used in condensation nuclei methods while a negative pressure-measuring device is used to determine volumetric leak rates. These tests require the use of relatively expensive equipment but follow similar protocols to those described in the qualitative fit test. OSHA requires that one of these test methods be used for fit testing of respirators with protection factors (OSHA calls this the "fit factor") required to exceed 100.

Respiratory Protection Programs

OSHA requires a written respiratory protection program. Key elements required in the program are:

- procedures for selecting respirators for use in the workplace
- medical evaluations (procedures and program description) of employees required to use respirators
- fit test procedures for tight-fitting respirators
- procedures for proper use of respirators in routine and emergency situations
- procedures and schedules for cleaning, disinfecting, storing, inspecting, repairing, discarding, and otherwise maintaining respirators
- procedures to ensure adequate air quality, quantity, and flow of breathing air for atmosphere-supplying respirators
- training of employees in the respiratory hazards to which they are potentially exposed during routine and emergency situations
- training employees in the proper use of respirators, including putting on and removing them, and any limitations on their use and maintenance
- procedures for regularly evaluating the effectiveness of the program

The procedures for selecting respirators should also include a written selection logic or procedure that bases respirator selection on the hazards to which the worker is exposed. This would need to include a hazard assessment with quantitative evaluation of the level of airborne hazards. This usually requires air sampling for time-weighted average concentrations, short-term peak exposures, and a determination of oxygen content if there is a potential for oxygen deficiency. The hazard assessment should also include odor thresholds or irritation levels and the service life evaluations or the availability of units with end-of-service-life indicators.

Instruction and training for employees using respirators should include:

- An explanation of the nature and extent of the hazard requiring the use of the respirators. This should include the information required under the Hazard Communication Standard (e.g., acute and chronic effects of exposure, warning signs, emergency procedures, etc.). An explanation of other control methods being considered or other efforts to reduce employee exposures should also be presented at this time.
- An explanation of the respirator selection process and OSHA requirements
- Demonstrations on the proper fitting, donning, wearing, and removing of the respirator including an opportunity for hands-on experience with the negative pressure test and other fit-testing procedures

- An explanation of the limitations, capabilities, and operation of the respirator
- An explanation and demonstration of the proper maintenance and storage of the respirators
- An explanation and demonstration on how to inspect the respirator for defects
- Information on how to recognize emergencies and what to do in these situations

This information and training should emphasize practical hands-on experience for the respirator user. It is also advisable to conduct a written and/or performance evaluation of the worker's knowledge. These records should be maintained with the employee's training records.

Additional information on the key elements of a respiratory protection program can be found in the OSHA standard for respiratory protection, the ANSI Z88.2 standard, in the model program for small business available from OSHA, and the NIOSH *Guide to Industrial Respiratory Protection*.

Protective Clothing

Hazards to the skin can result from unprotected exposures to harmful biological, physical, and chemical agents. Most of these injuries and disease risks can be prevented or reduced through the appropriate selection and use of protective clothing or other control methods. Hazards to the skin are common to many occupations in both the public and private sectors. There have been a number of serious accidents in laboratory personnel resulting from improper or inadequate dermal protection. Even though the quantity of chemicals being used in the laboratory tends to be small, the proper selection and use of protective clothing can be critical. Protective clothing is regulated by OSHA in the General Industry Standards under 1910.132 and for specific uses under various standards including the Construction Industry Standards and Maritime. For example, requirements for emergency response to chemical spills are listed in 1910. 120, while blood-borne pathogens requirements may be found under 1910.1030.

An Overview of Hazards Affecting the Skin

There are three major categories of hazards affecting the skin: biological, physical, and chemical hazards. Biological hazards include infection from agents and disease common to humans, animals, and the work environment. Biological hazards common to humans include pathogenic microbes such as those causing AIDS, hepatitis, tuberculosis, Legionnaires disease, and others. Infectious agents from animals include those such as Anthrax. Environments that can potentially represent a hazard from biological agents include sewage treatment plants, composting facilities, clinical and microbiological laboratories, genetic research facilities, as well as other special work environments.

Physical hazards can be categorized as those that are a result of trauma (produce injuries such as cuts), excessive heat and cold, vibration, and radiation. All of these are possible in the laboratory setting. Trauma to the skin from physical hazards (cuts, abrasions, etc.) is common for many laboratory tasks such as glass cleaning and in industrial sectors such as construction and meat cutting. Common to work environments are hazards from extreme heat, such as use of furnaces in labs, molten metals work, and fire fighting, while extreme cold is common to work tasks such as handling of cryogenic liquids. The protective attributes of clothing for these hazards are related to the insulation provided (generally increases with thickness), whereas protective clothing for flash fire and electric flashover requires flame-resistance properties. Protection from some forms of both ionizing and non-ionizing radiation can be achieved using protective clothing. Excessive vibration can have several adverse effects on body parts, primarily the hands. Occupations such as mining (hand-held drills) and road repair (pneumatic hammers or chisels) are two examples in which excessive vibration can lead to diseases such as Raynaud's phenomenon.

Chemicals may present more than one type of dermal risk (e.g., a substance such as benzene is both toxic and flammable). Adverse effects of skin contact with chemicals can be generally categorized as causing irritation, an allergic response, corrosion (chemical burns), skin toxicity, systemic toxicity (permeation through the skin), and promotion of cancer of the skin or other body cancers. Some chemicals may present more of a hazard from the dermal route than through inhalation. Chemicals that have the potential to significantly contribute to a worker's overall dose by the dermal route are identified by OSHA in their permissible exposure limits (PEL) and the American Conference of Governmental Industrial Hygienists (ACGIH) in their threshold limit values (TLV) by a "skin" notation. Not all chemicals that present dermal hazards are identified in these listings. For example, corrosives are not included.

Chemical Permeation of Barriers

The diffusion of chemicals through "liquid-proof" protective clothing barriers has been known for at least the last 30 years. For example, acetone has been shown to diffuse through (permeate) Neoprene gloves to the inside surface within 30 minutes of direct liquid contact. This movement of a chemical through a protective clothing barrier is called permeation. The permeation process consists of the diffusion of chemicals on a molecular level through protective clothing. The permeation process occurs in three steps.

1. absorption of the chemical at the barrier surface
2. diffusion through the barrier
3. desorption of the chemical on the normal inside surface of the barrier

The permeation of a barrier, since it is essentially diffusion, is directly related to temperature and inversely related to barrier thickness. The time elapsed from the initial contact of the chemical on the outside surface until detection on the inside surface is called the breakthrough time. The permeation rate is the steady-state rate of movement (mass flux) of the chemical through the barrier after equilibrium is reached. The permeation rate is normally reported in mass per unit area per unit time (e.g., $\mu g/cm^2/min$) and may be normalized for thickness. Most current laboratory testing done for permeation resistance is for periods of up to 8 hours to reflect normal work shifts.

Permeation Testing

A standard test method for determining permeation of protective clothing is F739–91, "Test Method for Resistance of Protective Clothing Materials to Permeation by Liquids or Gases Under Conditions of Continuous Contact," published by the American Society for Testing and Materials (ASTM). In brief, the test consists of placing the barrier material between a reservoir of the challenge chemical (liquid or gas) and a collection cell connected to an analytical detector, as shown in Figure 46.11. It should also be noted that the actual reported breakthrough time is also related to the sensitivity of both the analytical method and system (i.e., collection system). Variations on this method can also be performed for determining permeation from intermittent contact and a novel approach measuring evaporation through the barrier by weight loss for chemicals that have a vapor pressure of at least 10 mm Hg.

Aside from the permeation process just described, the two other chemical-resistance properties of concern are degradation and penetration. Degradation is a deleterious change in one or more physical properties of a protective material caused by contact with a chemical. For example, latex rubber gloves may go into solution with

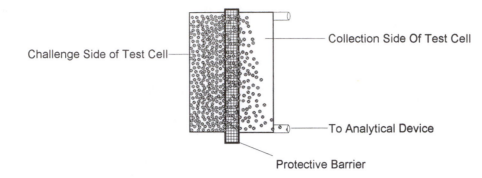

FIGURE 46.11 Representation of permeation testing.

contact by hexane or toluene. Many times degradation can be assumed if the barrier swells or has a change in physical appearance (wrinkles, burns, color changes, etc.). Penetration is the flow of a chemical through zippers, open seams, pinholes, cuts, or imperfections in the protective clothing on a nonmolecular level. Penetration protection is important when the exposure is unlikely or infrequent and the toxicity or hazard minimal. Penetration usually is a concern for splash protection garments.

Given the lack of published skin permeability and dermal toxicity data, the approach taken by most safety and health professionals is to select a barrier with no breakthrough for the duration of the job or task. However, it is important to recognize that there is no protective barrier currently available that provides permeation resistance to all chemicals. For situations where the breakthrough times are short, the safety and health professional should select the barrier(s) with the best performance (i.e., longest breakthrough time and/or lowest permeation rate) as well as considering other control measures (such as a clothing change). A recent case in which contact with methyl mercury resulted in the death of a "gloved" laboratory worker is an example of the importance of effective barrier resistance.

There are several guides that have been published listing chemical resistance data (many are also available in an electronic format). In addition to these guides, most manufacturers publish current chemical and physical resistance data for their products.

Types of Protective Clothing

Protective clothing includes all elements of a protective ensemble (e.g., garments, gloves, and boots). Protective clothing items can range in complexity from simple gloves to gas-tight, fully encapsulating suits. Protective clothing can be made of natural materials, synthetic fibers, or various polymers. Materials that are woven, stitched, or are otherwise porous (not resistant to liquid penetration or permeation) should not be used in situations where protection against a liquid or gas is required. Specially treated or inherently nonflammable porous fabrics and materials are commonly used for flash fire and electric arc protection. Protection against heat requires clothing that provides flame (burning) resistance and thermal insulation. Other specialized fire fighting applications may require aluminized coverings for protection against an infrared component and use of a vapor barrier for protection from steam production. Table 46.4 summarizes typical physical, chemical, and biological performance requirements and common protective materials used listed by hazard.

Gloves

Protective gloves are available in a wide variety of natural materials and fibers, synthetic fibers, polymers, and combinations such as polymer-coated natural fiber (manufactured using a dipping process). Four common examples of the different types of polymers or elastomers are shown in Figure 46.12. Gloves are probably the most common PPE used in laboratories. Polymer gloves are available in a wide variety of thicknesses ranging from very lightweight (<2 mm) to heavyweight (>5 mm) with and without inner liners or substrates (called scrims). Gloves are also commonly available in a variety of lengths ranging from approximately 30 cm for hand protection to gauntlets of approximately 80 cm extending from the worker's shoulder to the tip of the hand.

It should also be noted in this section that latex rubber gloves (also called natural rubber) can cause irritant contact dermatitis, chemical sensitivity dermatitis, or severe allergic reactions in some workers. Natural latex rubber gloves have poor chemical-resistance properties but are extensively used in laboratories and for biological hazards, especially in the health care fields. Options available for these situations include use of gloves made of other rubbers or plastics (such as thin nitrile), use of

TABLE 46.4 Summary of Common Protective Clothing Applications

Hazard	Performance Characteristic Required	Common Protective Clothing Materials
Chemical	Permeation, penetration, degradation resistance	Polymeric materials
Biological	Protection against microbes	Polymeric materials
Radiological	Provide shielding or not allow liquid penetration	Polymers and lead-lined protective clothing
Thermal	Insulation	Thick natural or man-made fabrics or with insulating layers
Fire	Insulation and fire resistance	Aluminized gloves; flame-resistant-treated gloves; aramid fiber and other special fabrics
Mechanical abrasion	Abrasion resistance; tensile strength	Heavy fabrics; leather with metal studding
Cuts and punctures	Cut resistance	Metal mesh; aromatic polyamide fiber, and other special fabrics

Adapted from *The Occupational Environment—Its Evaluation and Control*, published by AIHA.

hypoallergenic latex gloves (not effective against allergic reactions to latex), and use of nonpowdered latex gloves.

Boots

Boots are also available in many different natural materials and polymers (as noted in the section on foot protection), while chemically resistant boots are available in only a limited number of polymers. Chemically resistant boots may be obtained as boot covers (worn over work boots) or as an integral boot. Common polymers and rubbers used in chemically resistant boot construction include PVC, butyl rubber, nitrile, and neoprene rubber. Boots made with combinations of polymers for added chemical resistance are also commercially available.

Garments

Protective garment items can range from simple single components such as jackets, pants, and hoods to one-piece, fully encapsulating (gas-tight) polymer suits with attached gloves and boots. These are available in a number of styles and materials of construction ranging from natural fibers to polymer combinations. Natural fiber garments are not appropriate for chemical exposures. The newer nonwoven polyethylene disposable suits are very commonly used for clean room applications and protection against particulate contamination. These polymer garments, sometimes incorrectly called "paper suits," are made using a special process where the fibers are bonded or pressed and glued together rather than woven (see Fig. 46.13). Breathable (microporous) suits allow some water vapor transmission and are less heat stressful. Most of these suits are not chemical or liquid resistant. Spun-bonded garments with various coatings such as polyethylene, PVC, and Saran can provide chemical resistance. Some protective barriers will have multiple lay-

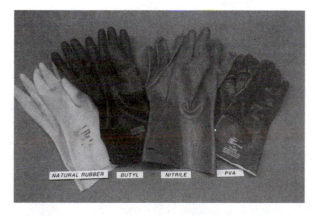

FIGURE 46.12 Examples of common polymer or elastomer gloves. (Courtesy of Zack Mansdorf.)

FIGURE 46.13 Example of a spun-bonded garment worn by a worker with hard hat and respirator. (Courtesy of MSA.)

ers or laminants. Layered materials are generally required for polymers that do not have good inherent physical integrity and abrasion resistance properties (e.g., Teflon) or to improve chemical resistance.

The fully encapsulating gas-tight suit of one-piece construction provides the highest level of protection provided by chemical protective clothing. These suits are used for emergency responses to large spills of hazardous chemicals. They are also sometimes used for work in extremely dangerous atmospheres, such as certain maintenance work in pilot plant operations. In the majority of these configurations, the suit also protects the respiratory protection device (air line or SCBA), since it is worn within the suit.

Hazard Assessments for Selection of Protective Clothing

OSHA also requires that a hazard assessment be performed before assigning protective clothing. The approach is the same as that used for other protective equipment. The emphasis in this hazard assessment would be the hazards presented by the dermal route of entry as well as special considerations unique to protective clothing, such as the need for decontamination and effects of any added heat stress. The laboratory situation offers a special challenge for conducting hazard assessments. This is because the chemical hazards encountered tend to be high but the volumes and frequency of contact tend to be low. There is usually also the need for high levels of dexterity (e.g., for synthesis work in a hood). Nevertheless, the basic tenets of hazard assessment still apply: an assessment of the risk to the user from contact and the likelihood of that contact.

Worker Education and Training

Worker education and training follows that required of all PPE and should include:

- the nature and extent of the hazard(s)
- when protective clothing should be worn
- what protective clothing is necessary
- use and limitations of the protective clothing to be assigned
- how properly to inspect, don, doff, adjust, and wear the protective clothing
- decontamination procedures, if necessary
- signs and symptoms of overexposure or clothing failure
- first-aid and emergency procedures

- the proper storage, useful life, care, and disposal of protective clothing

This training should incorporate all of the elements listed above and others of pertinence that have not already been provided to the worker through other programs. For those training subjects and topics already provided to the worker from other training programs, an assessment of the need for refresher training should be undertaken. Finally, the workers should have an opportunity to try out the protective clothing before a final selection decision is made and significant quantities ordered.

Protective Clothing Programs

A good professional practice would be to develop a written program. This will reduce the chance for error, increase worker protection, and establish a consistent approach to the selection and use of protective clothing. A model program should contain the following elements:

1. an organization scheme and administrative plan
2. a risk-assessment methodology
3. an evaluation of methods other than protective clothing to protect the worker
4. performance criteria for the protective clothing
5. selection criteria and procedures to determine the optimum choice
6. purchasing specifications for the protective clothing
7. a validation plan for the selection with medical surveillance, as appropriate
8. decontamination and reuse criteria, as applicable
9. a user training program
10. auditing plans to assure procedures are consistently followed

You will note that this program is somewhat parallel to that required for respiratory protection by OSHA; however, it is not an OSHA requirement.

General References

ACGIH. Dermal Absorption. In *Documentation of Threshold Limit Values and Biological Exposure Indices*; American Conference of Governmental Industrial Hygienists: Cincinnati, 1992.

ASTM. Test Method for Resistance of Protective Clothing Materials to Permeation by Liquids or Gases Under Conditions of Continuous Contact (Method F739); American Society for Testing and Materials: West Conshohocken, PA, 1991.

ASTM. Test Method for Resistance of Protective Clothing Materials to Permeation by Liquids or Gases Under Conditions of Intermittent Contact (Method F1383); American Society for Testing and Materials: West Conshohocken, PA, 1992.

ASTM. Test Method for Resistance of Protective Clothing Materials to Liquid Permeation—Permeation Cup Method (Method F1407); American Society for Testing and Materials: West Conshohocken, PA, 1995.

Brown, P. L. Protective Clothing for Health Care Workers: Liquid-proofness versus Microbiological Resistance. In *Performance of Protective Clothing*, STP 1133; American Society for Testing and Materials: West Conshohocken, PA, 1992; pp. 65–82.

Coletta, G. C.; Mansdorf, S. Z.; Berardinelli, S. P. Chemical Protective Clothing Test Method Development: Part II, Degradation Test Method. *Am. Ind. Hyg. Assoc. J. 41*, 26–33.

Colton, C. E.; Nelson, T. J. Respiratory Protection. In *The Occupational Environment—Its Evaluation and Control*; DiNardi, S. R., Ed.; American Industrial Hygiene Association: Fairfax, VA, 1997.

Davies, J. Conductive Clothing and Materials. In *Performance of Protective Clothing*, ASTM STP 989; American Society for Testing and Materials: West Conshohocken, PA, 1988; pp. 813–831.

Day, M. A Comparative Evaluation of Test Methods and Materials for Thermal Protective Performance. In *Performance of Protective Clothing*, ASTM STP 989; American Society for Testing and Materials: West Conshohocken, PA, 1988; pp. 108–120.

Forsberg, K.; Mansdorf, S. Z. *Quick Selection Guide to Chemical Protective Clothing*, 3rd ed.; Van Nostrand Reinhold: New York, 1997.

Furr, A. *Handbook of Laboratory Safety*, 3rd ed.; CRC Press: Boca Raton, FL, 1990.

Grandjean, P. *Skin Penetration: Hazardous Chemicals at Work*; Taylor and Francis: New York, 1990.

Henry, N.; Schlatter, N. Development of a Standard Method for Evaluating Chemical Protective Clothing to Permeation of Hazardous Liquids. *Am. Ind. Hyg. Assoc. J. 42*, 202–207.

Herrick, R. F., Ed. Personal Protection. In *Encyclopaedia of Occupational Health and Safety*, 4th ed.; Stellman, J. M., Ed.; International Labour Organization, 1997.

Johnson, J.; Anderson K., Eds. *Chemical Protective Clothing, Vol. II*; American Industrial Hygiene Association: Fairfax, VA, 1990.

Johnson, J.; Schwope, A.; Goydan, R.; Herman, D. *Guidelines for the Selection of Chemical Protective Clothing, 1991 Update*; Dept. Of Energy, Office of Environment, Safety and Health, National Technical Information Service: Springfield, 1992.

Johnson, J. S.; Stull, J. Measuring the Integrity of Totally Encapsulating Chemical Protective Suits. In *Performance of Protective Clothing*, ASTM STP 989; American Society for Testing and Materials: West Conshohocken, PA, 1988; pp. 525–534.

Linch, L. L. Protective Clothing. In *CRC Handbook of Laboratory Safety*; CRC Press: Boca Raton, FL, 1971; pp. 124–137.

Mansdorf, S. Z. Risk Assessment of Chemical Exposure Hazards in the Use of Protective Clothing—An Overview. In *Performance of Protective Clothing*, ASTM STP 900; American Society for Testing and Materials: Philadelphia, 1986; pp. 207–213.

Mansdorf, S. Z. Personal Protective Equipment. In *Complete Manual of Industrial Safety*; Prentice Hall: Englewood Cliffs, NJ, 1993; Chap. 12.

Mansdorf, S. Z. Industrial Hygiene Assessment for the Use of Protective Gloves. In *Protective Gloves for Occupational Use*; CRC Press: Boca Raton, FL, 1994.

Mansdorf, S. Z. Personal Protective Clothing. In *The Occupational Environment—Its Evaluation and Control*; DiNardi, S. R., Ed.; American Industrial Hygiene Association: Fairfax, VA, 1997.

Mansdorf, S. Z. Personal Protective Equipment. In *Applications and Computational Elements of Industrial Hygiene*; Stern, M., Mansdorf, S. Z., Eds.; CRC Press: Boca Raton, FL, 1998.

Mickelsen, R. L.; Hall, R. A Breakthrough Time Comparison of Nitrile and Neoprene Glove Materials Produced by Different Manufacturers. *Am. Ind. Hyg. Assoc. J.* **1985**, *48*, 941–947.

Mickelsen, R. L.; Roder, M.; Berardinelli, S. P. Permeation of Chemical Protective Clothing by Three Binary Solvent Mixtures. *Am. Ind. Hyg. Assoc. J.* **1986**, *47*, 189–194.

National Research Council. *Prudent Practices in the Laboratory*; National Academy Press: Washington, DC, 1995.

NIOSH. *Preventing Allergic Reactions to Natural Rubber Latex in the Workplace*, Publication 97-135; U.S. Department of Health and Human Services, Public Health Service, Centers for Disease Control, National Institute of Occupational Safety and Health: Atlanta, 1997.

NIOSH. *Proposed National Strategies for the Prevention of Leading Work-Related Diseases and Injuries*; U.S. Department of Health and Human Services, Public Health Service, Centers for Disease Control, National Institute of Occupational Safety and Health: Atlanta, 1988.

NIOSH. *Report to Congress on Workers' Home Contamination Conducted Under the Workers Family Protection Act*; DHHS, NIOSH: Cincinnati, 1995.

OSHA. General Industry Standards, 29 *CFR* 1910.132. Occupational Health and Safety Act of 1970 (84 Stat, 1593). Revised 1996.

OSHA. General Industry Standards, 29 *CFR* 1910.139. Occupational Health and Safety Act of 1970 (84 Stat, 1593). Revised 1997.

Perkins, J. L. Chemical Protective Clothing (Vol. I): Selection and Use. *J. Appl. Ind. Hyg.* **1987**, *2*, 222–230.

Perkins, J. L. Solvent-Polymer Interactions. In *Chemical Protective Clothing, Vol. I.* American Industrial Hygiene Association: Fairfax, VA, 1990.

Respiratory Protection. In *Fundamentals of Industrial Hygiene*; Plog, B., Niland, J., Quinlan, P., Eds.; 4th ed.; National Safety Council: Itasca, IL, 1996.

Respiratory Protection, A Manual and Guideline; Colton, C., Birkner, L., Brosseau, L., Eds.; 2nd ed.; American Industrial Hygiene Association: Fairfax, VA, 1991.

Sansone, E. B.; Tewori, Y. B. The Permeability of Laboratory Gloves to Selected Solvents. *Am. Ind. Hyg. Assoc. J.* **1978**, *39*, 169–174.

Slater, K. Comfort or Protection: The Clothing Dilemma. In *Performance of Protective Clothing*, 5th Vol., STP 1237; American Society for Testing and Materials: West Conshohocken, PA, 1996; pp. 486–497.

Stull, J. *PPE Made Easy: A Checklist Approach to Selecting and Using Personal Protective Equipment*; Government Institutes: Rockville, MD, 1998.

47

Incompatibles

LESLIE BRETHERICK

The dictionary defines incompatibles as "having opposing characteristics." In a chemical context this seems to imply that two or more reactive chemicals are essential for adverse effects to appear. Perversely, under some circumstances a single labile chemical may be able to decompose or react with itself to give similar effects of excessive energy release. The properly trained chemist knows (or will find out) how to control his chemical work so that such effects are not likely to occur, but if the unexpected happens, such effects will be contained or minimized by precautions.

This chapter tells why types of chemicals may react or interact adversely, how to foresee such problems, and where to find relevant details. It does not deal with imcompatibility of chemicals with people or with the environment. These topics appear elsewhere in this book. In the text below, the name of a single chemical a reference in parentheses, such as "(1/177)" following the name of a single chemical, or "(2/238)" after the name of a group of similar chemicals, gives the volume and page number(s) of the reference[1] wherein and whereon extensive detail will be found.

Factors for Self-Reactivity

Main factors influencing the stability of a chemical are its elemental composition, its structure and bond energies, and the conditions to which it is exposed. These are summarized here.

Labile Compounds

The properties of a particular chemical depend entirely on the number of elements present, and on the energy of the various bonds linking them together. Bond energies are expressed in terms of heat content, and most laboratory data handbooks have tables listing thermochemical data for a range of compounds. Most relevant here is the heat of formation (ΔH_f^0), given in kCal or kJ per mole. This shows the total heat content of a gram mole of a compound formed from its elements. The values are negative when the reaction is *exothermic*, but

when heat is absorbed in the reaction, the *endothermic* (energy-rich) product is assigned a positive value.

Endothermic compounds are thermodynamically unstable because their decomposition is exothermic and the heat accelerates the decomposition reaction. So except for (resonance-stabilized) benzene and toluene, any organic compound with a higher positive value for ΔH_f^0 is more likely than not to be unstable. Most but not all organic endothermics have been involved in violent decomposition incidents (2/119). Some typical endothermic structures can be identified in Table 47.1 by the + prefix in the ΔH_f^0 column, values being in kJ/g for ready comparison. Compounds in the table are classified on the basis of the elements present. Note that the energy of decomposition ($-\Delta U$) will be less than the ΔH_f^0 value if products other than the elements are formed on decomposition.

Exothermic compounds are thermodynamically stable, but they may be kinetically unstable under particular conditions. The usual approach adopted for judging the likely stability of an exothermic compound of known composition and structure is to relate it to known compounds of similar type. Table 47.1 also contains examples of exothermic compounds with their $-\Delta U$ values. Low stability in both types is often associated with structures containing mutiple bonding, small rings, high local or total proportions of nitrogen or oxygen, or with halogen to oxygen, or with nitrogen to sulfur bonds (2/129). It was proposed for industrial practice that open vessels (manhole-sized minimum vent) should not be used for exotherm potentials above 0.5 kJ/g of reactor contents. For closed vessels with pressure relief lines, a limit of 0.15 kJ/g was proposed.[2]

The proportion of oxygen in compounds containing it may have a large effect on stability. The effect is greatest when just enough oxygen is present to oxidize fully all the other elements that are present (except nitrogen and halogen). This is called zero balance point, and the potential for energy release is maximum (2/297). But the potential is still present to a considerable extent when a deficiency of oxygen (negative balance) exists: TNT has a −64 wt% balance. More detail on all these matters is in Stull[3] and Young.[4]

TABLE 47.1 Labile Structures and Energies

Elements, Compound Type	ΔH_f^0 (in kJ/g)	$-\Delta U$	Structure	Reference
C, H				
Alkyne				
Ethyne	+8.72		HC≡CH	(1/254)
Alkene				
Ethene	+2.18	−4.33[a]	$H_2C=CH_2$	(1/288)
Dienes				
Propadiene	+4.80		$H_2C=C=CH_2$	(1/401)
1,3-Butadiene	+2.07		$H_2C=CH-CH-CH=CH_2$	(1/502)
Akenyne				
1-Buten-3-yne	+2-76		$H_2C=CH-C≡CH$	(1/485)
C, H, N				
Diazoalkane				
Diazomethane	+4.58		$H_2C=N^+=N^-$	(1/157)
Nitriles				
Formonitrile	+4.83		HC≡N	(1/146)
Ethanedinitrile	+5.91		N≡CC≡N	(1/362)
Cyanamide	+1.40		N≡CNH_2	(1/156)
Arenediazonium salt				
Benzenediazonium chloride	−1.5		$C_6H_5-N^+≡N\ Cl^-$	(1/727)
Heterocycle				
Aziridine	−2.02		$\overline{CH_2NHCH_2}$	(1/321)
Tetrazole				
	+3.39		$N=NNHN=CH$	(1/159)
C, H, O				
Heterocycle				
Oxirane	−1.52		$H_2\overline{COCH_2}$	(1/306)
Peroxides				
Most low MW peroxides explore; few values.				(2/280)
C, H, N, O				
Nitro compounds				
Nitromethane	−3.92		H_3CNO_2	(1/175)
Nitrobenzene	−1.76		$C_6H_5NO_2$	(1/739)
Alkyl nitrates				
Methyl nitrate	−4.42		H_3CONO_2	(1/180)
H, N, O				
Hydrazine	+1.57		H_2NNH_2	(1/1672)
Nitrogen oxide	+3.01		NO	(1/1781)
Dinitrogen oxide	+1.85		N_2O	(1/1789)
O, X				
Chlorine dioxide	+1.52		ClO_2	(1/1397)
Perchloryl perchlorate	+1.45		$ClO_3^-\ ClO_4^+$	(1/1438)

[a]Determined at high temperature and pressure.

Figure 47.1 shows the effects of an unexpected detonation of a 4.5 M kg dump of the 2 : 1 double salt ammonium nitrate—ammonium sulfate. The salt, formerly used as a fertilizer, is, in fact, oxygen balanced (1/1678).

Redox Compounds

Another important cause of internal incompatibility is the simultaneous presence in the structure of reducing and oxidizing functions. Typical examples are the salts of a reducing base with an oxidizing acid, such as hydrazinium perchlorate (1/1368) and hydroxylaminium nitrate (1/1684). Their strongly exothermic gassing decomposition on ignition finds use in solid propellants.

The redox compounds formed by coordination of reducing bases to metal salts with oxidizing anions have the added possibility of (faster) metal-catalyzed decomposition of the complex. In dihydrazinecobalt(II) chlorate,

FIGURE 47.1 BASF Oppau factory site, September 1921. (Reproduced from *Chemical Health & Safety*, May/June 1995. Copyright American Chemical Society 1995.)

which explodes on the slightest impact or friction, the low-valent metal adds to the reducing function (1/1414). Redox compounds are rather numerous (2/350).

Monomers

So far, only decomposition reactions of labile compounds have been discussed, but monomers differ in that their polymerization reactions are synthetic reactions. Monomers are typically derivatives of an alkene or diene, often with a polar substituent group present. Initiation of a monomer by *in situ* generation of free radicals causes monomer molecules to react together exothermically to give long-chain polymeric species. These may have molecular weights exceeding 1 M (with tens or hundreds of thousands of monomer residues linked), so the total heat of reaction can be very large. This is also true if two or more monomers are being co-polymerized.

Increasing molecular weight of the polymerization mixture leads to an increase in its viscosity, which may in a concentrated solution lead to stirring and cooling problems. Measures to control polymerization reactions are therefore essential to prevent thermal runaway, which has led to fire and/or explosion on many occasions (2/324). Additionally, most monomers are subject

to autoxidation reactions and the peroxidic products may initiate polymerization unless careful attention is given to stabilization of monomers and to their storage conditions (2/308).

Mutual Reactivity

Air-Reactive Chemicals

A mixture of oxygen with about 4.5 volumes of nitrogen is the oxidant responsible for most fast reaction incidents (fires) involving mainly organic chemicals. With very reactive types such as alkylmetals (2/22), alkylphosphines (2/29), or some metal powders (2/344), ignition and fire occur after a few milliseconds delay (see Chapters 53 and 59).

With less reactive materials, an elevated temperature and/or ignition source may lead to fire (1/1937). Once a fire is established, the fast radical reactions involved in combustion must be interrupted to extinguish it. Choice of a preferred extinguishing agent must take account of various fuel incompatibilities.[5]

The very much slower reaction of air with a wide range of susceptible organic compounds to form hazardous peroxide species is well understood and has been reviewed comprehensively;[6] also see Chapter 52. Finally, remember that liquid air is a potent oxidant (24 vol% oxygen) for many organic substances, and that such oxidation reactions may become explosive (2/198).

Water-Reactive Chemicals

Along with air, water is the most common chemical likely to contact other chemicals either as vapor or liquid. There are several well-recognized groups of chemicals that are incompatible with water (and steam) owing to vigorous or violent reactions with it, especially if in restricted quantity. These include metals (groups IA and IIA, 2/14), metal hydrides (2/220), anhydrous metal halides of groups IIIB, IVA metals, halides of some group IVB, VB, and VIB elements, and some oxides thereof (2/264).

There is an unexpected incompatibility in aqueous solutions of low-valent transition metal chlorides in sealed containers at room temperature. Both chromium(II) and titanium(II) chlorides will slowly reduce the solvent water to liberate hydrogen gas, eventually bursting the bottles.[7]

Oxidant-Reactive Chemicals

Oxidation reactions are by far the most common and serious type to be involved in this context of incompatibility. The factor common to all oxidation reactions is that electrons are transferred from the substrate (material being oxidized, an electron source) to the oxidant, which functions as an electron sink. (Conversely, in reduction reactions electrons are transferred to the substrate from the reducing agent or electron source.) The mnemonic OIL RIG (Oxidation Is Loss, Reduction Is Gain) helps one recall electron flows in oxidation or reduction reactions.

Hazardous reactions or incidents have been recorded for some 170 individual oxidants and for some 70 groups of oxidants (2/287). Of these, nitric acid is the oxidant most frequently involved, incidents with nearly 200 different substrates or mixtures being detailed (1/1566–1601). This wide involvement of nitric acid is due to its ability to function as an oxidant in dilute or concentrated aqueous solutions and at normal or elevated temperatures. Oxidation is invariably accompanied by more or less gas evolution, usually capable of rupturing closed or near-closed reaction vessels.

Close behind nitric acid for hazardous involvement come perchloric acid (1/1352) and hydrogen peroxide (1/1624). Perchloric acid as a cold 70% aqueous solution behaves as a very strong but non-oxidizing acid, but when heated above 150°C, or when anhydrous, it be-

haves as an extreme oxidant to most susceptible materials, frequently leading to explosion. Its covalent derivatives, readily formed with alkanols or diols, are all treacherously explosive. Perchloric acid and its salts are best avoided, but any essential use commands great respect and care in handling (1/1352, 2/231, 305).

Hydrogen peroxide has the reputation of being a clean oxidant, as its major byproduct is water, but do not be misled into thinking it is gentle, too. Concentrated or "high-test" solutions contain up to 44 wt% of active oxygen and show a tendency with various organic compounds to form peroxidic species as intermediates or products. A good example is the explosive trimeric acetone peroxide, which when heated on a steel plate will perforate it (1/1052). Peroxy acids, derived from both organic and inorganic acids by replacing a hydroxyl by a hydroperoxy group, form an important group of powerful oxidants (2/312, 314).

Reduceant-Reactive Chemicals

The word "reduceant" is used here to mean reducing agent, and there are fewer powerful reduceants than there are oxidants. It appears in principle that there should be as many, for a reduction reaction seems just an oxidation viewed from the other end. In practice, not all reactions are reversible. A few groups of reduceants that may react with substrates vigorously or violently under unsuitable conditions include dissolving metals (hydrogen evolved), groups IA and IIA metals heated with higher group metal halides (1/1819) or nonmetal halides (1/1820), powdered aluminum or titanium heated or on impact with metal oxides (thermite reactions, 2/383), or metal hydrides or complex hydrides (2/74, 220).

Both hydrazine (1/1672) and hydroxylamine (1/1662) are powerful reduceants, and are extremely reactive or unstable when concentrated. All reactions of reduceants with known oxidants (redox reactions) tend to high-energy release (2/351), as evidenced by use of hydrazine as a rocket propellant fuel (2/355).

An extreme example of reduceant incompatibility exists for burning metals in a fire fighting context. Several common extinguishants contain combined oxygen (water 89%, carbon dioxide 72%, dry sand 53%) and are not inert to burning metals. In contact with these extinguishants, sodium, potassium, magnesium, titanium, or zirconium burn more fiercely (or explode) than in air. The three last-named metals also continue to burn in nitrogen, forming the nitrides. Powdered graphite will extinguish all five metals. Dry sodium chloride has been used to extinguish sodium fires, but if used on a lithium fire, the liberated sodium burns more fiercely. Halons are also unsuitable for metal fires. More relevant data are available.[8]

Other Reactant Pairs

Acid and base are an obvious pair with opposing characteristics, and if not diluted the neutralization exotherm may be very large. Syrupy orthophosphoric acid (90 wt% solution) is 50 N, 98% sulfuric acid is 36 N, and 50 w/v% sodium hydroxide solution is 19 N. All these also have a significant heat of dilution.

Developer and fixer residues from silver-based photography are often stored with a view to later recovery of the valuable silver content. Usual storage and recovery procedures run the risk of formation of explosive silver nitride, and many accidents have occurred. Use of zinc powder to precipitate metallic siver is a safer alternative (2/366).

Molecular sieves are widely used for drying gases and liquids, but they are selective adsorbents, and their intrapore exchangeable sodium ions are a hidden source of base in large-pore sieves. Drying nitromethane with the latter led to a violent explosion, ascribed to formation of explosive sodium *aci*-nitromethanide (1/179). Several other incidents with sieves are recorded (2/242). Many other unlikely events lurk within[1]; check there *before* you start your experiment; it could save your life!

Quantitative Pressure Effects

The general destructive effects of sudden exothermic decomposition of thermolabile compounds or mixtures have long been known, but quantitative studies of such reactions are more recent. Equipment was developed at Hoechst in Germany (2/338) to record the onset temperature (°C), maximum pressure attained (bar), and maximum rate of pressure rise (bar/s) for several compounds. Two compounds, azobenzene at 340°/13 bar/65 bar/s and 4-nitrosophenol at 120/23/5 were rated as low hazard. 4-Nitrophenol at 290/>199/1030 and 2-nitrobenzaldehyde at 200/945/8,700 were rated otherwise. The results for (oxygen balanced) ammonium dichromate at 227/510/68,000 explain why an attempt to dry a 1-ton batch of the salt in a heated vacuum drying oven led to demolition of the latter (1/1487). Other pressure effects of chemical reactions have been discussed.[7]

Unexpected Toxic Products

Any significant evolution of toxic volatile products from a reaction should be trapped or absorbed within a fume hood. But small amounts could be released during work-up of products or washup of equipment from contact of cleaning agents with residues. Avoid contact of:

- acids with bleach (Cl_2), azides (HN_3), cyanides (HCN), nitrites (nitrous fumes), or sulfides (H_2S)
- sulfuric acid with metal halides (HX), nitrates (NO_2), or formates or oxalates (CO)
- reducents with arsenic compounds (AsH_3), selenides (H_2Se), or tellurides (H_2Te); note that aluminum or galvanized vessels with non-neutral water content are reducing systems! (1/21)

Uncontrolled Incompatibles (Spills)

Murphy's law predicts that untoward events in a chemical context will occur when and where least expected, so be prepared in the chemical storeroom. Enough is known about incompatibles to ensure that accidental spillage or a shelf collapse will not lead to catastrophe. The two buzz words are *classification* and *segregation*, so that in the event of loss of containment, only compatibles are likely to make contact. Since the major hazard in a chemical store (or lab) is fire, it is logical to base the classification and segregation on fire-related criteria. A scheme capable of storing any likely chemicals and gas cylinders in up to eight separated areas, each with specific fire protection or ventilation, has been proposed.[9] Inventory control is also an essential requirement.

Information Sources and a Predictive Method

Additional to *Bretherick's Handbook*[1] are the *NFPA Manual 491M*,[10] *Prudent Practices*,[11] and Luxon,[12] and for emerging current information *CA Selects*[13] and *Laboratory Hazards Bulletin*[14] are reliable and comprehensive. Following an ASTM study of the characteristics, preparation, and use of chemical compatibility charts, an *ASTM Guide*[15] has been published.

When no factual information is available from literature resources, the ASTM PC-based CHETAH (CHEmical Thermodynamics And energy (Heat) release evaluation) program[16] can in most cases be used to calculate the heat-release potential for a given compound or reaction mixture and, based on several criteria, furnish an overall hazard rating (2/76).

References

1. *Bretherick's Handbook of Reactive Chemical Hazards*, 6th 2-vol. ed.; Urben, P. G., Ed.; Butterworth-Heinemann: Newton, MA, 1999 (also on disk/CD-ROM and online at www.bretherick. com).
2. Grewer, T.; Duch. E. *Proc. 4th Sympos. Loss Prev. Saf. Process Inds.*, Vol. 4; Rugby, IChE, 1983, A1–A11.

3. Stull, D. R. *Fundamentals of Fire and Explosion*, AIChE Monograph Series No. 10; AIChE: New York, 1977.
4. *Improving Safety in the Chemical Laboratory: A Practical Guide*, 2nd ed.; Young, J. A., Ed.; Wiley: New York, 1991.
5. *NFPA Principles of Fire Protection Chemistry*; National Fire Protection Association: Quincy, MA, 1989.
6. Kelly, R. J. *Chem. Health Saf.* **1996**, *3*(5), 28–36.
7. Bretherick, L. *Chem. Health Saf.* **1996**, *3*(3), 26.
8. *NFPA Fire Protection Guide to Hazardous Materials*; National Fire Protection Association: Quincy, MA, 1994.
9. *Safe Storage of Laboratory Chemicals*, 2nd ed.; Pipitone, D. A., Ed.; Wiley: New York, 1991; Chapter 4.
10. *NFPA 491M, Hazardous Chemical Reactions*; National Fire Protection Association: Quincy, MA, 1991.
11. *Prudent Practices in the Laboratory: Handling, and Disposal of Chemicals*; Committee on Prudent Practices, National Academy Press: Washington, DC, 1995.
12. *Hazards in the Chemical Laboratory*, 5th ed.; Luxon, S. G., Ed.; Royal Society of Chemistry: Cambridge, 1992.
13. *CA Selects: Chemical Heath and Safety*; ACS Chemical Abstracts Service: Columbus, OH; bimonthly (also online).
14. *Laboratory Hazards Bulletin*; Royal Society of Chemistry: Cambridge, UK; monthly (electronic versions also in *Chemical Safety NewsBase*).
15. *ASTM Guideline on Chemical Compatability Charts*; ASTM Committee E27.02: West Conshohocken, PA, due 1999.
16. *CHETAH version 7.2*; Committee E27.07, American Society for Testing and Materials: West Conshohocken, PA, 1998. An enhanced version is in preparation.

48

Corrosives and Irritants

JAY A. YOUNG

Both OSHA[1] and ANSI[2] define a corrosive substance as "[a] chemical that causes visible destruction of, or irreversible alterations in, living tissue by chemical action at the site of contact." The same references define an irritant as "[a] chemical, not a corrosive chemical, that causes a reversible inflammatory effect on living tissue at the site of contact."*

More to the point, corrosivity and irritability are determined by *in vitro* tests, typically those originally suggested by Draize[3,4] and adopted by various regulatory bodies.[5] However, note that whether or not determined by the Draize tests to be a corrosive or an irritant, a chemical that by chemical action at the site of contact causes visible destruction of, or an irreversible alterations in, or causes reversible inflammatory effects in, living tissue is a corrosive or irritant agent.

Examples

Solutions of all strong acids and bases at concentrations greater than 1 molar are corrosive. Be aware that in nonaqueous solvents, some chemicals not ordinarily considered to be a strong acid or base nevertheless are strong acids or bases in such solvents, but neither the label nor the MSDS for these substances will necessarily warn of this property.

Vapors, mists, and dusts (of the solid acids, bases, or their anhydrides) of all acids and bases should be considered corrosive or at least irritating to the delicate membranes of the respiratory, digestive, and other systems and organs (eyes and ears, e.g.).

For all corrosives and irritants, the degree of harm varies with the circumstances: the greater the concentration, the longer the duration before the agent is flushed off the skin; the higher the temperature, the larger the affected area of skin or eye, the greater is the degree of harm to the victim.

Ordinarily, weak acids and bases at any concentration and in any solvent can be considered to be contact irritants. However, there are exceptions, notably some weak acids such as hydrofluoric acid solutions at all concentrations greater than 0.01 molar and 100% (glacial) acetic acid are corrosive. Usually, but not always, when it is appropriate to do so, the label or the MSDS will warn that a weak acid or base is corrosive.

Some severe corrosives, particularly aqueous alkaline solutions, are insidious. They cause no immediate pain, and, depending on the concentration, no sensation may be felt for several hours. Aqueous solutions of hydroflu-

---

*Strictly, the irritant definitions are slightly different; the OSHA definition of an irritant violates Fowler's recommendations in that it confusingly uses "which" clauses in place of the definitory "that" usage. Further, the OSHA definition explicitly describes an irritant effect as due to chemical action at the site of contact.

oric acid that are less than approximately 20 molar (40%, wt/wt) are also insidious, remaining painless for as long as 24 hours after initial exposure.

Almost all organic liquids are irritants in that, penetrating through the skin, they dissolve the underlying fatty layer (dissolution is a chemical action) and consequently produce an inflammatory effect. Some organic liquids (e.g., phosgene) are corrosive. The vapors of most of the volatile organic compounds are irritants, and a few (phosgene vapors serve as an example) are corrosive. All but a few organic vapors also are toxic to a greater or lesser degree; see Chapter 20.

Two corrosives, phenol and aniline, require special mention. Collapse and death may occur a few minutes after massive skin exposure to phenol despite prompt emergency care. In addition, since it is a skin anesthetic, a victim may be unaware of phenol on the skin. Aniline does not cause pain but nevertheless is rapidly absorbed through the skin causing anoxia by the formation of methemoglobin.

Chemicals other than acids and bases can be either corrosive or irritating. Examples include both oxidizers (such as hydrogen peroxide, benzoyl peroxide, and the halogens) and reducing agents (such as metal hydrides and amides).

Identification

Obviously, the only sure way to identify a substance as a corrosive or irritant is to subject it to an *in vitro* test and assume that the skin, or eye, or mucous membrane of the test animal has the same susceptibility as one's own skin, eye, or mucous membrane. Since this is impractical, users must necessarily rely on their own, or others', experience or on the MSDS and/or label for such information.

Typical phrases in MSDSs and labels are "causes burns," "causes severe burns," "may cause burns," "may cause severe burns," "causes eye burns," "may cause blindness," or other similar statement for corrosives and "causes skin irritation," "may cause skin irritation," "causes eye irritation," "may cause eye irritation," or other similar statement for irritants.

Users should be aware that on some labels and in some MSDSs these phrases, although appropriate, may be absent or well hidden in small print.

Precautionary Measures

To prevent harm from exposure to corrosives and irritants, avoid contact. For eye protection, at least wear type G, H, or K safety goggles and for better protection also wear a Type N face shield.[6] Do not use a cream or lotion to protect the skin; wear impermeable gloves with gauntlets at least long enough to cover the wrist and lower forearm; and keep in mind that no glove is likely to remain indefinitely impermeable. See Chapter 46 for further information. Wear protective clothing. If ventilation is inadequate to protect respiratory systems from corrosive or irritant exposures, use a properly operating hood or local exhaust or, as a last resort and temporary remedy, wear a respirator (see Chapters 42 and 43).

Contact Lenses

Most authorities, but not all, caution against wearing contact lenses when working with corrosives, anticipating that a liquid corrosive can seep between the lens and eyeball by capillary action and that in the event of an accident, the delay in removing a contact lens from the eye could limit the effectiveness of first-aid treatment.

On the other hand, various authorities dispute the adverse effect of any delay, pointing to the absence of reliable information concerning the harm arising from any capillary action, and suggest that, in fact, the contact lens protects the cornea![7,8]

The American Chemical Society Committee on Chemical Safety recommends that in environments where exposure to corrosives and/or irritants is likely or possible, contact lenses be worn only for therapeutic (i.e., not cosmetic) reasons and that "When contact lenses must be worn, that fact must be made known so that in the event of a chemical splash into an eye, the lenses can be removed—as quickly as possible—before flushing out at an eyewash fountain."[9]

In this author's experience, few persons who have been splashed in the eyes by a corrosive chemical would have been emotionally or physically able to remove their own contact lenses, and it is rare in such an emergency to find other nearby persons with the knowledge and skill necessary to competently remove contact lenses from the eyes of a victim.

First Aid

In the event of exposure to a corrosive or irritant, the only acceptable and universally effective first-aid treatment is prompt and copious flushing with water continuously for at least 15 minutes followed by appropriate medical attention. See Chapter 30 ("Chemical First Aid") for further information.

References

1. OSHA Hazard Communication Standard, 29 *CFR* 1910.1200, Appendix A.
2. ANSI Z129.1-1994. Hazardous Industrial Chemicals—Precautionary Labeling; American National Standards Institute: New York, 1994.
3. Draize, J. H.; Woodward, G.; Calvary, H. O, *J. Pharm. and Exptl. Therapeutics* **1944**, *83*, 377–390.
4. Draize, J. H.; Woodward, G.; Calvary, H. O. *Appraisal of the Safety of Chemicals in Foods, Drugs, and Cosmetics*. Association of Food and Drug Officials of the United States: York, PA, 1959; 3rd printing 1975, pp 46–59.
5. 16 *CFR* 1500.41 and .42, 40 *CFR* 158.135 and 798 subparts B and E, and *Guidelines for Testing of Chemicals*, Sections 404–405; Organization for Economic Cooperation and Development: Paris, France, 1987.
6. ANSI Z87.1-1989, Practice for Occupational and Educational Eye and Face Protection. American National Standards Institute: New York, 1989.
7. Segal, E. B.; Cullen, A. P. *Chem. Health Saf.* **1995**, *2*(1), 16–24.
8. Segal, E. B.; Young, J. A. *Chem. Health Saf.* **1995**, *2*(4), 4.
9. *Safety in Academic Chemistry Laboratories*, 6th ed.; Committee on Chemical Safety, American Chemical Society: Washington, DC, 1995; p. 47.

49

Hazardous Catalysts in the Laboratory

FRANCIS P. DALY

Over the past several decades catalysis has had a significant impact on both the chemical and petroleum industries. In the second half of the twentieth century successes in catalysis have created new industries related to pollution abatement. Today catalysis-based chemical syntheses account for 60% of manufactured chemical products and 90% of current chemical processes.[1] It is estimated that by the year 2001, the total worldwide catalyst market will be $10.7 billion.

Catalysts are manufactured in a multitude of shapes and forms (e.g., monoliths) commonly used in pollution-abatement applications; tablets, extrudates, spheres, aggregates, and the like used in fixed-bed petroleum and chemical applications; and powders used predominately in stirred-tank reactors for the manufacture of specialty chemicals. Catalysts used in all these applications have their own special handling requirements. Such requirements can be designed to avoid exposure to hazardous dust[2] and toxic or poisonous chemicals; to reduce the risk of fire or explosion,[3] pollution of the environment, and so forth. Handling requirements are also dictated by the state of the catalyst, that is, fresh[4] versus spent[5] catalyst.

As fresh catalysts, supported precious metal catalysts (e.g., platinum, palladium, rhodium, ruthenium) and activated metal catalysts (e.g., nickel, cobalt, copper) used in a variety of commercial hydrogenation processes, can present significant hazards. Of greatest concern is the oxidation of combustible organic vapors, resulting in fire.

Such fires can be initiated via the heat generated by the air oxidation of a pyrophoric or self-heating catalyst.

Even fresh catalysts such as zinc oxide can be converted *in situ* to a pyrophoric compound (e.g., zinc sulfide), which if not handled properly can result in fire in the presence of a fuel source such as carbon. Nickel carbonyl poisoning has been reported when a nickel methanation catalyst was being cooled down and purged with nitrogen that contained a small amount of carbon monoxide.[3]

The intention of this chapter is to review some of the hazards associated with pyrophoric and self-heating catalysts and safe handling practices designed to reduce the associated risks. Considering the variety of catalysts being used commercially, it is prudent to consider all catalysts as posing some risk that can be magnified if not handled properly. Consequently, as a start in understanding safe catalyst handling practices, reference should be made to the catalyst's Material Safety Data Sheet.

Supported Precious Metal Catalysts

Precious metals, such as platinum, palladium, rhodium, ruthenium, and iridium, have long been recognized for their outstanding catalytic properties. At moderate temperatures, under reducing conditions, and in liquid medium, precious metals are of prime importance in the conversion of hydrocarbons. Their inability to form bulk carbides in the presence of hydrocarbons enhances their versatility.

Carbon-Supported Catalysts

As is well understood in catalysis, dispersing a metal on a support can provide a catalytic metal surface significantly greater than that available with the bulk metal. Typical supports include alumina, silica, and carbon. Carbon is most commonly used as a powder in hydrogenation applications. Such precious metal carbon-supported catalysts are usually sold as a water-wet solid with a moisture content typically in the range of 40–70%. In this form, the fire hazard of the catalyst is minimized. However, on storage in loosely sealed containers, wet catalysts can lose water by evaporation. In such cases the hazard associated with the catalyst can increase significantly. Therefore, it is suggested that in such cases the catalyst be handled as if it were a dry catalyst.

Some applications require the use of a dry catalyst whose form is most susceptible to rapid ignition of organic vapors. Dry platinum and palladium catalysts in the presence of air can easily ignite highly flammable alcohols such as methanol, ethanol, and propanol. Some ignitions can result in an explosion. Hydrocarbons heavier than air (e.g., propylene) can actually travel to a dry catalyst source, resulting in ignition and fire.

Handling Procedures

To avoid these hazards the following guidelines are recommended:

- Whenever possible, water-wet rather than dry catalysts should be used. A water-wet catalyst will reduce the risk of ignition.
- Prior to charging a reactor with catalyst, purge the reactor with nitrogen. Charge the catalyst into the reactor under a blanket of nitrogen. The absence of oxygen will eliminate any ignition potential.
- Both reactor and transfer vessels should be grounded to avoid ignition due to static discharge.
- Avoid drying used catalysts. Such catalysts may contain absorbed hydrogen and could ignite on drying.
- Used catalysts should be washed with water and kept water-wet and out of contact with combustible vapors or liquids.

After the reaction is complete and the contents of the reactor have been discharged, the reactor should be washed to ensure that no traces of catalyst remain on the walls. Such traces when dry could ignite hydrocarbons when added to the reactor in the presence of oxygen.

Raney and Activated Metal Catalysts

Raney and activated metal catalysts, commonly referred to as Raney-type catalysts, are prepared by leaching out a catalytically inactive metal from a coarse, powdered alloy of the active metal. The result of this process is a catalytic metal with a large surface area, a corresponding increased adsorption capacity to adsorb hydrogen, and adsorbed hydrogen itself (from the reaction with alkali used in the leaching process). Typical activated metal catalysts include nickel, cobalt, and copper. The catalysts are typically shipped under water at a pH >9.0.

Raney and activated metal catalysts are potentially pyrophoric if allowed to dry in air or any gas that contains oxygen. Reacting exothermically with oxygen to form the metal oxide or adsorbed hydrogen on the catalyst reacting exothermically with oxygen to form water, they quickly flare up to a red heat and will ignite combustible materials that come in contact with the catalyst.

Handling Procedures

To avoid potential hazards associated with these catalysts the following guidelines are provided:

- Personnel handling shipments of these catalysts should wear an easily removable coverall and flame-retardant boots. Goggles or face shields should also be worn.
- Some catalysts will contain carcinogenic compounds. Measures to avoid inhalation, ingestion, or skin absorption must be taken.
- Spilled catalyst should be immediately flushed from the area with water. Contaminated cleaning materials should be disposed of in a safe place where they cannot cause a fire.
- Small amounts of hydrogen may be evolved during storage. Open flames, smoking, and so on, must be avoided in areas where the catalyst containers are stored or are being opened.
- All catalyst washing and separation steps should be performed under inert gas conditions.
- Replace water with process solvent via repeated washing steps. If a solvent is desired that is not mixable with water, start with methanol or ethanol and replace alcohol with the required solvent (e.g., benzene, cyclohexane).

Typically, spent catalysts are still pyrophoric. For safe handling they may be rendered nonpyrophoric by oxidation under water. Oxidants such as sodium hypochlorite or sodium nitrite as well as air may be used. Before treatment the catalyst must be washed free of residual organic compounds and volatile solvents.

References

1. *Technology Vision 2020: The U.S. Chemical Industry*; American Chemical Society: Washington, DC, 1996.

2. Fulton, J. W. *Chem. Eng.* **1987**, 94(12), 99–101.
3. Roettger, K. D. *Ind. & Eng. Chem.* **1957**, 49(10), 1731–1733.
4. Reynolds, M. P.; Greenfield, H. *Chem. Ind.* **1996**, 68, 371–376.
5. Habermehl, R. *Chem. Eng. Prog.* **1988**, 84(2), 16–19.

50

Flammables and Combustibles

JAMES G. GALLUP

One important factor in evaluating the acceptability of a chemical for a specific application is whether a chemical will burn. A related element of the evaluation is how readily the chemical will ignite and the potential expected impact of the fire that is foreseen in the specific application. A similar evaluation is necessary if, rather than choosing a chemical for an application, the chemicals are already in use.

In either instance the analysis is useful as the basis for decisions in protecting personnel as well as the buildings and equipment from fire exposure.

In order to evaluate the fire properties of a chemical, a basic level of knowledge of the chemistry of fire is needed. Then the general means to control the fire hazards of the chemical can be explored. The remainder of this chapter follows this format. More exhaustive treatments are found in publications of the Factory Mutual Insurance Company,[1] the Industrial Risk Insurers,[2] and the NFPA.[3,4] See Table 50.1 for a list of applicable NFPA Standards.

Fire Chemistry

Definition of Burning

The chemical process of burning is a rapid oxidation, the combining of oxygen (or other oxidizer) with a material containing carbon or other combustible. An explosion could be viewed as very rapid oxidation.

In order for oxidation of a material to occur, the material normally must first be vaporized. The burning or oxidation then occurs in the vapor-air mixture above the material rather than at the surface of the liquid or solid. This concept is important in understanding the fire behavior of flammable and combustible chemicals. Solids seldom burn directly without first vaporizing. One common exception is the glowing form of burning such as the glowing of charcoal.

The rate of vaporization of a chemical can be increased by increasing the surface area exposed to the heat source. This is why a chemical such as kerosene may be difficult to ignite in a dish but is easily ignited as a spray using the same ignition source. The form of chemical can dictate whether the chemical will burn. Even materials which are normally considered noncombustible such as steel will burn if in the proper form, such as steel wool.

TABLE 50.1 Checklist for NFPA Standards

NFPA Standard	Applicability
NFPA 30	Flammable and Combustible Liquids Code
NFPA 30B	Manufacture and Storage of Aerosol Products
NFPA 33	Spray Application Using Flammable or Combustible Materials
NFPA 34	Dipping and Coating Processes Using Flammable or Combustible Liquids
NFPA 36	Solvent Extraction Plants
NFPA 45	Fire Protection for Laboratories Using Chemicals
NFPA 49	Hazardous Chemicals Data
NFPA 51B	Cutting and Welding Processes
NFPA 55	Compressed and Liquified Gases in Portable Cylinders
NFPA 68	Venting of Deflagrations
NFPA 69	Explosion Venting System
NFPA 70	National Electrical Code
NFPA 86	Ovens and Furnaces
NFPA 321	Basic Classification of Flammable and Combustible Liquids
NFPA 325	Fire Hazard Properties of Flammable Liquids, Gases, and Volatile Solvents
NFPA 430	Liquid Solid Oxidizers
NFPA 491M	Hazardous Chemical Reactions
NFPA 497A	Classification of Class I Hazardous Location for Electrical Installations in Chemical Process Areas
NFPA 497M	Classification of Gases, Vapors, and Dust for Electrical Equipment in Hazardous Locations
NFPA 704	Identification of the Fire Hazards of Materials

TABLE 50.2 Definition of Flammable and Combustible Liquids

General Classification	Specific Classification	Flash Point (°F)	Boiling Point (°F)	Vapor Pressure (PSIA)
Flammable Liquid	IA	<73	<100	<40
Flammable Liquid	IB	<73	>100	<40
Flammable Liquid	IC	>73 100		<40
Combustible Liquid	II	>100 <140		
Combustible Liquid	IIIA	>140 <200		
Combustible Liquid	IIIB	>200		

Flammable Liquid Terminology

Traditionally, flammable liquids have been loosely defined as liquid chemicals that give off a sufficient amount of vapors to burn at room temperature, when an ignition source is present. Other liquids that burn are termed combustible. Thus, gasoline is a flammable liquid; diesel fuel is a combustible liquid.

Table 50.2 breaks down flammable and combustible liquids into subcategories. Table 50.2 also provides a more rigorous definition of flammable liquids and combustible liquids.

A combustible liquid will act as a flammable liquid if the liquid is heated. Heating increases the rate of vapor generation. The lowest temperature under controlled laboratory conditions* at which a liquid gives off enough vapors to burn near the surface of that liquid is the flash point of the chemical. Table 50.3 lists the flash point of a few chemicals.

Flammable limits is an additional important factor in evaluating the hazards of flammable liquids. Given an ignition source, a small amount of flammable vapor mixed within a large volume of air will not ignite because the mixture is too lean. As more flammable vapor is added, the mixture will reach a point where it will ignite. As a percentage of vapor in air, this point is referred to as the lower flammable limit (sometimes called the lower explosive limit). If the mixture is enclosed, the mixture will explode at the lower flammable limit.

As more vapor is added to the mixture, the mixture remains flammable until the mixture is too rich in vapor to burn. This is known as the upper flammable limit (or upper explosive limit). Adjustment to the carburetor of an automobile keeps the mixture of gasoline and air between the upper and lower flammable limits.†

TABLE 50.3 Flash Point of Common Chemicals

Chemical	Flash Point (°F)
Gasoline	−50
Benzene	−4
Acetone	3
Methyl Ethyl Ketone	21
Ethyl Alcohol	54
Kerosene	110
Motor Oil	300

The upper and lower flammable limits vary with the ambient temperature. In general, an increase in the ambient temperature will lower the lower flammable limit and raise the upper flammable limit. If precise values are needed, the operating temperature must be specified. Table 50.4 shows the lower and upper flammable limits of four chemicals at 70°F.

Several other properties of flammable liquids and vapors from time to time are needed in order to evaluate the flammable hazards of the chemical. The density of the chemical will determine whether the liquid will float on water or sink. For example, most hydrocarbon liquids are less dense than water. This factor is important for fire fighting to determine whether the chemical will float on the surface of the suppression water and be spread during fire fighting efforts.

The density of the vapor is also often important. The vapor density is expressed as a ratio to the density of air,

---

*See for example, ANSI Z11.24-1979, ANSI Z11.7-1979, and ASTM D 3278-78 describing the Tagliabue determination, the Pensky-Martens procedure, and the Setaflash test for flash point, respectively.
†Typically, 7.6 and 1.4% by volume, respectively.

TABLE 50.4 Flammable Limits of Common Chemicals

Chemical	Lower Flammable Limit	Upper Flammable Limit
Benzene	1.3	7.1
Acetone	2.6	12.8
Ethyl Alcohol	3.3	19.0
Kerosene	0.7	5.0

with air being 1.0. The vapor density will determine if the vapor will rise from the surface of the source chemical and collect at upper levels or if it will sink in air and flow into low areas such as basements. For example, when the vapor and air are at the same temperature, all hydrocarbon vapors except methane are denser than air and will flow like a liquid into low areas. These features will have an impact on decisions regarding ignition source controls and ventilation.

Water miscibility has a large impact on fire-suppression planning for flammable liquids. If a flammable liquid is miscible with water, such as the lower-molecular-weight alcohols, the mixture of water and the chemical can remain flammable early in fire suppression. Although the mixture eventually will be diluted as more suppressing water is added, and thus the fire weakened, the mixture will continue to burn until the diluted mixture no longer produces enough vapors to burn.

The heat of combustion can be used to predict the impact of burning of a chemical or to analyze what occurred during a fire as a postmortem of the fire. For example, the heat of combustion of hydrocarbon flammable liquids are 2½ times as great as the heat of combustion of wood products. Heats of combustion are often expressed in terms of in MJ/kg.

A number of sources publish information on the flammability of chemicals. The sources include the NFPA[3,4] and Lewis.[5] Related safety issues are treated in the OSHA requirements.[6]

Boiling Liquid Expanding Vapor Explosion (BLEVE)

A final property of flammable and combustible liquids and gases is worth exploring. The property is the propensity to produce a Boiling Liquid Expanding Vapor Explosion (BLEVE). If a closed container of flammable liquid or combustible liquid or gas is heated, such as during a fire, some of the liquid inside boils and vaporizes and collects at the top of the container. The pressure inside the container increases due to the vaporized chemical. As the heating continues the pressure continues to increase.

If in addition the external heat source provides localized heating above the vapor/liquid level inside the tank, the container weakens at the point of heating. When the container strength is less than the vapor pressure, the container opens and the vapors are released in a rush and are ignited. The liquid boils and releases vapor quickly due to the decrease in pressure. An explosion occurs, often with serious destructive force. This type of explosion is referred to as a BLEVE.

Fire Spread Mechanisms

Fire will spread by three distinct heat transfer mechanisms, conduction, convection, and radiation. The two mechanisms most important to the fire safety of chemicals are convection and radiation.

Conduction is heat transfer by direct contact such as the means by which an electric stove element heats a sauce pan. Convection is heat transfer from the fire by means of air currents. Radiation is heat transfer in all directions by radiant energy such as the warmth one feels when sitting in front of a fireplace.

The spreading of fire by heat transfer is an important consideration when evaluating the fire safety of chemicals. Can the fire be conducted from one side of a wall to an unaffected area? Will convective forces spread the fire to the roof or floors above? Will the radiant energy spread the fire to combustible material adjacent to the fire?

Labeling

Usually, flammable and combustible liquids are labeled with the types of placards shown in Figures 50.1 and 50.2 in order to identify the broad category of chemicals and their hazards.

In Figure 50.2 the upper number designates the flammability hazard rating. The number one represents a low hazard rating, the number four represents a very high hazard rating.

Usually, information on the flammability of chemicals can be found in the material safety data sheets supplied by the manufacturer; also see NFPA 49[7] and NFPA 325.[8]

Flammable Liquids

Fire Control Overview

The chemistry of flammable liquids fires leads to several basic safety premises. The first is that whenever possible a noncombustible liquid should be substituted for a combustible or flammable liquid. Often substitution is impossible due to the properties of the chemicals that are needed. Sometimes substitution only appears to be impossible because the noncombustible product does not have the same solvent properties as the more flammable product. The noncombustible product may not work as well or may require additional human effort. The decision to use the more flammable product should include an analysis of the fire safety of the chemical.

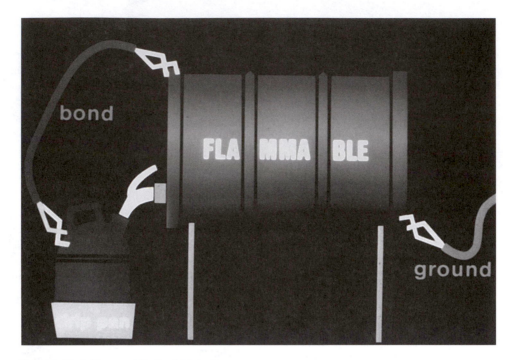

FIGURE 50.1 Flammable liquid label.

The second important safety premise is that when noncombustible substitution is not possible, a combustible chemical should be substituted for a flammable chemical wherever possible. A higher flash-point chemical is safer than a lower flash-point chemical.

A third premise is that when a flammable or combustible liquid must be used, limitations on quantities will limit the total fire exposure. This premise is widely used in all building codes and flammable liquid standards.

A fourth premise is that a small tight enclosure is more hazardous for a given quantity of flammable liquid use than a large open-air facility. A small tight enclosure will permit the vapors in the enclosed air to reach the lower flammable limit more rapidly than a larger open-air facility. An ignition source can ignite the vapors more readily in the smaller enclosure.

A fifth premise is that flammable liquids should not be stored or used in basements. Flammable vapors can col-

FIGURE 50.2 Typical hazard label.

lect at low levels and can be ignited by stray ignition sources.

Sixth, a precaution premise: It would seem at first thought that one way to prevent a fire or explosion of a flammable or combustible liquid vapor would be to so arrange conditions as to produce a vapor-air mixture that is above the upper flammable limit. Although it is true that such mixtures will neither burn nor explode, the only way to dispose of such mixtures is to first dilute them, whereupon the mixture must pass through the hazardous flammable range before it reaches the lower flammable limit.

Based on the density of hydrocarbon flammable liquids, water is sometimes viewed as an incompatible extinguishing agent because the liquid floats on water. This is a misconception. Water has tremendous cooling capabilities and in the form of mist, for example, is one of the finest extinguishing agents for flammable- or combustible-liquid fires. Water is readily available and is available in a continuous supply from municipal water systems. In liquid form, water will keep containers cool so that the container will not be as likely to BLEVE. Always, water must be applied so as to limit splash spreading of the fire, and water/oil or other resulting mixtures should be contained and controlled when using water to fight flammable-liquid and other fires.

Storage

Indoor versus Outdoor

Based strictly on fire chemistry and fire spread mechanisms, it is obvious that outdoor storage of flammable liquids is safer than indoor storage. With outdoor storage, vapors that may escape are carried away in the air. Ignition sources are fewer in number. Heat transfer from a fire to other stored materials is less likely as is heat transfer from fires of flammable liquids to the remainder of the building.

If a flammable chemical can be stored outside or in a detached facility, the consequences of a fire are less. This should be one of the first considerations. Although safest, outdoor detached storage is generally not practical nor is it economically feasible except for bulk storage. Environmental issues also limit the utility of outdoor storage.

Containers

Flammable and combustible liquids are shipped by truck or rail in bulk form or in containers. Bulk liquids are normally stored in tanks at the user's site. Refer to NFPA 30[4] for requirements for the unloading and storage facil-

ities. The larger industrial insurers also establish requirements for loading, unloading, and storage facilities.

The storage containers range in size from small laboratory size liter or smaller glass containers and safety cans to large portable metal totes that normally do not exceed 660 gallons. Container construction varies from glass to plastic, including steel, and fiber drums.

Each container type presents unique fire hazards. Glass is normally used in only the smallest containers due to the potential for breakage and release of the flammable liquid. Glass containers will resist fire spread during the early stages of a fire but will tend to break as the fire progresses. Once broken, the flammable liquid will contribute to fire growth.

Enclosed metal containers other than safety cans can BLEVE during fire exposure. BLEVEs are extremely hazardous to fire fighters. Recent advances in drum design include plastic bungs in the drums that can melt and relieve the excess pressure. The resulting fires tend to release more heat than the heat released from fires involving nonrelieving drums that BLEVE, but the overall storage is safer for fire fighters because the drums with the plastic plugs do not BLEVE.

Plastic containers tend to melt and burn early in a fire. The contents of the containers then feed the fire. Involvement of plastic containers can be slowed by packaging pallet loads of the plastic containers in corrugated enclosures.

Fiber drums are constructed of combustible material. The material burns more slowly than plastic and does not melt. The fiber drums restrict the involvement of the contents better than plastic containers.

Storage Rooms

In order to limit the total size of a fire, building codes and NFPA 30[4] limit the total quantity of flammable liquids that may be stored in any one location. Limiting the size of the fire also limits the fire from spreading to other properties or to other portions of the same property. The reader should reference the local codes for the maximum permissible quantities.

Except for remote detached flammable-liquids storage buildings, quantities of flammable and combustible liquids are stored properly only in rooms that are surrounded by fire walls. Openings in the walls are protected by fire doors and fire dampers. At least two exits are provided from the hazardous materials storage room. Generally the exits must be within 75 feet from any point in the room.

Spills are controlled by providing dikes to contain the spills. The floors are designed to be liquid-tight. As an alternative the floors may be sloped to drains which carry the spill to holding tanks. These holding tanks are normally designed large enough to hold not only the spilled liquid but also the sprinkler water. Sometimes the liquids are drained to oily water separators to separate the oil from the water mixture.

Storage rooms should have independent ventilation systems to exhaust flammable vapors from the rooms. The standards have specific requirements for the ratio of ventilation to room size. Since almost all flammable vapors are denser than air, the exhaust system should have intakes near the floor to pick up these vapors.

Fires are generally suppressed using an automatic sprinkler system. If materials are stored in racks, sprinklers also should be placed within the racks to cool the drums so as to reduce the potential for a BLEVE.

An alternative to standard automatic sprinklers is a foam water sprinkler system. Foam will suppress a fire quicker using less water. The traditional foams are not as environmentally friendly as standard sprinklers. Thus, the environmental impact of using fire fighting foam should be considered.

Manual fire fighting devices are normally provided for flammable-liquid storage facilities. Fire hoses and fire extinguishers rated for flammable-liquid hazards are used.

If flammable liquids are to be dispensed or transferred in the room, vapors will be generated during the transfer. Ventilation is needed to control the vapors. Controls are also necessary for electrical ignition sources.

Standard electrical equipment produces sparks in light switches and in electrical motors. The surface temperature of some high-wattage incandescent light bulbs is higher than the flash point of some flammable liquids. Lower wattage incandescent light bulbs that break can have sufficient residual energy to also ignite flammable liquids. Such electrical equipment is a potential ignition source in flammable-liquid storage rooms, particularly when vapors are present during dispensing. Only special explosion-resistant electrical equipment should be installed in the rooms to limit this ignition source. NFPA 30[4] and the National Electrical Code have detailed requirements for electrical equipment in flammable dispensing locations. The nearer to the source of the flammable vapors, the better the electrical equipment should be. The equipment nearest the source will be classified as Division 1. The next is Division 2 where fewer vapors are found.

The most flammable of the flammable liquids are classified as Class 1A and 1B liquids.[4] Vapors from these products are liberated freely in dispensing rooms. The potential for explosion is high. NFPA 30[4] requires explosion relief panels in the walls of the dispensing rooms to divert the force from the explosions and to reduce the

resulting damage to the structure. NFPA 68[9] specifies the design requirements for the explosion relief panels.

The design of flammable-liquid storage and dispensing rooms must eliminate ignition sources to the greatest extent possible. Smoking and open flames must be eliminated. Equipment with hot surfaces must be reduced. Friction sources and spark-producing equipment must be eliminated. The most common ignition sources in industrial occupancy arise from cutting and welding operations. These sources are eliminated by following the requirements of NFPA 51B[10] and those of the insurance company.

Flammable and Combustible Liquids— Use and Handling

Static potentials occur when dissimilar materials move past one another, such as during the pouring of a flammable liquid. A static discharge has sufficient energy to ignite flammable liquids. To eliminate this ignition source, the container of transfer should be firmly electrically bonded to the receiving container with a bonding wire. The system should be further firmly electrically grounded to the building grounding system using a grounding wire. Figure 50.3 shows the bonding and grounding of a flammable-liquid drum. The transfer of flammable liquids from drums or other large containers generally occurs in the flammable-liquid storage room

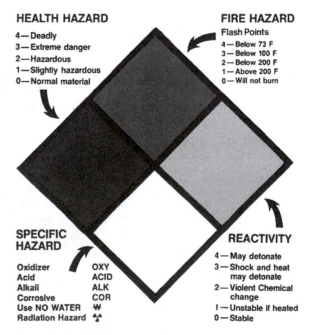

FIGURE 50.3 Bonding and grounding.

but may occur at the point of final use. The flammable-liquid storage room is often a safer location because ignition sources have been eliminated. In any event, the transfer-receiving system should be bonded and grounded.[6]

The safe transfer of flammable liquids generally involves hand pumps or self-closing valves to limit spills and vapor escape. Figures 50.4 and 50.5 show the two types of safe transfer equipment. When drums are placed on the side for dispensing, specially designed air vents are placed in the upper bung. The vents are designed not only to admit air during dispensing but will also relieve pressure during a fire exposure. Specially designed drip pans with flame arresters are placed under the self-closing nozzles. The flame arresters are a type of screen that limit fire transfer from the outside of a container into the container. Numerous other safety devices are available from suppliers for the safe use and handling of flammable liquids in specific applications.

FIGURE 50.4 Bonding of drum.

The location of ultimate use of flammable and combustible liquids is generally remote from the bulk storage rooms. Flammable liquid storage cabinets are employed at the location of use to store liquids in their containers before and after use. The cabinets are insulated to reduce fire exposure during the early stages of the fire. Figure 50.6 shows a flammable-liquid locker. The cabinets have ventilation ports to limit vapor buildup inside the cabinets. The ventilation ports should be kept closed unless the vents are hooked to a ventilation system or are vented to the outside.

NFPA 30[4] limits both the total quantities of materials that may be stored in any one cabinet and the number of cabinets that may be stored in any one fire area.

Whenever feasible, safety cans, as described in NFPA 30[4] and in the OSHA regulations[6] should be used to transport flammable liquids from bulk storage areas to the areas of final use. Safety cans are designed to limit flammable-liquids spillage and to provide for venting. The cans are also designed with flame arrestors. Figure 50.7 shows the cutaway of the safety can with a flame arrestor. The flame arrestors are a screen that prevents fire from outside the can from transferring to the inside by dissipating the heat from the exterior flame.

Because flammable liquids are often excellent solvents, flammable liquids are at times safely used in dip tanks with self-closing lids to safely wash parts. But flammable liquids should never be used in the open to clean parts, equipment, or floors. Such practices are not safe due to the nature of the flammable vapors. Use safety solvents instead of flammable liquids. These solvents have a higher flash point and are safer to use.

Generally, however, safety solvents require a higher level of human effort to complete the solvent tasks. An educational program is usually necessary to encourage workers to use the safety solvents.

Flammable Gases

Gases are products that would normally be found in gaseous form at normal atmospheric temperatures and pressures. These products are liquified and stored in liquid form through the use of high pressures. Containers are normally high-pressure tanks and bottles. The tanks and bottles can BLEVE during external fire exposure. Normally separation distances are required between storage structures or buildings and if the external exposures are great or the consequences of a BLEVE are high, the tanks are protected by fixed-water-spray systems.

Most flammable gases are heavier than air; very few are lighter than air. Fire protection systems, ventilation

FIGURE 50.5 Bonding and grounding of drums.

FIGURE 50.6 Flammable liquid cabinet.

FIGURE 50.7 Safety can cutaway.

systems, and ignition source controls should consider the nature of the gases.

Flammable Solids

Flammable solids are generally supplied in the form of coarse powders for industrial and laboratory usage. These powders are not as difficult to protect as flammable liquids because the solids do not produce vapors, are generally not stored in large quantities, and most do not flow when spilled.

The primary fire protection problem occurs when finely divided powders or dusts are suspended in air. An ignition source can then readily ignite combustible suspended solids causing them to explode violently. If combustible solids can be foreseeably suspended in air during processing, ignition source controls are needed, fire suppression is needed, and general explosion relief panels are necessary for the enclosure. Since finely divided powders, as distinct from coarse powders, do tend to flow and thus generate a static electrical charge, particular precautions to prevent the buildup of static charge should be employed when handling such powders. Refer to the MSDS for a finely divided powder and to NFPA 61A, 61B, 61C, and 61D* for further information.

---

*These four publications pertain to the prevention of fires and explosions for a variety of dusty agricultural products.

Flammables and Combustibles in Laboratories

Laboratories generally contain numerous flammable and combustible liquids, solids, and flammable gases. Many of the features described in this chapter apply to laboratories, including construction features, containment of spills, ventilation, storage limitations, bonding and grounding, and fire suppression. Refer to NFPA 45[11] for further information.

Special Products and Facilities

Aerosols

Aerosols present a special flammable-liquid problem. The containers are constructed of light metal and are prone to rapidly BLEVE during a fire. The cans become flying projectiles when they BLEVE. The danger to fire fighters is extreme. NFPA 30B[12] addresses storage of aerosols. The recommended storage includes chain-link barriers to limit the travel of the aerosol cans during BLEVEs.

Chemical Reactors

Medium- and larger-scale processes that involve chemical reactions are generally conducted in reactor vessels. The processes and vessels include safety practices and devices such as overpressure relief and containment. Refer to specific NFPA, Factory Mutual, or Industrial Risk Insurers requirements for specific types of reactors. These will include NFPA 36[13] for solvent extraction processes which are extremely hazardous.

Dipping/Coating/Ovens

Dipping and coating are two processes in which flammable paints are applied to other products. The processes are hazardous because flammable-liquid vapors are generated during the painting and drying processes. The drying of products is often completed in ovens, which liberate additional flammable vapors within the enclosure. NFPA 34[14] addresses dipping and drying processes.

Hazardous Materials and Storage Lockers

Prefabricated portable storage rooms or lockers have become available to store hazardous waste. These rooms have also been used to store flammable-liquid raw materials and finished manufactured flammable liquids when an adequate storage enclosure is not available. The storage lockers are generally fully self-contained and can be stored outside or inside a building. When used for the storage of flammable or combustible materials, the storage lockers should be provided with fire suppression and containment, proper ventilation, and spark-proof electrical equipment. If so equipped and constructed of fire-rated materials, the lockers can be located inside buildings and will meet most building code requirements for flammable liquid storage rooms.

Hydraulic Fluids

Hydraulic fluids are used in process equipment to operate mechanical functions. Often mineral oil is used because the lubricity values are attractive and the material does not degrade seals in the equipment. Mineral oil is a combustible liquid with a flash point of approximately 120°F. Under the high pressures needed for hydraulic equipment, a leak in the hydraulic system will create an atomized spray that is easily ignited by process heat sources or open flame.

The primary fire protection solution is to try to find a hydraulic fluid with a higher flash point. Fluids are also available with additives that reduce the ignition capabilities. Sometimes a water/oil emulsion is used as a hydraulic fluid.

Oxidizers

Oxidizers are chemicals such as oxygen, chlorine, perchlorates, and permanganates that support combustion but do not burn independently (see Chapter 51). Oxidizers can react violently with flammable and combustible materials.

Oxidizers should be stored separately from flammable and combustible materials in cool dry locations. Fire suppression systems should be provided in the storage areas. Since some oxidizers, for example, oleum, react exothermically with water, the selected fire-suppression system for such oxidizers should be nonaqueous.

FIGURE 50.8 Flammable liquid tank farm.

Spray Finishing

Spray finishing using flammable paints will liberate flammable vapors as well as create a flammable atmosphere of atomized paint droplets. The applicable NFPA standard is in NFPA 33.[15] Generally spray finishing is conducted in paint spray booths. The booths are constructed of noncombustible materials and are designed to be easily cleaned. The booths have ventilation with exhaust inlets that will maintain a flow of air through the spray area of the booth. The ventilation system generally includes a filter to remove paint particles from the exhaust stream. The filters are designed to be cleaned.

The electrical equipment in the booth will be listed for flammable atmosphere. The booth will have sprinklers in the spray areas as well as in the exhaust chambers and ducts where flammables may condense. The sprinklers will be controlled by a separate valve so that the booth sprinkler system may be shut down without shutting down the building sprinkler system.

Tank Farms

Tank farms are bulk flammable- and combustible-liquid storage facilities. Figure 50.8 shows a typical tank farm. Liquids are brought in by tanker truck or rail car. Transfer of liquids is generally conducted using loading/unloading racks. The racks have fire-suppression systems and have the ignition sources controlled. NFPA 30[4] addresses tank farms as well as loading and unloading facilities.

The installed electrical systems should be sparkproof. Tank spills should be contained by dikes and drained away from other tanks to reduce exposures. The tanks should have pressure relief vents. Many tanks are protected with foam systems for extinguishing purposes. Some tanks will have monitor nozzles from the water system that will spray the tanks for cooling purposes during fires or to cover the dike areas. Sometimes the monitor nozzles will use foam as the extinguishing agent.

References

1. *Loss Prevention Dam Sheets 7–29, Flammable Liquids In Drums and Small Containers*; Factory Mutual Insurance Company.
2. *Flammable and Combustible Liquids Storage*, IM. 10.2.4; Industrial Risk Insurers.
3. *Fire Protection Handbook*, 18th ed.; National Fire Protection Association: Quincy, MA, 1997.
4. *Flammable and Combustible Liquids Code*, NFPA 30; National Fire Protection Association: Quincy, MA, 1993.
5. Lewis, R. J. *Dangerous Properties of Industrial Materials*, 9th ed.; Van Nostrand Reinhold: New York, 1996.
6. OSHA, *Occupational Safety and Health Standards For General Industry*, Subpart H, 29 *CFR* 1910.106.
7. *Hazardous Chemicals Data*, NFPA 49; National Fire Protection Association: Quincy, MA, 1994.
8. *Fire Hazard Properties of Flammable Liquids, Gases, and Volatile Solvents*, NFPA 325; National Fire Protection Association: Quincy, MA, 1994.
9. *Venting of Deflagrations*, NFPA 68; National Fire Protection Association: Quincy, MA, 1994.
10. *Cutting and Welding Processes*, NFPA 51B; National Fire Protection Association: Quincy, MA, 1994.
11. *Fire Protection for Laboratories Using Chemicals*, NFPA 45; National Fire Protection Association: Quincy, MA, 1997,
12. *Manufacture and Storage of Aerosol Products*, NFPA 30B; National Fire Protection Association: Quincy, MA, 1994.
13. *Solvent Extraction Plants*, NFPA 36; National Fire Protection Association: Quincy, MA, 1993.
14. *Dipping and Coating Processes Using Flammable or Combustible Liquids*, NFPA 34; National Fire Protection Association: Quincy, MA, 1989.
15. *Spray Application Using Flammable or Combustible Materials*, NFPA 33; National Fire Protection Association: Quincy, MA, 1989.

51

Identifying Oxidizing and Reducing Agents

RUDY GERLACH

Reactions of active oxidizing agents with active reducing agents often are decidedly vigorous and can be dangerous. When the quantities are large and/or if either or both are strong oxidizing or reducing agents, the consequences can be disastrous. Any (male) chemist whose age in years is greater than the number of hairs on his head can recall from personal experience more than a few catastrophic accidents involving combinations of a few tens of pounds of strong oxidizers with strong reducers. Chapter 47 considers the incompatibility of oxidizing and reducing agents in some detail and Chapter 52 deals with a particular class of oxidizers, the peroxides. Here we concentrate on identifying elements and compounds as potential oxidizing and reducing agents.

Incomplete Lists

Indeed, lists of oxidizing and reducing agents could be used to identify these agents, but the lists necessarily would be incomplete. Nevertheless, see Table 51.1, our very own incomplete list.

TABLE 51.1 An Incomplete List of Common Oxidizing Agents

Solid	Liquid	Gas
azides	bromine	chlorine
bromates	chloric acid	fluorine
chlorates	chlorous acid	nitrogen tetraoxide
chlorites	hydrogen peroxide	nitrous oxide
chromates	hypochlorous acid	oxygen
dichromates	nitric acid	oxygen difluoride
hypochlorates	perchloric acid	ozone
hypochlorites	sulfuric acid	steam
iodates	water	
metallic peroxides		
nitrates		
nitrites		
organic peroxides		
perborates		
perchlorates		
permanganates		
persulfates		
picrates		

A similar list of common reducing agents would include several inorganic compounds and almost all organic compounds, and although also incomplete, it would be too lengthy to be useful.

Instead of relying on listings, we rely on the stability principle. In terms most useful to our purpose, the stability principle can be stated as: All elements tend to achieve the oxidation state in which they are the most stable under the conditions then existing.

For most elements under ordinary conditions, the most stable oxidation state (valence state) is either known to most readers or can be derived by reference to the Periodic Table.

The Periodic Table

Thus, under ordinary conditions the elements in each column of the periodic table have closely related stable oxidation states. For all the alkali metals, the stable oxidation state is +1, for every halogen, −1. For the chalcogens (Group VIA) although −2 is the stable oxidation state for oxygen, that of tellurium is best considered to be zero, with the other chalcogens perhaps −2, perhaps zero. Although the stable oxidation state of the alkaline earths can safely be assigned as +2, we cannot assign +3 to all of the Group IIIA elements, nor +4 to Group IVA, and certainly not +5, or −3 for that matter, to the Group VA elements. And, when it comes to the B Group elements the assignment of stable oxidation state numbers is less certain.

However, keeping in mind that the concept of oxidation state is arbitrary in the first place, and using a little common sense, it is possible to determine in many cases that a compound or element is an oxidizing agent or a reducing agent.

For example, to determine that an element is an oxidizing agent, refer to Figure 51.1 and compare the oxidation state of the element, zero, to the numbers assigned by custom to identify the oxidation states of the elements when they are combined.

Thus, is elemental iron an oxidizing agent? Since it is in the top row of Group VIIIB, when it reacts, its final

TABLE OF PERIODIC PROPERTIES OF THE ELEMENTS

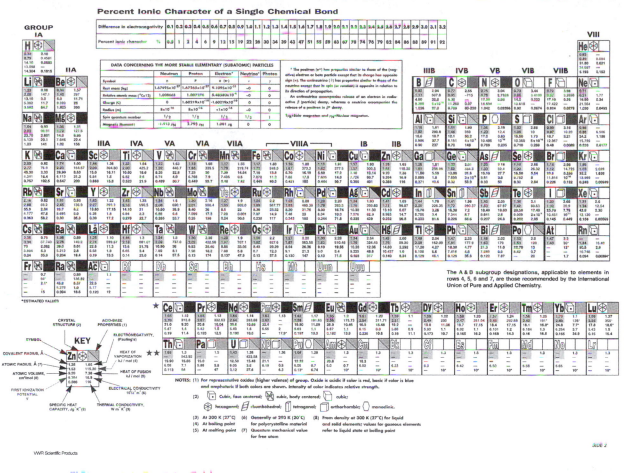

FIGURE 51.1 Periodic Table.

oxidation state will be either +2 or +3. Clearly, when it is in the oxidation state of zero, it can only be oxidized to the +2 or +3 state; it cannot be reduced. Therefore, elemental iron cannot be an oxidizing agent. Is elemental iron a reducing agent? Clearly, yes, since it can indeed itself be oxidized from zero to either +2 or +3. Could combined iron in the +3 (ferric) state be an oxidizing agent? Yes, because it can be reduced to +2 or zero. What about iron in the +2 state? It can go either way, oxidized to +3 in the presence of almost any oxidizing agent or reduced to zero in the presence of a sufficiently strong reducing agent.

Chromates contain chromium in a +6 oxidation state; since the +3 oxidation state for chromium is more stable, chromates are oxidizing agents. Manganese is more stable in the +2 or +4 state than in the +7 state; hence permanganates are oxidizers. Copper in the +1 state is a reducing agent and tends to disporportionate, to oxidize

and reduce itself so to speak, forming a mixture consisting of copper in the +2 state and elemental copper.

Similarly, elemental fluorine is an oxidizing agent, as are all the other halogens in the elemental state, readily achieving the −1 oxidation state. Indeed, one might conclude that because fluorine is the most active halogen, the other halogens would not be particularly strong oxidizing agents, compared to fluorine. The conclusion is correct, but beware: To conclude that elemental bromine, being a weaker oxidizing agent than fluorine, is relatively innocuous is decidedly incorrect. In the presence of a moderate to strong reducing agent, elemental bromine can be a tiger.

Stated differently, the application of the stability rule does not predict comparative strengths of agents as oxidizers and reducers. Therefore, until one's own personal experience demonstrates otherwise it is prudent to consider that any combination of an oxidizing agent and reducing agent is a combination of strong agents. Always take appropriate precautions, wear suitable protective clothing, and until it is certain that it is safe to do otherwise, mix very small amounts of reactants, a few milligrams preferably, heat cautiously, do not subject a mixture of oxidizer and reducer to frictional or impact shock.

Do not overlook the obvious: Water is an oxidizing agent since the hydrogen in water is readily reduced by alkali metals and the alkaline earths. Carbon dioxide is not as inert as some seem to think—the carbon in CO_2 is quickly reduced by aluminum and other active metallic reducing agents. At moderately high temperatures, gaseous nitrogen reacts readily with magnesium.

Prepare in advance: If the substance you intend to work with was purchased from a supplier, read the labels, refer to the Material Safety Data Sheet, remembering that MSDSs are not always reliable. Read the label. If these sources are not available or seem to be equivocal, refer to *Bretherick's Handbook*.[1,2] Since the information in Chemical Abstracts is available on any computer with a modem, you can easily determine whether the original literature describing the substances you intend to work with describes their oxidative and/or reductive potential.

Insidious Hazards

Finally, in the practice of chemical redox safety it is not sufficient to identify substances as oxidizing or reducing agents; it is also necessary to recognize their presence. For example, if a regulator has been inadvertently lubricated with oil, there is a resulting insidious hazard if that regulator is attached to a cylinder of compressed oxygen.

Although liquid nitrogen is generally inert itself, since it is colder than liquid oxygen under normal conditions, gaseous oxygen from the air condenses in the liquid nitrogen and soon enough one has a Dewar flask containing, instead of an inert liquid, a tiger at least as ferocious as liquid bromine. Liquid air presents the same insidious risk since from it the nitrogen evaporates more rapidly than the oxygen, resulting in an ever increasing concentration of oxygen in the remaining liquid.

Consider the storage of chemicals. Despite almost an overemphasis in the literature of chemical safety, in some places chemicals are still stored on the shelves in alphabetical order, with "Acetone" and "Acid, Perchloric" in close proximity. In some other facilities where it has been recognized that alphabetical storage is unsafe, the chemicals are stored by class; thus, all acids are stored together with the consequent proximity of bottles of nitric and glacial acetic acid.

Some spill clean-up materials such as sweeping compounds based on sawdust are reducing agents. They should not be used when cleaning up a spilled oxidizing agent.

Dusty atmospheres and the resulting deposits of dust are more likely in an industrial workplace than in, say, a laboratory. But since most dusts are reducing agents, oxidizing agents should be neither used nor handled in either location unless suitable precautions have been taken to eliminate or control possible dusty redox reactions.

In some cases the proximity of oxidizing agent to reducing agent cannot be avoided. The instability of ammonium nitrate, for example, is perhaps too well-known. There are several other compounds with similar propensities that need not be named here.[2]

References

1. Urben, P. G., Ed. *Bretherick's Handbook of Reactive Chemical Hazards*, Vol. 1, 5th ed.; Butterworth-Heinemann: Oxford, 1995.
2. Urben, P. G., Ed. *Bretherick's Handbook of Reactive Chemical Hazards*, Vol. 2, 5th ed.; Butterworth-Heinemann: Oxford, 1995.

52

Peroxidizable Organic Chemicals

RICHARD J. KELLY

Organic peroxides are carbon-based chemical compounds that contain the characteristic peroxide oxygen-oxygen bond. The primary types of organic peroxides are hydroperoxides (R—O—O—H) and diallyl peroxides (R—O—O—R′, where R and R′ are alkyl moieties). Several other types of peroxides exist, including acylperoxides, polyperoxides, per-oxyesters, alkylidene peroxides, peroxyacids, and cyclic peroxides.

Most organic peroxides are, to varying degrees, shock-, heat-, or friction-sensitive.[1] Shock, shaking, friction, or heating of the liquid may cause an explosion. Whereas the reactive hazard of organic peroxide reagents (i.e., purchased or synthesized organic peroxides) is usually well known, the accumulation of adventitious organic peroxides in common solvents may lead to an unrecognized explosion hazard.[2–4] Unexpected explosions involving peroxidized organic chemicals have been reported in the literature many times. Although peroxidizable chemical hazards are usually associated with labs, such chemicals are widely used in many types of industry and operations.

Peroxidation Chemistry

A wide variety of organic compounds spontaneously form peroxides by a free-radical reaction of the hydrocarbon with molecular oxygen in a process of autooxidation. The reaction can be initiated by light (photoperoxidation) or by a contaminant. Like other free-radical reactions, it is self-propagating, and one initiating event may form many peroxide molecules.[1]

Although ethers are the most notorious peroxide formers, other peroxidizable organic moieties include acetals, certain allylic alkenes (olefins), chloro- and fluoroalkenes, dienes, aldehydes, amides, lactams, ureas, some alkylarenes, ketones, vinyl monomers, and some alcohols. Figure 52.1 presents peroxidizable moieties in order of decreasing hazard.[2,3,5,6] Obviously, it is an extensive list, and not all chemicals that fall into these categories have been shown to form potentially dangerous peroxides. This is a part of the problem that must be addressed when developing a control program for peroxidizable chemicals.

Peroxidation is most likely to occur in compounds with activated hydrogen atoms in which the intermediate free radical can be stabilized through resonance or a similar mechanism.[1,6,7] For example, activated hydrogen atoms occur on carbon atoms linked by oxygen as ethers, on tertiary carbon atoms (e.g., the central carbon in isopropyl alcohol), on carbon atoms in vinyl structures, and in allylic and benzylic structures (i.e., on carbon atoms adjacent to vinyl or phenyl moieties). As a rule, a chemical containing more than one of these susceptible structures is at particular risk of peroxidation.[4] For example, vinyl groups that are further activated by the addition of an attached halogen atom, a phenyl or carbonyl moiety, or another unsaturation are very susceptible to peroxidation.[6]

There is some controversy about the hazardous peroxidizability of simple primary alcohols, which lack an activated hydrogen atom.[1,5,7–9] At present, most data do not suggest that such compounds, including ethanol, 1-propanol, 1-butanol, 1-pentanol, 1-heptanol, and 1-octanol, form hazardous peroxides. In contrast, the secondary alcohol 2-propanol (isopropanol), which possesses an activated hydrogen, will form concentrations of peroxides up to 4.20%, and 2-butanol will form a 12% peroxide solution. Explosions involving these peroxidized alcohols have caused serious injuries.[8–10] It is important to note, however, that some primary alcohols such as those with special structures that result in activation of hydrogen may be peroxidized. For example, 2-methylpropanol, 3-methylbutanol, 2-phenylethanol, and 2-ethylbutanol all have activated hydrogen, and all have been shown to be peroxidizable.[8]

Rate of Peroxidation

The rate of peroxidation is a function of the parent chemical. Of the common lab ethers, tetrahydrofuran (THF) probably has the highest rate of peroxide formation.[11] However, it is clear that the rate of peroxidation alone does not determine the degree of hazard posed by the chemical. A combination of factors, including the rate of peroxidation, the maximum peroxide concentration, volatility of the parent compound, and reactivity of the peroxides formed, determine the actual risk.

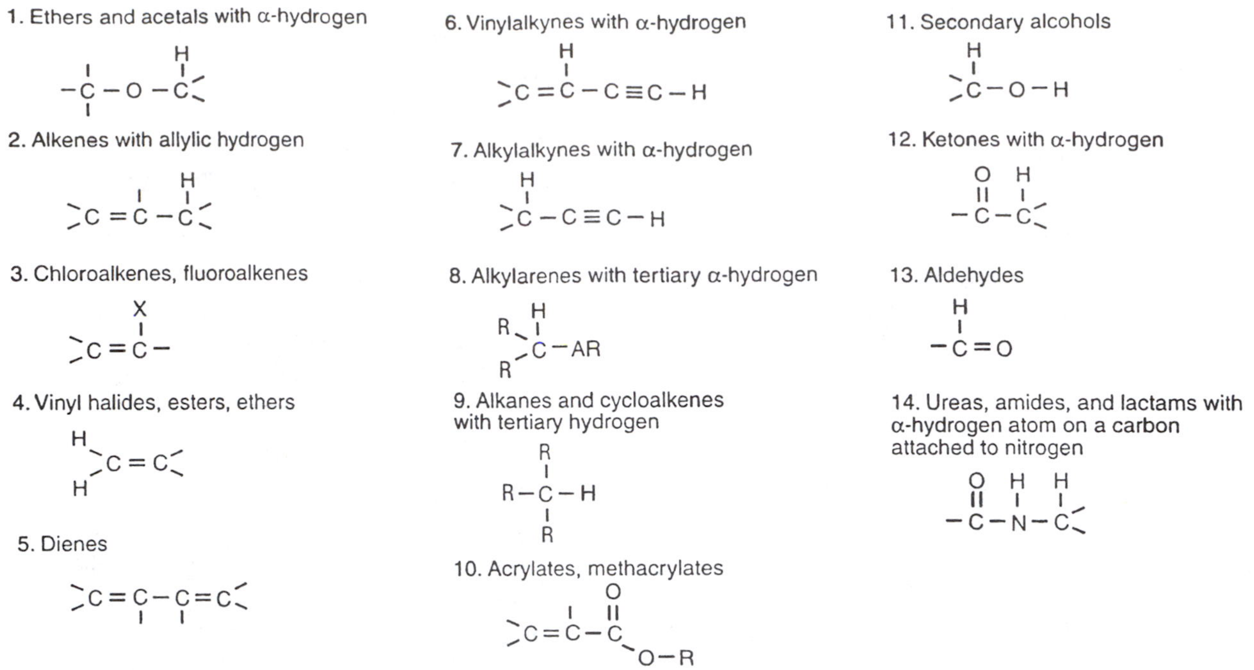

FIGURE 52.1 Peroxidizable organic moieties, numbered from most[1] to least[14] likely to form dangerous peroxides.

The risk of hazardous peroxidation generally decreases as the molecular weight of the compound increases. Ten or more carbon atoms at a peroxidizable site are considered low-risk systems.[3] Dialkyl ethers (R—O—R) are almost all susceptible to peroxidation, whereas diaryl ethers (AR—O—AR) are generally much less susceptible. The National Safety Council states that diaryl ethers with no hydrogen atoms on either oxygen-linked carbon atom need not be treated as peroxide formers.[3]

Peroxide accumulation is a balance between the formation of peroxide and degradation or further reaction of the peroxide. Peroxides are less volatile than the parent compound, and thus they concentrate as the parent chemical evaporates. Peroxidation may be accelerated by exposure to heat, light, and oxygen or air. Storing chemicals in open, partially empty, or transparent containers and at elevated temperatures may promote peroxidation. The effects of light and heat are somewhat variable and unpredictable, but the effect of exposure to oxygen is always to increase the formation of peroxides. Further, intentional or incidental contamination of peroxidizable alcohols with ketones may photosensitize the alcohol and exacerbate the effect of light exposure.[10,12] Similar interactions may occur that have not been documented.

Generally, it is recommended that volatile organic peroxidizables be refrigerated to slow oxidation. However, peroxide accumulation may actually be enhanced by refrigeration, as the rate of peroxide degradation is slowed more than is the rate of peroxide formation. This is known to be true for some peroxidizable organometallic compounds, which thus should not be refrigerated.[3] There is little or no evidence that refrigeration slows oxidation of diethyl ether, and the extreme volatility of this liquid means that a leak may form an explosive atmosphere, even in deep freeze.[6] Also, excess cooling, approaching the freezing point of the chemical, may cause the precipitation of peroxides from solution, which makes the container very shock-sensitive and dangerous. Thus, refrigeration of volatile peroxide formers is a double-edged sword. If they are to be so stored, only completely spark-proof refrigerators should be used to store volatile peroxide formers.

Secondary Reaction Products

Hydroperoxides are usually the initial product of autooxidation. However, over time, hydroperoxides may react further to form dialkyl, polymeric, cyclic, and other "higher" peroxides.[1] This result is important because these secondary reaction products are more difficult to detect and remove than simple hydroperoxides.[6,11] The common peroxide tests are only sensitive to hydroperoxides and may yield a false negative or low result when higher peroxides are present. Para-dioxane, in particular, has been reported to form very significant levels (more than 30% of total peroxide) of diperoxide products.[11] Isopropyl ether forms a variety of higher peroxides, including cyclic peroxides of acetone, which may be particularly explosive.[13]

Classes of Peroxide Formers

For some chemicals, the concentration of peroxide reaches dangerous, shock-sensitive levels without concentration. Other chemicals do not usually accumulate potentially explosive concentrations of peroxides unless the volatile organic material is reduced in volume. Such reduction may occur through incidental evaporation for very volatile compounds (e.g., diethyl ether) or through distillation.[2,3,8] Distillation of peroxide-containing organic compounds has resulted in many serious accidents.

Part A of Table 52.1[2,3] lists representative chemicals that may form explosive concentrations of peroxides without concentration by evaporation or distillation. Indeed, some of these may form explosive concentrations of peroxides even if never opened. Part B lists chemicals that accumulate peroxides but that usually become hazardous only if evaporated or distilled or otherwise treated to concentrate the peroxides. Note that the most common lab peroxidizable chemicals, including diethyl ether, THF, cyclohexene, the glycol ethers, and 2-propanol, usually require some concentration to generate hazardous concentrations of peroxides. However, many of these chemicals are quite volatile, and repeated small withdrawals from a container may allow enough evaporation to occur to concentrate peroxides to explosive levels. Part C lists certain peroxidizable vinyl monomers that may exothermically polymerize as a result of decomposition of accumulated peroxides; that is, the peroxides initiate very energetic polymerization of the bulk monomer. Part D lists other peroxidizable chemicals that have not been clearly characterized and thus have not been included in Parts A–C.

If properly packaged in manufacturing, peroxidizable pressurized gases such as butadiene, tetrafluoroethylene, vinyl acetylene, and vinyl chloride should be relatively resistant to autooxidation. However, these gases are sometimes transferred from the manufacturer's cylinder to another container in the lab, and it is difficult to completely eliminate residual air from the receiving vessel. An inhibitor should be placed in the receiving container before transfer is accomplished. The hazard of peroxidation may become much greater if these gases are condensed inside the cylinder or secondary vessel. All pro-

cesses involving these gases should be thoroughly evaluated to determine the likelihood of forming a liquid phase.[6]

Hazardous Levels of Peroxides

Various sources suggest that the minimum hazardous concentration of peroxides in solution in organic chemicals is in the range 0.005–1.0% (50–10,000 ppm) as hydrogen peroxide.[2,3,9,11,13] It has not been possible to reconcile this broad range (more than 2 orders of magnitude) because none of the authors cited provide any direct reference or data to support their statements. The high value (1.0%) is quoted from the National Safety Council,[3] but the council provides no references to support this statement and recommends an administrative control value of 100 ppm. A University of California at Davis chemical safety document states, "There is not even agreement as to what concentration of peroxides constitutes a hazard."[14] One reference states that, with respect to distillation of the easily oxidized isopropyl ether, "[t]he temperature and concentration at which explosion becomes probable has never been authoritatively stated. Even very small concentrations may be dangerous, since it is concentrated in the still system."[13] The Canadian Centre for Occupational Health MSDS for diethyl ether suggests that when the concentration of peroxide exceeds 100 ppm a hazard may arise if the solution is concentrated.[15] Presumably, as it concentrates it becomes more and more unstable. At some point, the solution spontaneously explodes.

Hazardous Peroxide Concentrations

In most peroxide safety programs, a concentration of 100 ppm of peroxides is used as a control point. However, this value has no scientific justification. This criterion seems to be based on the practical limit of detection of the potassium iodide method traditionally used to detect peroxides. It is likely that this value is at least 1 order of magnitude overly conservative in some cases.

Notably, this value is almost certainly quite overly conservative and excessively burdensome for the chemicals shown in Part B of Table 52.1 when they are not used in distillations or other similar processes. For example, it certainly seems excessive to limit the amount of peroxide in isopropanol used as a wipe cleaning compound to 100 ppm.

From a theoretical perspective, it should be impossible for most solutions of <1% peroxides to explode. However, the selection of a rational peroxide concentration control

TABLE 52.1 Classes of Peroxide-Forming Chemicals

A. Chemicals that form explosive levels of peroxides without concentration

Butadiene[a]	Divinylacetylene	Tetrafluoroethylene[a]	Vinylidene chloride
Chloroprene[a]	Isopropyl ether		

B. Chemicals that form explosive levels of peroxides on concentration

Acetal	Diacetylene	2-Hexanol	2-Phenylethanol
Acetaldehyde	Dicyclopentadiene	Methylacetylene	2-Propanol
Benzyl alcohol	Diethyl ether	3-Methyl-1-butanol	Tetrahydrofuran
2-Butanol	Diethylene glycol dimethyl	Methylcyclopentane	Tetrahydronaphthalene
Cumene	ether (diglyme)	Methyl isobutyl ketone	Vinyl ethers
Cyclohexanol	Dioxanes	4-Methyl-2-pentanol	Other secondary alcohols
2-Cyclohexen-1-ol	Ethylene glycol dimethyl ether	2-Pentanol	
Cyclohexene	(glyme)	4-Penten-1-ol	
Decahydronaphthalene	4-Heptanol	1-Phenylethanol	

C. Chemicals that may autopolymerize as a result of peroxide accumulation

Acrylic acid[b]	Chlorotrifluoroethylene	Vinyl acetate	Vinyladiene chloride
Acrylonitrile[b]	Methyl methacrylate[b]	Vinylacetylene	
Butadiene[c]	Styrene	Vinyl chloride	
Chloroprene[c]	Tetrafluoroethylene[c]	Vinylpyridine	

D. Chemicals that may form peroxides but cannot clearly be placed in sections A–C

Acrolein	tert-Butyl methyl ether	Di(1-propynyl) ether[f]	4-Methyl-2-pentanone
Allyl ether[d]	n-Butyl phenyl ether	Di(2-propynyl) ether	n-Methylphenetole
Allyl ethyl ether	n-Butyl vinyl ether	Di-n-propoxymethane[d]	2-Methyltetrahydrofuran
Allyl phenyl ether	Chloroacetaldehyde	1,2-Epoxy-3-isopropoxypropane[d]	3-Methoxy-1-butyl acetate
p-(n-Amyloxy)benzoyl chloride	diethylacetal[d]	1,2-Epoxy-3-phenoxypropane	2-Methoxyethanol
n-Amyl ether	2-Chlorobutadiene	p-Ethoxyacetophenone	3-Methoxyethyl acetate
Benzyl n-butyl ether[d]	1-(2-Chloroethoxy)-2-	1-(2-Ethoxyethoxy)ethyl acetate	2-Methoxyethyl vinyl ether
Benzyl ether[d]	phenoxyethane	2-Ethoxyethyl acetate	Methoxy-1,3,5,7-
Benzyl ethyl ether[d]	Chloroethylene	(2-Ethoxyethyl)-o-benzoyl	cyclooctatetraene
Benzyl methyl ether	Chloromethyl methyl ether[e]	benzoate	β-Methoxypropionitrile
Benzyl 1-napthyl ether[d]	β-Chlorophenetole	1-Ethoxynaphthalene	m-Nitrophenetole
1,2-Bis(2-chloroethoxy)ethane	o-Chlorophenetole	o,p-Ethoxyphenyl isocyanate	1-Octene
Bis(2 ethoxyethyl) ether	p-Chlorophenetole	1-Ethoxy-2-propyne	Oxybis(2-ethyl acetate)
Bis(2-(methoxyethoxy)ethyl]	Cyclooctene[d]	3-Ethoxyopropionitrile	Oxybis(2-ethyl benzoate)
ether	Cyclopropyl methyl ether	2-Ethylacrylaldehyde oxime	β,β-Oxydipropionitrile
Bis(2-chloroethyl) ether	Diallyl ether[d]	2-Ethylbutanol	1-Pentene
Bis(2-ethoxyethyl) adipate	p-Di-n-butoxybenzene	Ethyl β-ethoxypropionate	Phenoxyacetyl chloride
Bis(2-ethoxyethyl) phthalate	1,2-Dibenzyloxyethane[d]	2-Ethylhexanal	α-Phenoxypropionyl chloride
Bis(2-methoxyethyl) carbonate	p-Dibenzyloxybenzene[d]	Ethyl vinyl ether	Phenyl o-propyl ether
Bis(2-methoxyethyl) ether	1,2-Dichloroethyl ethyl ether	Furan	p-Phenylphenetone
Bis(2-methoxyethyl) phthalate	2,4-Dichlorophenetole	2,5-Hexadiyn-1-ol	p-Phenylphenetone
Bis(2-methoxymethyl) adipate	Diethoxymethane[d]	4,5-Hexadien-2-yn-1-ol	n-Propyl ether
Bis(2-n-butoxyethyl) phthalate	2,2-Diethoxypropane	n-Hexyl ether	n-Propyl isopropyl ether
Bis(2-phenoxyethyl) ether	Diethyl	o,p-Iodophenetole	Sodium 8,11,14-
Bis(4-chlorobutyl) ether	ethoxymethylenemalonate	Isoamyl benzyl ether[d]	eicosatetraenoate
Bis(chloromethyl) ether[e]	Diethyl fumarate[d]	Isoamyl ether[d]	Sodium ethoxyacetylide[f]
2-Bromomethyl ethyl ether	Diethyl acetal[d]	Isobutyl vinyl ether	Tetrahydropyran
β-Bromophenetole	Diethylketene[f]	Isophorone[d]	Triethylene glycol diacetate
o-Bromophenetole	m,o,p-Diethoxybenzene	β-Isopropoxypropionitrile[d]	Triethylene glycol dipropionate
p-Bromophenetole	1,2-Diethoxyethane	Isopropyl 2,4,5-	1,3,3-Trimethoxypropene[d]
3-Bromopropyl phenyl ether	Dimethoxymethane[d]	trichlorophenoxyacetate	1,1,2,3-Tetrachloro-1,3-
1,3-Butadiyne	1,1-Dimethoxyethane[d]	Limonene	butadiene
Buten-3-yne	Dimethylketene[f]	1,5-p-Methadiene	4-Vinyl cyclohexene
tert-Butyl ethyl ether	3,3-Dimethoxypropene	Methyl p-(n-amyloxy)benzoate	Vinylene carbonate
	2,4-Dinitrophenetole		Vinylidene chloride[d]
	1,3-Dioxepane[d]		

[a]When stored as a liquid monomer.

[b]Although these chemicals form peroxides, no explosions involving these monomers have been reported.

[c]When stored in liquid form, these chemicals form explosive levels of peroxides without concentration. They may also be stored as a gas in gas cylinders. When stored as a gas, these chemicals may autopolymerize as a result of peroxide accumulation.

[d]These chemicals easily form peroxides and should probably be considered under Part B.

[e]OSHA-regulated carcinogen.

[f]Extremely reactive and unstable compound.

Source: References 2 and 3.

value is complicated by several factors. For example, when peroxidized chemicals are dispensed from screw-cap bottles, some of the liquid may remain on the threads and cap. The liquid evaporates, leaving pure peroxide in the threads of the cap. Unscrewing the cap may initiate an explosion.[16,17] Thus, a solvent with modest peroxide contamination can explode because of peroxide concentration at the cap.

Dilute solutions of peroxidizable chemicals do not usually pose a peroxide hazard. For example, solutions of polyether nonionic surfactants usually are adequately dilute to prohibit the formation of dangerous concentrations of peroxides. Furthermore, they have a low volatility (boiling point above 300°C or a vapor pressure of <0.1 mmHg at 20°C) and are not likely to concentrate. In most cases, dilute solutions of low-volatility peroxidizables need not be treated as peroxidizables.[2,3]

Inhibitors

Many methods can be used to stabilize or inhibit peroxidizable organic chemicals and thus reduce oxidation. Use and storage under an inert atmosphere are helpful precautions and will greatly reduce peroxidation in most cases. Some manufacturers add hydroquinone, 2,6-di-tert-butyl-p-methylphenol (BHT), diphenylamine, or a similar compound to the chemical in trace quantities. Phenolic compounds are often added to commercial vinyl monomers. It is interesting that phenolic inhibitors are ineffective if some oxygen is not present; thus, these inhibited chemicals should not be stored under inert gas.[2]

Iron will inhibit the formation of peroxides in diethyl ether, which is one reason that this compound is usually sold in steel containers. However, iron or other metals will not inhibit isopropyl ether and are not known to be effective for other chemicals. In fact, iron may catalyze peroxidation in some chemicals. Despite the use of inhibitors, peroxide explosions have occurred. Inhibitors are only one part of a peroxide control program.

Inhibitors are depleted as peroxides are formed and degraded. Eventually, the inhibitor is totally depleted, and the peroxide-forming chemical will act as an uninhibited chemical. This may result in the rapid accumulation of peroxides in a chemical that has been stable for a long time. If inhibited chemicals are retained for extended periods, not only must the peroxide content be periodically evaluated, but the inhibitor level should be determined.[3] If the inhibitor is exhausted, new inhibitor must be added or the chemical must be treated as an uninhibited chemical.

Control of Peroxidizable Organic Chemicals

The control program set forth here represents common practice and generally accepted procedures. As noted above, some of these requirements may be excessively restrictive, reflecting the lack of a sound scientific basis. On the other hand, following these guidelines is no guarantee that an explosion will never occur.

Purchasing and Storage

Ideally, purchases of peroxidizable chemicals should be restricted to ensure that the chemical is used completely before it can become peroxidized. Table 52.2 shows some commonly recommended safe storage periods for peroxide formers.[2,3] It should be stressed that these are minimum criteria, and many authors suggest more frequent testing. In the past, the National Research Council recommended testing of chemicals in Part B of Table 52.1 every 6 months,[18] but it makes no specific recommendation in its most recent publication.[6] The ACS Committee on Chemical Safety recommends a blanket 3-month time limit for all peroxidizables unless tested and found to be peroxide-free.[19] The University of California at Davis places a 12-month limit on the storage of unopened containers and stipulates a maximum storage time of 3–6 months for opened containers.[14] Baumgarten[20] recommends that containers of uninhibited THF not be used for distillation 2–3 days after opening. The chemist or industrial hygienist should be alert for operations and uses of peroxidizable chemicals in which more frequent testing should be recommended, such as chemicals in frequent use or prolonged storage, use involving extensive evaporation or air exposure or light or heat exposure, and use in distillation without substantial bottoms.

Containers of peroxidizable chemicals must be dated when received and again when opened. Many commer-

TABLE 52.2 Safe Storage Period for Peroxide Formers[a]

Description	Period
Unopened chemicals from manufacturer	18 months
Opened containers	
Chemicals in Part A, Table 1	3 months
Chemicals in Parts B and D, Table 1	12 months
Uninhibited chemicals in Part C, Table 1	24 months
Inhibited chemicals in Part C, Table 1	12 months[b]

[a]Data from References 2 and 3.
[b]Do not store under inert atmosphere.

cially purchased ethers have a manufacturer's bottling date and expiration date. However, because many manufacturers do not put expiration dates on containers of other types of peroxide formers, the organization or person receiving the chemical must provide a receipt date. The person who opens the container should add an opening date. Figure 52.2 shows the label used at the Lawrence Livermore National Lab.

All peroxide formers should be stored in sealed, air-impermeable containers. In most cases, dark amber glass with a tight-fitting cap is appropriate. Diethyl ether should be stored in steel containers because the iron tends to neutralize peroxides. At least one manufacturer is now providing isopropanol in steel containers. Containers with loose-fitting lids and ground glass stoppers on glass bottles should not be used for storage. Generally, plastic containers should not be used, although an exception is often made for relatively short-term storage of isopropanol in plastic squeeze bottles smaller than 1 L. Experience has shown that such storage does not result in hazardous peroxide accumulation under normal use. However, 2-propanol that will be distilled should not be stored in plastic bottles. Chemicals that are listed in Part A of Table 52.1 should be stored under nitrogen if possible.

Surveillance

A responsible person should maintain an inventory of peroxidizable chemicals or annotate a general chemical inventory to indicate which chemicals are subject to peroxidation. By the expiration date, the person using the chemical or the person responsible for the chemical should either dispose of the chemical or test it for peroxide content using one of the procedures described below. Any chemical with a peroxide content >100 ppm should be disposed of or decontaminated.

Chemicals inhibited with low levels of antioxidants should be tested for antioxidant capacity (inhibitor level or function) if peroxide concentrations >25 ppm are detected during any test. If the inhibitor is exhausted, a new inhibitor of the type and concentration originally used by the manufacturer should be added, or the chemical should be treated as if it is uninhibited.

Old containers of peroxidizable chemicals, or containers of unknown age or history, must be handled very carefully. Any peroxidizable chemical with visible discoloration, crystallization, or liquid stratification should be treated as potentially explosive. If the container shows no outward sign of deterioration, it may be possible to test the chemical by using special precautions against agitation and providing extra protection to the tester. Typically, the unopened container should be carried to a nearby hood by a person wearing chemical goggles, a face shield, heavy gloves, ear muffs, and a thick wool or tough quilted cloth overgarment covering the body, arms, and legs from the chin to the wrists and ankles. A heavy rubber apron should be worn over the other protective clothing. In the hood, with secondary containment and using a blast shield, the person should test the chemical with the least possible disturbance, typi-

---

WARNING: MAY FORM EXPLOSIVE PEROXIDES

Store in tightly closed original container. Avoid exposure to light, air, and heat.
If crystals, discoloration, or layering are visible, contact your ES&H Team immediately.
Check for peroxides before distilling or concentrating.

THIS CHEMICAL HAS A LIMITED SHELF LIFE

Container received on _____ .   Container opened on _____ .

Test or dispose of _____ months after receipt or_____months after opening.
Do not use chemical if >100 ppm of peroxides are detected.

TESTING, DEPEROXIDATION, AND STABILIZATION RECORD

Test Date ____ Peroxides____Post-Treatment ____ Inhibitor Added ____

Test Date ____ Peroxides____Post-Treatment ____ Inhibitor Added ____

FIGURE 52.2 Label for peroxide forming chemicals.

cally by use of a calorimetric peroxide test strip. If high levels of peroxide are detected, the container should be handled as a potential bomb.

Use

Operations that result in evaporation of peroxide formers or extensive exposure to air or oxygen are particularly dangerous. Distillation is notable for the number of accidents it has caused. Any peroxide former that is to be distilled should be tested for peroxides before use, preferably immediately before use. The chemical should not be distilled if it contains >100 ppm of peroxides. Any distillation operation using peroxide formers should leave at least 10% bottoms, and 20% bottoms should be used until it is determined that the operation does not accumulate explosive levels of peroxides. If possible, add a nonvolatile organic compound such as mineral oil to the distillation. It will remain behind and dilute the remaining peroxides. During distillation, the solution must be stirred with a magnetic stirrer or an inert gas bleed. Air or other oxygen-containing mixtures should never be used to maintain mixing during distillation.

The peroxides in higher-boiling-point chemicals are usually degraded during distillation. Long-chain alkyl ethers and the glycol ethers are such chemicals. However, if they are distilled at reduced pressure, reducing the temperature of the process, decomposition may not occur and unstable levels of peroxide concentrations may result. Uninhibited chemicals listed in Part C of Table 52.1 should not be distilled.

Peroxide Detection Methods

Although there are numerous quantitative, semiquantitative, and qualitative methods to detect peroxides in organic and aqueous solutions, four are commonly used. They include two qualitative variations on the iodine detection method, the qualitative ferrous thiocyanate method, and the use of semiquantitative redox dip strips. Recently, the use of titanium sulfate has been suggested as a means to detect peroxides,[21] but it is not widely used.

The dip strip method has the advantage of being the most gentle test, an important consideration if the chemical is shock-sensitive. It also has another substantial advantage: It can detect, to some extent, dialkyl peroxides, polyperoxides, and cyclic peroxides, compounds that are not efficiently detected by the other methods (except, perhaps, the titanium sulfate method). Some solvents, notably isopropyl ether and dioxane, may form significant and hazardous levels of these higher reaction prod-

ucts. Furthermore, the standard peroxide removal procedures may remove all of the hydroperoxides but leave behind dangerous levels of alkyl peroxides, polyperoxides, and cyclic peroxides.[1,11] The routine ferrous thiocyanate and iodine methods may yield a false negative in this case, but the dip strip test would likely detect the remaining peroxides, although perhaps not quantitatively. The dip strip method, however, is difficult to use with water-immiscible, low-volatility chemicals.

Additional tests include the perchromate test and the mercury test. A comparatively complex reflux method is available for the detection of total peroxides, including alkyl peroxides and polyperoxides.[4,22]

Ferrous Thiocyanate Method

This qualitative method relies on the oxidation by peroxide of colorless ferrothiocyanate ($Fe2^+$) to the red ferrithiocyanate ($Fe3^+$).[2,3] One drop of reagent is added to 1 drop of the chemical to be tested. A barely discernible pink color indicates that peroxides are present at a concentration of about 10 ppm. A clear pink to cherry color suggests a concentration of about 20 ppm. A red color indicates a concentration of about 80 ppm, and a deep red indicates a concentration as high as 400 ppm.

The reagent is prepared by dissolving 9 g $FeSO_4 \cdot 7H_2O$ in 50 mL 18% HCl. A little granulated zinc is added, followed by 5 g NaSCN. When the transient red color fades, an additional 12 g NaSCN is added, and the liquid is decanted from the unused zinc into a clean, stoppered bottle. The shelf life of this reagent is very limited.

This method is sensitive only to hydroperoxides; it will not detect most other peroxides. However, this method is very sensitive and rapid for the detection of hydroperoxides, and a quantitative photometric version of this test has been developed.[23]

Iodide Tests

This is the most common qualitative test for adventitious peroxides. The two primary variations on this procedure involve the oxidation of iodide to iodine by the peroxide, with the resulting formation of a yellow to brown color. ASTM has published a version of this method,[24] and commercial test kits using this principle are available from many sources.

Method A

In this method, 1 mL of the material to be tested is added to an equal volume of glacial acetic acid, to which approximately 0.1 g NaI or KI (10% wt/vol) has been added. A yellow color indicates a low concentration of peroxides (40–100 ppm as hydrogen peroxide). A brown color indicates a higher concentration of peroxide. Blanks must always be prepared. The test solution has a very short shelf life and will naturally result in high blank values if stored for any length of time. Variations on this method include the use of a 20% wt/vol reagent and hydrochloric acid instead of glacial acetic acid[11] and the use of sulfuric acid.[13] Heating is recommended to enhance detection of higher peroxides that may not be detected by the routine process.[4]

In quantitative variations on this method, the liberated iodine is titrated with 0.1 M NaSCN.[4] This degree of precision is usually unnecessary for routine testing, and these methods do not increase the sensitivity of the method to higher peroxides.

Method B

In this method, 1 mL of a freshly prepared 10% KI solution is added to 10 mL of the organic liquid in a 25-mL glass-stoppered vial. A barely visible yellow color suggests a peroxide content of 10–50 ppm. A clear and definite yellow color indicates a concentration of about loo ppm, and a brown color indicates a higher concentration. Variations on this method include using a 20% KI solution and using a 1:1 solvent/reagent ratio.[8] Some authors recommend vigorous shaking or mixing.

Comparison of Methods A and B

Method A is often faster than Method B. Up to 15 min may be required for formation of color in Method B, whereas the color usually forms in <1 min for Method A.[2,3] For both procedures, the color formed is a function of the peroxide content and the chemical tested.

Some scientists disagree about the sensitivity of these methods to peroxides other than hydroperoxides. Burfield[11] states that the method is sensitive only to hydroperoxides. Noller and Bolton[4] indicate that the acidified Method A is sensitive to hydroperoxides, peroxyacids, diacylperoxides, and some peroxide esters, but not to dialkyl and alkyldiene peroxides. Davies[1] makes a similar statement. Noller and Bolton[4] suggest that heating, perhaps with the addition of hydroiodic acid, may be necessary to detect these compounds. Mair and Graupner[22] use a combination of glacial acetic acid and hydrochloric acid plus heat to detect all peroxides. These latter two procedures are complex reflux processes that require extensive experience in chemistry and substantial lab facilities.

Dip Strips

E. Merck and Aldrich Chemical make dip strips for the semiquantitative detection of peroxides in organic and inorganic solutions, and they are available from many suppliers. The strip incorporates the enzyme peroxidase, which transfers oxygen from peroxide to an incorporated organic redox indicator. The indicator turns blue in the presence of peroxides. Comparison color scales are provided for organic and aqueous tests. The range for organic chemicals varies, depending on the manufacturer: 0–25 ppm for the Merck product and 1–100 ppm for Aldrich. Higher ranges can be accommodated by dilution of the suspect chemical with a miscible, nonperoxidized chemical.

For volatile organic chemicals, the test strip is immersed in the chemical for 1 s; then the tester breathes slowly on the test strip for 15–30 s, or until the color stabilizes. Vapor in the breath provides water for the reaction to proceed. The color is then compared with the scale provided on the bottle.

Modifications of this procedure are needed to test nonvolatile organic compounds. The Aldrich strips have been successfully used as follows[8]: For water-miscible compounds, add 3 drops of water to 1 drop of chemical to be tested. Wet the dip strip in the mixture, wait 2–3 min or until the color stabilizes, and multiply the result by 4. For water-immiscible compounds, mix 3 drops of a volatile ether with 1 drop of the low-volatility compound to be tested. Wet the dip strip and breathe on the reaction zone for 30–60 s, or until the color stabilizes, and multiply the measured value by 4.

Bottles of 100 strips cost about $45. The article by Mirafzal and Baumgarten[8] compared the test strips (Aldrich version) with traditional methods and found good agreement. These strips have a limited shelf life, but refrigeration is not recommended once the container has been opened, because repeated cooling and heating cause condensation that will ruin the strips. Storage under a dry, inert atmosphere will prolong the shelf life.

Literature from E. Merck indicates that their test strips will detect hydroperoxides and most higher peroxides, but some polyperoxides may be poorly detected.[25]

Titanium Sulfate

Historically, the reaction between titanium and peroxide has been used for the detection of titanium[26] and the detection of organic and inorganic peroxides.[27,28] To detect organic peroxides, a solution of titanium sulfate in 50% sulfuric acid is used. When this reagent is added to a peroxidized solvent, a yellow-orange complex is formed. It has been stated that this method will detect higher peroxides, especially polyperoxides.[21,29] The higher peroxides are hydrolyzed by the strong acid and are thereby made detectable.

The test reagent is prepared by dissolving a small

amount of TiO_2 in hot concentrated sulfuric acid and adding this to an equal volume of water. Although it is a very old procedure, there is not a lot of modern experience or literature that discusses this method in any detail. Little seems to be readily available on the limit of detection of the method, stability of the reagent, specific procedures for its use, or the sensitivity to higher peroxides other than polyperoxides. As discussed previously, strongly acidified versions of more traditional methods have been outlined,[4,22] and it is unclear what advantage the titanium sulfate procedure has relative to these procedures. This method can only be recommended if it is compared by the user with other methods of known sensitivity and used accordingly.

Summary of Detection Methods

Dip strips provide the highest sensitivity and the most accurate quantitation of peroxide concentrations for routine testing. Furthermore, they are easier, faster, and safer to use than other methods, and they detect a wider range of peroxides than do other simple methods. They are, however, somewhat inconvenient to use when testing nonvolatile solvents, and they have a limited shelf life after the container is opened.

If it is important to detect all peroxides, including alkyl peroxides and polyperoxides, the acid reflux method should be used by a knowledgeable chemist in a lab hood. The titanium sulfate method may prove to be equivalent, but supporting data on this procedure are limited. Where precision is desired, the quantitative versions of the potassium iodide or ferrous thiocyanide tests may be used. If absolute precision and accuracy are required, the potassium iodide-acid reflux combination method should be considered.

Removal of Peroxides

In some cases, it might be desirable to remove peroxides from chemicals rather than dispose of the entire chemical. This can be done safely for relatively low levels of peroxides (<500–1000 ppm). Scrubbing of concentrations >1000 ppm may pose an unacceptable hazard, depending on the chemical involved. Scrubbing of discolored, crystallized, or layered peroxide formers is almost certainly too hazardous and should not be attempted. These severely peroxidized compounds should be treated as potential bombs.

Many methods and materials have been used to remove peroxides: activated alumina, ferrous salts, assorted amines, sodium metabisulfate, potassium permanganate, silver hydroxide, stannous chloride, sodium

hydroxide, cerous hydroxide, sodium, lithium, lithium aluminum hydride, sodium perborate, hydroquinone, ion-exchange resins, copper chloride, indicating molecular sieves, triphenylphosphene, lead oxide, acidified zinc, iron, mercury, sodium carbonate, ammonia, glacial acetic acid, and phosphoric acid esters.

All methods have some drawbacks. Some are potentially explosive, such as those involving the use of alkali metals, alkali metal hydrides, sodium borohydride, sodium hydroxide, and reflux over cuprous chloride or stannous chloride. Some methods entail heating, refluxing, or other operations that require special expertise and equipment. These include refluxing with indicating-activated molecular sieves; refluxing with hydrochloric acid, potassium iodide, and acetic acid; and refluxing with cuprous chloride or stannous chloride. Some methods, such as the use of mercury, ion-exchange resins, and molecular sieves, generate undesirable waste products. Others, including the use of ferrous sulfate or sodium bisulfite solutions, tend to introduce water into the solvent, which may need to be removed. Finally, many methods, such as the use of indicating molecular sieves, cuprous chloride, and ferrous sulfate, fail to remove dialkyl or higher peroxides.

Method 1

Hydroperoxides can be removed by passing the solvent through a column of activated alumina.[7] This method works for water-soluble and water-insoluble chemicals. The washed solvent should be retested to ensure that it has been cleaned adequately. The alumina apparently catalyzes the degradation of some peroxides, but in some cases the peroxide may remain intact on the alumina, making it potentially shock-sensitive. The alumina can be deactivated by flushing with a dilute acid solution of potassium iodide or ferrous sulfate.

The amount of alumina required depends on the quantity of peroxide. As a start, a column containing 100 g of alumina should be used for 100 mL of solvent.[8] More alumina or passage through a second column may be required to eliminate peroxides. This method is relatively slow and expensive, but it avoids shaking the solvent and does not add water. It will not reliably remove dialkyl peroxides, although there is some controversy about this.[1]

Method 2

Peroxides in water-insoluble chemicals can be removed by shaking with a concentrated solution of ferrous salt[2]; 60 g $FeSO_4$, 6 mL concentrated H_2SO_4, and 110 mL H_2O are a standard solution. Another formulation is 100 g

FeSO$_4$, 42 mL concentrated HCl, and 85 mL H$_2$O. The peroxide former is extracted two to three times with an equal volume of the reagent. Drying over sodium or magnesium sulfate can be used to remove dissolved water. Shaking should be very gentle for the first extraction. This method has been shown repeatedly to be quite effective for most peroxides, but it is not reliable for removing alkyl peroxides.

Method 3

Blue-indicating molecular sieve (4-8 mesh, type 4A) is added to containers of peroxidized chemicals and allowed to sit for 1–30 days.[11] An amount equivalent to about 5–10% (wt/vol) of the peroxidized liquid is used. Alternatively, the mixture can be refluxed, and the reaction occurs within 4 h. The peroxide is broken down, and the indicator in the sieve is consumed. When run at room temperature, this process is apparently safe, slow, and controlled. Dialkyl peroxides are not efficiently removed, especially from dioxane. This method may be particularly suited to treatment of THF, diisopropyl ethers, and diethyl ethers, which may be decontaminated at room temperature in a couple of days.

Other Methods of Removing Higher Peroxides, Including Dialkyl Peroxides

In one suggested procedure, a 10% molar excess of sodium or potassium iodide is dissolved in 70 mL glacial acetic acid. A small quantity of 36% (wt/vol) HC is added, followed by 0.01 mol of the dialkyl peroxide. The solution is heated to 90–100°C on a steam bath over the course of 30 min and held at that temperature for 5 h.[6] Zinc dissolved in acetic or hydrochloric acid has been recommended.[1,4] Prolonged treatment with ferrous sulfate in 50% sulfuric acid has also been recommended.[1]

Acknowledgment

Work performed under the auspices of U.S. Department of Energy by the Lawrence Livermore National Lab under contract W-7405-Eng-48 UCRL-JC125005.

References

1. Davies, A. G. *J. Roy. Inst. Chem.* **1956**, 80, 386–89.
2. Jackson, H. L.; McCormack, W. B.; Rondestvedt, C. S.; Smeltz, K. C.; Viele, I. E. *J. Chem. Educ.* **1970**, 47(3), A175–88.
3. *Recognition and Handling of Peroxidizable Compounds: Data Sheet 655*; National Safety Council: Chicago, IL, 1976, 1982, and 1987.
4. Noller, D. C.; Bolton, D. *J. Anal. Chem.* **1963**, 35(7), 887–92.
5. Urben, P. G., Ed. *Bretherick's Handbook of Reactive Chemical Hazards: An indexed Guide to Published Data*, 5th ed.; Butterworth-Heinemann: Newton, MA, 1995.
6. *Prudent Practices in the Lab: Handling and Disposal of Chemicals*; National Research Council, National Academy Press: Washington, DC, 1995; pp. 54–55, 99–101, 162–163.
7. Bretherick, L. *J. Chem. Educ.* **1990**, 67(9), A230.
8. Mirafzal, G.; Baumgarten, H. E. *J. Chem. Educ.* **1988**, 65(9), A226–29.
9. Sharpless, T. W. *J. Chem. Educ.* **1984**, 61(5), 476.
10. Bathe, F. M. *Chem. Br.* **1974**, 10, 143.
11. Burfield, D. R. *J. Org. Chem.* **1982**, 47, 3821–25.
12. Pitt, M. *J, Chem. Br.* **1974**, 10, 312.
13. Hamstead, A. C. *Ind. Eng. Chem.* **1964**, 56(6), 37–42.
14. Seabury, J. *UC Davis Safety Net.*, 5/85; University of California at Davis, Office of Environmental Health and Safety.
15. *MSDS for Diethyl Ether.* Canadian Centre for Occupational Health and Safety: Hamilton, Ontario, 1993.
16. Bretherick, L. In *Improving Safety in the Chemical Lab*, 2nd ed.; Young, J. A., Ed.; Wiley-Interscience: New York, 1991; p. 120.
17. Steere, N. V. *J. Chem. Educ.* **1964**, 41, A575–79.
18. *Prudent Practices for Disposal of Chemicals from Labs*; National Research Council, National Academy Press: Washington, DC, 1983; pp. 73–76 and 242–46.
19. *Safety in Academic Chemistry Labs*, 6th ed.; Committee on Chemical Safety, American Chemical Society: Washington, DC, 1995; p. 62.
20. Baumgarten, H. E. *Organic Synthesis*; Wiley and Sons: New York, 1973; Vol. 5, p. 796.
21. Walters, D. *Chem. Health Saf.* **1996**, 3(3), 35.
22. Mair, R. D.; Graupner, A. *J. Anal. Chem.* **1964**, 36, 194–204.
23. Wagner, C. D.; Clever, H. L.; Peters, E. D. *Anal. Chem.* **1947**, 19, 980–82.
24. ASTM, *Standard Test Methods for Assay of Organic Peroxides*: E 288–91, ASTM Subcommittee E15.22; ASTM: West Conshohocken, PA, May 1991.
25. *Peroxide Test Strip, Merckquant*; E. Merck/EM Science: Postfach 41 19, D-6100 Darmstadt, Germany; EM Science: 480 Democrat Rd., Gibbstown, NJ 08027; 609-354-9200; fax 609-423-4389.
26. Wilson, C. L., Wilson, D., Eds. *Comprehensive Analytical Chemistry*; Elsevier: New York, 1959; Vol. 1A, p. 392.
27. Meites, L. *Handbook of Analytical Chemistry*; McGraw-Hill: New York, 1963; p.12–113 (Table 12–43).
28. Jacobson, C. A., Ed. *Encyclopedia of Chemical Reactions*; Vol. 7, Reinhold: New York, 1958; p. 411.
29. Mahn, W. *J. Academic Lab Chemical Hazards Guide-book*; Van Nostrand Reinhold: New York, 1991; p. 10.

53

Disposal of Shock- and Water-Sensitive, Pyrophoric, and Explosive Materials

GEORGE C. WALTON

This chapter describes the process of disposing of shock-sensitive, water-sensitive, pyrophoric, and explosive materials. The process begins with the identification of the waste and includes such tasks as opening the waste storage container and withdrawing a sample for field or lab testing; physically or chemically treating the waste, transferring the material from the storage container to a treatment vessel as necessary; disposing of treatment residues; and cleaning all process or treatment equipment. Successful completion of these tasks requires a knowledge of the chemistry of the materials/wastes as well as a knowledge of the regulations governing waste management. This discussion will concentrate on the thought processes involved in reactive or energetic waste disposal rather than on specific technical subjects such as the molarity of reagent solutions and temperatures of reactions.

Definitions

Carefully defining terms is crucial in energy-intensive operations such as these. Small errors in concept or practice could have significant results in terms of property damage and human safety. To simplify matters, the terms energetic or reactive will be used to describe all shock-sensitive, explosive, water-sensitive, and pyrophoric materials.

Chemical Definitions

First, the hazards of the waste (or material) for disposal need to be defined. Table 53.1 lists selected definitions of materials that may be shock-sensitive, water-sensitive, or explosive. The definitions of ignitability and reactivity make all shock- and water-sensitive, pyrophoric, and explosive materials hazardous wastes. Often there is no clear difference between and among these materials. For example, a solution of dinitrophenol is regulated by DOT as a poison liquid. If some moist material (with more than 15% water) has collected on the cap of the container, that material is regulated as a flammable solid. If

the container has not been used for a long period of time and the deposits on the cap have completely dried, then the deposits are a DOT-regulated explosive.

Regulatory Definitions

The second set of definitions is essentially regulatory. Problems in the current waste-management system center around permit issues. By definition no one may treat a hazardous waste on-site or accept waste for treatment from off-site without a permit. Treatment, as defined in Table 53.2, consists of almost any operation that alters the physical state, chemical structure, or the general hazards of a material. Thus, almost any waste disposal process that laboratory or production chemists complete is treatment. Permitting is a personal, not corporate issue; individuals working for both private and governmental organizations have received felony convictions for improper waste treatment or disposal.

The concept of waste exemptions for laboratories, in fact, the definition of a "laboratory," is in flux. Local or state environmental regulatory agencies or EPA regional offices should be consulted to determine what, if any, permit is required. Complicating waste disposal operations are the states' rights under RCRA. As of 1997, approximately 30 states have adopted the federal standards on a state-by-state basis.

Safety

All energetic waste-disposal processes must begin with a thorough knowledge of the hazards to be faced. Even relatively well-known materials pose significant risks. Tollens Reagent consists of silver nitrate, sodium hydroxide, and dilute ammonia solution. It is used both to detect aldehydes and to silver mirrors, and may form silver fulminate on standing. The stability of Tollens Reagent was studied and reported[1]:

E. H. Barry of New York told to Mr. Cohen of the Franklin Institute that, in order to test the stability of

TABLE 53.1 Selected EPA and DOT Definitions of Air-Reactive, Water-Reactive, Shock-Sensitive, and Pyrophoric Materials

Agency/Term	Definition
EPA Ignitability Characteristic Hazardous Waste (40 *CFR* 261.21)	A liquid with a flash point less than 60°C (140°F)
	Not a liquid and is capable, under standard temperature and pressure, of causing fire through friction, absorption of moisture, or spontaneous chemical changes and, when ignited, burns so vigorously and persistently that it creates a hazard
EPA Reactivity Characteristic Hazardous Waste (40 *CFR* 261.23)	It is normally unstable and readily undergoes violent change without detonating
	It is capable of detonation or explosive reaction if it is subjected to a strong initiating source or if heated under confinement
	It is readily capable of detonation or explosive decomposition or reaction at standard temperature and pressure
	It is defined as a Forbidden Explosive or an Explosive A or Explosive B by DOT at 49 *CFR* 173
	It reacts violently with water or forms potentially explosive mixtures with water
	When mixed with water, it generates toxic gases, vapors, or fumes in a quantity sufficient to present a danger to human health or the environment
DOT Primary Explosive (49 *CFR* 173.59)	A substance that is manufactured with a view to producing a practical effect by explosion, is very sensitive to heat, impact, or friction, and even in very small quantities, detonates
DOT Organic Peroxide [49 *CFR* 173.128(a)]	Any organic compound containing oxygen in the bivalent –O–O– (O is oxygen) structure and that may be considered a derivative of hydrogen peroxide, where one or more of the hydrogen atoms have been replaced by organic radicals
DOT Oxidizer [49 *CFR* 173.127(a)]	A material that may, generally by yielding oxygen, cause or enhance the combustion of other materials
DOT Spontaneously Combustible [49 *CFR* 173.124(b)]	A liquid or solid pyrophoric material that, even in small quantities and without an external ignition source, can ignite within five (5) minutes after coming in contact with air
	A self-heating material that, when held in contact with air at 140°C for 24 hours and without an energy supply, is liable to self-heat to 200°C (392°F)
DOT Dangerous When Wet [49 *CFR* 173.124(c)]	A material that, by contact with water, is liable to become spontaneously flammable or to give off flammable or toxic gas at a rate greater than 1 liter per kilogram of materials per hour

silvering solution it was allowed to stand over the weekend in a closed room, in an open beaker. On Monday morning the contents and the beaker fragments were found to have been scattered all over the ceiling.

All aspects of the proposed energetic or reactive waste-disposal operation must be carefully studied and all questions carefully, completely, and truthfully answered. Engineering controls (structures and equipment), administrative controls and work practices (how equipment is utilized; how reagents are handled), and personal protective equipment must be used to protect human life, the environment, and property during energetic waste-disposal operations.

Disposal Operations

Energetic waste-disposal operations consists of answering five sets of questions:

1. Is the waste identified? Are we sure?
2. What are the waste-disposal process goals? How is "best" defined?
3. What are the consequences of errors? Can we manage mistakes?
4. Can the existing organization (personnel, equipment, technical resources) complete the proposed operation?
5. What are treatment-residue disposal requirements?

If chemical operations are begun before a plan is developed, permitting, process control, safety, and residue-management issues can force changes in concepts that decrease safety, increase costs, and delay disposal. Plan first, work second.

Waste Identification

As with any chemical reaction, the reactants must be known before the process can begin. With energetic materials, absolute identification of the waste may be difficult or impossible. Crystals from peroxide-forming compounds tend to form at the boundary between the highest oxygen level and the highest compound level. For liquids, this is generally in the threads of the container cap or bung. Turning the threaded closure impinges on the crystals which may result in a detonation. For any

TABLE 53.2 Selected Federal (U.S. EPA) Definitions and Classifications of Disposal and Treatment Operations

Term	Definition, Example, or Code
Disposal [40 *CFR* 260.10(a)]	Discharge, deposit, injection, dumping, spilling, leaking, or placing any solid waste or hazardous waste in or on any land or water so that such solid waste or hazardous waste or any constituent thereof may enter the environment or be emitted into the air or discharged into any waters, including ground waters
Treatment [40 *CFR* 260.10(a)]	Any method, technique, or process, including neutralization, designed to change the physical, chemical, or biological character or composition of any hazardous waste so as to: neutralize such waste recover energy or material resources from the waste render such waste nonhazardous or less hazardous render such waste safer to transport, store, or dispose of render such waste amenable for recovery, amenable for storage, or reduced in volume
Thermal Treatment (40 *CFR* 264, App I)	T07: Rotary Kiln Incinerator T11: Pyrolysis T13: Wet Air Oxidation T15: Microwave Discharge
Chemical Treatment (40 *CFR* 264, App I)	T22: Chemical Oxidation T23: Chemical Precipitation T24: Chemical Reduction T25: Chlorination T26: Chlorinolysis T31: Neutralization
Physical Treatment (40 *CFR* 264, App I)	T35: Centrifugation T36: Clarification T40: Filtration T46: Ultrafiltration
Removal of Specific Components (40 *CFR* 264, App I)	T48: Absorption-Molecular Sieve T51: Catalysis T60: Liquid-Ion Exchange T61: Liquid-Liquid Extraction
Boiler and Industrial Furnaces (40 *CFR* 264, App I)	T80: Boiler T81: Cement Kiln T82: Lime Kiln T86: Blast Furnace

planned process, generally the first operation is simply opening the waste container and withdrawing a sample. A variety of techniques using pneumatic or hydraulic apparatus can either penetrate or cut the top of the closure, allowing a sample to be withdrawn.

Waste identification poses risks. Organic peroxide solutions of unknown concentration should not be analyzed by IR spectrometry. The input of energy can cause deflagrations or detonations. Air-sensitive materials require special attention. Errors in interpretation of field screening data are easy to make. Chlorosulfonic acid produces volumes of dense white smoke, chars paper and most elastomers, and reacts violently with water. Hexane (or similar solvents) containing dissolved metal alkyls (diethyl zinc, triethyl aluminum) ignites spontaneously in moist air, producing smoke; chars virtually all paper and plastic; and reacts violently with water. A simple field test of the liquid or smoke with pH paper will separate the two. Without careful thought and detailed planning, it may be possible to confuse the white fumes from chlorosulfonic acid with iso-butyl aluminum in heptane. Plan first, work second.

For personal safety, it may be necessary to complete field screening or tentative waste identification under inert atmospheres in glove bags or boxes or while wearing personal protective equipment and air-supplying respirators. Again, the process poses risks. Can the field chemists actually use, and use safely, the equipment necessary to identify the waste? From personal observation, *most injuries in energetic waste disposal occur as a result of compromise of safety or operating rules because the operation simply cannot be done under the proposed conditions and with the proposed safety equipment.* Planning and practice are necessary to safely conduct waste identification procedures.

Waste-Disposal Goals

Obviously the "best" option will be selected to implement a waste disposal plan. But "best" is hard to define. Complete destruction of energetic materials with the residues disposed of locally eliminates transportation and RCRA/Superfund liability at a treatment, storage, or disposal facility (TSDF). But this "best" solution requires significant practical chemical knowledge, safety controls, and energetic reaction experience. On the other hand, simple dilution or stabilization followed by transportation and incineration at a TSDF is "best" in terms of limiting personal exposure but creates long-term liabilities at a TSDF. The "best" solution must consider the chemistry of the waste, the treatment capabilities and capacities of the generator, the costs of various options, and the generator's risk-management program.

The goals of any energetic waste disposal option define virtually all steps in the process, perhaps even the extent to which waste identification must be carried out. If the process will involve the chemical destruction of a specific molecular species, then care must be taken to positively identify the waste or specific isomers of the waste. For example, reverse nitration of nitro groups in the *ortho* position is affected by steric hinderance.[2] Contaminants in the waste that will interfere with the destruction process must be identified and controlled. If the process is essentially diluting or stabilizing the waste and packaging it for transportation to an off-site TSDF, then trace contaminants and specific conditions may not be so critical.

All TSDFs must have a waste analysis plan, defining the types and quantities of waste acceptable at the facility, and a system for approving a specific waste prior to delivery of the waste to the TSDF. Companies use various names—waste data sheets, waste profile sheets, authorization request forms—for the waste approval forms. Field sales staff and TSDF chemists review these forms to insure that a specific waste, in a specific form and volume, is acceptable at a given TSDF. This technical review of the waste is an asset for generators in planning the waste-disposal process by using the TSDF's practical experience to accomplish the most work with the least effort and cost, under the safest conditions. A reasonable goal for most low-volume energetic materials disposal plans is the proper DOT packaging in small quantities after dilution or stabilization for acceptance at a RCRA-regulated incinerator.

Consequences of Error

An iterative planning process,[3] as shown in Figure 53.1, should be used to ensure that all reasonable steps have been taken to identify and control mistakes in identification, process selection, process operation, safety and residue management before waste disposal begins. Some errors result in minimal effect. If a waste-disposal plan for

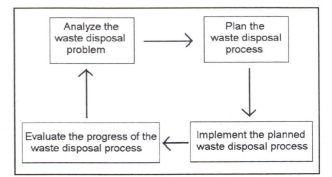

FIGURE 53.1 The iterative planning process for disposal of energetic wastes.

picric acid is based on a laboratory bottle containing 100 grams and if 10,000 grams of alcohol:water solution were selected as the diluent, the true mass of the picric acid could vary from near zero to slightly over 100 grams with little real effect. However, if a 10-gram vial of 1-methyl-3-nitro-1-nitroso-guanidine is mishandled during disposal, the detonation may cause significant physical damage. The release of such a potent carcinogen will definitely require significant decontamination to remove the biological hazard. Some silver salts, such as silver fulminate, are light-sensitive explosives. The consequences of errors can kill people and destroy property. As above, research and plan first, work second.

The consequences of error may be expensive. Incomplete or inaccurate research may indicate that wastes are soluble in specific solvents at specific levels. If a waste-disposal plan is based on this wrong assumption, multiple waste streams will be formed during attempted dilution. Each stream may then require a separate management process. This increases personnel exposure to hazardous materials and wastes, increases waste volumes, and greatly increases waste-disposal costs.

Organizational Assets

Can the proposed team and equipment actually complete the planned disposal process? Personnel must have the training and experience to recognize actual and potential hazards and manage them with engineering controls, work practices, and personnel protective equipment. Also, personnel must realize their own limitations in terms of training, experience, and knowledge. While TSDFs are tremendous assets in planning (or conducting) the disposal of energetic materials, manufacturers have a wealth of technical knowledge and experience in dealing with reactive wastes. This knowledge and experience is probably the greatest asset a generator of waste has, but the generator must ask for it.

During disposal operations of reactive and energetic wastes, forces and conditions not usually encountered under laboratory and production conditions will be encountered. Static loads on processing equipment may be converted to dynamic loads during exothermic reactions. Can the equipment—laboratory or production—sustain those loads? Corrosion-resistant containers may be filled with materials outside their design concentrations. For example, aluminum tanks for 98% nitric acid may not be able to contain dilute concentrations encountered during on-site treatment operations. Equally as important are the gaskets between flanges, even the bolts holding the flanges together. Can they contain the concentrations and pressures that will develop during energetic waste disposal? Does the "team"—people, process and safety equipment, technical resources—have the capability to complete all disposal tasks?

Residue Management

Once all aspects of a carefully and thoughtfully planned and correctly executed energetic waste-disposal operation have been completed, where do the residues go? Two examples illustrate why residue management is important. First, during emergency response to an aluminum phosphide fire, the "best" solution was to quench the reacting phosphide with water, creating several hundred gallons of phosphine-contaminated water. The local publicly owned treatment works (POTW) required such extensive treatment and testing that it was cheaper to package the water in 55-gallon drums and transport it off-site for hazardous wastewater treatment. Second, one gallon of red fuming nitric acid was chilled to reduce fuming, diluted with ice water, neutralized with sodium carbonate, solidified with oil dry (expanded clay), and buried in a municipal solid-waste landfill. The first question in the planning process was directed to the local landfill. Would they accept the neutralized, solidified acid? If yes, then all planning was directed to safely, efficiently, and cheaply creating a waste the landfill would accept. Backward planning—starting with the last step and going toward the first—based on the disposal of waste-treatment residues may indicate the "best" waste-disposal process.

Specific Operations

The following waste-disposal operations are provided as examples of the process of disposing of reactive and energetic materials that protect human safety and the environment, comply with local, state, and federal regulations, and manage money well.

Mercuric Oxycyanide

A rural hospital had a 20-year-old 2-pound bottle of mercuric oxycyanide (CAS 1335-31-5). Local sewer/landfill disposal was not considered because of the mercury content. Hazardous waste landfills would not accept the waste due to its strong oxidizing and explosive potentials. Stabilization with mercuric cyanide, a procedure in the literature, was not considered practical because no facility would accept the stabilized residue. The hospital did not have, and could not efficiently create and permit, the "team" necessary for complete chemical treatment. A hazardous waste incinerator would accept the material if the following procedure, described on a waste acceptance form, was followed:

1. Cut with a mixture of sodium chloride and soda ash. (The salt : ash mixture was listed as the best fire extinguishing media on the MSDS. The waste was cut to 10% of its initial strength based on U.S. Army Corps of Engineers and DOT standards of reclassifying Class 1.1 [or old Class A Explosives] to flammable solids.)
2. Pack 1 pound of the cut or diluted waste in a 1-gallon HDPE jug.
3. Fill each jug with approximately 1 gallon of tap water. After screwing the caps on securely and taping them, pack the jugs in a DOT-approved drum.
4. Describe the drum on a hazardous waste manifest as HAZARDOUS WASTE SOLID, N.O.S. (MERCURY CYANIDE, MERCURIC OXYCYANIDE, AND SODA ASH IN WATER), 9, NA3077, PG III.

The essential questions of energetic waste disposal were answered. The waste was identified in sufficient detail to allow the mixing process to be done safely. The goal of the waste disposal process was clearly stated—follow the procedure negotiated while completing the waste approval form. The consequences of error were minimized by creating a safe work environment for the mixing process and by selecting a technique that involved minimal manipulation of an explosive, reactive, toxic molecule. A team of explosive-handling chemists, in the proper respiratory and blast protective equipment, using the proper tools, was formed using contractor support. Finally, process residue disposal was assured because that was really the first critical step considered in the plan.

Silver Fulminate

A student in a college chemistry laboratory, in specific violation of oral and written safety rules, prepared approximately 1 liter of Tollens Reagent the second week of September in a glass-stoppered reagent bottle. As he was cleaning out his laboratory space in December, he reported the presence of the aged reagent to the teaching assistant. Some precipitate could be seen around the ground-glass stopper and on the bottom of the container. The reported composition of the solution in the bottle was considered sufficient information to identify the waste as silver fulminate. Due to the high probability of a detonation during any opening, transfer, or reaction process, it was decided the "best" option was to detonate the container. Two electric blasting caps were taped to the bottle, one each at the neck and across the bottom. The bottle was placed in a vermiculite-filled detonation chamber in a 55-gallon drum. Approximately 1 inch of perlite was placed on top of the vermiculite. The caps were detonated. As per the disposal plan approved by the state, postdetonation samples for silver (EPA Waste Code D011[4]) were collected. Surface soil sampling was biased toward any particles of perlite. The vermiculite in the pail/drums was excavated only to ensure the reagent bottle had been ruptured. This was considered proof that any silver fulminate had been destroyed. The vermiculite was replaced, the drum sealed and then sent to a hazardous waste landfill. As no silver was detected in surface soil samples, no further action was taken. The critical questions for energetic waste disposal had been asked and answered.

Methyl Ethyl Ketone Peroxide (MEKP)

Seventeen gallons of MEKP had been stored at elevated temperatures and no longer met the manufacturer's assay standards. Due to incomplete and improper waste-disposal process design, dilution with a mixture of diesel fuel and water was attempted. However, MEKP does not dissolve in diesel fuel and water. The original waste dilution process failed, more than doubling the volume of hazardous waste for disposal. The characteristics of the waste had not been identified and the consequences of failure not been included in the plan. Personal exposure, levels of effort, and costs all were needlessly increased. The important energetic-waste-disposal questions, discussed above, had not been asked or answered.

Other Materials

A variety of sources supply much useful information in identifying, treating, or disposing of potentially energetic or reactive materials. Some, but not all, of these references include the National Safety Council,[5] *Prudent Practices* series,[6] *Improving Safety in the Chemical Laboratory*,[7] the NIH Research Safety Symposium Series[8] (and especially publications by Dr. Eric Sansone[9]), and manufacturers' technical data bulletins and material safety data sheets.

References

1. Federoff, B. T.; Kaye, S. M., Eds.; *Encyclopedia of Explosives and Related Items*, 10 vols.; Dover, NJ: US Army Research and Development Command, 1960–1983.
2. Urbański, T. *Chemistry and Technology of Explosives*, 4 vols.; New York: Pergamon Press, 1964.
3. *Recognizing and Identifying Hazardous Materials*, 2nd ed.; United States Fire Administration, National Fire Academy: Emmitsburg, MD, 1991.
4. EPA Publication SW-846: *Test Methods for Evaluating Solid Waste—Physical/Chemical Methods*; U.S. Environmental Protection Agency: Washington, DC, 2000.
5. National Safety Council; *Recognition and Handling of Peroxidizable Compounds*, Data Sheet I-655, Rev. 87; National Safety Council: Chicago, 1987.
6. *Prudent Practices in the Laboratory: Handling and Disposal of Chemicals*; Committee on Prudent Practices for Handling, Storage, and Disposal of Chemicals in Laboratories; National Academy Press: Washington, DC, 1995.
7. *Improving Safety in the Chemical Laboratory*, 4th ed. Young, J., Ed.; New York: Wiley, 1991.
8. NIH Research Safety Symposium, Office of Research Safety; National Cancer Institute: Frederick, MD, 1990.
9. Lunn, G.; Sansone, E. B. *Destruction of Hazardous Chemical in the Laboratory*; New York: Wiley, 1994.

54

Compressed Gases

GEORGE WHITMYRE

When subjected to ballistic impact, the human body has a consistency of Jello. The hazard common to all compressed gas fittings, tubing, and pressure vessels is ballistic failure, the release of kinetic energy that was stored as pneumatic pressure.

The compressed gas checklist is offered here as a starting point to help compressed gas users analyze how process upsets and human error can affect their compressed gas system. It should be *adapted*, not used *verbatim* either in part or as a whole. Adapt it to accommodate the subtle variations in your workplace or laboratory process, layout, and local requirements. Adapt it for research group laboratory audits. Adapt it for compressed-gas safety training. Use it as a guide, adapt it to meet your needs.

Use your modified version of the checklist for your hazard review of your compressed-gas systems. Ensure that adequate safety factors and safety controls are installed, maintained, and operated so as to eliminate or mitigate the effects of upsets and errors.

Compressed Gas Checklist

Training and Hazard Review

1. All lab personnel have received low-pressure compressed-gas safety training in addition to Hazard Communications (Chemical Hazard Right-to-Know) and the OSHA Laboratory Standard training. Company or institution policy may require additional basic training elements.
2. A standard operating procedure is established by a job hazard analysis (JHA) for lab compressed gas usage.[1] This may meet in part OSHA requirements for selecting PPE.
3. Compressed gas operations that meet established prior approval criteria require a hazard review.[2,3] Triggers for hazard review include:
   a. pressures >150 psi
   b. glassware or plastic ware under vacuum or pressure
   c. >400 ft^3 flammable gas

   d. corrosive, pyrophoric, toxic, or other reactive gas
   e. unattended operations
   f. large-scale pressure operations
   g. a change in a previously reviewed compressed gas system, including a change of equipment operators

Ordering Cylinders

4. The smallest practical cylinder size is ordered to minimize the total compressed gas volume in the lab. Half-filled cylinders are ordered if small cylinders are unavailable and full cylinder pressure is unimportant.[4]
5. Toxic or reactive gases are ordered in a "balance inert" mixture whenever possible.
6. Critical orifice valves are ordered on flammable, corrosive, pyrophoric, or toxic gases.
7. Disposable cylinders are not refilled or used for other purposes. They are emptied, valve-removed, and cylinder-drilled before disposal.

Cylinder Storage

8. Cylinders are in 3-point contact, chained or strapped storage, secured from theft and tampering.[5] In earthquake zones, chain them individually at high and low points on the cylinder body with 3/16" chain and eye bolts anchored directly into a wall stud or into a similar structural member.[3]
9. Storage areas are dry and temperature does not exceed 125°F (50°C).
10. Cylinders are not stored in corridors, egress routes, labs, or other workplaces.
11. Either 20-ft distance or a fire wall at least 5 ft high separates oxygen cylinders from flammable gases and flammable-liquid hydrocarbons both in storage and in the lab.
12. Indoor storage rooms have an outside wall or other pressure-relief blow-out panel.
13. Lecture bottles and other small cylinders have

organized storage with the cylinder valves protected.

14. Continuous stainless steel tubing is used to deliver gases to the laboratory from a sheltered outdoor location to eliminate transporting cylinders.

15. Remote gas delivery tubing is valved at the terminations and has the gas type clearly labeled at required intervals.

16. Remote gas tubing systems for pyrophoric gases and flammable gases may require inerted or evacuated conduits with leak detection in the conduit or in the pipe chase and in the lab.

Moving Gas Cylinders

17. Industrial safety glasses with side shields are routinely used by all lab personnel when moving a cylinder, connecting a gas regulator, and when performing any lab activity with compressed gases.

18. A gas cylinder cart (with restraining chain) is used to move cylinders. Cylinders are not rolled, pushed, dragged, manually lifted, or hoisted by hooking the cylinder cap.[6]

19. Gas cylinder carts have a low-level cradle and restraining chains to secure small gas cylinders.[4]

20. The regulator is removed and cylinder cap installed when moving a cylinder.

21. Hand injuries are prevented when moving a cylinder by awareness of hand positions.

22. When shifting a cylinder from the gas cylinder cart to a chained restraint, use one hand on cylinder cap, the other hand on neck of cylinder to maintain control of cylinder.

23. Building elevators are considered a confined space when transporting gas cylinders. People do not ride with gas cylinders in case of discharge of the cylinder contents through the pressure-relief device.

24. Only the smallest sizes of gas cylinders are hand carried to the laboratory either singly in a rubber bucket or in the original DOT shipping box.

25. A 2-person carry cradle is used to manually lift and carry a standard cylinder over rough terrain for field laboratory use.

26. DOT-authorized vehicles, not passenger cars, transport gas cylinders on public roads.

Point-of-Use

27. Gas cylinder paint color is not standardized so it is not relied on for gas identification. Unla-

beled or unknown gases are returned to the gas vendor.[7]

28. A chain or strap is provided for each cylinder at point of use to prevent cylinders from falling over. Lecture bottles and other small cylinders are secured with a chain clamp, channel strut clamp, or in a bench stand.

29. In earthquake zones cylinders over 3 ft tall are chained individually at high and low points on the cylinder body, with 3/16″ chain and eye bolts anchored directly into a wall stud or into a similar structural member.

30. The gas cylinder cap is stored in low storage, below head level.

31 The cylinder valve is closed and regulator depressurized when the gas is not in use.

32. To reduce the gas cylinder count, manifolds are used to distribute gas from one cylinder to multiple-use points within one lab or to a group of labs.

33. Empty or unused gas cylinders are capped and promptly removed from the lab.

34. An emergency eyewash station with a 0.6 gal/min water flow and a safety shower delivering 15 gal/min deluge is available where reactive or corrosive gases are used.

35. Gas cylinders are returned to the gas vendor with 30 psi gas pressure to prevent contamination. Tag the cylinder and notify the gas vendor if drawdown occurs.

Preventing/Troubleshooting Gas Cylinder Leaks

36. A special hook wrench is used to free rusted or jammed cylinder caps, not a screwdriver that can engage and open the cylinder valve.[8]

37. Cylinders with leaks, tight valve action, or other problems are immediately returned to the gas vendor. It is prudent to inspect all gas cylinders for "fish egg" valve leaks with a polymer-soap solution before accepting them from the gas vendor. The gas user must not attempt to tighten valve packing nuts or leaking cylinder valve components; this is the gas vendor's responsibility.

38. Reactive, flammable, or toxic gas cylinders discovered leaking in the lab are not immediately transported out of the building by the lab user. Alert others in the vicinity to evacuate the hazard area. The gas cylinder valve is turned off if there is no imminent danger to the lab user. **Do not** turn off ignition sources or even light

switches when leaving the room. From a safe location call the local hazmat response team to evaluate the leak and take appropriate action. They may move it to a ventilated gas cabinet or a lab hood, bag-over or use a special pressure vessel overpack to remove it from the building.[9]

39. An inert, nontoxic gas cylinder with a minor leak can be contained in a ventilated gas storage cabinet or a lab hood until the gas vendor arrives.

40. To tighten or replace leaking gas system components, the lab user depressurizes the system before making repairs.

Regulator Installation

41. The valve on a gas cylinder is never "cracked open" to "blow out dust."

42. A regulator is always attached to the cylinder before opening the gas cylinder valve.

43. The CGA (Compressed Gas Association) regulator fitting exactly matches the cylinder valve CGA.[10]

44. Regulators are never modified by the gas user; a CGA fitting adaptor is never used.

45. Teflon thread tape is NOT used on any CGA cylinder valve fitting (parallel machine threads).[4]

46. Teflon thread tape is used on all NPT pipe thread fittings (tapered pipe threads).

47. The gasket washer on flat-faced CGA fitting is replaced at every cylinder change out.

48. Lab gas users understand and practice correct regulator installation and removal:

    Step 1: Inspection—The regulator and the cylinder valve CGA fittings are inspected for dent or scratch flaws across the mating surface before the regulator is attached. Use a finger to feel for these flaws or use a bright light and magnifying glass. Neither overtightening or Teflon thread tape will stop a leak caused by a fitting flaw. Return flawed regulator or flawed cylinder to the vendor. Attempting to tighten a flawed fitting will emboss the flaw on the mating surface.

    Step 2: Installation—The CGA fitting is hand tightened then snugged with a wrench. Avoid using excessive force. Never use a wrench extension lever; it will distort the machine threads. A CGA connection will go together with light wrench torque.

    Step 3: Isolate Diaphragm—Once the pressure-reducing regulator is attached, the pressure setting knob is rotated counterclockwise (CCW) to avoid a high-pressure gas surge on the diaphragm. The small regulator outlet valve is closed.

    Step 4: Open Cylinder Valve—The cylinder valve is opened slowly ½ to 1 turn; the operator facing away in case the pressure gauge or regulator diaphragm ruptures.

    Step 5: Set Pressure—The pressure knob is turned CW to increase the outlet pressure.

    Step 6: Leak Test—All fittings from the cylinder to the regulator are leak tested with a polymer-soap solution.

    Step 7: Bleed Off—To shut down, bleed off the regulator pressure and then rotate the pressure adjusting knob CCW. This avoids rupturing the regulator diaphragm when the main cylinder valve is again opened. The main cylinder valve is closed and the regulator is depressurized when the equipment is unattended or not operating.

49. The gas user stays alert for signs of regulator failure including regulator creep and gas leaking from a diaphragm rupture. The faulty regulator is returned for vendor repair.

Pressure Safety

50. ASTM pressure rating tables are used to select appropriate tubing diameter and wall thickness.[11]

51. ASME-approved pressure vessels, not makeshift vessels are used for gases.[9]

52. Pressure vessels and pressure relief devices are inspected on schedule and replaced when required.

53. Pressure vessels and other components are protected by quality pressure relief devices such as a rupture disk or a pressure-relief poppet valve.

54. Because tube fittings from different manufacturers are not interchangeable, they are not intermixed. Ballistic fitting failures are prevented by using compatible, correctly installed tube fittings from one source.

55. Tube fitting ferrules are the same material (same hardness) as the tubing to make a reliable seal.

56. Correct application of Teflon thread tape or anaerobic sealing compound is used on (tapered)

pipe threads, avoiding overlaying the first thread of the fitting to prevent a tape fragment from pinching off.

57. Teflon thread tape or sealing compounds are not used on (parallel) machine threads of tube fittings.

58. Domestic plumbing fittings have a low MAWP (maximum allowable working pressure) and are not used for lab compressed-gas service.

59. Tubing is prepared by cutting to length with a tubing cutter tool and ends deburred.

60. Cone-end fittings are selected for very high-pressure work (>5,000 psi).

61. Tubing is bent with a tube bending tool, allowing enough straight tubing for the fittings. Hand bending produces irregular sharp bends, constrictions, and tube stresses.

62. Continuous tubing runs are made where possible to reduce the fitting count, minimizing leak points.

63. All components of a pressurized system have MAWP above highest expected operating pressure (Fig. 54.1).

64. Pressure venting and overpressure relief devices protect components with low pressure ratings and prevent expanding gas from bursting tubing and vessels.

65. Uninterruptible (no shutoff valve) pressure-relief valve or rupture disc protects system components.

66. Pressure-relief device discharges to a safe area or blowdown vessel.

67. Vent discharge line is sized for 2-phase (liquid-gas) flow or for manifold vent lines.

68. Ordinary glassware or plastic ware are not used for pressurized (or vacuum) systems.

69. Lockout and pressure bleed are provided for pneumatic or hydraulic (in addition to electrical LOTO) systems subject to OSHA lockout regulations.

70. Prior approval requirements including an equipment and procedural hazard review for gas handling systems, reactors, and other apparatus was completed.

71. Gas handling equipment is designed to protect the lab users from high-pressure gas injection into the body.

72. Gas tubing and components are securely anchored at intervals with tube clips, clamps, or straps to prevent sagging or tube whipping if the tube bursts or a tube fitting fails.

73. Where flow reversal is possible, especially in manifolded gas systems, cylinder discharge lines are equipped with quality check valves to prevent cylinder contamination and inadvertent mixing of backflow of incompatible gases.

74. Lab building drinking water is protected from pressure injection of chemical reactor products in the event of a cooling coil leak. This is accomplished with a sanitary double-check valve or a makeup water air gap if a reservoir and pump system provide cooling.

Barricades and Shields

75. Permanent or portable shields cover all exposed sides to protect the lab worker from high-velocity projectiles, like tube fittings and other components that fail, especially in systems with high kinetic energy stored as compressed gas.

76. High-pressure systems operated in lab exhaust hoods have a supplemental operator shield; ordinary lab exhaust hood sashes composed of safety glass in sheet metal frames are not reliable as an explosion barrier.

77. High-pressure barricades are constructed of hot rolled steel panels bolted or welded to heavy-angle iron frames with polycarbonate (not acrylic) view windows.

78. Pressure venting of protective barricades is directed up or away from the user to an unoccupied vent-discharge area.

79. Hydrostatic (liquid) pressure testing is performed with a polycarbonate barrier protecting the operator. Pressure testing with a compressed gas is performed only with an adequate steel or concrete barrier between the gas systems and the equipment operator.

80. Portable safety shields designed for benchtop use have heavy base weights or are bolted down.

Precautions for Oxygen

81. Oxygen-enriched atmosphere (OEA) >22% is avoided by discharging waste oxygen to an exhaust vent that is free of organic materials and ignition sources or by venting the oxygen directly outside.

82. Precautions are taken to prevent OEA from contact with oils, grease, and other hydrocarbons. Oxygen is entrapped in hair and clothes where it can ignite by low-energy ignition sources or from contact with hydrocarbons.[8,12]

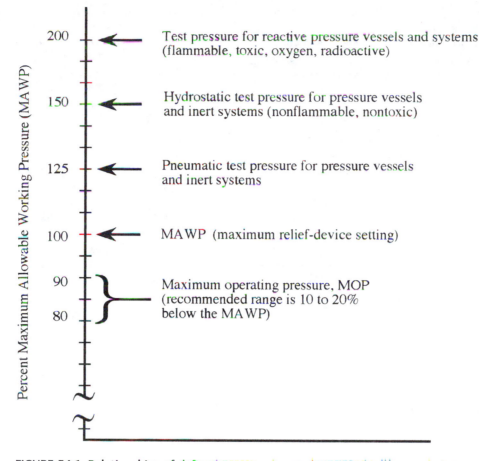

Percent Maximum Allowable Working Pressure (MAWP)

200 — Test pressure for reactive pressure vessels and systems (flammable, toxic, oxygen, radioactive)

150 — Hydrostatic test pressure for pressure vessels and inert systems (nonflammable, nontoxic)

125 — Pneumatic test pressure for pressure vessels and inert systems

100 — MAWP (maximum relief-device setting)

90
80 — Maximum operating pressure, MOP (recommended range is 10 to 20% below the MAWP)

FIGURE 54.1 Relationships of defined pressure terms. (Reprinted with permission from Lawrence Livermore National Laboratory Document M-010, *Health and Safety Manual*, 1997.)

83. Oxygen cylinders are used only with factory-clean oxygen regulators. Lubricants and other organic contaminants on regulators can cause a regulator explosion or a fire when regulator is adapted for oxygen.

84. Quick-action valves including ball valves, poppet, and plug valves are not used in oxygen systems since rapid oxygen compression can cause compression ignition of metal tubing, plastic tubing, and plastic components.

85. Particulates are eliminated from oxygen piping since friction of particles on tubing surfaces will ignite oxygen fires in metal tubing from particle impact ignition.

86. In use or in storage, oxygen cylinders (except in oxyacetylene welding kits) are separated from reducing gases by 20 ft or by a firewall at least 5 ft high).

87. Reactive gases (e.g., oxygen and hydrogen) are mixed or blended only at the actual point of use, not in a pipeline. Backflow-prevention devices and flashback arrestors are installed where required by hazard review of the gas system.

Inert/Asphyxiating Gases

88. Lab confined spaces are identified and posted. This includes lab tanks, bins, test chambers, dry ice storage chests, elevators, and other low elevation or enclosed areas where inert gases could collect, displacing breathing air.

89. Oxygen deficiency is monitored with a gas detector before entering any potential confined space. Procedures must conform to OSHA confined space entry requirements. Stratification of

oxygen-deficient and toxic gases may occur within a confined space.

90. The lab is not reentered following a large toxic gas or asphyxiant gas leak (e.g., liquid nitrogen spill) until air sampling determines it is safe.

Precautions for Flammable Gases

91. The count of flammable gas cylinders does not exceed the number allowed by local codes or life safety codes (e.g., a maximum of two hydrogen gas cylinders are permitted below grade by 29 *CFR* 1910.106).[6]

92. Metal tubing is used for flammable gases to dissipate flow-induced static charge.

93. Liquefied gas cylinders are kept upright so that only the vapor phase, not the liquified gas phase contacts the cylinder valve pressure relief device. Local codes may prevent piping liquefied gas into a building.[13]

94. A grounding clamp with a pointed set screw and clamp ridges is used with a bonding cable to equalize and dissipate static charge induced by gas flowing in tubing.

95. An earth electrical ground is connected to apparatus either insulated by the dielectric material of laboratory benchtop or the dielectric exhaust hood base.

96. Because hydrogen will heat and self-ignite upon rapid expansion (negative Joule-Thompson) a gas cylinder is never "cracked open" to "blow out dust." A regulator is always attached to the cylinder before opening the gas cylinder valve.

97. Flammable gases are ordered premixed from the gas vendor with an inert gas component when possible to achieve a nonflammable mixture or to reduce the required ignition energy.

98. Self-polymerizing gases like acetylene are never purified. Acetone is never used at more than 15 psi or mixed with metals that react to form unstable, explosive metal acetylides.

99. An absolute minimum number of flammable gas cylinders are used in the laboratory to reduce total fuel load in case of a building fire. Electrolytic hydrogen generators producing ultrapure hydrogen are good substitutes for hydrogen cylinders in many lab applications. Manifold distribution can also help minimize the cylinder count.

100. A lab hazard review determines if a flashback arrestor, check valve, or a critical orifice valve is appropriate for a particular flammable or toxic gas application.

101. Waste gas is vented into a non-recirculating exhaust duct or lab hood exhaust, provided that no ignition sources are in the ductwork and that an interlock stops gas flow in event of exhaust failure. Waste gas is not burned in a flare and an open flame is not used for leak testing.

102. Although classified electrical service is not required in labs using flammable gas, ignition sources (e.g., hot plates, stirrers, static discharge, electrical switches, motors) are eliminated from lab hoods and other areas where flammable gas is used. Classified wiring is used where flammable or explosive atmospheres are *normally* present and ignition sources cannot be removed.[6]

103. Routinely inert flammable gas systems before and after flowing reducing gas to purge air. Check valves isolate the inert from the flammable or reactive gas to prevent mixing in the regulator or gas cylinder.

Toxic, Corrosive, and Pyrophoric Gases

104. Only corrosion-resistant gas regulators, flow-limiting valves, flow-restrictor orifice valve, and other components designed for control of reactive gas are used.[14]

105. Reactive and corrosive gas cylinder change-out and purge procedures require a trained two-person, SCBA airpack-equipped team.

106. Pyrophoric gases and cross-purge system are located in the lab hood or in a standard ventilated gas cabinet with airflow monitor and ventilation-failure alarm.

107. Processes using pyrophoric gases are ventilated and the work area is monitored with fail-safe gas detector. Either detecting gas at the PEL or a ventilation failure should remotely shut down the pyrophoric gas flow by an air-operated valve system.

108. Toxic, corrosive, or reactive gas effluents are scrubbed, absorbed, or neutralized before venting to an exhaust system at negative pressure with respect to the lab.

109. The sense of smell is not relied on to determine gas concentration (e.g., sulfur gases deplete the olfactory receptors, hydrogen fluoride and carbon monoxide are odorless).

Compressed Air

110. Air used for drying lab glassware or in cleaning operations is regulated to less than 30 psi. Alternately, a safety air nozzle with pressure-relief side ports is used.

111. Tygon tubing and other plastic tubing is clamped to a hose barb fitting and used up to only a few psi pressure. Short runs of flexible tubing are used to prevent tube whipping in event of overpressure. Generally copper tubing, either soldered or connected with tube fittings, is used in small diameters at pressures under 150 psi.

112. PVC plastic pipe is not rated for compressed air service.

Breathing Air

113. Compressed air is not used as breathing air since it often has hydrocarbon contaminants.

114. Engineering and administrative controls are always used to eliminate the need for respirators or breathing air when possible. An OSHA respirator program that includes training, medical exams, equipment maintenance, documentation, and fit-testing elements is in place where SCBA or airline breathing air is used.

115. Breathing airline connections are labeled and unique so that asphyxiation death from inadvertent connection to inert gas or contaminated air is prevented.

References

1. *Job Hazard Analysis in an Industrial Laboratory Environment*; Industrial Safety Data Sheet 706; National Safety Council: Itasca, IL, 1992.
2. Hoffman, J. M., Master, D. C, Eds.; *Chemical Process Hazard Review*; ACS Symposium Series 274; American Chemical Society: Washington, DC, 1985.
3. Whitmyre, G. *Compressed Gases: Compressed Hazards*; American Chemical Society: Washington, DC, 1995.
4. Whitmyre, G. *Compressed Gases: Safe Handling Procedures*; American Chemical Society: Washington, DC, 1995.
5. *Storage, Use and Handling of Compressed and Liquefied Gases in Portable Cylinders: NFPA 55*; National Fire Protection Association: Quincy, MA, 1998.
6. *Safe Handling of Compressed Gases in Cylinders, P-1*; Compressed Gas Association, Inc: Arlington, VA, 2000.
7. *Fire Protection for Laboratories Using Chemicals: NFPA 45*; National Fire Protection Association: Quincy, MA, 2000.
8. *Safe Handling of Compressed Gases*; Matheson Gas Products: East Rutherford, NJ, 1983.
9. Ernest Orlando Lawrence Berkeley National Laboratory, *Health & Safety Manual*; http://www.lbl.gov.ehs/pub3000, 1999.
10. *American National, Canadian and Compressed Gas Association Standard for Compressed Gas Cylinder Valve Outlet and Inlet Connections; ANSI.CSG/CGA V-1*; Compressed Gas Association, Inc: Arlington, VA, 1994.
11. Callahan, F. J. *SWAGELOK^{TM} Tube Fitting and Installation Manual*; The SWAGELOK Companies: Solon, OH, 1998.
12. *Oxygen Systems*; Technical Bulletin No. 5; The SWAGELOK Companies: Solon, OH, 1992.
13. *National Fuel Gas Code; NFPA 54*; National Fire Protection Association: Quincy, MA, 1999.
14. *Stanford University Toxic Gas User's Handbook*; http://www.stanford.edu/dept/EHS/lab/tgo/manual/index.html, 1997.

55

Hydrogenations

J. M. LAMBERT, JR.
W. S. HAMEL

Hydrogenation Techniques

Hydrogenations are a class of chemical reactions in which hydrogen is added to the molecular structure of organic compounds being hydrogenated. Typically, these compounds are in a liquid state during the hydrogenation. Hydrogenation reactions are normally catalyzed and carried out at elevated temperatures and pressures, and in some cases at atmospheric pressure and room temperature. The reactions can be conducted in batch or continuous-flow operations.

Batch Reactors

Batch reactions are those in which a specified volume of the liquid compound to be hydrogenated, often in a solvent, is added to a reactor, hydrogenated, and removed from the reactor before the sequence is repeated. Catalysts are added to the reactor to increase the rate and, in certain cases, the selectivity of the hydrogenation. It is common to use catalysts that are in a powdered or finely ground state.

It is advantageous to provide enhanced contacting of the liquid reactant and the hydrogen with the catalyst surface. This is accomplished by (1) using reactors that are capable of operating at elevated pressures to allow more hydrogen to dissolve in the liquid and adsorb on the catalyst surface; (2) using reactors that provide increased agitation to increase the gas/liquid and the liquid/solid mixing; and (3) using elevated temperatures to increase the overall reaction rate.

The hydrogenation apparatus shown in Figure 55.1 is of a style that has been used for many years.[1] This unit provides a glass vessel into which the liquid reactant and catalyst are added. The vessel is then connected to a pressurized reservoir of hydrogen, heated to the desired reaction temperature, and shaken to provide increased mixing. Monitoring the decrease in reservoir pressure allows the reaction progress to be measured. Units commercially available today are capable of pressures to 60 psig and temperatures to 80°C in vessels having volumes from 250 to 2000 mL. The oscillating motion of the vessel has led this unit to become known as a "shaker."

A similar type of unit, known as a "rocker" is used for hydrogenation reactions at much higher pressures and temperatures. Rockers, such as the one shown in Figure 55.2, are capable of pressures to 6300 psig and temperatures to 400°C in vessels having volumes from 500 to

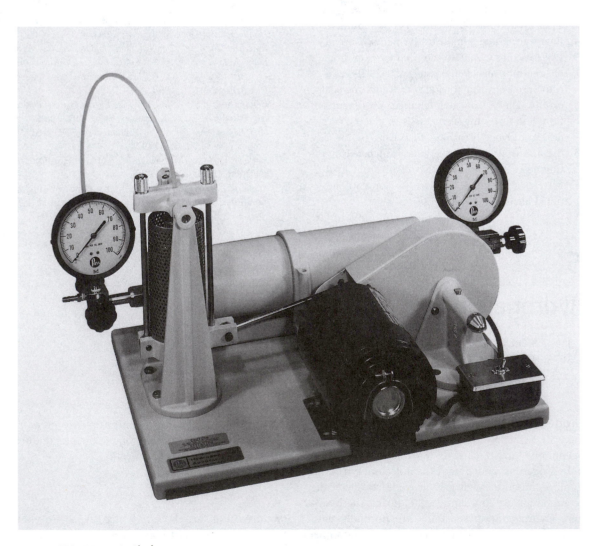

FIGURE 55.1 Shaker reactor.

FIGURE 55.2 Rocker reactor.

1000 mL. Agitation in a rocker is achieved by pivoting the reactor up to 45° above and below the horizontal position.

For reactions requiring increased agitation and mixing, it is common to use a stirred reactor apparatus, such as the unit shown in Figure 55.3. Here the agitation is provided by the high-speed rotation of impellers magnetically coupled through the pressure-containing wall to a variable-speed electric or air motor. Various types of impellers are employed, for example, gas entrainment impellers for increased gas/liquid mixing and pitched-blade turbine impellers for increased liquid/solid mixing. Laboratory-scale stirred reactors can range in size from 25 mL to 5 gal, while commercial-scale reactors can be 1000 gal, 5000 gal, or larger.

Most hydrogenations in stirred reactors are performed with powdered catalysts. A slurry is formed when the catalyst is mixed with the liquid reactant; therefore, these reactions are known as slurry reactions. It is also possible to conduct fixed-bed reactions in a stirred reactor by placing the catalyst, usually in the form of pellets or extrudates, in an annular basket. This also eliminates the need for a filtration step during the recovery of the product liquid and allows the catalyst to be recycled.

One type of fixed-bed reactor uses a *dynamic* basket. In this case, the basket of catalyst is rotated inside the reactor so as to move the catalyst through the fluid. A second type of fixed-bed reactor holds the catalyst basket stationary and uses a specially designed impeller system to move fluids through the basket. This style is referred to as a *static* basket reactor. Figure 55.4 shows the two styles of baskets.

Continuous-Flow Reactors

Stirred vessels, configured for slurry or fixed-bed use, are not always operated as batch reactors. They can also be operated as continuous-flow reactors, particularly on a

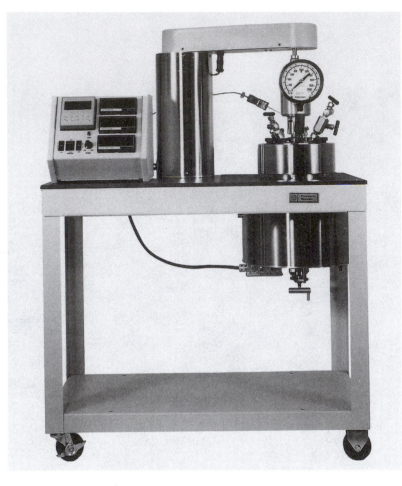

FIGURE 55.3 Stirred reactor.

laboratory scale. In a continuous-flow reactor, both the liquid reactant and the hydrogen are continuously fed to the reactor. Product liquid and gases are removed at a rate that maintains a constant liquid level and a constant pressure within the reactor. On a commercial scale, the most common continuous-flow reactor is a fixed-bed tubular reactor, operating in a co-current, downflow mode.

Properties of Hydrogen

Hydrogen is a colorless, odorless gas having a density[2] at STP of 0.090 g/L. As this density is nearly 15 times less than that of air, hydrogen will rise above the air in a quiescent environment. This is particularly important to recognize when there is a potential for hydrogen to escape into a hood or room that is not provided with adequate ventilation. Dropped ceilings provide an exceptionally hazardous place to trap hydrogen.

Hydrogen will burn with a source of oxygen and does so with a flame that is colorless. It is often the case that there is no carbon or nitrogen present to provide the characteristic yellow or blue color often associated with flames.

Although an energy source, such as a spark, is required to initiate a hydrogen-oxygen flame at lower temperatures, auto-ignition (also called spontaneous ignition) will occur at temperatures exceeding 550°C, provided the concentration of hydrogen is not too rich or too lean.[3]

If the concentration of hydrogen is too high, there will not be enough oxygen to allow combustion to occur. Similarly, if the concentration of oxygen is too high, there will not be enough hydrogen to allow combustion to occur. In between these extremes combustion is supported and the concentrations are said to be within the *flammability limits*. The flammability limits for hydrogen in air range from 4 to 75% hydrogen.[2] It is worth noting that the flame speed of hydrogen[3] in air is 291 cm/s. This is nearly eight times faster than the flame speed of methane and nearly twice as fast as an acetylene flame.

One of the least remembered facts about hydrogen is that its temperature increases upon expansion. When we consider the frost that accumulates on an opened cylinder of nitrogen or the effect of aftershaves and astringents evaporating from our skin, we recognize that most gaseous expansions are accompanied by cooling. The thermodynamic explanation for these effects are embodied in the Joule-Thompson effect.[4] The rate of heating or cooling during an adiabatic expansion is characterized by the Joule-Thompson coefficient. As the temperature increases, the value of the coefficient decreases and eventually becomes negative at the so-called inversion temperature. Above this temperature, an adiabatic expansion will cause an increase in temperature. The inversion temperature for most gases is well above room temperature, so we are accustomed to witnessing a cooling effect. The inversion temperature for expansion of gases from one atmosphere to vacuum can be approximated as six times the critical temperature.[5] The values for

TABLE 55.1 Joule-Thompson Inversion Points[4,5]

Gas	T (°C)
Hydrogen	−75
Nitrogen	485
Oxygen	655
Helium	−240
Argon	630

some typical gases arc listed in Table 55.1. Note that the inversion temperatures of hydrogen and helium are well below room temperature.

Hazard Identification

The hazards of working with hydrogen and performing hydrogenations can be broadly grouped into three categories: chemistry, pressure, and temperature. Prior to beginning any reactions involving the use of hydrogen

FIGURE 55.4 Catalyst baskets.

it is important that the implications of each of these categories be understood. The safety of the persons in the laboratory rests in the hands of the operators and it is their responsibility to fully understand the ramifications of their experimentation, processes, and reactions.

Chemistry

Understanding the chemistry is the first category. One should be familiar with the intended chemistry, as well as the possible side reactions. Of great importance is to understand the stoichiometry of the reaction. Most hydrogenation reactions consume hydrogen from the gas phase leading to a reduction in the total number of moles present in a batch reaction, but such is not always the case. As shown in Table 55.2, some reactions with hydrogen can lead to an increase in the number of gas phase molecules present. When this occurs there will be an increase in pressure. Without prior expectation or knowledge of the possible stoichiometry, hazardous situations can occur, and appropriate steps to prevent catastrophic failure should be taken.

Care must also be taken that reactions are conducted in vessels made from the appropriate metals. Corrosion of the vessel can lead to a weakening of its pressure-containment ability. The use of strong acids, such as hydrochloric acid, provide an interesting example of metal corrosion. Stainless steel is not corroded by anhydrous HCl; however, it is rapidly corroded by wet HCl. Other material choices for the use of HCl include Alloy B-2 and titanium. The extent to which these metals are resistant to corrosion by HCl depends on the concentration of oxidizing ions. With as little as 50 ppm of Fe^{+3} present, the corrosion resistance of Alloy B-2 is greatly diminished. When no oxidizing ions are present, the corrosion resistance of titanium is greatly reduced.

It should also be recognized that catalysts can become contaminated if any of the reagents present react with the wall of the vessel. It is possible, for example, to leach iron or nickel from the wall of a stainless steel vessel and deposit it on the surface of a catalyst. Such depositions may act as a promoter or as a poison for the desired reactions or undesired side reactions.

TABLE 55.2 Stoichiometry of Three Hydrogenation Reactions

Reaction	Example	Molar Change
Hydrogenation	$C_4H_8=C_4H_8 + H_2 \rightarrow C_8H_{18}$	$2 \Rightarrow 1$
Hydrogenolysis	$C_4H_8=C_4H_8 + 2H_2 \rightarrow 2C_4H_{10}$	$3 \Rightarrow 2$
Hydrocracking	$C_4H_8=C_4H_8 + H_2 \rightarrow 4C_2H_4 + H_2$	$2 \Rightarrow 5$

An operator should also be cognizant of the expected rate of reaction. Many reactions proceed at a rate which can be easily monitored and controlled. Some reactions, such as the hydrogenation of oxygen, proceed much more quickly than others. Explosive reactions cannot be contained in pressure vessels designed for stresses which build up much more slowly. The shock wave from an explosion can damage a pressure vessel before a relief device has had time to actuate. Experienced chemists associate the following chemical groups with the possibility of explosions: acetylide, amine oxide, azide, chlorate, diazo, diazonium, fulminate, N-haloamine, hydroperoxide, hypohalite, nitrate, nitrite, nitro, nitroso, ozonide, peracid, perchlorate, and peroxide.[6]

The hydrogenation of a double bond in hydrocarbons is an exothermic reaction, with approximately 30 kcal/mole of heat being liberated. Conjugated double bonds, such as those found in dienes, are somewhat more stable and liberate heat on the order of 57 kcal/mole, rather than the expected 60 kcal/mole. The added resonance energy of aromatic rings allows benzene to liberate only 50 kcal/mole when hydrogenated to cyclohexane.[7]

It should be remembered that reaction rates are exponentially dependent on the temperature. Faster reactions produce more heat per unit time, which raises the temperature, which, in turn, raises the rate, and so on. Thermal runaway of reactions can occur when the temperature is allowed to rise uncontrolled. Higher temperatures also equate to higher pressures in closed vessels, so that thermal runaways become undesirable pressure excursions.

Pressure

Any of the above examples can lead to pressure excursions which, when alleviated, can discharge materials from the reactor vessel. Additional care should be exercised when these materials are toxic, mutagenic, or carcinogenic. Secondary enclosures should be mandated.

Pressure excursions are sometimes caused by overfilling a batch reactor with liquid. Liquids expand upon heating and they can eventually fill the vessel completely with liquid. In this state, small changes in the volume of liquid create very large hydraulic pressure forces. These forces can easily create a condition of overpressurization.

This condition can occur when heating any fluid, but water is particularly worthy of note. As shown in Table 55.3, the expansion of water is minimal below a temperature of 100°C, but expands 115% at temperatures around 200°C, 140% around 300°C, and over 300% before reaching its critical temperature at 374°C.[8] The liquid volume multiplier is the ratio of the specific volume

TABLE 55.3 Water Properties at Elevated Temperatures and Pressures[8]

Temp. (°C)	Liquid Volume Multiplier	Vapor Pressure (psig)
25	1.00	—
100	1.04	0
200	1.15	211
250	1.25	562
285	1.35	990
300	1.40	1230
325	1.53	1735
350	1.74	2385
371	2.28	3080
372	2.45	3110
373	2.75	3150
374	3.17	3195

of water at the listed temperature to its specific volume at 25°C.

To prevent potential hazards, a maximum allowable water level (MAWL) is defined as:

$$MAWL = \frac{(0.9)\ (Vessel\ Volume)}{(Volume\ Multiplier\ at\ Max.\ Temp.)}$$

Thus, a 1000-mL vessel to be operated at 300°C should not be filled to more than 643 mL. A good rule of thumb is the "2/3 Rule." Vessels should not be filled to more than two thirds of their capacity at room temperature.

American-made pressure vessels are designed in accordance with the standards of the American Society of Mechanical Engineers (ASME). Similar organizations produce similar standards in other countries. Each vessel is then rated depending on the strength of the metal from which it is made, the temperature of use, and the thickness of the metal. The maximum allowable working pressure (MAWP) is stamped on the vessel. A safety rupture disk is typically installed on all pressure vessels and is designed to rupture before reaching the MAWP. Should a vessel be pressurized to its maximum, the operator should be aware of certain consequences:

- The vessel will have expanded, but will generally return to its original dimensions when the pressure is relieved.
- Pressure gauges can generally be taken to pressures 10% greater than the dial maximum without significant loss of accuracy.
- Rupture disks will experience fatigue whenever they are stressed to more than 70% of their rated burst pressure. Repeated excursions will lead to premature failure.

Should a rupture disk fail during a hydrogenation, the gaseous and vaporous contents of the reactor will be expelled from the vessel. As the vessel is now open to the ambient atmosphere, precautions should be taken to prevent the continued addition of hydrogen. This is most easily accomplished by installing an excess-flow-check valve between the hydrogen source and the inlet to the reactor.

It is also important to consider the location to which the contents of a reactor will be expelled when a safety disk ruptures. It is common to vent the material into a hood, although the nature of the material being vented should be considered. Toxins, mutagens, and carcinogens will require special handling.

One should also consider the effects of the loud noise caused by the rapid release of high-pressure gases. In addition, operators should be aware that a possibility exists for the reactor to move because of the escaping gases. It is best to vent the gases vertically or be sure that the reactor is securely fastened to the bench or floor.

Escaping hydrogen may also pose a fire or explosion potential as it mixes with the ambient air. The most common means of addressing this potential are to remove any potential sparking sources, work in a barricaded room, and provide adequate ventilation. The removal of sparking sources implies the use of intrinsically safe, explosion-proof equipment. This subject is addressed in a separate chapter of this book. Barricades are generally designed with reinforced concrete on at least three walls. The fourth is sometimes designed to freely move so as to direct the expanding gases in a safe direction. It should be noted, however, that the amount of hydrogen is typically quite small, particularly when protected with an excess-flow-check valve. For example, in a 1-liter reactor filled 2/3 with liquid, there are 333 cc of hydrogen. If the reaction were taking place at a pressure of 1500 psig and a temperature of 150°C, this would amount to 0.8 cubic feet of hydrogen at ambient conditions. If a typical hood has 48 cubic feet of space, the concentration of hydrogen would be on the order of 1.7%. This is well below the flammability limits and exemplifies the need for and value of providing ventilation with a high turnover of air within the hood.

Temperature

The effects of temperature are the third general category of hazards to be identified. Much of today's equipment is electronically controlled and is generally considered to be fault-protected. Care should be taken to understand the limitations of all equipment. Thermocouples, for example, can become shorted and give erroneously low outputs. Temperature controllers, if not properly equipped with short-protection circuitry, interpret this as a low temperature and continue to apply maximum power to a heater. It is therefore, advisable to have a second thermocouple, in parallel circuit, to detect high-temperature excursions and terminate power to the heater before safe limits are exceeded.

It is also advantageous to use controllers that minimize temperature overshoots when heating to a new setpoint. This will prevent pressure excursions to levels that may cause fatigue in sensitive components. It will also assist in keeping the reaction rate under control and reduce the possibility of thermal runaway.

General Precautions

Given the multitude of things that could go wrong, even with the best planning and execution of experimental chemistry, there are a number of rules that should be followed as a minimum level of safety:

- Allow only trained operators to handle, adjust, or use the equipment.
- Operate hydrogenations in an area that will minimize the number of people affected if something does go awry.
- Minimize the amount of equipment located in the area that might be affected.
- Store unnecessary chemicals away from the reaction site.
- Monitor, maintain, and maximize the amount of ventilation in the area.
- Select appropriate hardware by ensuring that all components used in a hydrogenation are rated for the temperature, pressure, and corrosion resistance required.
- Insist on equipment that passes third party certification requirements.
  - Ensure that pressure vessels conform to ASME, TÜV, or other applicable code.
  - Ensure that all electrical devices conform to CSA, CE, NEC, or other applicable code.
  - Ensure that all equipment is manufactured by reputable companies conforming to ISO 9001 standards and are not homemade.
- Ensure that all equipment is installed and operated in a safe, approved manner, making sure that all electrical and local safety codes are met.
- Use explosion-proof equipment when dictated by code or when in doubt about the safety of a proposed operation.
- Ensure that all equipment is properly maintained.

References

1. Voorhees, V.; Adams, R. *J. Am. Chem. Soc.* **1922**, 44, 1397–1405.
2. Weast, R. D., Ed. *CRC Handbook of Chemistry and Physics*, 57th ed.; CRC Press: Boca Raton, FL, 1976.
3. Kanury, A. M. *Introduction to Combustion Phenomena*; Gordon & Breach: New York, 1975.
4. Perry, R. H., Chilton, C. H., Eds. *Chemical Engineers' Handbook*, 5th ed.; McGraw-Hill: New York, 1973.
5. Sienko, M. J.; Plane, R. A. *Chemistry: Principles and Properties*; McGraw-Hill: New York, 1966.
6. Livingston, H. K. *Chemistry* **May 1968**, 38.
7. Solomons, T. W. G. *Organic Chemistry*, 6th ed.; Wiley: New York, 1996.
8. *Steam*, 39th ed.; Babcock and Wilcox: New York, 1978.
9. Smith, D. T. *J. Chem. Educ.* **1964**, 41, A520–A527.

56

Chemical Inventory Control and Methods

SHARON E. STASKO

Chemical management and control are the responsibility of all persons involved in chemical acquisition, use, and disposition. Strong controls enhance safety programs, support regulatory compliance, reduce cost, and lower risk. Such controls require a detailed chemical inventory and policies and procedures for *maintaining and using* inventory data. A computerized system that uses bar coding and container-level monitoring can deliver the detailed inventory.

Getting Started

Implementing a chemical inventory program requires management to commit people and money, and to support change. It requires collaboration among many organizational groups, which may be asked to change how they order, receive, store, monitor use of, and dispose of chemicals. Throughout the process, they will want to know how their workload will be affected and what new tasks they will be expected to complete.

Communicating the purpose of the inventory and the expected benefits will generate support. Similarly, soliciting end-user input throughout the planning stages will help to avoid the "you didn't ask us how we do it" objections during implementation.

Planning and communication will ensure benefits to the organization and its employees, and to the community.

Define the Scope of the Inventory Project

The first step in inventory management is to define the scope of the proposed inventory. The scope of the project will limit the objectives that can be achieved. It will also identify who should be included in the planning stages and what resources will be required throughout the project.

- Will it apply only to one group or department, or will it apply to the organization?
- Will all chemicals be included regardless of source, use, or hazard ratings?
- Is this to be a stopgap program or a best practices effort delivering a sustainable program?

Document Objectives and Set Goals

Step two is establishing objectives. The objectives will define the type of system required and the data elements that should be recorded. In addition, they will identify the policies and procedures needed to implement and maintain a system. Sample objectives include the following:

- comply with federal, state, and local regulations for record keeping and reporting
- reduce liability exposure
- control chemical purchase and disposal costs
- improve MSDS management

Identify the objectives and then list the specific goals that must be achieved to signal successful inventory implementation. For example, measurable goals include:

- reduce the number and size of like chemicals stored throughout the facility
- prepare a list of all on site chemicals by location and user group
- identify training, storage, transportation, and safety requirements for each item
- provide inventory data in a form that can be accessed by all employees
- document MSDS availability for each item
- improve cost control by permitting direct charge for chemicals to projects, accounts, and so forth

Prepare Site Policies and Procedures

Step three is defining site policies and procedures required for chemical inventory. These guidelines will grow out of the scope and objectives, and an honest assessment of the need and ability to modify current operating procedures. Identify what substances will be permitted on site; what will be included in the inventory; how chemicals will be ordered, received, and recorded in the inventory; what transactions will be monitored (transfers, refills, etc.), and by whom.

Who Is in Charge

Step four is identifying who will sponsor and ultimately manage the program. Although a group may review, evaluate, and select an inventory system, and the group may collaborate throughout the implementation stage and play a role in maintenance, it will not retain responsibility for the inventory performance and data.

Do not invest in an inventory if a manager has not been identified. The manager must be accountable, responsible, and empowered to make the system work.

Building Inventory Records

Historically, numerous tools including card files and spreadsheets have been used to list chemicals. These tools serve a limited group and require labor-intensive updating. A computerized chemical management system, however, provides a dynamic resource to the entire user community. Not merely a list of items, such a system includes relational databases that minimize data entry, ensure consistent data, and track every chemical by container. Ideally, the system monitors chemicals from requisition until the container and any remaining chemical are properly disposed. Used in a network environment—and

perhaps on the organization's Intranet, the data are available to everyone who works with the chemicals.

The data included in the inventory must respond to the stated management objectives. *Minimum* data for a meaningful inventory record include:

- the container identification—a number applied to individual containers
- the chemical identification—Chemical Abstracts Service (CAS) number and description
- the size of the container—unit and unit of measure stored in different fields
- the location to the building and room level

A more robust and comprehensive inventory will include the information listed in Table 56.1.

Container Labeling

Numbered labels for chemical containers are central to a computerized system. Attached to each container at receipt, the label identifies the chemical and points to acquisition, storage, safety, regulatory, and disposal data.

TABLE 56.1 Data Needed for a Complete Chemical Inventory Record

Field Name	Definition
ID number	Bar coded number identifying specific inventory item
Description	Chemical name
CAS number	Chemical Abstracts Service number identifying chemical
Vendor ID	Code identifying the chemical vendor
Vendor's product ID	Vendor product number; permits vendor-specific MSDS control
Grade	Grade designation for the chemical
Size	Amount of chemical in the container, e.g., 500
Unit of measure	Standard unit of measure; for example, grams, liters, gallons
Original amount	Original amount received
Amount remaining	Current amount; original amount less consumption
Location	Code identifying the chemical storage area
Lot number	Vendor's lot number
Container	EPA container type or size code
Acquisition method	User-defined code that identifies purchases, samples, etc.
Purchase order number	Purchase order number
Order date	Purchase order date
Date in	Date received
Expiration/reevaluation date	Provided by the vendor or calculated using shelf life and received date
Bottle open date	Date container opened
Cost	Purchase cost, total and unit
Surplus	Flag to identify surplus chemical available for sharing
Inventory date	Date of last inventory
Last change	Date of last change to record
Notes	Editable data specific to individual chemical record
MSDS reference	MSDS location, date
Safety data	Multiple fields to monitor storage, hazards, personal protective equipment requirements
Regulatory data	Regulated status, permitted quantities

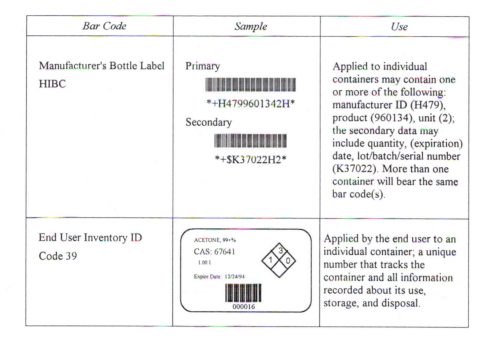

Bar Code	Sample	Use
Manufacturer's Bottle Label HIBC	Primary *+H4799601342H* Secondary *+$K37022H2*	Applied to individual containers may contain one or more of the following: manufacturer ID (H479), product (960134), unit (2); the secondary data may include quantity, (expiration) date, lot/batch/serial number (K37022). More than one container will bear the same bar code(s).
End User Inventory ID Code 39	ACETONE, 99+% CAS: 67641 1.00 l Expire Date: 12/24/94 000016	Applied by the end user to an individual container; a unique number that tracks the container and all information recorded about its use, storage, and disposal.

FIGURE 56.1 Bar code use in chemical identification.

Although inventory systems without individual container controls are common, the benefits of container labeling are numerous, for example:

- easy correlation of chemical to location and to end user exposure
- clear identification of shelf life and reevaluation requirements
- specific chemical and location data to facilitate chemical sharing
- detailed, auditable disposal records for individual items
- quick response to recall notifications

Labeling systems should be consistent and easy to use, and labels must be appropriate for the chemical environment. The label and adhesive must be able to withstand extremes in temperature, be applicable to glass, metal, plastic, and other surfaces. The label ink must resist spills, fluorescent and ultraviolet lighting, and moisture. Label size may vary according to container size and environment, from 0.25″ × 0.75″ for lab use to 4″ × 10″ or larger for use in production storage areas.

Providing the identification number in bar code format will permit scanning and reduce time required to record routine inventory transactions. The standard bar code symbology for chemical inventory purposes is code 39 or the Health Industry Bar Code (HIBC), a code 39 variation. Code 39 permits use of numbers, letters, and a limited character set. Printing labels as they are

needed allows the user to adjust the label size to the container and include other information on the label such as chemical name, CAS number, receipt date, original amount, expiration date, NFPA codes, or location.

The inventory bar code differs distinctly from bar codes applied to containers by manufacturers or suppliers. Manufacturer bar codes provide useful data but because identical bar codes can be used on more than one container, they cannot be used for inventory control. Figure 56.1 summarizes the differences between manufacturer and inventory bar codes.

Chemical Identification

An inventory system requires a standardized method such as a "catalog" or master file to describe all chemicals. The chemical identification can be entered into the master file one time and used repeatedly to prevent corrupted data—including odd synonyms, spelling errors, and general inconsistency. Once created, a master file containing 5000 records, for example, provides a tool to speed data entry for an inventory that may total 25-, 50-, or 100,000 container records.

The master record should contain the chemical name, the CAS number, vendor identification, and vendor product number. The CAS number or chemical name alone will not provide adequate detail for researchers, production managers, or health and environmental safety staff. The vendor and product number will speed the initial inventory, facilitate future purchasing, promote shared use, and allow vendor-specific MSDS management.

The master file can relate to synonyms, structures, safety data, regulatory requirements, personal protective equipment (PPE), NFPA guidelines, or mixture constituents. Some purchased inventory systems include a startup listing of chemicals provided by one or more vendors; if not, many vendors will provide the information in database format.

Storage Locations

By regulation, the physical location of hazardous chemicals must be tracked at the building or lot level.[1] Meeting other objectives may require tracking at the room level. Room-level tracking will enable the organization to monitor storage practices, employee and visitor exposure to chemicals, training requirements, PPE availability, and MSDS distribution.

Compliance with building code standards or with local, state, or federal guidelines requires control by zones, which may be composed of several rooms on a floor. The total amount of substances by classifications within each zone must be calculated according to the building construction and the floor.[2]

Associating chemicals with their actual storage locations within the rooms will allow quick assessment of compatibility and compliance with storage guidelines. In addition, knowing the specific bench, hood, corrosive cabinet, flammable cabinet, freezer, shelf, or bin will simplify shared use. A location hierarchy includes buildings at a site, rooms within buildings, benches or aisles within rooms, cabinets within an aisle, and so forth.

Each location requires a unique identifier. For example, assigning hood 1 in Smith Hall, Room 101 the random number "H1445" reduces the amount of information that must be entered for any chemical stored in that location. In addition, the identifier can be printed as a bar code and applied to the location for scanning in transactions such as periodic inventory that use location information.

Accountability

A complete inventory record will identify who purchases, manages, uses, or is potentially exposed to the chemicals. Associate chemicals with organizational units, projects, customers, or employees if your objectives include training, safety, and liability and cost reduction.

Acquisition Data

The site guidelines for allocating chemical costs and for chemical sharing will drive the requirements for acquisition data. The source of the item (new purchase, vendor sample, etc.), its cost, purchase order number, receiving and expiration dates, and lot number may be useful.

Safety Information and MSDS

The relational database format for inventory control will improve tracking of hazard and safety information. The databases can manage OSHA, EPA, and state regulatory lists; identify carcinogens and other specific hazards; and track PPE, safe storage guidelines, reporting requirements, and MSDS availability for each master record.

Users should be able to move easily from any chemical record to the MSDS, whether it is stored electronically or in paper form. A database of on-site MSDSs will reduce the manual labor required to maintain data sheets and provide quick access to them. If vendor-specific sheets are maintained, the MSDS database should include the following fields:

- CAS
- vendor
- product number
- chemical name
- MSDS original preparation date
- received date
- MSDS location
- update date(s)

The CAS/vendor/product number provides a direct link to the inventory record and to the master chemical record. The location field points to the data sheet whether it is stored on a CD, in a file on the server, or in the Environmental Health and Safety paper files.

Conducting the Initial Inventory

When the physical inventory begins, the inventory system and procedures must be in place. As the inventory proceeds, the site must be prepared to record new acquisitions and document disposal of items that have been included in the inventory.

Who Should Conduct the Inventory?

The individuals who will perform the inventory are collectively "the team." At least one individual should become the coordinator. This person will arrange schedules, ensure access to all locations, maintain the resources needed by the team, and provide inventory reports to individual locations and groups, as well as to management, throughout the process. The coordinator will give clear information requirements to the team whether it is composed of in-house staff or is outsourced.

Team members must demonstrate a good knowledge of chemicals and lab safety guidelines. They must be computer literate.

In each location, a responsible person should be available to answer questions and provide access to locked areas. After the inventory, provide a listing of the chemicals to the person responsible for the location and request a review of accuracy and completeness.

Scheduling

The schedule for the inventory must take into consideration the normal work being conducted in each area, the volume of chemicals within the area, and the resources available. Find out if laboratories are being moved, if new spaces are being opened, or if large quantities of chemicals are to be delivered for new projects. Such activities will affect the schedule.

The initial inventory will be labor-intensive. Plan for 2–3 minutes per container if you have a good master chemical file; schedule more time if the entire master file must be created during the inventory. This time estimate includes removing the container from storage, wiping it clean, entering data into the computer, printing and/or applying the label, and returning the container to its proper storage location.

Include all inventory activities in the schedule: a walkthrough to assess each area, training, the roundup, and the initial inventory itself.

No matter how the schedule is defined, let everyone know the schedule. Circulate the schedule so laboratories will be ready for arrival of the inventory team.

Roundup

Prior to the initial physical inventory, conduct an inspection and roundup of the chemicals stored in each location. The goal of the roundup is to remove items that should not be inventoried. The roundup will reduce the time required for inventory, but it may cause a one-time increase in disposal costs.

Evaluate age, usefulness, packaging, labeling, storage conditions, and current need for the chemical. As appropriate, segregate chemicals for disposal or transfer to other groups.

Roundup also affords an opportunity to identify and bar code all storage locations.

What to Include

When defining what will be included in the inventory, use language that limits interpretation. Review the project's objectives.

Because the term "hazardous" can be debated among different groups, the EH&S department should clearly define what substances meet the organization's "hazardous" definition.

Provide guidelines to the team for recording quantity. Rather than an estimate of the amount remaining in a container, the inventory record usually documents the original amount.

On-Hand Substances

All hazardous substances on-hand at the start of the inventory process should be labeled and listed in the inventory database. These would include research chemicals, and possibly cleaning and maintenance materials, agricultural supplies such as fertilizer, and office supplies such as toner.

Some organizations include all purchased substances, whether included on current OSHA, EPA, or local-regulated substance lists or not. Some substances may be listed but not labeled, for example, bleach, rock salt, and correction ink. Such meticulous application provides thorough records of exposure in the event classifications change; it also represents best practices in risk-minimization.

Samples or Demonstration Materials

Samples received for analysis, evaluation, trial, or other use from vendors, trade groups, colleagues, and competitors should be included in the inventory if they meet the guidelines proposed for other substances in the site policies and procedures.

Untagged Items

Some chemicals may be inappropriate for tagging due to container size, storage requirements, or inaccessibility. Such items should be assigned a number and marked in your database records as "untagged." A list of untagged items for each location can contain the bar codes that would otherwise be on the containers. The list can be taped to a freezer, for example, for ready access.

Noninventoried Items

Because omitting substances from the inventory calls for individual judgments relating to chemical hazards or reporting requirements, a list of "excluded items" will simplify compliance. Be specific about excluded items, for example, "Exclude work in process." Leaving the decision about what should be included to the end-user can result in inconsistency or even liability. The list should be prepared by the EH&S staff.

Nonallowed Items

Identify any items that are not permitted on site in any quantity. Document procedures for the inventory team to use with these items.

Safety Issues

The chemical inventory process requires specific training if the task is to be completed without incident. The inventory team will move every container in the facility at least twice. *Emphasize safe handling procedures.*

Standard laboratory safety guidelines must be followed whether the inventory is being conducted in a laboratory, in a tool shed, or in a basement storage area. The team will need:

- safety glasses
- lab coat with cuffs secured to prevent catching on bottles
- appropriate gloves
- flat, rubber-soled shoes
- a step stool to enable access to high areas, and comfortable seating at the low areas

Provide written, detailed guidance for handling the following:

- open, leaking, or bulging containers
- unlabeled containers
- research in process
- unventilated spaces

Ensure that each person has the emergency response team number printed and taped on the clipboard or computer, for example. A person with responsibility for the location being inventoried should be available to answer questions.

Inventory Tools

To complete the inventory with as few delays as possible, the team will need the right tools:

- a lab cart for easily moving the inventory equipment around the facility
- reference books to look up CAS numbers or chemical names
- clean wipes for removing dust from containers before labels are applied
- alcohol or cleanser for cleaning in dirty environments, such as shop maintenance areas where oily surfaces will be encountered
- bar code printer or preprinted bar code labels

- a tabletop computer, laptop computer, portable data collector, and power supply, or preprinted forms (if computers or electrical power are not available)
- preprinted forms for recording exceptions such as locked and inaccessible storage areas
- an extension cord

Bar Codes

Bar codes must be applied to containers in a way that makes them scannable. The beam emitted by the bar code reader must touch the entire length of the bar code. For that reason, it is recommended that bar coded labels be applied vertically, not horizontally. Consistent placement will eliminate judgment errors and simplify future inventory scanning tasks.

Maintaining the Inventory

Maintaining inventory records after the initial inventory is complete will ensure that the organization achieves the best return on its investment. At a minimum, routine maintenance includes adding new items and marking items as "disposed." Budget about two minutes per item for adding new items, less for documenting disposal.

Purchase

The organization's purchasing and receiving procedures present the greatest challenges to maintaining a chemical inventory system. If the organization allows end-users to order via electronic commerce or telephone or allows delivery directly to the end-user, special procedures or programs will be needed to maintain inventory records. When practical, the chemical system should be integrated with other systems to overcome these obstacles.

Chemical purchases can be reduced if users have access to the inventory system and sharing is encouraged. They can determine chemical availability before they initiate a purchase requisition.

Orders should be for the smallest amount required for the task, even if the per unit purchase cost is greater. Disposal costs typically exceed the savings in purchasing costs.

Receipt

A single point of chemical entry simplifies inventory control. Receiving staff record new items in the database, print and apply labels, and verify MSDS status. When labeling at the receiving point is impractical because of packaging or hazard concerns, the label can be taped to the exterior container and applied to the chemical container or the "untagged" list by the end-user.

If multiple entry points are permitted in the organization, inventory procedures must be defined and adhered to rigidly. The burden will rest with the receiving end-user.

Use

To improve day-to-day management and to maintain database integrity, procedures should be in place for changes such as transfers from one location or user to another, consumption, refills, vendor returns, or disposals. These transactions can be recorded at the PC or in portable data collectors that take advantage of the bar codes used throughout the facility. The data can be transferred electronically to the PC where records are updated.

To reduce disposal costs, the site may want to identify chemicals as "surplus" or "available for sharing" and circulate a list of these items to facilitate further use.

Periodic Physical Inventory

Sites should schedule periodic physical inventories for one or more locations or responsible groups. An inventory updates location records and validates the accuracy of transaction records for transfers, use and disposal. It can be used to evaluate chemical storage and disposal practices, to document losses following disasters, or to audit substances following personnel changes.

If bar codes have been used to identify locations and inventory items, the periodic inventory will proceed rapidly, permitting as many as 500 items to be scanned in a single morning. The user will scan the location, then scan the chemical bar codes within each location. The collected data, once transferred to the PC, will permit quick reporting and identification of missing or mislocated items.

Disposal

Marking chemicals as "Disposed" is necessary to maintain accurate records of on-site quantities and to document past activity. The inventory system should include on-site and disposed inventory records and all should be available for viewing and reporting at any time. The disposal record will document how both the container and any remaining chemical are disposed, as well as how empty containers are disposed. It may include the lab pack, waste carrier, disposal method or site, as well as manifest and certificate of disposal numbers.

Scanning the container bar code is the most accurate way to document disposal transactions. The bar code reader can be used in any location whether a lab or a waste accumulation facility. Be sure that items are not counted as "disposed" if they have been transferred to an accumulation area. They remain on-site and the inventory system should monitor the time items have been in storage.

References

1. The BOCA National Building Code, 1993, p. 15ff. www.bocai.org.
2. The Health Industry Bar Code (HIBC) Supplier Labeling Standard, www.hibcc.org/sls_sec2.html.

57

Chemical Storage

LYLE H. PHIFER

Storing chemicals safely is largely a matter of simply being logical. The individual who is responsible must have a reasonable chemical knowledge, access to Material Safety Data Sheets (MSDSs) on all the products under his control, and a knowledge of the OSHA rules relating to chemical storage.

OSHA Rules

There are surprisingly few specific OSHA regulations associated with chemical storage. Nearly every reference to storage involves flammable and combustible liquids. A flammable liquid is defined by OSHA as having a flash point below 100°F (29 *CFR* 1910.106(a) (19)) and is designated as Class 1. This is subdivided into Class 1A—having a flash point below 73°F and boiling point below 100°F (29 *CFR* 1910.106(a) (19) (I)); Class 1B—having a flash point below 73°F and boiling point at or above 100°F (29 *CFR* 1910.106(a) (19) (ii)); and Class 1C—having a flash point at or above 73°F and below 100°F (29 *CFR* 1910.106.(a) (19) (iii)). Combustible liquid is defined as having a flash point at or above 100°F (29 *CFR* 1910.106(a) (18)). This is subdivided into Class II—having a flash point at or above 100°F and below 140°F (29 *CFR* 1910.106(a) (18) (I)); and Class III—having a flash point at or above 140°F and below 200°F (29 *CFR* 1910.106(a) (18) (ii)).

Container Size Restrictions

Container size restrictions are imposed depending on the class of the liquid and the location of the storage areas. Table 57.1 gives the maximum allowable size of the containers. As an exception, Class 1A and Class 1B materials can be stored in up to 1 gallon glass or plastic containers if exposure to metal would render them unfit for its intended use (29 *CFR* 1910.106(d) (2) (iii) (a) (1)).

In general, laboratory quantities of Class I liquids should be transferred to safety cans. Fifty-five-gallon drums should be kept in a secure, isolated, preferably outdoor, location. It should be noted that transferring Class I liquids from 55-gallon drums to a metal safety can should be performed with a metal connector to avoid static electrical problems. The drum should also be grounded. This can be done with a metal grounded plate or direct ground.

Specifications for Solvent Storage Cabinets

Specifications for solvent storage cabinets are also included in the OSHA rules. Curiously, there is no specification for venting of vapors. Even a minor leakage of vapors from Class I liquids can easily result in an explosive air mixture in the storage cabinet; adequate venting can only be accomplished with some type of fan. This presents a potential explosion problem if corrosive effects over time result in overheating of the fan motor or bearing and result in igniting vapors. Even though OSHA approves of storage cabinets for flammable liquids, the question can be raised as to how safe they really are. Properly maintained safety cans are very logical replacements for storage cabinets.

Hazard Communication

The OSHA Hazard Communication Standard specifies that the label on chemical containers must include the identity of the chemical, appropriate hazard warnings, and the name and address of the manufacturer, importer, or other responsible party. If this label is damaged or otherwise defaced, it should be immediately replaced with a new label containing the appropriate information.

OSHA rules require that a MSDS be available for every chemical on the premises. It is highly recommended that the MSDS be available or readily accessible on a computer, that all employees having access to chemicals be familiar with access to the appropriate MSDS, and that the facility have a requirement that any individual using the chemical read and understand the MSDS before handling the chemical.

Inside Storage Rooms

Inside storage room specifications are included in the OSHA rules. The important aspects include:

1. Construction must meet the required fire resistance rating for their use.

TABLE 57.1 Maximum Allowable Size of Containers and Portable Tanks (29 *CFR* 1910.106, H-12)

Container Type	Flammable Liquids			Combustible Liquids	
	Class 1A	Class 1B	Class 1C	Class II	Class III
Glass or approved plastic	1 pt	1 qt	1 gal	1 gal	1 gal
Metal (other than DOT drums)	1 gal	5 gal	5 gal	5 gal	5 gal
Safety cans	2 gal	5 gal	5 gal	5 gal	5 gal
Metal drums (DOT specification)	60 gal	60 gal	60 gal	60 gal	60 gal
Approved portable tanks	660 gal	660 gal	660 gal	660 gal	660 gal

2. Openings to other rooms must be provided with liquid-tight raised sills or ramps at least 4 in. high. As an alternative, the floor can be at least 4 in. lower than the surrounding area, or an open grated trench can be in the room, provided it drains to a safe location.

3. Doors must be self-closing fire doors.

4. If Class I flammable liquids are to be stored, the room cannot be on a basement level.

5. Ventilation can be either gravity or mechanical, but it must provide complete change of air at least six times per hour. Makeup air and exhaust must be from and to the outside. For a mechanical system, the control switch must be outside the room and the same switch must control the lighting in the room. A portable light adjacent to the switch must be available if Class I flammable liquids are dispensed in the room.

6. There are specific limits on the total volume of each class of flammable liquid stored in one room.

7. Containers over 30 gal in size cannot be stacked on top of each other.

8. If piles or pallets of containers are stored in the room, there are restrictions that apply. Distance limits are applied depending on what class is being stored. No pile may be closer than 3 ft to the nearest beam, chord, girder, or other obstruction and 3 ft from sprinkler heads or any overhead fire protection system.

Overhead Storage

Another potential storage problem is overhead storage access. Any area over a work area must house a posted sign of the maximum storage weight and be surrounded by at least a 4-ft-high barrier capable of withstanding 200 lbs of side pressure and a kick plate. This applies regardless of what is being stored in that area, including items other than chemicals.

General Rules for Chemical Storage

There are several generalizations that should be applied to the storage of chemicals:

1. Purchase the smallest quantity appropriate for the expected use. The large economy size may look to be a bargain; however, disposal costs when the chemical is outdated or decomposed can run the cost to many times more than the savings. In addition, a spill or accidental breakage will result in a much easier and safer cleanup.

It should be noted that very small vials seldom break even if dropped onto a concrete floor.

2. Put the date of purchase on every container before placing in storage. This is particularly valuable in the case of unstable compounds or peroxide formers. Several producers, in particular for small quantities, are putting expiration dates on their products. These are not necessarily reflections of the expected stability, but rather are warranty dates or suggested time frames for replacement if stability problems may be encountered. In any case, no one should use products after the warranty date and the materials should be disposed of as waste. Maintaining inventory data on a computer makes it fairly easy to keep track of this information.

3. Isolation of dissimilar reactive materials is considered to be very important, particularly where larger quantities (over 1 L) are involved. This should include strong acid-base groups, oxidizers, and strong reducing agents. It is recommended that peroxide-formers be isolated from other chemicals primarily to ease the monitoring for peroxides.

4. A chemical hood should never be used for chemical storage. If something happens like an explosion or fire, everything in the hood will be involved in the accident. A good rule is to have only one product in the hood at one time.

5. The floor, either in a warehouse or laboratory, is not the place to store chemicals, even temporarily. In an emergency, someone may stumble over the containers and create even more of an emergency.

6. Laboratory versus Storeroom Storage: Laboratory storage of chemicals is usually different from storeroom or warehouse storage. Storerooms can effectively control and manage inventory by any of a number of systems, and usually there is a storeroom manager with specific responsibilities for safe storage. In the lab, the odds are that the container has been opened and exposed to air and moisture. There is frequently no order as to how they are stored (i.e., incompatible materials in same area). Rarely does anyone keep track of a chemical's age during laboratory storage. Transferring of chemicals to another container invites inadequate labeling of the new container. Frequently no adequate record is available as to which chemicals are present in the area. With time, one is likely to find old and outdated products which should be disposed.

All of these items should be addressed. Some of the potential problems can be solved by a regular, scheduled inspection of each lab area. Having standardized labels available can help solve the labeling problem. A small-quantity approach to chemical purchasing and inventory management can also contribute to safe storage of chemicals in the laboratory.

Special Situations

There are specific rules for storing certain types of materials, including:

1. *Controlled substances, including drugs and drug precursors*: It is not legal to have certain chemicals on your premises without a special permit.
2. *PCBs*: Purchases of these materials are reported by the suppliers to the U.S. Environmental Protection Agency (EPA). There are special rules for labeling, including the location of the materials. Disposal also requires permits.
3. *Ethanol*: Tax-free ethanol is available for laboratory use. The ethanol must be kept under lock and key and its use reported.
4. *Carcinogens or highly toxic materials*: These materials should be kept in a separate storage area, preferably with a separate ventilation system and limited access.

Emergency Planning

No matter how safe and efficient the chemical storage area may appear, there are potential emergencies that should be addressed. The following checklist can help in developing an emergency preparedness and prevention plan.

Fire

- Should a fire be fought by employees? Under what circumstances?
- What information should be available for the fire department?
- Are there ample alarms and sprinklers present in the storage area?
- Are emergency exits available and known to everyone who may be in the area?
- Is automatic call equipment available for hours when no one is on the premises?
- Are there other flammables in the area that need not be there?
- Are fire extinguishers available and checked on a regular basis?
- Are they the appropriate extinguishers for the types of material being stored?
- Do employees know how to shut off sprinklers?
- Are there procedures in place for reporting a fire?

Earthquakes

- Is shelving firmly secured so no possibility of turning over exists?
- Are there restrainer strips in the front of the shelves?

- Is the order of the inventory adjusted so that large containers are on the lower part of the shelving?
- In the event of the collapse of the area, what preparations for handling the situation are in place?

Flooding

- What is the possibility of the area being flooded?
- Are there empty or nearly empty storage tanks or containers that will float in the area?
- Are there water-reactive chemicals stored in the area?
- Are there water-resistant labels on all the containers?

Spills

- Is there a written spill-handling procedure in place, and have the employees been trained in spill cleanup?
- Is proper respiratory protection equipment available, and have employees received the required training in its use?
- Are appropriate absorbents available for liquid spills?
- Are emergency evacuation procedures in place and exercises scheduled on a regular basis?
- Are there adequate containment curbs in place to prevent the chemical from geting outside and contaminating soil or surface waters?
- Are spill kits available where appropriate?

Miscellaneous

- Are there any water-reactive chemicals stored on shelving over a sink?
- Are waste storage cans corroded or leaking?
- Is accumulated waste being disposed of within the required time limits?
- Is there a regularly scheduled inspection of the storage area for loose labels and caps?
- Are there chemicals which are no longer needed or out of date in the storage area?
- Does the facility have a policy on what constitutes an emergency for reporting purposes?
- If employees are to respond to emergency situations, have they had the required OSHA Hazardous Waste Operations and Emergency Response training?
- Does the facility have a written emergency-response plan, and does it address chemical storage area emergencies?

58

Cryogenic Safety

GEORGE WHITMYRE

Cryogenic Gases

By definition, cryogens are gases, liquids, and solids at or below 200 K (−73°C, −100°F). Substances at these low temperatures no longer exhibit their familiar characteristics. As one writer succinctly puts it[1]: "The main cause of cryogenic accidents is that working with or around cryogens is alien to the normal exerience of most people and consequently introduces unexpected hazards."

Engineering and administrative controls are essential to the safe use and handling of cryogens. To that end the checklist below is presented as a basis for readers' *adaptation* to their local environs. It should not be used *verbatim*, either in part or as a whole.

Of equal, if not superseding, importance, administrators of workplaces where cryogens are used and handled and the users themselves should attend and participate in training courses dealing with the safe use and handling of cryogens. Such courses are available from federal agency labs and several industrial and academic engineering departments. Typically, the courses emphasize the ongoing development in safety technology driven by consensus standards of recognized prudent practices along with local, state, and federal codes.

Cryogenics Safety Checklist

Freezing and Frostbite

Full-body protection including hands, eyes, and foot coverage is essential to prevent a cryogen spill or splash from freezing body tissues. Upon thawing, severe tissue destruction may occur, with edema and blood clots blocking local blood circulation.

1. The user is aware that hand and body contact with cryogenic-cooled surfaces, especially moist skin, will cause almost instant frostbite. Expanding gas, except hydrogen and helium (negative Joule-Thompson coefficient), may produce skin freeze burns from adiabatic cooling.
2. Splash goggles are used at all times when using liquid cryogens. Full-length face shield protection is required when using a wide-mouth Dewar flask

and when making pressurized liquid cryogen transfers.
3. Loose, insulating gloves with a gauntlet cuff are worn to protect the cryogen user from cold surfaces; they are quickly thrown off if soaked with a cryogen.
4. Workers handling large volumes of cryogens wear a leather apron, have cuffless trousers that extend over boots, and use a face shield, splash goggles, and gloves.
5. Frostbite first-aid procedures are included in cryogen emergency procedures training (see Box 1).[1,2]

Preventing Injuries When Moving Dewars

A 160-L Dewars weighs over 500 lbs when full and presents a serious potential for crushing injuries to hands, arms, legs, and feet. Awareness of hand and foot positions relative to the Dewar cart and using only equipment designed for Dewars can help prevent accidents.

6. Small, capped Dewar flasks of up to about 10 L volume with a handle can be hand carried with adequate personal protective equipment. Dewars

---

Box 1
Cryogen Frostbite First Aid

- Immerse with water at 100–110°F to warm rapidly. Use body heat or warm air if a water bath or shower are unavailable.
- Do not expose the wound to an open fire or flame.
- Do not rub or massage the affected parts of the body.
- Patients should not walk on frostbitten feet.
- Do not use medications. Do not apply dressings on intact skin.
- Keep the patient calm and at normal body temperature until medical help arrives.

Source: Jefferson Lab EH&S Manual-6500 Cryogenic and ODH Safety.

---

at or above 25 L volume require a rolling dolly base; 80- to 160-L and larger sizes require a wheeled base or a 4-wheel, pneumatic tire cart designed for moving large Dewars. Except for tip-back Dewar moving carts, cryogen containers are kept vertical.[3]

7. Transporting and using cryogenics from uninsulated containers (e.g., ordinary lab glassware and plastic ware) may cause frostbite burns, container embrittlement, and breakage. Liquid oxygen may possibly react with some polymeric materials.

8. Cryogen users are aware of their hand and feet positions to prevent crushing injuries when transport dollies are rolled, loaded, and unloaded. Bystanders are alerted when they are in the vicinity of a moving cart.

9. Where colleagues may step through doorways into the path of a moving Dewar cart, a spotter walks a few steps ahead of the loaded cart. The Dewar is pulled when rolling through a doorway, over a threshold, or loading on or off an elevator.

10. Extra clearance is provided around the Dewar to keep hands and feet clear of walls and lab equipment when tipping or setting a transport dolly upright.

Bulk Storage

Stationary 500-gal and larger Dewars provide liquid cryogen for transfer to smaller Dewars as needed. The addition of an external vaporizer coil provides a pure gas source for lab use.[6]

11. Bulk cryogen tanks are sited away from basements, depressions, and tunnels.

12. Inside piping from the bulk vessel vaporizer is clearly labeled along the pipe run and at all terminations in the building.

13. Oxygen meters monitor areas where inert gas from high volumes of cryogens are piped to rooms below grade.

Preventing Pressure Explosions

Cryogenic fluids trapped in closed systems will build up pressure rapidly upon heating. Vacuum-jacketed Dewars and other insulation systems for cryogenic fluids will reduce heat flow into the cryogen, but some heat flux is unavoidable. A pressure-relief poppet valve must constantly vent the gas pressure to prevent sudden rupture of a sealed vessel or pipeline. This normal venting produces a disconcerting hiss that may alarm inexperienced lab personnel.

14. Users are aware that the rapid expansion of cryogenic liquid-to-gas in sealed, nonvented containers (e.g., cold traps) can cause the vessel to burst. Glass sample tubes, cold traps, and other fragile components of a cryogenic system are shielded in the event that thermal shock or a pressure buildup causes a pressure explosion.

15. Glass Dewars are either contained in a Dewar shield or wrapped with fiber-reinforced tape. Plastic mesh over glass Dewars improves the hand grip but will not reliably stop glass fragments from an implosion.

16. Backup pressure relief is installed to vent 2-phase (liquid and vapor) flow caused by unusual system heating (fire or exothermic events). Uninterrupted (no valves) rupture disc venting is sized to relieve the maximum 2-phase flow from worst-case conditions.

17. Every pipeline or other part of the cryogenic system that can be isolated by valving off has a separate pressure relief. In addition to pipelines, pressurized supply Dewars and cryogen transfer hoses connected point-to-point have overpressure protection.[5]

18. Pressure relief is installed on sealed cooling bath space, cold fingers, external cryogen cooling jackets, or evacuated space contacting a cryogen. High-pressure buildup can result from cryogens trapped in these areas.

19. The pressure relief is set at or below the MAWP of the lowest-rated system component. All cryogen contact surfaces, valves, piping, and other devices are rated for cryogenic service.

20. Low-temperature material strength, material shrinkage, and cryogenic embrittlement are considered. For example, leaks undetected at room temperature can develop at cryogenic temperatures when cold shrinkage exceeds normally encountered tolerances. Most plastics become brittle when exposed to cryogenic fluids and some metal alloys develop stress cracks from cold shrinkage.

21. Liquid helium and hydrogen cryostats and other systems are pressurized so condensing air will not block pressure-relief devices.

22. Vent and relief valves that dump large quantities of gas or cryogenic liquids in normal or emergency modes are not vented into a building.

23. The retaining chain or safety cable is kept in good condition on pressurized Dewars with a re-

Dewar Leaks and Tip-Overs

24. The user is aware that internal pressure-building coils in larger Dewars produce a normal frost line on the external Dewar shell. Also, hissing of vapor venting from the pressure relief valve is a normal sound.

25. In the event of a tip-over, the area is evacuated. A pressurized Dewars that suddenly develops a horizonal frost line may have a ruptured inner vessel leaking into the vacuum jacket. The area is kept clear until the rupture disc discharges the contents and vapors dissipate as shown by oxygen meter readout.

26. Gentle handling of Dewars is practiced on the loading dock and in the lab. Hand-rolling 80-L and larger Dewars, tilted on the bottom edge, can result in tip-over. Also, fluid inertia from rolling can transmit enough torque to the inner vessel to snap off the internal retaining stem, producing a cryogen leak into the vacuum jacket.

Precautions for Liquid Oxygen

Oxygen enrichment (>22% by volume) accelerates burning of ordinary materials and lowers ignition energy of materials normally considered noncombustible.[6]

27. Organic materials including chemicals, fabrics, lubricants, and asphalt are excluded from liquid oxygen work areas.[2]

28. Air-freezing cryogens are handled in Dewars and vacuum-jacketed piping to prevent air freezing on the surfaces. When air that is frozen between the pipe and insulation vaporizes it can blow off the insulation explosively. This is prevented either with low-strength insulation that yields at low pressure or with air-tight insulation to prevent condensation. Liquid oxygen dripping off cryogenic piping colder than 90°K can cause combustion or oxygen enrichment.

29. Liquid nitrogen cold traps are prone to liquid oxygen buildup from freezing air.

30. The liquid air user is aware that nitrogen boils off liquidfied air at a lower temperature than liquid oxygen, causing oxygen enrichment (see Table 58.1).

Asphyxiation Hazards

Air with less than 19.5% oxygen by volume (partial pressure less than 135 mm Hg) is an oxygen-deficient hazard (ODH). An oxygen detector is required to assess if a depressed oxygen level exists in the workplace (see Box 2).

Cryogens and gases with density > 1.0 (e.g., carbon dioxide from vaporized dry ice, argon, cold nitrogen vapor, Freon, SF_6) accumulate in low areas, tunnels, trenches, and confined spaces. The lab air-exchange rate is considered when calculating worst-case release from a cryogenic fluid spill or from a catastrophic release of bulk cryogen storage.[7]

31. Cryogens are used only in ventilated areas because they have a high volume expansion when vaporized.

32. The area is evacuated immediately if there is an uncontrolled release of a cryogenic liquid or vapor. Air monitoring is required when reoccupying areas inerted by a release.

TABLE 58.1 Physical Properties of Common Cryogens

Cryogen	B.P. (°K)	Liquid-to-Gas Expansion Ratio	Gas Density (lb/ft³)	Critical Temperature (°K)	Critical Pressure (atm)	Liquid Density (g/l)
He	4.2	780	0.14	5.2	2.2	125
H₂	20.3	865	0.07	33.0	12.8	71
Ne	27.1	1470	0.70	44.4	26.2	1206
N₂	77.3	710	0.97	126.3	33.5	808
air	—	—	1.00	—	—	—
Ar	87.3	860	1.39	150.9	48.3	1402
O₂	90.2	875	1.11	154.8	50.1	1410
CO₂	194.7	790	1.70	304.2	72.8	1560
R-12	243.4	294	4.35	385	40.6	1487

Source: Jefferson Lab EH&S Manual Appendix 6500-R1.

33. Never enter a cryogenic vapor cloud or an area with a known or suspected ODH; exit away from the cloud and hold your breath until you are out of the vapor.
34. Building elevators are considered a confined space when transporting cryogenics. People do not ride with pressurized Dewars in case of discharge of the cylinder contents through the pressure-relief device.
35. Dry ice users are aware of the asphyxiation hazard of CO_2 and are careful to keep their faces out of dry ice storage chests.

References

1. *DOE Jefferson Lab EH&S Manual*, Cryogenic and ODH Safety; http://www.jlab.org/ehs/manual/EHSbook-434.html; 2000.
2. NASA, Glenn Research Center. *Safety Manual*; http://osat.gre.nasa.gov/safety; 2000.
3. *Safe Handling of Cryogenic Liquids*, Publication P-12, Compressed Gas Association, Inc.: Arlington, VA, 1993.
4. Cryogenic Engineering and Safe Practices Bibliography, 14 Oct 1994 post to Safety List Server by Edward C. Connors in safety archives: gopher://siri.uvm.edu.
5. *Health & Safety Manual*, Ernest Orlando Lawrence Berkeley National Laboratory; http://www.lbl.gov.ehs/pub3000, chapter 7.8, Cryogenic Systems; 2000.
6. *Guide to Safe Handling of Compressed Gases*; Matheson Gas Products: East Rutherford, NJ, 1983.
7. *Cryogenic Fluids in the Laboratory*, Data Sheet 688; National Safety Council: Itasca, IL, 1980.

59

Explosive and Reactive Chemicals

IRV KRAUT

This chapter deals with the hazards presented by a cross-section of reactive and potentially explosive chemicals with basic technologies proven by field experience to reduce the threat these compounds pose to the individuals involved.

The problems we face today are inherited. The bottles, cans, flasks, cylinders, drums, and vials that contain compounds such as azides, organic peroxides, ethers, and nitro-aromatics are legacies from our predecessors. In some cases, these gifts from the past were merely forgotten at the back of a chemical storage cabinet. Others were known to be difficult or impossible to dispose of safely and thus stored for future consideration. In some cases, the materials were too expensive to manage and thus left for better budgetary times. Whatever the reason or excuse, someone today is reading through an MSDS or a chemical dictionary in hopes that the chemical that he/she found is not as bad as feared.

In truth, the individual will be lucky to find sufficient information in current texts to render a sound conclusion. With little information in hand, the investigator will either make a wrong decision, spend far too much time searching the Internet for an answer, or, if fortunate, find someone who has had sufficient experience with the chemical to accurately identify characteristics of concern and thus allow the investigator to take proper steps to remediate the problem.

Common Reactive Chemicals

Alkali Metals

Sodium metal, lithium metal, potassium metal, and NaK compose a good cross-section for review. One of the primary users of alkali metals is the nuclear industry. Both sodium metal and NaK (sodium/potassium alloy) are used as heat-exchange media and thus are often the subject of waste-management concerns when vessels containing these reactive metals are identified for waste management. It is important to note that small laboratory containers of alkali metals need only be packaged in appropriate DOT containers for shipment to various EPA-permitted treatment facilities in the United States. The problems connected with these materials are managing sealed tanks or drums often involving bulk quantities.

Alkali metals are water reactive! Alkali metals are not listed as pyrophoric; however, they may react with moist air or wet gases (CO_2). Explosive hazards connected with alkali metals are associated with the liberation of hydrogen. Lithium may react with nitrogen gas forming unstable nitrides; therefore, noble gases such as helium or argon should be used with lithium as a cover gas. Alkali metals may form peroxide super oxides often associated with the metal being exposed to air. Super oxides may appear as a crust covering the compound and are to be considered shock sensitive.

Case History

A contractor is hired to download a 200-gallon stainless steel tank of sodium metal. The generator is a U.S. Government facility. The scope of work requires the contractor to remove all sodium from the tank, package the sodium in gallon containers under a nitrogen cover gas, and decontaminate the tank to RCRA "empty" status.

The contractor wraps the tank with heat tape and increases the temperature of the sodium to a molten state. The contractor then places a glove box assembly over a tank inlet and accesses the tank. A new dip tube is installed (bypassing the original dip tube possibly plugged), and a nitrogen gas stream is introduced into the tank to force the sodium up and through the dip tube.

The sodium is transferred through a dry, inert transfer system into containers set up to receive the material. Once the tank is empty, the contractor utilizes a steaming operation to decontaminate the tank. The steam system may include an entrained combination of steam and nitrogen. Close attention is required to ensure that there are no pockets of sodium left in the tank that may have survived the initial transfer. A fiberoptic examination of the tank is recommended to ensure complete decontamination.

Peroxides

Diethyl ether, isopropyl ether, dioxane, tetrahydrofuran, dicyclopentadine—these are examples of organic solvents that may present the formation of explosive peroxides under specific conditions. The following sections discuss the formation of peroxide compounds, procedures for inspecting and testing for the presence of peroxides, and options for treating and disposing of peroxides.

Identifying the Problem

A chemist is reaching for an amber, one-pint glass bottle of tetrahydrofuran (THF). The bottle is dusty and has sat in the back of the flammable storage cabinet for many years. The chemist stops midway through his/her reach and stares intently at the container. Adhering to the inside of the bottle at the midway level is a crystalline structure that looks like an ice formation on a window in January. The chemist considers the structure and leaves the bottle alone.

Peroxide-forming solvents such as THF come from the manufacturer with an inhibitor. Hydroquinone or tertiary butylcatechol are the inhibitors of choice depending on the solvent in question. The purpose of the inhibitor is (as the name implies) to reduce, prevent, or inhibit the formation of peroxides. These inhibitors do their job, but various factors can surface that deplete the inhibitor and leave the solvent free to form peroxides.

Peroxide Formation

Peroxides form inorganic solvents as a result of autoxidation. Common peroxide-forming solvents can be divided into the following groups:

- ethers, including open chain and cyclic ethers, acetals, and ketals (e.g., ethyl ether, isopropyl ether)
- hydrocarbons with allylic, benzylic, or propaglic hydrogen (e.g., cumene, cyclohexane)
- conjugated dienes, eneynes, and diynes (e.g., butadiene, furans)

Autoxidation in solvents is facilitated by three factors:

1. exposure to oxygen
2. exposure to light, including sunlight
3. storage time

Oxygen is a necessary ingredient for peroxide formation. A cap or bung left off a container or drum, or a loose-fitting seal, may supply sufficient oxygen to support peroxide formation by eliminating the inhibitor and supporting the initiation of the autoxidation process.

Light, including sunlight, also promotes the elimination of inhibitors and stimulates the autoxidation process. Light, however, cannot promote the autoxidation process unless sufficient oxygen is present in the container. Once formed, peroxides can, in direct sunlight, undergo autodetonation. Storage time simply allows peroxides to develop and form structures. Since autoxidation is a self-sustaining reaction, the rate of peroxide formation increases with time.

The development of peroxide formation is initially a reaction that is visible only through analytical testing. Established methods such as potassium iodide testing or peroxide test strips will provide a positive response to the presence of peroxides in the parts per million (ppm) range. An interesting field observation from this writer is that a container free of an inhibitor and left alone will begin and continue the formation of peroxides unabated until visible, hard peroxide structures develop. In other words, the formation of peroxides appears to be self-sustaining.

An examination of a bottle (amber or brown glass is used to prevent light penetration into the bottle) by the chemist can be enhanced by the use of a flashlight (which will not worsen a problem) to side- or back-light the container in question. Once the container is illuminated, the chemist should look initially for the presence of any substance other than liquid. If nothing is observed, then the chemist should concentrate on the liquid phase and discern if any hair-like structures are seen floating in suspension. If such structures are observed, then peroxide precautions should be initiated until testing proves or disproves the concern.

The historical viewpoint is that peroxides will most certainly be found in the cap threads of the container. While this is possible, the vast majority of peroxide formations known to this writer were found to have descended to the bottom of the container through their own weight or adhered to the inside of the container. In truth, each container is unique and presents its own characteristic. Peroxide formation has appeared to viewers as "chipped ice, cut diamonds, moths balls, Styrofoam, cotton candy, or lumpy silica." Whatever material you observe should be considered a peroxide until proven otherwise.

Metal cans and drums cannot be inspected visually and must be opened to allow appropriate testing. Opening such containers is a delicate procedure due to the possibility of peroxide formation in the cap or bung threads. While peroxide formation tends to occur less frequently in the cap area than in other container areas, metal cans and drums should only be opened by trained individuals, and the application of remote opening equipment should be considered. Metal containers are believed to accelerate the rate of peroxide formation. The scientific documentation supporting this belief, however, is largely anecdotal.

The formation of peroxides in an organic solvent can be inhibited (1) by adding an inhibiting compound to the solvent or (2) by purging the oxygen from the free space in the solvent container. Chemical manufacturers add inhibitors to this class of solvents with the exception of those solvents used for HPLC. These are specifically manufactured without inhibitors, because inhibitors interfere with the UV detection process.

Explosive Hazard

In the early 1980s, an East Coast environmental company performed a trazl test on 5 grams of hard peroxide removed from a desiccated bottle of isopropyl ether. The trazl block weighed 4 pounds and was hollow in the center to about half its depth. The peroxide was shocked by a #6 electric blasting cap that was placed against the peroxide inside the block. The electric blasting cap by itself would not produce much effect on the block other than a slight expansion of the center hole. The combination of the blasting cap with the peroxide crystal essentially destroyed the block. A witness at the test site told this writer that the blast produced a shock wave not associated with a small blasting cap.

Peroxides formed inside an uninhibited organic solvent container must be considered both explosive and shock sensitive. In the presence of a solvent, the peroxide may be slightly less sensitive to shock than if the peroxide were inside a dry bottle or can and thus did not have the motion-absorbing effect of a liquid. Consequently, a container of solvent with peroxides is an explosive/flammable hazard, while a container with only peroxide crystals (the solvent having evaporated) is a nonflammable explosive hazard.

Laboratory Testing

Several methods are utilized to test for the presence of peroxides. The following two tests are among the most common:

- *Commercially available peroxide test strips.* These test strips provide quantitative results and are simple to use. The test strip is saturated with a representative sample of the liquid in question. A section of the strip changes color if peroxides are present; this color change is compared to a graph that indicates the peroxide concentration in the ppm range. Test strips typically register as high as 100 ppm.
- *Potassium iodide (KI) test.* In this test, 100 mg of potassium iodide is dissolved in 1 ml of glacial acetic acid. Then 1 ml of suspected solvent is added.

TABLE 59.1 Compounds that May Form Peroxides during Storage

Compound	Test Cycle in Storage	Special Handling
Isopropyl ether	Every 3 months	Consume or discard within 3 days of opening
Divinyl acetylene	Every 3 months	Consume or discard within 3 days of opening
Vinylidene chloride	Every 3 months	—
Potassium metal	Every 3 months	Avoid oil, hydrocarbons if KO_2 is present
Sodium amide	Every 3 months	—

A pale yellow color indicates a low concentration of peroxides; a bright yellow or brown color indicates a higher concentration of peroxides. This is a preferred method for testing di-isopropyl ether.

A peroxide test should be performed each time material is removed from a container. If the material is removed on a daily basis, tests should be done every other day. Containers of peroxide-forming compounds should be marked with the date the container was first received and first opened, the results of the first peroxide test, and the results of the last peroxide test before disposal. Tables 59.1, 59.2, and 59.3 show the testing requirements for common peroxidizable compounds during storage as well as for compounds while in use.

The results of peroxide testing dictate how the material should be handled. The following are general levels of risk association with various concentrations of peroxides:

* *Between 5 and 30 ppm:* Expired compounds testing within this range pose little or no threat of violent reaction on the given test date. For compounds testing in this range the investigator should consider adding fresh inhibitor to retard the autoxidation process, and the container should be tightly sealed in preparation for off-site waste disposal.
* *Between 30 and 80 ppm:* Expired or mismanaged compounds that test within this range may pose a threat to operations in the laboratory or facility.
* *Greater than 80 ppm:* Any solvent testing in excess of the maximum quantifiable limits of standard peroxide test strips must be considered potentially shock sensitive.

Treatment and Disposal Options

Deactivation Most, if not all, peroxide-forming chemicals are regulated as hazardous wastes. The BDAT for peroxides is deactivation to eliminate the ignitability characteristic (55 CFR 22546). Technologies that may be used to deactivate peroxide formers (D001 oxidizers) include chemical oxidation, chemical reduction, incineration, and recovery. Any of these technologies is acceptable

TABLE 59.2 Compounds that Readily Form Peroxides in Storage through Evaporation or Distillation[a]

Compound	Test Cycle in Storage	Special Handling/Testing
Diethyl ether	Every 12 months	HPLC grades of this compound are normally packaged without inhibitors. Test uninhibited compound.
Tetrahydrofuran	Every 12 months	Every 3 months, if uninhibited
Dioxane	Every 12 months	Every 3 months, if uninhibited
Diclopentadiene	Every 12 months	Every 3 months, if uninhibited
Isoprene	Every 12 months	Every 3 months, if uninhibited
Grignard reagents	Every 12 months	Every 3 months, if uninhibited
Methyl acetylene	Every 12 months	Every 3 months, ff uninhibited
Cumene	Every 12 months	Every 3 months, if uninhibited
Acetaldehyde	Every 12 months	Anhydrous acetaldehyde will autoxidize at 0°C or below under ultraviolet light catalysis to form paracetic acid, which may react with more acetaldehyde to produce explosive acetaldehyde monoperacetate.

[a]Concentration processes (evaporation or distillation) defeat the action of most autoxidation inhibitors. Special handling and accountability are required of those compounds offered at HPLC grade, because HPLC-grade materials are packaged without inhibitors.

TABLE 59.3 Compounds that Pose Hazards Due to Peroxide Initiation of Polymerization

Compound	Test Cycle in Storage	Special Handling/Testing
Butadiene	Every 12 months	Every 3 months, if stored as liquid
Styrene	Every 12 months	Every 3 months, if stored as liquid
Vinyl acetylene	Every 12 months	Every 3 months, if stored as liquid
Vinyl pyridine	Every 12 months	Every 3 months, if stored as liquid
Vinyl chloride	Every 12 months	Every 3 months, if stored as liquid

Note. When stored in the liquid state, the peroxide-forming potential dramatically increases.

provided it eliminates the ignitability characteristic. Most, if not all, off-site EPA-permitted treatment and disposal facilities will require that containers be peroxide free and do not pose an explosive hazard.

Stabilization/Reduction Peroxides within a container can be chemically stabilized. The following describes one chemical procedure that has been used to stabilize peroxides. (The reader is cautioned that any procedure used to handle a sensitive chemical or eliminate peroxides should be undertaken only by very experienced personnel who understand the potential for uncontrolled exothermic reactions during the procedure.) The solvent container is accessed through its cap by a remotely operated nonsparking drill. A Teflon catheter is then inserted through the access point to draw a 1 cm^3 sample of solvent for testing. Three standard peroxide strips are used to measure the sample's peroxide concentration. All negative indications are verified by adding a drop of sample solvent to a 10% potassium iodide solution for calorimetric evaluation.

If the container is found to contain peroxides, a buffered solution of ferrous ammonium sulfate is injected into the container. This produces an oxidation-reduction reaction that is often very exothermic and proven to be successful in eliminating peroxides. The container is re-tested continuously until all peroxides have been dissolved and peroxide tests are shown to be negative. Hydroquinone (an inhibitor) is then added to stabilize the container and guard against an immediate recurrence of peroxidation. Finally, the container is resealed and placed in a safe storage area pending off-site disposal.

Open Detonation Open-air detonation or burning of peroxide-forming compounds has been used by police bomb squads and government explosive technicians in an effort to assist the private sector. This practice was found to have two major disadvantages:

1. Potentially shock-sensitive materials were subjected to extensive movement prior to disposal.
2. Numerous EPA and DOT regulations were circumvented in the interest of public safety.

Nitro Compounds

The word "nitro" conjures up an image in the minds of many people of a compound that is explosive. This is not always the case and, in fact, many nitro compounds have few or no explosive properties at all. It is important that the chemist fully understand the characteristics of the specific material before making a judgment based on superstition.

Nitro compounds most commonly utilized in a laboratory setting are:

nitrobenzene
nitroguanidine
dinitrophenylhydrazine
trinitrofluoronone
trinitrophenol (picric acid)
picryl chloride
hexanitrodiphenylamine

When purchased from the chemical supplier, these compounds arrive through common carrier (UPS) and have Department of Transportation descriptions ranging from flammable solids to explosives. While the container may have a label that reads "explosive when dry," the chemicals are not shipped as explosive materials. In truth, these materials do not present explosive characteristics when originally manufactured but can and do present an explosive hazard under specific conditions.

Contamination and Desiccation

The chemical that has garnished the most interest over the past twenty years is trinitrophenol (picric acid). Historically, picric acid was discovered in 1771 by Peter

Woulfe, a British chemist. Its name is derived from the Greek word "pikos," meaning bitter, due to the intensely bitter and persistent taste of its yellow aqueous solution. In the past, this strong acid was used as a fast dye for silk and wool and in aqueous solutions to reduce the pain of burns and scalds.

When dry, picric acid has explosive characteristics similar to those of TNT. Table 59.4 summarizes the explosive characteristics of picric acid. The first experiment to use picric acid as an explosive bursting charge was conducted in the town of Lydd, England, in 1885. In 1888, the British adopted picric acid as a military explosive under the name Lyddite. Since that time, picric acid has been used by many countries as a bursting charge under the names Pertite (Italy), Melinite (French), Shimose (Japan), and Granatfulung 88 (Germany).

Today, the use of picric acid as a military explosive has been largely discontinued due to its toxicity and sensitivity to metal and concrete shell linings.

The toxicity of nitro compounds, in general, is an important consideration for chemists and laboratory technicians. Absorbed through the skin and through inhalation, acute nitro exposures can depress the central nervous system and reduce the body's ability to carry oxygen through the bloodstream. Prolonged exposure may result in chronic liver and kidney damage. OSHA's permissible exposure level (PEL) for picric acid, for example, has a time-weighted average (TWA) of 100 μ/m^3, with a "skin" notation to indicate the possibility of dermal absorption.

Proper personal protective equipment, such as gloves and respirators, should be worn when handling nitro compounds in other than a controlled environment such as a fume hood where gloves alone may be sufficient.

As a rule of thumb, the higher the nitrated level of the compound, the more potential for explosive characteristics. As such, trinitro and hexanitro compounds, such as trinitrobenzene and hexanitrophenylamine, are characteristically more sensitive to detonation than nitrobenzene or dinitrophenylhydrazine.

In reviewing literature on various nitro compounds, it becomes apparent that the more desiccated the material, the more dangerous the compound. In continuing the discussion on trinitrophenol/picric acid, this author has had substantial experience in the 1970s and 1980s when the methodology for disposal of nitro compounds such as picric acid was open-air detonation. It becomes apparent that bottles containing picric acid, dinitrophenylhydrazine, picryl chloride, and nitro guanidine only underwent true detonation when the substances inside the container were desiccated. In all cases, when the containers contained free liquids or had the appearance of wet sand (moisture laden), the bottles failed to detonate. It should be noted that the procedure used to detonate the containers was always the same, and involved the wrapping of the bottle with a donor charge of 100-grain detonating cord. The use of "det-cord" produced an enormous shock to the containers and revealed that wet nitro compounds failed to sympathetically detonate. It should be noted at this juncture that containers that have formed explosive structures due to contamination (metal picrates) may detonate even in the presence of moisture. Containers of this type are fortunately the exception and not the rule.

TABLE 59.4 Picric Acid

Picric Acid (Trinitrophenol, Lyddite, Melinite, Shimose, Pertite, Picronitric Acid, Carbazotic Acid, etc.)
yellow crystals: colorant
gross formula: $C_6H_3N_3O_7$
MP: 122.5°C
Autoignition temp.: 572°F
MW: 229.1
Oxygen balance: −45.4%
Heat of explosion: 1080 kcal/kg
Density: 1.767 g/cm^3
Lead block test: 315 cm/10g
Detonation velocity (when confined): 7350 m/s
Deflagration point: 570°F = 300°C
CAS: 88-89-1
UN 0154 dry or wetted with less than 30% water, by weight
UN 1344 with 30% or more water, by weights
Toxicity
OSHA PEL: TWA 100 μ/m^3 (skin)
ACGIH TLV: TWA 0.1 mg/m^3
DFG MAK: 0.1 mg/m^3

Note. Picric acid, like many trinitrocompounds, is absorbed through the skin and through inhalation. Acute picric acid exposure can produce a depression of the central nervous system and a reduction in the body's ability to carry oxygen through the bloodstream. Prolonged exposure may result in (chronic) kidney and liver damage. Percutaneous absorption may cause vomiting, nausea, abdominal pain, staining of the skin, convulsions, or death.

Proper personal protective equipment (gloves, respirator, SCBA, Level B attire) should be donned when handling picric acid in other than a laboratory setting. Chemists and technicians in the lab should always wear gloves and utilize a fume hood for safe handling procedures.

Perchloric Acid

The most persistent problem connected with this compound is the contamination of perchloric acid fume hood systems. Several factors contribute to a buildup of perchloric acid salts and to the manner in which such a system should be addressed:

- The fume hood does not have a washdown system.
- The fume hood has a washdown system that is inoperable.

- The fume hood is left unattended for many years and little, if any, reliable information is available as to its usage.

Perchloric acid salts are commonly found within a system at elbows, bends, joints, or fan blade housing. These salt deposits are to be considered shock sensitive, and the removal of a perchloric acid system should be preceded by a thorough hot water wash or steaming of the system followed immediately by a chemical decontamination of the system utilizing a suitable reducing agent such as aqueous sodium hydroxide. Testing of various surface areas within the system should be employed as a final step to ensure the elimination of perchloric acid salts prior to the dismantling and removal of the system.

Contractors selected to perform these tasks should have skills relative to the chemistry of the materials in lieu of a standard construction contractor.

General References

USEPA Handbook, *Approaches for the Remediation of Federal Facility Sites Contaminated with Explosive or Radioactive Wastes*; Office of Research and Development, EPA/625/R-93/0113: Washington, DC, September 1993.

Standard Operating Procedures, Peroxide Forming Solvents; ETSC Government Services, Inc.: Schaumberg, IL, rev. 1994.

SECTION VI

RADIOLOGICAL AND BIOLOGICAL SAFETY

NELSON COUCH

SECTION EDITOR

60

Non-Ionizing Radiation

JOSEPH M. GRECO

Optical Radiation

Optical radiation is comprised of the ultraviolet (UV), visible, and infared (IR) portions of the electromagnetic spectrum. The International Commission on Illumination (the CIE—Commission International d'Eclairage) Committee on Photobiology has designated optical radiation spectral bands,[1] summarized in Table 60.1. While ionizing radiation is described in terms of photon energy (electron volts), and radiofrequency radiation in terms of frequency (Hz), optical radiation is normally described in terms of wavelength (nm, µm, or mm). Exposure to optical radiation is normally expressed in units of W/cm^2 (for continuous sources) or J/cm^2 (for pulsed or short-term exposure).

This chapter addresses only noncoherent (i.e., non-laser) sources of optical radiation.

Ultraviolet Radiation

Ultraviolet radiation resides between ionizing (gamma and X) radiation and the visible light region in the electromagnetic spectrum. The CIE defines UV wavelengths as 100 to 400 nanometers (nm) and divides it into three regions (UV-A, UV-B, and UV-C), as listed in Table 60.1. (The UV-C and UV-B wavelengths are collectively called the *actinic* region.) These three regions vary widely with respect to penetration and potential for causing biological effects. For example, the shorter UV wavelengths (below 200 nm) are highly attenuated by most materials (including the atmosphere) and are not considered an exposure problem because sources are uncommon. Prolonged exposure to other UV wavelengths (especially 270 to 320 nm) can cause deleterious bioeffects as well as degradation of many materials such as plastic and rubber.

Sources

Although the earth's ozone layer absorbs most of the UV produced by the sun, the remaining unattenuated solar UV reaching the earth is by far the largest source of exposure for the general population. Exposure to UV varies and is dependent on time of day, latitude, elevation, cloud cover, and presence of reflective surfaces. As an example, Machta measured the daily variation in UV energy (erythemal effective irradiance) in Denver, CO, on a clear summer day. The maximum value (11.3 µW/cm^2) was reached at approximately noon, while low dense cloud cover somewhat reduced this intensity.[2]

UV-emitting equipment in the work/laboratory setting is common and is used for a wide variety of purposes. The radiation emitted from UV sources is either deliberate (e.g., killing bacteria, ink curing) or incidental (un-

TABLE 60.1 Divisions of the Optical Spectrum According to the International Commission on Illumination (CIE)

Region	Wavelength Range	Other Terminology
Ultraviolet (UV)	100 to 400 nm	
UV-C	100 to 280 nm	Far Ultraviolet
UV-B	280 to 315–320 nm	Middle Ultraviolet
UV-A	315 to 380–400 nm	Near Ultraviolet
Light	380–400 to 760–780 nm	Visible
Infrared (IR)	760–780 nm to 1 mm	
IR-A	760–780 to 1400 nm	Near Infrared
IR-B	1400 to 3000 nm	Middle Infrared
IR-C	3000 nm to 1 mm	Far Infrared

wanted UV from arc welding). Box 1 lists some processes and devices found in the workplace and other locations that emit noncoherent UV radiation.[3-5]

Bioeffects

Effects from exposure to UV radiation are very dependent on wavelength. The organs that are most vulnerable are the eyes (especially the cornea and lens) and skin. In addition, photosensitization to UV can occur from various substances. It should be noted that a small amount of UV exposure is beneficial, in that it promotes Vitamin D production.

Effects on the skin can be divided into acute and chronic. The most common acute effect from overexposure to UV is erythema (sunburn), which is a photochemical response of the skin resulting from overexposure to the UV-C and UV-B regions (esp. 200–315 nm). The severity of the effect depends on the duration and intensity of the exposure. Exposure to UV-A alone requires far greater levels to induce erythema; the UV-A dose required to produce erythema is 800–1000 times that of the UV-B dose. Maximum sensitivity of the skin occurs at 295 nm.[6]

Chronic exposure to sunlight (especially wavelengths shorter than 315 nm) accelerates skin aging and increases the risk of developing skin cancer. These effects may not manifest themselves for years or even decades. A causal relationship is most clearly established between UV-B exposure and squamous cell carcinoma, and less so for basal cell carcinoma. The relationship between malignant melanoma and UV exposure is complex. The risk attributable to UV exposure after age 30 is uncertain; there may be a critical time during which exposure incurs increased risk of malignant melanoma.

The primary effect on the eye from UV overexposure is keratoconjunctivitis (also known as welder's flash, snow blindness, and arc eye), which involves the cornea. Symptoms include a "ground-glass" sensation in the eye, aversion to bright light, and pain/discomfort within hours of exposure. The maximum effect on the eye for a given exposure is produced at 270 nm. The cornea is considered the "skin" of the eye, and the corneal cells have a very short lifespan—only about 48 hr. Damage to this outermost layer, then, is usually repaired within two days by normal cell turnover. Permanent corneal damage is rare.

Photosensitivity, described as hypersensitivity to certain wavelengths in the optical spectrum, may be genetic, associated with various immune states, or induced. Certain creams, lotions, shampoos, and drugs (such as tetracycline) may induce hypersensitivity to UV radiation in many individuals.[5] Fitzpatrick et al. lists many common photosensitizers.[7]

Exposure Guidelines

ACGIH The ACGIH annually publishes Threshold Limit Values for Chemicals and Physical Agents.[6] UV radiation Threshold Limit Values (TLVs) "refer to UV radiation in the spectral region between 180 and 400 nm and represent conditions under which it is believed that nearly all workers may be repeatedly exposed without adverse health effects."

The UV TLV is highly dependent on the wavelength. The TLV is given for an 8-hr day in units of J/m^2 (and mJ/cm^2) for specific wavelengths. These TLVs may be used directly for narrowband sources; however, for broadband sources, an effective irradiance, E_{eff}, in units of $\mu W/cm^2$, should be calculated. The TLV for effective irradiance is an accumulation of 3 mJ/cm^2 for an 8-hr exposure duration.*

ICNIRP The International Non-Ionizing Radiation Committee of the International Radiation Protection Association (INIRC/IRPA) published guidelines for exposure limits for UV radiation in 1985. These guidelines were amended in 1989, and published in 1991.[8] In 1996, the International Commission on Non-Ionizing Radiation

---

*Some UV radiometer manufacturers manufacture detectors that respond to UV radiation according to the relative spectral effectiveness. Since the measurement result using this detector is in units of effective W/cm^2, comparison of the reading to the permissible UV exposure table will readily provide a determination of the allowable exposure duration per day.

Box 1
Processes and Devices
that Emit UV Radiation

Germicidal lamps	Tanning beds
Curing equipment	Welding equipment
Fluorescent lamps and equipment	Ink curing equipment
Mercury lamps	Halogen lamps
Xenon arcs	Dermatology lamps ("Wood's lamps")
Metal smelting	Solar lamps
Glass processing	Projection systems
"Black lights"	

Protection (ICNIRP) reviewed these guidelines, and no substantial changes were made.[9] These guidelines are very similar to those published by the ACGIH.

Best Safe Practices

Although many sources of UV radiation also emit some visible light, some do not. A germicidal lamp, for example, may only emit a faint visible glow while emitting intense UV radiation. Therefore, the amount of visible light should not be used as an indicator of the UV intensity. A combination of engineering/administrative controls and personal protection should be used to limit UV exposures. Engineering controls are preferred due to their broader effectiveness.

Shielding UV leakage should be contained and shielded whenever possible. If viewing the process is required, observations should take place through materials such as polycarbonate plastics that attenuate UV-C and UV-B wavelengths while remaining transparent to visible light.

Interlocks If access to equipment that houses UV radiation is required, the housing should be interlocked such that the UV source will be turned off when the housing or cover is removed. Interlocks should be tested regularly to ensure proper operation.

Nonreflective Materials If possible, reflective materials such as shiny metals (e.g., aluminum), glossy light-colored paint, and paint containing titanium should be avoided. It should also be noted that materials that appear dark and nonreflective in the visible portion of the spectrum may not be the same in the UV portion.

Awareness/Warning Signs Be aware of any UV radiation sources in your work area and inform those who have potential for exposure: new employees, maintenance personnel, visitors, and so on. For example, some fume hoods are equipped with normal lighting as well as UV germicidal lamps. Ensure that workers using the hood (or performing maintenance) are aware of the UV lamps and protect themselves accordingly. Labeling UV sources with an appropriate warning sign, in combination with an indicator light, is recommended.

If UV radiation exposure cannot be avoided, the following will help to minimize the exposure.

Personal Protection All skin with potential for exposure (especially hands and face) should be protected. Gies et al. measured effectiveness of various types of clothing and fabrics, and determined that clothing can provide significant protection against UV exposures.[10] In general, dense, heavy fabrics with no direct transmission paths are recommended for UV protection during welding operations. Most gloves found in a laboratory environment will protect hands from UV exposure. Safety glasses with a UV-protective coating are available that will protect against UV wavelengths and still afford adequate visibility (they sometimes are light yellow in color). Moseley measured the UV and visible light transmission properties of several types of protective eyewear used in medical, industrial, and welding applications.[11]

Time and Distance Decrease the amount of time spent near UV sources and maintain a safe distance whenever possible.

Survey/Measurements If exposures to UV light cannot be avoided, the effective UV irradiance should be measured, and a maximum permissible exposure duration should be determined, based on accepted standards.

Ozone Production Ozone (O_3) is a colorless, toxic gas, irritating to the lungs and mucous membranes, having a characteristic pungent odor. It can be formed by the photochemical reaction between short-wavelength UV radiation and oxygen molecules present in air. The threshold for odor detection ranges from 0.005 ppm to ? ppm. The TLV (2000) for ozone ranges from 0.05 to 0.2 ppm, depending on workload (and resultant inhalation rate). Odor detection, therefore, may not provide adequate protection from occupational exposures. Exposure to 50 ppm for 0.5 hr may be fatal. Diffey and Hughes provided a way to estimate the concentration of ozone that may exist in the vicinity of UV-emitting lamps.[12]

General room ventilation should provide adequate protection from the hazards of ozone in most cases. Very intense, short-wavelength sources of UV may require additional localized ventilation, which should be interlocked with the UV source.

Visible and Infrared Radiation

Incoherent sources of visible light (400 to 760–780 nm) and infrared (i.e., "heat") radiation (760–780 nm to 1 mm) are common, and most do not present a hazard. This section addresses the relatively few sources of intense visible/IR radiation that should be handled with care, but does not cover conditions such as heat stress or conductive heat. Visible and IR sources are discussed together, because most (but not all) IR sources will also have a visible component.

Sources

Like UV radiation, the single largest source of visible/IR radiation is the sun. Occupational sources include lamps and lighting systems (flashbulbs, photocopiers, etc.), space heaters, and industrial settings, such as glass works and iron foundries. Some devices, such as remote control units and IR loops for the hearing-impaired use IR signals, but the output power level is insignificant.

Bioeffects/Exposure Standards

Skin injury is usually not a concern from most conventional sources of IR radiation (such as industrial furnaces and open arcs) because the natural aversion response—moving away from the source of heat—will normally occur before injury can take place.[4] The eye does not have adequate thermal warning properties and is the organ most susceptible to visible/IR overexposure. Middle IR (1400 to 3000 nm) and far IR (3000 nm to 1 mm) is absorbed in the cornea. Visible (400 to 700 nm) and near IR (700 to 1400 nm) passes through the cornea and is absorbed by the retina. It should be noted that glare can overwhelm adaptive capacity of the eye, causing visual fatigue and reduced visual perception, which can make a worker more prone to accidents.

Injury to the retina can be caused by two mechanisms—thermal injury (focusing thermal energy onto the retina) and a photochemical reaction to chronic exposures to visible light (especially blue wavelengths).

The ACGIH provides threshold limit values for visible and near-infrared radiation. Biological weighting factors for both the Blue-Light Hazard and Retinal Thermal Hazard are assigned to wavelengths between 400 and 1400 nm. The ACGIH lists a 10 mW/cm^2 limit for IR exposure for periods greater than 1000 seconds in hot environments. Whillock et al. measured the radiance levels at four industrial sites (a glass works, an iron foundry, a large-scale float glass facility, and a small food preparation facility). They found that the radiance did not indicate potential photochemical or burn hazards to the retina from these operations. However, the authors did note that a potential for lenticular injury (cataracts) exists for long-term exposures (10 to 15 years) to IR radiation.[13]

Best Safe Practices

In areas where intense IR radiation may be encountered, such as steel mills and other industrial facilities, infra-red-absorbing glasses should be used for shielding. Often, an electrically conductive coating is placed on the front of the glasses to reflect IR radiation to avoid excessive temperature rises in the glass.[5]

Electric and Magnetic Fields

Static Magnetic Fields

Magnetic fields result from movement of electric charge and will always exist when there is an electric current. The fields may either be time varying, as the case of 50/60 Hertz (Hz) magnetic fields emitted from power lines and electrical appliances, or they may be static, meaning they have a frequency of 0 Hz.* Static magnetic fields, also called direct current (DC) fields, are treated differently from time-varying fields in terms of measurement, biological effects, and exposure standards.

Magnetic flux density is considered the most relevant quantity for expressing magnetic fields associated with biological effect.[14] The most widely used units for this parameter are the tesla (T) and the gauss (G). One tesla equals 10,000 gauss. Another quantity used is magnetic field strength, measured in units of amperes per meter (A/m) and the oersted (Oe). Magnetic field strength and magnetic flux density can be considered equivalent when measuring non-ferromagnetic material, such as cells or tissues. Common static magnetic field quantities and their conversions are listed in Table 60.2.

Sources

Earth's Magnetic Field All persons are exposed to the earth's natural magnetic field, which has a magnetic flux density range of 0.3 to 0.7 G, depending on geographic location.[14]

MRI and NMR Magnetic resonance imaging (MRI) has become increasingly common in clinical settings for diagnosis of disease. In addition to a static magnetic field, these imaging devices also use radio frequency fields (1 to 100 MHz range), which remain very local to the magnet enclosure. Stuchly and Lecuyer measured stray static magnetic fields around four types of MRI devices. At a 1-m distance on axis with the device, magnetic flux densities ranged from 100 to 750 G and were dependent on the nominal magnet strength as well as magnet

---

*All magnetic fields may be seen as having oscillations if enough time elapses. For the purpose of this section, static magnetic fields do not vary with time, and have an infinitely long wavelength.

TABLE 60.2 Common Static Magnetic Field Quantities and Conversions

Unit	Quantity	Conversion
Tesla (T)	Magnetic flux density	1 T = 10,000 G 1 mT = 10 G 1 T = 8×10^5 A/m 1 T = 10,000 Oe
Gauss (G)	Magnetic flux density	1 G = 0.0001 T 1 G = 0.1 mT 1 G = 80 A/m 1 G = 1 Oe
Amperes per meter (A/m)	Magnetic field strength	1 A/m = 1.28×10^{-6} T 1 A/m = 1.28×10^{-2} G 1 A/m = 1.28×10^{-2} Oe
Oersted (Oe)	Magnetic field strength	1 Oe = 0.0001 T 1 Oe = 80 A/m 1 Oe = 1 G

housing design. Isoresponse lines were generally circular around the device.[15]

Nuclear magnetic resonance (NMR) devices used in the analytical lab are similar in design and field strength to MRI units, although the center bore is typically only several inches in diameter, as required for sample placement.

Other Sources Strong magnetic fields are also found in high-energy technologies (fusion reactors, DC power generation/distribution, superconducting generators); research facilities (bubble chambers, particle accelerators, isotope separation units); the aluminum industry (electrolytic processes); and transportation systems (magnetically levitated trains). In addition, static fields may be used in theft detection and airport security systems.

Biological Effects and Exposure Standards

Biological Effects Available reports suggest that there are no deleterious effects conclusively demonstrated for static magnetic fields up to 20,000 G.[16] Current epidemiological studies show only a weak-to-nonexistent relationship between static magnetic field exposures and cancer.[17]

Pacemaker Implants Cardiac pacemakers and other metallic devices (defibrillators, aneurysm clips, prosthetic devices, etc.) may be sensitive to elevated static magnetic fields. Pacemakers that were programmed to provide a predetermined electrical pulse to the heart may revert to an asynchronous pacing mode when subjected to fields as low as 17 G. They returned to their normal operation when removed from the elevated field.[18] For this reason,

current guidelines restrict persons implanted with pacemakers to flux densities of 5 G.

Exposure Standards Table 60.3 lists current exposure standards for the ACGIH, the International Commission on Non-Ionizing Radiation Protection (ICNIRP), the World Health Organization (WHO), and the United Kingdom's National Radiation Protection Board (NRPB).[16,19–21]

Safety Considerations

Potential risks to workers and the general public from static magnetic fields should be addressed. For MRI/NMR units, safety issues include personnel exposure to magnetic fields, cryogenic liquids (used for cooling), hazards from unsecured ferromagnetic objects, and pacemaker/implant concerns.

Personnel Exposure Plots of the surrounding magnetic field should be available from the manufacturers of MRI/NMR units. If unavailable, the vicinity of MRI/NMR should be surveyed with an appropriate gaussmeter to determine magnetic flux density. Proper choice of the survey instrument is essential—a meter that is made to characterize only extremely low frequencies (ELF) will not properly measure a static magnetic field.

Various boundaries should be established to warn personnel. For example, a 5-G line can be established around an NMR unit with warning tape and/or a sign for the person implanted with a cardiac pacemaker or other metallic device, and a 10-G line for access of NMR staff and use of ferromagnetic materials. It is important to remember to survey all rooms adjacent to the MRI/NMR as well as those on floors above and below, since the stray field extends in all three axes.

TABLE 60.3 Summary of Static Magnetic Field Exposure Standards

Organization	Remarks	Exposure Standard (G)	Exposure Standard (mT)	Pacemakers/ Devices
ACGIH (2000)	Whole body (time-weighted average)	600	60	5 G (0.5 mT)
	Extremities (time-weighted average)	6000	600	
	Ceiling value—whole body	20,000	2000	
ICNIRP (1994)	Continuous occupational exposure— whole body	2000	200	
	Short-term occupational exposure— whole body	20,000	2000	5 G (0.5 mT)
	Extremities	50,000	5000	
	General public	400	40	
WHO (1989)		20,000	2000	
NRPB (1993)	Averaged over 24 hr	2000	200	
	Max. to whole body	20,000	2000	
	Max. to arms and legs	50,000	5000	

Cryogenic Liquids Most MRI and NMR superconducting magnets are cooled by liquid nitrogen or helium. Normally nontoxic, these gases can become an asphyxiant if adequate ventilation is not provided, especially if the magnet undergoes a rapid quench (loses its superconductivity and boils off coolant). MRI/NMR units should never be located in an airtight room. In addition, the units should be situated in the room in a way that facilitates easy transfer of cryogenic liquid, in order to reduce the chance of a spill.

Unsecured Objects Ferromagnetic tools and objects are potential missiles and should not be allowed near strong magnetic fields. Attractive forces depend on the mass of the object and increase rapidly close to the magnet. An attractive force that initially may be slightly noticeable may become uncontrollable in a very short distance. Gas cylinders are especially hazardous. All ferromagnetic materials should be kept away from fields greater than 10 G.[22]

Magnetic-Sensitive Materials Besides unsecured objects, other materials and devices can be influenced (normally detrimentally) by strong magnetic fields. Some effects, such as disturbance of computer monitors, can be remedied by either degaussing the screen, moving the monitor away from the sources, or more expensively, shielding the monitor by a material such as mu metal. Cards that have a magnetic strip on the back, such as credit cards, identification passes, and so on, may be adversely affected by strong magnetic fields. It is useful to remove these items and any other personal items (watches, glasses with metallic frames, pens, etc.) before approaching an MRI/NMR unit. A list of deleterious effects to devices from various flux densities are listed in Table 60.4.

Sub Radio-Frequency Magnetic and Electric Fields

Magnetic and electric fields with a frequency of 30 kilohertz (kHz) and below are referred to as sub radio-frequency (sub-RF). This includes:

Extremely Low Frequency (ELF): 30 to 300 Hz (including 50/60 Hz power frequencies)
Voice Frequency (VF): 300 to 3,000 Hz
Very Low Frequency (VLF): 3 to 30 kHz

This section addresses primarily ELF fields. The acronym "EMF"—electromagnetic fields or electric and magnetic fields—is a broad term often seen in the popular press. EMF has been used to describe 50/60 Hz (power frequencies) as well as an all-encompassing term for any time-varying field.

TABLE 60.4 Deleterious Effects from Static Magnetic Fields

Effect	Magnetic Field Density (G)	Magnetic Field Density (mT)
Effects on electron microscopes	1	0.1
Disturbance of color computer displays	1 to 3	0.1 to 0.3
Disturbance of monochrome computer displays	3 to 5	0.3 to 0.5
Erasing credit cards and bank cards	10	1
Effects on watches	10	1
Erasing magnetic tapes	20	2
Erasing floppy disks	350	35

Electric field strength, measured in V/m, is determined by the level of applied voltage and is fairly constant over time. Magnetic flux density, measured in gauss or tesla, is produced by flowing electrical current and will vary over time. Field intensities decrease rapidly with distance.

Sources

The primary source of ELF fields in the workplace, home, and elsewhere is from the generation, transmission, distribution, and use of electricity. Anything that plugs into an electrical outlet will emit ELF electric and/or magnetic fields. Table 60.5 lists various ELF and VLF sources and their flux densities and/or electric field strength.[23–26]

Biological Considerations

Increased electric power usage worldwide exposes virtually all persons in modernized countries to low levels of ELF electromagnetic fields. Because of the size of the exposed population, interest in possible adverse biological effects is proportionately large. Current knowledge of the effects on humans is not complete, even though there have been a multitude of studies, including *in vitro, in vivo*, animal, epidemiological, and human. While some epidemiological studies show an association between exposure to ELF fields and an adverse health effect, many others show no association.

Many mutlidisciplinary bodies have reviewed the available data and reports. The World Health Organization in 1984 raised questions regarding reproducibility, validity, confounding factors, and other factors of epidemiological studies.[27]

Oak Ridge Associated Universities (ORAU), at the request of the U.S. Department of Labor and the Committee on Interagency Radiation Research and Policy Coordination (CIRRPC), evaluated the available literature in 1993 (about 1000 journal articles over the past 15

years). They concluded that there was no convincing evidence to support the contention that ELF exposure produces ill health effects.[28]

In 1996, the National Research Council of the National Academy of Sciences (NRC/NAS) examined more than 500 studies spanning 17 years of research, and concluded that no clear, convincing evidence exists to show exposure is a threat to human health. Cancer development, reproductive and developmental abnormalities, and learning and behavioral problems were examined in particular.[29]

Concerns about reproductive effects, especially miscarriages, on women using video display terminals have been raised. A study jointly sponsored by the National Institute for Occupational Safety and Health and the American Cancer Society found no excess risk of spontaneous abortion among women who used VDTs.[30] Fertility and developmental effects, if any, are unknown.

Exposure Guidelines

Table 60.6 summarizes sub-RF electric and magnetic field exposure guidelines for the ACGIH and the International Non-Ionizing Radiation Committee of the International Radiation Protection Association (IRPA/INIRC) [now known as the International Commission on Non-Ionizing Radiation Protection (ICNIRP)].

Pacemakers and other medical devices may be susceptible to sub-RF electromagnetic fields. The ACGIH recommends that, lacking specific information from the manufacturer, persons implanted with pacemakers and other devices should not be exposed to greater than 1 G and 1 kV/m at power frequencies.

Best Safe Practices

Measurements/Surveys There are various survey devices, from inexpensive to sophisticated, that will measure sub-RF electric and magnetic fields, especially 50/60-Hz

TABLE 60.5 ELF and VLF Sources and Their Magnetic Flux Densities and/or Electric Field Strength

	Magnetic Flux Density (μT)	Magnetic Flux Density (mG)	Electric Field Strength (V/m)
Directly below transmission lines	5 to 80	50 to 800	10,000
Transmission line right-of-way	2 to 5	20 to 50	1000 to 2000
Directly below distribution lines	0.1 to 3	1 to 30	up to 100
Building wiring, electrical appliances, power tools (distance-dependent)	0.1 to 150	1 to 1500	up to 100
Electric arc welding machines (operator position)	1 to 10,000	10 to 100,000	1 to 300
Electrolytic processes	1000 to 10,000	10,000 to 100,000	—
Video display terminals/computer monitors (operator position)	0.1 to 1	1 to 10	2 to 5

TABLE 60.6 Exposure Limits for Sub-RF Fields.

Organization	Frequency	Remarks	Magnetic Flux Density (mT)	Electric Field Strength
ICNIRP	50/60 Hz	Occupational—whole day	0.5	10 kV/m
	50/60 Hz	Occupational—short term (2 hr max)	5	30 kV/m
	50/60 Hz	Limbs	25	—
ACGIH	1 to 300 Hz	Occupational	$60/f^a$ (1 mT for 60 Hz)	—
	300 Hz to 30 kHz	Occupational (partial and whole body ceiling value)	0.2	—
	0 Hz to 100 Hz	Occupational	—	25 kV/m
	100 Hz to 4 kHz	Occupational (partial and whole body ceiling value)	—	$2.5 \times 10^6/f^a$ V/m
	4 to 30 kHz	Occupational (partial and whole body ceiling value)	—	625 V/m
	50/60 Hz	Pacemakers and medical devices	0.1	1 kV/m

$^a f$ = frequency in Hz.

magnetic fields. Also available are personal magnetic field monitors that are worn by a person over a short time to determine personal exposure (as opposed to field measurements).

Prudent Avoidance   Some individuals advocate the policy of "prudent avoidance," which suggests taking steps to avoid uncertain risks, when the cost is modest. Examples of prudent avoidance in the home include moving the electric alarm clock away from the bed (where a person sleeps for 8 hours), or preheating the bed with an electric blanket, rather than leaving it on all night while sleeping.[31]

References

1. Commission International de l'Eclairage (International Commission on Illumination), *International Lighting Vocabulary*, Publication CIE No. 17 (E1.1); Paris, 1970.

2. Machta, L. *CIAP Measurements of Solar Ultraviolet Radiation*; Cotton, L. G., Hass, W., Komhyr, W. Eds.; U.S. Department of Transportation Final Report, Interagency Agreement DOT-A5-20082; Washington, DC, 1975.

3. National Radiation Protection Board, *Protection Against Ultraviolet Radiation in the Workplace*; Harwell, UK, 1977.

4. Cremonese, M.; Mariutti, G.; Matzeu, M. Irradiance Measurements of Some UV Sources. *Health Phys.* **1982**, *42*(2), 179–185.

5. Sliney, D.; Wolbarsht, M. *Safety with Lasers and Other Optical Sources*; Plenum Press: New York, 1980.

6. American Conference of Governmental Industrial Hygienists. *Threshold Limit Values for Chemical Substances and Physical Agents and Biological Exposure Indices*; Cincinnati, OH, 2000.

7. Fitzpatrick, T. B.; Pathak, M. A.; Harber, L. C.; Seiji, M.; Kukita, A. *Sunlight in Man, Normal and Abnormal Photobiologic Responses*; University of Tokyo Press: Tokyo, 1974.

8. International Non-Ionizing Radiation Committee of the International Radiation Protection Association (INIRC/IRPA), Guidelines on Limits of Exposure to Ultraviolet Radiations of Wavelengths Between 180 nm and 400 nm. In *IRPA Guidelines on Protection Against Non-Ionizing Radiation*; Duchene, A. S.; Lakey, J. R. A.; Repacholi, M. H. Eds.; Pergamon Press: Oxford, UK, 1991.

9. International Commission on Non-Ionizing Radiation Protection (ICNIRP), Guidelines on UV Radiation Exposure Limits, Matthes, R., Ed. *Health Phys.* **1996**, *71*(6), 978.

10. Gies, H. P.; Roy, C. R.; Elliott, G.; Zongli, W. Ultraviolet Radiation Protection Factors for Clothing. *Health Phys.* **1994**, *67*(2), 131–139.

11. Moseley, H. Ultraviolet and Visible Radiation Transmission Properties of Some Types of Protective Eyewear. *Phys. Med. Biol.* **1985**, *30*(2), 177–181.

12. Diffey, B. L.; Hughes, D. Estimates of Ozone Concentrations in the Vicinity of Ultraviolet Emitting Lamps. *Phys. Med. Biol.* **1980**, *25*(3), 559–561.

13. Whillock, M. J.; Bandle, A. M.; Todd, C. D.; Driscoll, C. M. H. Measurements and Hazard Assessment of the Optical Emissions from Various Industrial Infrared Sources. *J. Radiol. Prot.* **1990**, *10*(1), 43–46.

14. *IRPA Guidelines on Protection Against Non-Ionizing Radiation*; Duchene, A. S.; Lakey, J. R. A.; Repacholi, M. H., Eds.; Pergamon Press: New York, 1991.

15. Stuchly, M. A.; Lecuyer, D. W. Survey of Static Magnetic Fields Around Magnetic Resonance Imaging Devices. *Health Phys.* **1987**, *53*(3), 321–324.

16. American Conference of Governmental Industrial Hygienists, *Documentation of the Threshold Limit Values and Biological Exposure Indices*, 6th ed., Vol. III, 1991; pp. PA-51–54.

17. Moulder, J. Static Electric and Magnetic Fields and Cancer FAQs, Medical College of Wisconsin, http://www.mcw.edu/gcrc/cop/static-fields-cancer-FAQ/QandA.html

18. Pavlicek, W.; Geisinger, M.; Castile, L. et al., The Effects of Nuclear Magnetic Resonance on Patients with Cardiac Pacemakers. *Radiology* **1983**, *147*, 149–153.

19. International Commission on Non-Ionizing Radiation Protec-

tion, Guidelines on Limits of Exposure to Static Magnetic Fields. *Health Phys.* **1994**, *66*(1), 100–106.

20. World Health Organization, *Magnetic Fields Health and Safety Guide*, Health and Safety Guide No. 27; Geneva, Switzerland, 1989.

21. Kowalczuk, C. I. et al. *Biological Effects of Exposure to Non-Ionizing Electromagnetic Fields and Radiation. I. Static Electric and Magnetic Fields* (NRPB-R238). National Radiation Protection Board: Chilton, UK, 1991.

22. *Site Planning Guide for Superconducting NMR Systems.* Bruker Instruments: Bellerica, MA, 1991.

23. Hitchcock, R. T.; Patterson, R. M. *Radio-Frequency and ELF Electromagnetic Energies.* Van Nostrand Reinhold: New York, 1995.

24. U.S. Environmental Protection Agency, *Magnetic Field Measurements of Everyday Electrical Devices*; EPA/402-R-92-008, 1992.

25. *EPRI Slide Library on EMF Health Effects*; Electric Power Research Institute: Palo Alto, CA, 1993.

26. International Radiation Protection Association/International Labour Organization, *Protection of Workers from Power Frequency Electric and Magnetic Fields*, Occupational Safety and Health Series Number 69; International Labor Office: Geneva, 1994.

27. World Health Organization. Extremely Low Frequency (ELF) Fields, *Environmental Health Criteria 35*; Geneva, Switzerland, 1984.

28. *Health Effects of Low-Frequency Electric and Magnetic Fields*, Report prepared by an Oak Ridge Associated Universities Panel for the Committee on Interagency Radiation Research and Policy Coordination, 1992.

29. *Possible Health Effects from Exposure to Residential Electric and Magnetic Fields*, Committee on the Possible Effects of Electromagnetic Fields on Biologic Systems; National Research Council, National Academy Press: Washington, DC, 1996.

30. Schnorr, T. M.; Grajewski, B. A.; Hornung, R. W.; Thun, M. J.; Egeland, G. M.; Murray, W. E.; Conover, D. L.; Halperin, W. E. Video Display Terminals and the Risk of Spontaneous Abortion. *New Engl. J. Med.* **1991**, *324*(11), 727–733.

31. Morgan, M. G. *Electric and Magnetic Fields from 60 Hertz Electric Power: What Do We Know About Possible Health Risks.* Department of Engineering and Public Policy; Carnegie Mellon University: Pittsburgh, PA, 1989.

61

Non-Ionizing Radiation: Radio-Frequency and Microwave Radiation

R. TIMOTHY HITCHCOCK

For the purposes of this section, radio-frequency (RF) radiation is electromagnetic radiation between the frequencies of 300 gigahertz (GHz) and 3 kilohertz (kHz). Microwave radiation is a subsection of this spectral region, and includes frequencies between 300 GHz and 300 megahertz (MHz).

Sources

Sources of RF energies in the laboratory may include plasma processes such as chemical vapor deposition (CVD), chemical analysis (nuclear magnetic resonance, NMR), cathode-ray-tube devices such as oscilloscopes and video display terminals (VDTs), microwave ovens, induction processes such as induction heaters, energy pumps and pulse-forming networks for some lasers, and communications equipment (transceivers and cellular devices). A more complete listing is included in Table 61.1. A number of useful reviews and articles are available.[1–18]

Biological Effects

Useful data concerning human exposure are limited and demonstrate no clear trends. Therefore, health science has relied on animal studies to establish biological effects. Animal studies have established effects in many major systems and organs including nervous, reproductive, neuroendocrine, sensory, and immune.[19–25] Combined interactions of radio-frequency fields with neuroactive drugs and chemicals have been reported.[26–27]

Cataracts

Cataracts have been demonstrated in test animals exposed to microwave radiation, but not in man. The most

TABLE 61.1 Sources of RF Radiation in the Laboratory

Source	Comment
CVD reactors	Typically operate between 100 kHz and 27 MHz; used in synthesis, hardening, etching, cleaning, or stripping processes
Chemical analysis	NMR RF radiation around 1 to 100 MHz, and used to change energy state in carbon-hydrogen framework; glow-discharge for elemental analysis and glow-discharge spectrometry; microwave-induced plasma spectrometry; excite discharge from specific lamps; monitor moisture; from about 900 MHz to 30 GHz
CRT devices	Broadband devices with most RF energy around 10 to 35 kHz and some higher harmonics; used in information processing technologies
Microwave reactors, pressure vessels, and ovens	915 and 2450 MHz; used as heat source
Induction heaters	Low-frequency devices used to heat conductive materials; energy source for inductively coupled plasma units operating at 27 MHz
Some lasers	Energy pump for some carbon monoxide and carbon dioxide lasers around 1 to 27 MHz; produced during normal operation of some Q-switched lasers; generated by accelerator for electron lasers
Communications	Cellular devices around 800 MHz to 6 GHz; citizens band around 27 MHz

effective frequencies are around 1 to 10 GHz. Thresholds have been demonstrated for rabbits receiving ocular exposure (versus whole-body exposure) at 2450 MHz.[28,29] No cataracts were found when animals were unrestrained and received whole-body exposures,[30,31] even when exposures were almost lethal. A number of epidemiological investigations and clinical evaluations have been performed, but none has found an excess of cataracts in populations purported to be RF-exposed.[32–36]

Behavioral Effects

Reversible behavioral disruption in short-term studies of operant behavior in test animals is an effect often cited as the biological basis of human exposure guidelines. This endpoint has been found to be a sensitive measure of RF exposure and has been described by various researchers at differing RF frequencies and with more than one study species.[37–40] Generally, behavioral effects are a thermal effect related to significant increases in body temperature.[40,41] A few studies have demonstrated effects on cognitive abilities in animals at low levels of exposure, relative to effects on operant behavior.[42] However, more study is necessary to produce a cohesive body of definitive research.

Although early reports from workplace evaluations of East European countries and Russia discuss nonspecific symptoms associated with the nervous system (headache, fatigue, insomnia)[43] these have never been conclusively linked to RF exposure. In the United States, there have been reports of nervous system effects in two case reports of apparently high, acute overexposure to microwaves.[44,45]

Reproductive and Developmental Effects

Reproductive and developmental effects have been established in test animals exposed to relatively high levels of RF. For example, reproducible teratogenic effects at 27.12 MHz were observed in rats when the whole-body average (WBA) specific absorption rate (SAR, RF dose rate) was around 10–11 W/kg.[46–48] These effects appear to be thermal in nature, as determined from measurement of rectal temperatures.[47] Other researchers found developmental abnormalities in rodents exposed at 2450 MHz. These effects also occurred at high values of WBA SAR.[49,50]

Human studies have not demonstrated any trends.[51–57] Two studies have reported effects on semen,[58,59] but the small sample sizes involved and lack of exposure data make interpretation difficult. A few early studies of operators of cathode-ray-tube type video display terminals suggested a link between VDT use and some reproductive endpoints, but when the study included field measurements, no statistically significant increase in risk was observed.[55] In a hypothesis-generating study, a difference in the sex ratio of the offspring was reported in female physical therapists who had higher indicators of RF exposure.[54]

Cancer

Some studies have suggested that microwaves may be a possible tumor promoter in test animals,[60,61] while other studies did not find significant differences between RF-exposed groups and groups of control animals.[62–66] In one study, no significant differences were observed in rats by tumor type, but when the data on all primary malignancies were combined, there was a statistically significant finding. However, this was not consistent with other study measures, where the animals were generally healthy and disease free in more than 100 other endpoints.[67,68]

A few studies have examined the cancer question in test animals that are susceptible to certain types of tumors. Toler and colleagues found no statistically significant increase in mammary tumors in mice susceptible to that tumor, but did observe an increase in bilateral ovarian epithelial stromal tumor. However, the numbers of epithelial stromal tumors and total epithelial stromal tumors were not significantly different between exposed and control animals.[69] Other studies also reported no statistically significant increase in mammary tumors in susceptible mice.[70,71] No significant differences were found between RF-exposed and control animals in brain cancer for rats inoculated with glioma cells.[72] One group of researchers did report a statistically significant difference in the incidence of total lymphoma and nonlymphoblastic lymphoma in transgenic mice.[73]

With regard to animal studies, there are both positive and negative findings, but, in general, no trends have been observed. Hence, the animal data must be viewed as inconclusive.

In studies of U.S. military peronnel[74,75] and the radar development staff at the MIT Rad Lab,[76] no meaningful differences were observed in cancer mortality. A marginally significant difference in cancer of the gall bladder and bile ducts was reported when the Rad Lab workers were compared to a group of physician specialists. No other differences were reported.[76] In a study of Polish military personnel, statistically significant increases were reported in the mortality rates for all malignancies and cancers of the alimentary canal, nervous system, hematopoietic system, and lymphatic organs.[77] In a study of Motorola, Inc., employees, no association was established between lymphatic/hematopoietic cancers and RF exposure.[78]

Brain cancer has become the focus of study for users of cellular telephones. In general, no statistically significant associations have been reported between cell phone use and overall mortality[79] or brain cancer in users[80] or workers.[78] However, there have been few studies reported and these are limited, in general, by latency considerations where the user population is young and has used cellular phones for a relatively brief time period.

A hypothesis involving testicular cancer and the use of police radar was developed in one study, which found an excess of this cancer in a group of Washington State policemen who used radar guns.[81] In a study of cancer incidence for police officers in Ontario, Canada, the risk of melanoma was statistically significantly increased, but the anatomical distribution of melanoma was similar to the general public. The risk of testicular cancer was nonsignificantly elevated.[82]

In conclusion, although there have been a few positive findings, these studies have not provided conclusive evidence that RF radiation is carcinogenic to humans.[1,83,84] Furthermore, these studies have not identified trends and have not included useful information concerning measures of exposure or dose received by the study populations.

Other Hazards

Kingston and his colleagues have reported that, during chemical sample preparation, microwaves may superheat solutions, which may lead to melting of reaction vessels or explosions.[85] Metallic implants (e.g., metal staples, cochlear implants) or metallic objects that are worn (e.g., jewelry, watches, metal-framed spectacles) may perturb RF fields. RF fields may also interfere with electronic devices, medical electronics (e.g., cardiac pacemakers), or electro-explosive devices.

Exposure Guidelines

In the United States, the foremost guidelines are those published by the Institute of Electrical and Electronics Engineers, Inc. (IEEE)[86] and the American Conference of Governmental Industrial Hygienists (ACGIH).[87] Both guidelines are frequency dependent and draw an envelope around the human whole-body absorption spectrum, as shown in Figure 61.1. The plateau region between about 30 and 300 MHz, where the exposure limits are lowest, is called the whole-body resonance region. This is where the human body absorbs RF energy best because the wavelength-to-body size is optimum, which is called geometrical resonance. At lower frequencies, in the so-called subresonant range, absorption is lowest. At higher frequencies, in the microwave region, absorption is intermediate, and this region is called the quasi-optical region.

Geometrical resonance occurs both when the body is not grounded, called the free-space condition, and when the body is grounded. In general, the resonance fre-

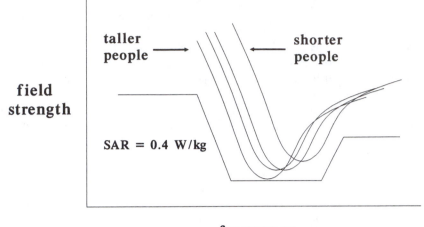

frequency

FIGURE 61.1 A family of whole-body average absorption spectra is displayed. These vary with RF frequency, as discussed in the text. An envelope exposure curve, composed of three plateau regions and two transition (sloping) regions, is positioned around the spectra. The exposure curve may be in terms of field strength, as it is here, or power density.

quency is lower by about a factor of two when the body is grounded. For example, a 5 ft 9 in. person will have a free-space resonant frequency around 70 MHz, while the grounded-resonance frequency is around 35 MHz. Both conditions are covered in the envelope approach utilized in the exposure guidelines.

The IEEE guideline is two-tiered and has a frequency range of 300 GHz to 3 kHz. The two-tier limits are based on whether or not individuals are in a controlled or uncontrolled environment. A key difference between the two environments is that people in the controlled environment are aware of the potential for RF exposure, while individuals in the uncontrolled environment have no knowledge or control of their exposure. In general, the exposure limits in the controlled environment are five times those in the uncontrolled environment. This difference does not have a biological basis but is due to the application of an additional fivefold safety factor to derive the levels for the uncontrolled environment. The ACGIH guidelines are for occupational exposures between 300 GHz and 30 kHz.

The fundamental quantity of the exposure criteria is the specific absorption rate (SAR), or RF dose rate. The whole-body SAR is 0.4 watts per kilogram (W/kg) in the controlled environment and in the ACGIH limits. In the uncontrolled environment, it is 0.08 W/kg. The SAR does not vary with frequency.

The guidelines include three frequency-dependent, free-field quantities that may be measured or calculated to determine if an overexposure exists. These are the electric-field strength (E), magnetic-field strength (H), and power density (W or S). The exposure limits, in terms of these quantities, are called derived limits and may be viewed as surrogate measures for the SAR. In application, if the measured free-field value is less than the exposure limit, then the whole-body average SAR is less than the threshold limit, that is, 0.4 W/kg or 0.08 W/kg, and the exposure is acceptable.

At frequencies less than 100 MHz, both IEEE and ACGIH guidelines require the evaluation of induced and contact currents. Low-frequency RF fields can induce currents to flow within the body, and these currents may flow out through the bottom of the foot as a short-circuit current. Contact currents are RF currents that are generated or stored in a conductive material, with the potential for transmission to the hand when grasping contact is made. The exposure limits for induced and contact currents are shown in Figure 61.2.

The Occupational Safety and Health Administration does address RF radiation in 29 *CFR* 1910.97; however, this section is dated and generally is not used. More recent federal exposure guidelines have been developed by the Federal Communications Commission,[88] but these apply only to communications and broadcasting. The Department of Defense has a standard for RF radiation, DOD Instruction 6055.11, that is based on the IEEE guideline.

Although not an exposure guideline, the Food and Drug Administration has published a product performance standard for microwave ovens. This standard re-

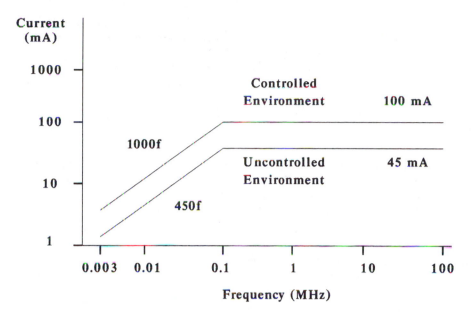

FIGURE 61.2 Exposure limits for induced and contact currents are displayed for the controlled and uncontrolled environments. The limits for induced currents are for a single foot and must be doubled for both feet. When used for contact currents, the limits are for grasping contact with the hand.

quires manufacturers to comply with certain performance and labeling specifications, including criteria for acceptable levels of microwave leakage.[89]

Evaluation

Hazard evaluation of RF radiation usually involves measurement with broadband measurement equipment.[1,25] These instruments require no tuning and have a calibrated, frequency-independent response. Instruments to evaluate free fields include antenna arrays that are sensitive to E or H fields. E-field elements are small, orthogonal monopole or dipole antennas, while H-field elements are small, orthogonal loop antennas. The antennas are linked to detectors, with diode or thermocouples being most common. The output from the detector is directed to the metering instrumentation via a conductive lead or fiberoptic cable.

Instruments for induced currents are either stand-on or clamp-on devices. Stand-on units are a type of parallel-plate capacitor that resemble a bathroom scale. Clamp-on devices measure current flowing through the body or an appendage with a current transformer. Contact current instruments are hand-held units that are connected to a conductive plate that is placed on the ground. A sample probe is used to make contact with the surface to be sampled. The current flows through a

circuit, where it is measured, and then out the plate to ground.

The most important rule in measurement is that the operational frequency of the source must be within the calibrated frequency response of the instrument being used. Knowledge of the operational frequency will also allow the selection of the appropriate exposure limit, since the guidelines are frequency dependent. Also, it will determine what quantities must be measured, as compiled in Table 61.2.

Measurement involves positioning a "stickman" made of a low-permittivity material (e.g., polyvinyl chloride tubing) at the location of interest. The stickman should be 200 cm tall and marked in 20-cm increments. The evaluator collects a minimum of 10 measurements at the designated points on the stickman, then calculates an arithmetic average.[86] (Note: The squares of E and H

TABLE 61.2 Measurement Quantities as a Function of Frequency

Frequency Range	Quantity
300 GHz to 300 MHz	E or H or S as spatial average
300 MHz to 100 MHz	E and H as spatial average
100 MHz to 3 kHz	E and H as spatial average induced currents; contact currents

field measurements must be averaged, while values in power density do not require squaring prior to averaging.) The average value is compared to the exposure limits, since the limits are based on whole-body average SARs.

Control Measures

Shielding effectiveness (SE), or attenuation, is used to describe the effectiveness of a material as an RF shield. SE, in decibels, is determined from the contributions of reflection, absorption, and internal reflection of a material or materials. A suitable shielding material may be used to make an enclosure to shield workers or to shield components that may be susceptible to an applied RF E or H field. Particular attention must be paid to conductive gasketing, ventilation openings, access ports, bonding, and grounding when designing and constructing an enclosure. Therefore, it is important to involve a knowledgeable RF engineer in the process to ensure that the total enclosure provides the necessary attenuation to reduce the intensity of the RF field outside the enclosure.

Shielding materials for electric fields include aluminum, copper, brass, bronze, silver, gold, tin, lead, and conductive polymers. Materials may be combined or machined to produce composite or plated plastics, laminates and film coatings, clad metals, conductive paints, and arc-sprayed metals. Meshes, other woven textiles, and perforated materials may be used. Materials that are used for magnetic fields are iron, some stainless steels (430), mild steel, and nickel-iron and cobalt-iron alloys.[1,25]

A control measure that is useful when conveyor or ventilation openings are required in an enclosure, is the waveguide-beyond-cutoff. A waveguide is a metal tube (circular, rectangular, or square) that is used to confine and guide electromagnetic waves. Although most waveguides are designed to allow propagation of RF radiation, a waveguide-beyond-cutoff is designed to make use of the cutoff frequency characteristic of a waveguide, thereby minimizing propagation of RF waves that are below a specific frequency cutoff.[90] This is effective because RF waves are unable to propagate efficiently through an opening that is smaller than 1/2 the wavelength. As such, conveyor or ventilation openings that have openings smaller than 1/2 the RF wavelength will not permit leakage outside the enclosure. This is the same principle that prevents RF energy from leaking outside of a microwave oven door.

For frequencies near the human, whole-body, grounded-resonance frequencies, about 10 to 40 MHz, it has been shown that the SAR may be reduced by separating the body from the ground by a small space. For a person exposed at these low frequencies, this effect occurs because the ground connection is broken, which shifts the body into the "free-space" absorption-rate mode. The SAR is reduced since exposure remains at low resonant frequencies, and the body is effectively in free space. This has been demonstrated by simulating an air gap between a human subject and the ground with expanded polystyrene and hydrocarbon resin foam.[91] The same effect has been observed with common footwear and socks, where the SAR has been modified when human volunteers were exposed at frequencies between 10 and 40 MHz. The use of shoes and socks reduced the SAR in comparison to barefoot volunteers by the following: 15 to 45% for nylon socks and thin-leather-soled shoes and 35 to 75% for wool socks and rubber soles.[91]

References

1. Hitchcock, R. T.; Paterson, R. M. *Radio-Frequency and ELF Electromagnetic Energies*; Van Nostrand Reinhold: New York, 1995.
2. Girshick, S. L.; Lindsay, J. W. *Mat. Res. Soc. Symp. Proc.* **1994**, *416*, 3–11.
3. Murata, M.; Takeuchi, Y.; Sasagawa, E.; Hamamoto, K. *Rev. Sci. Instrum.* **1996**, *67*, 1542–1545.
4. Hosokawa, N. *Thin Solid Films* **1996**, 136–142, 281–282.
5. Stanley, J. L.; Bentley, H. W.; Denton, M. B. *Appl. Spectrosc.* **1973**, *27*, 265–267.
6. Alder, J. F.; Brennan, M. F.; Clegg, I. M.; Drew, P. K. P.; Thirup, G. *Trans. Inst. Meas. Control* **1983**, *5*, 99–111.
7. Meinel, H.; Rembold, B. *Radio Electron. Eng.* **1979**, *49*, 351–360.
8. Shick, C. R.; DePalma, P. A. *Anal. Chem.* **1996**, *68*, 2113–2121.
9. Marcus, R. K.; Harville, T. R.; Mei, Y. *Anal. Chem.* **1994**, *66*, 902–911.
10. Jones, D. G.; Payling, R.; Gower, S. A.; Boge, E. M. *J. Anal. At. Spectrom.* **1994**, *9*, 369–373.
11. Sheppard, B. S.; Caruso, J. A. *J. Anal. At. Spectrom.* **1994**, *9*, 145–149.
12. Majetich, G.; Hicks, R. *J. Microw. Power Electromagn. Energy* **1995**, *30*, 27–45.
13. Bennett, B. *Process Cont. Eng.* **1994**, *47*(12), 56–57.
14. Bond, G. M.; Moyes, R. B.; Pollington, S. D.; Whan, D. A. *Chem. Ind.* **1991**, *18*, 686–687.
15. Pugnet, M.; Downing, B.; Michelson, S. *J. Microw. Power Electromagn. Energy* **1993**, *28*, 18–24.
16. Lee, J. H.; Kim, D. S.; Lee, Y. H.; Farouk, B. *J. Electrochem. Soc.* **1996**, *143*, 1451–1458.
17. Rhoades, C. B., Jr. *J. Anal. At. Spectrom.* **1996**, *11*, 751–757.
18. Stuerga, D. A. C.; Gaillard, P. *J. Microw. Power Electromagn. Energy* **1996**, *31*, 87–100.
19. National Council on Radiation Protection and Measurements, *Biological Effects and Exposure Criteria for Radiofrequency Electromagnetic Fields* (NCRP Report No. 86); Bethesda, MD, 1986.
20. Environmental Protection Agency, *Biological Effects of Radiofrequency Radiation* (Report No. EPA-600/8-83-026F). Elder,

J. A.; Cahill, D. F. Eds.; National Technical Information Service: Springfield, VA: 1984.

21. Heynick, L. *Critique of the Literature on Bioeffects of Radiofrequency Radiation: A Comprehensive Review Pertinent to Air Force Operations* (Report No. USAFSAM-TR-87-3); USAF School of Aerospace Medicine: Brooks Air Force Base, TX, 1987.

22. Saunders, R. D.; Sienkiewicz, Z. J.; Kowalczuk, C. I. *J. Radiol. Prot.* **1991**, *11*, 27–42.

23. Wilkening, G. M. Nonionizing Radiation, In *Patty's Industrial Hygiene and Toxicology General Principles*, 4th ed.; Clayton, G. D., Clayton, F. E., Eds.; Wiley & Sons: New York, 1991, pp. 657–742.

24. Murray, W. E.; Hitchcock, R. T.; Patterson, R. M.; Michaelson, S. M. Nonionizing Electromagnetic Energies. In *Patty's Industrial Hygiene & Toxicology*, 3rd ed., Vol. 3, Part B; Cralley, L., Cralley, L., Bus, J. Eds.; Wiley & Sons: New York, 1995; pp. 623–727.

25. Hitchcock, R. T. *Radio-Frequency and Microwave Radiation*; American Industrial Hygiene Association: Fairfax, VA, 1994.

26. Lai, H.; Horita, A.; Chou, C. K.; Guy, A. W. *IEEE Engineering in Medicine and Biology Magazine* **1987**, *6*, 31–36.

27. Frey, A. H.; Wesler, L. S. *J. Bioelect.* **1990**, *9*, 187–196.

28. Carpenter, R.; Van Ummersen, C. *J. Microwave Power* **1968**, *3*, 3–19.

29. Guy, A. W.; Lin, J. C.; Kramar, P. O.; Emery, A. F. *IEEE Trans. Microwave Theory Tech.* **1975**, MTT-23, 492–498.

30. Michaelson, S. M.; Howland, J. W.; Deichmann, W. B. *Ind. Med.* **1971**, *40*, 18–23.

31. Appleton, B.; Hirsch, S. E.; Brown, P. V. K. *Ann. N.Y. Acad. Sci.* **1975**, *247*, 125–134.

32. Cleary, S. F.; Pasternack, B. S.; Beebe, G. W. *Arch. Environ. Health* **1965**, *11*, 179–182.

33. Cleary, S. F.; Pasternack, B. S. *Arch. Environ. Health* **1966**, *12*, 23–29.

34. Majewska, K. *Pol. Med. J.* **1968**, *38*, 989–994.

35. Siekierzynski, M.; Czerski, P.; Gidynski, A.; Zydecki, S.; Czarnecki, C.; Dziuk, E.; Jedrzejczak, W. *Aerospace Med.* **1974**, *45*, 1146–1148.

36. Bonomi, L.; Bellucci, R. *Bollettion Di Oculistica* **1989**, *68*(S7), 85–98.

37. D'Andrea, J.; Gandhi, O.; Lords, J. L. *Radio Sci.* **1977**, *12*(6S), 251–256.

38. de Lorge, J. O. The Effects of Microwave Radiation on Behavior and Temperature in Rhesus Monkeys. In *Biological Effects of Electromagnetic Waves* (HEW Publication (FDA) 77-8010; Johnson, C., Shore, M. L., Eds., Bureau of Radiological Health: Rockville, MD, 1976, pp. 158–174.

39. de Lorge, J. O. Disruption of Behavior in Mammals of Three Different Sizes Exposed to Microwaves: Extrapolation to Larger Mammals. In *Proceedings of the 1978 Symposium on Electromagnetic Fields in Biological Systems*, International Microwave Power Institute: Edmonton, Canada, 1978, pp. 215–228.

40. de Lorge, J. O. The Thermal Basis for Disruption of Operant Behavior by Microwaves in Three Animal Species. In *Microwaves and Thermoregulation*; Adair, E. R., Ed.; Academic Press: New York, 1983, pp. 379–399.

41. D'Andrea, J. A. *Health Phys.* **1991**, *61*, 29–40.

42. D'Andrea, J. A. *Bioelectromagnetics* **1999**, *20*, 64–74.

43. Silverman, C. *Am. J. Epidemiol.* **1973**, *97*, 219–224.

44. Forman, S. A.; Holmes, C. K.; McManamon, T. V.; Wedding, W. R. *J. Occup. Med.* **1982**, *24*, 932–934.

45. Williams, R. A.; Webb, T. S. *Aviat. Space Environ. Med.* **1980**, *51*, 1243–1244.

46. Lary, J. M.; Conover, D. L.; Foley, E. D.; Hanser, P. L. *Teratology* **1982**, *26*, 299–309.

47. Lary, J. M.; Conover, D. L.; Johnson, P. H.; Burg, J. R. *Bioelectromagnetics* **1983**, *4*, 249–255.

48. Lary, J. M.; Conover, D. L.; Johnson, P. H.; Hornung, R. W. *Bioelectromagnetics* **1986**, *7*, 141–149.

48. Berman, E.; Carter, H.; House, D. *J. Microwave Power* **1982**, *17*, 107–112.

50. Berman, E.; Carter, H.; House, D. *Bioelectromagnetics* **1982**, *3*, 285–291.

51. Kallen, B.; Malmquist, G.; Moritz, U. *Arch. Environ. Health* **1982**, *37*, 81–84.

52. Kolmodin-Hedman, B.; Mild, K. H.; Hagberg, M.; Jonsson, E.; Anderson, M. C.; Eriksson, A. *Int. Arch. Occup. Environ. Health* **1988**, *60*, 243–247.

53. Taskinen, H.; Kyyronen, P.; Hemminki, K. *J. Epidemiol. Comm. Health* **1990**, *44*, 196–201.

54. Larsen, A. I.; Olsen, J.; Svane, O. *Scand. J. Work Environ. Health* **1991**, *17*, 324–329.

55. Schnorr, T. M.; Grajewski, B. A.; Hornung, R. W.; Thun, M. J.; Egeland, G. M.; Murray, W. E.; Conover, D. L.; Halperin, W. E. *New Engl. J. Med.* **1991**, *324*, 727–733.

56. Nielsen, C. V.; Brandt, L. *Scand. J. Work Environ. Health* **1990**, *16*, 323–328.

57. Brandt, L.; Nielsen, C. V. *Scand. J. Work Environ. Health* **1990**, *16*, 329–333.

58. Lancranjan, I.; Maicanescu, M.; Rafaila, E.; Klepsch, I.; Popescu, H. I. *Health Phys.* **1975**, *29*, 381–383.

59. Weyandt, T. B. *Evaluation of Biological and Male Reproductive Function Responses to Potential Lead Exposures in 155 MM Howitzer Crewmen*; National Technical Information Service (AD-A247 384); Springfield, VA, 1992.

60. Szudzinski, A.; Pietraszek, A.; Janiak, M.; Wrembel, J.; Kalczak, M.; Szmigielski, S. *Arch. Dermatol. Res.* **1982**, *274*, 303–312.

61. Szmigielski, S. A.; Szudzinski, A.; Pietraszek, A.; Bielec, M.; Janiak, M.; Wrembel, J. *Bioelectromagnetics* **1982**, *3*, 179–191.

62. Santini, R.; Hosni, M.; Deschaux, P.; Pacheco, H. *Bioelectromagnetics* **1988**, *9*, 105–107.

63. Wu, R. Y.; Chiang, H.; Shao, B. J.; Li, N. G.; Fu, Y. D. *Bioelectromagnetics* **1994**, *15*, 531–538.

64. Svedenstal, B. M.; Holmberg, B. *Int. J. Radiat. Biol.* **1993**, *64*, 119–125.

65. Imaida, K.; Taki, M.; Yamaguchi, T.; Ito, T.; Watanabe, S.; Wake, K.; Aimoto, A.; Kamimura, Y.; Ito, N.; Shirai, T. *Carcinogenesis* **1998**, *19*, 311–314.

66. Imaida, K.; Taki, M.; Watanabe, S.; Kamimura, Y.; Ito, T.; Yamaguchi, T.; Ito, N.; Shirai, T. *Jap. J. Cancer Res.* **1998**, *89*, 995–1002.

67. Guy, A. W.; Chou, C.-K.; Kunz, L. L.; Crowley, J.; Krupp, J. *Effects of Long-Term Low-Level Radiofrequency Radiation Exposure on Rats*, Vol. 9. Summary (Report USAFSAM-TR-85-64), United States Air Force, School of Aerospace Medicine: Brooks Air Force Base, TX, 1985.

68. Chou, C.-K.; Guy, A. W.; Kunz, L. L.; Johnson, R. B.; Crowley, J. J.; Krupp, J. H. *Bioelectromagnetics* **1992**, *13*, 469–496.

69. Toler, J. C.; Shelton, W. W.; Frei, M. R.; Merritt, J. H.; Stedham, M. A. *Radiat. Res.* **1997**, *148*, 227–234.

70. Frei, M. R.; Berger, R. E.; Dusch, S. J.; Guel, V.; Jauchem, J. R.;

Merritt, J. H.; Stedham, M. A. *Bioelectromagnetics* **1998**, *19*, 20–31.

71. Frei, M. R.; Jauchem, J. R.; Dusch, S. J.; Merritt, J. H.; Berger, R. E.; Stedham, M. A. *Radiat. Res.* **1998**, *150*, 568–576.

72. Salford, L.; Brun, A.; Persson, B. *Wirel. Netw.* **1997**, *3*, 463–469.

73. Repacholi, M. H.; Basten, A.; Gebski, V.; Noonan, D.; Finnie, J.; Harris, A. W. *Radiat. Res.* **1997**, *147*, 631–640.

74. Robinette, D.; Silverman, C.; Jablon, S. *Am. J. Epidemiol.* **1980**, *112*, 39–53.

75. Garland, F. C.; Shaw, E.; Gorham, E. D.; Garland, C. F.; White, M. R.; Sinsheimer, P. J. *Am. J. Epidemiol.* **1990**, *132*, 293–303.

76. Hill, D. G. A Longitudinal Study of a Cohort with Past Exposure to Radar: The MIT Radiation Laboratory Follow-up Study, (Ph.D. dissertation), Johns Hopkins University, University Microfilms International: Ann Arbor, MI, 1988.

77. Szmigielski, S. *Sci. Total Environ.* **1996**, *180*, 9–17.

78. Morgan, R. W.; Kelsh, M. A.; Ahao, K.; Exuzides, K. A.; Heringer, S.; Negrete, W. *Epidemiology* **2000**, *11*, 118–127.

79. Rothman, K. J.; Loughlin, J. E.; Funch, D. P.; Dreyer, N. A. *Epidemiology* **1996**, *7*, 303–305.

80. Hardell, L.; Nasman, A.; Pahlson, A.; Hallquist, A.; Mild, K. H. *Int. J. Oncology* **1999**, *15*, 113–116.

81. Davis, R. L.; Mostofi, F. K. *Am. J. Ind. Med.* **1993**, *24*, 231–233.

82. Finkelstein, M. *Am. J. Ind. Med.* **1998**, *34*, 157–162.

83. Silverman, C. Epidemiology of Microwave Radiation Effects in Humans. In *Epidemiology and Quantitation of Environmental Risk in Humans from Radiation and Other Agents*; A. Castellani, Ed.; Plenum Press: New York, 1985; pp. 433–458.

84. Elmwood, J. M. *Environ. Health Perspect.* **1999**, *107* (Suppl. 1), 155–168.

85. Kingston, H. M.; Walter, P. J.; Engelhart, W. G.; Parsons, P. J. Chemical Laboratory Microwave Safety. In *Microwave Enhanced Chemistry: Fundamentals, Sample Preparation, and Applications* (ACS Professional Reference Book Series); American Chemical Society: Washington, D.C., 1997.

86. IEEE, *Safety Levels with Respect to Human Exposure to Radio Frequency Electromagnetic Fields, 3 kHz to 300 GHz* (IEEE C95.1); IEEE: New York, 1999.

87. American Conference of Governmental Industrial Hygienists, *1996 TLVs and BEIs*; ACGIH: Cincinnati, OH, 1996, 113–116.

88. Federal Communications Commission, *Guidelines for Evaluating Effects of Radiofrequency Radiation* (47 *CFR* Parts 1, 2, 15, 24, and 97), *Fed. Regis.* **1996**, *61*(153), 41006.

89. Food and Drug Administration, Subchapter J—Radiological Health (21 *CFR* Part 1030); U.S. Government Printing Office: Washington, DC, 1971.

90. Ruggera, P. S.; Schaubert, D. H. *Concepts and Approaches for Minimizing Excessive Exposure to Electromagnetic Radiation from RF Sealers* (HHS Publication (FDA) 82-8192); U.S. Government Printing Office: Washington, DC, 1982.

91. Hill, D. A. *IEEE Trans. Microwave Theory Tech.* **1984**, MTT-32, 772–778.

62

Ionizing Radiation: Radiation Safety Program Elements

PHILIP E. HAMRICK

Biomedical and other research and teaching organizations often must address a wide variety of procedures using radioactive material, some of which may be unique. To meet the challenges, evaluate risks, and provide the safest practical working environment requires the development of a program[1,2] with necessary controls and flexibility to address new procedures and minimize unnecessary and burdensome regulations. Each organization has program requirements that are specific to its operations, but there are a number of elements that are common to many programs. The program elements that are listed or discussed are not intended to be exhaustive, but should prove helpful in the design or review of a radiation safety program.

The major program areas addressed are as follows:

regulations, licensing, radiation safety manual, inspections, audits, records, and environmental.

Regulations and Guidelines

The Nuclear Regulatory Commission (NRC) is the primary federal agency that establishes and enforces laws and regulations[3] pertaining to the use of radioactive material. The agency was established by the Energy Reorganization Act of 1974 and became effective January 19, 1975 (10 *CFR* 1.1). The principal NRC offices are located in Washington, DC, with five regional offices. The primary mailing address for the Washington, DC, office is U.S. Nuclear Regulatory Commission, Washington, DC 20555-0001. The addresses for the four regional offices are (10 *CFR* 1.5):

Region I, USNRC, 475 Allendale Road, King of Prussia, PA 19406-1415

Region II, USNRC, 81 Forsyth Street, SW, Suite 23T85, Atlanta, GA 30303

Region III, USNRC, 801 Warrenville Road, Lisle, IL 60532-4351

Region IV, USNRC, 611 Ryan Plaza Drive, Suite 400, Arlington, TX 76011-8064

The principal regulations for the establishment of a radioactive materials program are given in Box 1. The Occupational Safety and Health Administration regulations found in 29 *CFR* are based on the regulations in 10 *CFR* Part 20. The regulations given in 49 *CFR* 171–180 include regulations applicable to the shipment and packaging of nonradioactive material as well as those applicable to radioactive material. Environmental Protection Agency regulations in 40 *CFR* 60 relate to incineration and air monitoring and are not specific for radioactive material. However, 40 *CFR* 61 gives the national

standards for radionuclide emissions. The regulations given in 40 *CFR* 260–262 are not specific for radioactive material, but deal with other hazardous waste that may be a component of the radioactive waste stream.

Many states have an agreement with the NRC to oversee the use of radioactive material within the state. These "Agreement States" have conforming regulations that have been approved by the NRC and that are at least as stringent as NRC regulations. The agreement states establish programs for licensing, inspection, and enforcement that are similar to the NRC programs in nonagreement states. The jurisdiction of the state does not extend to federal facilities or nuclear power plants, which are regulated directly by the NRC. Many states also regulate naturally occurring and accelerator-produced radioactive materials (NARM),[4] whereas the NRC generally regulates "byproduct" material. Byproduct material is any radioactive material produced or made radioactive "by exposure to radiation incident to the process of producing or utilizing special nuclear material" (10 *CFR* 20). This includes the tailings or waste produced by the extraction of uranium or thorium from

---

Box 1
Principal Regulations

Principal Regulations for the Establishment of
a Radioactive Materials Program

10 *CFR* 20	Standards for Protection Against Radiation
10 *CFR* 19	Notices, Instructions, and Reports to Workers
10 *CFR* 31	General Domestic Licenses of Broad Scope for Byproduct Material
10 *CFR* 33	Specific Domestic Licenses of Broad Scope for Byproduct Material
10 *CFR* 35	Medical Use of Byproduct Material
10 *CFR* 71	Packaging and Transportation of Radioactive Material

Additional Important Federal Regulations

21 *CFR* 1020	Performance Standards for Ionizing-Radiation-Emitting Products
29 *CFR* 1910.96	Occupational Safety and Health Standards—Toxic and Hazardous Substances
40 *CFR* 60	Standards for Performance of New Stationary Sources
40 *CFR* 260-262	Hazardous Waste
40 *CFR* 141	Drinking Water Standards
49 *CFR* 171-180	Transportation of Hazardous Materials

its ore for use as source material. Those states having agreement status with the NRC are: Alabama, Arkansas, California, Kansas, Kentucky, Florida, Maine, Massachusetts, Mississippi, New Hampshire, New York, North Carolina, Texas, Tennessee, Oregon, Iowa, Illinois, Arizona, Colorado, New Mexico, Louisiana, Nebraska, Washington, Maryland, North Dakota, South Carolina, Georgia, Utah, Nevada, and Rhode Island. All states regardless of NRC agreement status will normally have some regulations affecting the use of ionizing radiation.

In addition to federal and state regulations, counties and cities may have regulations. These regulations often address transportation of radioactive material through the city or county, sewer concentrations, or incineration/air concentration limits.

A number of voluntary organizations publish standards, reports, and other valuable information on radiation, radiation protection, and radiation protection procedures. Some of the more notable of these are: National Council of Radiation Protection (NCRP), International Commission of Radiation Protection (ICRP), American National Standards Institute (ANSI), and International Atomic Energy Agency (IAEA). The primary professional organization in the United States concerned with radiation protection is the Health Physics Society. This society publishes a journal, *Health Physics*, a newsletter, and sponsors annual and semiannual scientific meetings. Other important professional organizations with information on radiation safety are American Association of Physicists in Medicine, American Conference of Governmental Industrial Hygienists, and American Industrial Hygiene Association.

Licensing and Registration

The NRC and/or states control the use of radioactive material through the process of licensing. Since state regulations will vary from state to state, the discussion will be restricted to the NRC licensing process.

Some widely used sources of radiation or devices containing sealed radioactive material do not require the user to apply for a license but are under a "general" license issued by the NRC (10 *CFR* 31). Table 62.1 lists some devices and sources along with any applicable radionuclide and activity limits falling in this category.

Although a general license does not require an application by the licensee, certain regulations must be followed and records kept for these sources. Some states require that the sources be registered. Those sources that require leak testing and proper operation of the on/off mechanism must be tested at no longer than six-month intervals. States will also normally require that ionizing-

radiation-producing machines such as X-ray devices and linear accelerators be licensed or registered. The user must check with the appropriate state agency for applicable requirements.

Specific licenses require the submission of an application to the NRC or appropriate state agency. The licenses can vary in scope from a license to use one particular radionuclide, such as ^{14}C with a stated maximum activity for biological studies, to specific licenses of broad scope in which the licensee is authorized to use any of a wide range of radionuclides, up to a stated maximum activity. The broader the scope of the license, the greater the number of requirements and safeguards that must be met. Therefore it is to the advantage of the licensee when applying for a license to restrict the application to those items likely to be needed. However, in a research or hospital setting, where the requirements for a particular radionuclide may not be recognized until a short time before it is needed, a license of broad scope is desirable. Valuable research time or treatment options may be lost while an application for license amendment is being processed.

The NRC has established three types of licenses of broad scope, Type A, Type B, and Type C. The Type A license is the most flexible and authorizes the "receipt, acquisition, ownership, possession, use, and transfer of any chemical or physical form of the byproduct material specified in the license, but not exceeding quantities specified in the license" (10 *CFR* 33). The Type A license also requires more administrative controls than either Type B or Type C requires. In particular, a radiation safety committee must be established for review and approval of proposed uses. Factors that must be considered are adequacy of facilities, equipment, training and experience of the user, and operating or handling procedures. Most major universities, leading research and development laboratories, and hospitals involved in biomedical research require a Type A license. For smaller organizations that have known and stable radionuclide requirements, a Type B license is more appropriate, and for an individual physician or small clinic, a Type C license is sufficient.

The three license types require that the licensee establish administrative procedures to completely control the procurement, use, and disposal of radioactive material and maintain associated records. Type A and B licenses require the appointment of a radiation safety officer who is qualified by training and experience in radiation protection and who is available when advice or assistance is needed on safety procedures. Normally, the radiation safety officer will have a college degree or equivalent at the master level in health physics. For a Type C license, the applicant must have a college degree or equivalent

TABLE 62.1 General License Material

Item	Isotope(s)	Applicable Activity Limit per Item (mCi)
Static eliminator	Po-210	0.5
Ion-generating tube	Po-210	0.5
	H-3	50
Luminous safety devices for aircraft	H-3	10,000
	Pm-147	300
Ice detection device	Sr-90	0.05
In vitro tests by physicians or veterinarians	I-125, I-131, C-14	0.01
	H-3	0.05
	Fe-59	0.02
Measuring,[a] gauging, or controlling device	Various	Depends on specific license conditions required of the manufacturer

[a]Requires periodic leak testing.

at the bachelor level in the physical or biological sciences or in engineering and at least 40 hours of health physics training.

One of the major differences in the license types is that the maximum activity for one radionuclide allowed under Types B and C is restricted to the activity listed in 10 *CFR* 33.100 of columns I and II, respectively, of schedule A. If more than one radionuclide is used, the sum of the fractional limits must not exceed one. Table 62.2 summarizes the major differences.

Radiation Safety Manual

Each organization that uses radioactive material or sources of ionizing radiation should develop policies for radiation workers that are specific to the types of work and procedures being performed. These policies should be made available to employees in a convenient, readable format such as a radiation safety manual or manual of policies and procedures.

Some areas that may be covered in a typical manual are presented in Table 62.3.

Inspections

A necessary component of any radiation safety program is the establishment of regular and thorough inspections by the licensee. Inspections should not be viewed solely as a means of identifying violations of safety policies and procedures but can be an important part of the ongoing training of radiation workers. Review of work procedures or working conditions and discussion with workers can be valuable in identifying potential problems and hazards. Also, workers can be introduced to current and emerging information and technology with which they may be unfamiliar.

The frequency of inspections will vary with each organization and with specific license conditions. However, it is desirable for inspection personnel to visit work areas often and become familiar with the personnel and work

TABLE 62.2 License Types

Type of License	Activity Limits	Requirements
A	As specified in the license	Radiation Safety Committee and Radiation Safety Officer
B	10 *CFR* 33.100 Column I of Schedule A[a]	Radiation Safety Officer
C	10 *CFR* 33.100 Column II of Schedule A[a]	B.S. degree or equivalent and 40 hr training and experience in safe use of radioactive material

[a]If more than one isotope is used, the sum of the fractional limits for each isotope must be less than one.

TABLE 62.3 Radiation Safety Manual Topics

Area	Some Possible Topics
Emergencies[5,6]	Phone numbers, spill and cleanup procedures, personnel decontamination procedures
Organization and authority	Type of NRC or state license, organizational hierarchy and authority, role of radiation safety committee, enforcement policies
Responsibilities	Radiation Safety Committee, Radiation Safety Officer, Principal User (person in charge of use of radioactive material for a laboratory or group), other radiation workers
Facilities	Procedure and activity limits as determined by available facilities and equipment
Safety	General, and isotope- or source-specific safety procedures
Monitoring	Requirements for monitoring badges, bioassay procedures, personnel and room monitoring, exposure limits
Radioactive material	Security, inventory, procurement, receipt, transfer, transport, shipping, disposal, labeling, associated records
Radiation-producing devices	Room design, shielding, security, filtration, collimation, interlocks
Animals	Special procedures related to procurement, use, and disposal
Forms	Copies and examples of forms such as: application to use radioactive material with proposed procedures, application for monitoring, statement of training, procurement, request for radioactive waste disposal, inventory, clearance for departing personnel, declaration of pregnancy

that is being performed. Many of these visits could be short and informal with more formal inspections limited to one or two times per year. As radiation workers become more familiar with inspection personnel, they will more likely seek advice or consultation on safety problems and concerns.

Formal inspections can be announced or unannounced. However, some unannounced inspections must be made each year to ensure that legal requirements are being followed. If key personnel are not present during an unannounced inspection, it may be desirable to announce the next inspection and request that key personnel be present. If the inspection has revealed hazards or hazardous practices that require immediate attention, a meeting of key personnel must be quickly called to review the situation. The meeting should provide time for full discussion of any problems areas detected, possible solutions, and implementation plans. An enforcement policy must be established so that users know the consequences of initial and repeated violations. Depending on the nature of the violation, the enforcement may be of an escalating type.

One of the most important elements in any inspection program is monitoring for surface and air contamination. Radiation workers as part of normal safety operating procedures will monitor radiation levels, test for contamination, and monitor air concentrations if applicable. The Radiation Safety Office must confirm that monitoring is being done by inspection of records and by inde-

pendent monitoring. The presence of contamination is an indication that safety procedures are inadequate or not being followed. The type and frequency of independent monitoring required will depend on the activity used, the volatility of the material, and other potential hazardous conditions. If contamination is often detected in a particular area, work procedures must be discussed with the personnel to explore possible ways in which the contamination can be reduced. In some biomedical organizations, independent swipe tests and monitoring surveys are made of all laboratories each month. This is one of the best quantitative indicators of the effectiveness of the safety program.

In addition to monitoring surveys, the following items should be included in a typical inspection of a biomedical laboratory:

- condition of monitoring and inventory records
- proper posting of areas where radioactive material is used or stored
- presence or availability of reference material such as a radiation safety manual and experimental protocol procedures
- presence of required-personnel-monitoring dosimeters
- verification that all personnel are properly trained and that the training is up to date
- compliance with security policies
- use of proper shielding

- condition of waste collection and storage
- presence and currency of emergency information
- evidence of food or drinks in work areas
- overall cleanliness

Audits

Whether it is called an audit or an annual report, each program should be reviewed at least once per year. The review can be "in house" or by a group external to the program. For those programs with a Radiation Safety Committee, the committee may serve as the annual audit or review group. The Radiation Safety Officer normally will have the responsibility for collecting and recording much of the data used in an annual report. The report should cover all aspects of the radiation safety program and comparisons made with previous years to identify trends, improvements, or problem areas.

Table 62.4 lists some elements that should be considered for inclusion in an annual program review. Some of these items are not applicable to all programs, while other programs should include additional elements specific to those programs.

Records

Records are important to almost all aspects of a radiation safety program. In the listing of possible audit procedures in Table 62.4, a direct or implied reference is made to records in almost every case. Only through accurate and thorough records can comparisons be made and trends detected. Records[14] are vital to document events that may have future impact or legal implications. Although it may seem desirable to make a record of everything that happens, this is clearly impractical. A balance must be reached where only the most important events, or those that are likely to be perceived as important in the future, are recorded. It is evident that events in the past, once considered unimportant or minor, now are viewed as very important. An example is the recent congressional concern over the use of radioactive tracers in experiments using uninformed or "captive" (prisoners) people. In many of the experiments being investigated, it appears that guidelines and regulations that were in effect at the time of the experiments were followed. How well one can predict or anticipate these changes in point of view may determine the difference between an adequate program and an excellent program.

Some of the most important records that must be maintained for long periods of time (license termination) are individual dose records (10 *CFR* 20.2102). If possible, the records should be kept permanently. Recent regulatory changes require that records be kept of any activities that would affect future decommissioning. These records must be kept until the license is terminated or the records transferred to the proper regulatory authority. Many of the records of the day-to-day operation, such as swipe or monitoring records, must be kept for up to 3 years (10 *CFR* 20.2102) or since the last inspection by a regulatory body such as the NRC. If there are events in which personnel are exposed, even if the dose was not measurable, a record of the event and the dose evaluation should be filed with the dose record.

Other important records that should be maintained until license termination, even if not required, are: training records, annual received activity by radionuclide and annual inventory by radionuclide, annual activity by radionuclide of generated waste by disposal method (especially for any waste treated on site), any known environmental releases or spills, and remediation success.

Environmental Concerns

As environmental concerns continue to increase, it is desirable that all radiation safety programs have the ability not only to address issues as they arise, but also to establish background levels and to demonstrate the ability to monitor the environment for low levels of radioactive material used in the program. For those institutions or organizations that have not used any radioactive material in the past, but anticipate using radionuclides, the opportunity to gather background data for air, soil, and buildings should not be missed. For those who are already using radioactive material, buildings on the same site that have never used radioactive material may serve as good background. It may be necessary to find similar buildings or land off-site for the establishment of background levels. Background will vary considerably with building material, soil type, location, and time. If possible, sampling should be done over a long period to account for seasonal and other temporal fluctuations.

The level of effort put into environmental monitoring will depend on many factors, such as the size of the program, the activity, the presence of volatile material, on-site waste treatment, and the proximity of the site to urban or densely populated areas.

Summary

Each institution that uses radioactive material or ionizing radiation must establish a program that ensures that legal requirements are met and that personnel exposures

TABLE 62.4 Audit Elements

Element	Regulatory Reference	Possible Procedures
Training	10 *CFR* 19.12, 10 *CFR* 33.15	1. List number of people trained previous year by category. 2. Check that procedures are in place to assure all radiation workers are receiving required training. 3. List number of workers removed from radiation work due to failure to attend required training.
Procurement and receipt of radioactive material	10 *CFR* 20.1906, 10 *CFR* 33.13	1. Review procurement procedures. 2. List any instances of unauthorized procurement or receipt of any radioactive material that did not go through proper procedures. 3. List total number of packages received for the year and the total activity received by isotope. 4. List any cases where packages were contaminated and the actions taken.
Radioisotope inventory	10 *CFR* 33.13, 10 *CFR* 30.35	1. Compare present total activity on site by isotope with license limits. 2. Compare with previous years. Is there any indication that the license limits should be modified?
Leak testing of sealed sources	License conditions	1. Check records of sealed sources to determine that all have been checked as required (normally every 6 months). 2. List any instances in which contamination was detected and which required notification to the regulatory authority.
Personnel monitoring[7]	10 *CFR* 20.1502, 10 *CFR* 20.1201–08, 10 *CFR* 20.2106–7	1. List number of people being monitored by monitoring method, whole-body dosimeter, extremity monitors, urine assay, thyroid, whole-body counting, and so on. 2. List number of people with doses greater than an established investigational limit or list number by dose categories and compare with previous years. Any trends? 3. List any actions taken during the year to reduce the dose for specific individuals.
Surveys for contamination[8]	10 *CFR* 20.1101	1. Review records of tests for contamination. Arrange the records by location and by user to identify problem areas and continuing problems. 2. List number of contaminations detected and compare with previous years.
Laboratory inspections	10 *CFR* 20.1101, 10 *CFR* 33.13–.14	1. Review laboratory inspection reports. 2. List any laboratories not inspected at least once during the year and reason. 3. List results of inspections by type and number of violations to identify areas that need attention and to keep potential doses as low as reasonably achievable (ALARA). 4. List any laboratories with continuing problems.
Waste disposal	10 *CFR* 20.2001–.2007, license conditions	1. Review waste disposal records. 2. List activity by isotope picked up for disposal. 3. List for each disposal method the volume and the total activity by isotope. Methods of disposal include: storage for decay, release to sanitary sewer, incineration, other on-site treatment, off-site treatment, shipment for land burial and indefinite storage. 4. Compare with previous years and review the program for volume reduction and waste minimization.
Waste storage and shipment	10 *CFR* 20.2001–.2007, 10 *CFR* 71.5, 71.10, 49 *CFR* 170–177	1. Review physical storage location, condition of storage, number of drums/containers, and space for future storage. 2. Review packing procedures for proper waste classification, labeling and manifest completion. 3. Insure that the following are on file: shipping papers, certifications, notifications to state agencies, and manifests.
Shipment of radioactive material other than waste	10 *CFR* 20.2001–.2007, 10 *CFR* 71.5, 71.10, 49 *CFR* 170–177	1. List number of shipments and total activity by isotope. Compare with previous years. 2. List number of exempt, White I, Yellow II, and Yellow III shipments 3. Determine that copies of receivers' licenses are on file.

TABLE 62.4 *Continued*

Element	Regulatory Reference	Possible Procedures
Environmental air releases[9]	10 *CFR* 20.1101	1. Using "Comply" code, determine dose to maximally exposed person and compare with regulatory limit (10 mrem/y). 2. Insure any required reports to NRC have been filed.
Calibration[10,11,12]	10 *CFR* 20.1501	1. Review calibration procedures and records. 2. List number of instruments calibrated and frequency of calibration for each. 3. List any periods in which scheduled calibrations were not performed and the reason.
Facilities[13]	10 *CFR* 30.36, 33.13, 40.42	1. Review present facilities, compare with previous year. 2. Are there program changes or increases in activity that indicate the need for improved facilities? 3. Have any facilities been decommissioned?
Emergency procedures	10 *CFR* 30.32(i), license conditions	1. Review any required emergency plans. Is all applicable information up to date?

are kept as low as reasonably achievable. These goals can be met by following required regulations, obtaining appropriate licensing, establishing operating policies and procedures, performing thorough inspections and audits, and keeping associated records. Depending on the proposed usage, environmental concerns should be addressed early in program development. Although individual requirements will vary, inclusion of these principal program elements should provide the basis for the development of an excellent radiation protection program.

References

1. NCRP 59, Operational Radiation Safety Program: Washington, DC, 1980.
2. *CRC Handbook of Management of Radiation Protection Programs*; Miller, K. L., Weidner, W. A., Eds.; CRC Press: Boca Raton, FL, 1986.
3. NRC Standards for Protection Against Radiation, 10 *CFR* 1, 19, 20, 31, 31, 33, 35 and 71.
4. *Guides for Naturally Occurring and Accelerator-Produced Radioactive Materials (NARM)*; Conference of Radiation Control Program Directors (HHS Publication FDA 81-8025): Washington, DC, 1981.
5. NCRP 111, Developing Radiation Emergency Plans for Academic, Medical or Industrial Facilities: Bethesda, 1991.
6. IAEA 152, Evaluation of Radiation Emergencies and Accidents, Technical Report: Vienna, 1974.
7. NCRP 115, Risk Estimates for Radiation Protection: Bethesda, 1993.
8. NRC Guidelines for Decontamination of Facilities and Equipment Prior to Release for Unrestricted Use; U.S. Nuclear Regulatory Commission: Washington, DC, 1993.
9. EPA 520/1-89-002, A Guide for Determining Compliance with the Clean Air Act Standard for Radionuclide Emissions from NRC-Licensed and Non-DOE Federal Facilities: Washington, DC, 1989.
10. NCRP 112, Calibration of Survey Instruments used in Radiation Protection for the Assessment of Ionizing Radiation Fields and Radioactive Surface Contamination: Bethesda, 1991.
11. IAEA 133, Handbook on Calibration of Radiation Protection Monitoring Instruments, Technical Report, Vienna, 1971.
12. ANSI N323, Radiation Protection Instrumentation Test and Calibration, New York, 1978.
13. IAEA 1, Safe Handling of Radionuclides, Safety Series, Vienna, 1973.
14. NCRP 114, Maintaining Radiation Protection Records, Bethesda, 1992.

63

Ionizing Radiation: Fundamentals

DANIEL D. SPRAU
PHILIP E. HAMRICK
FREDERICK L. VAN SWEARINGEN

Radiation and radioactive materials are very helpful tools used in research, medicine, and industry. When misused, risks must be carefully evaluated. Ionizing radiation is unique in the environmental health and safety field. This is because humans are unable to detect it with their physical senses. No one can see, feel, smell, hear, or even taste ionizing radiation! People cannot detect it without instrumentation. Coupled with a prevalent fear about anything "nuclear," a mystique has arisen regarding radiation sources. Individuals often view radiation and radiation protection issues as being very complex and not easily understood. This results in radiation health effects being somewhat exaggerated (when compared to other complex environmental health and safety issues). It also leads to the assumption that simple radiation protection techniques are difficult to achieve.

One reason radiation is viewed as being very complex is that, in several ways, it has a dual nature. Radiation can be ionizing or non-ionizing, behave like a wave or a particle, come from either radioactive material or electronic devices, be internal or external to the body, and be naturally occurring or man-made. All of these dual concepts promote the idea of nuclear complexity when, in fact, the basics of ionizing radiation and radiation safety are relatively simple.

This chapter will briefly review the basics of ionizing radiation. Non-ionizing radiation is covered elsewhere in this handbook. This chapter covers radiation sources, biological effects, and detection methods. Finally, it reviews some common radiation protection principles. Only those elements are emphasized that are critical to understanding and working safely with radioactive material, since many references are available that treat this topic extensively.

Common Terms

The following are common terms essential to understanding ionizing radiation fundamentals. These are grouped by category:

Radiation—Energy emitted and traveling through space or a medium as waves. By extension, this includes particulate emissions, such as alpha or beta particles or those found in cosmic radiation.

Electromagnetic Radiation—Energy emitted in small discrete pulses called photons that travel with associated wave motion from changing electric or magnetic fields. Electromagnetic radiation ranges from X-rays (and gamma rays) of short wavelength, through the ultraviolet, visible, and infrared regions, to radar and radio waves of relatively long wavelength. All electromagnetic radiation travels in a vacuum at the speed of light. This definition does not include particulate emissions, such as alpha, beta, and neutrons.

Ionizing Radiation—Any radiation capable of displacing electrons from atoms or molecules, thereby producing ions. Examples include alpha and beta particles, gamma and X-rays, and neutrons.

Non-ionizing Radiation—Any electromagnetic radiation not capable of displacing electrons from atoms or molecules. Examples include ultraviolet, visible light, infrared, microwave, and radio frequency waves.

Radioactivity—The spontaneous process by which unstable atoms or nuclei emit radiation, either directly when unstable atomic nuclei disintegrate (undergo nuclear transformation) or as the result of a nuclear reaction. Sometimes called radioactive decay.

Atomic Number—The number of protons in the nucleus of an atom. This equals the number of electrons in an electrically neutral atom. All atoms of an element have the same atomic number.

Atomic Mass—The mass of an atom, usually given as the sum of the number of protons and neutrons in the atom's nucleus. Alternatively, the atomic mass may be expressed in atomic mass units. Similar to atomic weight.

Half-life—The time in which half the atoms of a particular radioactive substance disintegrate to another nuclear form. Half-lives vary from millionths of a second to billions of years.

Isotopes—Atoms having the same number of protons

(same atomic number) but different numbers of neutrons in their nuclei (different atomic masses). For example, the most abundant isotope of carbon has 6 protons and 6 neutrons for an atomic mass of 12. Another naturally occurring isotope of carbon has 6 protons and 8 neutrons for an atomic mass of 14. This isotope of carbon is unstable (radioactive) due to having too many neutrons and decays by the emission of a beta particle.

Activity—the number of nuclear transformations per unit time.

Becquerel (Bq)—The International System of Units (SI) unit of measurement of activity, equal to one nuclear transformation per second (s^{-1}).

Curie (Ci)—The unit frequently used in the United States to measure activity. One curie equals 3.7×10^{10} nuclear transformations per second (Bq), which is approximately the activity of 1 gram of radium.

Dose—The amount of all types of radiation that is actually absorbed in any material. Dose usually refers to *absorbed dose*, the energy imparted to matter by ionizing radiation per unit mass of irradiated material at the point of interest. This unit does not account for the biological effects of the particular radiation or how it is absorbed, only the amount absorbed.

Gray (Gy)—The SI unit of absorbed dose. One gray is equal to the absorption of 1 Joule of energy per kilogram of material (Jkg^{-1}). One gray equals 100 rads.

Rad—The unit for absorbed dose frequently used in the United States. One rad is equal to the absorption of 0.01 Joule of energy per kilogram (or 100 ergs per gram) of material.

Dose Equivalent—A quantity used in radiation protection to express the biological effects of all radiations on a common scale for calculating the effective absorbed dose. It is defined by multiplying the absorbed dose by certain modifying factors to arrive at a quantity that is essentially independent of the type of radiation.

Sievert (Sv)—The SI unit of dose equivalent. It is equal to multiplying the absorbed dose (in grays) by modifying factors. One sievert equals 100 rem.

Rem—The unit of dose equivalent frequently used in the United States. It is equal to the absorbed dose (in rads) multiplied by the quality factors (QF) that account for the type of radiation encountered and how the radiation is absorbed. (For example: QF = 1 for beta, gamma, and X-ray; QF = 20 for alpha; QF ranges from 3–10 for neutrons, depending on neutron energy.)

Basics

Although there are several different types of radiation, they can be divided into two physical forms: small particles and electromagnetic waves. Particles include alphas, betas, and neutrons. Electromagnetic waves include visible light, microwaves, gamma rays, and X-rays. The term *nuclear radiation* refers to where much of the radiation originates—the nucleus of an atom.

Sometimes an atom or its nucleus is in an "excited" state. To return to a more normal or "stable" state, the atom or nucleus must release excess energy or particles. It does so by emitting radiation. The radiation emitted depends on several factors, including the size of the atom, the final or "daughter" atom, and the excess energy (the amount greater than that required for stability). Table 63.1 lists and describes the primary types of ionizing radiation.

The physical properties and the hazard of a source of radioactive material depend on several factors: the amount of activity, the type of radiation emitted (alpha, beta, etc.), the energy of the radiation, and the half-life of the material. This information is available in various references and charts for a large number of isotopes.[1,2,3]

Even though any given atom releases radiation spontaneously and randomly, the probability of its occurring for a large number of atoms can be accurately predicted. The following equation describes the change in activity as a function of time:

$$A = A_0 e^{-(0.693 \, *t/T)},$$

where A is the activity as a function of time, A_0 is the activity at some initial time $t = 0$, t is the time that activity A is measured, and T is the half-life. The factor of 0.693 represents the ln(2), and the negative sign in the exponent indicates a decreasing activity with time.

Radiation energy is usually given in units of thousands of electron volts (keV) or millions of electron volts (MeV). An electron volt (eV) is the energy needed to move one electric charge through a potential of 1 V. For example, consider the picture tube in a color television set. Electrons are accelerated in the tube through a potential of about 30 kV, giving them an energy of about 30 keV. The inside surface of the front of the tube is coated with a phosphor that emits light when struck by the electrons. By controlling the movement of the electrons, a picture is produced. These electrons are indistinguishable from beta particles.

Radiation energies vary widely following radioactive decay. Isotopes may emit beta particles of lower energy (<250 keV—only penetrate the superficial layer of skin) to higher energy (1–2 MeV—will penetrate a centimeter or more in tissue). Alpha particles usually have energies of 1–10 MeV, but they will not penetrate the outer layer of skin. However, both alpha and beta particles can damage cells if the material is internally deposited (discussed later). Isotopes emit gamma rays and X-rays, which

TABLE 63.1 Primary Types of Ionizing Radiation

Type of Radiation	Description
Alpha particle	A positively charged particle ejected spontaneously from the nuclei of some radioactive elements of higher atomic mass. It is identical to a helium nucleus, having a mass number of 4 and an electrostatic charge of +2. (The magnitude of one electrostatic charge is equal to the charge on an electron or 1.602×10^{-19} coulomb.) Alpha particles may be stopped by thin sheets of paper or plastic, or by a few centimeters of air. Alpha particles are an internal hazard but not an external hazard.
Beta particle	A charged particle emitted from a nucleus during radioactive decay, with a mass number equal to 1/1836 that of a proton. A negatively charged beta particle is identical to an electron. A positively charged beta particle is called a positron. Large amounts of beta radiation may cause skin burns, and beta emitters are harmful if they enter the body. Beta particles are easily stopped by a thin sheet of metal or plastic.
Gamma ray	High-energy, short-wavelength electromagnetic radiation emitted from the nucleus. Gamma radiation frequently accompanies alpha and beta emissions and always accompanies fission. Gamma rays are very penetrating. They are best shielded against by dense materials, such as lead or uranium, or thicker layers of less dense materials, such as concrete, soil, or water. Gamma rays are similar to X-rays, but are usually more energetic. Gamma rays and X-rays are deeply penetrating and present an external hazard.
X-ray	Electromagnetic radiation similar to gamma radiation, but produced in the outer orbital electron shells. X-rays can also be produced by sudden acceleration or deceleration of a charged particle, such as occurs in a medical X-ray machine.
Neutron	An uncharged elementary particle found in the nucleus of the atom with a mass slightly greater than that of the proton. It is only emitted from atoms having a relatively high atomic mass, such as uranium. Neutrons can be an external hazard.

range from approximately 10 keV to 5 MeV. The higher energy forms are highly penetrating and may require thick shielding.

When radiation interacts with matter, positive or negative ions may be produced. Charged particles interact by coulombic repulsion and attraction to strip electrons from atoms. The large doubly-charged alpha particles produce a higher concentration of ions per unit path length in matter than the lighter singly-charged beta particles.

For photons, three principal ionization mechanisms are the photoelectric effect, Compton scattering, and pair production. In the photoelectric effect, the photon is completely absorbed by an atom, and an electron is ejected. In Compton scattering, a photon may be thought of as colliding with an electron. Part of the photon's energy is transferred to the electron, and the remaining energy is retained by the scattered photon. Depending on the photon's initial energy and the energy transferred to the electron, the scattered photon may interact several times

by additional Compton collisions. For high-energy photons (>1.02 MeV), pair production occurs. The first 1.02 MeV of photon energy is converted to form an electron-positron pair, with remaining energy contributing to the motion of the pair. This pair interacts by coulombic forces just as for beta particles. The positron and an electron will come together and annihilate each other—their rest masses form two 0.511 MeV photons that can also contribute to ionization. The mechanisms of interaction are dependent on incident photon energies and atomic number of the material. The photoelectric effect is the principal means of interaction for photons at lower energies (<100 keV), Compton scattering predominates at mid-energies, and pair production is dominant at high energies.

Sources

Radiation sources can be classified as naturally occurring or man-made. Figure 63.1 (based on data from

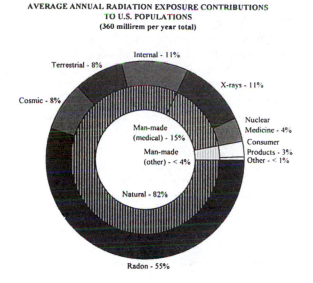

AVERAGE ANNUAL RADIATION EXPOSURE CONTRIBUTIONS
TO U.S. POPULATIONS
(360 millirem per year total)

Internal - 11%
Terrestrial - 8%
Cosmic - 8%
X-rays - 11%
Man-made (medical) - 15%
Man-made (other) - < 4%
Nuclear Medicine - 4%
Consumer Products - 3%
Other - < 1%
Natural - 82%
Radon - 55%

FIGURE 63.1 Average annual radiation exposure contributions to U.S. populations.

NCRP 93[4]) shows approximate contributions from the major sources to average annual radiation dose in the United States. Actual doses to individuals will differ based on their circumstances, such as if they are radiation workers or if they have had many medical X-ray procedures.

The United States population receives over 80% of its radiation dose from natural sources (called background radiation). Background radiation comes from cosmic radiation and naturally occurring radioactive material (NORM). Cosmic radiation originates in outer space. Some of it penetrates the earth's atmosphere and reaches the surface. NORM is present in the rocks, soil, and all living materials. Many radioactive isotopes compose NORM, such as isotopes from elements of uranium, radium, radon, and potassium. Exposure to radon gas alone contributes about 55% of the average citizen's annual radiation dose. NORM found in the ground (terrestrial) adds various amounts of radiation depending on where one lives and the construction materials (stone, brick, concrete) of one's home and office. Everyone is born with a small amount of NORM, and additional small levels are found in food and water (internal sources).

Man-made sources account for about 20% of the U.S. population's dose, mostly from X-ray and nuclear medicine procedures. Some comes from consumer products, such as smoke detectors and some gas lantern mantles. Man-made sources and uses of radioactive material found in the "other" category include:

- occupational exposure to radiation sources
- nuclear fuel cycle, including reactors
- electronic equipment, such as X-ray luggage inspection equipment used in airports
- general and biomedical research applications
- industrial applications
- fallout from nuclear weapons tests
- military uses

Biological Effects

Following their discoveries in the late 1890s, X-rays and radioactive materials were soon used in research and medical applications. It was quickly evident that radiation could damage biological tissues. Hair loss, skin reddening, and worse tissue-damaging effects were seen in some early radiation workers, including researchers, early radiologists, and radium dial painters. Following World War II, the effects of radiation on atomic bomb survivors were studied. Studies were also done on patient groups who received radiation as part of their treatments. Finally, numerous radiological studies were done on animals, plants, cells, and tissue cultures.

Biological effects of radiation can be classified as follows:

Somatic Effects Effects occurring in the person exposed to radiation. These may be further categorized. Prompt effects are observed soon after a large acute dose. Delayed effects occur many years after exposure to radiation. Stochastic effects are those whose probabilities of occurring increase with dose but whose effect is the same (cancer). Nonstochastic effects are those whose severities get worse at higher doses (a sunburnlike skin reddening leads to burns at higher doses).

Genetic Effects Abnormalities that may occur in future children of exposed individuals and in subsequent generations (although this has not yet been seen in humans).

Teratogenic Effects Effects observed in children who were exposed during the fetal and embryonic stages of development.

Radiation Models

Over the past 50 years, more time and money has been spent determining the health effects of radiation than of any other toxic or hazardous agent. All types of radiation have been studied for their internal and external hazards. Effects at high and moderate doses and dose rates of radiation are well known. However, effects at low doses and low-dose rates are uncertain because the risk levels are very small and there are competing nonradiation causes, including natural variations between human groups. There is general agreement that risks are smaller at lower doses. But, extrapolations from results at higher doses and rates are highly dependent on the extrapolation methods or models used. This makes it difficult to predict health risks of deleterious effects for any given individual exposed to a low radiation dose (i.e., <0.1 Sv).

Because of these difficulties, several hypotheses have developed. The *linear hypothesis* extrapolates linearly the risk at higher doses, which are known, down to lower doses, where there is uncertainty. It predicts that no matter how small the dose, there is always some corresponding risk to human health. The *threshold hypothesis* predicts that there are virtually no health effects from small doses of radiation until a certain threshold dose is reached. Beyond that point, harmful health effects increase until they match the known risks at higher doses. A third hypothesis, the *hormesis hypothesis*, proposes that low doses have a beneficial effect on humans. At high dose, detrimental risks are still predicted. Finally, there are other variations of these, such as the quadratic, linear-quadratic, and supralinear hypotheses.

There is a large body of research data to support each theory. For example, some cancers follow variations of the linear hypothesis. Some effects, such as skin reddening and cataract formation, more closely follow the threshold hypothesis. There are many studies, such as on radiation effects on plant growth, that support the hormesis hypothesis. However, the most widely used at this time is the linear hypothesis. It is used by most regulatory bodies to set radiation protection standards.

Acute Radiation Effects

Exposure to radiation sources, either internal or external, normally results in deposition of energy in tissue. For external alpha and low-energy beta radiation, energy may be deposited only in the dead layer of skin with little or no health effect. For living tissue, biological effects are directly related to the amount of energy deposited and how it is deposited. For example, alpha particles are highly ionizing and deposit their energy over a short path length of tissue. The energy of beta and gamma radiation is deposited over a longer path length. Thus, for the same amount of energy deposited per unit mass of tissue (dose in rad or gray), alpha radiation will produce more biological damage than do either beta or gamma radiation. The dose equivalent, measured in rem or sieverts, accounts for this difference, so that identical dose equivalents produce similar biological effects regardless of the type of radiation.

Some other factors are known. Cells tend to be more sensitive to radiation if they are rapidly dividing, have a long dividing future, and are relatively unspecialized. Dose, dose rate, type of radiation, exposure to other hazardous materials, health, and age all contribute to effects on health. Some effects will be delayed and not observed until much later—possibly 20 to 40 years later. Usually, though, higher doses and dose rates result in quicker and more noticeable results. For the same total dose equivalent, an acute exposure (short time period) gives a greater biological effect than a chronic exposure (long time period). Acute effects would only be likely in an accident situation or in some high-radiation-dose medical treatments. Table 63.2 summarizes what is known about high doses of acute radiation.

Chronic Radiation Effects

Chronic radiation doses are usually spread over many days or years. These protracted doses allow the body to repair some of the damage. Fortunately, most worker doses are usually so small that it is not possible to notice any effect, or effects are masked by exposure to other hazardous materials.

The International Commission on Radiation Protection (ICRP60)[5] has estimated (based on the linear hypothesis) that the total risk of somatic and genetic effects is approximately 7.2×10^{-4} per person-rem. [Due to a dose rate reduction factor, the NCRP (NCRP 115)[6] more recently estimated risks about half this level.] Thus, if 10,000 people each received a dose of 1 rem, about 7 would be expected to develop a radiation-induced cancer during their lifetimes or pass on some genetic defect as a result of the dose. This number can be compared to current U.S. cancer risk estimates, where about one-third of all individuals get cancer during their lives and about 1 in 6 will eventually die from cancer. Under the linear hypothesis, if the number of people were 1000 and the dose to each were 10 rem, about 7 people (0.7% of the total) would still be expected to be affected.

Effects on the Unborn

When radiation leaves energy in a germ cell, the damage can result in mutation or chromosome breaks. This defect can then be passed on to the offspring. The radiation-induced mutation does not differ from a spontaneous mutation. Genetic effects from radiation have been detected in animal studies. However, even for Japanese atomic bomb survivors exposed to relatively high levels

TABLE 63.2 Acute Whole Body Radiation Dose Effects

Whole Body	Dose	Acute Dose Effects (assumes no medical treatment)
<0.25 Gy	(25 rads)	No evident physical changes
0.25–1 Gy	(25–100 rads)	Subclinical range, minor blood chemistry changes
1–2 Gy	(100–200 rads)	White blood cell (leukocyte) loss
>2.5 Gy	(>250 rads)	Acute Radiation Syndrome
		Nausea
		Chills
		Epilation (hair loss)
		Erythema (skin reddening)
>3.5 Gy	(>350 rads)	Hematopoietic Syndrome
		Decrease in red blood cell production
4.50 Gy	(450 rads)	LD 50/60
		50 percent will die within 60 days
>6 Gy	(>600 rads)	Gastrointestinal Syndrome
		Death of epithelial cells
		Blood infection
		Fluid loss
10 Gy	(1000 rads)	LD 100/60
		100 percent will die within 60 days
>10 Gy	(>1000 rads)	Central Nervous System Syndrome

of radiation, no significant excesses of genetic effects have been observed. Thus, the probability of passing on a genetic effect due to radiation is thought to be very small.

As the embryo and fetus are made up of large numbers of rapidly dividing and radiosensitive cells, they are considered to be more sensitive to radiation. The amount and type of damage that could result depend on the absorbed dose and the gestation stage.

Radiation received during the first 2 weeks after conception (preimplantation stage) can result in spontaneous abortion or resorption of the embryo. At this time, the woman probably does not know she is pregnant. If the embryo continues to develop, no other significant effects have been detected.

Radiation received during weeks 2–8 postconception (organogenesis stage) can result in developmental abnormalities. The organ system(s) under development when the radiation is received will determine the type of abnormality that develops. In children exposed *in utero* from the Hiroshima atomic bomb, smaller head size was detected at this stage.[7] Animal data suggest that other malformations are likely, but a statistical increase has not yet been detected in people at lower radiation doses.

Based on Japanese data, children exposed to radiation 8–15 weeks after conception (synaptogenesis period— rapid neuron development and migration) have a slightly increased risk of mental retardation. The NCRP[6] provides a risk estimate of 4.0×10^{-5} per rem for severe

mental retardation during this period of development. Radiation also may have a general impact on intelligence and other neurological functions.

From week 16 on, the developmental risk factors decrease. During the third trimester, the risks appears to be no greater than for adults.

During all stages before birth, there is a slightly increased risk that the child might later develop childhood cancer from the prenatal radiation exposure. This risk is greatest during the first trimester, and it reduces during the second and third trimesters. Even if prenatal exposure were as much as 10 rads, the risk of childhood cancer is less than 2% if the radiation were received during the first trimester and 0.5% if received during the second and third trimesters.

ALARA

Projected biological effects from using radiation sources should always be weighed against expected benefits from using the sources. Benefits include excellent diagnostic and treatment tools for patients and medical personnel, electrical power generation, industrial applications, and effective research tools for scientific investigators. Using radiation sources provides employment for many thousands of workers. These benefits, coupled with the fact that most doses to workers are very small when compared to background levels, make using radiation sources practical and desirable. In the workplace, all hazards must be considered when evaluating health risks. Many other physical, chemical, and biological hazards to which workers are exposed are much more hazardous than working with small amounts of radioactive material.

Though risks may be small, the goal of any radiation protection program is to keep doses as low as reasonably achievable (ALARA) and as close to background levels as possible. It is impossible to reduce all exposures if useful and practical work with radioactive material is to be accomplished. However, by following good radiation safety procedures, doses can be kept low.

Detection

Since the human body cannot directly sense ionizing radiation, other methods must be used to detect and quantify it. Detection can be achieved by several different means. In the early 1900s, X-ray tube exposures were sometimes checked through use of skin reddening. Fortunately, this is no longer done. There are five primary methods currently being used to detect and measure ionizing radiation. They are:

1. ionization in gas-filled chambers
2. scintillation
3. photographic films
4. thermoluminesense
5. semiconductors

Gas-filled chambers or tubes come in many configurations depending on their intended use. The Geiger counter is probably the most widely recognized and is a very sensitve detector of radiation. It is designed to detect and display or "count" the number of ionizing events occurring per unit time in the chamber. It does not discriminate between types or energy of the radiation and is limited to detecting only those radiations having enough energy to penetrate the window of the detector tube. Even with a thin mylar window, the Geiger counter is very ineffective for detecting alpha and low-energy beta radiation. Other gas-filled detectors are used to measure the charge produced by ionization in the gas (called ionization chambers). Still others, such as gas flow proportional counters, can be used to discriminate to some extent between energies and between alpha and beta radiation.

Scintillation detectors are very versatile and are used to detect and discriminate between various types and energies of radiation. When ionizing radiation interacts with scintillation materials, light is produced that is proportional to the energy of the radiation. Solid crystals of scintillants, such as sodium iodide or germanium, cou-

pled to photomultiplier tubes and a multichannel analyzer, can be used to discriminate between gamma and X-ray energies. This produces an energy spectrum of the radiation source. Alpha and low-energy beta particles can be detected using liquid scintillation counters in which the radioactive substance to be detected is mixed with a solution containing scintillants. The light emitted is detected by photomultiplier tubes, and spectra can be generated to help discriminate between alpha and beta particles. Since beta particles from an isotope exhibit a spread of energies, discrimination between betas of different maximum energy is not as good as with gamma rays. Liquid scintillation counters are the primary instruments used in biomedical research for detecting radioactive material used as tracers in various biological systems.

One principal use of photographic film is to produce images from X-ray procedures. Sometimes scintillation material ("screen") is used in conjunction with film to reduce the amount of radiation required to produce the image. Film is also used as a dosimeter for measuring exposure to radiation. One example is the common "film badge." The greater the radiation dose, the darker the film gets. Film is more sensitive to radiation at some energies, so calibration curves must be used to correlate the density of the film with the exposure.

Thermoluminescence refers to a property exhibited by some materials. When exposed to radiation, they trap some radiation energy in metastable states. This trapped energy is released as light when the material is later heated. The amount of light released is proportional to the amount of radiation absorbed. Some thermoluminescent materials make excellent dosimeters and have replaced a large percentage of film dosimeters. These dosimeters are very sensitive and are linear with dose over a wide range of doses and dose rates.

Semiconductors, although not as widely used as the other detectors discussed above, are becoming more prevalent and can be made quite sensitive to radiation. Depending on the configuration, the detectors can discriminate between types of particles and energies.

Protection Principles

Radiation protection principles are based on the philosophy of keeping doses as low as reasonably achievable (ALARA). *Fundamental objectives* of any radiation protection program are:

- Limit exposures to all radiation to as low levels as feasible and always within the limits set by the regulatory agencies (Nuclear Regulatory Commission or state agencies).

- Limit entry of radionuclides into the human body by ingestion, inhalation, absorption, or through open wounds from unconfined radioactive material, and always remain below the regulatory limits.
- Limit release of radioactive material to the environment.

An *important secondary objective* is to obtain desired and reliable results from industrial processes, tests, experiments, and medical procedures. To accomplish this objective, preplanning and following the above principles are required. It is necessary to analyze in advance the hazards of each job and to plan emergency procedures for possible accidents.

Protection principles may be divided into two general categories: those for external sources, such as radioactive material or radiation-producing machines; and those for potential internal sources, such as radioactive material inside the body.

External

External radiation dose can come from electronic equipment, gamma emitters, and high-energy beta emitters. The degree of external exposure hazard depends on increasing or decreasing the "big four" radiation protection factors: amount, distance, time, shielding.

Amount

Decrease the amount. The external exposure hazard can depend on the amount of radioactive material being used. To reduce exposure, use the smallest amount of activity needed to perform an experiment, industrial application, or clinical procedure.

Box 1
Guidelines for Using Radiation Producing Machines

- Before working with any radiation-producing machine, train personnel adequately to know exactly what work is to be done and the safety precautions to be taken.
- Provide written operating and safety procedures to personnel before they operate the equipment.
- Have equipment operator supervise visitors and students in areas where radiation-producing machines are used.
- Do not leave radiation-producing machines unattended in an operational mode.
- Before using equipment, have a qualified expert review the structural shielding requirements for any new unit or for an existing one in which changes are made.
- Follow restrictions rigidly if the safe use of the equipment depends on the mechanical setup of the unit or on technique factors.
- Do not tamper with or defeat shutter mechanisms and interlocks.
- All warning lights should be "fail safe."
- Use a manual-reset cumulative-timing device that either indicates elapsed time or turns off the unit when the total exposure reaches a previously determined limit.
- Be careful around X-ray diffraction equipment, which can be particularly hazardous because of high-exposure rates in the primary beam (e.g., in excess of 500,000 rads per minute at the X-ray tube port).
- For larger irradiators and accelerators that are separately licensed, post and follow detailed operating and emergency procedures.
- Properly maintain all radiation-producing equipment. Repairs should only be made by properly trained technical staff.

Distance

Increase distance. Distance alone can reduce the dose rate from alpha and beta emitters to background levels. They do not travel far in air (a few centimeters for alphas and a few meters for even the most energetic betas). An X-ray or gamma radiation source whose size is small compared to the distance from the source may be considered a point source. Radiation from a point source obeys the inverse square law:

$$I_1/I_2 = (d_2)^2/(d_1)^2 \rightarrow I_2 = I_1 * (d_1)^2/(d_2)^2$$

where I_1 is the dose rate at a distance d_1 from the source and I_2 is the dose rate at a distance d_2 from the source. Thus, if the distance to a source doubles, the dose rate is reduced by a factor of four.

Time

Decrease time. The total dose received from a radiation source depends on the total time spent near the source. Spend as little time near a source as possible and use it as effectively as feasible.

Shielding

Increase shielding. Absorbing-material is needed around a radiation source when using significant levels of gamma or high-energy beta emitters. Shielding material and thickness depend on the amount and type of radiation. Many times, lead is used around X-ray and gamma sources, while plastics are used around high-beta sources.

Box 1 lists some guidelines for using radiation-producing machines.

Internal

Internal exposure can come from any radioactive material. Although low-energy beta emitters and alpha emitters are slight external hazards (other than to the skin), they may be serious internal hazards. The primary goal is to limit the amount of radioactive material entering the body by practicing good contamination control. Materials can enter the body by:

- breathing radioactive vapor, aerosols, or dust
- consuming radioisotopes in food, water, from contaminated hands, or from smoking

---

Box 2
Guidelines for Using Radioactive Material

- Before starting any work with radioactive material, personnel must understand the work to be done and the safety precautions to be taken.
- Develop written procedures that personnel are to do for each project or process. Necessary equipment, waste containers, and survey instruments must be present.
- Characteristics of the radioactive material should be known, such as type of radiation, significant and typical amounts, and chemical form.
- Perform a "dry run" of the procedure or process before it is actually performed with radioactive material. This will test for potential problems and help to minimize time spent using radioactive materials.
- Wear proper laboratory coats, other protective clothing, and personnel dosimeters when handling radioactive material.
- Wear gloves when hand contamination is likely and unsealed sources are used, or when there is a break in the skin on the hand. Do not use the telephone, counting equipment, handle books, open cabinets, and so on, wearing contaminated gloves.
- When possible, increase distance and shielding to reduce radiation exposure.

(continued)

Box 2
Continued

- A radiation worker from the area should supervise visitors and students who visit an area where radioactive material is used.
- Do not leave radioactive material unsupervised or unsecured. Unauthorized persons may handle or remove it, particularly without realizing it is radioactive.
- As a general practice, confine work with radioactive material to only the area necessary. This simplifies the problem of confinement and shielding, and aids in limiting the affected area in case of an accident.
- Monitor work area frequently for contamination control. Monitor personnel after they finish work and before they leave the work area.
- When skin contamination or radioactivity uptake is suspected, contact the Radiation Safety Office at once.
- Properly cover all work surfaces and storage areas (table top, hood, floor, etc.). Some facilities, especially in older buildings, are very difficult to decontaminate.
- Use absorbent mats or paper, such as those having a plastic back and absorbent paper front. If contaminated, they may simply be discarded as radioactive waste.
- Place plastic or metal trays (stainless steel washes easily) on the surface when liquids are to be used. The lip of the tray serves to confine a spill.
- Conduct experiments or procedures that might produce airborne contamination (volatile isotopes, dust, or gases) in a hood, dry box, or other suitable closed system.
- Practice good housekeeping. If an area is kept neat, clean, and free from equipment and materials not required for the immediate procedure, the likelihood of accidental contamination or exposure is reduced.
- Minimize the accumulation of radioactive wastes and dispose of them properly.
- Open shipments of radioactive material and check them for contamination in a properly equipped laboratory.
- When feasible, keep radioactive material, particularly liquids, in unbreakable containers. Use a secondary container as backup in case the primary container leaks.
- NEVER PIPETTE RADIOACTIVE SOLUTIONS BY MOUTH. Always use a pipette filling device.
- Do not smoke in areas where work with unsealed radioactive sources is in progress or where contamination may exist. Under no circumstances should cigarettes, cigars, or pipes be laid on tables or benches where radioactive work has been or is in progress.
- Personnel working in areas containing radioactive material must wash their hands thoroughly before eating, smoking, applying cosmetics or ChapStick, or leaving the work area.

- entering through a wound
- absorption through the skin

Box 2 lists some guidelines for using radioactive material.

Summary

Ionizing radiation and radioactive materials are very useful tools of medicine, research, industry, and government. There are many benefits to their use. However, they can be harmful if used improperly. The goal of a good radiation safety program is to minimize the risks while retaining the benefits.

Although radiation is often thought of as a complex topic, there are a few simple safety concepts. To reduce exposures to radiation and radioactive materials, one can:

- decrease the amount(s) used
- decrease the time spent around the source(s)
- increase the distance from the source(s)
- increase the shielding around the source(s)
- practice good contamination control

There are many specific procedures that can be followed to incorporate these fundamental ideas. And, there are detection methods to help track your progress. The main goal is not only to have a program that is in regulatory compliance, but to also have one that maintains exposures ALARA.

References

1. Golnick, D. J. *Basic Radiation Protection Technology*, 3rd ed.; Pacific Radiation Corporation: Altadena, CA, 1994.
2. *NCRP Operational Radiation Safety Program*, Report No. 59; National Council on Radiation Protection and Measurements: Washington, DC, 1978.
3. IAEA 1, *Safe Handling of Radionuclides*, Safety Series #1; International Atomic Energy Agency: Vienna, 1973.
4. *NCRP Ionizing Radiation Exposure of the Population of the United States*, Report No. 93; National Council on Radiation Protection and Measurements: Bethesda, MD, 1987.
5. *ICRP 1990 Recommendations of the International Commission on Radiological Protection*; ICRP Publication 60, Annals of the ICRP 21, International Commission on Radiological Protection: Pergamon Press, Elmsford, NY, 1991.
6. *NCRP Risk Estimates for Radiation Protection*, Report No. 115; National Council on Radiation Protection and Measurements: Bethesda, MD, 1993.
7. Wagner, L. K.; Lester, R. G.; Saldana, L. R. *Exposure of the Pregnant Patient to Diagnostic Radiations: A Guide to Medical Management*; Lippincott: Philadelphia, 1985.

64

Radiation Emergency Response, Decontamination, PPE

BOB WILSON

Modern research laboratories use a wide variety of radiation sources, including radioactive materials (as solids, liquids, and gases) and X-ray machines. These sources usually span a wide range of quantities, energies, and intensities. Radioactive materials will generally include a variety of low-energy beta emitters (H-3, C-14, S-35), medium-energy beta emitters (Ca-45, Sr-89, P-33, etc.), high-energy beta emitters (P-32, Sr-90, etc.), and gamma emitters (I-125, I-131, Na-22, Cr-51, etc.). Alpha emitters are currently more seldom used, but will be present with certain static eliminators and the low-energy beta emitter Pb-210. Analytical X-ray machines and electron-beam devices are also commonly present in a laboratory complex. Any of these radiation sources can be a hazardous agent directly or indirectly involved in an emergency situation. It is prudent to plan ahead and be prepared to respond properly to a radiation emergency of any scope.

Definition

A radiation emergency may be defined as the loss of control or containment of a radiation source resulting in an actual or potential threat sufficient to require an immediate response.[1] Each facility should establish its own interpretation of potential radiation emergencies. It may benefit one facility to be very conservative in its definition, while another may be better served with a broader interpretation. The actual scope of operation and safety response resources should play a role in an operational definition. Regulatory requirements will also influence a specific definition.

As a guide, radiation emergencies may be further defined as minor and major incidents. Certainly, an uncontrolled exposure, contamination, or any source loss are events that require a prompt and effective response.

Precautions

A well-managed laboratory work environment is a safe workplace. Good management includes a commitment to communication and training. This continually reduces the probability of poor work habits and human error, the precursors to many emergencies. Other, specific beneficial laboratory safety precautions include[2]:

1. use of the minimum necessary radioactivity
2. establishment of segregated work areas
3. careful containment of radioactivity at the work station and in storage
4. good ventilation
5. easily decontaminated surfaces in working area
6. good housekeeping (this is very important and often overlooked)
7. consistent use of protective clothing, including lab coat, gloves, safety glasses
8. thorough, frequent hand washing, especially at breaks and lunch
9. conscientious and timely secured storage of sources
10. monitoring of personnel and work areas at appropriate frequencies
11. training and supervision of workers

Scope

The range of laboratory radiation emergencies includes spills, facility contamination, personnel contamination, personnel intakes, fire, source loss, and significant doses (whole body, extremity, skin, or organ).

Radioiodines are usually the most radiotoxic of commonly used radioactive materials. Iodination procedures can release radioiodine vapor as an airborne inhalation hazard in addition to spills and contamination. Thyroid radioiodine uptakes can be readily measured and any thyroid dose evaluated for significance and further ac-

tion.[3] If there is a significant uptake, thyroid blocking by potassium iodide, as prescribed by a physician, can help reduce the organ dose even when taken after the exposure.[4] Phosphorus-32, a common high-energy beta emitter, is metabolized as calcium and concentrates in the bone when ingested. Significant skin dose is possible from microcurie quantities of P-32 unless the skin is promptly decontaminated. Most other commonly used laboratory radionuclides are less radiotoxic. Skin wipes, direct measurements, and bioassay procedures (urinalysis or external organ counting) are adequate for at least a preliminary assessment. In the case of any actual or suspected intake, contact your Radiation Safety Officer at once.

Analytical X-ray machines are one type of radiation source that can deliver an acute dose large enough to cause an overt radiation injury. Any exposure to the X-ray beam from one of these devices should be considered as a severe emergency.[5] The most serious injuries have usually occurred when the operator did not notice the beam was on or that his or her hand or fingers were exposed. Such accidents have happened with both open-beam and enclosed systems when safety shutters were dismantled, interlocks defeated, or malfunctions occurred.

Response Planning

With planning, the laboratory staff can readily be prepared to respond promptly and effectively to an incident. An in-lab incident response plan should be in place even when a good central safety service is provided. In many institutions, a plan is part of the lab's authorization requirement. A competent radiation emergency response is based on several fundamentals:

1. Planning/Preparation
2. Training/Experience
3. Equipment
4. Follow-up

These fundamentals provide an overview of laboratory-response expectations and procedures. The laboratory staff, especially management/supervisory personnel, should critically review existing institutional plans. Feedback to the central safety service should be included. Each laboratory can then build on the institutional plans by adding its own experiences and expectations relating to the radiation sources in use, type of use, risk of incidents, and the potential of significant incidents. The individual lab should develop its own specific incident response procedures, designed to support the institutional plan. An example emergency management procedure is shown in Table 64.1. An example analytical X-ray emergency procedure is shown in Table 64.2. An example radioactive material source security procedure is shown in Table 64.3. Much emergency-response training can be acquired through the plan development process. Routine initial and refresher response training is done through periodic reviews of the laboratory and institutional emergency-response plans.

Most radiation incident response equipment and sup-plies are readily available in the laboratory. For example, one or more radiation survey instruments should routinely be at hand. Disposable gloves, lab coats, absorbent materials, warning tape, and so forth, are commonly available in quantity. However, a distinctive, customized laboratory response kit is also recommended. This ensures readiness of essentials and easily overlooked special items. The kit should be well maintained and kept close to, but not directly in, the areas of potential need.

TABLE 64.1 Emergency Management Procedures for Radiation

Equipment Needed	
Description	**Location**
(Make, Model) Survey Meter (or equivalent)	Lab ____, Cabinet (Drawer) Posted
Emergency Response Pack (post a list of contents on the pack)	Lab ____, Posted Cabinet
Supplies (quantities of absorbents, gloves, etc.)	Lab ____, Dept. Stores, etc.)

Response

1. a. Secure control of the area(s) involved.
   b. Contain spills, liquid or powder.
   c. Identify injured and potentially exposed, contaminated persons and remove from the area as feasible.
   d. Secure medical attention as necessary.
   e. Shut down ventilation system as necessary.

2. *Confirm radiation status.* Confirm the radiation source involved, physical and chemical forms, maximum quantities or levels. *Be alert for other hazards.*
   a. Note warning signs, labels, posted instructions.
   b. Conduct initial radiation survey of the facility and personnel.
   c. Liaison with Radiation Safety Officer, as appropriate.

3. *Establish the "Clean Line"*
   a. Articles or persons crossing the line are contaminated.
   b. Articles coming out must be sealed in clean, surveyed, tagged bags.
   c. Dress in and out.
   d. Set up to survey, bag, and tag all items removed.

4. *Conduct surveys*
   a. External radiation levels: record external radiation survey results as counts per minute (CPM) or mrem/hr, location, distance from source, date, time.
   b. Removable contamination levels: begin wipe survey at clean line. Record results by CPM or disintegrations per minute (DPM), location, date, time.

5. *Confine visible spills.* Use available absorbent materials (vermiculite, bench paper, paper towels, etc.).

6. *Decontamination*
   a. Review information, survey results.
   b. Maintain site security pending an action plan by experienced, trained personnel.

7. *Personnel inhalation, ingestion, or absorption*
   a. Identify persons, collect skin and nasal wipes.
   b. Begin urine sample collection schedule.
   c. Consider body counting techniques.

8. *High external dose*
   a. Process available dosimeters on an emergency basis.
   b. Perform dosimetry mockup.
   c. Consider medical follow-up.

9. *Notification to the regulatory agency*
   Notify the applicable (state or NRC) regulatory agency as required. This will be when a source is lost or stolen, effluent limits are exceeded, or dose limits are exceeded. Keep in mind that the notification may need to be made promptly.

TABLE 64.2 Analytical X-Ray Machines

Radiation Emergency Procedures

IF YOU ARE EXPOSED TO THE DIRECT X-RAY BEAM, OR SUSPECT AN EXPOSURE, *IMMEDIATELY* FOLLOW THESE STEPS:

1. Shut off the X-ray beam.
2. Remain calm. Call these contacts until (1) medical advice is obtained and (2) the incident is reported.

Medical Advice/Incident Reporting

_____, M.D., Director, Emergency Medicine. (Phone No., Pager, etc.)
Institution Safety Office . . . . (Phone No., Pager, etc.)
_____, (Other) . . . . (Phone No., Pager, etc.)
AFTER HOURS CONTACT, SECURITY (Phone No., Pager). Ask for Radiation Safety Assistance

Safety Procedures

X-ray diffraction and spectrographic devices generate in-beam radiation dose rates of 30 to 7000 rads/sec. Severe tissue damage can be inflicted by very brief exposures to these high-dose rates. Surgical treatment or amputation may be required when small body parts, such as fingers, receive greater than 1000 rads.

It is imperative that stringent safety precautions be applied when using these devices. Safety precautions include mechanical and electrical guards as well as proper training and instruction. The following safety procedures have been established to help prevent accidents. Adherence to these rules is mandatory.

1. NO PERSON SHALL BE PERMITTED TO OPERATE ANALYTICAL X-RAY MACHINES UNTIL THEY HAVE:
   a. received instruction in relevant radiation hazards and safety
   b. received instruction in the theory and proper use of the machine
   c. demonstrated competence, under supervision, to safely use the machine
2. RADIATION EXPOSURE TO THE OPERATOR AND OTHERS SHALL BE KEPT AS LOW AS PRACTICABLE. RADIATION SAFETY SURVEYS SHALL BE CONDUCTED PERIODICALLY.
3. OPERATORS SHALL WEAR RADIATION BADGES AS ASSIGNED.
4. OPERATORS SHALL REMAIN IN CONSTANT ATTENDANCE WHILE THE X-RAY BEAM IS ON, OR THE DEVICE SHALL BE SECURED AGAINST ACCESS BY UNAUTHORIZED PERSONS.
5. *ANY* CHANGES IN THE STATUS OR LOCATION OF A DEVICE SHALL BE REFERRED TO THE RADIATION SAFETY OFFICER FOR PRIOR APPROVAL.

The kit should be small and simple. Multiple kits in strategic locations can be useful. Make sure a kit can be reached without having to enter an incident area to get it. This is an advantage of having more than one kit location. Each kit should contain, as a minimum:

1. disposal impervious gloves, several pairs
2. shoe covers, several pairs (segregated by size and well marked)
3. plastic waste bags, large, one or more, with several zippered bags for samples
4. absorbent materials, such as plastic-backed diaper paper
5. barrier ribbon or rope to cordon off the area
6. note paper and pens, warning signs, warning tape, survey plan guides, indelible black markers
7. filter papers for wipe-test surveys (plan for approximately 200 wipe tests)
8. decontamination supplies (detergent, white vinegar, etc.)
9. disposable protective coveralls, two pairs (appropriate sizes)
10. 6-in. forceps, tweezers, and/or tongs
11. 6-in. cotton swabs (a few dozen)

A kit such as this can readily fit in a small bag or spill bucket. Replenish promptly after each use. Check the contents periodically. Items can be misplaced or deteriorate.

Response Procedures

When an incident occurs, prompt initiation of the planned response is a critical factor. Failure to immediately recognize an incident and notify lab personnel can greatly complicate the situation. Evacuate all those who have not been affected by the incident to an uninvolved area. Get medical aid for any injured. Send all potentially contaminated personnel to a separate area for individual monitoring. These people should not eat, drink, smoke, or apply cosmetics (all avenues for ingestion of radioactive contamination) until they have been surveyed and released.[6]

As judged appropriate, contain the spill or situation to the immediate area. This can be done quickly, by throwing paper towels or bench paper on a liquid spill. For a powder spill, moisten the spill and cover, taking great care to avoid fanning the spill area. Close windows and shut off room ventilation if possible. Leave hoods running. Post lookouts or close doors and post signs as appropriate. In accordance with institution policy, report the incident to the central safety service. Conduct a thor-

TABLE 64.3 Radiation Source Security Procedures

1. Missing Radiation Source Emergency Procedures:

STEP 1: Stop ALL Building Trash Pickups IMMEDIATELY. POST A GUARD AT THE (Building) DUMPSTER!!

STEP 2: Contact Your Building Housekeeping Supervisor for Assistance.

Housekeeping Supervisor: Name _____   Phone No. _____

STEP 3: As Soon As Steps 1 and 2 Are Underway, Immediately Notify the Following:

Institution Safety Office: (Phone, Pager)

Security (if after 5:00 p.m.):   (Phone, Pager)

PRINCIPAL INVESTIGATORS ARE DIRECTLY RESPONSIBLE FOR THE COMPLETE SECURITY OF ALL RADIATION SOURCES AND RADIOACTIVE WASTES UNTIL TRANSFER OR PICKUP. ANY LOSS OF RADIOACTIVE MATERIALS SERIOUSLY JEOPARDIZES YOUR USE PRIVILEGE. ASSURE PHYSICAL SECURITY OF ALL SOURCES BY ATTENDANCE OR LOCK AT ALL TIMES.

2. Security of New Radiation Sources:

A. Do not leave radioactive materials in a shipping box after receipt. Remove the radioactive material and put it into proper storage at once.

B. Always confirm that a radionuclide shipping box is empty, uncontaminated, and shipping labels are obliterated before releasing the box for disposal in the ordinary trash or recycling.

C. Be acutely aware of your radioactive inventory at all times.

3. Security of Radioactive Waste:

A. Use only standard, posted containers for waste storage. Nonstandard containers are prohibited.

B. Use only approved intermediate radioactive waste containers. Never leave any containers unsecured. Empty intermediate containers at the end of each day.

C. Remove radioactive waste on a regular schedule.

DO NOT ALLOW RADIOACTIVE WASTE TO REMAIN UNDULY IN THE LABS!!

D. Radiation label tape is inadequate as a warning to lay persons. DO NOT DEPEND ON IT AS AN ADEQUATE WARNING!! USE FULL SIZE STICKERS AND LABELS.

ough radiation survey of all personnel, starting with those least likely to be contaminated. This allows for the early release of people from the area, an advantage in itself, and provides a body of helpers. Next, survey those who were actually or potentially involved. Identify every individual and record survey findings for each. Any contamination on skin, clothing, or shoes calls for that person to be scheduled for a bioassay. Any contaminated clothing should be removed and bagged, with owner identification, radionuclide(s), activity, date, and so forth. Conduct a thorough survey of the incident area. Begin the survey well away from the incident area and work inward toward the area. Wipe-test surveys are the most sensitive and also provide a measure of the contamination transferability. Instrument surveys are faster and may be adequate, depending on the radionuclide. When the scope and nature of the incident have been determined, pause and plan how to conduct the recovery (cleanup, search, reenactment, etc.).

Closure

After securing and controlling the incident site and taking care of involved individuals, recovery operations are begun. For most incidents, this is an informal action to put routine activities back in place. In some cases, specialized radiation safety assistance may be needed. Recovery operations may include:

Decontamination Any residual radioactivity must be removed to acceptable guideline levels (200 DPM per 100 cm^2 or other in-house guide) before routine activities can resume.

Search Any radiation source that is missing constitutes an emergency. All control is lost. The source may have been moved to a public area or uncontrolled area. The source must be found or confirmed as not being an exposure risk.

Dose Assessment The radiation safety officer or other specialized assistance should be contacted to perform bioassays and dose assessments as necessary.

Investigation Any incident should be reviewed to discover the root cause of the event. For more serious events, a formal investigation may be conducted. Each review or investigation should be documented for further review and joint critique.

Reenactment As part of an incident-root-cause investigation and especially for some dose-assessment purposes, a reenactment of the initial event may be useful.

Decontamination

Radioactivity decontamination is a common aspect of radioactive materials use. Equipment, apparatus, glassware, and associated elements in routine use will become contaminated. Most laboratory radiation emergencies are due to overt contamination events. Personnel contamination may also occur. This includes an intake, skin, or clothing contamination. Wearing protective shoecovers, gloves, and, as necessary, anticontamination clothing, workers can safely carry out common decontamination procedures. Decontamination efforts begin after an incident has been brought under control and a complete radiation survey is done. A decontamination plan is made and followed:

1. Collect materials and supplies.
2. Delineate the affected area, beginning well to the "clean" side.
3. Use appropriate PPE.
4. Proceed with selected methods working from the outside inward toward the center.
5. Begin with standard materials. Monitor frequently as the work progresses.
6. Progress to more aggressive methods only as necessary.
7. Keep cleanup waste and liquids to a minimum.
8. Prevent any additional spread of the contamination.
9. Continue decontamination efforts until monitoring confirms that established release limits have been achieved.
10. Record the final monitoring results to document adequate cleanup.

Establish a "Clean Line," or delineated border for the decontamination work area. An individual should be stationed at the Clean Line to monitor everything passing out of the area, including workers. All items passed out must be monitored, bagged, and tagged. Helpers at the Clean Line can assist in passing in supplies, and handling materials being passed out. The less traffic across the Clean Line, the better. The initial incident survey serves as a starting guide. Soap and water scrubbing are normally adequate. Work from the outside of the contamination inward. Use as little liquid as possible to prevent spreading. Keep close control of the decontamination process. Further spread of contamination can otherwise result. Continue to conduct wipe-test surveys behind the work to monitor progress. Record the final survey results to document adequate cleanup.

For personal (body, skin) decontamination: (a) remove contaminated clothing; (b) wash contaminated skin with soap and water without vigorous scrubbing; (c) carefully shampoo or clip contaminated hair. Do not progress to more aggressive methods without medical guidance.[7] If the skin is abraded, radioactivity may be absorbed through the skin. Household vinegar, an effective skin decontaminate for phosphates, should be kept on hand when radiophosphates (P-32, P-33) are used.[8] In addition, the use of such an ordinary household product is reassuring to people. A *freshly prepared* solution (10% by weight in water) of sodium thiosulfate anhydrous is effective for skin (or equipment, floors, etc.) decontamination of radioiodines. A 1-lb container of granular stock material will suffice for months or years. Be thoughtful, patient, and empathetic during the process. Individuals may become upset when they discover they have radioactivity on their body. Part of the incident response is to keep everyone calm, reassured, and informed. Achieving this is a major part of the job.

Beta-emitting skin contamination can give a significant dose to the skin.[9] Normalized dose conversion factors have been calculated and are available for several nuclides, including most of those used in research labs.[10]

PPE

All laboratory workers should wear proper personal protective equipment (PPE). This includes a lab coat with full length sleeves, safety glasses or goggles, fully enclosed footwear, and appropriate protective gloves. At times, PPE may include full face shield, splash apron, and protective gloves especially selected for specific materials. For radiation workers, the assigned personnel monitoring device(s) is always worn. Wear the extremity monitoring finger ring *under* gloves. Body badge should also be worn under the protective clothing. Otherwise, the monitoring devices are subject to contamination themselves, giving extremely erroneous results.

PPE should only be worn at the work station or in the work area. Lab coats should not be worn out of the lab. Protective gloves are worn both to keep hazardous material off the worker and to keep the material within the work area. The gloves must be removed at the work station when leaving. This closes one common route for the spread of contamination.

Summary

Every laboratory that uses radioactive materials or X-ray devices should design and establish a plan for promptly responding to radiation emergencies. Emergency response can be facilitated by developing an appropriately equipped kit and having decontamination supplies on hand. The consequences of a lab mishap are minimized by the selection and use of appropriate Personal Protective Equipment.

References

1. NCRP Report No. 111, *Developing Radiation Emergency Plans for Academic, Medical or Industrial Facilities*, August 30, 1991.
2. Martin, E. B. M. *Health Physics Aspects of the Use of Tritium*; Science Reviews Ltd, and H & H Scientific Consultants Ltd, Occupational Hygiene Monograph No. 6, 1982.
3. Nishizawa, K.; Maekoshi, H. *Health Physics* **1990**, *58*, 165–169.
4. Prime, D. *Health Physics Aspects of the Use of Radioiodines*; Science Reviews Ltd, and H & H Scientific Consultants Ltd, Occupational Hygiene Monograph No. 13, 1985.
5. Martin, E. B. M. *A Guide to the Safe Use of X-Ray Diffraction and Spectrometry Equipment*; Science Reviews Ltd, and H & H Scientific Consultants Ltd, Occupational Hygiene Monograph No. 8, 1983.
6. Fritz, R. *RSO Magazine* **1997**, *2*, 15–18.
7. Shleien, B.; Slaback, Jr., L. A.; Birky, B. K. *The Health Physics and Radiological Health Handbook*, 3rd ed.; Williams & Wilkins: Baltimore, 1998.
8. Party, E. *Health Physics* **1991**, *60*, 458.
9. Faw, R. E. *Health Physics* **1992**, *63*, 443–448.
10. McGuire, E. L.; Dalrymple, G. V.; McClellan, J. L. *Health Physics* **1990**, *58*, 399–401.

65

Dosimetry

CARMINE M. PLOTT

Radiation dosimetry (or simply *dosimetry*) is the measurement of absorbed dose or dose rate resulting from the interaction of ionizing radiation with matter. This branch of science attempts to relate quantitatively specific measurements made in a radiation field to any chemical or biological changes produced by the radiation. It therefore provides the basis for establishing dose-effect relationships and is essential for monitoring radiation exposure to individuals. It is an essential component of any radiation protection program.

Types of Exposure

An individual's exposure includes exposure to radiation emitted by radioactive material that has been taken into the body, called *internal exposure*, or by radiation that originates outside of the body, called *external exposure*. External exposure may come from either radioactive material or electronic devices. Regardless of whether the exposure is from an internal or external source, radiation dose has the same biological risk. Regulations require that dose from internal and external exposures be added together if each exceeds 10% of the annual occupational limit and that the total be within occupational limits.[2] The sum of the internal and external exposures is called

the total effective dose equivalent (TEDE) and is expressed in units of rems or Sieverts (Sv).

Internal Exposures

Radioactive materials may enter the body through breathing, eating, drinking, or open wounds, or they may be absorbed through the skin. The most common route of entry is breathing contaminated air. Radioactive materials may be present as fine dust or gases in the workplace. They may also be present as a result of resuspended particles from working on contaminated surfaces.

Once inside the body, the radioactive material localizes in various organs or is excreted, depending on the metabolism of the material by the body. Bioassay is the determination of the kinds, quantities, or concentrations, and in some cases, the locations, of radioactive materials in the body. It may be completed by direct, *in vivo* measurement of these materials in the body, as with a thyroid count following an intake of iodine-131. Bioassay may also be completed via *in vitro* analysis and evaluation of materials excreted or removed from the body, such as urinalysis for hydrogen-3. Because most radionuclides are excreted in a few days, bioassays at routine intervals are as critical to internal dose assessment as immediate bioassays following suspected internal exposures.

To limit the radiation risk to specific organs and the total body, an annual limit on intake (ALI) has been established for each radionuclide.[2] This is the amount of a specific radioactive material that, if taken into the body by inhalation or ingestion, would result in the maximum allowed dose to an individual organ or to the worker's whole body. The actual intake is computed using bioassay data.[1] The dose is then calculated by determining what fraction of the ALI was taken into the body; the resultant dose is that same fraction of the maximum allowed dose. Regulations[2] require monitoring of intakes of radioactive materials for adults likely to receive an intake in excess of 10% of the applicable ALI.

Regulations also specify the concentrations of radioactive material in the air to which a worker may be exposed for 2000 working hours per year. These concentrations are termed the derived air concentrations (DAC) and represent total air concentrations that workers are allowed to breathe if no external radiation dose is received. The result of breathing air at 1 DAC (for 2000 working hours) would be an intake of one ALI.

External Exposures

Sources of external exposure include the betas and gammas emitted from radioactive materials, X-rays from radiographic equipment and accelerators, and possibly neutrons from high-energy accelerators. Short working times and maximum distances between the source of exposure and the worker are effective in reducing dose from external sources. Shielding is also an effective means to reduce exposure.

Dose Limits

For radiation protection purposes, there are several general categories of exposure to consider: (1) occupational exposure to radiation workers (including pregnant women); (2) exposure to members of the general public; (3) medical exposure that deals with the intentional exposure of patients undergoing diagnostic or therapeutic procedures; (4) exposure to natural background radiation; and (5) exposure to consumer products. There are specific annual dose limits for only the first two categories.

To comply with these dose limits, area surveys and monitoring must be completed to evaluate the magnitude of radiation levels, concentrations or quantities of radioactive materials, and the potential radiological hazards. In addition, workers may be required to participate in routine bioassay programs to assess internal exposures or may be issued individual radiation monitoring devices to assess external exposures.

Occupational Dose Limits

Dose resulting from exposure of an individual to radiation in a restricted area or in the course of employment in which the individual's job duties involve exposure to radiation is termed *occupational dose*. Dose limits are defined for adult workers, minors, and the embryo/fetus of women who have declared pregnancy.

Adult Workers

For adult workers, the annual dose limits include:

- 5 rems (0.05 Sv) for the total effective dose equivalent (TEDE), which is the sum of the deep dose equivalent (DDE) from external exposure to the whole body and the committed effective dose equivalent (CEDE) from intakes of radioactive material
- 50 rems (0.5 Sv) for the total organ dose equivalent (TODE), which is the sum of the DDE from external exposure to the whole body and the committed dose equivalent (CDE) from intakes of radioactive material to any individual organ or tissue, other than the lens of the eye
- 15 rems (0.15 Sv) for the lens dose equivalent (LDE), which is the external dose to the lens of the eye
- 50 rems (0.5 Sv) for the shallow dose equivalent (SDE), which is the external dose to the skin or to any extremity

Minors

For minor workers, the annual occupational dose limits are 10% of the dose limits for adult workers.

Embryo/Fetus

Once an occupationally exposed woman declares her pregnancy by informing her employer in writing, she is subject to more restrictive dose limits for the embryo/fetus. For the entire pregnancy, the embryo/fetus dose limit is 0.5 rem (5 mSv) and is controlled by restricting the exposure to the declared pregnant woman. The embryo/fetus dose is the sum of the DDE from external exposure to the pregnant woman plus the internal dose from radionuclides within the embryo/fetus and radionuclides within the declared pregnant woman. Efforts must be made to avoid substantial variation above a uniform monthly exposure rate to the declared pregnant woman. If the dose to the embryo/fetus has exceeded 0.5 rem (5 mSv) or is within 0.05 rem (0.5 mSv) of this dose by the time the woman declares her pregnancy, the additional dose to the embryo/fetus cannot exceed 0.05 rem (0.5 mSv) during the remainder of the pregnancy.

Restricting the occupational exposure to the woman who declares her pregnancy raises questions about privacy rights, equal employment opportunities, and the possible loss of income. Accordingly, declaration of pregnancy by a female radiation worker is completely voluntary. Also, she may withdraw her declaration of pregnancy for any reason.[3] For example, she may believe that her benefits from receiving the occupational exposure would outweigh the risk to her embryo/fetus from the radiation exposure.

Dose Limit to Members of the General Public

The annual dose limit from licensed or registered operations to individual members of the general public is 0.1 rem (1 mSv). This does not include dose from disposal of radioactive material into sanitary sewerage. In addition, the dose in any unrestricted area from external sources may not exceed 0.002 rem (0.02 mSv) in any one hour.

Personnel Dosimeters

Regulations require monitoring of all individuals likely to receive 10% of the dose limits[2] in one year. Issuing dosimeters, or devices designed to measure radiation dose, to workers is the easiest way to demonstrate compliance. Table 65.1 shows characteristics of an ideal dosimeter.

A dosimeter records the radiation dose to the area on the body in which it is worn. Workers should wear the device where it provides an indication of the exposure received by the trunk of the body. It is acceptable to wear the device on the front of the body at waist level or at chest level facing forward. When a protective lead-equivalent apron is worn (as with X-ray equipment), the dosimeter should be placed outside of the apron at collar level as the unprotected head, neck, and eyes receive 10 to 20 times more exposure than the protected body trunk.[4]

There are three main types of personnel monitoring devices used for measuring external exposure: pocket dosimeters, thermoluminescent dosimeters (TLDs), and film badges. Note that an *accredited dosimetry processor or provider* must supply the latter two. This is an organization with current accreditation from the National Voluntary Laboratory Accreditation Program (NVLAP) of the National Institute of Standards and Technology for the types of radiation for which the dosimeter is designed. For each approved radiation type, the processor reports the *deep-dose equivalent, H_d; shallow dose equivalent, H_s; or lens-dose equivalent, H_l,* as appropriate.

Pocket Dosimeters

The self-reading pocket ionization chamber (about the size of a fountain pen) is a simple capacitor and is used to measure exposure from X-rays or gammas. It contains one positively charged (central) electrode and one negatively charged (outer) electrode. When the charged electrodes are exposed to ionizing radiation, the air surrounding the central electrode becomes ionized and discharges the device in direct proportion to the radiation exposure. The advantages and disadvantages of pocket dosimeters are shown in Table 65.2. These devices are generally sensitive to exposures ranging from 0 to 200 mR (0 to 5.2×10^{-5} C kg^{-1}).

Pocket dosimeters afford workers immediate read-out of exposure results. This is particularly useful for tracking exposure while working in high-dose-rate areas. However, they do not provide a permanent legal record of exposure unless the results are documented.

Thermoluminescent Dosimeters (TLDs)

When thermoluminescent (TL) material is exposed to ionizing radiation, electrons are excited to higher energy levels and are trapped in lattice imperfections or impurity sites. The higher the radiation dose, the more electrons will be trapped. Subsequent heating of the TL material will release the electrons from their traps, causing light emission as they return to their original or normal state. A TL analyzer or reader is used to measure the amount of radiation by first heating the materials and

TABLE 65.1 Characteristics of Ideal Dosimeter

Adequate sensitivity
Sufficient accuracy
Adequate reproducibility
Energy independent response
Linear relationship between response and exposure
Dose rate independent response
Angular independent response
Cost effective
No response to unwanted radiations
Easy identification
No interference with worker
Minimum sensitivity to environmental factors
Simple but rugged construction

TABLE 65.2 Advantages and Disadvantages of Pocket Dosimeters

Advantages	Disadvantages
Small, compact	Not cost effective for monitoring many employees
Reasonably accurate and sensitive	Must be recharged or "zeroed" prior to next use
Can monitor exposure for short periods of time	Subject to mechanical shock, which may discharge electrode
Immediate readout	No permanent, legal record of exposure

then recording the amount of light emitted by them. The total light emitted is proportional to the radiation energy absorbed in the material and hence is proportional to the radiation dose.

Various chemical compounds, almost exclusively in crystalline form, exhibit properties of thermoluminescence. These materials are available as powders or solid dosimeters in the form of small ribbons or chips as

shown in the badge in Figure 65.1. TL material is reusable upon heating to a high temperature. This process, called *annealing*, is similar to the heat cycle used in analyzing the material except that the temperature is raised higher and sustained for a longer period of time. For example, lithium fluoride (LiF), known as TLD-100, is the most commonly used TL material. TLD-100s are normally read while heating to 255°C over a time span of 20 seconds. These dosimeters are then annealed at 400°C for 1 hour. Tests have shown that a given dosimeter can be used and annealed over 1000 times without loss of radiation sensitivity.[5]

Advantages and disadvantages of TLDs are shown in Table 65.3. The small chip size makes them particularly useful for monitoring extremity doses. Because of its effective atomic number, LiF (Z_{eff} = 8.3) is sufficiently near soft tissue (Z_{eff} = 7.4) and is considered tissue equivalent. From this standpoint, its dosimetric characteristics are ideal. For X-rays and gammas, its dose response is linear over the useful radiation protection range of 10 mR to 1000 R (2.58×10^{-6} to 0.258 C kg^{-1}). However, at low photon energies, LiF over-responds by about 30%[5]; corrections must be applied to account for this effect.

FIGURE 65.1 Thermoluminescent dosimeter (TLD) badge containing lithium fluoride chips. (Courtesy Landauer, Inc., Glenwood, IL.)

TABLE 65.3 Advantages and Disadvantages of Thermoluminescent Dosimeters

Advantages	Disadvantages
Light, durable	Cannot determine day of exposure
Tissue equivalent for accurate dose assessment	One-time readout of crystal (glow curve data must be saved on computer)
Not affected by normal temperature changes or humidity (so exchanged quarterly)	Calibration tedious
After processing or reading, crystals may be reused	Crystals maintain exposure "history" when exposed to high doses
Ideal for extremity monitoring	

Although TLDs respond quantitatively to betas, the sensitivity is dependent on the beta energy; the beta energy must exceed approximately 250 keV to penetrate the badge "window" and contribute to dose. Note that TLDs are generally calibrated by the dosimetry provider with a 2.27 MeV beta source held perpendicular to the badge. Measurements[5] show that LiF TLDs indicate only 28% of the correct response based on the 2.27 MeV standard. In addition, angular response curves show a 49% under-response at 45° incident to the badge.

Film Badges

Photographic emulsions are widely used in radiation dosimetry. Shown in Figure 65.2, film badges are the most commonly used and economical personnel monitoring devices. Advantages and disadvantages are shown in Table 65.4. Typically, this badge consists of two pieces of dental film covered by light-tight paper in a compact plastic container. Ionizing radiation darkens the film and the degree of darkening is proportional to the radiation exposure.

Due to the sharpness of images cast by the filters on the film, the filters in the badge holder actually allow the measurement of the approximate energy of the radiation reaching the film. In addition, the filters provide information regarding the direction from which the radiation reached the film and whether the exposure was from a

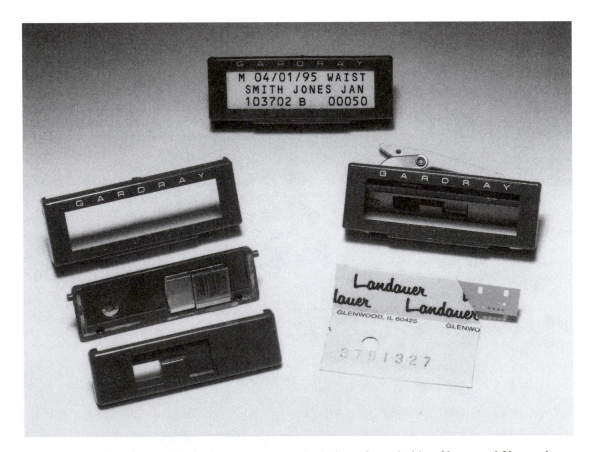

FIGURE 65.2 Film badge components, including plastic holder, filters, and film pack. (Courtesy Landauer, Inc., Glenwood, IL.)

TABLE 65.4 Advantages and Disadvantages of Film Badges

Advantages	Disadvantages
Light, durable	Damaged film pack light-sensitive
Reasonably priced (cost-efficient for monitoring large numbers of employees)	Temperature and humidity may fog film over long periods of time (so exchanged monthly)
Permanent record of exposure	Cannot determine day of exposure
Film can be reanalyzed	
Filters allow estimate of direction from which exposure came	
Filters can distinguish exposure from scatter versus primary	
Filters can discriminate between X-rays and gamma rays	

primary source of radiation or from a source of scatter radiation.

The film packet contains a special film that is sensitive to doses ranging from 10 mrem (0.1 mSv) to 500 rem (5 Sv). At photon energies above 200 keV, film demonstrates an energy-independent response. But at low energies, the film over-responds by as much as 2000 to 4000%.[5] Therefore, it is essential that the film be placed correctly in the holder.

As with TLDs, the detection efficiency for film is strongly dependent on beta energy and the angle of incidence into the badge. The chief problem with film is the absorption of the beta particles by the paper wrapping around the film emulsion before betas reach the sensitive layer. Published response curves show the film emulsion can detect only betas above 0.2 to 0.3 MeV in energy.[5] In addition, published angular response curves for film badges show that betas incident at 45° rather than perpendicular to the badge will be under-reported by 58%.[5] Commercial dosimetry providers recognize some of these difficulties and *generally* do not report any doses from betas below 1 to 1.5 MeV.

Personnel Monitoring Program

Personnel monitoring is the continuous measurement of occupational radiation exposure; it does not include exposure to natural background or exposure incurred during diagnostic or therapeutic medical procedures. Results of such a program will verify that occupational doses are below regulatory limits, provide means to en-

sure that doses are being maintained as low as reasonably achievable (ALARA), provide measures of radiation-protection-program effectiveness, and provide legal records of occupational dose. In addition, because monitoring devices are worn or carried by workers at all times, they provide some indication of the work environment and habits of the individuals included in the program.

The type of work performed at the facility will determine what types of dosimeters are provided to the employees. For example, workers with high potential for X-ray or gamma exposure are issued film badges that are exchanged monthly. The monthly reports provided by the commercial dosimetry service allow for timely review of cumulated doses; the monthly reports also allow more timely assessment of administrative and engineering controls to reduce personnel exposures.

Types and quantities of radioactive materials will determine the types of dosimeters needed for lab workers. For example, many radionuclides commonly used in basic science research include low-energy beta or low-energy gamma emitters. As a result, external exposure is minimal and dosimeters are not required. Because the materials are generally "unsealed," potential for internal exposure exists, particularly by route of hand-to-mouth contamination. Good radiation safety techniques (wearing personnel protective equipment such as lab coats and gloves and never pipetting by mouth), security of radioactive materials (to ensure accountability of such materials), and routine area monitoring (smear surveys) will reduce potential for internal exposures.

External exposures to laboratory workers will unlikely approach 10% of the occupational limit because stock solutions are generally well shielded. However, these individuals, particularly those working with phosphorus-32 or iodine-131, may be issued TLD badges that are exchanged quarterly. Ring dosimeters may also be issued to these workers as well as to those working with iodine-125 to monitor extremity doses incurred while pipetting solutions, handling test tubes, and so forth.

Female employees who have declared their pregnancy should be issued a separate, monthly film badge to monitor dose to the embryo/fetus. The results provide a legal document to demonstrate compliance with the 0.5 rem (5 mSv) during the entire pregnancy. This exchange frequency will demonstrate that there is no significant variation of exposure from month to month.

Quarterly TLDs may also be used to verify compliance with the dose limit to members of the general public. Although compliance may be demonstrated with area survey data, these dosimeters are easily deployed in waiting rooms, corridors, and so on. By exchanging them quarterly, annual doses at a variety of locations may be determined.

References

1. Lessard, E. T.; Yihua, X.; Skrable, K. W.; Chabot, G. E.; French, C. S.; Labone, T. R.; Johnson, J. R.; Fisher, D. R.; Belanger, R.; Lipsztein, J. L. *Interpretation of Bioassay Measurements*; NUREG/CR-4884 (BNL-NUREG-52063), Brookhaven National Laboratory: Upton, NY, 1990.

2. NRC Standards for Protection Against Radiation, 10 *CFR* 20.

3. NRC. *Instruction Concerning Risks from Occupational Radiation Exposure*; Regulatory Guide 8.29, U.S. Nuclear Regulatory Commission: Washington, DC, 1996.

4. Statkiewicz-Sherer, M. A.; Visconti, P. J.; Ritenour, E. R. *Radiation Protection in Medical Radiography*; Mosby: St. Louis, MO, 1993.

5. Gollnick, D. A. *Basic Radiation Protection Technology*, 3rd ed.; Pacific Radiation Corporation: Altadena, CA, 1994.

66

Lasers

WESLEY J. MARSHALL*

Lasers are a wonderful technological breakthrough, and new uses are being discovered all the time. It may seem strange, however, that even a one-milliwatt laser requires a warning label, and yet a 100-watt light bulb does not have one. A laser produces radiant energy in the optical portion of the electromagnetic spectrum (wavelengths between 180 nm and 1 mm). The individual photons (particles or bundles of energy) of laser light are the same as ordinary light, but a laser produces a photon beam through stimulated emission, with all photons of the same wavelength and in phase (known as coherence), and traveling in the same direction. Laser stands for light amplification by the stimulated emission of radiation (see Box 1).

In contrast, a light bulb produces light of all colors, which propagates in all directions. Even when lenses are used with the bulb, such as in a flashlight, the produced beam is fairly broad. Laser light is different in that it can be highly concentrated in both space and time. Although everyone is exposed to low-intensity optical radiation every day, when a concentrated laser beam interacts with human tissue, the tissue may be heated several degrees centigrade, or photochemical changes may take place due to the high concentration of photons. The lens and cornea concentrate laser energy by 100,000 times so that a retinal burn or lesion may result.

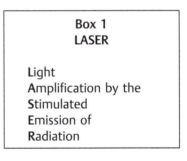

Box 1
LASER

Light
Amplification by the
Stimulated
Emission of
Radiation

Figure 66.1 illustrates the difference between viewing a conventional light source and intrabeam viewing of a laser beam. It is this tissue interaction that the laser safety professional wishes to prevent. Since the product of the wavelength and frequency of optical radiation is equal to the speed of light, either quantity may be used to specify the laser energy. Since the frequencies involved are quite high, optical radiation is identified by its wavelength. Lasers usually emit optical radiation of one wavelength or frequency, although some lasers emit several wavelengths simultaneously. The energy may be emitted in a continuous wave (CW), in individual pulses, in Q-switched pulses (duration of a few nanoseconds), or as a continuous stream of pulses. Laser theory was developed by Charles Townes and Arthur Schawlow in 1958. The first actual device was a pulsed ruby laser produced by Theodore Maiman of the Hughes Research Laboratories in 1960.[1] The CW helium-neon gas laser, which is used in many grocery store sales counters, was developed in 1961. The first diode laser, which is now

---

*The opinions or assertions herein are those of the author and should not be construed as official positions of the U.S. Department of the Army or the U.S. Department of Defense.

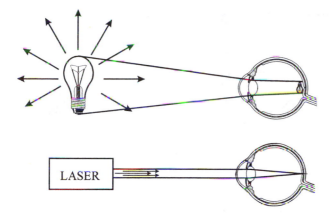

FIGURE 66.1 Comparison of the retinal image size produced by a conventional light source to that produced by a collimated laser source.

used in pocket pointers, was developed in 1962. Lasers in all parts of the optical spectrum are now available (see Box 2). Visible lasers with an output power of tens of watts are often used in laser light shows. Laser weapons which produce beams of light that can destroy vehicles or bridges require further development.

Biological Effects and Exposure Limits

Although skin injury is possible from some types of lasers, eye injury from lasers is usually of most concern. The injury mechanisms are mostly either thermal or photochemical. In experiments with animals (normally rhesus monkeys), thermal injury is immediately apparent and is usually evaluated 1 hr after exposure. Photochemical injury takes much longer to become evident. This injury is usually evaluated 24 hr after exposure. Both mechanisms are incorporated into maximum permissible exposure (MPE) limits for lasers. Often, two methods of evaluation must be compared to determine the dominant effect. Exposure limits are listed in a variety of published standards.[2,3,4]

Retinal Hazard Region (400 to 1400 nm)

Wavelength Correction Factors

Lasers with wavelengths between 400 and 1400 nm are the most hazardous because they can cause retinal injury. Near-infrared energy (700 to 1400 nm) has less injury potential than light (400 to 700 nm) due to reduced transmission in the eye and reduced absorption at the retina. As shown in Figure 66.2, correction factors C_A and C_C account for these differences.

The MPE for a Nd:YAG, Q-switched pulsed laser (wavelength of 1064 nm) is 10 times higher than for a visible Q-switched pulsed laser. The MPE for the eye in the retinal hazard region is (for exposure durations between 1 ns and 10 s):

$$MPE = 0.5 \times C_A \ \mu J \cdot cm^{-2} \quad \text{for } \lambda < 1050 \text{ nm and } t < 18 \ \mu s$$

$$MPE = 5.0 \times C_C \ \mu J \cdot cm^{-2} \quad \text{for } \lambda \geq 1050 \text{ nm and } t < 50 \ \mu s$$

otherwise: \hfill (1)

$$MPE = 1.8 \times C_A \times C_C \times t^{0.75} \ mJ \cdot cm^{-2} \quad \text{for } t \leq 10 \text{ s}$$

where t represents the pulse width for pulsed lasers or the anticipated exposure duration for CW lasers. Computing MPEs for exposure durations longer than 10 s can be much more involved; however, unintentional exposure is usually limited to 0.25 s for visible wavelengths (400 to 700 nm) and to 10 s for longer wavelengths. Since skin is highly reflective in the retinal hazard region, skin MPEs in this region are two C_A times higher than skin MPEs for the far-infrared.

Exposure to Multiple Pulses

When an individual is exposed to a continuous stream of pulses from a repetitively pulsed laser, more injury potential exists than either exposure to a single pulse of the same pulse energy or exposure to a CW source with the same average power. When an individual can be exposed to n pulses, the MPE is adjusted with a correction factor, C_p, equal to $n^{-0.25}$. For example, an exposure to 10,000 pulses results in a C_p of 0.1. In addition, the average power must be checked against the MPE for a CW

Box 2 Common Laser Wavelengths (nm)	
Nitrogen	337
Argon	457, 476, 488, 514
Krypton	351, 356, 530, 647
Copper vapor	511, 578
Helium–Neon	633
Ruby	694
Alexandrite	755
Gallium–Arsenide	905
Neodymium:YAG	1064
Erbium:Glass	1540
Carbon dioxide	10,600

Data from Reference 1.

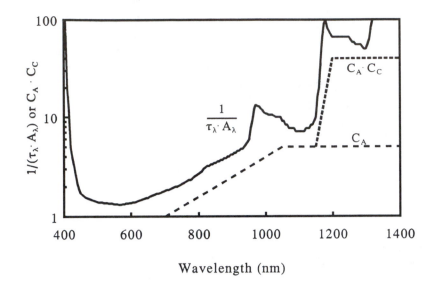

FIGURE 66.2 Correction factors C_A and C_C compared to the reciprocal of the product of the transmission of the ocular media and absorption by the retina. (Data from Reference 5.)

exposure, and the most conservative evaluation is used. In the retinal hazard region, the CW limit dominates only when the laser is pulsed at several kilohertz or more.

Extended Sources

A laser beam striking a matte surface produces a diffuse reflection, which is often an extended source when viewed at a close distance. The laser energy from a diffuse reflection is less hazardous than that from the direct beam. Not all the reflected energy enters a person's eye, and the portion that does impact the eye is not focused onto a diffraction-limited retinal area, as shown in Figure 66.3. Still, some lasers can pose a serious hazard for a person observing a diffuse surface, such as a wall or target board, used to stop a laser beam. In addition, some lasers (mostly diode projection systems) have a larger source than the usual small source. The MPEs for extended sources are higher because the laser energy or power entering the pupil is distributed over a larger portion of the retina, thus lowering the potential for retinal heating. However, a conservative approach to hazard evaluation is to assume that all laser sources are small sources.

Optically Aided Viewing

To match measured and calculated radiometric quantities to the potential for biological injury, limiting apertures and measurement apertures are used. For possible

retinal injury to the unaided eye, a 7-mm limiting aperture simulates the dark-adapted pupil. Only that laser power or energy that is transmitted by this aperture contributes to the injury potential. The potential use of optical viewing aids must also be considered, such as eye loops or binoculars. Normal eye-glasses do not affect the hazard evaluation. A 50-mm measurement aperture (simulating 7 × 50 binoculars) is used when the radiant

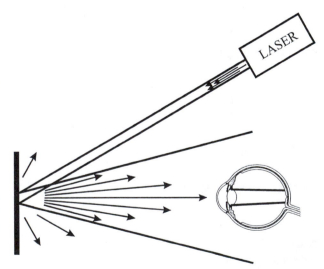

FIGURE 66.3 Retinal image produced by a diffuse reflection of a collimated laser beam. Only a portion of the reflected energy impacts the eye, and the portion that enters the eye does not form a diffraction-limited image on the retina.

energy can be transmitted by common optics (wavelengths between 302 nm and 2.8 μm), and use of these devices is likely. In this case all the energy focused by the optics into the limiting aperture is considered as contributing to the retinal injury potential. Generally, a hazard analysis of lasers intended for outdoor use will include the potential use of magnifying optics.

Middle and Far-Infrared Exposure (1400 nm to 1 mm)

Eye exposure to wavelengths greater than 1400 nm poses a potential for injury to the cornea based on thermal heating. Small beam diameters have a more injurious effect for middle and far-infrared wavelengths since the laser energy is not transmitted to the retina. Except for wavelengths greater than 100 μm, the limiting aperture D_f is reduced from the 7 mm used to represent the pupil, for visible and near-infrared wavelengths, to 1 mm for corneal exposure to pulsed infrared lasers. For skin exposure or for eye exposure for more than 10 s, D_f is increased to 3.5 mm to account for thermal diffusion and natural eye movements which increase the exposed area. For middle and far-infrared wavelengths, the MPEs are shown in Figure 66.4. The MPEs for the eye, in particular wavelength bands (wavelengths less than 2600 nm), have been set much higher than for wavelengths farther into the infrared, since the laser energy is absorbed within a volume of tissue rather than just at the corneal surface. For example, the MPE for an erbium: glass laser (1540 nm) is 1 J · cm^{-2} for any accumulation of radiant exposure up to 10 s. In contrast, the MPE for a CO_2 laser (10,600 nm) is 10 mJ · cm^{-2} for a single 100-ns pulse. Each of these lasers has the same MPE for a CW exposure of 10 s, since the local temperature rises in tissue from the absorbed energy have had time to equalize.

Multiple-pulse exposures are limited both by an equivalent CW exposure and C_P times the single-pulse MPE. The equivalent CW-exposure MPE is usually dominant for lasers pulsed at more than a few hertz. For example, all pulses within a 10-s exposure to a 1540-nm laser are treated as one pulse with the combined energy from all the individual pulses.

The skin MPE for both middle and far-infrared wavelengths is about the same as that for the eye, as shown in Figure 66.4. However, differences exist in the various standards.[2,3,4,6]

Ultraviolet Exposure (180 to 400 nm)

Exposure to ultraviolet laser radiation for the eye or skin is considered additive over at least an entire workday (30 ks), as the biological effect is usually from photochemical injury and not thermal heating. For wavelengths between 180 and 302 nm, the MPE is 3 mJ · cm^{-2}. For wavelengths between 315 and 400 nm, the MPE is 1.0 J · cm^{-2}, and the MPE varies between 302 and 315 nm. The radiant exposure is added from repeated exposures, even if one exposure occurs in the morning and the other in the afternoon. In addition, the MPE is limited to 0.56 $t^{0.25}$ J · cm^{-2}, where t represents the pulse length for pulsed lasers or the exposure duration for CW lasers. The limiting aperture for the eye is 1 mm for short exposures (less than 0.3 s) and 3.5 mm for

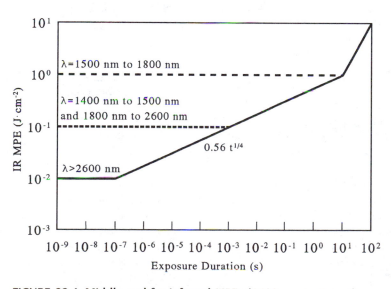

FIGURE 66.4 Middle and far-infrared MPEs (1400 nm to 1 mm).

exposures of 10 s or more. For skin exposure, the limiting aperture is 3.5 mm for all exposure durations.

Hazard Classification and Evaluation

Although laser hazard classification is based strictly on optical radiation hazards, other hazards are associated with a laser besides the laser beam. Many lasers use potentially hazardous mixtures of gases and have power supplies that use a combination of cooling water and high voltage, creating a potential shock hazard. Proper ventilation and mounting brackets must be considered when gas cylinders are used. Gas cylinders of different types must be stored separately to comply with OSHA requirements. Ventilation is also important when products of combustion caused by laser beams are toxic. Some lasers can produce ultraviolet radiation hazards when covers are removed.

The beams from high-power laser light displays have caused glare and flashblindness to pilots operating aircraft near airports in critical phases of flight. Safety zones now exist near airport runways: the sensitive zone, the critical zone, and the laser-free zone.[7] Visible laser beams, even less than the MPE in these zones, could interfere with safe aircraft operation. Coordination with both the FDA and the FAA is required before staging an outdoor laser light show.

Hazard Classification

Lasers are classified according to the optical hazards that they produce. The idea is to first classify the laser and then apply the control measures for that class. The best practice is to build the safety into a product, rather than to force the user to use personal protection or operate the laser in a prescribed manner. Lasers are usually classified according to the federal laser product performance standard.[6] These lasers are required to have a warning label that states the maximum output power, or energy for pulsed lasers, and the hazard class. Safety features are required to be built into the laser product by the federal standard for each laser class, such as a beam attenuator, a remote interlock connector, and an emission indicator.

When lasers have been modified by the user, such as adding a beam expander or limiting the power output, they may need to be reclassified according to the ANSI standard.[2] The ANSI standard provides guidance for the safe use of the laser product. The hazard classifications by ANSI, as listed below, are similar, but not identical, to the federal standard.

Class 1

The Class 1 accessible emission limit (AEL) is the MPE multiplied by the area of the limiting aperture (0.385 cm^2 for a 7-mm pupil). Lasers emitting less than this level do not present a hazard under normal use, but may enclose a more powerful beam. When a Class 3 or Class 4 laser is embedded in a Class 1 laser system, interlocks are required to prevent hazardous exposure to the operator.

Class 2

Visible lasers that do not exceed the Class 1 AEL within 0.25 s (less than 1-mW output power) are Class 2. Some laser pointers are Class 2. A yellow caution label warns people not to stare into the beam. Class 2a lasers, such as many supermarket checkout lasers, emit less power, do not exceed the Class 1 AEL for a 1000-s exposure duration, and have a simpler label.

Class 3

Class 3 lasers present an instantaneous direct beam hazard, and the class contains two subcategories. Class 3a includes visible lasers less than 5 mW, such as pointers. ANSI also includes invisible lasers less than 5 times the Class 1 AEL in this category, although the federal standard does not. Many laboratory lasers are Class 3b. Class 3b includes lasers emitting less than 500-mW average power and less than 30 C_A mJ per pulse for pulsed visible and near-infrared lasers.

Class 4

More powerful lasers are Class 4. Many industrial lasers emit much more than 500 mW and are used in precision welding and cutting operations. Class 4 lasers may produce a potential fire hazard, skin hazard, and a hazard from viewing a diffuse reflection of the beam.

Hazard Evaluation

Irradiance and Radiant Exposure

Lasers, other than Class 1, produce a potential hazard. The laser's irradiance (power per unit area), E, for CW lasers, or radiant exposure (energy per unit area), H, for pulsed lasers, is averaged over the limiting aperture before comparison is made to the MPE. Otherwise, the hazards are often perceived as greater than they really are.

For CW lasers, the MPE in joules is divided by the exposure duration to correct the units for comparison. The average irradiance or average radiant exposure can be calculated by:

$$E_{avg} = \frac{4\Phi}{\pi[\text{Max}(D_L, D_f)]^2} \quad \text{and} \quad H_{avg} = \frac{4Q}{\pi[\text{Max}(D_L, D_f)]^2} \quad (2)$$

where Φ represents the CW output power, Q is the output energy per pulse, and $\text{Max}(D_L, D_f)$ represents the maximum of the limiting aperture D_f and the beam diameter D_L.

A Gaussian beam has a maximum irradiance in the center of the beam and the irradiance gradually decreases at the edges. For laser safety calculations, the beam diameter and beam divergence ϕ are at $1/e$ of peak irradiance points, which, for a truly Gaussian beam, is about 71% of that usually specified (at $1/e^2$ points).

Optically aided viewing causes a variety of effects related to the magnifying power: the laser's beam diameter is reduced, the viewer is effectively moved closer to the laser source, and the laser's source size is increased. Optically aided viewing is approximated by the effects of 7×50 binoculars. Figure 66.5 is a plot of the radiant exposure from a diode laser, averaged over a 7-mm aperture, versus viewing distance for both unaided viewing and viewing with 7×50 binoculars. The laser emits 1 µJ per pulse at 10 kHz, has a 1-cm exit beam diameter, and a beam divergence of 0.5 mrad (both at $1/e$ points),

and is an extended source when directly viewed within a few meters through optics. The distance where the radiant exposure no longer exceeds the MPE is termed the nominal ocular hazard distance (NOHD).

Nominal Hazard Zone

The nominal hazard zone (NHZ) defines the hazardous area created by a laser beam: from the direct beam, pointing inaccuracies, and reflections of the beam. When the laser beam is used in a laboratory or used with a suitable backstop, the direct beam is hazardous only between the laser source and the backstop. The NHZ defines where control measures are needed to prevent hazardous personnel exposure. If an entrance door is within the NHZ, a door interlock is necessary. If the NHZ extends to the eyes of personnel, then eye protection is needed.

When the beam is used outdoors and unterminated, it can be hazardous at a considerable distance from the laser. For operation outdoors, both unaided and optically aided conditions need to be considered. For example, a ruby laser with a 50-mJ output, a 5-mm beam diameter, and a beam divergence of 0.5 mrad would have an MPE of 0.5 µJ · cm^{-2}. The beam diameter D_L at a distance r may be approximated by the product of r and ϕ. At a distance of 7 km (700,000 cm) from the laser, the beam would be about 3.5 m (350 cm) across. The beam radi-

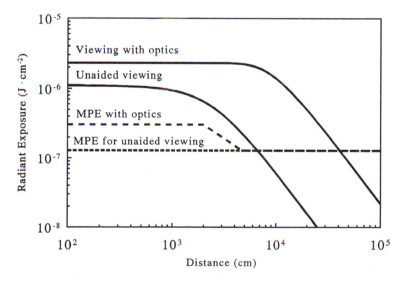

FIGURE 66.5 Corneal average radiant exposure versus distance from a diode laser source when viewed with the unaided eye or for optically aided viewing. Two MPEs are shown because this laser appears as a large source within a few meters when viewed with magnifying optics.

ant exposure is then found from Equation 2 to be about $0.52 \, \mu J \cdot cm^{-2}$. The NOHD is then about 7 km for unaided viewing. However, this distance would normally be reduced somewhat due to atmospheric absorption and scattering of the beam.

A simplified method for calculating the irradiance E or radiant exposure H from a diffuse reflection is:

$$E = \frac{\Phi}{\pi r_1^2} \text{ and } H = \frac{Q}{\pi r_1^2} \quad (3)$$

where r_1 is the distance from the observer to the surface. As a worst-case approximation, these calculated irradiances or radiant exposures may be compared directly to the MPEs. If the same ruby laser, from the example above, were aimed at a white diffuse target within a few meters, an observer located 50 cm from the target would receive a radiant exposure of $6.4 \, \mu J \cdot cm^{-2}$, or about 13 times the MPE. Since the laser spot on the target would be an extended source at this distance, the MPE would be increased somewhat. However, the above analysis would provide a conservative estimate of the hazards. Table 3 of ANSI Z136.1-1993 provides laser energies for Q-switched lasers that will produce a diffuse reflection hazard at various distances based on the actual source size.[2]

Optical Density

The necessary optical density, OD, considering use with or without magnifying optics is

$$OD = \log_{10} \left(\frac{\Phi \text{ or } Q}{\text{Class 1 AEL}} \right). \quad (4)$$

For example, the necessary OD for a 50-watt CO_2 laser would be 3.7 because the Class 1 AEL is about 10 mW. The irradiance behind the eye protectors would be reduced by 5000 times.

Best Safe Practices

Laser control measures, which apply mostly to Class 3b and Class 4 lasers, are usually divided into the categories of engineering, administrative, and personal protection.

Engineering

Control measures are built into laser systems as required by the federal standard,[6] such as an interlocked case, an emission indicator, scanning safeguards, a beam attenuator, and a key switch. Other control measures, such as door interlocks, warning signs, and key control, must be

established for each laboratory. Lasers should be secured against unauthorized use by either using the built-in key lock or by restricting access to the device. When an individual could be injured by entering the laboratory, a door interlock should be connected to the laser's interlock connector. A beam shutter connected to a door interlock can also be used as an alternative to interrupting the laser's power supply. A well-lit laboratory with lightly colored walls helps offset reduced visibility when laser goggles are sometimes necessary.

Laser beams should not be positioned so that a person's eye would enter the beam either in a standing or sitting position. Beams should not be directed toward a doorway. By positioning equipment in this way, the probability of injury is lessened even when required eye protection is inappropriately removed during a testing procedure. Beams should be terminated on the optical table by something other than a removable detector or mirror. A beam directed upward from under an optical table has caused retinal injury when the beam missed the turning mirror and struck the operator in the eye. Common causes of accidents are listed in Box 3.

Administrative

Administrative controls are usually preferred to door interlocks in medical treatment facilities due to patient risk caused by a beam interruption. A nonionizing radiation control program should be established by the designated Laser Safety Officer (LSO). The LSO should keep an inventory of laser devices as well as pertinent laser safety guidance for workers. Laser workers should be given safety training commensurate with the hazards they encounter. A standard operating procedure (SOP) should

---

Box 3
Common Causes of Laser Accidents

- Exposure during beam alignment
- Misaligned optics
- Equipment malfunction
- Upwardly directed beams
- Eye protection not used
- High voltage
- Intentional exposure
- Untrained operators
- Ancillary hazards
- Exposure after service procedure
- Failure to follow SOP

Data from Reference 2.

be written for specific laser devices and locations. It should be brief, contain little background information, an authorized list of users, and necessary specific safeguards to protect personnel from injury. Laser eye exams are required for laser workers who use Class 3b and Class 4 lasers routinely.

Personal Protection

Although gloves and skin protection are sometimes necessary for lasers, most personal protection devices are laser eye protectors. Eye protection for the specific laser should be available before potentially hazardous situations arise. Eye protection designed for one wavelength may provide little or no protection at a different wavelength. The OD printed on the goggle at the specific laser wavelength should be either equal to or higher than that required. In addition, the damage level of the goggle needs to be considered for high-power lasers. Visual transmission is important since accidents are more likely when an individual is hampered visually. Goggles that have darkened with age provide little protection, since the operator cannot wear the goggles and still perform required tasks.

References

1. *Encyclopedia of Lasers and Optical Technology*; Meyers, R. A., Ed.; Academic Press: New York, 1991.
2. ANSI Z136.1-1993, American National Standard for the Safe Use of Lasers; American National Standards Institute: New York, 1996.
3. ACGIH, Threshold Limit Values for Chemical Substances and Physical Agents, Biological Exposure Indices; American Conference of Government Industrial Hygienists: Cincinnati, OH, 1996.
4. International Electrotechnical Commission Standard IEC 825-1. Radiation Safety of Laser Products—Part I: Equipment Classification, Requirements and User's Guide; IEC: Geneva, 1993.
5. CHPPM, Technical Guide No. 085, Hazard Analysis of Broadband Optical Sources; U.S. Army Center for Health Promotion and Preventive Medicine: Aberdeen Proving Ground, MD, 1981.
6. FDA. 21 *CFR* 1040.10 and 1040.11.
7. FAA Order 7400.2D, Part 8, Chapter 34, 1996.

67

Biological Safety: Program Elements

DEBRA L. HUNT

Biological safety programs are developed to prevent or minimize (1) the risk of occupational infections, (2) adverse reactions to biological products in the workplace, and (3) release of organisms to the environment. Typically, biosafety programs apply to clinical or infectious disease research laboratories. The need for biological safety principles extends to related workplaces, such as tissue culture labs, research or production labs using recombinant DNA (rDNA) or other biotechnology, animal labs, environmental water-quality labs, or any lab or work area in which human blood, other body fluids, tissues, or infectious agents are handled or present.

The elements of a biosafety program include containment principles and practices that have evolved from the knowledge and experiences of microbiologists over the past century and parallel other laboratory safety programs. Although many references exist that summarize biosafety principles, the two primary resources for control of biological hazards are (1) the CDC/NIH document,

Biosafety in Microbiological and Biomedical Laboratories[1] (BMBL), which discusses containment when working with specific microorganisms, and (2) the OSHA Bloodborne Pathogen Standard[2] for workplace exposure control when work with human blood or other potentially infectious materials is anticipated.

Methods of Biological Containment

The purpose of containment is to reduce or eliminate the exposures to the worker, other people in the work area, and the outside environment. The most common means of exposure to biological hazards include respiratory exposure to biological aerosols, inoculation via needles or other sharp instruments, mucous membrane exposure of the eyes, nose, or mouth via hand contact or droplet exposure from procedures, and ingestion of contaminated material. The elements of containment reduce the risks for these exposures and include work practices and techniques, primary barriers (safety equipment), and secondary barriers (facility design).

Standard Work Practices

The most important means of containment is the consistent use of standard safe work practices that form the foundation for national safety guidelines and regulations. All laboratory workers handling hazardous materials (chemical, radiological, or biological) must make an assumption of potential exposure hazards and use basic precautions consistently. Standard safe microbiological practices (see Box 1) will minimize most exposures except biological aerosols, and are consistent with the prudent practices used for safe handling of chemicals in labs.

Primary Barriers

Primary barriers protect lab workers in the immediate area from exposures to biological materials. Alteration in the techniques used to safely handle this material is a type of "primary barrier"; however, the term is generally used for safety equipment that can reduce specific types of exposure. Safety equipment includes biological safety cabinets (BSCs),[3] splash shields, safety centrifuge cups, and personal protective equipment such as respirators, face protection, gloves, and gowns. Examples of primary barriers and the biological exposures they are designed to prevent are included in Table 67.1.

TABLE 67.1 Application of Primary Barriers to Biological Containment

Potential Route of Exposure	Primary Barriers (examples)
Aerosol	Biological safety cabinets (BSC) Safety centrifuge cups HEPA-filtered equipment Respirators
Inoculation	Self-sheathing needles and sharps Blunted instruments Puncture-resistant sharps containers *Work Practices:* 　Do not recap needles 　Find alternatives for sharps in labs
Mucous Membrane (face)	BSC or splash shield Mask plus eye protection (goggles/shield) Chin-length faceshield *Work Practices:* 　Wash hands
Nonintact Skin	Personal protective equipment: 　gloves, gowns, lab coats, boots, caps
Ingestion	Standard microbiological practices: 　Wash hands 　Do not mouth pipette 　Do not eat, drink, smoke in lab 　Do not store food in lab 　Decontaminate work surfaces

Secondary Barriers

Secondary barriers refer to methods to protect the environment outside of the immediate laboratory or work area, and generally refer to the facility design. For most labs handling biological materials, such barriers include separation of the work area from public access, handwashing facilities, and the availability of an autoclave. Labs handling higher risk agents or agents that may be transmitted by the aerosol route may require additional containment such as double-door entry, dedicated exhaust, and negative-flow air pressure into the work area.

Assignment of Biosafety Levels

The national guidelines published by the CDC and NIH[1] are based on historical data and experience that describe four levels of containment for safe management of infectious agents. Low-risk organisms (i.e., those that do not cause disease in immunocompetent humans) are assigned to biosafety level one (BSL-1), with higher levels of containment (BSL-2 through BSL-4) recommended for organisms with increasingly higher risks of transmission to personnel, the environment, or the community. The

---

Box 1
Standard Safe Microbiological Practices

- Wash hands after handling biological materials, after removing gloves, and before leaving the lab.
- Do not eat, drink, smoke, store food, handle contact lenses, or apply cosmetics in the lab work area.
- Do not mouth pipette.
- Manipulate biological materials carefully to minimize the production of aerosols or droplets.
- Use extreme precautions with needles and other sharp instruments, avoiding their use if possible.
- Use protective equipment (gloves, faceshields, etc.) as needed to protect the mucous membranes of the face or nonintact skin.
- Decontaminate work surfaces at least once a day and after any spill of biological material.

biosafety levels consist of combinations of microbiological practices and primary and secondary barriers to reduce the potential hazards identified in the risk assessment of the work in the lab.

Risk Assessment

As with chemical or radiological safety planning, a risk assessment is done to determine the types of agents to be handled and their hazard potentials (e.g., chemical assessments review characteristics such as flammability, corrosiveness, toxicity, or route of personal exposure). A similar risk evaluation is done for biological hazards by reviewing (1) the hazards associated with the microorganism such as the pathogenicity, the consequences of infection, and the routes of exposure; (2) the types of manipulations expected to occur in the lab, such as procedures that result in aerosolization or splashes; (3) the potential "dose" of the exposure (i.e., propagation of cultures to high titers); and (4) the availability of preventive immunizations or treatment (see Box 2). Appropriate levels of containment and practices can then be applied to minimize the risks.

Biosafety Levels

The guidelines for combinations of work practices and primary and secondary barriers that define each biosafety level can be found in the BMBL[1] and are summarized in Table 67.2. In addition, general recommendations for biosafety levels for specific organisms that have caused occupational infections are summarized in the BMBL. The laboratory director is ultimately responsible for appropriately applying the recommended biosafety level for the lab. If additional information is obtained during the risk assessment, the level of containment may be increased or decreased. For example, if new procedures are performed that increase the potential for aerosol production, the organism has developed multiple resistance to antibiotics, or the amount of material handled increases, more stringent precautions might be instituted. A combination of levels of containment may be appropriate. For example, work with clinical levels of HIV is safely conducted in BSL-2 facilities using BSL-2 work practices and safety equipment; but propagation of HIV in culture or in concentrated amounts should be conducted in BSL-2 facilities with BSL-3 work practices and safety equipment.

Biosafety Level 1 (BSL-1)

BSL-1 containment is appropriate for work with well-characterized microorganisms not known to cause disease in immunocompetent human adults. Typically, this applies to undergraduate and secondary school teaching laboratories. BSL-1 represents a basic level of containment that relies on standard microbiological work practices, with no special primary or secondary barriers other than a sink for handwashing. Examples of BSL-1 organisms include *Bacillus subtilis*, *Naegleria gruberi*, and infectious canine hepatitis virus.

Biosafety Level 2 (BSL-2)

BSL-2 containment is appropriate for work with indigenous, moderate-risk organisms that are associated with human disease and generally present in the community. BSL-2 represents the level of containment that is appropriate for most laboratories, including clinical, diagnostic, and medical teaching labs, as well as any lab using human blood or other body fluids. Organisms classified as needing BSL-2 containment are generally transmitted by percutaneous or mucous membrane exposures or by ingestion of biological material. Therefore, in addition to the BSL-1 requirements, extreme precautions must be taken with sharps, primary barriers such as BSCs or splash shields should be used to prevent splashes, and access to the lab should be limited. Examples of BSL-2 organisms include *Toxoplasma sp.*, *Salmonella sp.*, human hepatitis viruses, and human immunodeficiency viruses (HIV). Also include any work with human blood or

---

Box 2
Risk Assessment for Biological Hazards

- Does the organism cause infection in healthy humans? How severe is the disease if acquired?
- Have there been documented occupational infections with this organism?
- What is the route of exposure? Can it be spread through aerosols, or does it require inoculation or direct contact to cause infection?
- What is going to be done with the organism? Is there a potential for aerosol production, such as sonication, centrifugation, or opening lyophilized preparations?
- What is the relative quantity of the organism to be handled? Will clinical specimens from infected patients be handled, or is the organism to be propagated in cultures to high titers?
- Is there a vaccine available for lab workers? Is there treatment available should an infection occur?

TABLE 67.2 Summary of Biosafety Levels

Biosafety Level	Agents	Work Practices	Primary Barriers (Safety Equipment)	Secondary Barriers (Facilities)
1	Not known to cause disease in healthy adults	Standard microbiological practices	None required	Handwashing sink
2	Associated with human disease; Indigenous; *Hazards*: inoculation, ingestion, mucous membrane exposure	BSL-1 plus: —limited access —"sharps" precautions —warning signs —safety manual	BSC or physical containment equipment for procedures causing splashes or aerosols PPE (gloves, gowns, face protection) as needed	BSL-1 plus: —autoclave available
3	Cause human disease with serious or lethal consequences; Indigenous *or* exotic; *Hazards*: aerosol exposure plus those above	BSL-2 plus: —controlled access —decontamination of all lab wastes and lab clothing before laundering —baseline serum	BSC or physical containment equipment for *all* manipulations PPE (gloves, gowns, face protection, *respirators*) as needed	BSL-2 plus: —physical separation from public access —self-closing, double door access —dedicated exhaust —negative airflow into lab
4	Cause life-threatening disease; Dangerous/exotic; *Hazards*: aerosol exposure plus those above	BSL-3 plus: —clothing change before entering —shower on exit —decontamination of all material before leaving lab	*All* manipulations are done in Class III BSC *or* other BSCs in combination with full-body, air-supplied, positive pressure suit	BSL-3 plus: —separate building or isolated zone —dedicated supply/ exhaust, vacuum, and waste system —other requirements in BMBL

Adapted from Reference 1.

other potentially infectious material as defined in the OSHA Bloodborne Pathogen Standard.[2]

Biosafety Level 3 (BSL-3)

BSL-3 containment is appropriate for work with indigenous or exotic microorganisms that may cause serious or lethal infection and can be transmitted via biological aerosols. This level of containment may be appropriate for some clinical, research, diagnostic, and medical teaching labs, as well as for some production facilities. Because aerosol transmission is a major concern, additional primary and secondary barriers are recommended, such as BSCs for all manipulations of material, double-door entry for controlled access to the lab, and a ventilation system to minimize the release of biological aerosols from the lab. Examples of BSL-3 organisms include *Mycobacterium tuberculosis, Coxiella burnetii,* or St. Louis encephalitis virus.

Biosafety Level 4 (BSL-4)

BSL-4 containment is appropriate for facilities in which work is done with organisms that are exotic and pose a high risk for life-threatening disease, are transmitted by

the aerosol route, and for which there is no available vaccine or treatment. Very few BSL-4 labs exist in the United States, and any facility with the need for this level of containment should contact the CDC for advice. All manipulations of biological material at this level must be done in a maximum-containment BSC (Class III) or by personnel in a full-body, air-supplied positive-pressure suit. The BSL-4 facility is isolated, such as in a separate building, with complex, specialized ventilation and waste management systems to protect the environment and community. Examples of BSL-4 organisms include Marburg or Congo-Crimean hemorrhagic fever, Ebola virus.

Bloodborne Pathogens

The biosafety levels of containment discussed above are applicable to laboratories working with specific agents, including hepatitis viruses or HIV. Other recommendations have been published for work with HIV in research and production settings.[4] Blood from sources known to be infected with bloodborne agents has typically been handled according to the recommended biosafety level for the agents, that is, BSL-2 for HIV- or HBV-contaminated blood.

In 1987, the CDC issued new guidelines for handling human blood and body fluids.[5] Recognizing that human infections with HIV, hepatitis viruses, and other blood-borne pathogens may not be clinically apparent and the infectivity of blood is not always known, the CDC recommended the consistent use of standard precautions for *all* human blood and body fluids. This strategy has been referred to as "universal precautions."

The universal precautions recommendations have formed the basis for federal regulations from OSHA in the Bloodborne Pathogen Standard.[2] OSHA has essentially mandated that human blood and other potentially infectious material (PIM) are hazardous materials in the workplace. ("Potentially infectious material" includes semen, vaginal secretions, cerebrospinal, synovial, pleural, pericardial, peritoneal, and amniotic fluids, saliva in dental procedures, any body fluid visibly contaminated with blood, or all body fluids in situations where it is difficult to differentiate between fluids; any unfixed human tissue or organs; HIV-containing cell or tissue cultures, organ cultures, and HIV- or HBV-containing culture medium or solutions; and blood, organs, or other tissues from HIV- or HBV-infected animals.[2]) The Bloodborne Pathogen Standard and its requirements apply to any work setting where occupational exposure to human blood and other PIM can be expected. An outline of the required elements of the standard is found in Box 3.

Methods of Exposure Control

The Bloodborne Pathogen Standard requires the use of universal precautions, along with combinations of engineering controls, work practices, and personal protective equipment (PPE) in order to control blood exposures in the workplace. These requirements are consistent with the methods of containment outlined for BSL-2 recommendations for labs. A review of the specific requirements of the Bloodborne Pathogen Standard and their applications in laboratory environments can be found in *Biological Safety: Principles and Practices*, 3rd edition.[6]

Box 3
Required Elements of the OSHA Bloodborne Pathogen Standard

I. Development and Distribution of an Institution-Specific Exposure Control Plan
II. Exposure Risk Determination for all Employees
III. Control Methods:
  A. Universal Precautions
  B. Engineering Controls
  C. Work Practice Controls
  D. Personal Protective Equipment
  E. Additional requirements for HIV and HBV Research Laboratories/Production Facilities
  F. Housekeeping Practices
  G. Laundry Practices
  H. Regulated Waste Disposal
  I. Warning Signs
IV. Training
  A. All Affected Employees
  B. Additional Training/Demonstration of Proficiency for Research Labs
V. Employee Health Program
  A. Hepatitis B Vaccination
  B. Postexposure Evaluation and Follow-up
VI. Recordkeeping
  A. Training
  B. Employee Medical Records

Data from Reference 2.

---

the requirement for an Exposure Control Plan for labs covered under the Bloodborne Pathogen Standard. The need to post warning signs in a BSL-2 workplace corresponds to the OSHA requirement for biohazard warning signs to mark contaminated equipment or areas where blood or other PIMs may be stored.

Universal Precautions/BSL-2 Work Practices

Standard work practices are emphasized in both the OSHA standard and BSL-2 and are consistent in the control of hazards from environmental or unapparent exposures. Examples include handwashing, routine housekeeping, spill decontamination, prohibition of eating, smoking, and so forth, in the lab, and no mouth pipetting. The OSHA standard also emphasizes the safe handling of sharps, with specific requirements for a "no recapping" work practice policy for needle use. The safety manual recommended for BSL-2 labs is consistent with

Engineering Controls

The engineering controls defined by OSHA are equivalent to the safety equipment of BSL-2. In fact, OSHA advocates the use of available technology and devices to isolate the hazard from the worker. Examples of appropriate safety equipment when working with human blood or PIM include self-sheathing needles for prevention of needlesticks, conveniently located, puncture-resistant containers for sharps disposal, BSCs or splash shields to protect the face, or sealed safety cups or rotors in centrifuges.

Personal Protective Equipment (PPE)

Another strategy to minimize worker exposure to blood or other PIM is the appropriate use of PPEs. The combination of OSHA's engineering controls and PPE represent the primary barriers of BSL-2. OSHA requires the use of gloves when contact with infectious material is anticipated—that is, when performing phlebotomy, handling clinical specimens, infected animals, or soiled equipment, and cleaning spills. Face protection is required when splashes to the face are expected and equipment such as a BSC is unavailable. Likewise, gowns, aprons, or lab coats are needed for expected splatter to protect street clothes or uniforms.

Other OSHA Requirements

In addition to the requirements to implement the above safety precautions, OSHA requires that institutions make all necessary safety equipment available to affected employees in order to comply with these requirements. Institutions must develop an Exposure Control Plan, make it accessible to all employees, and review it at least annually for applicability and effectiveness. Other issues covered under the Bloodborne Pathogen Standard include specific requirements for HIV or HBV research laboratories, training, and employee health concerns.

HIV and HBV Research Laboratories and Production Facilities

The recommendations from the CDC for work with HIV in research laboratories[4] have been adopted by OSHA in the Bloodborne Standard. Work with HIV or HBV in research quantities may be done in BSL-2 facilities. Because of the higher titers of viruses in research labs, BSL-3 special work practices and containment equipment must be used. Work with HIV or HBV in production facilities must be done under BSL-3 facility design criteria and using BSL-3 work practices and equipment. Employees in both research labs and production facilities must demonstrate proficiency in working with human pathogens before they may work with HIV or HBV.

Training

Institutions must formally identify those employees with potential occupational exposures to blood and other PIM, and provide required safety-training information to them before they begin their work, annually, and any time their work responsibilities may change. The elements of this training are clearly defined in the standard,[2] and must be provided by a trainer knowledgeable in the subject.

Employee Health Concerns

One of the best strategies for prevention of occupational infection is an applicable vaccination program. The OSHA Bloodborne Pathogen Standard mandates that institutions provide the hepatitis B vaccine free of charge to all employees covered under the standard. To emphasize the importance of the vaccine, OSHA requires that employees who refuse the vaccine sign a declination form indicating they have been informed of the effectiveness of the vaccine and they may receive the vaccine at any time in the future if they change their minds.

Adherence to safety precautions will reduce, but not eliminate, occupational blood or body fluid exposures. Therefore, a postexposure evaluation program is a necessary component of a biological safety program. The OSHA Bloodborne Pathogen Standard does not require routine, serologic surveillance, but does mandate medical evaluation, follow-up, and documentation of any "exposure incident." (An exposure incident is a specific workplace incident involving parenteral exposure, splash to the mucous membranes of the face, or nonintact skin exposure to blood or other PIMs.[2]) Follow-up and treatment of exposed employees must comply with current Public Health Service recommendations for bloodborne pathogen exposure protocols.[7,8,9]

References

1. *Biosafety in Microbiological and Biomedical Laboratories*, 4th ed.; Richmond, J. Y., McKinney, R. W., Eds.; U.S. Department of Health and Human Services, CDC and NIH, U.S. Government Printing Office: Washington, DC, 1999. (also found: http://www.cdc.gov/od/ohs/biosfty/bmbl4/bmbl4toc.htm)
2. OSHA Occupational Exposure to Bloodborne Pathogens Standard; 29 *CFR* 1910.1030, 1991. (also found: http://www.osha-slc.gov/OshStd_data/1910_1030.html)
3. *Primary Containment for Biohazards: Selection, Installation, and Use of Biological Safety Cabinets*; U.S. Department of Health and Human Services, CDC and NIH: Washington, DC, 1995. (also found: http://www.cdc.gov/od/ohs/biosfty/bsc/bsc.htm)
4. Centers for Disease Control. Agent Summary Statement for Human Immunodeficiency Virus and Report on Laboratory-Acquired Infection with Human Immunodeficiency Virus. *Morbid Mortal Weekly Rep.* **1998**, *37*(suppl. no. S-4), 1–22.
5. Centers for Disease Control. Recommendations for Prevention of HIV Transmission in Health-Care Settings. *Morbid Mortal Weekly Rep.* **1987**, *36* (suppl. no. 2-S), 3S–18S.
6. Hunt, D. L. HIV-1 and Other Blood-Borne Pathogens. In *Biological Safety: Principles and Practices*, 3rd ed.; Fleming, D. O., Hunt,

D. L., Eds.; American Society for Microbiology: Washington, DC, 2000.

7. Centers for Disease Control and Prevention. Public Health Service Guidelines for the Management of Health-Care Worker Exposure to HIV and Recommendations for Postexposure Prophylaxis. *Morbid Mortal Weekly Rep.* **1998**, 47(No RR-7), 1–34.

8. Centers for Disease Control and Prevention. Recommendations for Follow-up of Health-Care Workers after Occupational Exposure to Hepatitis C Virus. *Morbid Mortal Weekly Rep.* **1997**, 46, 603–606.

9. Centers for Disease Control. Immunization of Health-Care Workers. *Morbid Mortal Weekly Rep.* **1997**, 46(RR-18), 1–42.

68

Research Animal Biosafety

SCOTT E. MERKLE

The research use of animals poses a variety of specialized health and safety risks to research and animal-care staff. Occupational health and safety programs must address the risks posed by the species of animal (e.g., zoonotic diseases and animal allergens) as well as those intentionally introduced into the experimental project (e.g., test chemicals, infectious agents, anesthetic agents, disinfectants, and pest control chemicals). Effective control of these risks requires a multidisciplinary team approach, involving health and safety staff, facility veterinarians, research investigators and technicians, engineering, maintenance and housekeeping staff, and animal care personnel. This chapter provides a basic overview of animal biosafety guidelines and standards, the principal types of hazards that may be encountered in the research use of animals, and the commonly employed methods to control these hazards.

Guidelines and Standards

As workplaces, animal housing and research facilities must be in compliance with all applicable standards established by OSHA and state and local jurisdictions. In addition, there are federal guidelines and standards that are especially applicable to the use of animals in research (Table 68.1). Even though some of the guidance is focused on the care and well-being of the laboratory animal, aspects of occupational health and safety are often addressed.

Certain animal-research procedures, especially those involving small animals and hazardous biological agents, often will be conducted within Class II biosafety cabinets. In such cases, the biosafety cabinets should be constructed, tested, and certified according to the NSF Standard 49, Class II (Laminar Flow) Biohazard Cabinetry.[5]

The Association for the Assessment and Accreditation of Laboratory Animal Care, International (AAALAC) is a voluntary, nonregulatory organization that provides independent evaluations of animal care and use programs. Accreditation by AAALAC involves initial and periodic on-site reviews of the animal care and use program, including management practices and facilities. The institution's occupational health and safety program is one of the key elements in the accreditation review process.

Hazard Recognition

Consideration for employee health and safety must be an integral part of the research project's design, and each project must be individually evaluated for possible physical and exposure hazards. The physical hazards associated with animal research will vary depending on the type of animal involved—small animals (mice, rats, rabbits, birds, etc.), nonhuman primates, and large animals (e.g., livestock and farm animals). Each species will pose a different potential risk of causing physical injury based on their size and strength. Likewise, there are differences among species in their potential to carry infectious agents that can infect and cause disease in humans (i.e., zoonoses). Examples of some hazards associated with animal research include:

- Physical hazards
  Animal bites and scratches
  Puncture wounds from needles and other sharps
  Burns during cage washing or autoclaving

TABLE 68.1 Pertinent Federal Guidelines and Standards

Guidelines	Description
Biosafety in Microbiological and Biomedical Laboratories[1]	Jointly published by CDC and NIH; defines recommended biosafety practices and facilities for experiments on animals potentially infected with agents which may cause human infection
Guide for the Care and Use of Laboratory Animals[2]	Published by the National Research Council; provides recommendations for institutional policies and practices for the humane care and use of animals in research
Occupational Health and Safety in the Care and Use of Research Animals[3]	Published by the National Research Council; contains guidelines for ensuring effective occupational health and safety in animal research involving hazardous agents

Standards	Description
Guidelines for Research Involving Recombinant DNA Molecules (NIH Guidelines)[4]	Specifies biosafety practices for experiments constructing or using recombinant DNA, including their introduction into whole animals. Adherence to the NIH guidelines is mandatory for all NIH funded projects
Occupational Safety and Health Act	Provides regulatory authority to the Department of Labor for workplace health and safety. Regulations are published in 29 *CFR* 1910. Standards on occupational exposure to bloodborne pathogens (1910.1030), formaldehyde (1910.1048), and hazardous chemicals in laboratories (1910.1450) may be particularly relevant to the animal-research setting
Animal Welfare Act and Amendments	Provides regulatory authority to the Department of Agriculture for the humane care and treatment of animals, including those used in research. Regulations are published in 9 *CFR* 93
Foreign quarantine and importation of etiologic agents	CDC regulations, at 42 *CFR* 71, requiring permits for the importation of any animal capable of being a host or vector of human disease

Musculo-skeletal injuries from moving or transporting cages
- Chemical hazards
  Toxicological test agents (e.g., carcinogens, reproductive toxins)
  Chemical disinfectants and cleaning agents (e.g., sodium hypochlorite)
  Anesthetic agents (e.g., ether, isoflurane)
  Chemical sterilants (e.g., ethylene oxide)
  Formalin in necropsy procedures
- Biological hazards[6]
  Bacterial diseases (e.g., salmonellosis and brucellosis)
  Chlamydial diseases (e.g., psittacosis from birds)
  Fungal diseases (e.g., ringworm from *Microsporum* in dogs and cats)
  Parasites (e.g., mites, fleas from dogs, cats, rabbits, and rodents)
  Protozoa (e.g., toxoplasmosis from cats)
  Rickettsial diseases (e.g., Q fever from sheep)
  Viral diseases (e.g., rabies, hepatitis A, and Cercopithicene herpesvirus 1)

- Radiological hazards
  Radioisotopes
  X-ray equipment
- Allergens
  Animal dander and hair
  Airborne fecal or urinary proteins
  Latex allergy from gloves

The development of a proposed animal-research protocol is a crucial step in hazard recognition and assessment. Typically, the research protocol is reviewed by the institution's Animal Care and Use Committee to ensure that experimental procedures are appropriate and include adequate provisions for the humane treatment of the animals. The appropriate level of containment, personal protective equipment, and safety procedures should be specified during the protocol review and approval process. For projects involving hazardous agents, the proposed research protocol should provide sufficient information to evaluate the potential for exposure at all stages of the project. Some of the issues that should be considered when hazardous chemical agents are involved are presented in Box 1. While chemical exposure is always possible during dose preparation, exposure from dosing will depend on the method of dose administration and potential release of aerosols or vapors (Table 68.2).

Risk Reduction Approaches

The basic principals of biological containment apply to the animal facility setting. Primary barriers, in the form of safety equipment (e.g., microisolation cages) and personal protective devices, are employed to prevent the accidental dispersion of biohazards within the immediate animal laboratory environment. Secondary barriers, typically incorporated into the facility design, are provided to prevent the spread of infectious agents outside the laboratory. Examples of secondary barriers include decontamination procedures and equipment (e.g., autoclaves), handwashing stations, restricted or controlled access to animal areas, use of ventilation systems or air locks to prevent air flow from biohazardous areas, and separate buildings or isolation rooms within the animal facility.

The CDC and NIH[1] have prescribed four levels of increasingly protective practices, equipment, and facilities for research using animals that may be infected with agents capable of causing disease in humans. These levels, designated as Animal Biosafety Levels 1–4, are based on the assigned Biosafety Level of the infectious agent. The Animal Biosafety Levels, summarized in Table 68.3, are cumulative in that additional precautions are prescribed for each successive level. Conducting the animal research project at the appropriate biosafety level should provide an adequate degree of employee protection from potential exposure to pathogenic agents under ordinary conditions.

Engineering Controls

These controls are the preferred approach to reducing potential exposures to biological, chemical, and radioactive agents. Engineering controls include ventilated enclosures (e.g., Class IIA, IIB, or III biological safety cabinets[7] or laboratory hoods) and specialized ventilation systems providing controlled air movement between different areas of the facility to achieve a "clean/dirty" concept. Generally recommended ventilation rates in animal holding and procedure rooms range from 10 to 15 air changes per hour.[2] The minimum ventilation rate to accommodate the heat load generated by the animals can also be calculated.[8] However, the calculated ventilation rate may need to be adjusted to control for other factors besides heat buildup, such as toxic gases or vapors, particulates, odors, and animal allergens. Air exhausted from animal research areas and rooms where toxic chemicals or pathogenic agents are present should not be recirculated.

Personal Protective Clothing

Clothing requirements for entering and working in animal facilities are usually specified for the protection of the animals (e.g., jumpsuit, head and shoe covers, surgical mask, and latex gloves). Where hazardous chemical or biological agents may be present, clothing must be carefully selected for worker protection, and the requirements for their use must be specified in the animal research protocol. For example, disposable full-body lab garments, chemical-resistant gloves, face shields or safety glasses/goggles, and NIOSH-approved respiratory protection may be needed when animal-handling procedures also involve hazardous agents. In addition, animal-care technicians and support personnel may need safety shoes (with attention to nonslip treads for working in wet areas) and hearing protection from noise produced by cage washers and other equipment.

Box 1
Health and Safety Considerations in Animal-Research Protocols Involving Hazardous Agents

- Route and method of dose administration
- Type and duration of containment needed during and post dose administration
- Pharmacokinetics of hazardous agent (i.e., urinary or fecal excretion of unchanged agent and hazardous metabolites)
- The potential for exposure during tissue harvesting and necropsy procedures
- Procedures for cage changing and cleaning when excretion products and bedding are considered hazardous
- Type of cage to be used (e.g., disposable cages for mice or rats)
- Necessary safety and health precautions for the specific hazards of the agent (e.g., posting of warning signs and labels, types of personal protective clothing and equipment, and the need for worker or workplace monitoring)
- Regulatory status of waste products and animal carcasses and appropriate methods of disposal

TABLE 68.2 Methods of Dose Administration and Potential for Exposure

Method of Administration	Potential for Exposure	Nature of Exposure[a,b]
Inhalation	High	Aerosols and vapors Spillage during preparation of dose concentrations
Feeding	High	Aerosols Cage and animal contamination Cross contamination
Water	High	Aerosols Cage and animal contamination Spillage Generally less hazardous than feeding
Skin painting	Moderate	Cage and animal contamination Spillage during painting
Gavage	Moderate[a]	Regurgitation Passage of unaltered dose Spillage during dosing
Injection	Moderate to Low[a]	Spillage Aerosol formation during dosing Accidental injection

[a]Assumes that dosing compound has minimal vapor pressure under normal animal room conditions. Materials with elevated vapor pressure represent an additional vapor hazard requiring ventilated primary containment.
[b]Risk will be increased if toxic metabolites or excretion products are associated with the dosing compound.

Training

Education programs are critical to ensure that employees learn and maintain the special skills needed for safe and humane animal care. Training in procedures and techniques in animal handling and restraint can minimize the risk of animal bites and scratches. Information must be provided to employees (in compliance with OSHA's Hazard Communication Standard, 29 *CFR* 1910.1200) so they can recognize potential physical, chemical, and biological hazards in their work environments. Employees should periodically review SOPs and the safety provisions of research protocols (e.g., emergency actions to take for wound care from animal bites or needlesticks).

Signage

Appropriate hazard information and warnings should be posted on each animal holding and procedure room. A responsible person to contact in the event of emergencies should also be listed. Hazard information on entrance signs and placards and posted safety protocols must be kept current because activities in animal procedure and holding rooms can frequently change. Signage often becomes the most immediate means of communication and coordination between the research investigators and animal-care staff.

Medical Surveillance

The employee's health status should be evaluated upon initial assignment and at periodic intervals. These evaluations serve several purposes such as assessing the employee's fitness to wear respiratory protection, ensuring that necessary immunizations are obtained (e.g., tetanus and hepatitis B), and monitoring for early symptoms of disease (e.g., lab animal allergies) when intervention is possible. Surveillance for diseases that can be transmitted from the worker to the research animal should also be included in the medical evaluation (e.g., periodic testing for tuberculosis in those working with primates). Specific components of the medical examinations should be based on potential exposures and the individual's occupational history.

This chapter has outlined the basic considerations necessary to ensure an adequate level of employee health and safety in animal research. Other specialized precautions may be needed due to the extremely varied nature of animal-research activities and the settings in which they take place. The appropriate level of containment and preventive measures must be determined for the specific circumstances of each research project. The importance of careful preplanning and coordination between the research investigator, facility veterinarian, and health and safety professional cannot be overemphasized. Each

TABLE 68.3 Summary of Animal Biosafety Levels

Animal Biosafety Level	Agents	Practices	Equipment	Facilities
1	Not known to cause disease in healthy human adults	Standard animal-care practices	As required for normal care of each species	Standard animal facility —no recirculation of exhaust air —directional air flow recommended
2	Associated with human disease *Hazard:* percutaneous exposure, ingestion, mucous membrane exposure	Level 1 practices plus: —limited access —biohazard warning signs —sharps precautions —biosafety manual —decontamination of all infectious wastes and animal cages prior to washing	Level 1 equipment plus: —primary barriers: containment equipment appropriate for animal species —PPEs: lab coats, gloves, face, and respiratory protection as needed	Level 1 facility plus: —autoclave available —handwashing sink available in animal room
3	Indigenous or exotic agents with potential for aerosol transmission; disease may have serious health effects	Level 2 practices plus: —controlled access —decontamination of clothing before laundering —cages decontaminated before bedding removed —disinfectant foot bath as needed	Level 2 equipment plus: —containment equipment for housing animals and cage dumping activities —Class I or II BSCs available for manipulative procedures (inoculation, necropsy) that may create infectious aerosols —PPEs: appropriate respiratory protection	Level 2 facility plus: —physical separation from access corridors —self-closing, double-door access —sealed penetrations —sealed windows —autoclave available in facility
4	Dangerous or exotic agents that pose high risk of life-threatening disease; aerosol transmission, or related agents with unknown risk of transmission	Level 3 practices plus: —entrance through change room where personal clothing is removed and lab clothing is put on —shower on exiting —all wastes are decontaminated before removal from facility	Level 3 equipment plus: —maximum containment equipment (Class III BSC or partial containment equipment in combination with full body air-supplied positive-pressure personnel suit) used for all procedures and activities	Level 3 facility plus: —separate building or isolated zone —dedicated supply and exhaust, vacuum and decontamination systems —other requirements as outlined in Reference 1

Reproduced from Reference 1.

plays a vital role in achieving the goals of the research effort—quality research, humane animal care, and protection from workplace hazards.

References

1. CDC/NIH. *Biosafety in Microbiological and Biomedical Laboratories*, 4th ed.; Richmond, J. Y., McKinney, R. W., Eds.; DHHS Publication No. (CDC) 93-8395, 1999.
2. *Guide for the Care and Use of Laboratory Animals*; National Research Council, Commission on Life Sciences, Institute of Laboratory Animal Resources, National Academy Press: Washington, DC, 1996.
3. *Occupational Health and Safety in the Care and Use of Research Animals*; National Research Council, Commission on Life Sciences, Institute of Laboratory Animal Resources, National Academy Press: Washington, DC, 1997.
4. NIH. Guidelines for Research Involving Recombinant DNA Molecules, *Fed. Regist.* **1994**, 59, 34496–34547.
5. NSF. Standard 49, Class II (Laminar Flow) Biohazard Cabinetry; Ann Arbor, MI: 1992.
6. Fox, J. G.; Lipman, N. S. Infections Transmitted by Large and

Small Laboratory Animals. *Infect. Dis. Clin. North Am.* **1991**, 5(1), 131–163.

7. CDC/NIH. *Primary Containment for Biohazards; Selection, Installation and Use of Biological Safety Cabinets*; Richmond, J. Y., McKinney, R. W., Eds.; DHHS, 1995.

8. ASHRAE. *Handbook: Fundamentals*, Chapter 9, Environmental Control for Animals and Plants; American Society of Heating, Refrigerating, and Air Conditioning Engineers: Atlanta, GA, 1993.

69

Biological Safety: Emergency Response and Decontamination Procedures

FREDERICK L. VAN SWEARINGEN

ANTONY R. SHOAF

Management and staff of a facility establish procedures to ensure that operations are safe for employees, students, and visitors. These include daily operations, such as proper decontamination of laboratory areas and equipment. However, procedures must also be developed to ensure that personnel can handle foreseeable emergencies.

There are two goals of this chapter. The first goal is to consider what emergencies could happen to a facility that uses biological materials and how to start planning for them. The second goal is to briefly detail some precautions and procedures to take for the most common occurrences: decontamination following a spill and some routine decontamination needs.

Emergency Response Considerations

Any facility may eventually have to deal with an emergency situation. Although steps can be taken to prevent problems such as accidents and spills, other emergencies may be unavoidable. However, anyone can plan to deal with an emergency. Know what goes on at the facility. Evaluate possible emergency scenarios. Then, establish appropriate plans of action. Plans should not only cover the facility as a whole, but also unusual laboratory situations.[1]

Spills

Historical information suggests that human error is the most common cause of spills and personal exposure to infectious materials in the laboratory.[2] Fortunately, most spills are small. This might be the reason why many in-fected individuals followed in case studies do not know when or where they acquired their infections. Equipment and facility failures occur, but not as often as spills. Thus, spills are the most likely biological emergency that laboratory personnel will face.

In any emergency, the first concerns are to prevent unnecessary exposure and help those who are hurt or contaminated. A biohazard spill should only be handled by properly trained personnel. Decontaminate the spill when it is safe to do so after putting on the appropriate protective clothing. Information on proper spill response is given later in this chapter. Although housekeeping personnel may not know how to safely handle a biohazard spill, they can clean the area after the infectious hazards have been eliminated.

A "Biohazard Spill Kit" should be available in areas where spills are likely to occur. This kit should contain materials and equipment needed to handle spills likely to happen in that area. Specific written plans should be developed for spill cleanup. Personnel should be trained in the methods to safely use the biohazard spill kit.

Other Emergency Situations to Consider

Emergency planning may also be needed for the following situations:

• *Fires*—Many facilities establish general procedures on how to deal with a fire. These plans should include ways to limit the spread of potentially infectious materials. Although heat will inactivate any infectious agents that are consumed in the fire, fire also creates drafts. Fires may create aerosols of infectious agents as well as toxic fumes or smoke.[1] These aerosols could result in personnel exposure. However, the self-contained breathing apparatus worn by fire fighters will protect them from aerosols generated by the fire.

- *Weather*—Adverse weather can also create emergencies for a lab using biological materials. Frozen water pipes burst, causing water damage or flooding problems. Destructive weather, such as tornadoes and hurricanes, causes physical destruction to containment systems and power loss. Storms cause power loss, resulting in failure of containment systems, loss of ventilation to rooms or hoods (if not on emergency power), and loss of lighting or refrigeration. All hazardous laboratory work should cease until the safety systems are again operational. A serious situation might require evacuation of the premises.

- *Natural disasters*—Some areas may have to deal with natural disasters, such as earthquakes or floods. Generally, these affect a larger geographical area than just the facility, so outside assistance may be limited at first. Disasters can cause some of the same problems as weather (loss of power, flooding, destruction of containment systems).

- *Other emergencies*—All facilities might have to plan for theft, terrorist acts, or bomb threats. Many college and university campuses are relatively open, so security of laboratory contents, including biohazardous materials, is a concern. The authors were involved with a sewage leak. Sewage contains many potentially infectious agents that must be considered. In this case, the contaminated office held records that needed to be decontaminated and retained. Several treatment options, including radiation, were investigated for some items.

General Goals

The first goal in any emergency response is to prevent or limit personnel exposure. A second consideration is to minimize damage to the facility. Other factors to consider are the cost of each response option and whether that option presents additional concerns, such as availability of disposal sites for wastes generated in the cleanup process.

These general guidelines should be considered when planning any emergency response:

- Warn individuals in adjacent areas of potential hazards to their safety.
- Restrict unnecessary movement into or through the area.
- If possible, limit the spread of any contamination or spill.
- Summon professional help. In a fire, call the fire department or the facility's fire brigade. In a medical emergency, summon medical help. As needed,

notify appropriate facility personnel (security, biosafety officer, infection control specialist, chemical safety officer, radiation safety officer, risk manager, management supervisor).

- Render assistance and first aid to those involved. Remove individuals from an area if they could be further injured by staying. However, injured people not in danger of further harm should only be moved by qualified medical personnel or trained emergency team members. Remove any contaminated clothing. Wash contaminated skin with mild soap and water.

Decontamination

Decontamination is a process or treatment that reduces or eliminates microbial contamination to an acceptable level to minimize the probability of infection. Decontamination can range from sterilization of an item to simple cleaning with soap and water. Sterilants and disinfectants are used for decontamination of potentially infectious materials.

Types of Decontaminants

Choice of proper decontaminant depends on many factors: (1) nature of the item to be decontaminated, (2) number of microorganisms present, (3) resistance of microorganisms to decontamination method, (4) amount of organic soil present, and (5) characteristics of the decontamination method (concentration, contact time, temperature, etc.). Decontaminants can be categorized both by their pathogen-killing ability and by their physical properties.

Several federal agencies (EPA, FDA, CDC) have developed guidelines on effectiveness of decontamination agents. EPA registers chemical agents submitted to it. Some common chemical decontaminants are given in Table 69.1. A description of a common classification system follows.

Sterilant

A sterilant is an agent intended to destroy all microorganisms (viruses, bacteria, pathogenic fungi) and their spores on inanimate surfaces. Sterilants can be both physical and chemical agents. A common use of a physical agent is the steam autoclave, which uses moist heat under pressure. Many steam autoclaves operate at 121°C, usually for at least 60 minutes. Dry heat alone (171°C for 1 hour, 121°C for 16 hours) is adequate for sterilization. Chemicals are also used as sterilants. Ethylene ox-

TABLE 69.1 Some Common Chemical Decontaminants

Chemical Compound	Typical Use Conditions	Classification	Used to Decontaminate
Alcohols (ethyl or isopropyl)	70–85%, 10–30 min 70%	Intermediate Antiseptic	Glassware, surfaces Skin
Chlorine compounds (bleach)	0.05–5%, 10–30 min (500–50,000 mg free Cl/L)	Intermediate	Glassware, liquid discard, surfaces
Ethylene oxide (gas)	450–500 mg/L, 55–60°C, 105–240 min	Sterilant	Equipment, glassware, instruments
Formaldehyde (liquid—Formalin)	6–8%, many hours 1–8%, 10–30 min	Sterilant Low to high	Equipment Glassware, instruments, liquid discard
Glutaraldehyde	Variable 2%, 10–600 min	Sterilant Intermediate to high	Instruments Glassware, instruments, surfaces
Hexachlorophene	1–3%	Antiseptic	Skin
Hydrogen peroxide	6–30% 3–6%	Sterilant Intermediate to high	Equipment Surfaces
Iodophor compounds	0.5–2% (30–50 mg/L free iodine) 1–2% avail. (1–2 mg/L free iodine)	Low to high Antiseptic	Instruments, surfaces Skin
Paraformaldehyde (heated)	0.3 g/ft^3, > 23°C, 60–180 min	Sterilant	Equipment
Phenolic compounds	0.2–3%, 10–30 min	Low to intermediate	Glassware, instruments
Quaternary ammonium compounds	0.1–2%, 10–30 min	Low	Glassware, surfaces

Note. Some disinfectants (e.g., ethylene oxide) leave harmful residues. Precautions should be taken to ensure complete removal when used in reusable lab equipment used for research with viable organisms or tissue culture work.
Adapted from References 1 and 3.

ide gas (450–500 mg/L at 55–60°C) is used in ethylene oxide sterilizers. Some liquid chemicals, such as a 6–8% solution of formaldehyde, can also be effective sterilants.

Disinfectant

A disinfectant is an agent intended to destroy or irreversibly inactivate specific pathogenic microorganisms on inanimate surfaces. However, many disinfectants are not effective sterilants. They will not kill bacterial spores at the lower concentration levels and shorter contact times commonly used. Disinfectants are further categorized as high level, intermediate level, or low level. This is based on the fact that microorganisms can be grouped according to their resistance levels to a range of physical and chemical agents. A high-level disinfectant will kill some microorganisms that are resistant to lower-level disinfectants. Figure 69.1 shows the descending order of resistance by microorganisms to different disinfection levels.

A high-level disinfectant may be the same agent used as a sterilant except under different use conditions. For example, hot water pasteurization (using moist heat at 75–100°C) is a high-level disinfection method. Lower concentrations of formaldehyde are high-level disinfectants, rather than sterilants. Some high-level disinfectants are sterilants if a long contact time (e.g., 10 hours) is used.

Intermediate-level disinfectants do not kill most bacterial spores. However, they do kill more resistant forms of bacteria, such as the tubercle bacillus. They correspond to the EPA designations of "hospital disinfectant" and "tuberculocidal." Commonly used dilutions of household bleach are intermediate-level disinfectants. Some alcohols (70% ethyl or isopropyl) can be intermediate-level disinfectants if they do not evaporate (resulting in short contact time).

Low-level disinfectants are generally effective against a range of viruses, fungi, and vegetative bacteria. Within a practical contact time, though, they may not kill bacterial spores, mycobacteria, all fungi, or all small or nonlipid viruses. They may be registered by the EPA as "hospital disinfectants." However, they are not effective tuberculocides. Common low-level disinfectants are quaternary ammonium compounds (sometimes called "quats").

LEVELS OF DECONTAMINATION

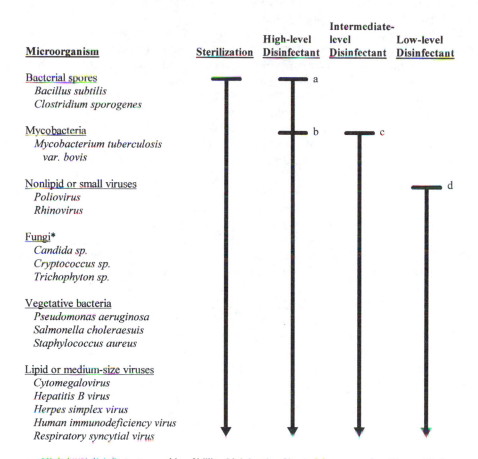

a: High-level disinfectants capable of killing high levels of bacterial spores only with extended contact times

b: High-level disinfectant action at routine contact times, although with some spore killing

c: Hypochlorites exhibit some spore killing, whereas other intermediate-level disinfectants failed to kill some nonlipid and small viruses. Phenolics failed to kill some rhinoviruses and enteroviruses, and ethanol failed to kill a test virus in this category.

d: Low-level disinfectants do not kill all nonlipid or small viruses at routine contact times, nor do they kill all fungi.

* Includes asexual fungi spores, but not necessarily chlamydospores or sexual spores

FIGURE 69.1 Descending order of resistance of microorganisms to different decontamination levels. (Data are from References 3 and 8.)

Antiseptic

An antiseptic is a chemical that kills pathogenic microorganisms and that is formulated to be used on the skin or tissue. Antiseptics are not designed to be used as disinfectants, although most antiseptic formulations have active ingredients that would be classified as low-level disinfectants. Alcohols (ethyl, isopropyl) and hexachlorophene are examples of antiseptics.

Use of Bleach as a Common Disinfectant

The most commonly used disinfectant is household bleach (sodium hypochlorite). It has many advantages: it is inexpensive, readily available, rapid acting, and a broad spectrum of microorganisms are killed. But, it also has disadvantages: it is extremely corrosive at higher concentrations (e.g., aluminum), irritating at higher concentrations to personnel, and inappropriate for some materials (will take color out of carpets and fabrics). Dilution of fresh commercial household bleach (a 5.25% solution of sodium hypochlorite contains 50,000 mg/L of free available chlorine) provides chlorine concentrations appropriate for uses requiring an intermediate level of decontamination.[3] For large spills of cultured or concentrated infectious agents, a 1 : 5 dilution ratio (1 part bleach in 4 parts water) and contact time of 20–30 minutes at room temperature is recommended. For spills involving lower concentrations or when used on porous surfaces, a 1 : 10 dilution (1 part bleach in 9 parts water) is recommended. For hard, smooth surfaces that have been cleaned, a 1 : 100 dilution may be sufficient. It was reported that a 1 : 100 dilution inactivated HIV within 2 minutes and HBV within 10 minutes. Two points of caution are necessary. First, these dilutions are for fresh bleach; "old" bleach has a lower concentration as it loses strength. Second, putting bleach (or any disinfectant) on a liquid spill will lower the concentration of the bleach in the spill.

Bleach, like other liquid disinfectants, is routinely used to decontaminate a wide range of items, such as work surfaces, discarded liquids, instruments and glassware, and contaminated waste generated during spills. Other decontamination methods are not as versatile. For example, autoclaves can decontaminate instruments, glassware, and some wastes, but not work surfaces.

Cleaning of Spills

Most laboratory spills are small and can be easily cleaned. Every worker should know how to clean spills of the materials with which they work. They should know where the biohazard spill kit is located and how to use it. The following procedures can guide in safely cleaning spills.

Liquid Spills

Proper response to a liquid biohazard spill depends on several factors: (1) the hazard of the agent, (2) the volume spilled, and (3) whether aerosols or droplets of infectious material were generated. There are two ways to manage a spill.[2] In the first method, the spill is cleaned up, then the area is decontaminated. This reduces the amount of organic matter in the spill, enhancing the effectiveness of the disinfectant. However, personnel must handle materials (possibly broken glassware) that contain active biohazards. This should only be done by a knowledgeable person who uses appropriate protective clothing and safety equipment. In the second method, the spill is decontaminated first, then the area is cleaned.

Personnel are handling less biohazardous materials. This is considered a more conservative approach, and it will be discussed further.

If significant aerosols or droplets are generated, first evacuate the room or area for about 30 minutes. This allows droplets to settle and aerosols to be reduced by the ventilation system. Most labs in recently constructed or renovated buildings are under negative pressure with respect to the hallway, so this inhibits aerosols from moving into the hallway. If respiratory protection is needed, individuals who have received OSHA-required training in respiratory protection should provide the primary response.

If a small volume of a minimally hazardous agent is spilled, such as a test tube or Petri dish containing an organism that requires only Biosafety Level 1 precautions,[4] it can be cleaned easily. In most cases, use a paper towel soaked with an appropriate disinfectant. The spill site should be left wet for the disinfectant's contact time as recommended by the manufacturer. The individual only needs to wear gloves and minimal protective clothing (lab coat or gown, shoe covers).

If a larger spill, especially for an agent requiring Biosafety Level 2 (or higher) precautions, use more care. For aerosols, evacuate the room and close the doors for about 30 minutes. Remove or decontaminate clothing of affected individuals. Before entering the room, put on appropriate protective clothing and respiratory protection. If high concentrations of protein (e.g., blood, serum) are involved, absorb the bulk of the spilled liquid prior to disinfection.[5] While confining the spill, flood the spill site with an appropriate intermediate- or high-level disinfectant (for examples, see Table 69.1). Keep disinfectant on the spill for the manufacturer's recommended contact time. Remove spilled material and disinfectant. Rinse the spill site with water or a detergent solution. Dry a floor site to prevent slipping. Place spill material into biohazard container for decontamination or disposal. Personnel must wash their hands following cleanup of the area. Clean other potentially contaminated areas of the body (hair, face).

If the spill contains broken glass or other sharp objects, remove these using remote handling procedures (tongs, forceps), not the hands. If the skin is punctured or contaminated, start routine first aid. Wash the skin site with soap and water, encouraging bleeding of the wound. If appropriate, bandage the wound site. If contamination gets in the eyes, nose, or mouth, wash with large quantities of water. Seek prompt medical attention as needed.

Figure 69.2 is a flowchart of how these procedures could be incorporated into a comprehensive liquid-spill response plan.

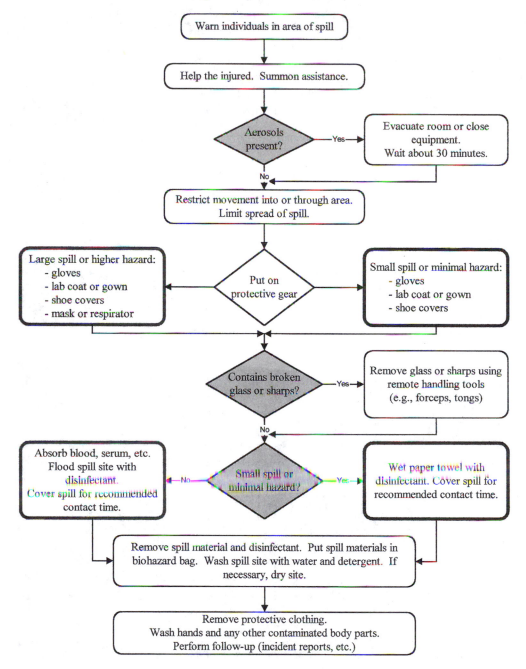

FIGURE 69.2 Flowchart of a comprehensive response to a liquid biohazard spill.

Solid Spills

To reduce aerosols from dry dusts, slightly wet spilled solids. Carefully sweep up spilled solids of low-virulent agents and place them into appropriate disposal containers. Biohazardous solids of higher virulence may require additional precautions, such as using a vacuum equipped with a HEPA filter. Decontaminate the spill site using an appropriate disinfectant.

Spills Involving a Biological Safety Cabinet

Start cleanup of a spill in a biological safety cabinet (laminar flow hood) while the cabinet is still operating. Remove contaminated absorbent paper and place it into a biohazard bag. Work in a way that limits further aerosol generation in the cabinet or escape of contaminants from the cabinet. Using a towel soaked in disinfectant, wipe off material splattered onto the cabinet's interior or on items within the cabinet. Choose proper disinfectant carefully to limit problems. Flammable organic solvents (e.g., alcohol) may reach dangerous concentrations in a biological safety cabinet if too much is used. (However, alcohol wipes are safely used for routine cleanup in biological safety cabinets.) Finally, remove contaminated gloves and wash hands after the spill has been cleaned and before placing clean absorbent paper into the cabinet.

Large spills (where liquid flows through the front or rear grille) require more extensive decontamination.[6] Decontaminate surfaces of items within the cabinet, then remove those items. Close the drain valve to the cabinet. Pour a suitable liquid disinfectant onto the work surface and through the grilles to the drain pan. Leave disinfectant on the spill for the manufacturer's recommended contact time (20–30 minutes is usually sufficient). Absorb liquid on the work surface with paper towels, and discard them into a biohazard bag. Attach a flexible tube to the drain valve. Carefully empty the drain pan into a collection vessel that contains more disinfectant. The collection vessel should be large enough to contain all of the spilled liquid, disinfectant, and wash water. To minimize aerosols, submerge the unattached end of the tube into the disinfectant in the collection vessel. Finally, flush the drain pan with water, then remove the drain tube.

Spills Involving Equipment

A centrifuge can become contaminated when a tube breaks within it. If an infectious agent was in the tube, leave the centrifuge closed for at least 30 minutes to allow droplets to settle. Open the lid and carefully remove any broken glass using a hemostat or other remote handling device. Then, decontaminate the centrifuge as if it were a liquid spill.[7]

Other equipment (refrigerators, freezers, incubators, shaker baths) are more prone to becoming contaminated, but they might also contain aerosols. Use the same procedures used to decontaminate liquid spills. If aerosols are formed, leave area until aerosols are reduced. Put on protective equipment, clean spill, then disinfect the area. Personnel should then wash their hands following these procedures.

Spills Containing Multiple Hazards

Spills containing multiple hazards have one or more hazards (chemical, infectious, radioactive) in addition to the biohazard. Personnel should respond to the primary hazard first, then follow through for the lesser hazards. The low levels of radioactivity used in many labs will usually be of lesser concern until after the primary infectious agent has been inactivated. These spills are normally handled initially similar to handling a biohazard spill. After the biological hazard is removed, the remaining waste is treated as radioactive waste. The area is also checked for radioactive contamination.

If a spill contains multiple hazards, notify the personnel who also have direct responsibility for responding to those hazards. The laboratory supervisor can provide specific information about the materials involved in the spill. A chemical safety officer or industrial hygienist can help on chemical incompatibilities and hazardous air concentrations. An infection control or biosafety officer can offer information about appropriate disinfectants to use for specific infectious agents. Immediately notify the radiation safety officer if radioactive material is involved in a spill.

Because multiple hazards may be present, care is needed in the choice of decontaminant. The disinfectant must be compatible with other chemicals that are involved. Do not use combinations that are incompatible or that release toxic fumes. This is also true for radioisotopes. Hypochlorite solutions are commonly used for decontamination of infectious hazards, but the chlorine volatilizes radioactive iodine. Radioiodine spills are usually treated with an alkaline sodium thiosulfate solution (a 10% solution is often used) to stabilize the iodine prior to decontamination. Do not use chlorine disinfectants when sodium thiosulfate has been used to stabilize the iodine. Other potentially incompatible combinations exist, such as using chlorine disinfectants with ammonium-containing materials.

Cleaning of Surfaces and Equipment

To reduce personnel exposure to biohazardous agents, routine decontamination of contaminated work surfaces and equipment is necessary. Following are general guidelines regarding items that need decontamination:

- work surfaces daily and after any spill of viable material[4]
- initial decontamination of some wastes when stored or transported
- HEPA filters in biological safety cabinets prior to replacement
- contaminated equipment prior to removal for discard or repair
- contaminated work areas before any maintenance work or inspections are done

Surfaces

Trained laboratory personnel should clean surfaces requiring effective decontamination. These surfaces include countertops, shelves, refrigerators, and other lab equipment where biohazards were used. Housekeepers are generally not trained to decontaminate biohazardous surfaces. Their cleaning duties usually require only low-level disinfectants. Trained individuals should also decontaminate surfaces in specialized facilities, such as cold rooms and sterile rooms.

Although carpeting is not normally found in laboratories, it is found in hospital areas through which blood samples and specimens are transported. Hypochlorite solutions may bleach the color out of carpets, so another disinfectant should be considered.

Wastes

Potentially infectious wastes generated routinely or from accidents should be decontaminated prior to disposal. Due to state regulations or waste facility requirements, some infectious wastes may need to be decontaminated at the generating facility prior to shipment to the disposal site. Decontamination protects those who handle the wastes (laboratory workers, housekeeping staff, waste haulers) from the risk of infection. However, other wastes may remain infectious until they are sent for treatment. The outer containers must be free of significant contamination.

Biological Safety Cabinets

At the end of each workday, decontaminate accessible surfaces of the biological safety cabinet. Wipe interior surfaces (work surface, interior of front sash glass, sides, and back) with a suitable disinfectant. Decontaminate containers and equipment used within the cabinet when work is completed.

A trained individual must decontaminate the biological safety cabinet prior to filter removal or internal repair of contaminated portions of the unit. A gas decontamination agent is used. The most common method is to heat paraformaldehyde flakes in the presence of high humidity, subliming off formaldehyde. Entire labs and ductwork have been decontaminated by this method. However, formaldehyde is considered a potential occupational carcinogen (thus the OSHA formaldehyde standard). Hydrogen peroxide vapor has also been used successfully,[6] but is still somewhat experimental.

Other Equipment

Disinfect contaminated surfaces of equipment to be removed from the lab for repair or discard. This includes large items (e.g., freezers used to store biohazards). Some disinfectants can be corrosive to sensitive parts. Choose a disinfectant that has low corrosivity but is effective against the biohazards used in the equipment. For example, formaldehyde and glutaraldehyde are used as high-level disinfectants for certain surgical instruments.[8]

Notes About Personal Protective Equipment (PPE)

Decontamination agents are used to kill microorganisms and, in high enough concentrations, they can also harm people. For example, the aldehydes (formaldehyde, glutaraldehyde) can be irritating and potentially hazardous (toxic, mutagenic, or carcinogenic).[3,5] They should be used only with the proper protective gear. A manufacturer will give some information about PPE on its MSDS. Additional information can be found in this and other books, in some safety supply catalogs (chemical resistance and corrosion resistance guides), and on some Web sites. It should be noted that the PPE recommendations found in many MSDSs are conservative, as they are generally for larger exposure situations than will usually occur in the laboratory.

Also, check the OSHA requirements, as they can influence the choice of decontaminant used. If respiratory protection must be worn, individuals must be properly trained in its use and fit tested. OSHA requires long-term medical surveillance when using some chemicals. In either case, another decontaminating agent might be considered.

References

1. *Biosafety in the Laboratory: Prudent Practices for the Handling and Disposal of Infectious Materials*; Committee on Hazardous Biological Substances in the Laboratory; National Academy Press: Washington, DC, 1989.

2. *Managing Hazardous Materials Emergencies in Biomedical Research Facilities*; Proceedings of the 12th NIH Research Safety Symposium; NIH Publication No. 91-3200; National Institutes of Health: Washington, DC, March 1990.

3. Favero, M. S.; Bond, W. W. Chapter 24: Sterilization, Disinfection, and Antisepsis in the Hospital. In *Manual of Clinical Microbiology*; American Society for Microbiology: Washington, DC, 1991; pp. 183–200.

4. DHHS/CDC/NIH. *Biosafety in Microbiological and Biomedical Laboratories*, Richmond, J. Y.; McKinney, R. W., Eds.; DHHS (CDC) Pub. No. 93-8395; U.S. Department of Health and Human Services: Washington, DC, May 1993.

5. National Committee for Clinical Laboratory Standards. *Protection of Laboratory Workers from Infectious Disease Transmitted by Blood, Body Fluids, and Tissue*, 2nd ed. (tentative); NCCLS Document M29-T2, The National Committee for Clinical Laboratory Standards: Villanova, PA, September 1991.

6. DHHS/CDC/NIH. *Primary Containment for Biohazards: Selection, Installation and Use of Biological Safety Cabinets*; Richmond, J. Y., McKinney, R. W., Eds.; U.S. Department of Health and Human Services: Washington, DC, September 1995.

7. National Committee for Clinical Laboratory Standards. *Protection of Laboratory Workers from Instrument Biohazards* (tentative); NCCLS Document 117-P, The National Committee for Clinical Laboratory Standards: Villanova, PA, September 1991.

8. Bond, W. W.; Ott, B. J.; Franke, K. A.; McCracken, J. E. Chapter 64: Effective Use of Liquid Chemical Germicides on Medical Devices: Instrument Design Problems. In *Disinfection, Sterilization, and Preservation*, 4th ed.; Block, S. S., Ed.; Lea & Febiger: Philadelphia & London, 1991.

SECTION VII

LABORATORY DESIGN

STEPHEN SZABO

SECTION EDITOR

70

Isolation Technology

DENNIS EAGLESON

Increasing use of chemicals of high hazard to personnel handling them and better awareness of health and safety issues causes the technology for isolating these substances from the work environment to improve rapidly. Generally, these high-hazardous substances are found in laboratories and pilot plants in industries such as pharmaceutical, agricultural, biotechnology, and other chemical research organizations. A basic principle for protecting people from exposure is to limit the volume and concentration of the substance. Risk should be assessed, and the use of proper protective equipment, procedures, and engineering controls should be prescribed.[14] Isolation technology deals with the knowledge and application of engineering controls to separate or isolate the people and their environment from the hazardous chemical substances they are using.

Principles

There are a few major principles for the technology of isolation to understand in order to select or design appropriate equipment solutions for the containment of hazardous chemicals.

Creating Pressure Differentials

The first principle is the creation of pressure differentials in order to ensure that air will flow in the desired direction. In isolating a chemical from people or the surrounding area, the objective is to have air moving in toward the source of the hazard rather than away from it. There is a relationship between the size and characteristics of the openings in an enclosure and the amount of air that will flow at a given velocity ($Q = VA$). The velocity at which the air will flow is a function of the pressure differential.[10] The principle is simple: Air will flow from one area of higher pressure to an adjacent area where the pressure is lower.

Sealing and Fastening

The technology involved in physically sealing rooms, equipment, and materials is dependent on knowledge of a number of techniques. Items like gaskets, fasteners, fittings, welding, and leak testing are all employed in utilizing isolation technology.

Entrance and Exit from Isolation Space

Entry to and exit from the isolation space is another technology that needs careful thought and planning. In order for all materials and personnel to pass into and out of the isolation space without loss of the hazardous chemical, the designer must choose the appropriate strategy, procedures, and equipment. The use of airlocks that allow only one door to be opened at a time is usually the best solution. It does not, however, lend itself to a continuous-flow operation. An airlock may be equipped with a cycle to put items into the air lock, evacuate it, and refill it with treated atmosphere or decontaminate it before opening the inner door to the isolation space. The process is reversed when exiting.

Ergonomics of Isolation Space

Ergonomics of isolation spaces is extremely important. The design of equipment should be such that people are able to perform the work safely with minimum fatigue and maximum productivity. The space should be planned to allow easy and convenient access to all machinery, instrumentation, supplies, and product that will be needed in the isolation space. Serviceability of process equipment should also be considered early in the design phase to allow safe and easy access to areas that will routinely need attention.

Definitions

Since isolation technology is changing rapidly to reflect various applications, names of devices may differ in various disciplines. The following definitions are applied for this chapter.

High Hazard These are particularly hazardous substances. They include select carcinogens, reproductive toxins, and substances that have a high degree of acute or chronic toxicity.[2] A determination of the hazard of the substance and of the exposure levels must be made based on reliable criteria.[1,3,7]

Glovebox A box-like structure that creates a physical barrier that is totally enclosed during use. It is often provided with exhaust ventilation so that virtually no contaminant within the box can escape into the room. It permits manual manipulations within the box through armholes provided with impervious gloves sealed to the box at the armholes.[5,13]

Isolator (or Barrier Isolator) An enclosure system that protects the product and the work environment by physically separating them from the surrounding environment. Prevents any direct contact between human operators and the product being processed. It is generally provided with a treated air supply as well as exhaust.[13] This term is often used in relation to clean or sterile processing.

Isolation Technology Specific techniques used to provide an element of separation of work environment, process, and product from surrounding environment.[13]

Equipment Alternatives for Chemicals of High Hazard

Fume hoods with face velocities of 100 feet per minute (fpm) and higher, special purpose ventilation hoods and devices, personnel protective devices, and gloveboxes are all isolation equipment solutions to protecting personnel from hazardous chemical exposure.

Fume Hoods

Chemical fume hoods are dealt with in depth in another chapter of this handbook and will not be extensively covered in this section. Often, the hazard of materials to be used in the hood will determine the appropriate face velocity for the hood. Generally, the higher the risk, the higher the face velocity that will be required.[14] There still is one open plane of the fume hood subject to drafts and other disturbances that reduce the effectiveness of the airflow barrier. Usually, face velocity is specified as being 100 to 120 fpm for fume hoods when using chemicals of higher toxicity. This face velocity value may be con-

sidered too high for safest hood performance. If fume hoods are not considered adequate when operating at reasonable face velocities, gloveboxes should be considered instead.[7,14]

If an application also has a requirement for clean air (free of particulates) as well as a need for containment of small quantities of chemicals, the use of total exhausting laminar flow hoods may be an option. This design is a variation of the more common Biological Safety Cabinet, which is used for microbiological work but permits no recirculation of the air containing chemicals. Its containment ability is roughly equivalent to the high-performance fume hoods, but care should taken in selecting the correct hood. [11]

When using fume hoods for high-hazard chemicals, related equipment and facilities may need to be upgraded to the additional risk. Careful selection of the materials used for the fume hood and ductwork should be done so as to maximize resistance to the chemicals in use. The exhaust fan itself may have to be explosion-proof or to have a non-sparking fan wheel material. A velocity sensor with an alarm will be needed to indicate if the face velocity drops below a preset value. If there are lights in the hood, they may need to be explosion-proof. Electrical outlets may also need to be explosion-proof. A fire suppression system may be an additional requirement.[7]

Modifications of other types of equipment may be useful in order to improve on the open-face fume hood. An enclosed box with ports as openings for operators' arms allows for the increased face velocity requirement to be met with less loss due to turbulence. Keep two things in mind. First, the operator's arms and hands and any other implements that are in use will be removed from the enclosure at some time after completion of the work and will invariably have the contamination of the chemical on their surfaces, thereby spreading the substance into adjacent rooms and being carried out of the facility with the operator. Second, turbulence still will be a factor and loss may occur even though it may be an improvement over an open-face fume hood.

There are many special adaptations of fume hoods that limit the access and openness of the hood to do a better job at containing hazardous materials than a standard fume hood. Many are designed with specific applications in mind and should be used expressly for that purpose. The use of spot ventilation for controlling fumes close to the source is also a good technique when combined with other devices and facilities.[7,10]

Personal Protective Equipment (PPE) in an Isolation Facility

Personal protective equipment is used to protect workers from exposure in the absence of effective engineering control. Engineering control is used to capture the chemical at its source and is the preferred good industrial hygiene practice. Personal protective equipment is employed only when engineering controls and work practices and procedures are inadequate to protect the workers.

Complete suiting up of the individual is an option that is available for protecting the personnel from exposure to highly hazardous chemicals over and above the use of normal PPE (gloves, gowns, faceshields, filter and gas masks). These suits are ventilated with treated air supplied for breathing either from a central source or from a self-contained breathing apparatus (SCBA) (Fig. 70.1). When pressurized and intact, they will do an excellent job of preventing exposure. However, to be truly effective in preventing release to the building and surrounding environment, a special isolation-type facility must be used.

This facility should allow access only to personnel with PPE. It is designed and built to be easily cleanable and is equipped with a HEPA filtered vacuum cleaner. There will be ventilation requirements for it and its antechambers utilizing pressure differentials to direct the airflow into the contaminated area instead of allowing it to flow out into the adjoining spaces. The room should be designed to have a high number of air changes per hour (ACH), usually falling in the range of 6 to 20 ACH. The air supplied should be 100% fresh air. An isolation room uses the same principles of isolation as other equipment, except the worker is inside the isolation space. The facility includes airlocks in order for personnel to change clothes and put on PPE. A shower facility will be included. The facility layout will be designed so as to enable the personnel to gown and to decontaminate properly to prevent release of chemical into the environment. Air supply and exhaust lines are separate from the building services and are protected with HEPA and/or chemical removal or neutralization.[14] The exhaust flow from the room should be monitored to determine when to change the filters.

The facility itself should be specially designed for the specific research program and the hazards associated with it. This includes denying access to nonauthorized personnel. A safety plan will be in place and people trained to carry out the plan. Adequate and appropriate storage will be available for the chemicals, including flammable storage cabinets, refrigerators, and freezers appropriate for flammable materials storage. A safety deluge shower will be installed along with eyewash stations. The facility will be administered in accordance with all regulations dealing with chemicals and related to the health and safety of the personnel, hazardous waste disposal, and inventory control and tracking.[7]

While this design approach is effective in accomplishing the objective of isolation and preventing contamination escape, it is also very capital and personnel intensive and thus very costly to operate.

Simple Gloveboxes

Gloveboxes can provide an "absolute barrier" to contain the hazardous material within the workspace, close to the source (Fig. 70.2). They allow people to work safely in a more normal environment with fewer special facility requirements. A glovebox is preferred when the materials and procedures being used are of high or unknown risk. The risk should be considered greater when the planned manipulation or accident would generate hazardous personnel exposures if the work was done in an ordinary fume hood.[5] The term "glovebox" is used in the generic sense, and the user should make sure that the particular design of the glovebox is suitable for high hazard chemical use.

Trends

For a number of reasons the use of gloveboxes for highly hazardous materials is increasing. The chemical professional should be aware of the availability of this type of equipment in order to ensure appropriate health and

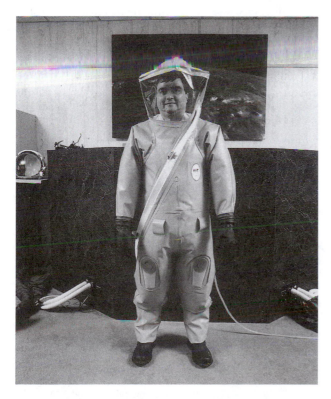

FIGURE 70.1 Chemical/biological ventilated protective suit. (Courtesy of ILC Dover.)

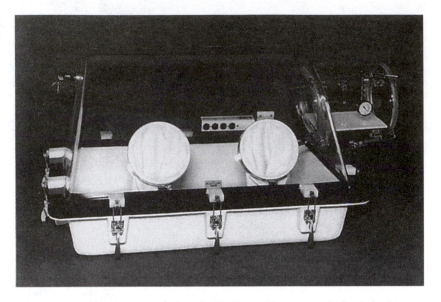

FIGURE 70.2 Simple nonventilated glovebox. (Courtesy of Plas-Labs.)

safety as well as to provide a working environment that is cost effective.

An understanding of the following trends is iseful for planning facilities.

Trends That are Driving the
Move to Gloveboxes

- Greater number of more toxic compounds
- Greater potency of new compounds being synthesized
- More chemicals of unknown toxicity; new chemical compounds where there is little or no information of effects of long-term exposure
- Speed at which new chemical compounds are being created
- "Microchemistry" enables the use of smaller containment devices
- Need to reduce the dependence on large energy-consuming ventilation hoods, and the lack of adequate HVAC capacity in old labs
- Dual requirements for both toxic chemical containment and contamination-free processing
- Increasing use of automation
- Increasing awareness of health and safety regarding occupational exposures
- Less expensive alternative to designated rooms and controlled work spaces
- Products that are sensitive to environmental conditions: temperature, humidity, oxygen, and so on
- Confinement to a small area allows the process to be controlled more readily

Specialized Gloveboxes

Glovebox is a generic term for a type of equipment. It includes a variety of design types, each with its own sets of criteria, specification, and design features. The types have been classified in a number of ways by different organizations. Generally, the types can be differentiated by the intended use of the equipment as seen in Table 70.1.

While there is an appeal to conduct work in a glovebox because of an increased need for worker and/or product protection, the specifier should be realistic in assessing the trade-offs involved with this type of equipment. Table 70.2 indicates advantages and disadvantages.

In order to most effectively design and specify glovebox configuration, the designer must make a careful assessment of the procedures to be used. There are a number of features available for glovebox design that enable

TABLE 70.1 Types of Gloveboxes Based on Application

Glovebox Type (Use)	Application
Nuclear	Radioactive materials[7, 8]
Chemical	High acute or chronic toxicity[5]
Class III Biological Safety Cabinet	High-risk biological agents; P4 containment level[9]
Isolator Barrier System	Sterile processing[13]
Dry Box	Controlled atmosphere for air-sensitive materials (inert gas, humidity)

TABLE 70.2 Glovebox Advantages and Disadvantages

Advantages	Disadvantages
Contains spills, powders; allows for clean-up inside containment device	Restricts operator movement, reach
Less noise	Ergonomically difficult design: vision, dexterity, comfort
Lower cost than a controlled room	Difficult to handle large objects
Less area required than a controlled room	Need for rationalization of the complete process outside the box: handling, preparation, transport, waste removal
Quick user access unlike controlled rooms or PPE suits	
Very high level of safety (and sterility) achievable due to the physical barrier	

them to be more "user friendly." It is extremely important to plan ahead in order to ensure that the user can operate the process safely and efficiently. Calculation of protection factors is a technique that is useful, especially for a more complicated or multistep process.

The glovebox is an "absolute" barrier (that is, it is a sealed enclosure). The weakest link of this system is the gloves themselves. They are easily punctured, torn, cracked, will deteriorate, and have selective permeability for various chemicals. Careful testing and inspection of the gloves before each use is necessary. Usually, frequent routine replacement of gloves is necessary and desirable. There are a wide variety of materials and styles of gloves. Some of the differences include attributes relating to materials, configuration, durability, flexibility, chemical resistance, and permeability.

The following list includes some design criteria that can be used when determining specifications for a glovebox.

Some Design Suggestions for Gloveboxes

- Adequate access to the interior: using passthrough boxes, doors, air locks, removable panels, service fixtures
- Design for cleanability: specify radius corners, same criteria that are used for radioactive particulate materials
- Differential pressure: use of a "pressure cascade" from one containment area to the next; will be a conflict of pressure requirements for containment and for aseptic or clean processing that will need to be resolved (positive versus negative pressure).
- Choice of materials of construction: use materials that are impermeable, nonabsorbent, resistant to

corrosion from chemicals to be used, resistant to disinfectants, easily cleanable, and resistant to abrasion caused by cleaning.

- Durability: especially important in areas where usage is heavy, such as production and scale-up activities
- Gloved arms should be able to reach all areas within the box, or special tools will have to be available.
- Determine if the HVAC system should be designed as an "on-board" system or part of the building systems.
- Gauge or monitor for measuring pressure
- Inlet and outlet valves, if a gaseous environment
- Viewing windows of adequate size and location
- Adequate lighting levels inside box

Glovebox with a Controlled Environment

Some chemical reactions and activities require operations to take place in an oxygen-free and/or low-humidity environment. Certain chemical synthesis and hazardous reactions can be done in the type of glovebox that provides an inert-gas atmosphere (Fig. 70.3). In addition to the design details of all other gloveboxes, these systems include additional features. They will have a recirculation system to continue to purify the air and to keep humidity at the desired level. They may be equipped with a freezer to store the chemical. They are usually fitted with air lock passthrough boxes that are evacuated after materials have been placed into them and then the evacuated air replaced with the atmosphere found in the box. Once that cycle is complete, it allows for opening the inner door to complete the passage of materials into the glovebox. The cycle is reversed to remove items and waste materials.

Aseptic Processing

A growing use for glovebox technology is in the processing of sterile products (Fig. 70.4).[12] While the trend recently has been in the pharmaceutical and biotechnology industries, the design features and approaches can be applied to other uses that require a high degree of contamination control. These applications require additional technologies related to the world of microbes. Interiors should be sealed to create smooth surfaces that do not harbor microbial growth. Places where water can collect should not be allowed, as the water supports microbiological bloom.

Aseptic processing requires equipment designed specifically for this purpose in order to be successful. Some of these considerations are listed below.[6,13]

FIGURE 70.3 Controlled atmosphere glovebox with freezer. (Courtesy of Innovative Technology, Inc.)

FIGURE 70.4 Aseptic filling isolator. (Courtesy of The Baker Company, Inc.)

Special Design Considerations for
Aseptic Processing

- CIP/RIP (Clean-in-Place, Rinse-in-Place): for automatic cleaning by flushing the interior of the isolator
- Sterilization: choice of methods for rendering the interior of the isolator free of microorganisms
- RTP (Rapid Transfer Port): to ensure entry into the isolator without breaking the physical barrier
- UV (ultraviolet) tunnel: to decontaminate outside of materials while bringing them into the isolator
- Special HVAC requirements: for example, increased drying capacity to aid in sterilization process
- Control, monitoring, and documentation of the process: to ensure lot traceability and sterility assurance

- Additional environmental conditions: product may be sensitive to temperature, humidity
- Treatment of the access to and from the system: In a continuous process there is a need to provide reliable protection to the entry and exit openings
- Absence of marks, crevices, and cracks that would harbor microbes

While this industry is concerned with ensuring the sterility and purity of the product, many of the chemicals used are potent and can be harmful to the operators. Consequently they require containment or limited release to the room environment.[8] It is helpful to study how both of these needs are addressed in design and design verification in order to accomplish the same objectives for other uses (Fig. 70.5).

Many of the same issues apply to the semiconductor industry. Here, there also are needs for protecting the

FIGURE 70.5 Containment and product protection enclosure. (Courtesy of The Baker Company, Inc.)

product from contaminants found in the normal room environment. In addition, there are chemicals used in the process of manufacturing that may be harmful to the operators. A similar approach in this industry is evolving through the use of what are called "mini-environments," which enclose the process, keeping people out of the process area.

Exhaust Requirements and Room Environment

There may be specific requirements for exhaust flow from a glovebox or other containment device based on the materials used in the process, local and federal requirements for release of the material, and any unique facility requirements. Generally, if the hazardous material is particulate, it will be filtered first by a high-efficiency filter (HEPA) before being released to a fume hood or chemical exhaust system. If a gas or a vapor is generated, additional scrubbing or filtration may be necessary.[1] There should be a minimum exhaust flow rate capability of 50-cfm/sq. ft. of door area.[7,10] OSHA requires a ventilation rate of at least two volume changes per hour and a pressure (negative) of at least 0.5″ water column (WC).[1] (Note ACGIH manual says 0.25″ WC static pressure in a closed system.[10]) Planning should be done to sequence the chemical removal devices based on the material to be moved. For example, if a HEPA filter is placed first and a corrosive chemical is present, it could cause the filter media to deteriorate.

Equipment Design Specifications

When selecting a glovebox or other containment equipment for purchase, the buyer should carefully select the specification criteria that will suit the application or intended use. Since there is a wide variety of equipment types available, there is no one specification that will be suitable for all uses. The American Glovebox Society publication, "Guideline for Gloveboxes,"[4] serves as a useful template for selecting design specifications. ANSI/AIHA Z9.5 is also a good source from which to develop specifications.[5] The buyer and supplier should work together to have a firm understanding of the requirement to ensure a good result. The specification should form part of the contractual agreement between vendor and supplier. Common accepted practice for work with hazardous chemicals is for the glovebox to operate under negative pressure, so that if there is a leak in the box or glove it will be inward into the hazardous area. For certain applications, gloveboxes are designed to operate under positive pressure because of the need for humidity or oxygen control. This design requires routine testing for leaks and continuous monitoring of the system.[2,14]

Special consideration should be given when changing out filters in order to prevent loss to the environment of particulates trapped on the filter. The procedure for glove changing should be worked out, validated, and people trained to demonstrate proficiency. *Radiation Hygiene Handbook* contains a good description and illustration of the glove changing methods.[15] Design should allow for decontamination of interiors, service work to be done, and removal of contaminated waste without undue exposure of residues to the service personnel.

Materials should be nonporous and corrosion resistant to chemicals that will be used. Interior corners should be rounded and joints and seams minimized and sealed. The box should be designed to anticipate a chemical spill with a practical procedure for cleaning.

Design of gloveboxes requires much thought and planning in order to allow workers to conveniently reach all areas where their access is needed through the gloves available. There are some basic principles of ergonomic design to do this: matching glove port location to a model of typical human proportion, slanting the face of the box for ease of viewing and reach, and providing lifting and tooling aids inside the box. The best approach during the design phase is for the end users of the equipment to work closely with the designer in order to have the design reflect the projected actual use and movements of the operators. It may be desirable to mock up sections of the box and evaluate the ergonomics of the design during the design process.

Exhaust system design should be specific to the application and will be similar to the requirements for a facility dedicated to chemical containment. ANSI/AIHA Z9.5 describes some exhaust system recommendations.[5]

The design of the glovebox should accommodate the method to be used for decontamination of the workspace. In some instances, this may be just ensuring that a liquid can be drained or mopped up adequately. Neutralizing agents can be selected to cover the spill in the box before removal. In other situations, such as with a gaseous environment, the box may need a vent to the exhaust or to a chemical removal system in order to purge the box and replace the gas or vapor. Much thought should go into this aspect of the design, and a combination of these approaches may be employed. As well as the interior of the box, determine what other related parts of the system (like the exhaust) will be in need of decontamination. Even after decontamination, personnel working inside the enclosure system should wear protective equipment and assume there is some risk.

Performance Criteria

Another key part of the contractual agreement should be the specific design and manufacturing criteria that are agreed on in order to prove that the contract has been fulfilled based on successful acceptance requirements of performance and construction details.

It is useful to express the performance criteria as a checklist.

Suggested Checklist of Field Performance Tests for Gloveboxes

- airflow volume and pattern
- box internal differential pressure
- filter leak test
- pressure decay and/or box leak test
- electrical safety and integrity
- alarm condition verification and calibration
- illumination

The need for documented performance, both prior to shipment to the customer and after installation, is critical to a successful installation, especially if there is a requirement for documentation, such as for achievement of FDA approvals. Development of the test protocols and acceptance criteria should be agreed on by both parties as early as possible in the life of the design phase. Any test and inspection criteria must relate directly to the design document specification in order to have meaning.

References

1. OSHA Occupational Exposure to Hazardous Chemicals in Laboratories, 29 *CFR* 1910.1450.
2. OSHA Toxic and Hazardous Substances Standard, 29 *CFR* 1910.1000.
3. OSHA Health Hazard Definitions Standard, 29 *CFR* 1910.1200, Appendix A.
4. AGS Guideline for Gloveboxes; Publication No. AGS-G001-1994; American Glovebox Society: Denver, CO, 1994.
5. ANSI/AIHA, Z9.5 American National Standard for Laboratory Ventilation; American Industrial Hygiene Association: Fairfax, VA, 1992.
6. *Barrier Isolation Technology Conference*; International Society for Pharmaceutical Engineering: Tampa, FL, 1995.
7. *CRC Handbook of Laboratory Safety*, 4th ed.; Furr, A. K., Ed.; CRC Press: Boca Raton, FL, 1995.
8. *Containment of Potent Compounds*; International Society for Pharmaceutical Engineering: Tampa, FL, 1995.
9. *Handbook of Facilities Planning*, Vol. 1, Laboratory Facilities; Ruys, T., Ed.; Van Nostrand Reinhold: New York, 1990; p. 339.
10. *Industrial Ventilation: A Manual of Recommended Practice*, 22nd ed.; Committee on Industrial Ventilation, American Conference of Governmental Industrial Hygienists, Inc.: Cincinnati, OH, 1995.
11. *Introduction to Laminar Flow Biological Safety Cabinets*; The Baker Company Inc.: Sanford, ME, 1995.
12. *Isolated Pharmaceutical Applications*, Lee and Midcalf, HMSO Publications Centre, 1994.
13. Wagner, M.; Akers, J. E. *Isolator Technology: Applications in the Pharmaceutical and Biotechnology Industries*; Interpharm Press: Buffalo Grove, IL, 1995.
14. *Prudent Practices for Handling Hazardous Chemicals in Laboratories*; Committee on Prudent Practices for Handling, Storage, and Disposal of Chemicals in Laboratories; Commission on Physical Sciences, Mathematics, and Applications, National Research Council, National Academy Press: Washington, DC, 1995.
15. *Radiation Hygiene Handbook*; Blatz, H., Editor-in-Chief; McGraw-Hill: New York, 1959.

71

Lab Design/Radiosynthesis Lab Design

NELSON W. COUCH
JOHN J. NICHOLSON
SHIMOGA R. PRAKASH

The purpose of this chapter is to provide insights on the subtle differences between a general lab and the one designed for preparation of compounds labeled with beta- and gamma-emitting isotopes, that is, compounds containing long-lived radionuclides like carbon-14, tritium, and sulfur-35.

Basic Considerations for Radioisotope Lab Design

The basic difference between a lab designed for use of radioactive materials (RAM) and those designed for work with chemical or biological agents is the control of radioactive material contamination. The term "contamination" is used to mean the presence of radioactive material where it is not wanted. It can cause unnecessary exposure of personnel and can lead to expensive clean-up or disposal of equipment. Another difference is that labs designed for work with radioisotopes must also meet licensing requirements outlined by the regulatory agencies.

Basic Concepts of Design

A concept that should always be kept in mind when developing an environment where radioisotopes will be used is that of contamination "zones."[1] An area where radioisotopes are not present and therefore no chance for contamination exists would be the first or lowest contamination zone and would allow unlimited or unrestricted access. Examples of such an area would be offices, break rooms, or cafeterias. The second level might be where contamination is highly unlikely, but possible and access could be somewhat limited. An example of such an area might be a room containing a scintillation or gamma counter or entryway. The third contamination zone would be one in which work with radioisotopes occurs and contamination is always possible. Access to such areas is limited; examples of such areas are the radiation room or the entryway. The fourth and highest level of consideration would be an area where contamination levels are or could be high. Access to these areas is highly controlled. Examples of such areas are the interior of radioactive hoods or gloveboxes.

In considering adjacencies, thought must be given to the potential for radioisotope movement. Whenever a sample is moved there is potential for the spread of contamination or a spill. The need for transporting radioactive samples or the distance over which they must be moved should be minimized.

Ventilation Systems

The lab ventilation system offers the most important protection feature for lab workers. Usually the air is exhausted from the lab through the lab hood, but an adequate supply of air must be provided. Use of recirculated air in any lab situation is not recommended, and this is especially true for radioisotope labs. Therefore, once-through is the method of choice.

The hoods may be exhausted through multiple stacks or a single stack. However, care must be exercised to exhaust stack emissions high enough above the building to prevent entrainment into the intake system or that of an adjacent building.

Another consideration is the ventilation flow pattern. The airflow should be from areas of lowest contamination toward areas of highest potential contamination, thus reducing the opportunity to spread airborne contamination.[2]

Separate Services and Systems

The utilities required for a radioisotope lab are no different from those of any other lab. The lab needs supplies of hot and cold water, distilled water, air, nitrogen, a vacuum system, and special gases like P-10 (argon—10% methane). However, in the case of the vacuum system it might be advisable to consider a separate system for radioisotope labs should the system become contaminated by aspiring radioactive material. Use of water aspirators as a source of vacuum should be minimized. Likewise, the cylinder change area should not be within the laboratory.

Storage

Adequate storage for radioactive materials and samples must be provided. Also adequate space for waste collection and segregation must be thought through, as lab personnel will be handling both radioactive and chemical wastes. Due to the difficulty and expense involved in disposal, generation of mixed waste should be minimized.

Additional Considerations for a Radiosynthesis Lab Design

Basic Thoughts for Lab Layout

To prevent cross-contamination, the radiosynthesis lab should have modules that are earmarked for working with a specific isotope, a carbon-14 module, a tritium module, iodination module, and an analytical module. These modules could be rooms or zones with varying amounts of radioactivity. A darkroom for autoradiographic work is beneficial, but with the advent of newer techniques like phosphor imaging, it may not be required. Island benches with appropriate data lines for housing necessary chromatographic equipment (HPLC, GC, TLC scanners), gloveboxes, or space for mobile carts containing equipment are a requirement. These should be positioned to suit convenient maneuvering around the hoods. Glassware, chemicals, and solvents required for working in the hoods should be readily accessible and preferably stored in cabinets underneath the hoods (a diagram of a typical radiosynthesis laboratory is shown in Fig. 71.1).

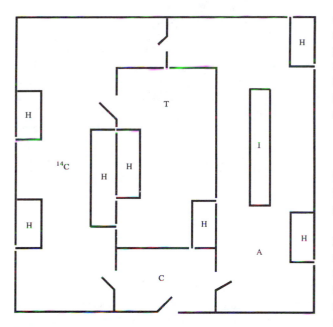

FIGURE 71.1 Diagram of a typical radiosynthesis laboratory. Several variations of this design are possible. Location of emergency exit doors, glass windows, and laboratory benches are not shown. Key: H is radioisotope hoods, C is changeover room, I is island bench, ^{14}C is carbon-14 laboratory, T is tritium laboratory, and A is analytical laboratory.

The location of the radiosynthesis lab should be carefully thought out. Since the radiolabeled compounds created by the lab will be used by other labs, they should be located as close as possible to their primary customers.

Access and Security

Due to the nature of the radioisotope work and the activities used within a radiosynthesis lab, access should be restricted and limited only to those highly trained professionals who have a need to be present.

Entry Area

Entry to the radiosynthesis lab must be through a controlled access point. This space will provide a buffer zone as a person enters to don coveralls, shoe covers, and head cover (if necessary), or other form of outer garments. Persons will then be allowed to cross over and enter the radiosynthesis lab itself. When they are ready to exit they will re-enter the entry area (or change over room) and remove the coveralls, shoe covers, and head cover and dispose of them, wash their hands, and check for contamination before crossing over to the other side

of the entry area and exiting. In addition to providing a controlled entrance and a space for donning or doffing protective clothing, the entryway provides a means to maintain the lab negative to the remainder of the building. Thus if there is a spill or airborne contamination release in the lab, it cannot spread to outside the lab space.

Emergency Exits and Emergency Response

The radiosynthesis lab, like any other space, must have two escape routes. In addition, emergency eyewash and safety showers must be provided.

Another aspect of emergency response is containment. Each entryway (including emergency exits) should be constructed with a slightly raised threshold to facilitate spill containment. Also, each penetration of the floor or walls must be carefully sealed to prevent leakage into adjacent spaces. The bottom line is that the spill must be contained within the lab space and not allowed to leak out.

Waste Water Collection and Discharge

Local agencies may set more strict sanitary sewer disposal limits than those established by the Nuclear Regulatory Commission (NRC). Sink drains from a radiosynthesis lab should preferably drain to a pair of hold-up tanks, which will allow the contents to be checked and documented prior to discharge to the sanitary sewer. If the radioactive content of the waste water is above the limits, the water can be pumped out of the holding tank and sent for treatment or alternate disposal. It is a good idea to work with a pair of holding tanks since one can be filling while the other is being checked, pumped out, or discharged. Although there are no NRC requirements for intermediate holding tanks, local considerations may require them. In practice the concept appears sound.

Cooling water for use in experiments may be provided via a separate system that discharges directly to the sewer, provided there are no chances of contamination. If possible, it is a good idea to drain cooling water used for experiments through a separate system. For example, draining cooling water through the hold-up tanks will fill them up quickly, putting undue stress on the people who must assay the contents and discharge it to the sewer.

Hood Exhaust Monitoring

License conditions may require demonstration that the total effective dose equivalent to personnel likely to receive the highest dose from air effluents of a radiosynthesis lab not exceed some established limits. There may also be restrictions on the average concentration of air effluents and potential doses to members of the general public. In addition, it is prudent to maintain air emissions of radioactive material as low as reasonably achievable. Some method of determining potential concentrations and doses must be used. Potential concentrations of air emissions and doses to personnel may be calculated or can be measured directly by means of air sampling devices. The method of determining these values should take into account all the radionuclides that may be present in the air effluent.

If a method of direct collection and measurement is selected, the performance of the effluent collection and sample analysis systems should be verified by regular calibration to ensure reliable and accurate results. Monitoring should take place after the air effluent has passed through any filtration systems. In some instances samples may be collected, before and after filters, to monitor filter efficiency. There are several NCRP and ANSI documents that provide in-depth information on designing and operating air effluent monitoring systems.[3,4,5,6]

There are several computer codes available that can be used to model public dose from emissions.[7] However, if a computer code is selected, one has to be sure that it can be validated and uses methods approved by regulatory agencies applicable to the license covering the radiosynthesis lab.[8]

Analytical Lab Requirement and Access

The radiosynthesis lab will need access to an analytical lab and radioactive counting equipment. It may be preferable to locate some of this equipment within the lab itself to reduce the potential for cross-contamination with other labs within the facility. However, one must also consider the possibility of high background counts if the equipment is located within the radiosynthesis lab itself. Therefore, a separate analytical lab and counting room located in the lowest contamination zone would eliminate possibilities of high background counts.

Storage

A radiosynthesis lab will be required to work with reasonably high specific-activity starting materials. Therefore, they will need adequate storage space for several mCi to Ci quantities of precursor materials. In addition, it will be necessary to store and perhaps dispense compounds prepared in the radiosynthesis lab. Use of a −70°C freezer may be required.

Decommissioning Considerations

Plans for decommissioning a radiosynthesis lab should start when the lab is being designed. Adequate pre-planning, design, selection of materials and equipment, and construction with an eye on decommissioning issues will make the task much easier and save money.

The next section of this chapter provides guidance on the construction materials for a radiosynthesis lab, including floor, wall, and ceiling coverings, furniture and bench tops, and chemical fume hoods. When considering options during the design and construction phase, consideration must be given for each option as to how the long-term decommissioning costs may be impacted. Selection of materials that may be less costly, but can be cleaned as easily as some others, may save money in the short term, but any savings may be lost once the decontamination and disposal costs are figured in.

In general, materials should be impervious, nonporous, chemically resistant, and easily cleanable. Air and water effluent lines should be maintained as short and straight as possible. Administrative controls for contamination control can go a long way to reducing decommissioning costs. Regular inspection and auditing of the radiosynthesis lab for compliance with established contamination control procedures and guidelines can also help reduce decommissioning costs.

When designing the radiosynthesis lab, attention should be given to the characteristics of the radionuclides to be used, the types of labeling procedures that may be employed, and the expected length of time the lab will be used.

There may be other decommissioning considerations besides radioactive materials. It can be expected that a radiosynthesis lab may be working with chemicals, potent compounds, or biological materials that have their own hazards and decommissioning issues.

Considerations for Construction Materials

Wall, Ceiling, and Floor Coverings

Floors are more likely to become contaminated than walls or ceilings are—in that order. However, the construction materials are important for all three areas from the standpoint of contamination control. In general, floor, wall, and ceiling coverings must be impervious, nonporous, chemically resistant, and easily cleanable. Seams and cracks in floor or wall coverings are difficult to decontaminate and must be avoided. Tile floor coverings are often used in lab areas, but there are several considerations. Use of tile floor coverings will introduce cracks that are difficult to decontaminate. Should a major spill occur, the tiles might need to be disposed of as radioactive waste and replaced with new tiles. But, at least, tiles are easy to replace should the need arise. If a tile floor is chosen, the largest size tiles practical should be selected to reduce the number of cracks. In addition to reducing the number of cracks, the tile floor must be kept waxed to fill the cracks and thus help reduce contamination problems. But the wax must not create a slippery surface.

It is possible to choose continuous floor coverings with heat-welded seams or epoxy-sealed floors and thus eliminate cracks completely. These choices make the most sense in a radiosynthesis lab. Tile floor coverings may make more sense in a tracer lab where the quantity of RAM used is much smaller than in a radiosynthesis lab.

Where the floor meets the wall or furniture, coving should be used to create a smooth transition. The joints between the coving, floor, and walls or furniture should be sealed to reduce the chance of contamination working into them.

Wall covering can be epoxy-painted sheetrock or a laminated-sheet product such as vinyl. The sheet vinyl product can be thermally sealed along the seams to create a seamless wall covering. By use of vinyl floor and ceiling materials, the space can be totally encapsulated, minimizing the number of seams that can become contamination sites. Should an area need to be replaced, it can simply be cut out and a new piece thermally welded into place as a replacement.

Ceilings may be constructed of drop-in materials, since the chance of them becoming contaminated is rare unless a major incident occurs. Use of softer ceiling materials will reduce reflected noise in the lab, making the space more comfortable to work in. Materials should be easily cleanable or else considered expendable should they become contaminated.

Furniture and Bench Tops

A major element in furniture selection must be contamination control. Use of wood or wood-laminated furniture will provide absorbing surfaces for RAM contamination. Decontamination will be nearly impossible, thus creating the necessity for expensive disposal. Metal with appropriate resistant coatings is the best choice of furniture construction materials. In addition, consideration should be given to selecting only necessary furniture.

As with any horizontal surface, the bench tops are prone to becoming contaminated. Therefore, they must be constructed from smooth, impervious, nonporous ma-

terials that are easy to clean. For example, stone, impregnated stone, resin, stainless steel, or various laminated plastics have been used successfully. The bottom line in furniture consideration is that no unnecessary furniture should be present.

Hoods

The same criteria hold for hood construction materials as those mentioned above. In general, hoods for use with radioisotopes are constructed of 304 or 316 stainless steel. However, the literature does mention the use of fiberglass-reinforced polyester resin as a possible alternative. Again, the ease of decontamination is the key. Therefore, coving the corners and minimizing the seams is a necessity.

In addition, there may be occasions where temporary shielding must be constructed inside a hood. Therefore, the ability for the hood structure to support sheet lead or lead bricks must be considered.

Considerations for Radioisotope Hoods

The radioisotope hood is one of the most important safety protection devices available to the lab worker. Therefore, much consideration must be given to its proper operation and maintenance. There are several critical elements to consider. The first is hood placement within the lab. Being located in heavy traffic patterns or near room supply diffusers can disturb the hood airflow. For best operation the hoods should be placed in low-traffic areas and as far away from supply air as possible. Second, the hood face velocity should be set for 100 feet per minute (fpm) ±20%. Since the face velocity will change over time due to fan maintenance, it is best to design for 120 fpm so that enough air capacity is available, but then to balance all hoods to run at 100 fpm. Since the hoods provide the first line of defense for the user, they should have some form of backup and a visible or audible alarm system should the power go off. This can be provided by UPS, battery, or emergency generator even if they have to operate at reduced volume. They should not be able to go positive with respect to the lab and discharge any contents into the breathing zone of the worker. Lab employees must be trained that a hood under emergency power is not for normal routine use: It is simply to keep them negative with respect to the lab. A full treatment of the various types of hoods is beyond the scope of this chapter. A full description of lab hood design and selection may be found in the ANSI guide on lab hoods.[9,10]

The ductwork connecting the hoods to the fans is obviously very important. The ductwork should be constructed of materials that resist corrosion, dissolution, or a tendency to melt or catch fire under normal or accidental conditions. Materials such as stainless steel, epoxy-coated steel, or various plastic materials have been used successfully. The ductwork should be kept as short as possible to reduce the potential for corrosion, condensation, and future decontamination or disposal issues. Hoods may be grouped into plenum chambers or ducted separately to dedicated fans. In all cases, cost, ease of fabrication, installation, maintenance, and eventually dismantlement must be considered.

The inside of the lab hood should be considered as potentially contaminated. Whenever work must be done on the hoods, the potential contamination must be assessed and decontamination accomplished prior to allowing repair or maintenance work to be done. Depending on the radioisotopes used, the hood exhaust air may require filtration prior to emission to the atmosphere. For example, working with radioactive iodine would require use of a charcoal filter in the hood exhaust to adsorb any release of radioiodine. Use of a HEPA filter may also be required to capture radioactive particulate that might be released. HEPA filters are of little use if the emission is in gaseous form such as radioactive iodine or tritium gas. However, occasionally tritium water vapor may attach to dust particles, which can be captured by the HEPA filter or plate out on the inside surfaces of the hood or duct system. Whenever filters are used, consideration should be given to the "bag in/out" type with the filter housings mounted such that they are easily accessible to maintenance staff. In addition, the exterior of the housings should be clearly marked with the radiation symbol.

Special Licensing Concerns for Radiosynthesis Labs

If the facility is not specifically licensed to conduct radiosynthesis procedures, the radioactive materials license will have to be amended. Federal and/or state regulatory agencies may ask for details regarding the following aspects (among others) of the radiation safety program when reviewing a license application or amendment for a radiosynthesis laboratory.

Increased Activity A radiosynthesis lab will be required to work with high-activity materials. Thus the facility license must be able to handle Ci quantities of various isotopes. For example, a facility may be licensed for 100 to 200 Ci of tritium, of which most of its activity will be required for use in the radioisotope lab. In addition, there will be special waste-disposal considerations due to the radiosynthesis lab activities. The radiosynthesis lab will create mixed waste that will have to be dealt with in some way or stored on-site until disposal can be arranged. The licensee should be able to handle on-site storage should that be necessary.

Bioassay Program The increased amounts of radioactive material routinely used in a radiosynthesis laboratory may require more frequent or different bioassays compared with those conducted in a routine radioactive material laboratory. Environmental monitoring of air and water emissions from the laboratory may have to be performed. Regulatory agencies may ask for details on the proposed monitoring program.

Radioactive Waste A radiosynthesis laboratory will produce some waste that is much higher in activity than that of a routine lab. A "mixed waste" consisting of hazardous chemical and radioactive waste will most likely be generated. This waste can be expensive and difficult or even impossible to dispose of. Some description of the plans to handle and dispose of the new types of waste will have to be included with the amendment or application.

Decommissioning Plans A review of the decommissioning plan and the estimated costs will be in order when considering adding a radiosynthesis laboratory. Most likely a revised decommissioning plan will have to be included with the license amendment.

Training of Users Due to the increased activity handled in a radiosynthesis laboratory and that some of the radioactive material may be in a gaseous form, the regulatory agencies may require specific training for radiosynthesis laboratory staff.

Security and Access Control This is always an issue with radioactive material laboratories; however, a radiosynthesis laboratory may warrant additional measures to secure material and limit access to unauthorized personnel.

A careful review of the current radioactive material license is in order when planning to add a radiosynthesis laboratory to an existing license. Regulatory agencies should be contacted in advance to discuss what issues might have to be dealt with.

User Considerations in Designing a Lab for Using Radioisotopes

A radiosynthesis lab, as mentioned before, is no different from a regular organic synthesis lab. However, caution must be exercised and every operation must be carefully thought of from a contamination-control point of view. Accounting and monitoring of radioactivity is of utmost importance. In general, hoods should be deep enough to accommodate spill trays (also containers for liquid radioactive waste and sharps) and tall enough to accommodate housing of vacuum manifolds, rotary evaporators, and chromatographic columns. It is also not advisable to have too large a hood (commonly seen in pharmaceutical and agrochemical industrial labs) with which decontamination and decommissioning issues become critical. If there are no space limitations, a walk-up hood for chromatographic work would be extremely useful.

References

1. Stewart, D. C. Radioisotope Lab Design. In *Handling Radioactivity: A Practical Approach for Scientists and Engineers*; Stewart, D. C., Ed., Wiley & Sons: New York, 1981; pp. 33–86.

2. Joseph, U. P. et al. Renovation of a Radiosynthesis Lab with High Airflow Requirements; *Pharmaceutical Engineering* **1994**, 64–70.

3. NCRP Commentary No. 3, "Screening Techniques for Determining Compliance with Environmental Standards" Jan. 1989, addendum Oct 1989.

4. ANSI N13.1(1969), "Guide to Sampling Airborne Radioactive Materials in Nuclear Facilities."

5. ANSI N42.18, "Specifications and Performance of On-Site Instrumentation for Continuously Monitoring Radioactive Effluents."

6. USNRC Regulatory Guide 4.20, "Constraint on the Release of Airborne Radioactive Materials to the Environment for Licensees Other Than Power Reactors," December 1996.

7. NCRP (1996), National Council on Radiation Protection and Measurements, *Screening Models for Release of Radionuclides to Atmosphere, Surface, Water, and Ground*, NCRP Report No. 123.

8. EPA 520/1-89-002, "A Guide for Determining Compliance with the Clean Air Act Standards for Radionuclide Emissions from NRC Licensed and Non-DOE Federal Facilities."

9. *Guidelines for Laboratory Design*, 2nd ed.; Diberardinis, L. J. et al., Eds.; Wiley & Sons: New York, 1993.

10. *Prudent Practices in the Laboratory*; National Academy Press: Washington, D.C., 1995.

72

High-Pressure Test Cells and Barriers

GEORGE W. MONCRIEF

High-pressure test cells and barriers are a necessary complement to chemical research and development laboratories when experimentation is performed on exothermic chemical reactions for which the chemistry is not fully understood. When the chemistry is fully understood, process control systems and pressure-relief devices reasonably ensure that a unit will operate within its safe operating envelope and, therefore, be safe to operate without a protective barrier. Fully understanding reaction chemistry or the thermal stability of chemical compounds requires having knowledge not only about the desired chemistry but also the undesired chemistry leading to runaway reactions. Chemical research facilities need access to small-scale thermochemical testing equipment such as Differential Thermal Analyzers (DTA), Differential Scanning Calorimetry (DSC), and Accelerated Rate Calorimeters (ARC). Sometimes larger scale pressure bombs and autoclaves require barricaded test cells. Often, for economic or propriety information protection

reasons, it is desired to have in-house capability to perform these tests. Some nonchemical research facilities also have need for such facilities, for example, to perform tests on equipment operating at extreme pressures, temperatures, or speeds.

Test cells allow potentially dangerous experiments to be remotely conducted from the other side of a protective barrier. A viewing port or remote camcorder system is usually available for visual observation. Figure 72.1 shows such a cell from the operator's side. A viewing port, composed of two 1/2-in. polycarbonate discs, and a control panel for the operator to remotely monitor temperatures, pressures, flow rates, agitation speeds, and so forth, can be seen.

Figure 72.2 shows a typical high-pressure autoclave equipment set inside the cell. The cell is a reinforced concrete total containment barrier rated for 1/4 lb trinitrotoluene (TNT). The cells have a high ventilation rate to prevent vapor-air explosions.

FIGURE 72.1 Operator's side of containment cell.

Design

The Departments of the Army, the Navy, and the Air Force manual entitled "Structures to Resist the Effects of Accidental Explosions"[1] offers a comprehensive approach to designing high-pressure cell barriers. Design details for both shelters (blast originating external to the structure) and barriers (blast originating inside the structure) constructed of reinforced concrete, steel, or a combination of concrete and steel are presented in this Army/Navy/Air Force manual. Concepts and some selected details from this manual are presented here in this chapter. The Army/Navy/Air Force manual makes some simplifying assumptions but with the assurance that such assumptions will lead to conservative design. However, specific designs should only be attempted by qualified design engineers. Historically, barriers have been constructed after cursory review based on past experience and limited engineering design with no assurances that they will perform as designed.[2]

Also it should be noted that unknown factors (e.g.,

unexpected shock wave reflections, construction methods, and quality of construction materials) can cause an overestimation of a barrier's ability to resist the effects of an explosion. To compensate for such unknowns, the Army/Navy/Air Force design manual recommends that the TNT equivalent weight, that is, "effective charge weight" used in design formulas, be increased by 20% over actual weight. The design process is iterative, meaning that a construction element with known dimensions and properties is selected, then several design parameters are calculated, ultimately solving for the amount of deflection. This is then compared to a resistance-deflection curve (Fig. 72.3) and if the deflection is not within the design objective, a different size or material is selected and the process repeated. Similarly for steel barriers, strain on a stress-strain curve is calculated for selected elements having different thicknesses and the process repeated until the required strain is achieved.

Design of a protective barrier requires an understanding of donor and receiver systems as well as the protective system. The donor system is the type and amount of

FIGURE 72.2 Interior of containment cell.

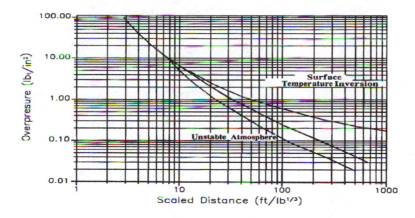

FIGURE 72.3 Correlation between overpressure and scaled distance. (Adapted from Reference 5.)

potentially explosive chemical and/or the physical equipment from which an expected blast pressure can be generated. The receiver system is the personnel and facilities that would receive the impact of an explosion if the protective system (i.e., barrier) were not present.

Donor Systems

Before barrier design can begin, an effective charge weight, W_E, in pounds of donor material must be determined in terms of its equivalent weight of TNT. TNT equivalency can be calculated from thermochemical data using computer programs such as ASTM's CHETAH[3] (Chemical Thermodynamic Energy Release Evaluation Program), or it can be determined empirically by standardized test methods such as the Trauzl lead block test. In a French variation of the Trauzl lead block test, a "coefficient of practical use" or c.p.u. is used to determine relative explosive power.[4] However, in France and the United Kingdom, picric acid is the reference standard for this test, while in Germany and the United States, TNT is the reference. TNT's explosive power is 0.94 times that of picric acid. Therefore, when using relative power data from the literature, be sure you know if picric acid or TNT is used as the reference standard.

TNT equivalency not only can be used for chemical energy but it can be extended to other forms of energy such as physical energy from equipment bursting at high pressure as will be discussed later.

For chemical explosions, the equivalent TNT charge can be determined from heats of detonation, H^d, as follows:

$$W_E = \frac{H_{EXP}^d}{H_{TNT}^d} W_{EXP} \times 1.2 \ (Safety \ factor) \quad (1)$$

where

W_E = effective charge weight (lbs) or equivalent TNT charge

W_{EXP} = weight of explosive in question (lbs)

H_{EXP}^d = heat of explosion (detonation or deflagration) of explosive in use (ft-lb$_f$/lb$_m$)

H_{TNT}^d = heat of explosion (detonation) of TNT (ft-lb$_f$/lb$_m$)
= 1.57E6 ft-lb$_f$/lb$_m$ (1120 cal/gm)

For nonchemical explosions such as rupture of a high-pressure autoclave, the effective equivalent TNT charge can be calculated from the following isentropic (i.e., constant entropy) expansion equation for an ideal gas[5]:

$$W_e = \left(\frac{P_1 V_1}{\gamma - 1}\right) \left| 1 - \left(\frac{P_2}{P_1}\right)^{(\gamma-1)/\gamma} \right| \times 1.2 \ \text{(Safety factor)} \quad (2)$$

where, temporarily switching from common engineering units to SI units,

W_e = expansion work (liter-atm)
P_1 = initial pressure of exploding vessel (atm)
P_2 = final pressure (1 atm)
V_1 = initial volume of exploding vessel (liters)
γ = heat capacity ratio (C_p/C_v) (dimensionless)

The liter-atm units of W_e can be converted to pounds of TNT (W_E) by multiplying by 4.85E-5 lbs TNT per liter-atm.

Knowing the energy of the explosion in terms of its TNT equivalency in pounds, one must next determine the force, or overpressure, on barrier walls at known distances from the explosion center. Overpressure on a wall can be calculated from the following scaling-law equation[5]:

$$Z = \frac{r}{W_{\text{TNT}}^{1/3}} \qquad (3)$$

where

Z   = scaling distance (ft/lb$^{1/3}$)
r   = distance from explosion to barrier wall (ft)
W_{TNT} = effective TNT equivalency weight of explosive (lbs)

Overpressure at the barrier wall is then determined from Figure 72.3. Weather conditions affect overpressure. The middle curve is for normal weather conditions, while the upper and lower curves are for abnormal weather conditions.

Missiles and shock waves must also be considered as donor system elements. Determining the worst-case scenario with respect to missile size, velocity, and scaled impulse on the barrier wall is extremely complex and not within the scope of this chapter. The Army/Navy/Air Force manual offers a rigorous approach to making this determination, which is summarized as follows:

• Primary fragment velocities are calculated using the Gurney[6,7] method. Primary fragments have very high initial velocities in the order of thousands of feet per second. They are relatively small in size compared to secondary fragments and are large in number.

• Likely sources of secondary fragments such as valves, piping, accessories, and their kinetic energy are determined next. The interaction of a blast wave with various-shaped objects is examined, and the specific impulse of objects impacting the barrier wall are calculated from the area of the objects facing the blast wave, the mass of the objects, and the pressure of the blast wave. The worst-case secondary missile scenario is selected for inclusion in the design. When protective barriers are located within 20 ft of a detonation, the velocity of an object striking the barrier can be assumed to be its initial velocity, imparted by the blast wave. Beyond 20 ft, scaling distances should be calculated using Equation 3.

• Reinforced concrete barriers are sandwiched between steel plates to provide best protection against missile penetration. Missiles can cause hazardous spalling of the exterior wall, exposing personnel to large chunks of concrete.

• The horizontal motion that shock waves impart to a barrier's foundation must be considered. The depth of the foundation and methods of reinforcing against shear forces is critical.

Receiver or Acceptor System

The receiver or acceptor system consists of personnel and facilities that would receive the impact of an explosion if a protective barrier were not present. The highest level of protection is needed for personnel. While human tolerance to blast overpressure is relatively high, the orientation (sitting, standing, face-on, or side-on) of the person, the rate of pressure rise, and the duration of the overpressure are very important factors. Lungs are considered the critical organ for survival. The threshold for lung hemorrhage is 30 to 40 psi for fast-rise, short-duration blast pressures and for severe lung hemorrhage, up to 80 psi.[1] Other potentially serious effects include eardrum rupture (Fig. 72.4); skin bruises (at overpressures

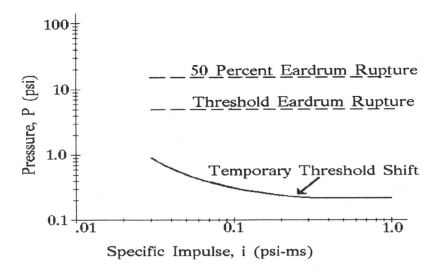

FIGURE 72.4 Human ear damage due to blast pressure. (Adapted from Reference 1.)

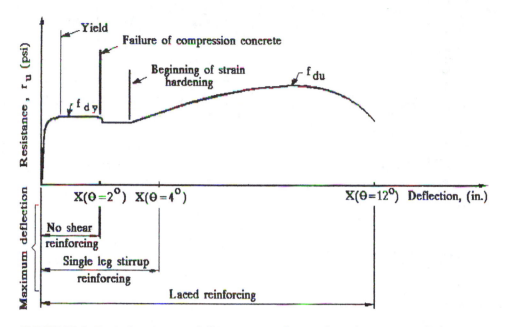

FIGURE 72.5 Typical resistance-deflection curve for reinforced concrete cell.

as low as 10–15 psi for long-duration loads); and being thrown off-balance (2–3 psi for persons located in the open).[1]

Protective System

In the case of reinforced concrete barriers, protective systems are designed from resistance-deflection curves. For steel barriers, stress-strain curves are used. These curves determine the response of a barrier to the force of an explosion. Barrier failure occurs at the point of maximum deflection. However, permanent deformation occurs at the yield point. Strain hardening occurs with rapid stress loads. This additional strength is indicated by the dynamic yield points, f_{dy} and f_{du}, shown in Figure 72.5 for reinforced concrete, and Figure 72.8 for steel. Considerable additional energy can be absorbed under conditions of dynamic loading. Below the static yield point, f_y, no permanent damage occurs and the barrier could be used again. When designing test cells that are expected to contain repeated explosions, the barrier's response should be in this elastic range. If, on the other hand, the purpose of the barrier is to protect personnel and facilities from an unlikely one-time event, it can be designed to protect out to its dynamic ultimate stress point, f_{du}.

Steel barriers are commonly designed out to the steel's maximum dynamic deformation limit for economy rea-

sons but, as with reinforced concrete, test cells that undergo repeated explosions need to be designed within the steel's elastic range. Otherwise, extensive repairs or complete replacement may be necessary after a barrier has been strained beyond its yield point. The dynamic yield, f_{dy}, on the "dynamic strength curve" shown as the dashed curve in the stress-strain diagram, may be a more appropriate design objective for test cells experiencing repeated blasts. Designing for the static yield, f_y, would be more conservative.

Reinforced Concrete

For reinforced concrete barriers, the resistance-deflection curve is as shown in Figure 72.5. The type of reinforcement used is extremely important to the strength of reinforced concrete barrier cells. Lacing reinforcing increases the strength of the barrier three times over conventional stirrup reinforcing and six times over no reinforcing as shown by the deflection values, X, in the resistance-deflection curve. This added strength is particularly important when potentially explosive equipment is situated "in-close" to the barrier wall, even if the cell has a high degree of explosion venting. Ideally, the explosion should be centered in the middle of the cell, which is where the main autoclave or test unit should be placed, but sometimes explosions occur in peripheral equipment close to the walls. Conventional and laced reinforcements are illustrated in Figures 72.6 and 72.7.

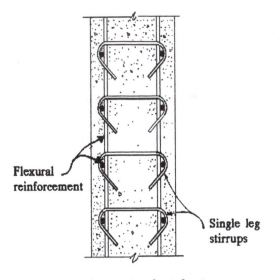

FIGURE 72.6 Conventional reinforcing. (Reproduced from Reference 1.)

Steel

Steel barriers use stress-strain curves instead of resistance-deflection curves (see Fig. 72.8). Steel exhibits a linear stress-strain relationship up to its yield point. Rapid stress loads result in "strain hardening" in steel as well. After reaching a maximum stress, called "the tensile strength," fracture occurs at an elongation amounting to 20 to 30% of the steel's original length. It is this ability of structural steel to undergo sizable "plastic" (permanent) deformations before fracturing that makes steel a desired barrier material for high-pressure cells.

Use

Due to the high cost of construction, usually only a small fraction of a research organization's total laboratory space is devoted to high-pressure cells. Hard decisions have to be made concerning what operations should or should not be in a barricade. Reserving this space for only the most hazardous operations opens the door to the possibility that an operation with underestimated hazards will not be allotted needed barricade protection. Conversely, allotting this space to operations not perceived as hazardous is inefficient use of facilities, but it can also lead to a breakdown in safe practices, particularly with respect to safety rules controlling when persons can and cannot enter a cell. For continued safe operation of a high-pressure laboratory, two administrative areas, that is, allotment of space and enforcement of safety rules, need to be discussed in more detail.

Space Allotment

Allotment of cell space should be based on decisions about how much is *not* known about the chemistry of an experiment. To be commercially successful, processes must be scaled up to a size that would make barricading prohibitively expensive. Therefore, risk assessments, which are made on virtually every chemical process today, should be asking the question: Is the process ready to forever shed its barricade or not? If the answer to that question is that there is not enough thermodynamic and kinetic information to ensure safe control, then it should be allocated space in a remote control cell. Computer-based thermodynamic calculations like CHETAH are useful for quantifying an equivalent TNT rating of a chemical system but should not be used as a go or no-go condition for allocating space. Once a rational decision is made to require remote operation, thermodynamic calculations like CHETAH, oxygen-balance calculations, literature values on TNT equivalency, and so on, should be used to quantify how much explosion energy should be in the cell, or, if cell space is not available, how much should be permitted in a specially shielded laboratory hood. This presupposes that all pressure cells and specially shielded alternatives to pressure cells have a TNT rating. If not, high priority should be given to getting all such facilities rated for a maximum allowable equivalent

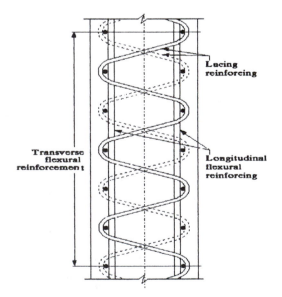

FIGURE 72.7 Laced reinforcing. (Reproduced from Reference 1.)

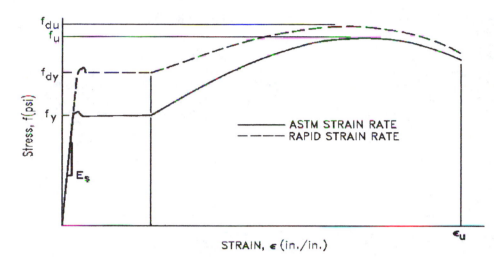

FIGURE 72.8 Typical stress-strain curves for steel. (Reproduced from Reference 1.)

charge of TNT. Other thermochemical data such as DTA, DSC, and ARC calorimetry data should be obtained as early as possible because these data are extremely useful for conducting risk assessments.[8]

Safety Rules

Safety rules for high-pressure cells are extremely important. Safe-operating procedures for a high-pressure cell should be based on the premise that every experiment in a pressure cell requires remote control. Because it must be operated under remote control, under no circumstances shall anyone enter a cell that is operating. Remotely operating an experiment is difficult and frustrating at best but it does not justify entering a cell to make quick adjustments to equipment. Also, there is never any reason that a door to a barricade should be left open when equipment in a cell is operating. If a researcher can justify relaxing either of these two rules, then his or her experiment should not be taking up valuable cell space that someone else could be putting to better use.

A rule to not lean against, touch, or in any way have a part of your body in contact with a barrier wall that is protecting against explosion is also needed. In an explosion, there can be enough energy imparted to a cell wall to seriously injure anyone leaning against the wall. Also, loose objects should never be left on beams, ledges, or other niches. They can be violently propelled into the operating area. Safety glasses should be required at all times, even in the so-called safe or protected operator area, because loose objects inadvertently left against a barrier, concrete spalling, or building motion could propel objects at personnel.

Conclusion

The services of a consultant or design engineering firm that has experience with explosion barrier design and demonstrated ability to fathom all of the demanding intricacies of barrier design should be engaged when building high-pressure cells. All design premises and assumptions affecting the final TNT equivalency rating of each cell should be documented. What is the resistance-deflection (or stress-strain) curve for the final design? What basis was used for primary and secondary fragment energy? What are the limitations of the explosion venting, if any? Is the TNT rating of a cell applicable to the elastic yield points, f_y or f_{dy}, or is it intended for ultimate stresses, f_u or f_{du}?

This information should be part of a documented record for each cell and made available for process safety reviews when new operations are proposed for placement in existing cells. This is to ensure that the original design limitations of a cell are taken into consideration.

References

1. *Structures to Resist the Effects of Accidental Explosions*; TM 5-1300/NAVFAC P-397/AFR 88-22; Joint Departments of the Army, the Navy and the Air Force, 1990.
2. Palluzi, R. P. *Pilot Plant and Laboratory Safety*; McGraw-Hill, Inc.: New York, 1994.
3. ASTM DS 51, CHETAH—the Chemical Thermodynamic Energy Release Evaluation Program; American Society of Testing and Materials: Philadelphia, Version 7.0, 1996.
4. Medard, L. *Accidental Explosions, Vol 1, Physical and Chemical Properties*; Ellis Horwood Ltd: Chichester, England, 1989.

5. Crowl, D. A.; Louvar, J. F. *Chemical Process Safety: Fundamentals with Applications*; Prentice-Hall: Englewood Cliffs, NJ, 1990.

6. *The Initial Velocities of Fragments from Bombs, Shells and Grenades*; Report No. 405; Gurney, R. W., Ed.; Ballistic Research Laboratories: Aberdeen Proving Ground, MD, 1943.

7. *The Mass Distribution of Fragments from Bombs, Shells and Grenades*; Report No. 448; Gurney, R. W., Ed.; Ballistic Research Laboratories: Aberdeen Proving Ground, MD, 1944.

8. Loving. F. A. *Ind. Eng. Chem.* **1957**, 49(10), 1744–1746.

General References

Browne, H. C.; Hileman, H.; Weger, C. *Ind. Eng. Chem.* **1961**, 53(10), 52A–58A.

Manning, W. R. D.; Labrow, S. *High Pressure Engineering*; CRC Press: Cleveland, OH, 1971.

Tunkel, S. J. *Chem. Eng. Prog.* **September 1983**, 50–55.

73

Clean Rooms for Semiconductors

JOHN A. MOSOVSKY

Clean rooms are facilities that use HEPA (High Efficiency Particulate Air) or ULPA (Ultra Low Penetration Air) filters for removing particles at efficiencies up to 99.999% down to a size of 0.3 μ in order that manufacturing and/or research and development operations that are extremely sensitive to particle contamination can take place. Clean room facilities are classified in FED-STD-209E, "Airborne Particulate Cleanliness in Clean Rooms and Clean Zones,"[1] according to the number of particles counted (≥ 0.5 μ) per cubic foot of air. For example, a Class 100 clean room would contain 100 particles ≥ 0.5 μ per cubic foot of air. Figure 73.1[2] shows a diagram of airflow in a conventional laminar flow clean room. Because of its superior particle contamination control, clean room technology has been applied to a number of industries, including semiconductor, aerospace, pharmaceutical, flat panel display, and photovoltaics. The electronics or semiconductor industry has been particularly proactive in safe clean room design. Because of this and because of its tremendous growth, this chapter will concentrate on facility design, engineering controls, and equipment as they apply to the semiconductor industry.

Semiconductor Loss History

Through risk management, an organization is able to prioritize identified risks so that limited resources that are used for reducing or eliminating these risks can be applied most effectively. Loss statistics are just one of the many inputs required for effective risk management. Figure 73.2 shows semiconductor industry losses prepared by Factory Mutual Engineering and Research Corpora-

tion and Allendale Insurance, major underwriters of the industry.[3] According to these data, which represent a large segment of the world's semiconductor manufacturing, liquid damage (from chemical spills and heating, ventilating, and air conditioning equipment), fire damage (which includes water damage), and service interruption account for almost 75% of the losses experienced between 1986 and 1995. From a risk-management point of view, a clear benefit can be derived from considering these statistics during the design phase of a clean room construction project.

Clean Room Design and Construction

Various codes, standards, and guidelines exist that directly apply to clean room design and construction. Building and fire codes become law when they are adopted by local governments. Because of this legal stature, considerable attention must be paid to specific code requirements that apply to clean room facilities handling hazardous materials. Building and fire codes are, however, considered to be minimum design requirements and should therefore be complimented with other standards and guidelines. The NFPA and SEMI (Semiconductor Equipment and Materials International) provide valuable companion documents that specifically address clean rooms. These standards and guidelines will be discussed under a separate heading.

Building and Fire Codes

There presently exist three major building and fire code organizations that publish codes that apply to clean

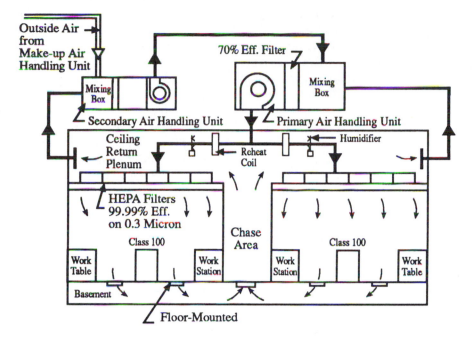

FIGURE 73.1 Laminar flow clean room. (Reproduced with permission from Factory Mutual Engineering Corporation, 1997.)

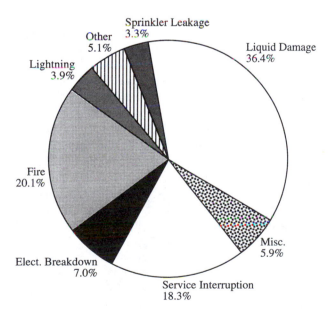

FIGURE 73.2 Semiconductor plant losses (percentage of total amount) by loss type, 1986–1995.

rooms. The International Conference of Building Officials publishes the Uniform Building and Fire Codes (UBC and UFC); the Building Officials and Code Administrators publishes a "national" building code (BOCA) and the National Fire Prevention Code (NFPC); and the Southern Building Code Congress publishes a "standard" building code (SBCC). Traditionally, each organization's codes have been adopted regionally by local governments throughout the United States—that is, UBC/UFC on the West Coast; BOCA on the East Coast and in the Midwest; and SBCC in southern United States and Midwest. The inconsistencies created as a result of this regional influence are currently being eliminated through a cooperative effort to develop one national code. Because the UBC/UFC codes have been in existence the longest, the focus of clean room facility design, construction, and hazardous materials in the next section will emphasize these code provisions. Code requirements can be complicated. The reader is urged to obtain a copy of the specific code that has jurisdiction over the facility being considered.

UBC/UFC Clean Room Provisions

The following information represents a summary of some of the code provisions applying to design and construction. Because codes apply to many different types of industrial and commercial facilities, classification of a proposed construction project is required. The classification considers the type, size, and location of the facility to be built and its proposed occupancy. Under Chapter 3 ("Use or Occupancy") of the UBC, semiconductor clean rooms are classified as H-6 occupancies. The H-6 designation is not reserved solely for semiconductor clean rooms but could apply to other facilities meeting the H-6 occupancy specifications:

> Group H occupancies shall include buildings or structures, or portions thereof, that involve the manufacturing, processing, generation or storage of materials that constitute a fire, explosion, or health hazard. . . . [4]

Under the Group H occupancy designation, seven divisions further define the occupancy. Division 6 covers:

> Semiconductor fabrication facilities and comparable research and development areas in which hazardous production materials (HPM) are used and the aggregate quantity of materials are in excess of those listed in Table 3-D or 3-E. Such facilities and areas shall be designed and constructed in accordance with Section 307.11. [4]

For further clarification, an HPM is defined in the code as:

> a solid, liquid or gas that has a degree of hazard rating in health, flammability or reactivity of 3 or 4 (*according to NFPA 704 "Standard System for the Identification of the Fire Hazards of Materials"*) and which is used directly in research, laboratory or production processes which have, as their end product, materials which are not hazardous. [4]

Tables 3-D and 3-E referenced in the Division 6 definition are extensive listings delineating allowable quantities of hazardous materials in "control areas," which are defined as occupied areas separated by one-hour fire-resistive construction. Section 307.11 of the code calls for compliance with the UFC and its provisions concerning separation, floors, ventilation, electrical design, shafts and enclosures, hazardous materials transport and storage, exits and service corridors, and piping and tubing. The UFC sections of greatest interest are Article 51, "Semiconductor Fabrication Facilities," Article 79, "Flammable and Combustible Liquids," and Article 80, "Hazardous Materials." [5] These articles provide specific requirements on the handling, use, storage, transport, dispensing, allowable quantities (which may differ from UBC requirements) and classification of hazardous materials, and emergency control and alarms.

Standards and Guidelines

Standards and guidelines used in conjunction with building and fire codes offer a comprehensive approach to safe clean room design and construction. NFPA 318, "Standard for the Protection of Clean Rooms," [6] provides descriptions of safeguards against fire and related hazards by addressing design considerations for ventilation, chemical storage and handling, hazardous gases, fire protection, and production and support equipment. NFPA 70, "National Electrical Code," [7] and NFPA 101, "Life Safety Code," [8] also have many provisions that directly apply to clean rooms. Factory Mutual Engineering Corporation's Loss Prevention Data sheets titled "Semiconductor Fabrication Facilities" [9] and "Clean Rooms" [2] address construction and design recommendations on facility systems and process equipment installation. SEMI's "Facilities Standards and Safety Guidelines" [10] provides information on clean room utilities as well as piping systems, effluent handling, gas enclosures, and process equipment. These guidelines also include test methods and specifications for ensuring safety and purity of materials and system components.

Materials Processing in Semiconductor Manufacturing

Knowledge of the chemicals and gases used and the process steps for manufacturing a semiconductor is needed in order to properly address clean room facility design and engineering controls. Comprehensive discussions of chemical and gas hazards can be found in this chapter's general references. The following discussion of process steps and the associated potential hazards of fabricating an integrated circuit is intended to provide a basic summary and is included here only to help clarify the next section on control technology.

The manufacture of a semiconductor can be divided into three broad operations: crystal growth, integrated circuit (IC) fabrication, and packaging. Although specific potential hazards are associated with the crystal growth and packaging operations, this discussion will focus on the clean room chemical- and gas-intensive IC fabrication process. This process can be further broken down into four specific operations: patterning, junction formation, deposition, and metallization. Using single-crystal silicon wafer substrates, a circuit pattern for a specific device is transferred to the wafer surface using photolithographic techniques. Electrical junctions are then formed in the silicon crystal lattice by diffusion and/or ion implantation of dopant species, that is, arsenic, boron, phosphorous, or antimony. Additional layers of an integrated circuit can be added through the deposition process in order to fabricate additional semiconductor components. This is achieved through epitaxial or chemical vapor deposition of single-crystal films. Once the circuit is completed, individual components are interconnected by the deposition of conductive metals or alloys through the metallization process. Chemicals and gases used throughout the fabrication process include mineral and organic acids; alkalis; peroxides; halogenated, aromatic, and polyaromatic hydrocarbons; metal hydrides; organometallic hydrides; ketones; metals; and corrosive and flammable gases. Potential health risks associated with these materials include acute and chronic inhalation toxicity, reproductive and carcinogenic risks, acute skin burns and ulcerations, and bone damage. Because of these risks and other potential physical hazards, clean room designs include extensive point source removal of contaminants and containment of high-hazard activities.

Control Technology

Engineering controls in semiconductor clean rooms are primarily concerned with the safe transfer of chemicals and gases from remote containers and bulk supplies to process equipment reaction chambers. Leak integrity and materials compatibility are the highest priorities for safety as well as purity. Once the source materials reach process equipment, enclosures and local exhaust ventilation represent secondary protection systems. Extensive monitoring and alarm systems are used to alert personnel to unsafe conditions. Although potential physical hazards such as ionizing and nonionizing radiation and noise associated with support equipment also exist, the following discussion on engineering controls will concentrate on gas transfer systems, process equipment and its support, and local exhaust ventilation.

Toxic, flammable, and/or pyrophoric gas leaks in an IC fabrication clean room most often involve the piping systems and gas cabinets containing compressed gas cylinders associated with process-equipment such as diffusion and deposition furnaces, epitaxial reactors, etching equipment, and ion implanters. As IC fabrication clean rooms become larger and the number of pieces of process-equipment increases, compressed gas cylinders are being replaced with bulk gas supplies. The advantages and disadvantages associated with this trend for silane gas have been well documented.[11] Storage of compressed gas cylinders should be kept to a minimum by practicing just-in-time delivery schemes. Storage areas must be well ventilated and clearly labeled and allow for segregation of gases according to hazard class. Fire detection and suppression systems should also be included. All gas cylinders should be leak checked on delivery, and cylinder valve flow-limiting devices should be specified from the gas supplier whenever possible, especially for metal hydrides. In addition to minimizing the consequences of toxic gas releases, flow-limiting devices can also reduce the area of evacuation zones needed in the event of a gas release, prevent excessive flow rates to process equipment, and allow for construction of smaller sized effluent treatment facilities. For example, at 200 psig, an arsine cylinder valve release without a flow-limiting device can be as high as 1085 liters/minute. With a 6-mil flow-limiting device, this same cylinder valve release can be reduced to 1.5 liters/minute.

Hazardous compressed gas cylinders are usually staged inside exhausted gas cabinets located inside gas houses or gas rooms attached to IC fab areas. Gas houses/rooms often include gas cabinet exhaust scrubbers, National Electrical Code hazardous location design requirements, fire detection and suppression systems, explosion venting, temperature control, continuous on-line gas leak monitoring, and security access systems. Because of silane's unique pyrophoric and potentially explosive properties, specific cylinder staging design recommendations should be followed.[12]

Gas cabinets serve to isolate hazardous compressed gas cylinders, provide controlled local exhaust ventilation, offer a means for safe purging before and after cylinder changes, and provide emergency and process gas control. Automatic, microprocessor-controlled gas panels are common design features, which are instrumental in reducing the human error so often experienced with manually operated panels. In addition, gas cabinets and panels include software and components specifically engineered to enhance safety. Some of these features include excess-flow switches, tied-diaphragm single-stage regulators, isolation and check valves, leak check ports, pressure and temperature sensors, exhaust failure monitors, and gas leak detectors. Many of the signals from these sensors cause automatic shutdown of process gas flow and activate local and remote alarms.

Gas delivery piping systems must be constructed from compatible materials and be of adequate strength and durability to withstand pressures and structural stresses. Welding is the preferred method for connecting piping lengths. The use of fittings should be kept to a minimum. Hazardous gas piping should be coaxial and be designed with a pressure monitoring system capable of detecting gas leaks into the annular space. Construction and leak integrity should follow SEMI specifications.[10]

Purchase orders for process equipment should include minimum equipment safety specifications. Because of the many inconsistencies among equipment suppliers, SEMI has published a standard intended to represent a minimum set of performance-based environmental health and safety considerations for equipment used in semiconductor manufacturing. SEMI standard S2-93, "Safety Guidelines for Semiconductor Manufacturing Equipment," defines safety considerations for equipment operation and maintenance dealing with chemical, gas, radiation, electrical, physical, mechanical, environmental, fire, explosion, seismic, ventilation, and ergonomic hazards.[10]

Vacuum pumps are common clean room support equipment because many semiconductor processes require a vacuum for proper reaction chemistry. Many safety issues exist with the pumping of hazardous gases. Harsh environments require special lubricants, flammable and pyrophoric gases require purging, pump oils may be contaminated and require special hazardous waste-disposal attention, and repair and maintenance operations require extreme caution. These issues are compounded by the fact that vacuum pumps are high-maintenance pieces of equipment but are located in service areas remote from process-equipment, which often leads to neglect. Specific controls for pumping hazardous gases can be found in American Vacuum Society Recommended Practices for Pumping Hazardous Gases.[13]

Many semiconductor processes are highly inefficient in using reaction source materials. Unused reactants from product formation must be scrubbed by proper environmental control equipment. Because a single process may use several classes of gases and chemicals, an integrated approach combining two or more treatment techniques may be needed. Important considerations for proper equipment selection include definition of effluent streams in terms of composition, flow rate, impurities, flow variations, reaction chemistries, and waste-disposal requirements. Once these parameters have been defined, an evaluation should consider available scrubber types, methods and techniques, limitations, collection efficiencies, and maintenance issues.

A safety status monitoring system is an important part of material flow in the IC fabrication process. Each piece of equipment using hazardous materials has a number of sensors that are designed to alert personnel to unsafe conditions. Gas leak detection, fire, exhaust failure, high temperature/pressure, purge failure, and so on, are some of the signals used, which in many cases are highly dependent on the type of process equipment used, its location, and its supporting equipment. These sensors must be tested on a periodic basis and tied into a central command center that is capable of initiating emergency actions, recording data, and monitoring critical on-line functions. Sensors are designed to warn of potentially hazardous events and minimize their consequences by initiating emergency actions such as automatic shutdowns, purging, ventilating, and so forth. Because of the number of sensors, alarms, and emergency actions involved, signal management systems should be used to monitor status. Systems are commercially available that tie in production operations and facilities maintenance monitoring as well as safety functions.

Many different types of local exhaust ventilation systems are used throughout the material flow process. Equipment exhaust systems are used for gas cabinets, wet process stations (sinks), photolithography equipment, equipment reaction chambers and valve enclosures, chemical storage cabinets, vacuum pumps, chemistry fume hoods, and support equipment such as cleaning-hoods and sinks, bead blasters, and ovens. Because standard exhaust designs and flow rates do not exist for some equipment, tracer gas containment studies are sometimes conducted in order to ensure personnel protection. References exist that address common exhaust systems as well as those systems unique to semiconductor manufacturing.[14,15]

Because of the complexity of semiconductor processes, formal process safety reviews should be conducted on equipment using hazardous materials. OSHA's Standard for Process Safety Management of Highly Hazardous Chemicals includes the elements necessary for a process safety management program; however, based on the standard's specified threshold quantities, many semiconductor processes are exempt from the standard's requirements. Process safety management is necessary, however, in order to evaluate and strengthen a process's lines of defense for preventing hazardous chemical/gas releases and for identifying, evaluating, and preventing failures in processes, procedures, and equipment that could result in chemical/gas releases. The American Institute of Chemical Engineers has developed guidelines for hazard evaluations, which include details for conducting What-if analyses, hazard and operability studies (HAZOP), failure mode and effects analyses (FMEA), and fault tree analyses (FTA).[16] Computer software is also available for helping to coordinate such analyses.

Future Trends

As semiconductor features become smaller, stricter contamination controls, clean room design flexibility, and cost reduction become drivers for "cluster tools" and mini-environments that isolate process operators from product, equipment, and hazardous chemicals. In addition to these drivers, customer demands for "green" products and processes are being translated into competitive advantage. Design for environment health and safety is being introduced throughout the product and process value chain. Organizations such as Sematech, the Semiconductor Industry Association, and the Semiconductor Safety Association are leading the efforts in research- and information-sharing of process and environmental health and safety improvements.

References

1. FED-STD-209E, Airborne Particulate Cleanliness in Clean Rooms and Clean Zones; Institute of Environmental Sciences: Mount Prospect, IL, 1995.
2. FMEC, Loss Prevention Data, Clean Rooms; Factory Mutual Engineering Corporation: Norwood, MA, 1988.
3. Semiconductor Fab Room Seminar; Factory Mutual Engineering and Research and Allendale Insurance: Berkeley Heights, NJ, 1996.
4. ICBO, Uniform Building Code; International Conference of Building Officials: Whittier, CA, 1994.
5. IFCI, Uniform Fire Code; International Fire Code Institute: Austin, TX, 1994.
6. NFPA 318, Standard for the Protection of Clean Rooms; National Fire Protection Association: Quincy, MA, 1995.
7. NFPA 70, National Fire Protection Association: Quincy, MA, 1999.
8. NFPA 101, Life Safety Code; National Fire Protection Association: Quincy, MA, 2000.
9. FMEC, Loss Prevention Data, Semiconductor Fabrication Facilities; Factory Mutual Engineering Corporation: Norwood, MA, 1991.
10. SEMI, Facility Standards and Safety Guidelines; Semiconductor Equipment and Materials International: Mountain View, CA, 1995.
11. Comparative Analysis of a Silane Cylinder Delivery System and a Bulk Silane Installation, Technology Transfer #95092976A-ENG; Sematech, Austin, TX, 1995.
12. Effects of Leak Size and Geometry on Releases of 100% Silane, Technology Transfer #96083168A-ENG; Sematech: Austin, TX, 1996.
13. O'Hanlon, J.; Fraser, D. *J. Vac. Sci. Technol.* **1988**, 6, 1226–1254.
14. ACGIH, *Industrial Ventilation Manual, A Manual of Recommended Practice*, 21st ed.; American Conference of Governmental Industrial Hygienists: Cincinnati, OH, 1992.
15. Williams, M. E.; Baldwin, D. G.; Manz, P. C. *Semiconductor Industrial Hygiene Handbook*; Noyes Publications: Park Ridge, NJ, 1995.
16. *Guidelines for Hazard Evaluation Procedures*, 2nd ed.; Center for Chemical Process Safety, American Institute of Chemical Engineers: New York, 1992.

General References

ACGIH, *Hazard Assessment and Control Technology in Semiconductor Manufacturing II*; American Conference of Governmental Industrial Hygienists: Cincinnati, OH, 1993.

Acorn, W. R. *Code Compliance for Advanced Technology Facilities*; Noyes Publications: Westwood, NJ, 1993.

Hazard Assessment and Control Technology in Semiconductor Manufacturing; American Conference of Governmental Industrial Hygienists; Lewis Publishers: Chelsea, MI, 1987.

The National Technology Roadmap for Semiconductors; Semiconductor Industry Association: San Jose, CA, 1994.

74

Ergonomics in Design

LAURA S. NYSTROM

Ergonomics is the study of work, derived from the Greek "ergon" meaning work, and "nomos" meaning laws of.[1] The recognition of ergonomics dates back to the 18th century with the observations of an Italian physician, Bernandino Ramazinni, who described various afflictions due to "certain violent and irregular motions and unnatural postures of the body."[2] Today, ergonomics focuses on the design of the jobs, tools, equipment, environment, and the overall organization of tasks and work flow. Through this focus, improvements can be realized in the health and safety of workers, morale, job and task performance, productivity, and quality.

Ergonomics Program Approach

The overall goal of a health and safety program is to maintain the health and well-being of the workers by avoiding accidents, injuries, and illnesses. Ergonomics is another facet of this health effort and ultimately should meld into an existing health and safety program. Ergonomics in the design of workplaces can be a two-pronged approach with both a reactive and proactive process. Initially, it will be reactive as you seek to solve existing problems. The long-term outlook is one in which ergonomics considerations are an integral aspect of the front-end design of equipment, office layout, process, or laboratory. If you are starting from ground zero, here is a suggested list of steps in achieving a proactive ergonomics program:

- Determine if ergonomics problems exist by review of:
  - occupational illness and injury logs
  - OSHA 200 logs
  - Workers Compensation data
- Gain management commitment by:
  - presenting the effect of ergonomic illnesses/injuries on the business
- Develop a policy and goals
- Assign resources
  - medical
  - safety and health
  - operations
  - engineering and maintenance
  - supervision
- Train the resources
- Develop a medical case-management program
- Provide employee with general ergonomic training
- Run a pilot program
  - locate problem areas
    - audit for ergonomic risk factors
    - examine worker complaints
    - evaluate jobs with high turnover
  - select the appropriate solutions
  - prioritize solutions
  - reexamine the solutions to ensure new problems haven't surfaced
- Refine the ergonomics program with learnings from pilot
- Proactive ergonomic design considerations
  - new equipment and facilities
  - upgraded equipment

Both the OSHA ergonomics guideline for meatpacking plants[3] and the latest draft of the ANSI proposed stan-dard on controlling work-related cumulative trauma disorders (CTDs)[4] highlight these elements as the necessary components of a program.

Matching Task Requirements with Worker Capability

Humans have evolved with specific capabilities and requiring the employee to adapt to the workplace imposes a cost to conforming. When the adaptations are too great, a mismatch occurs and can manifest itself as overexertion, strain or sprain, back injuries, and repetitive-motion disorders. The responses to the mismatches include injuries, illnesses, failure to perform, increased scrap and rework, job stress, absenteeism, and high job turnover.

Jobs, tools, equipment, work environment, and facilities should be able to accommodate 95% of the workers. To do this, six aspects of the human system need to be considered.[5]

Physical Size (Anthropometry)

Anthropometry is the science of human measurement. People's size affects their reach, clearances around equipment, and their ability to work standing or sitting. Ideally, workplaces should be adjustable to fit a variety of users.

Endurance (Physiology)

Endurance affects a worker's ability to perform such jobs as heavy jobs, because these demand excess oxygen, and place a strain on the cardiovascular system; jobs under high temperature conditions, because they may result in heat stress and illness; and static work or work involving heavy muscular contractions with little movement, because this results in rapid muscle fatigue.

Strength (Biomechanics)

Different amounts of strength are required to lift, push, carry, pull, maneuver, or move objects. Some people can lift twice as much as other people, but jobs should be designed to allow all employees to work at the tasks.

Movement/Posture (Kinesiology)

Many tasks require workers to use their hands. Some demand strength to grasp, move, or lift while others require dexterity or highly repetitive movement. Overuse of the hand and other parts of the body held in unnatural positions to perform highly repetitive motions can result in CTDs. Tools enhance the versatility of hands and should be used where appropriate, but with sound ergonomic design.

Working Environment (Noise, Light, Vibration, Temperature)

The work environment can affect the performance of the employees. The following elements require monitoring for both comfort and safety: thermal element—temperature, humidity, and airflow; lighting—adequate lighting levels, glare, reflections, shadows, and undue contrast; sound—excessive noise can damage hearing, high noise can impair performance and interfere with oral communications; vibration—hand/arm or whole-body vibration can impair performance and cause other forms of CTDs.

Information Processing (Cognition)

People are required to perform mental functions as part of nearly every job. Although some tasks may be merely repetitive, others require rapid processing of information to perform sequential tasks. The limit of cognitive process is influenced by factors such as training, memory, goals, intentions, personality, attitude, motivation, mood, stress, and knowledge but the ultimate result is to decide on a course of action.[6]

The checklists at the end of the chapter provide guidance on the application of these various systems in the office, laboratory, or manufacturing environment.

Revised NIOSH Lifting Equation

The 1991 Revised NIOSH Lifting Equation is a tool that can determine the maximum two-handed lift under specific lifting conditions. It derives a Recommended Weight Limit (RWL) that nearly all healthy workers could perform over a substantial period of time (e.g., up to 8 hours) without an increased risk of developing lifting-related low-back pain. The equation is devised so that the weight is acceptable for the majority of the population (90% of men and 75% of women.)[7]

The concept behind the revised NIOSH Lifting Equation is to start with a recommended weight that is considered safe for an "ideal" lift (i.e., load constant equal to 51 lb) and then reduce the weight as the lifting conditions worsen. The equation for calculating the RWL is based on a model that provides a multiplier for each of six task variables:

- H, horizontal distance of the hands away from the midpoint between the ankles. Measure at the origin and destination of the lift (cm or in.).
- V, vertical distance of hands above the floor. Measure at the origin and destination of the lift (cm or in.).
- D, vertical travel distance between the origin and destination of the lift (cm or in.).
- A, angle of asymmetry, how far the object is displaced from the front (mid-sagittal plane) of the worker's body. Measure at the origin and destination of the lift (degrees).
- F, frequency, average number of lifts per minute and duration of lifting.
- C, quality of the hand-to-object coupling, either good, fair, or poor.

The RWL is defined by the following equation:

$$RWL = LC \times HM \times VM \times DM \times AM \times FM \times CM$$

where

	Metric	U.S. Customary
LC = load constant	23 kg	51 lb
HM = horizontal multiplier	(25/H)	(10/H)
VM = vertical multiplier	$1 - (0.003 \lvert V - 75 \rvert)$	$1 - (0.0075 \lvert V - 30 \rvert)$
DM = distance multiplier	$0.82 + (4.5/D)$	$0.82 + (1.8/D)$
AM = asymmetric multiplier	$1 - (0.0032A)$	$1 - (0.0032A)$
FM = frequency multiplier	From Table 74.1	From Table 74.1
CM = coupling multiplier	From Table 74.2	From Table 74.2

The lifting index (LI) is a term that provides a relative estimate of the level of physical stress associated with a particular manual lifting task. The estimate of the level of physical stress is defined by the relationship of the weight of the load lifted and the recommended weight limit. The LI is defined by the following equation:

$$LI = \text{Load Weight/Recommended Weight Limit}$$
$$= L/RWL$$

where L = the weight of the object lifted (kg or lb). The goal LI is less than or equal to one.

TABLE 74.1 Frequency Multiplier (FM) Table

Frequency[a] (lifts/min) (F)	Work Duration					
	≤1 hr		−1 but ≤ 2 hr		>2 but ≤ 8 hr	
	V[b] < 30	V ≥ 30	V < 30	V ≥ 30	V < 30	V ≥ 30
≤0.2	1.00	1.00	0.95	0.95	0.85	0.85
0.5	0.97	0.97	0.92	0.92	0.81	0.81
1	0.94	0.94	0.88	0.88	0.75	0.75
2	0.91	0.91	0.84	0.84	0.65	0.65
3	0.88	0.88	0.79	0.79	0.55	0.55
4	0.84	0.84	0.72	0.72	0.45	0.45
5	0.80	0.80	0.60	0.60	0.35	0.35
6	0.75	0.75	0.50	0.50	0.27	0.27
7	0.70	0.70	0.42	0.42	0.22	0.22
8	0.60	0.60	0.35	0.35	0.18	0.18
9	0.52	0.52	0.30	0.30	0.00	0.15
10	0.45	0.45	0.26	0.26	0.00	0.13
11	0.41	0.41	0.00	0.23	0.00	0.00
12	0.37	0.37	0.00	0.21	0.00	0.00
13	0.00	0.34	0.00	0.00	0.00	0.00
14	0.00	0.31	0.00	0.00	0.00	0.00
15	0.00	0.28	0.00	0.00	0.00	0.00
>15	0.00	0.00	0.00	0.00	0.00	0.00

[a]For lifting less frequently than once per 5 min, set F = 0.2 lift/min.
[b]V is expressed in inches as measured from the floor.

TABLE 74.2 Coupling Multiplier (CM) Table

Coupling Type	Coupling multiplier	
	V < 30 in. (75 cm)	V ≥ 30 in. (75 cm)
Good	1.00	1.00
Fair	0.95	1.00
Poor	0.90	0.90

> Good—Optimal container with optimal handles or cut-outs (see notes 1, 2, 3). Loose objects must conform to note 6.
> Fair—Optimal container with handles or cut-outs that are nonoptimal (see notes 1, 2, 3). Optimal container with no handles or cut-outs must conform to note 4.
> Poor—Nonoptimal containers with no handles or cut-outs (see note 5). Loose objects that do not conform to note 6.

Notes.
1. An optimal handle design has 0.75–1.5 in. (1.9–3.8 cm) diameter, ≥4.5 in. (11.5 cm) length, 2 in. (5 cm) clearance, cylindrical shape, and a smooth nonslip surface.
2. An optimal hand-hold cut-out has these approximate characteristics: ≥1.5 in. (3.8 cm) height, 4.5 in. (11.5 cm) length, semi-oval shape ≥2 in. (5 cm) clearance, smooth nonslip surface, and ≥0.25 in. (0.6 cm) container thickness (e.g., double thickness cardboard).
3. An optimal container design has ≤16 in. (40 cm) frontal length, ≤12 in. (30 cm) height, and a smooth nonslip surface.
4. A worker should be capable of clamping the fingers at nearly 90 degrees under the container, such as required when lifting a box from the floor.
5. A container is considered less than optimal if it has a frontal length >16 in. (40 cm), height >12 in. (30 cm), rough or slippery surfaces, sharp edges, asymmetric center of mass, unstable contents, or requires the use of gloves. A loose object is considered bulky if the load cannot easily be balanced between the hand-grasps.
6. A worker should be able to comfortably wrap the hand around the object without causing excessive wrist deviation or awkward postures, and the grip should not require excessive force.

APPENDIX: Ergonomic Checklists

General Workstation Design Principles

Use of the following general workstation design principles will help achieve an optimum match between the work requirements and operator capabilities. This, in turn, will maximize the performance of the total system while maintaining human comfort, well-being, efficiency, and safety.

1. Make the workstation adjustable so that the tall person will fit, and the small person can reach easily.

2. Provide materials and tools in front of the worker to reduce twisting.

3. Avoid static work postures. Avoid jobs requiring operator to:
   - lean to the front or the sides
   - hold an extremity in a bent or extended position
   - bend the torso or head forward/backward more than 15 degrees

4. Set the work height at 2 inches below the elbows for most work and 6 inches above the elbows for precision work.

5. Provide adjustable, properly designed chairs.

6. Allow the workers, at their discretion, to alternate between sitting and standing. Providing floor mats/padded surfaces for prolonged standing

7. Support the limbs. Provide wrist-, arm-, foot-, and backrests as needed.

8. Use gravity to move materials.

9. Design the workstation so arm movements are continuous and curved. Avoid straight-line, jerking arm movements.

10. Keep arm movements in the normal area (limit reaches to 16 inches).

11. Provide dials and displays that are simple, logical, and easy to read and operate.

12. Eliminate excessive noise, heat, humidity, cold, and poor illumination.

Ergonomic Guidelines for Laboratory Employees

1. Position equipment and tools directly in front of your body to reduce twisting motions and reach distances.

2. Eliminate excessive reaches over 16 inches.

3. Set the work height (hand height) at 2 inches below the elbows to minimize static muscle fatigue of the shoulders, neck, and back.

4. Maintain good seated posture. Adjust the backrest so that it provides firm support to your lower back. Do not sit on the edge of the chair.

5. Minimize static loads and fixed work postures. Avoid:
   - leaning to the front or sides
   - holding an extremity in a bent or extended position
   - tilting your head forward more than 15 degrees
   - bending your body forward or backward more than 15 degrees
   - supporting your weight with one leg

6. Support the limbs. Use elbow-, wrist-, arm-, and backrests when needed.

7. Reduce the number of repetitive hand, wrist, and finger motions. The use of the thumb is preferred to the use of a single finger for trigger action. Motorized pipettes eliminate repetitive, forceful trigger-finger motions.

8. Maintain neutral (handshake) wrist postures. Design experiments so that the wrist does not need to be flexed forward, extended backward, or bent from side to side.

9. Reduce grip force requirements. Select pipettes with a compressible, rather than hard plastic-gripping surface. Use tools that utilize a full-hand power grip, rather than a more forceful precision finger grip.

10. Avoid pounding with the pipette to pick up tips.

11. Select tools which minimize stress on soft tissues. Stress concentrations result from tool handles that exert pressure on the palms or fingers. Examples include short-handle pliers and tools with finger grooves that do not fit the specific employee's hand.

Ergonomic Guidelines for Computer Operators

The simple adjustments outlined below may increase the comfort of your computer workstation. Consider the following to prevent musculoskeletal and visual fatigue:

Reprinted with permission from David Ridyard, CIH, CSP, CPE, Applied Ergonomics Technology, 270 Mather Rd., Jenkintown, Pennsylvania 19046.

_____ 1. *Adjust the height of your work surface and the height of your chair* so that your keyboard is at elbow height and your feet are flat on the floor. If the work surface is not adjustable, a footrest should be used when the feet do not rest flat on the floor.

_____ 2. *Adjust the backrest of the chair* so that it provides support to your lower back. Do not sit on the edges of the chair; rest your back against the backrest.

_____ 3. *Position the screen* directly in front of you. The distance between your eyes and the screen should be approximately 18 to 20 inches.

_____ 4. *Adjust the height of the screen* so that your eyes are level with the top of the screen.

_____ 5. *Tilt the screen to minimize glare.* Tilting the screen will help reduce glare caused by bright overhead lights.

_____ 6. *Draw drapes or shades and utilize task lighting* rather than bright overhead lighting when working at the computer to reduce glare.

_____ 7. *Use a document holder.* Documents placed flat on the desk will cause you to lean forward and flex your neck, leading to fatigue and discomfort. The document and screen should be located at approximately the same distance to eliminate constant eye refocusing at varying viewing distances.

_____ 8. *Keep the area under your desk clear* for adequate leg and knee room.

_____ 9. *When keying, rest your wrists and/or forearms, and keep the upper arms nearly vertical* to prevent fatigue. Use a wrist rest, if necessary, to maintain your wrists, hands, and arms in a straight horizontal line.

_____ 10. *Take frequent micro-breaks and stretch periodically* to reduce the soreness and stiffness related to fixed, static work postures.

Selection of Ergonomic Hand Tools

Poorly designed hand tools that combine repetitive forceful grip exertions with bent wrist postures can cause carpal tunnel syndrome and other cumulative trauma disorders. The following checklist can be used as a guide for selecting hand tools:

_____ 1. Maintain straight wrists. Avoid bending or rotating the wrists. Remember, bend the tool, not the wrist. A variety of bent-handle tools are commercially available.

_____ 2. Avoid static muscle loading. Reduce both the weight and size of the tool. Do not raise or extend elbows when working with heavy tools. Provide counterbalanced support devices for heavier tools.

_____ 3. Avoid stress on soft tissues. Stresses result from poorly designed tools that exert pressure on the palms or fingers. Examples include short-handled pliers and tools with finger grooves.

_____ 4. Reduce grip force requirements. A compressible gripping surface rather than hard plastic is best.

_____ 5. Whenever possible, select tools that utilize a full-hand power grip rather than a precision finger grip.

_____ 6. Maintain optimal grip span. Optimal grip spans for pliers, scissors, or tongs, measured from the fingers to the base of the thumb, range from 2.5 to 3.5 inches. Recommended handle diameters for circular-handle tools such as screwdrivers are 1.5 to 2 inches when a power grip is required, and 0.3 to 0.6 inches when a precision finger grip is needed.

_____ 7. Avoid sharp edges and pinch points. Select tools that will not cut or pinch the hands even when gloves are not worn.

_____ 8. Avoid repetitive trigger-finger actions. Select tools with large switches that can be operated with all four fingers. Also, use of the thumb is preferred to using a single finger for trigger action. Proximity switches are the most desirable triggering mechanism.

_____ 9. Protect hands from excessive heat, cold, and vibration.

_____ 10. Wear gloves that fit properly to enhance strength and dexterity.

Lifting and Lowering Tasks: Ergonomic Design

The following checklist should be used to eliminate the need to manually lift heavy or bulky materials, and reduce unnecessary bending, twisting and reaching when lifting materials:

1. *Optimize Material Flow Through the Workplace*:

_____ Reduce manual handling of materials to a minimum.

_____ Establish adequate receiving, storage, and shipping areas.

_____ Maintain adequate aisle and access areas.

2. *Eliminate the Need to Lift or Lower Manually*:

____ Lift Tables and Platforms	____ Elevated Pallets
____ Lift Trucks	____ Gravity Dump Systems
____ Cranes and Hoists	____ Elevating Conveyors
____ Drum and Barrel Dumpers	____ Vacuum Systems

3. *Increase the Weight So It Must Be Handled Mechanically*:

____ Palletized handling of raw materials and products.

____ Unit load concept (bulk handling in large bins).

4. *Reduce the Weight of the Object*:

____ Reduce the weight and capacity of the container.

____ Reduce the load in the container.

____ Specify the quantity per container to suppliers.

5. *Reduce the Hand Distance From the Body*:

____ Change the shape of the object or container.

____ Provide grips or handles.

6. *Convert Lift/Lower Combined with a Carry to a Push or Pull*:

____ Conveyors, Hand Trucks, Carts	____ Ball Caster Tables

Pushing and Pulling Tasks: Ergonomic Design

The following checklist should be used to eliminate manually pushing or pulling materials, or to reduce the exertion hazard when materials are pushed or pulled:

1. *Eliminate the Need to Push or Pull*:

____ Conveyors	____ Lift Tables
____ Slides or Chutes	____ Powered Trucks

2. *Reduce Force Required to Push or Pull*:

____ Reduce size and/or weight of load.

____ Utilize four-wheel trucks or dollies.

____ Utilize non-powered conveyors.

____ Require that wheels and casters on hand-trucks and dollies have:
   • Periodic Lubrications of Bearings
   • Adequate maintenance
   • Proper sizing (provide larger diameter wheels & casters)

____ Maintain floors to eliminate holes and bumps.

____ Require surface treatment of floors to reduce friction.

3. *Reduce the Distance of the Push or Pull*:

____ Relocate receiving, storage, production, or shipping areas.

____ Improve production to eliminate unnecessary material handling.

4. *Optimize Technique of the Push or Pull*:

____ Replace pull with a push whenever possible.

____ Use ramps with slope less than 10%.

Repetitive Hand Tasks: Ergonomic Design

Workers may experience pain, discomfort, and disabling hand and wrist disorders such as carpal tunnel syndrome if the high number of repetitions are combined with abnormal wrist postures and excessive forces. The following checklist should be used as a guide for designing safe hand and wrist activities:

____ 1. Reduce the number of repetitions per shift. Automated systems should be used whenever possible.

____ 2. Maintain neutral (handshake) wrist positions. Design jobs and tools so that the wrist is not flexed forward, extended backward, or bent from side to side. Avoid inward and outward rotation of the forearm when the wrist is bent to minimize stress to the elbow.

____ 3. Reduce the force or pressure on the wrists and hands. Reduce the weight and size of objects that must be handled repetitively. Avoid using the hand as a "hammer."

____ 4. Design tasks so a power grip rather than a finger-pinch grip can be used to grasp materials. Note that a pinch grip is five times more stressful than a power grip.

____ 5. Avoid reaching more than 16 inches in front of the body. To minimize shoulder disorders, avoid reaching above shoulder height, below waist level, or behind the body.

____ 6. Provide support devices where awkward body postures (elevated hands, elbows, and extended arms) must be maintained.

____ 7. Avoid tools and equipment that transmit vibration to the hands.

____ 8. Avoid exposure of the hands to cold, hot, and vibration.

____ 9. Wear gloves that fit properly to enhance grip strength and manual dexterity.

____ 10. Select and use properly designed hand tools.

Computer Workstation: Ergonomic Design

Use of the following set of computer workstation design guidelines will maximize the performance of the computer operator while maintaining human comfort, well-being, efficiency, and safety. Note that *adjustability* of the workstation is the key.

_____ 1. The *keyboard and mouse surface* should be height adjustable between 23 and 28 inches. Both keyboard and mouse should be positioned at elbow height.

_____ 2. *Workstation width* should be at least 30 inches.

_____ 3. *Depth of the workstation* should allow for screen, keyboard, and approximately 3 inches to serve as a wrist rest area.

_____ 4. *Edges of the work surface* must be rounded and at least 1 inch thick to prevent stress on the arms and wrists.

_____ 5. A *wrist rest* should be provided to enhance workstation adjustability.

_____ 6. *Knee room design considerations*:
- Knee-room height should be a minimum of 26 inches.
- Knee-room width should be a minimum of 20 inches.
- Knee-room depth should be a minimum of 15 inches.

_____ 7. *Chair design considerations*:
- Chair base should be five-point, on casters. Chair should swivel.
- Chair should have an adjustable seat height.
- Seat size between 15 and 17 inches deep, and 18 inches wide with a "waterfall" front edge.
- Seat slope would be between 10 degrees forward and 10 degrees backward.
- Seat pan and backrest should be upholstered. When seated, the seat pan and backrest should not compress more than 3/4 inch.
- Backrest size should be 7 inches high and 13 inches wide; backrest height should be adjustable between 3 and 6 inches; backrest depth should be adjustable between 14 and 17 inches; and backrest tilt should be adjustable 15 degrees forward and backward.
- Removable, height-adjustable armrests should be incorporated.
- Backrest, if adjustable, should have a locking mechanism.

_____ 8. *Screens* should be located directly in front of the operator at a distance of 18 to 20 inches.

Screens should be located at right angle to windows.

_____ 9. *Task lighting*, rather than overhead lighting, should be provided to minimize glare. Work surfaces and walls should be furnished with nonglare materials.

Audit of Materials-Handling Risk Factors

The following items should be considered when evaluating new or existing materials-handling equipment, processes, work activities, or workstations:

1. Does the task require any of the following activities?
   _____ Lifting, lowering, or carrying more than 25 pounds.
   _____ Lifting, lowering, or carrying objects that are too bulky to easily grip and hold close to the body.
   _____ Lifting, lowering, or carrying materials more than 50 times per shift.
   _____ Lifting above shoulder height or below waist level.
   _____ Lifting in cramped areas resulting in bending, reaching, twisting.
   _____ Pushing/pulling carts, and so on, that require large forces to get moving.
   _____ Maintaining a fixed or awkward work posture (e.g., overhead work, twisted or bent back, kneeling, stooping, or squatting).

2. Do any of the following unsafe practices or conditions exist?
   _____ Employees not following procedures (taking short-cuts, etc.).
   _____ Employees not using appropriate materials-handling equipment.
   _____ Materials-handling equipment inadequate or damaged.
   _____ Materials improperly stacked, loaded, or banded on fork trucks.

3. Do any of the following unsafe work area conditions exist?
   _____ Unsafe (cracked or broken pavement, etc.) walking surfaces.
   _____ Poor housekeeping (wet, oily floors; debris; clutter).
   _____ Poor layout of work area—crowding, congestion, excessive traffic.
   _____ Excessive noise, heat, humidity, cold, or poor illumination.

AREA AUDITED: _____

AUDITED BY: _____

Audit of Repetitive Hand Tasks

The following items should be examined when evaluating new or existing work activities requiring highly repetitive motions with the hands or wrists (i.e., manual packaging or inspection activities, and the use of hand tools):

1. Does the task require any of the following activities?
   _____ Performing the same motion every few seconds with the hands or wrists.
   _____ Repetitively shaking cartons, bottles, or other materials (repetitive bending of the wrists).
   _____ Bending the wrists when working. The wrists should be maintained in a neutral (handshake) position. The job task should not require the wrists to be flexed forward, extended backward, or bent from side to side.
   _____ Working for extended periods of time with awkward body postures (elevated hands and elbows; extended arms; reaching behind body).
   _____ Exerting high grip forces with the hands.
   _____ Using pinch grip rather than a power (curled-finger) grip.
   _____ Using the hand as a "hammer."
   _____ Repetitively reaching more than 16 inches in front of the body.

2. Do any of the following unsafe hand tool practices or conditions exist?
   _____ Using vibrating or impact tools or equipment.
   _____ Using tools that require bending or rotating the wrists.
   _____ Raising or extending the elbows when working with heavy tools.
   _____ Using tools requiring repetitive trigger-finger

actions, excessive grip forces, or excessive forceful exertions.

AREA AUDITED: _____

AUDITED BY: _____

References

1. Grandjean, E. *Fitting the Task to the Man—A Textbook of Occupational Ergonomics*, 4th ed.; Taylor & Francis: London, 1988.
2. Ramazinni, B. *De morbis artificum distraba*, 1717. (Translated by Wright, W. C., University of Chicago Press: Chicago, 1940.)
3. DOL, *Ergonomics Program Management Guidelines for Meatpacking Plants*; OSHA, Publication 3123, 1990.
4. *Control of Work-Related Cumulative Trauma Disorders*; Working Draft of ANSI Accredited Standards Committee Z-365, New York, June 10, 1997.
5. Alexander, D. P.; Alexander, J. P. *Ergonomics Design Guidelines*; Auburn Engineers, Inc.; 123 North College St., Auburn, AL, 1992.
6. Chemical Manufacturers Association. *A Manager's Guide to Reducing Human Errors*: Washington, D.C., July 1990.
7. Waters, T. R.; Putz-Anderson, V.; Garg A. *Applications Manual for the Revised NIOSH Lifting Equation*; DHHS(NIOSH) Pub. No. 94-110; U.S. Department of Commerce, NTIS: Springfield, VA, January 1994.

General References

Eastman Kodak Co. *Ergonomic Design for People at Work, Vol. 1*; Van Nostrand Reinhold: New York, 1983.
Eastman Kodak Co. *Ergonomic Design for People at Work, Vol. 2*; Van Nostrand Reinhold: New York, 1986.
Putz-Anderson, V. *Cumulative Trauma Disorders—A Manual for Musculoskeletal Diseases of the Upper Extremity*; Taylor & Francis: London, 1988.

75

Ergonomic Factors in Laboratory Design

JOSHUA O. KERST

The High-Performance Laboratory

Human Factors Engineering or ergonomics is an applied engineering discipline. The general philosophy of ergonomics is to design tools, workstations, work methods, and the ambient environments to meet and enhance the capabilities of people. Although ergonomics is an engineering discipline, it is also a philosophy. Its value is in allowing people to look at things from a different perspective; to break convention; to escape from stereotypes; to improve. Fundamental to understanding ergonomics is understanding the role of the laboratory and its evolution.

When the laboratory introduced its way into the

world of research, the research activities centered around a single table or countertop. Since then, this work area has evolved into lab benches. Drawers were added underneath and cupboards were installed overhead. Technology advanced to include automated processes for centrifugation, chemical assays, and processing. Computers were added for microprocessing, image analysis, and robotic control. Regulations were established for exhaust and fume control leading to fume hood and ventilation systems. With all of these advancements, and subsequent space requirements, there has been a concurrent development of occupationally induced degenerative trauma injuries in the laboratory.

Attention to ergonomic issues can help prevent the development of these injuries and illnesses and can result in a happier and more productive workplace. This chapter begins by explaining some background information on ergonomic concerns and their effects on the human body. Risk areas are targeted and explained. Guidelines for the design of sitting and standing laboratory workstations are outlined with detailed problems and solutions. Finally, a nine-step procedure is presented to effectively deploy ergonomics in the laboratory environment.

Ergonomics in the Laboratory: Is There a Need to Proceed?

Laboratory design plays a major role in the work performance of individuals. The lab environment can affect the functioning of the organization by influencing employee health and safety, product quality, and production efficiency, thereby influencing the effectiveness of the organization as a whole.

The laboratory affects people both directly and indirectly. A direct effect of lab environment design is the prevalence of *human error*. Human error affects hazardous accidents, quality of products, rework rate, and overall efficiency. Another direct effect is the lab environment contribution to *degenerative disease* through requirement of difficult postures or repetitive movements that are biologically intolerable.

Indirect impacts may include *reduced quality of work life*. This dissatisfaction is manifested in tardiness, absenteeism, unwillingness to cooperate and contribute, job turnover, and so on. Another factor is *fatigue*. Fatigue takes problems in the lab environment and increases their likelihood or exacerbates their consequences.

In the past, productivity was increased through improvements in the automation, components, and testing processes. At that time, people were relatively inexpensive and automated products were relatively expensive. Concerns for human performance and well-being were limited to rudimentary safety precautions related to machine guarding and other factors dictated by the minimum legislated standards.

Today, people are relatively expensive and machinery relatively inexpensive, due to increases in costs for benefits, wages, and injury compensation. It is becoming increasingly apparent that the human plays a significant role in determining the effectiveness of the lab. Another irony of automation is that the more automated a laboratory is, the more important people become. Human performance in decision making, programming, quality control, and so forth, impacts the entire automated process. It is clearer now that engineering and design programs that emphasize the machine side of the human-machine interface, to the exclusion of other factors in the system, have limited viability and result in limited improvements. The human component must be accommodated and integrated with all other system components so that the relative attributes of the individual components as well as the synergy of the entire system can be realized. This approach to integrating all components has been referred to as *system design*.

The performance of your people follows a very simple formula, which is summarized in one of W. Edwards Deming's 14 Points of Quality. Point #12 states: *"Nothing should stand between a worker and their right to pride of workmanship."* In continued efforts to achieve world-class quality performance, efforts have been made to remove all of the barriers that "stand between" people and success in the lab. This chapter focuses on the often-overlooked area of person-equipment layout needs for the high-performance laboratory.

Explanation of Ergonomic Concerns in Laboratories

Enhancing human performance is fundamental to enhancing laboratory performance. The better people perform, the more productive and profitable a lab will be. These problems range from poor morale and reduced productivity to injury and illness. A primary challenge to the health and safety of laboratory personnel is the development of cumulative trauma disorders or CTDs. CTDs are a class of illness that is the result of months and years of overuse of human joints and connective tissues to the point that they become sore and sometimes unusable.

The best way to visualize cumulative trauma is to think of one's joint structures as a bucket with fixed capacity. Microtrauma from the job and from non-job activities drips into each joint's trauma bucket. Fortunately, the body can heal with time and safely absorb a certain amount of trauma. But if we place more trauma into the bucket than can be absorbed by the natural healing process, the result is impaired movement.

Cumulative Trauma Disorders

Consequently, cumulative trauma is a disorder that is based on a dose exposure. Similar to hearing loss, cumulative trauma disorder occurs gradually over a long period of exposure to a low-level harmful agent. A brief exposure to these agents would not cause harm; however, prolonged exposure results in reduced ability to function. To prevent the development of cumulative trauma disorders there must be a reduction in the microtrauma to the joints. Three factors contribute to this microtrauma. They are the force applied by the person, the frequency of the force application, and the posture assumed during this activity. An investigation into these factors will unveil the root causes of CTDs in the laboratory.

Force

There exists a clear understanding of how an acute injury occurs. If a knife is pressed against the skin with a small amount of pressure, nothing happens. If the pressure increases, the skin may dent, but eventually the tissue will return to normal. Pressing harder can bruise the skin, and, with even more force, break the skin. Therefore, an "acute" or short-term injury depends on the transfer of energy to the human tissue. When the energy transferred exceeds the energy-absorbing capacity of the tissue, an injury occurs. Acute injuries have very short cause/effect relationships: A misguided hammer transmits excessive energy to a bone and breaks a finger; a shard of broken glass puts a large amount of force onto a small surface of the foot and penetrates the foot. Acute injury is easily understood when viewed from an energy-transfer perspective. However, what if force is not excessive? Can it still cause injury over the long term? Absolutely!

Frequency

If a soft drink can is lightly squeezed with the hands the sides of the can initially crinkle, but it regains its shape. The force applied was not strong enough to immediately destroy the can. However, if the force is repeatedly applied in this same manner, say 15 or 20 times, the can develops a *fatigue debt* and progressively tears apart. The can *wears* out and it *tears* apart.

It is the same for the human body, but instead of 15 or 20 repetitions, the frequency is measured in the hundreds and thousands of repetitions. The repeated applica-

tion of a force that is not strong enough to cause immediate damage can over the long term induce fatigue in tissues. Over time, fatigue develops in the skin, muscles, tendons, ligaments, nerves, and blood vessels and they wear out. Force can also be created without external movements. For example, if we bend halfway to the floor, stop, and hold this posture, we are creating a static force load even though there is no external movement. Both high-frequency movements and sustained postures contribute to the development of fatigue debt.

Posture

A third consideration is the posture in which the body is exposed to forces. Because the skeleton is essentially a lever system, there are certain postures in which force can be absorbed easier than in others (see Fig. 75.1). Phrased another way, there are certain postures that are more susceptible to injury. Typically, the closer to the extremes of a joint's range of movement, the less capable the joint is.

If the forces are extreme, an immediate injury will occur. If the force is below the threshold of immediate damage, the onset of a "wear and tear" injury will depend on the number of times per day the person is exposed to the force. And, if the postures are extreme, the combination of force and frequency will cause damage quicker than if the postures were more natural and neutral. It is the combination of force, frequency, and posture that causes wear and tear injuries.

Ergonomic Intervention Strategies

Although the injuries have complicated names like carpal tunnel syndrome, tendinitis, epicondylitis, and so on, they all result from a mismatch between the force applied, the frequency of the application, and the posture in which the person was working. It also follows that

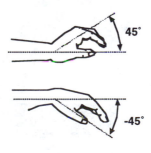

FIGURE 75.1 Joints lose strength as range of motion increases.

the higher the force, the lower the frequency or the better the posture must be. The higher the frequency, the lower the force or the better the posture must be. The poorer the posture, the lower the force or the lower the frequency must be. These three factors are permanently linked (see Fig. 75.2). Consequently, to prevent "wear-and-tear" injuries there must be removal of at least one, preferably two, and ideally all three of the variables that contribute to repetitive motion injury.

Laboratory-Specific Ergonomics Issues

Mismatches between force, frequency, and posture in the laboratory are outlined in Table 75.1. The more non-neutral postures observed, the more challenging the workplace and the more opportunities for improvement exist. An outline of ergonomic stressors and solutions is provided to aid in the identification of ergonomic mismatches.

Wrist: Avoid the Washrag

Bent-wrist postures can lead to tendon overloading. Any posture that requires the use of the hands to squeeze out a rag is one to be avoided. A straight wrist is best. Common tasks that may run the risk of a "washrag" or bent-wrist-type injuries are pouring material into test tubes, inserting pipette tips, drawing samples, ejecting samples, ejecting tips, typing on keyboards, and opening chemical jars and tubes. Some suggestions for avoiding wrist problems include the following:

- pump-activated, definite-measure dispensers for bottles rather than pouring
- thin-walled tips for pipettes to lower tip insertion and ejection forces

FIGURE 75.2 The force-frequency-posture triangle.

- pipetting tools such as multichannel pipettes, repeating micropipette dispensers, or motorized pipette dispensers to reduce activation forces
- pipetting carousels for easy storage and workstation access
- squeeze bottles for commonly used liquids to reduce the amount of bottle opening
- proper fitting gloves to increase friction with the hand and bottles to reduce required grip
- height adjustable keyboard trays for neutral keying wrist positions
- fixed jar and tube openers for twist-on caps

Elbow: Eliminate Elbows Out

The body knows that working with a bent wrist is bad. When confronted with a poor lab process design that would require bending of the wrist, the body subconsciously transfers the stress to the elbows by "winging" the elbows out to the side of the body. Elbow disorders have been associated with thumb-activated pipetting, working around large chemical protection shields, and wearing loose sleeves during chemical use. Elbows may also be extended to protect the body or sleeves from a chemical splash. Some suggestions for avoiding this concern include the following:

- pump-activated, definite-measure dispensers for bottles rather than pouring
- face shields or reduced width (max. 250 mm) reach around chemical shields
- safety glasses, face shields, and shields and proper-fitting long gloves with elastic cuffs to allow closer work

Vibration: Bad Vibes

The vibration from lab tools can permanently damage small blood vessels and nerves in the fingers. Vibration also induces muscle fatigue because we have to grip a vibrating tool harder to maintain control. A single test tube centrifuge with the vibrating rubber cup is a commonly used laboratory item that may put the technician at risk. Two additional stressors—mechanical stress and temperature extremes—can negatively affect the body. Mechanical stress is observed during excessive physical contact with a solid object, for example, leaning on lab benches or pipetting; temperature extremes involve contact with either hot or cold items, for example, autoclaved or frozen items. Avoid vibration and stressor concerns by implementing the following:

- vibration-absorbing grips for tools
- gel-based glove inserts to reduce vibration

TABLE 75.1 Ergonomic Interventions

Hit List Item	Problem	Solution
Washrag	1. Pouring into test tubes 2. Opening chemical jars/test tubes 3. Manual pipetting	1. Use pump, definite-measure pipettes and dispensers 2. Use good fitting gloves 3. Use jar-opener assists Use squeeze bottles
Elbows Out	1. Thumb activated pipetting 2. Large protection shields 3. Raising elbows to protect hanging sleeves	1. Use pump, definite-measure pipettes and dispensers 2. Face shields or protection shield 3. Lab coats with elastic cuffs
Bad Vibes	1. Test tube vibratory mixers 2. Vibration from centrifuges butting up to lab benches 3. Extreme temperatures 4. Mechanical stress	1. Vibration-absorbing grips 2. Leave a 50 mm gap between machine and lab benches 3. Temperature-resistant gloves 4. Padded or rounded edges
Shoulder Too High/Too Low	1. Low tables or benches 2. Reaching into fume hood 3. Pulling down hood face	1. Adjustable-height tables 2. Rolling step stools 3. Power-assist/angled hood face
Hungry Head	1. Looking into microscope or at computer monitors 2. Bent neck doing precision work 3. Looking in and under fume hoods 4. Reading dials or displays 5. Glare from windows on monitors	1. Increased focal length microscope lenses 2. Self-supporting magnifying glass with task lighting 3. Transparent hood with full-opening range 4. Proper height and size of dials and displays 5. Height/tilt surface for microscopes
Twist and Shout	1. Retrieval from bins 2. Reaching for drums 3. Reaching into chest freezers 4. Lifting heavy objects 5. Unprotected chemical work	1. Tilting or gravity feed bins 2. Load levelers 3. Upright freezers 4. Dollies and lifts to move boxes 5. Safety glasses or a face screen

(continued)

TABLE 75.1 *Continued*

Hit List Item	Problem	Solution
Don't Give Me Static	1. Sustained posture 2. Prolonged standing	1. Periodic stretching 2. Automated equipment Anti-fatigue matting

- magnetic stirrer bars for beakers and tubes rather than single-tube vibration mixers
- separation of any floorstanding centrifuges and adjacent lab benches by 50 mm
- rounded or padded edges on all work surfaces
- counterbalancing all containers in the centrifuge; use water-filled tubes if necessary

Shoulder: Design for the Comfort Zone

Ergonomics seeks to fit the heights of workstations and the sizes of tools to the different sizes of people. To achieve this the rule of thumb is: If the shoulder is too high, then the job is too high; if the shoulder is too low, then the job is too low. Work is also best in the area directly in front of our torso, where joints are the strongest, have the most control, and the eyes have the best visual acuity. The most common areas of shoulder risk are found during unnecessary overhead reaching such as closing a fume hood or storing chemicals too high or too low. The following suggestions will minimize occurrences of shoulders being too high or too low and help keep work in the comfort zone.

- adjustable-height tables based on the type of work performed (light, heavy, precision)
- standing precision work surfaces should be height adjustable from 910 to 1070 mm
- standing hand height of 825 to 980 mm for light work
- standing hand height of 725 to 870 mm for heavy work >5 kg
- adjustable height chairs and stools, with footrests
- rolling step stools with casters that lock with applied weight
- power-assisted hood faces, angled hood faces

Neck: Prevent the Hungry Head

Most of the value-added operations performed in labs are visually intensive. The positioning of viewing equipment and the amount of lighting is therefore crucial in eliminating poor neck, or "hungry head," postures. Satisfy the hungry head by including the subsequent suggestions where applicable:

- task lighting up to 1500 lux
- height- and angle-adjustable platforms under microscopes and monitors
- adjustable forearm support with microscope stages
- self-supporting magnifying glasses or lens extenders
- transparent fume hood faces with full-opening range to 2 m
- display height of 1200 to 500 mm (from floor) for standing workstations
- 200 to 675 mm (from seat pan) for seated workstations
- window blinds, computer glare screens

Back Problems: Twist and Shout

The back is constructed to provide a substantial amount of mobility. However, in return for this mobility our body must accept the negatives associated with an unstable and curved column of support. Loads are unequally distributed in a curved spine, and high muscle forces are required to stabilize it. Consequently, the lower back (lumbar area) is a target for injury because of its pronounced curvature. When bending down to pick up an object, the upper body is extended out over the floor. To keep from falling, muscles in the torso transform the spine into a rigid cantilever. This action generates even higher forces in the back muscles and the spine. The farther a load is from the spine, the larger the amount of force that is required to manipulate it. To reduce the stress on the back and shoulders, design the workplace so materials will be retrieved and discharged no more than 500 mm horizontally from the shoulders, ideally within 370 mm of the shoulders. Reaching and bending into low areas are the most common causes of back concerns. This often occurs with accessing carboys, chemical drums, chest freezers, and floor bins. Improved access can be achieved by providing cut-out access at lab benches and using tilting stands, fold-down bins, and rotating turntables so that the researcher or technician can rotate the equipment/material instead of reaching across it. Additional strategies include the following suggestions:

- use spring-loaded load leveler carts to maintain consistent retrieval heights
- upright refrigerator/freezer depth less than 500 mm
- placement of commonly used and heavy items within 370 mm of front of fume hood
- individualized work areas to limit cross-reaching for common-use items
- proper placement of instruments, for example, if used by right hand, place on right side
- equipment shared across benches no more than 750 mm apart
- cupboard or hood depth-access no more than 500 mm
- height- and lumbar angle-adjustable chairs where sitting is feasible
- 150 mm high footrests on lab stools and benches
- 150 mm high toe clearance at lab benches
- antifatigue matting/shoe inserts or padded insoles
- dollies and lifts for box or heavy-object handling

Static Muscle Loading

Static work postures and restricted workstations place extra stress on people. Just as too much movement can be fatiguing, not enough movement is fatiguing as well. To reduce the negative effects of static loading consider the following suggestions:

- periodic stretching of muscles, especially for precision work and repetitive jobs
- vary work patterns and tasks or adjust seating every 20 to 30 minutes
- periodic refocusing of eyes at various distances, especially during microscope work
- automated tasks where possible (e.g., pipetting, washing)

Steps for Implementing Ergonomic Design for the Laboratory

A nine-step process-improvement problem-solving model is outlined for the implementation of an effective laboratory ergonomics agenda. The basic framework of the process is provided. Customization and the development of a methodology that best supports your lab and operational characteristics is recommended.

1. Identify Your Opportunity

There are three important data sources that you can use to identify process improvement opportunities.

A. *Injury and Illness Data*—A review of the injury/illness log at your facility can help identify where mismatches have occurred in the past.
B. *Technician Feedback*—Gather information describing potential lab workstation challenges as seen from the employee's perspective. Document all feedback.
C. *Testing/Quality Data*—Evaluate testing and process quality data to identify which quality faults are derived from human performance problems. Then, see if there are specific barriers in the existing job/process design that contribute to these problems.

2. Form a Cross-Functional Team

A general understanding of where the opportunities exist should be the result of step 1. Deploy a cross-functional team to address these issues. A team could include technicians, supervisors, maintenance personnel, engineering, and health and safety personnel.

3. Define Specific Problems within the Current Process

Apply this chapter to investigate the possible ergonomics issues and problems to particular laboratory-based ergonomics challenges.

4. Define Desired Outcomes

Agree on the goal—is it to eliminate the identified problem or to limit employee exposure to an acceptable level? Once the goal is established, a measure for performance can be agreed on.

5. Define Root Causes and Solutions

Develop a list of the solutions to the specific problems you have identified. Refer to the solutions provided within this chapter.

6. Evaluate the Solutions for Feasibility

Develop an approximate cost for the problems identified. Using existing analysis tools, define which solutions are easy to implement and which are challenging. Develop an estimate of the resource allocation to solve the problem, and rank the recommendations on the basis of severity of the problem and the cost to fix it.

7. Implement Solutions and Track Projects to Completion

Using existing project management systems to follow these projects through to completion. Document the process for future reference.

8. Measure Progress

Track the project's progress using the performance measures agreed on in step 4. Additional measures to track are the number of job improvements completed, percentage of jobs improved, number of injuries/illnesses, and percentage of employees experiencing pain and discomfort from their job tasks.

9. Reward and Recognize

Tremendous benefits in the areas of improved employee morale, productivity gains, and improved product quality can result from the application of ergonomic principles. Recognizing the contributions of teams with public praise and appropriate rewards can maintain the momentum and keep the continuous improvement process going.

Discussion

The lab environment can affect the functioning of the organization by influencing employee health and safety, product quality, and production efficiency, and thereby influencing the effectiveness of the organization as a whole. Redefining laboratory activities in terms of ergonomic acceptability and prudently investing in laboratory equipment will support human performance.

General References

Brogmus, G; Sorock, G.; Webster, B. Recent Trends in Work-Related Cumulative Trauma Disorders of the Upper Extremities in the United States, An Evaluation of Possible Reasons. *J. Occup. Environ. Med.* **1996**, *38*(4).

Corlett, E. N.; Clark, T. S. *The Ergonomics of Workspaces and Machines, A Design Manual*; Taylor & Francis: London, 1995.

Deming, W. E. *Out of the Crisis*; W. Edwards Deming Institute, MIT-CAES: Cambridge, 1986.

Fredriksson, K. Laboratory Work with Automated Pipettes: A Study on How Pipetting Affects the Thumb, *Ergonomics* **1995**, *38*(5), 1067–1073.

Mond, M. C.; Walters, B. D.; Stricoff, S. R. Human Factors in Chemical Containment Laboratory Design; *Am. Ind. Hyg. Assoc. J.* **1987**, *48*(10), 823–829.

Pheasant, S. *Bodyspace*; Taylor & Francis: London, 1988.

Woodson, W. E.; Tilman, B.; Tilman, P. *Human Factors Design Handbook*, 2nd ed.; McGraw-Hill: New York, 1992.

76

Design Criteria for New Laboratories and Renovations

JANET BAUM

Programming

Programming is the very important starting point of the laboratory design process. The program process determines the needs of the owner and future occupants. The program process can also investigate and document building or renovations performance requirements. There are approximately 12 tasks that need to be completed to accurately document these requirements to the laboratory building design team. A program process may be led by and produced by an owner's representative, by a laboratory consultant, and/or by the design architect, with the participation of the owner and future occupants.

The first task is analysis of the existing facility, if one exists. The purpose is to understand the occupancy patterns of the scientists who will move into the new building or renovation. Factors that are measured and analyzed include equivalent linear measure of bench, equipment wall, chemical hood, safety station, chemical storage, and laboratory waste collection. Area calculations should include the population density in net assignable square feet per researcher and percentages of existing area used for the following functions: laboratories, laboratory support (including chemical storage), offices, and building support. If no existing facility exists, then analysis of a laboratory building that is comparable to the one proposed is still an informative task.

The second task is to interview future occupants, laboratory health and safety professionals, chemical hygiene and radiation safety officers, owner's representatives and administrators to investigate and document their goals, work processes and flows, relationships of program functions, current area or functional deficits, and level of flexibility, safety, and security required. Focus group interviews are particularly effective in discussing goals and revealing expectations. Group interactions help to build consensus. All interviews must be well documented and results distributed to those interviewed for review and corrections.

It is very helpful to gain data, from all the researchers interviewed, on existing equipment that will be moved and future equipment that will occupy space in the new building or renovation. In addition, tabulate the current and expected quantities of chemicals, listed by hazard category, to be stored and used in the new facility.

With the data collected from the existing area analysis and interviews, and with the participation of the owner and representative(s) of the future occupants, the program should establish adequate and safe space standards for the new laboratory. The program recommends the equivalent linear feet for bench, equipment, support functions, and area adjustments in net assignable area per full-time research worker.

At this point in the program process a room-type list can be compiled. The list should include the quantity of rooms and estimate of area for research laboratory modules, specialized laboratories, laboratory support functions, administrative areas, and laboratory building support. These room types can be diagrammed to test area estimates. Diagrams should be to scale with benches and/or major equipment. They should include recommendations of entry and egress and location of safety equipment and ventilation devices, such as chemical hoods (Fig. 76.1). Diagram the functional relationships and positions of the spaces on the room list. These can be a series of simple bubble diagrams, floor layouts to-scale, or just a location matrix (Fig. 76.2).

If the proposed project is a renovation, the next task is to compare the program area requirements with the area available for renovation. If the program area is significantly larger than the area available, the program must be reconciled to the area (and budget) available.

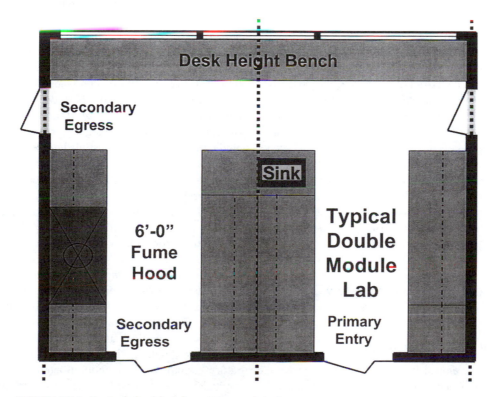

FIGURE 76.1 Typical double laboratory module layout.

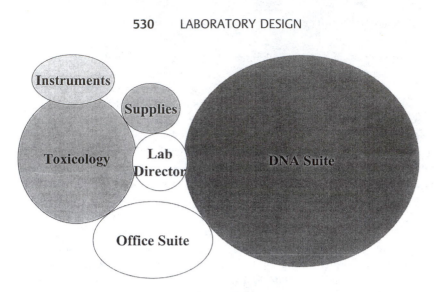

FIGURE 76.2 Adjacency diagram.

The owner and representative(s) of the future occupants need to decide on the priorities for the project to inform reconciliation of the area and budget.

If the program process extends into detailed programming, the next task is to develop laboratory design criteria. These are performance specifications that describe architectural, structural, mechanical, and electrical systems; fire, chemical, and laboratory safety equipment; and chemical and general storage and security issues (Table 76.1). Compute the net area and estimate the gross area of construction with data on and understanding of utility systems required for the new building or laboratory renovation.

The last task is to estimate cost of actual construction for the new building or renovation. This estimate should be provided by a construction cost specialist who has extensive credible experience with laboratory construction. With the assistance of the financial officer, estimate other project costs (Table 76.2). Again, this task offers the opportunity to reconcile the program area, scope, and quality with the proposed budget. The project should not proceed into the design phase until the estimated cost is within the budget. This is particularly important for renovations, because changes made later during either the design or construction phase can compromise safe performance of the laboratory.

Design Criteria for a New Laboratory Building or Building Renovation

Building and Fire Code Factors

Building codes developed from the need to improve life safety for building occupants and owners. Because the nature of processes and materials used in research, local and national codes that regulate construction of laboratory buildings for academic or medical institutions and industry have recently become more specific to actual laboratory conditions and life safety. Earlier versions of national codes and standards did not recognize research and teaching laboratories—they only regulated industrial processes and large-volume users of hazardous materials. National Fire Protection Association (NFPA) Standard 45, "Laboratories Using Chemicals," was the only nationally recognized consensus standard that laboratory designers could use. Current editions of national building codes now have adopted NFPA 45, or developed their own regulations, on the total amount of hazardous materials that laboratory buildings are permitted to contain in storage and in open and closed systems. National model building codes and NFPA define the criteria for low-hazard, intermediate-hazard, and high-hazard laboratories. The materials categories regulated by code include the following: combustible liquids, fibers and dusts, flammable or oxidizing cryogenics, explosives, flammable solids, liquids and gases, organic peroxides, oxidizers, pyrophoric materials, and unstable and water-reactive materials. Current national building codes also regulate quantities of corrosive, highly toxic, irritant, radioactive, sensitizing, toxic, and other materials that pose health hazards in solid, liquid, and gaseous form.

Building and fire code regulations not only specify the fire rating of construction materials and assemblies and fire protection systems; these codes regulate laboratory planning factors such as permitted occupancy, egress pathway, fire control areas, and laboratory unit size. Zoning codes regulate location and construction of laboratory buildings. Other national agencies, or their local equivalents, that may affect design of certain laboratory

TABLE 76.1 Room Design Performance Criteria Sheet

Identification No.	_____	**Classification**	Client Name	_____
		BSL-1,2,3 ____		
Department	_____	BSL-LS ____	Project Name	_____
		GMP ____		
Room Assignment	_____	GLP ____	Date	_____
		Clean Room		
Room Name	_____	Class		

ROOM DESIGN CRITERIA SHEET

OCCUPANCY

Hrs. Occupancy	_____	Room Quantity	_____
No. Occupants	_____	Room Area	_____
No. Animals	_____	Species	_____

FUNCTIONAL RELATIONSHIPS

Primary Room Activity _____
Secondary Activity _____
Room Adjacencies _____
Floor Adjacencies _____

ARCHITECTURAL

Floor Material	_____	Seamless	_____
Base Material	_____	Height	_____
Wall Material	_____	Finish	_____
Ceiling Material	_____	Height	_____
Door Material	_____	Hardware	_____
Door Width	Height ____	Type	_____
Hardware	_____		
Acoustic Criteria	_____		
Window Treatment	_____		
Floor Load	_____		

LAB CASEWORK

Standing Bench	LF ____	Type ____	Depth ____
Seated Bench	LF ____	Type ____	Depth ____
Wall Cupboard	LF ____	Type ____	Depth ____
Tall Cupboard	LF ____	Type ____	Depth ____
Mobile Bench	LF ____	Type ____	Depth ____
Tables	LF ____	Type ____	Depth ____
Wall Shelving	LF ____	Type ____	Depth ____
Reagent Shelf	LF ____	Type ____	Depth ____
Other		_____	
Countertop Material		_____	
Cage System	Size ____	Type ____	

LIGHTING

Fixture	Qty ____	Type	_____
Fixture	Qty ____	Type	_____
Switching		_____	
Timer		_____	

MECHANICAL

			Summer	Winter
Temperature	Set ____	Range ____		
Humidity	Set ____	Range ____		
Filtered Supply	YES ____	Type ____		
Fume Hood	Qty ____	Type ____	Size ____
Point Exhaust	Qty ____	Type ____	Size ____
Canopy Hood	Qty ____	Type ____	Size ____
Other Local Exhaust	Qty ____	Type ____	Size ____
Filtered Exhaust	Qty ____	Type ____	
Recirculate	NO ____	YES ____	
Air Exchanges	Hr ____	CFM ____	

SAFETY EQUIPMENT

Eyewash	Qty ____	Safety station	Qty ____
Deluge shower	Qty ____	Fire blanket	Qty ____
Protective garments	Type ____	Safety glasses	Qty ____
Chemical spill kit	Qty ____	Type ____	
Fire extinguisher	Qty ____	Type ____	
Fire suppression		Type ____	
Smoke detector	Qty ____	Type ____	
Other device	Qty ____	Type ____	

COMMENTS &
MAJOR EQUIPMENT _____

PLUMBING

		Bench	Hood	Description		Bench	Hood
Lab Sink	Qty ____			Local Polisher	Qty ____		
Cup Sink	Qty ____			Process Cooling	Qty ____		
Other Sink	Qty ____			Compressed Air	Qty ____		
Open Drain	Qty ____			Vacuum	Qty ____		
Cold Water	Qty ____			Nitrogen	Qty ____		
Hot Water	Qty ____			Steam	Qty ____		
RO/DI Water	Qty ____			Gas	Qty ____		
Floor or Trench Drain	Qty ____			Other	Qty ____		

ELECTRIC/TELECOMMUNICATION

		Volt	Amperage & Phases	UPS	Emerg.	Isolated Circuit
Bench	Qty ____					
Bench	Qty ____					
Wall	Qty ____					
Wall	Qty ____					
Wall	Qty ____					
Hood	Qty ____					
Hood	Qty ____					
Phone	Qty ____	Type ____				
Data Port	Qty ____	Type ____				

GAS CYLINDERS

					Liquid
Combustible	Qty ____		Inert	Qty ____	
Corrosive	Qty ____		Oxidizer	Qty ____	
Cryogenic	Qty ____		Unstable Reactive	Qty ____	
Flammable I-A	Qty ____		Toxic	Qty ____	
Flammable I-B	Qty ____		Highly Toxic	Qty ____	
Flammable I-C	Qty ____			Qty ____	

CHEMICALS

		in Safety Cabinet			in Safety Cabinet
Combustible Liquid II	Qty ____		Carcinogens	Qty ____	
Combustible Liquid III-A	Qty ____		Teratogens	Qty ____	
Combustible Liquid III-B	Qty ____		Mutagens	Qty ____	
Combustible Dust or Fiber	Qty ____		Alcohols	Qty ____	
Cryogenic Liquid	Qty ____		Amines	Qty ____	
Explosives	Qty ____		Anhydrides	Qty ____	
Flammable Solid	Qty ____		Benzene	Qty ____	
Organic Peroxide I	Qty ____		Borates	Qty ____	
Organic Peroxide II	Qty ____		Formaldehyde/formalin	Qty ____	
Organic Peroxide III	Qty ____		Fluoride compounds	Qty ____	
Organic Peroxide IV	Qty ____		Imides	Qty ____	
Organic Peroxide V	Qty ____		Isocyanate	Qty ____	
Peroxide detonatable	Qty ____		Ketones	Qty ____	
Oxidizer 4	Qty ____		Mercury	Qty ____	
Oxidizer 3	Qty ____		Monomers	Qty ____	
Oxidizer 2	Qty ____		Nickel compounds	Qty ____	
Oxidizer 1	Qty ____		Nitriles	Qty ____	
Pyrophoric	Qty ____		Perchloric acid	Qty ____	
Unstable Reactive 4	Qty ____		Phenol	Qty ____	
Unstable Reactive 3	Qty ____		Phosphorus	Qty ____	
Unstable Reactive 2	Qty ____		Petroleum oils	Qty ____	
Unstable Reactive 1	Qty ____		Olefins	Qty ____	
Water Reactive 3	Qty ____		Radioisotopes	Qty ____	
Water Reactive 2	Qty ____		Silver compounds	Qty ____	
Water Reactive 1	Qty ____		Stilbene	Qty ____	
Toxic	Qty ____		Tolulene	Qty ____	
Highly Toxic	Qty ____		Vinyl ester/sulfones	Qty ____	
Irritant	Qty ____		Xylene	Qty ____	

TABLE 76.2 Project Cost Components

Building construction and renovation costs	General, laboratory specialities, structural, mechanical, electrical, plumbing, fire protection, central utility plant, other
Non-building construction	Demolition, land, relocate/reconnect building utilities, site utilities, foundations, underpinning of adjacent buildings, site work, landscape, building permits, owner supervision, parking, other
Fees	Architect and engineers basic services, additional services, reimbursable expenses, planning and programming, economic feasibility study, interior design, landscape design, construction supervision, energy audits, equipment planning, legal, testing and inspection, local/regional agency approvals
Surveys	Land, soils, traffic, building conditions analysis, environmental impact statement, other
Equipment	Fixed (in addition to that in the construction contract), moveable, furnishings, fixtures, window treatment, graphics, telecommunications, data and computer systems, other
Financial	Interim financing, bid bonds, performance bonds
Insurance	Public liability, vehicle liability, property damage, fire and extended coverage, worker's compensation, employees' liability, other
Contingency	Construction, program scope, consultants, equipment, financing, other

buildings include the Food and Drug Administration (FDA), the U.S. Department of Agriculture (USDA), and the Environmental Protection Agency (EPA).

Floor Size and Number of Floors

Laboratory buildings obviously can be any size, height, or shape. However, building and zoning codes regulate these parameters. Laboratory buildings that are over 75 feet above the lowest level that can be accessed by fire department vehicles are considered to be "high-rise" under most building and zoning codes. Quantities of hazardous chemicals are more restricted on floor levels above the third floor. Careful analysis must be made to determine the best location within the building for scientific departments and individual laboratories that use large volumes of hazardous chemicals.

Research laboratory floors that are very large are sometimes subdivided into zones that function in two specific ways. The first relates to the building code and are referred to as control areas or laboratory units. Control areas are contiguous spaces within buildings where limited quantities of hazardous materials are stored, used, and do not exceed the allowable exempt amounts. The number of control areas allowed varies from four on the first floor to two on the third floor and above.

The second reason for subdivision of large floor plates is social. Informal, productive interactions between research scientists are enhanced when the size of the laboratory floor, or "neighborhood," is between 18,000 and 25,000 square feet. A strong sense of community supports positive behaviors such as open communications, and personal responsibility and safety.

Other important considerations to determine the size of the floor plate include site influences and constraints, campus configuration, zoning regulations, and the requirement for future expansion or connection to other buildings. In a laboratory building renovation, the floor plate size is normally fixed. Changing an existing floor size requires constructing an addition or demolishing part of the building.

Modular Design Approach

Laboratory design has used a modular approach for several decades. Modules are simply a unit of area used for planning certain types of buildings. Laboratory modules may have standardized layout of benches and equipment as well as standardized distribution of utilities. Modules in current design practice vary according to the requirements of the project. Scientific disciplines influence the module size, but there is benefit to standardize the size for long-term flexibility. The modular design approach helps retain safe egress aisle widths and egress pathways through years of renovation and change.

Module width is determined by the dimension of the minimum safe and functional aisle and the width required for work surfaces and equipment on both sides of the aisle. Width for a standard laboratory partition is usually included in the module width. Current good research laboratory practice recommends a minimum 5 ft aisle width. Module widths range between 10 ft 6 in.

and 11 ft 6 in. for most scientific disciplines. This width allows two zones of 30 in. to 36 in. wide for benches and equipment. If equipment is typically wider than 36 in. or narrow access aisles are required behind instrument benches to access connections behind, the module may need to be wider than 11 ft 6 in. Teaching laboratories too may require even wider laboratory aisles to allow the instructors safe and convenient access to each student bench.

Module length is determined by the needs of the laboratory. The range of module length typically is between 20 and 35 ft. Lengths outside this range should be carefully reviewed for safety and functional performance. Laboratory modules can be subdivided to provide smaller sized rooms for special laboratory support functions or for offices. However, safe egress must be maintained from all spaces.

Structural Strategies

Structural design criteria in laboratory buildings that influence laboratory safety center on flexibility and seismic performance. There will always be the need to change parts of laboratory buildings, including the structure. Laboratory building structures can be designed to facilitate minor changes. If the structural column line is offset from the laboratory module, structural members will not interfere with utility penetrations that normally occur on or near the module line.

The most common change is to make penetrations through the floor structure for new pipes and duct risers. Careful study of structural design documents is essential prior to making new openings so that structural components, such as reinforcing bars and beams, are not compromised. This is particularly important for building renovations. Sometimes additional steel is required to reinforce the sides of new openings. Another common change is installation of scientific equipment that is too heavy for the existing floor to support. In this case, too, study of structural design documents is required to determine the safe and effective modifications needed. Sometimes a thick steel plate placed beneath equipment can distribute heavy loads sufficiently to avoid structural modification. A structural engineer should be consulted for design of any structural modifications.

Laboratory building structures are designed for seismic loads in regions that historically experience earthquakes. This is a building code requirement. However, additional structural blocking should be included in all laboratory walls to allow earthquake restraints to be attached. In active seismic zones, all equipment and flammable liquid and chemical storage cabinets should

be fixed to the wall with angle iron and bolts or by strapping, made for this purpose, attached to the wall. Equipment and loose casework can become very destructive projectiles during an earthquake if they are not adequately restrained. Refer to the last paragraph of the subsection "Nonhazardous Laboratory Supplies," p. 544, for a description of earthquake bars on shelving.

Vibration isolation is a design factor that is often critical for scientific analytic equipment performance, but does not relate directly to laboratory safety. Equipment installation manuals document limits of vibration tolerance. This information should be gathered, if possible, during programming when major equipment lists are gathered. Some instruments can be supplied with integral vibration isolation platforms. Very sensitive instrumentation may perform best if located on grade level with the supporting floor slab isolated from the building structure. In this case, vibration caused by flow of fluids in pipes and air in ducts may be disruptive. Reroute pipes, air ducts, and other building-utility-distribution components away from the immediate environment of the instrument.

Mechanical equipment rooms and other building machine rooms should be structurally isolated so that vibration is not transmitted through the structure. These rooms should be enclosed with solid, sound-absorbing walls to diminish airborne noise and vibration.

Other structural design criteria are cost-effectiveness and speed of erection. These, too, are not directly related to laboratory safety, but must meet all life safety considerations by code and good practice.

Personnel and Materials Circulation

Laboratory circulation may be characterized in three primary types: life safety exit pathways, laboratory personnel internal circulation, and transport of laboratory materials and supplies. All have specific safety and performance considerations. Life safety exit pathway requirements covered by building codes, OSHA, and NFPA include public and fire-rated corridors, stairways, elevators, and building exits. In addition, the Americans with Disabilities Act of 1990 (ADA) recommends other design criteria for accessibility and safety of disabled workers. Please refer to the section on ADA compliance (p. 551).

Building codes are more liberal on certain exiting requirements if buildings are fully sprinklered. Sprinklering is highly recommended in buildings with laboratories. This should be one of the first considerations in planning for major renovation of laboratory buildings. Type of storage and location of water-sensitive and water-reactive chemicals should be given careful consideration.

Laboratory building egress corridors can be double or single loaded. They can run down the center of the building floor and have short secondary hallways that branch off the egress pathway like fingers. Corridors can form a racetrack around shared support or loop all around the exterior. Sometimes the corridors form a grid pattern. Each pattern of circulation has a different impact on building efficiency and performance. Often the shape of the building or site influences the choice. Taking into consideration all of these factors, the primary design criteria for corridor pattern is safe egress in emergency conditions. The circulation pattern should be easily recognized and understood by occupants, so they can safely evacuate the building. Signage and emergency lighting aid emergency procedures, but do not make up for disorientation difficulties caused by a confusing layout.

Building egress corridors can double as internal circulation for laboratory personnel. However, sometimes separate and parallel circulation is designed into laboratory floors to facilitate communication and sharing between adjacent laboratories. They are contiguous laboratory aisles and referred to as "ghost" corridors. They allow transfer of small volumes of experimental materials and supplies between laboratories without entering the public corridor.

Materials handling for research laboratory operations puts a great burden on public circulation corridors, because of the volume, frequency, and, often, the hazards involved. Separate, secure corridors for safe transport of laboratory chemicals, equipment, and supplies can perform very well. These "service" corridors also can be used effectively for internal laboratory circulation. Often, laboratory utility mains and ventilation systems are located in the ceiling space above (and below) service corridors. This allows operations and construction personnel to do preventive maintenance and make modifications without disrupting laboratory functions and without blocking public egress corridors. Service corridors can be designed to function as secondary fire egress pathways. Sometimes service corridors are designed to support laboratory equipment, such as freezers and ice machines, and supply storage, such as glassware and chemicals. There are many approaches to service corridor layouts. Each improves research laboratory performance and often is an added bonus for personnel safety and safe operation of laboratories.

Safety Systems and Distribution

Laboratory safety systems include emergency eyewash fountains and deluge showers, safety equipment, and first-aid kits. Minimum distribution requirements of some laboratory safety equipment and fire protection systems are included in most building codes and national laboratory standards. Design criteria above the minimum requirements should be based on discussions with the chemical hygiene officer and/or laboratory safety officer on recommended safety practices for specific hazard(s) present in individual laboratories. Please refer to the section, "Fire Protection Systems," below for discussion of detection, alarm, and suppression performance criteria.

Providing a safety station in each laboratory is a good general approach that enables health and safety staff to maintain up-to-date emergency supplies and equipment appropriate to each laboratory. Safety stations should be in the same location in every laboratory, where possible. Safety stations include storage for chemical spill kits, fire blanket, personal protective equipment (including the correct safety glasses and gloves for specific laboratory processes), first-aid kit, and a bulletin board to post health and safety announcements and information. Safety stations may also have fire extinguishers selected for the specific hazard(s) in each laboratory. Fire extinguishers are used by laboratory occupants only to assist in their escape from laboratories during a fire emergency. Extinguishers should be located in the rear of the laboratory or other locations where fire might block the egress route to laboratory exit doors.

Safety stations can be in close proximity to an emergency deluge shower or eyewash/shower combination units. Provide emergency showers and eyewash fountains in all laboratories and locations that need them, even if the number is above the minimum required by code and national standards. Eyewash fountains must have tempered water. Cold water cannot be tolerated for sufficient time to dilute and rinse away chemicals that cause eye damage. ANSI-approved eyewash fountains that are installed at laboratory sinks work well for able-bodied occupants, but are difficult, if not impossible, to use for disabled persons. Separate installations are required to meet the specific needs and disabilities of these laboratory occupants.

Emergency deluge showers can be installed within laboratories if there is no high-voltage equipment consideration. Locate emergency showers in laboratories where equipment or storage will not obstruct the shower or impede others to assist the person using the shower. An area above or beside the laboratory entry is often a good location, because the entry must be kept clear at all times. If there is high-voltage equipment or other processes that can cause hazardous reactions to water in laboratories, emergency showers can be located outside laboratories in corridors at no more than 75 ft apart. Emergency deluge showers installed within shallow alcoves in the corridors can be equipped with a simple shower curtain that draws around the zone where the user can undress and shower with some privacy. Lack of privacy is a major deterrent to laboratory occupants using the shower for the period required. Water to emergency deluge showers should be tempered. Most users are not be able to tolerate 15 minutes drenching with cold water.

Laboratory Safety and Performance Factors

Vibration adversely affects sensitive instruments and many laboratory animal species. Instrument manufacturers publish vibration criteria in installation instructions and openly offer information on building criteria and effective vibration isolation devices.

Noise adversely affects people who are trying to work in offices and in laboratories. High noise levels in chemical and biochemical laboratories have a long history. Ventilation air system and chemical hood noise added to typical equipment noise can rise to levels that border on or are unhealthy. Chronic exposure to high noise levels can, over time, diminish hearing acuity. Noise can also be very annoying and add to physical stress. No national building code regulates noise levels in buildings, but OSHA has established acceptable limits in the workplace (29 *CFR* 1910.95). The new National Institutes of Health Guide for Laboratory Design also recommends a noise limit of NC 60 in the laboratory generated by ventilation systems. Building owners and occupants need to consider what level of quiet they want to achieve in their ventilation systems and specify this for the architect and mechanical engineer.

Sustainable design embodies the goal for construction process and buildings to do no harm to the environment and, where possible, to give back in order to sustain the environment. Sustainability includes goals for low-embodied energy, biodegradation or recycling, use of renewable resources, not polluting, waste reduction, indoor air quality, and energy conservation.

Design Criteria for Engineering Systems

Utility Distribution Strategies

Laboratory utilities include gases, liquids, electric power, and telecommunications. Utilities service research processes and laboratory safety equipment. These utilities are distributed in pipes, conduit, and wires that must be brought to each laboratory and, in many buildings, to each laboratory bench. Two primary modes of distribution are horizontal and vertical. Horizontal distribution can be arranged in the space below the structure and above the finished ceiling or in an interstitial floor. Vertical distribution can be arranged in enclosed shafts or utility chases.

The primary safety design criteria in utility distribution are accessibility and flexibility. If utilities required for scientific processes are not accessible, often scientists will rig temporary and sometimes very unsafe connections to the nearest source. Extension cords are the most common makeshift approaches that can cause fires and tripping accidents. Rubber tubing and hoses that carry laboratory gases and water can be seen strung over light fixtures and ceiling grids. If tubing or hoses fail, flooding and accidents can happen. Flexibility in utility distribution is important for the same reasons as accessibility is.

HVAC

The purpose of this section is to outline some of the major issues of HVAC systems in laboratories and to give ranges of accepted good practice. Mechanical engineers and building owners hold widely divergent opinions on basic HVAC criteria. As in any laboratory design issue, the specific conditions and objectives of the project informs the process to determine the criteria to be adopted. There is no design formula that can automatically meet the required health and safety criteria in the highly varied conditions in research laboratory construction or renovation.

Human comfort conditions criteria in buildings are generally covered in national building codes. Specific and detailed criteria are published in American Society of Heating, Refrigerating and Air-Conditioning Engineers' (ASHRAE) *Guide* and in specific ASHRAE Standards. Criteria that pertain to laboratory ventilation systems are published in the following documents:

- ASHRAE *Applications*, 1999, Chapter 14, "Laboratories"
- ASHRAE Standard 110-1985, "Method of Testing Performance of Laboratory Chemical Hoods"
- ASHRAE Standard 90.1-1989, "Energy Efficient Design of New Buildings"
- ASHRAE Standards 100.4-1989 and 100.5-1991, "Energy Conservation in Existing Facilities," Industrial and Institutional, respectively
- Standard Z9.5, "Laboratory Ventilation" of the American National Standards Institute (ANSI) with American Industrial Hygiene Association (AIHA)
- OSHA 29 *CFR* 1910.1450, "Laboratory Standard"
- *Industrial Ventilation, A Manual of Recommended Prac-*

tices, 22nd Edition, 1992, by American Conference of Governmental Industrial Hygienists (ACGIH)

- ANSI/ASHRAE Standard 62-1989, "Indoor Air Quality Criteria"

This list of documents is representative of the breadth and detail of heating, ventilating, and air conditioning criteria that are generally accepted in the United States. Standards and criteria have been developed on many detailed aspects of HVAC design and equipment by authorities and associations cited above, and others, including the following:

- Air Moving and Conditioning Association (AMCA)
- American Society of Mechanical Engineers (ASME)
- Centers for Disease Control and Prevention (CDC)
- Environmental Protection Agency (EPA)
- Health and Human Services (HHS)
- National Fire Protection Association (NFPA)
- National Institutes of Health (NIH)
- National Institute for Occupational Safety and Health (NIOSH)
- Nuclear Regulatory Commission (NRC)
- National Safety Council (NSC)
- National Sanitation Foundation (NSF)
- Scientific Equipment and Furnishings Association (SEFA, formerly SAMA)
- Sheet Metal and Air Conditioning Contractors National Association (SMACNA)
- Underwriters Laboratory (UL)

Ventilation Criteria

Good laboratory practice has traditionally focused on the principle of contaminant containment by dilution or capture and removal from the building as the preferred method to maintain air quality within laboratories. To achieve this, typically 100% of air supplied to laboratories is tempered and conditioned outside air. Contaminated air is removed from laboratories and ejected outdoors by general exhaust systems and specific laboratory exhaust devices. Air, in laboratories that use and store chemicals, is normally not recirculated. The amount of supply air required to provide safe working conditions in laboratories varies depending on these factors:

- maintaining the required temperature
- minimum requirement by building code or national standard, ANSI/ASHRAE Standard 62
- heat loads produced within the laboratory by people, equipment, interior lights, and sunlight
- quantities and nature of odorous or hazardous materials used in the laboratory
- minimum or desired ventilation requirements

- maintenance of differential pressurization relationships between laboratory and nonlaboratory spaces
- make-up air for laboratory exhaust devices
- additional make-up air capacity for installation of future laboratory exhaust devices

The measure of fresh outdoor ventilation air is typically done in three ways: by cubic feet per minute per net square foot (cfm/nsf) of laboratory area; by cfm per occupant in some occupancies; or by air changes per hour for the enclosed laboratory volume. Minimum general laboratory criteria provide from slightly less than 1 cfm/nsf to 2 cfm/nsf of outside supply air. This is roughly equivalent to a range of from 6 to 12 air changes per hour for ventilation. Owners and engineers should carefully investigate the adequacy of laboratory supply air values that fall under the lowest values of these ranges for normal operating hours and conditions.

With very careful review, owners and engineers can set lower design criteria for air supply in truly unoccupied operating conditions. However, academic research laboratories often operate nearly 24 hours per day. Unless there is a lockout policy, owners and engineers should assume some laboratory operations and occupancy may continue after normal working hours. Design and operating engineers should make safe, reliable accommodation for this.

There are two basic systems to control the balance of supply and exhaust air volumes, constant volume (CV), and variable air volume (VAV). Each system has advantages and disadvantages (Table 76.3). The owner and engineer must evaluate and come to a joint decision on choice of ventilation system because the owner will inherit the wisdom, or lack thereof, and operating cost of that choice for many years to come.

Contamination Control

Contamination control is accomplished by dilution; by capture, containment, and removal; or by filtration. Ventilation criteria, just discussed, address dilution. Laboratory exhaust devices capture and contain contaminants. These devices are described and discussed in the section, "Design Criteria for Laboratory Fitout," below.

Exhaust systems remove contaminants. Exhaust fans discharge contaminated laboratory air at very high velocities through exhaust stacks, typically in the range of 3000 to 4000 linear feet per minute. The exhaust should enter the air stream high above the laboratory and surrounding buildings. Design engineers and architects should carefully design laboratory and other exhaust stacks so that contaminated air cannot re-enter through air intakes for the laboratory building air sup-

TABLE 76.3 Laboratory Ventilation Control Systems: Comparison of Constant Volume (CV) and Variable Air Volume (VAV) Systems

System	Relative Energy Savings	Pros	Cons
Constant volume Installation	Moderate	Good for one fan and stack per hood installation; less complicated to install and balance; constant volume exhaust through hoods	No energy savings when hood sash closed
Pattern of use	Moderate	Low flow or off setting when lab unoccupied	Always on when lab occupied; no energy savings even when hood sash closed
Operation and maintenance	Moderate	Less complicated to maintain and balance air in laboratory space	On-off mode of chemical hoods can throw air balance off between laboratories
Flexibility	Low	Good when few chemical hoods in lab space	Adding chemical hoods to laboratory space may exceed required exhaust volume
Variable air volume Installation	Moderate	Good for plenum and manifolded chemical fume hood Installation	More complicated to install and balance
Pattern of use	High	Good for continuous operation in either occupied or unoccupied modes	Chemical hood users must close sash for chemical fume hood to operate efficiently
Operation and maintenance	Low	Systems work well when operated properly and maintained	Controls are more complex and requires higher training and experience level of operations staff
Flexibility	High	Controls allow for change and rebalancing	

ply, or that of another building, even under adverse weather and wind conditions.

Exhaust from individual laboratory contaminant exhaust devices can be removed safely from the building in four ways:

- Individual exhaust ducts are routed directly to individual exhaust fans on the roof.
- Individual exhaust ducts are routed to a plenum and then discharged by a few high-volume high-velocity fans on the roof.
- Series of individual exhaust ducts are manifolded, by building zone, into a riser, routed to the roof, and discharged by an exhaust fan on the roof. Each zone has a fan.
- Series of individual exhaust ducts are manifolded into a riser, routed to a plenum, and then discharged by a few high-volume high-velocity fans on the roof.

Please note that the exhaust from specific devices, such as a radioisotope, perchloric acid, and acid-digestion chemical hoods, require dedicated, individual exhaust risers and fans for safe operation. Radioisotope hoods require activated charcoal and high-efficiency particulate air (HEPA) filters. Perchloric acid residue poses an explosion hazard. Ducts from these hoods must be designed as straight up as possible for thorough washdown provided by internal water spray heads. The design should also provide protection of water lines and heads from freezing. In addition to these features, chemical hoods dedicated to acid-digestion processes may require exhaust scrubbers and plastic, instead of metal, ductwork.

Key criteria for all exhaust systems are flexibility and reliability. Exhaust systems should have spare capacity to add additional laboratory exhaust devices as required for ever-changing operations with hazardous materials. Industrial grade electronic control systems are required to safely operate manifolded and plenum exhaust systems, to maintain negative pressurization of ducts that are inside the building, and to provide safe, efficient airflow.

In manifolded and plenum exhaust systems, reliability can be achieved by providing redundant exhaust fans. If one fan is under repair or shut down for preventative maintenance, the other(s) are designed to support 100%

of the load. In addition, installation of standby power allows continuous fan(s) operation during power outages. This maintains negative pressurization of laboratories relative to nonlaboratory spaces. In stand-alone individual fan systems, reliability is only achieved by having sash alarms that notify laboratory occupants and chemical hood users that the chemical hood is not operating safely. The benefit is that rarely more than just a few chemical hoods are out of service at any one time, except during power outages.

Differential Pressure

Laboratory buildings should have pressure positive to the outside atmosphere. Total building supply should be greater than total exhaust to prevent infiltration of water and air pollutants from the outside environment. Chronic air and water infiltration can cause serious building deterioration in just a few years. Infiltration also contributes to poor indoor air quality.

Nonlaboratory spaces within laboratory buildings should be pressurized more positively than laboratory spaces. Differential pressurization creates air movement from zones of higher to zones of lower pressure. This approach is used to support contamination containment along with other ventilation methods. In addition, certain laboratory spaces can be pressurized positive relative to other laboratory spaces. Directional airflow is required in high-hazard laboratory suites. Clean air flows from the access corridor to air lock to outer laboratory to the inner laboratory. Contaminated air is exhausted as described in the previous paragraph.

Plumbing

The primary laboratory safety issues in piped utilities are maintaining purity and adequate flow of the medium. Plumbing systems must protect hot and cold water supply systems for potable use from potential contamination by laboratory processes. Piping systems must protect laboratory gases from contamination and degradation by laboratory processes. Waste water must be monitored and treated before discharge, if needed, to protect municipal and well water supplies.

Domestic, potable water (drinking quality) can be protected from contamination by laboratory processes in three ways. If maintained well, coarse filters on the building inlet remove visible particulate from all incoming water. In some regions and municipalities, simple filtration brings noticeable improvement to general water quality. The design engineer can split the building water supply, from the point of entry to the building, into two parallel but separate systems—one for potable use and one for laboratory processes and building equipment use. The design engineer can protect the potable water riser by specifying a building backflow preventer. If there is not separation of potable and nonpotable water risers, the design engineer can specify backflow preventers on all laboratory water faucets, so that unclean water or chemicals cannot be accidentally siphoned back up into the supply stream through tubing attached to taps or from full containers placed under taps.

Laboratory process water is technically not potable. It is used primarily for washing at laboratory sinks and cooling for laboratory processes and equipment. Laboratory process water may be chilled and recirculated in closed systems. Closed, recirculating water systems can conserve hundreds of thousands of gallons of water per year that would otherwise go down the drain in typical research laboratory buildings.

Most academic, medical, and government research institutions and corporations do not permit hazardous chemicals to be poured down the drain. Laboratory water is still used to rinse off small amounts of common chemicals. This waste water should be monitored for pH level and neutralized before it is emptied into the municipal or owner's sewer system.

Purified Water

Purified water, that is, water that is filtered to remove particulate and/or organic materials or processed through an ion exchanger to remove ions, is used for many purposes. The level of purity depends very directly on the research process requirements. These requirements may vary widely within a research building. Low-level purified water can be economically generated and distributed from central equipment installation. High-level purified water can be generated centrally or locally. Because highly deionized water (10 megohms and greater) is extremely aggressive against typical piping materials and installation methods (welded stainless steel alloys and special plastics are exceptions), these systems are expensive to install and to maintain. Local water purifiers or polishers, which can be installed at laboratory sinks, are an economical and effective method to deliver high-purity water that can meet individual laboratory specifications. These devices, however, produce a significant flow of water to waste, more than do central systems. If water conservation is a goal for building operation, carefully evaluate local polisher options.

Laboratory central vacuum and compressed air are common utilities distributed to benches throughout laboratory buildings. Central vacuum, if it is appropriately sized and well maintained, meets most laboratory vacuum requirements, except for synthetic, physical, and other special chemistry activities that require high vacuum and/or greater control. Central vacuum systems eliminate the need for most water aspiration processes that send large volumes of water to drain. Here again, if water conservation is a goal, central vacuum is a good option. Central-building compressed air generally is not maintained pure or dry enough for use in experiments. It must be filtered and dehumidified. Researchers often use cylinder-supplied air instead and use building compressed air to run simple pneumatic equipment. Normal compressed air pressures of 40 psi and above can cause eye damage. Compressed air nozzles should not be mounted above or near eye level unless the nozzle is angled away from the operator's face.

The types of gases used in laboratories that are either fed from a central supply or tank farm range from flammable to explosive, from inert to toxic. Advantages of central gas supply are the ability to reduce the quantity of gas cylinders in laboratories, thereby reducing the hazard associated with moving them in and out. Long-term economy may be achieved with bulk tanks. Disadvantages of central supply are cost of initial installation of piping, gauges, and outlets as well as incurring the risk of leaks in hazardous gas lines or at outlets.

If large volumes of organic solvents are used in routine laboratory processes, the building owner and laboratory managers should consider the design, installation, and operation of solvent recovery equipment. This and other chemical waste-reduction technologies, such as silver recovery in photographic processing laboratories, greatly reduce polluting emissions.

Fire Protection Systems

Fire protection, detection, alarm, and suppression systems are covered by state and local building and fire codes. The National Fire Protection Association, a consensus organization, publishes many standards on fire detection and suppression systems. National Electric Code has standards on fire alarm systems. Fire alarm systems must meet the requirements of the Americans with Disabilities Act of 1990 (ADA). Provisions are made in systems for persons with hearing, visual, mobility, and some neurological disorders.

Sprinkler systems are recommended for laboratory buildings. Researchers sometimes offer arguments that very expensive research instruments are ruined by water released during false alarms. Properly installed and maintained sprinkler systems have a spectacular record of protecting lives of building occupants and property. Life safety considerations should be the first priority. Improved smoke- and fire-detection systems, as well as fire-suppression systems, including dry pipe and delayed-

action heads, diminish losses due to false alarms. If instrumentation or special processes must be protected, alternate fire-suppression methods and media can be engineered and used. Halon, a CFC, can no longer be installed, but new halon substitutes are available.

Laboratory building owners and laboratory managers can reference national codes such as the Uniform Code, BOCA National Fire Prevention Code, or the National Fire Protection Association standards on all aspects of fire protection including:

- Standard 10, Portable Fire Extinguishers
- Standard 11, Low Expansion Foam Extinguishing Systems
- Standard 12, Carbon Dioxide Extinguishing Systems
- Standard 12A, Halon 1301 Fire Extinguishing Systems
- Standard 13, Installation of Sprinklers
- Standard 14, Standpipes and Hose Systems
- Standard 15 Water Spray Fixed Systems for Fire Protection
- Standard 17, Dry Chemical Extinguishers
- Standard 22, Water Tanks for Private Fire Protection
- Standard 30, Flammable and Combustible Liquids Code
- Standard 68, Emergency Venting
- Standard 69, Explosion Prevention Systems
- Standard 70, National Electric Code
- Standard 78, Lightning Protection Code
- Standard 318, Clean Rooms (for application to this special laboratory type)

Electrical

Normal Power

Local and regional building codes and the NFPA National Electric Code proscribe design requirements for distribution of normal electric power in laboratory buildings. In the United States, laboratories typically need two voltages, 120 and 208 or 220 volts, 3-phase, 4-wire service. European Union and many Asian countries typically provide 220-volt service. Some laboratories with extremely sensitive electronic recording devices may require 5-wire installations. Most analytic instrument laboratories need exceptional amounts of power and dedicated circuits. For good performance, many analytical instruments require dedicated and/or isolated circuits.

To discourage laboratory occupants from using extension cords and local power strips, provide sufficient circuits and outlets in laboratories. Three to four outlets per circuit provide reliable, safe power for laboratory equipment. To reduce risk of electric shorts, ground fault interrupters are needed on laboratory bench circuits and at all outlets near sinks or other water sources.

Flexibility, stability, and reliability are important design criteria for electrical systems. Flexibility starts with space power capacity for the building and at laboratory panels. Provide access to both 120 and 208 or 220 volts power for each laboratory. A surface-mounted electric raceway allows easy changes and additions to outlets in laboratories. Limit duplex outlets to three or four on each circuit to avoid overloading.

Stability is achieved by good engineering design of distribution equipment and systems. Engineers can specify power conditioners to reduce line-voltage fluctuations. Many analytic instruments provide their own transformers, voltage conditioners, and uninterruptable power supplies to protect computers and electronic performance during experiments.

Emergency Power

Local and state building codes include emergency power requirements and design criteria. Emergency power is dedicated to operating life safety equipment and systems, such as fire pumps, fire alarm systems, egress lighting and signs, emergency communications systems, and smoke evacuation systems. Discuss with the chemical hygiene and laboratory safety officer if there are any critical exhaust fans and make-up air systems for chemical hoods and laboratories with particularly hazardous materials and processes, or heating systems and controls that need to be added to the emergency power system. Standby power, on the other hand, maintains critical laboratory processes and instrumentation. Clinical laboratories in health care institutions and forensic laboratories require power reliability for scientific testing. This extra power capacity can be supplied by increasing the size of the life safety generator or by installing a supplemental generator. In some regions of the United States, electric service reliability can be improved by providing separate power feeds from different power distributors to a double-ended transformer(s) or to two separate transformers. Investigate with the local power service provider(s) the quantity and duration of brownouts and power shutdowns during the year in the immediate region to evaluate whether a standby power system is needed and the design criteria for this system.

Uninterruptable Power Supply (UPS)

UPS are often supplied integrated into critical scientific instruments. If not, consider installing local or central UPS systems. Here, too, life and laboratory safety are normally not the driving issues, but rather loss of computer-controlled instruments and data acquisition that shut down with any power loss.

Lighting and Signage

Lighting design criteria for normal and emergency conditions are covered in local and state building codes. Laboratory building occupants are very sensitive to lighting quality. Reduction of glare, shadows, and contrast are important design criteria in laboratory settings because of typical visual tasks of scientific investigation and extensive use of video data terminals. Natural sunlight provides an ideal spectrum of light in the laboratory, but natural light must be controlled to reduce glare and supplemented for dark conditions outdoors. Light fixtures should provide sufficient brightness to the space and the appropriate light intensity for tasks conducted at the laboratory bench. Light intensity at the bench, recommended for typical chemistry laboratories, ranges from 80 to 100 foot-candles. Light fixtures that provide both direct and indirect light aid scientists' depth perception and 3-dimensional perception. For laboratories where fine motor tasks require precision movement, combined direct/indirect lighting works well. Some laboratories require special lighting fixtures for darkened, or blackout, conditions or for reduction of electromagnetic force and/or radio-frequency interference.

Data and Telecommunications

Emergency communications for fire department use are required by local and regional building codes and by the NFPA National Electric Code. Emergency communications assist fire fighters to investigate and respond to fires. Communications are particularly critical in laboratory buildings to ensure evacuation of the building.

Laboratories operate with extensive voice, data, and video communications modalities. These technologies and systems are the backbone of rapid information exchange that is the business of science.

Design Criteria for Laboratory Fitout

Workflow

Laboratory environments vary widely according to the nature of the process and product. The process ranges from education, to basic research, to product testing. Products likewise range from students' exposure to scientific principles, to publication of peer-reviewed scientific papers, to new discoveries, to simply better-quality products. It is difficult to generalize about activities in a laboratory. However, understanding the workflow, type of research, and range of laboratory activity is very important in the design of laboratories and selection of appropriate laboratory furniture and equipment. Placement of chemical hoods or other containment equipment, work surfaces, utility outlets including laboratory sinks, and storage of laboratory daily supplies are the key relationships to investigate to analyze workflow. The relationship of these four key components to laboratory occupants contributes to safe and efficient workflow patterns.

For example, if two people use a chemical hood at the same time, it is important to provide a work surface on both sides of the chemical hood (Fig. 76.3). Each user can set up materials without interfering with the other. Work surfaces directly adjacent to chemical hoods allow researchers to transfer materials easily from one place to the other. Turning and moving across the aisle to reach materials on a bench opposite the hood opens up more

possibility for accidents and spills than does a lateral shift of position.

Another example is the laboratory sink. The workflow relationship depends on the use and frequency of use by laboratory occupants. If laboratory occupants use the sink several times a day, and the sink has an emergency eyewash fountain, the sink needs to be centrally located in the laboratory (Fig. 76.4). If laboratory occupants use the sink infrequently or primarily for hand washing, the sink will get more use if it is near the laboratory exit door. That location prompts laboratory occupants to wash up before leaving the laboratory. On the other hand, if a laboratory process needs a water source and drain, locate the appropriate sink or cup-sink fixtures near the process.

Ideally, to sustain optimal efficiency, laboratory furniture and fixture configuration could change as processes and workflow change. However, due to hard connections to mechanical, electrical, and/or plumbing services, chemical hoods, sinks, some scientific equipment, and, to a lesser extent, utility outlets, are usually fixed

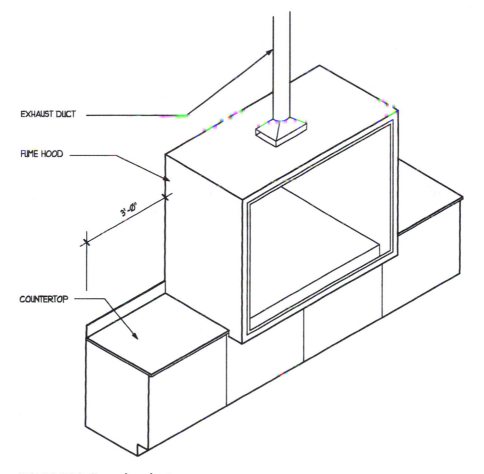

EXHAUST DUCT

FUME HOOD

3'-0"

COUNTERTOP

FIGURE 76.3 Fume hood setup area.

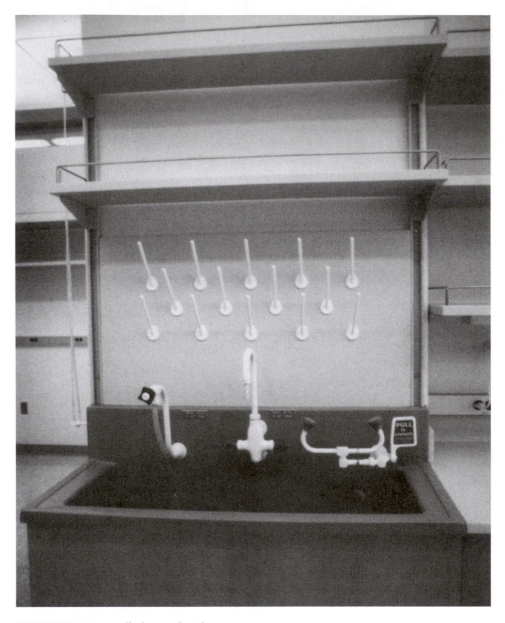

FIGURE 76.4 Centrally located sink.

in a laboratory. There are costs associated in moving them. Hard connections do not just plug in; they are directly wired, soldered, welded, or otherwise sealed connections of utility services to equipment and fixtures. Experienced, licensed, and bonded construction workers have the skills and equipment to relocate chemical hoods, sinks, and utility outlets. Often, maintenance and laboratory staff are not fully trained to do these jobs. On the other hand, maintenance staff may be able to reconfigure work surfaces and benches. Laboratory staff can move some equipment, such as carts and tables, safely with the aid of appropriate materials handling equipment.

Determining the level of flexibility required in a laboratory is often the most critical decision for selection of laboratory furniture systems and the cost of the system. Scientific research industries and institutions increasingly demand greater flexibility to reconfigure laboratory furniture to continually improve workflow and to safely accommodate new instruments and processes into their laboratories.

There is a range of ways to provide flexibility. One fitout option is a laboratory with 100% open floor area and utility racks suspended from the ceiling structure above to provide electric and piped services to an array of equipment and moveable tables below. A variation of this strategy is to add one wall of fixed components, such as sink, chemical hood, and adjacent benches. At the other extreme, a laboratory full of fixed benches has no free area to install floor-mounted instruments and new work processes. This chapter will describe the opportunities to provide flexibility in each section.

Materials Handling

Materials handling encompasses not only transport, but also storage of laboratory materials. The key to safe transport of materials within the laboratory is maintaining clear pathways of adequate width for movement of people as well as materials. A minimum recommended aisle is 5 ft clear (1.52 m). This distance allows a laboratory cart or person carrying materials to pass behind persons working at the bench (Fig. 76.5). Under other laboratory conditions, laboratory workers need wider aisles to move large cryogenic tanks or equipment. In some laboratories, wider aisles are needed for equipment servicing and operating clearance. Laboratory designers

should investigate all of these considerations in the design of each laboratory.

Gas Cylinders and Cryogenic Liquid Tanks

Cabinets, alcoves, or closets outside laboratories and specially designed for storage of hazardous gases, compressed gas cylinders, or cryogenic containers are safer than storing these materials within laboratories. Support personnel exchange spent cylinders from corridors and they do not have to enter the laboratory. In addition, compressed-gas cylinders and cryogenic liquid tanks clutter a laboratory, often obstruct laboratory aisles, and increase risks to laboratory occupants. If laboratory workers cannot find a safe location for compressed-gas cylinders outside the laboratory, they should dedicate a specific location in the laboratory for safe cylinder storage. Restraints should have lower as well as upper level limits in regions where there is earthquake potential.

Storage

Laboratory staff and researchers often store a full year's supplies within the confines of the laboratory because of funding and economies of bulk purchases. A large volume of bulk materials puts a tremendous burden on the

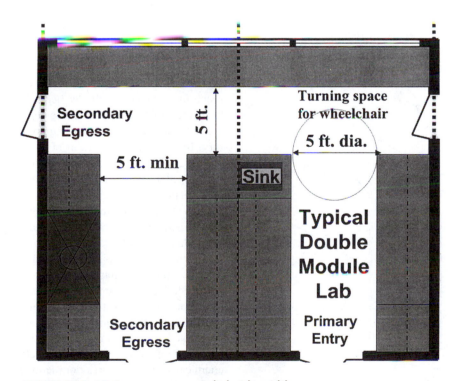

FIGURE 76.5 Minimum recommended aisle widths.

capacity of laboratory furniture to safely store them. Bulk storage hampers flexibility if it is awkwardly accommodated in base cupboards and drawers. For instance, benches could be relocated easily if they were not already stuffed to the brim and tables moved if the area below were not filled with boxes. Here are some problem areas and strategies to better accommodate materials storage.

The best solution is to not store any bulk materials in laboratories at all. This can be done with "just in time" shipments via overnight carriers from most scientific materials and chemical supply houses. Another solution is to establish a central storeroom that has sufficient area to provide a safe, secure, locked storage cage for each laboratory unit in the building. Laboratory staff can bring up a week's supply of materials or equipment that is not used frequently on an as-needed basis. In both ways, valuable laboratory space is not used for warehousing.

Nonhazardous Laboratory Supplies

There is a wide variety of storage units available from laboratory furnishings manufacturers for general nonhazardous supplies. Tall cupboard units are moveable and commodious and come in several sizes. Units are floor standing and have doors to keep contents clean, protected, and secure, if they have locks. Tall glassware and large items often fit in these cabinets. Storage shelves and racks can be made within cupboards to support unusually shaped items. In regions that experience earthquakes, tops of tall cupboards should be attached to the wall behind them to prevent them from tipping over or sliding across the floor during a seismic event (Fig. 76.6).

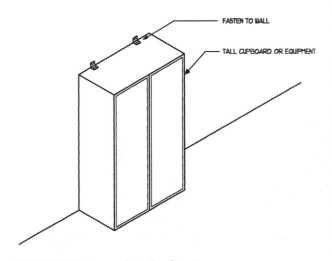

FIGURE 76.6 Secured wall cabinets.

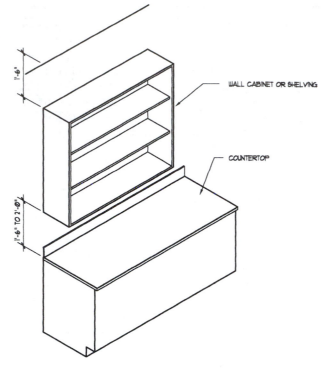

FIGURE 76.7 Countertop clearance for wall cabinets.

Wall cupboards, mounted on walls above laboratory benches and equipment, can hold supplies and keep them clean if cabinet doors are closed. Depths of wall cupboards limit the sizes of items that can be stored. Also, the height of top shelves is out of reach for most staff except those who are tall. Laboratory workers should use a step stool or stepladder to reach high shelves. It is unsafe to stand on open bench drawers to reach high shelves. Sprinkler heads need 1.5 ft (0.46 m) from the finished ceiling to the top of wall cabinets in order to provide proper fire suppression in the laboratory. In addition, the bottom of wall cabinets should be 1.5 to 2 ft (0.46 to 0.61 m) above bench countertops, so that cabinets will not obstruct experiment set-ups or equipment access on the bench (Fig. 76.7).

Open adjustable shelves supported by heavy-duty standards and brackets are installed directly on walls or on industrial slotted steel channels or posts. Open shelves are very flexible and limited only by the strength of the support systems and wall to which they are attached. Because open shelves have no enclosure, items stored on them collect dust and can fall off by accident or during an earthquake. For this reason, good laboratory practice requires restraints mounted on shelf edges in earthquake zones (Fig. 76.8). Metal shelving used for chemical storage requires nonmetallic, leakproof secondary containers for chemicals. Metal shelves and supports

FIGURE 76.8 Earthquake zone shelving.

can rust and fail in corrosive environments. Wood shelving is less vulnerable to failure in a corrosive atmosphere.

Chemical Supplies

Laboratories should store a maximum of one week's chemical supplies in the laboratory or follow a prudent limit recommended by the environmental health and safety officer. Laboratory staff should store supplies in excess of one week in a central chemical stockroom. Then staff can retrieve chemicals on an as-needed basis. Chemicals kept in the laboratory should be stored appropriately and separated according to class. Refer to the chemical storage chapters of this manual for further information regarding this topic.

There are laboratory furnishings specifically designed for chemical storage and approved by Underwriters Laboratory or Factory Mutual. These include flammable liquid storage safety cabinets and corrosive chemical storage cabinets that can be freestanding or installed beneath chemical hood superstructures. These storage units come with venting options. Contact your environmental health and safety officer to discuss what option is best for your use. Venting reduces staff exposure to fumes from bottles that are not tightly capped or leaking when the door is opened. Unvented flammable liquid storage cabinets provide better fire protection for contents than do vented cabinets. Other secured cabinets for toxic chemicals or restricted drugs are available from laboratory safety equipment manufacturers. Selection of the storage cabinet or safety device must be based on a full understanding of the hazard potential of the chemical.

Wall cupboards and even open shelves can be suitable for storing chemicals if the cabinet's recommended weight capacity is adequate. Reagent shelves are open shelves supported on or above island and peninsula benches (Fig. 76.9). Reagent shelves are frequently used to store chemicals as well as small instruments and supplies. Install earthquake restraint rods and raised front edges on all shelves that store chemicals, whether the shelves are in cupboards or are open. Bottles and jars that have a potential for leaking or spillage should have secondary containment. Secondary containment can be basins for individual bottles or deep trays for a group of bottles. Secondary containers should be made of chemical-resistant materials and be leak proof.

Shelves are made of several materials: finished wood, painted metal, sheet stainless steel, chemical-resistant

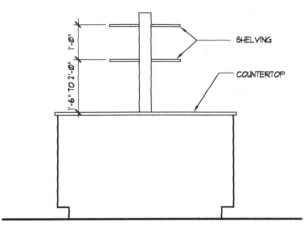

FIGURE 76.9 Reagent shelves.

plastic laminate on particle board, cast monolithic plastic, phenolic laminate, cast epoxy resin, stainless steel, and nickel-plated chrome wire. The most chemical-resistant material is epoxy. Epoxy shelves are strong, stain resistant, and do not burn or corrode. Epoxy, however, is very heavy and requires extra heavy-duty shelf-support systems. Carefully assess other shelving materials for safety or chemical reaction qualities before selection for chemical storage. Earthquake restraint bars are recommended for these shelves as well, in areas that have risk of this natural disaster.

Metal hardware and fittings will corrode rapidly in a corrosive atmosphere of chemical storage. Purchase storage units that have epoxy-coated or powder-coated hardware and fittings to prolong their use. Check hardware, especially shelf supports, on a biannual basis or more frequently, to make sure no support hardware shows significant corrosion.

Hazardous Waste Storage

Laboratory workers should keep quantities of containers and total volume of chemical waste at a minimum in the laboratory. If quantities of flammable liquids and petrochemicals are small, use approved standard safety cans and a designated shelf with secondary containment for small bottles of nonflammable chemicals. Laboratory aisles must stay free of obstructions. Mark the designated floor area for safety-can storage with yellow-striped caution tape. Cabinets designed for new chemical storage are also suitable for temporary storage of weekly volumes of laboratory chemical waste.

Researchers and staff should collect biological waste in red bags. Waste bin cabinets that hold red bags are very convenient for solid waste collection, especially if they are located beside the laboratory sink area where a lot of solid waste is generated. Sharps waste containers are small and are mounted on the wall near the work stations that use sharps.

Radioisotope waste storage methods vary according to the strength and half-lives of the isotopes. Plexiglas and lead brick shields and sheet lead-lined cabinets are common safe methods to temporarily store radioisotope waste materials in the laboratory. Staff trained to transport radioactive waste should remove it from laboratories upon notification by the laboratory. It may be stored in a licensed waste-collection depot and allowed to decay, then packaged and properly disposed of.

Ordinary Trash Storage

Requirements for recycling ordinary waste put more demands on precious laboratory floor area and laboratory furnishings. Recyclable waste streams include: plastics, glass and metal containers, white paper, computer printer paper, and cardboard and packaging. Waste containers need to be large and strong enough to hold contents between collections. Again, laboratory floor area must be set aside for waste containers. It is more convenient for both laboratory and housekeeping staff if waste containers are kept beneath a bench in the laboratory.

Bench and Major Equipment Locations

Egress

Egress is the pathway out of an occupied space or building. Local and state building codes such as the Uniform Building Code, BOCA Code, and the Southern Building Code regulate the means and design of egress out of buildings. National Fire Protection Association (NFPA) Standard 45, Laboratories Using Chemicals, establishes guidelines for egress out of laboratories that have been adopted as the code in some states.

Safe and direct egress is especially important during an emergency. Generally, fully sprinklered laboratories that are 300 net square feet or less and contain no hazard, such as a chemical hood or solvent storage, need only one exit. If all the conditions listed are not met, then two exits are required. Approved exitways for laboratory buildings lead directly to outdoors or to a passageway that leads to outdoors. Many building codes limit to one the number of rooms that occupants can pass through to reach an approved exit pathway.

Laboratory Aisles

Bench arrangements influence the egress path out of laboratories. Major laboratory aisles aligned with the exit door decrease confusion and, therefore, the time necessary to leave the laboratory during an emergency.

Aisles in research and development laboratories must remain clear, at the minimum width of 5 ft (1.52 m), with no obstructions so that occupants can get out from either end of the lab safely and quickly in an emergency. This width was recommended in an earlier paragraph on material handling in the laboratory. Common obstacles in laboratory aisles are chairs, stools, carts, trash receptacles, and equipment that protrude beyond the bench. These items can be pushed out of the aisle into knee spaces that are not routinely used for sitting. Aisles behind benches and worktables that are split, in order to provide access to services, should be 2 ft (0.60 m) wide to work most effectively.

In instructional laboratories, particularly at secondary school and undergraduate levels, laboratory aisles should be a minimum of 6 ft (1.83 m) to 7 ft (2.13 m) clear. This width accommodates increased traffic of greater numbers of people in teaching laboratories. Laboratory instructors must quickly and safely pass behind rows of students to respond to questions and potential accidents. Aisles must be wide enough to permit instructors to respond, without causing another accident.

Chemical Hoods

Placement of chemical hoods in a research and development environment influences workflow, egress pathway, and bench configuration. Laboratory design architects and engineers specify chemical hoods. Casework vendors furnish and install chemical hood cabinets as part of the casework system. Mechanical contractors install the ducts and ventilation equipment that operate chemical hoods. Finally, electrical engineers connect power to all of the equipment. Simply put, specification and installation of chemical hoods are the responsibility of many people. Coordination of the chemical hood type, specification, and its connection to ventilation systems and controls are essential to good performance. Designers who specify chemical hood type must request the recommendation of the laboratory safety officer early in the design process.

Laboratory designers investigate materials, methods, and safety considerations in the laboratory to determine the type, number, and location of chemical hoods needed. It is very important to locate chemical hoods away from the primary exit door and egress pathway from laboratories. Persons exiting the laboratory should not have to pass near any chemical hood because some processes conducted in chemical hoods and runaway reactions in hoods can be very hazardous.

Airflow patterns to the face opening improve if chemical hoods are away from high-traffic areas of the laboratory. Passersby disrupt airflow and reduce chemical hood performance. Provide two directions for operators to leave the immediate area of the chemical hood if possible. Dead-end aisles still pose a risk to laboratory workers who must rapidly escape the immediate area of the chemical hood. As discussed in the Workflow section, chemical hoods normally occupy a fixed position in the laboratory because of hard connections to plumbing, electrical, and building exhaust systems.

Recent National Institutes of Health (NIH) studies used computational fluid dynamics technology to model and assess chemical hood performance. Guidelines for fume cupboard installation from the British Safety Council support the NIH research results. These sources indicate that chemical hoods should be located a minimum of 1 ft away from any side wall. Airflow deflected off of walls perpendicular and next to the chemical hood cause turbulence. Chemical hoods that face each other across a laboratory aisle may also interfere with containment performance. Air transfer grilles, and distribution and design of air supply grilles can affect chemical hoods' contaminate capture ability. Please refer to Chapter 43 for further information on safety features and performance requirements for chemical hoods.

Biological Safety Cabinets

As with chemical hoods, biological safety cabinets are sensitive to interference in airflow to the face opening caused by traffic in laboratories. Biosafety cabinets should be installed and operated away from major laboratory circulation. If biological safety cabinets are not rated for chemical use and are used in the recommended manner, potential explosion and fire hazards are normally lower than those for chemical hoods. If the laboratory safety officer regards these hazards as low risk in a particular laboratory, biological safety cabinets can be placed along dead-end aisles. If, however, biological safety cabinets are rated for chemical use and are fully exhausted, they should be positioned with similar considerations as chemical hoods.

Gloveboxes

Gloveboxes vary in size and require adequate clearance at the input port. Gloveboxes, operated properly, provide total containment for a wide variety of hazardous and nonhazardous laboratory procedures. Because the chamber is sealed and closed to the laboratory atmosphere, airflow is not as critical an operating factor as it is for chemical hoods.

Laboratory workers often use relatively thick gloves to manipulate materials inside the chamber. This effort makes it more difficult for them to be physically comfortable and to maintain fine hand-to-eye coordination and dexterity. As with other containment devices, gloveboxes should be located away from high-traffic areas, in quiet zones of the laboratory. Compressed-gas-cylinder storage with manifolded lines to gloveboxes, or cylinder racks nearby, are required.

Casework

Materials, Properties, and Standard Systems

There are four basic materials used for construction of laboratory casework: metal, wood, phenolic resins, and plastic laminate. There are eight materials used for laboratory countertops: wood, stainless steel, plastic laminate, molded monolithic plastic, cast epoxy resin, ceramic tile, natural stone, and phenolic resin. Each material has advantages and disadvantages. Table 76.4 gives a comparison of these factors for casework, countertop, shelving, and sink applications.

There is always a heated discussion about casework materials concerning wood versus metal. Although wood and metal have very different aesthetic qualities, they have many other characteristics that are quite comparable including cost, availability, durability, and cleanability. In a fire, each has a characteristic disadvantage: Wood is combustible, and metal finishes release toxic fumes. If the laboratory routinely uses significant amounts of corrosive chemicals, wood casework will hold up longer and better than metal will. Wood, painted metal, and high-pressure plastic laminate do not do well in extreme humidity and wet laboratory environments, such as in cold rooms. Only stainless steel casework, while not perfect, performs better than wood in those conditions. Stainless steel performs very well in areas that must be sterile or extremely clean, such as surgeries, animal facilities, and microelectronics laboratories.

Wood casework manufacturers adhere to the standards of the Architectural Woodwork Institute's *Quality Standards Illustrated*, 7th ed., in the United States and the Architectural Woodwork Manufacturer's Association of Canada's *Quality Standards for Architectural Woodwork* in Canada. Standards for plastic laminate laboratory casework are also covered in the AWI Quality Standard. There is no current national standard for metal casework. Some of the quality characteristics covered by these standards are the strength and thickness of materials, method and visual quality of the finish, and very

important, the method of assembly, fasteners, and hardware. If roughly handled during shipment and delivery, poorly constructed casework can fall apart before it is installed. If well constructed and maintained, casework can endure for decades.

Casework Systems

Four standard casework systems are normally available in any of the materials listed in Table 76.2. Table 76.3 gives a comparison of the advantages and disadvantages of the systems. As discussed in the Workflow section, flexibility can be the key to productivity in the laboratory. Laboratory designers can plan each system to be flexible. It is a matter of exploring these options, because each comes with a cost attached.

One way to ensure long-term flexibility is to select only a few basic widths of casework components from the variety offered. Standard casework is manufactured in six increments starting at 1.5 ft (0.46 m) up to 4 ft (1.22 m), then in 1-ft increments up to 6 ft (1.83 m). If the combination of 2 and 4 ft, or 2.5 and 5 ft, wide components are selected, then components are interchangeable from one bench array to another. This selection strategy offers very great benefits for long-term flexibility with no cost premium for any of the casework systems listed. Laboratory designers can further control costs by selecting a handful of standard base cabinet designs from dozens offered in casework catalogs.

Countertops

Different performance demands for countertops suggest different materials to meet those demands. Table 76.5 shows performance criteria for common laboratory countertop materials. Sometimes no single material meets all the performance requirements of a laboratory. Countertops then may be selected for specific applications.

Clean rooms, operating rooms, cold rooms, and radioisotope laboratories normally use stainless steel, the easiest material to decontaminate. Heavy chemistry laboratories use epoxy or phenolic tops for chemical and heat resistance. Phenolic material is good for physics laboratories because of its true, flat surface. Both phenolic resin material (trade name is Trespa) and monolithic plastic material (trade name is Corian) are relatively new to laboratory applications. Their long-term performance in laboratory conditions is not well known.

As mentioned in the next section on ADA compliance, countertops with marine edges are preferred in many applications. A marine edge is continuous, smooth, and rounded with a raised bump along the edge of a countertop or shelf.

TABLE 76.4 Laboratory Casework Materials

Item	Material	Relative Cost	Pros	Cons
Casework	Wood, oak and maple species	Moderate	Durable, very chemical resistant finish, quiet, can be repaired and refinished, warm to touch	Combustible in lab fire, hardware subject to rust
	Metal, wide range of colors	Moderate	Durable, chemical resistant finish, can be repainted	Toxic vapors released in lab fire, metal subject to rust, dents, can be noisy
	Stainless steel, cabinets and countertop	High	Durable, very chemical-resistant finish, rust resistant, can be repaired and buffed	Expensive
	Plastic laminate, chemical-resistant available	Low	Slightly chemical-resistant finish, cheap	Scratches, cannot be repaired and refinished, hardware subject to rust, delaminates
Mounting Method	Floor mounted	Low	Very stable, vibration dampening, maximize storage volume, can be moved, sealed at floor	Countertop supported by base cabinet, need workman to move it
	Mobile	Moderate	Easy to move around on heavy-duty casters, no tools needed	Must be restrained in earthquake zone
	C-frame (cantilevered)	Moderate	Cupboard can be moved, counter is self supported, floor cleanable beneath	Amplifies vibration, need workman to move it, limited storage volume
	Lab frame or wall system	High	Cupboards can be moved, counter is self supported and adjustable height, floor cleanable beneath	Amplifies vibration, need workman to move it, limited storage volume
Countertop	Wood	Low	Natural material, durable, good impact resistance, long spans	Combustible in lab fire, not heat resistant, deflects, warps, cuts, not chemical or water proof
	Plastic laminate	Low	Inexpensive, easy to replace, color available, local fabrication and installation, cut and drill	Limited chemical resistance, scratches, delaminates, no heat or impact resistance
	Phenolic resin (Trespa)	Moderate	Light weight, chemical resistant, can cut and drill, doesn't deflect, stable, long spans	New material, life-span unknown in labs
	Ceramic tile	High	Very stable, heavy, vibration dampening durable, chemical and heat resistant, marine edge possible, color available	Heavy, cracks with dry ice and impact, small size pieces, lot of mortar joints that are hard to clean, not flat, dimensions vary
	Monolithic plastic (Corian)	Moderate	Moderate weight, cold and chemical resistant, can cut and drill on site, marine edge possible	New material, life-span unknown in labs
	Natural stone	Moderate	Very stable, heavy, vibration dampening, durable, chemical and heat resistant	Heavy, cannot cut or drill easily on-site, limited size pieces, lots of joints, not flat
	Epoxy resin	Moderate	Stable, heavy, vibration dampening, very durable, very chemical and heat resistant, marine edge possible	Heavy, cracks with dry ice, cannot cut or drill on site, limited size pieces, must joint, not flat, dimensions may vary

(*continued*)

TABLE 76.4 *Continued*

Item	Material	Relative Cost	Pros	Cons
	Stainless steel	High	Stable, durable, chemical, heat and cold resistant, marine edge possible, long spans, easy to decontaminate	Can stain, dent, expensive
Shelving	Wood	Low	Durable, chemical resistant finish, can be refinished, repaired, cut, drilled, cheap	Combustible in lab fire, scratches, deflects
	Plastic laminate	Low	Limited chemical resistant finish, cheap	Combustible in lab fire, deflects, delaminates
	Metal	Moderate	Durable, chemical resistant finish, can be refinished, moderate cost	Toxic vapors released in lab fire, all metal subject to rust, dents
	Phenolic resin (Trespa)	High	Light weight, chemical resistant, can cut and drill, doesn't deflect, stable, long spans	New material, life-span unknown in labs
	Epoxy resin	Moderate	Stable, heavy, very durable, chemical resistant, marine edge possible	Heavy, needs extra heavy-duty support, cannot cut or drill on site
	Monolithic plastic (Corian)	High	Moderate weight, some chemical resistance, can cut and drill on site, marine edge possible	New material, life-span unknown in labs
	Stainless steel solid	High	Durable, chemical resistant finish, can be decontaminated, marine edge possible	Expensive
	Nickel plated chrome wire	Moderate	Durable, chemical resistant finish, allows air circulation, can view contents from below	Moderate cost, needs plexiglass liner for chemical storage
Sinks	Epoxy resin	Moderate	Very stable, durable, very chemical and heat resistant, self-rimming set in possible	Heavy, cracks with dry ice, cannot cut or drill on site, must joint and seal
	Monolithic plastic (Corian)		Moderate weight, some chemical resistance, heat resistant, self-rimming set in possible	New material, life-span unknown in labs
	Stainless steel	Moderate	Stable, durable, chemical, heat and cold resistant, seamless, no joints, easy to decontaminate	Can stain, dent, expensive
Pegboards	Phenolic resin (Trespa)	High	Light weight, chemical and cold resistant, doesn't deflect, stable	New material, life-span unknown
	Epoxy resin	Moderate	Very stable, durable, very chemical and heat resistant	Heavy, cannot cut or drill on site
	Stainless steel	High	Stable, durable, light weight, chemical, heat and cold resistant, easy to decontaminate	Can stain, dent, expensive
	Plastic	Low	Light weight, slightly chemical resistant, stable, inexpensive	

TABLE 76.5 Laboratory Countertop Performance

Quality or Characteristic	Best Performer	Next Best
Flat surface	Phenolic	—
Dimensional consistency	Phenolic	Stainless steel
Resists chemical stain and deterioration	Epoxy, stainless steel	—
Heat resistance	Epoxy, stainless steel	Stone, ceramic tile
Cold tolerance	Wood, stainless steel	Plastic laminate
Low conductivity	Wood, plastic laminate	Ceramic tile, stone
Compressive strength	Epoxy, phenolic, stone	—
Finish scratch-resistant	Epoxy, stone	Stainless steel, ceramic tile
Weight limit	Stainless steel, phenolic	Epoxy, wood
Length of span	Stainless steel, phenolic	Wood
Least number of joints	Stainless steel, Epoxy	Plastic laminate, wood

ADA Compliance

Since the Americans with Disabilities Act of 1990, there has been concerted action in laboratories to accommodate people with a very wide range of disabilities. If a current laboratory occupant has a disability, his/her workplace must be modified so that the person can perform his/her job. If a future laboratory occupant is hired with a disability, the laboratory can be modified at that time in a manner appropriate to that person's disability. Often laboratory designers and owners interpret ADA compliance to mean simply to adjust the height of laboratory counters, chemical hoods, sinks, and primary laboratory safety equipment for individuals in wheelchairs (Fig. 76.10).

Floor-mounted, mobile, and C-frame systems benches are manufactured in normal-standing-height benches and normal-seated-height benches, 36 to 37 in. (0.91 to 0.94 m) and 29 to 31 in. (0.74 to 0.79 m), respectively. In advanced C-frame systems, bench heights are adjustable in 1-in. (2.5-cm) increments. This level of flexibility can accommodate every person in the laboratory, from the very tall to very short mobility-impaired persons and wheelchair users. Laboratory designers can design standard floor-mounted benches with removable drawers 6 in. (15 cm) high that can be taken out to convert a standard high bench to a low one, for a modest cost. If

required for the job, a laboratory sink, chemical hood, biosafety cabinet, and other primary workstation equipment should be provided at the appropriate height for persons in wheelchairs or with other mobility disabilities.

However, many other sensory and mobility disabilities of current laboratory occupants also influence selection and use of laboratory furnishings. Accommodation, by law, includes a full range of workplace adaptations or adjustments that enable handicapped workers to work more easily or more productively or that expand the range of jobs individuals can perform. These are some of the considerations:

- Provide emergency equipment at heights that are easily reached by disabled persons, including those in wheelchairs or with a limited range of arm movement or control.
- Avoid protruding or hanging objects in aisles lower than 7 ft (2.13 m).
- Maintain a minimum clear opening of 32 in. (0.81 m) at a 90° door swing. Provide lever-operated handles to operate doors. Provide large-vision panels in upper half of doors when allowed by building and fire codes.
- Provide cabinet hardware that can be easily operated by persons with limited arm movement or weakness or impaired sensation in fingers or hands. Cabinet and drawer pulls should not have sharp edges that can scratch people when doors or drawers are open. Contrast in color or surface reflectance helps laboratory workers with limited sight to see pulls. Label cabinet contents clearly and simply to minimize the need for verbal assistance. Use high-contrast marks and familiar symbols where possible. For persons with very limited sight, reinforce labels with tactile markings.
- Avoid frequent reorganization of storage areas or relocation of workstation.
- Contrast floor with walls especially at changes in floor level, such as at ramps to environmental rooms. This can be done with color, light to dark, or surface reflectance, shiny to dull.
- Provide visual and tactile backup for all auditory alarm systems and emergency equipment. Contrast these signals with environmental sounds. Exceed ambient noise level by at least 10 decibels for voice signals. However, understand that sudden loud noise or flashing lights of some emergency systems can trigger seizures.
- Provide good and appropriate lighting. Glare from windows and light fixtures must be controlled throughout the laboratory for all occupants. Provide task lighting that is adjustable for intensity,

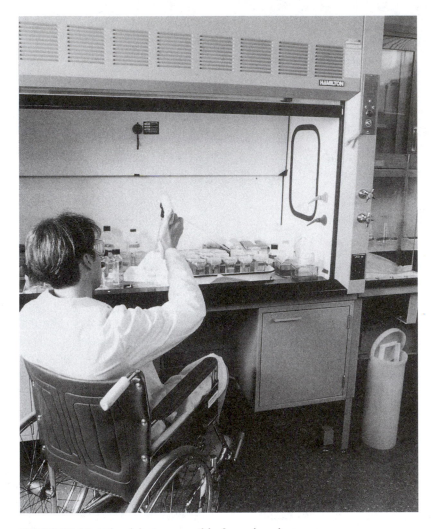

FIGURE 76.10 Wheelchair-accessible fume hood.

color, and angle. Undercabinet strip lighting often only produces glare and cannot be directed where light is actually needed.

- Provide countertops with marine edges (see definition, p. 548) and shelves with raised edges to give a tactile clue to the end of the surface. Edges keep spills from dripping and materials from falling. If the edge has color or brightness that contrasts with the countertop, the edge is easier to see.
- Provide a swivel base on all seating to enable face-to-face communication.
- Minimize ambient noise and vibration in the laboratory, so that individuals with limited hearing range or a speech disability are better able to communicate.
- For other motor disabilities, provide a wheeled cart or other aid for carrying materials. Carts should be fully enclosed for messy or hazardous items.

- Avoid slippery surfaces that might throw a person who is unsteady on his/her feet.

Many guides have been published since 1990 to aid building owners and designers to understand ways to adapt the laboratory workplace for their disabled employees.

Ergonomic Design

The purpose of ergonomic design is to improve productivity by reducing work process stress and on-the-job injury for all laboratory occupants—those who are able and those who are disabled. For the past decade, in corporate offices there has been emphasis on ergonomically designed workstations for white-collar workers. This phenomenon directly relates to the importance and proliferation of computer technology and recognition of repetitive stress injuries. Furniture manufacturers and designers responded by providing a wide range of ergonomically designed products for seating, workstations, and data and telecommunications devices.

The first principal of ergonomic design is that all people are not created the same, in size or proportions. Persons on the tall, short, and heavy end of the human spectrum are sometimes left uncomfortable with standard laboratory furniture. Without height accommodation, drawers become convenient but unsafe step stools for short researchers to reach utility outlets, reagent shelves, and wall cabinets. Tall persons can suffer chronic back pain from stooping for hours over a countertop or at a chemical hood that is too low. Heavy persons may need wider clearances to safely negotiate around the laboratory. Safe laboratory design includes the recognition that all laboratory workers are not the same.

Adjustable Casework

As discussed in an earlier paragraph on casework, there are several systems on the market that allow laboratory occupants to readily adjust countertop height to suit the process and the person. Many systems simply have two alternatives: standing-height and seated-height benches that support countertops or supports with 6-in. (15-cm) adjustments. Laboratories that have both normal seated and standing-height work surfaces, with knee spaces for chairs or laboratory stools, will go a long way toward improving ergonomic conditions in the laboratory.

Knee Spaces

Knee spaces for laboratory chairs should be provided at each bench. A seated position, with good support for feet either flat on the floor or by a foot ring, improves fine motor coordination and reduces fatigue. Knee spaces should be a minimum of 3 ft (0.91 m) wide so that laboratory stools with a wide-diameter footrest can be fully pushed up to or under the countertop. Knee spaces are not wasted space. If the knee space is not used for sitting,

it can be used for undercounter equipment or for storage, as described earlier.

Countertops

When laboratory workers use microscopes and other instruments for hours at a time, they lean against the countertop with their elbows or forearms. Sharp or hard edges can produce pressure injuries to elbows and loss of hand mobility or sensation. Countertops with marine edges remove any sharp condition, but are still hard. Special armrest pads should be provided for these workers. At laboratory benchtop instruments that have computer keyboards attached, provide convenient undercounter keyboard within the knee space. The keyboard tray provides a better height for use and puts the keyboard out of harm's way when not in use.

Laboratory Seating

High stools and chairs for all uses should be selected for postural stability and comfort. Laboratory seating should be adjustable, within certain limits, so that individuals of different body types can adjust the seat to their needs. Laboratory seating should swivel, so that balance can be maintained while moving into position at the bench. Persons with mobility and certain sensory disabilities may need seating that provides for their special needs.

Storage

Storage components in laboratory casework can enhance ergonomic quality of the laboratory. Drawer slides should be heavy duty and full extension, where possible, to assist staff in reaching and lifting drawer contents. Self-closing features are available in drawer hardware. Self-closing hardware is also available for swinging doors on wall and other cupboards. This feature improves the safety of persons moving in and around benches. Upper cupboard doors that are made of glass enhance safety. Contents are visible, which reduces the need to open doors and grope to find hard-to-reach items.

77

Moving and Decommissioning Laboratories

RICHARD FINK

Laboratories are constantly adapting to small changes and renovations to accommodate their ever-changing technical needs. Some industrial research and development labs, such as pharmaceuticals, renovate as frequently as every 5 years, while others, such as universities and government laboratories, may not renovate for 30 years or more. A recent survey of laboratory owners revealed that over 50% of their laboratories were 20 years old, about 25% were between 11 and 20 years old, and the rest were less than 10 years old.[1]

In a few cases, large-scale renovations may involve an entire laboratory, suite of laboratories, or entire buildings. Most laboratories are similar both in types of agents and the equipment that may be present. Thus, any move will involve:

- transportation of chemicals, radioactive materials, biological agents
- equipment (such as lab benches, sinks, shelves, cabinets, fume hoods, centrifuges, etc.)
- decontamination of equipment and building structures
- unpacking

It is helpful to think of these activities in several phases: planning, decontamination and moving, construction and renovation, and occupancy. The hazards present will depend on the types of laboratories involved; however, the approach applies to all kinds of laboratories including chemistry, biology, engineering, physics, geology, clinical, and radiation. Previous work has described specific decontamination procedures.[2-5]

Moving, renovating, or demolishing laboratories requires careful planning to account for the safety and health of all those involved. These include the laboratory occupants, moving personnel, contractors, and occupants of surrounding areas not involved in the renovation. Health and safety personnel and relevant facilities personnel must maintain continuous and open communication among these groups.

Planning

In order to facilitate communication, it is important that a project or move coordinator be appointed to obtain all the required facts related to the move or renovation and to coordinate the various activities. For large projects, such as whole floors or buildings, a committee that includes representatives of all involved parties and chaired by the move coordinator is desirable.

During the planning phase, the potential hazards must be identified and mitigation techniques specified. The current laboratory occupants can identify current and possibly past practices that may reveal potential problems. For example, use of perchloric acid at elevated temperatures without a scrubbing system may lead to the formation of explosive perchlorates in the exhaust ducts or vacuum systems.[2,4] Use of *Aspergillus flavus* or *parasiticus* may result in surfaces contaminated with aflatoxin. A variety of toxic materials such as acrylamide, metals, and arsenic may contaminate surfaces and lead to skin contact problems for the movers and construction workers. Mercury spills that are not properly remediated may result in the presence of mercury in the plumbing systems or under floors or cabinets. The use of azides in plumbing systems may lead to explosions. Improper disposal of radioactive materials can result in radioactive contamination of the plumbing or surfaces in the lab, resulting in exposure to nonlaboratory personnel. Finally, the presence of animals may result in allergic reactions.

The occupants must also identify what equipment and chemicals are to be moved to their new location and the appropriate handling procedures. They may also be responsible for decontaminating equipment and surfaces.

Health and safety professionals will need to ensure that all the right questions are asked and provide advice and assistance in assessing potentially hazardous conditions (perform surface or air sampling) and specifying appropriate decontamination procedures.

Contractors need to identify *how* they perform their work so that potential hazards that they may create can be identified. Examples of this include potential disturbance of asbestos-containing material, lead-containing paint, and surfaces contaminated with fixed radioactive material.

The move coordinator needs to act as the integrator of these related activities to make sure the parties involved fully interact with each other. Only in this way will the potential risks be minimized.

There are several important steps to incorporate in the planning phase:

- Identify potential hazards (see Table 77.1) and specify mitigation techniques.
- Initiate a program of minimizing the amount of hazardous materials to be moved. Reducing the volume and weight of materials will reduce the potential for a disastrous accident.
- Anticipate finding unknown and unlabeled materials. If possible, arrange to have one of your own labs or a contract lab perform identification.
- Identify who has to be notified about your move, that is, internal safety office(s) such as radiation and biological safety, outside agencies such as the U.S. Department of Agriculture (USDA) for veterinary service or plant service permits that are laboratory site specific, or the Nuclear Regulatory Commission (or a similar state agency).
- Moving or decommissioning a licensed radioactive laboratory may require prior approval of the regulators, and this process can be time consuming. Thus, in order that this does not cause a delay in the planned move schedule, it must be included early in the planning stage. In addition, if the laboratory is moving to an entirely new facility, that facility will need appropriate licensing and permitting. A record of the license and material shipped must be maintained for inspection by the regulatory agencies.
- Determine what must be decontaminated and make plans as to how to decontaminate the facility

that is being vacated. Equipment that has been chemically, radioactively, or biologically contaminated must be decontaminated prior to the move. For example, animal cages may be contaminated with excrements or secretions containing oncogenic chemicals, radioisotopes, and/or infectious agents.

- To minimize pest transfer, request pesticidal treatment of equipment prior to moving it to the new location.
- Complete a thorough radiation survey to determine contamination levels. The radiation safety office is a good starting place to examine records of prior contamination incidents, radiation survey results, and so forth.
- Decide whether to hire an outside expert company to move chemicals or to do it using in-house expertise.

Note that some radioactive materials do not require the user to be licensed or permitted for use. Naturally occurring radioactive materials such as uranium and thorium are two examples of this. Many chemistry, biology, and materials science laboratories use these materials on a routine basis without the knowledge of the radiation safety office. It is very important to determine whether a laboratory uses forms of these radionuclides before they become mixed with other chemical wastes.

Collect all packaging items needed prior to the move date. Carts, plastic bags, toweling or other cushioning absorbent materials, sealable-plastic or plastic-lined boxes, labels (fragile, universal biohazard, identification, location, caution radioactive material), sturdy tape, and spill kit should be readily available.

TABLE 77.1 Laboratory Activities Critical to Planning for Safe Renovations

Practice	Possible Effect
Use of perchloric acid at elevated temperatures without a scrubbing system	Formation of explosive perchlorates in exhaust duct or vacuum systems; use methylene blue test for evaluating the presence of perchlorates
Use of toxic materials that may contaminate surfaces	Agents such as aflatoxin, acrylamide, metals, arsenic, etc., may pose a skin contact problem to movers and/or construction workers
Hazardous chemical spills that were not properly remediated	Mercury may remain in plumbing systems or under floors or cabinets
Azides used in plumbing systems	Explosions
Use of pathogenic agents that may have contaminated surfaces	May pose an infection hazard to movers and/or construction workers
Radiation/ Radioactive materials	May pose a radioactive exposure hazard to movers and/or construction workers
Animals	Allergic reactions

Establish safety and emergency procedures for all phases of your move. For chemicals, biologicals, and radioactive materials, it is important that all handlers be trained in proper handling techniques and given awareness training in the potential health hazards following exposure. Potential emergencies include material spills, fires, slips and falls, and cuts. Protective clothing and spill-absorbent material should be readily available during packing, moving, and unpacking. Ample fire extinguishers should be available and personnel trained to use them. Each container or piece of equipment must have adequate labels identifying the agent, hazards, and precautions. Health and safety professionals should remain on-call until the unpacking phase. The types of hazards that may be identified in the planning phase are listed in Table 77.2. Although this list is not exhaustive, it may act as an initial checklist.

Any move or relocation should be viewed as an opportunity to review equipment and material needs and discard those that are not needed. This is the time to perform major "house cleaning."

Finally, plan on providing orientation and training for personnel to cover any safety changes in the new facility (equipment, procedures, etc.).

Decontamination and Moving

This phase involves the moving of chemicals, biologicals, and radioactive materials along with equipment to either a new laboratory facility or a holding area until the laboratory renovation is completed. The components of the move involve:

- packaging of chemicals, biologicals, and radioactive materials
- preparing the equipment that will be moved (potentially contaminated equipment must be decontaminated; however, exceptions may be made for radioactive material contamination)
- decontamination of remaining equipment and building structures, that is, benches, fume hoods, BSCs, and so forth
- transportation of materials
- unpacking

All equipment, apparatus, and fixed structures must be decontaminated and prepared for moving as necessary. Most of the preparation can be done just a day prior to the move, thus minimizing disruption of the laboratory work. Once decontamination has been performed, any work that could potentially recontaminate the premises must be prohibited.

Moving Biological Materials

Many laboratory materials are regulated. Regulated biological materials include all genetically engineered microorganisms; recombinant plants and seeds; organisms requiring USDA permits; human tissues, blood, or body fluids[6]; and certain human pathogens and toxins.[7] Care must still be taken to ensure that nonregulated materials are properly contained during the move.

All biological materials must be packaged prior to the move. Proper packaging consists of a primary sealed container placed within a secondary sealed unbreakable container, with enough absorbent material in between to contain and absorb any spill. Some examples are petri dishes in a plastic sleeve surrounded by paper towels, within a plastic lined box; stabs in a sealed plastic container with paper towels to cushion vials; and tissue culture dishes taped closed and placed in a sealed waterproof container with an absorbent.

TABLE 77.2
Potential Hazards Associated with the Decommissioning of Laboratories

Items Related to Occupant Activities

Chemicals used that may remain on surfaces or equipment or building systems
    Perchloric acid
    Mercury
    Nonvolatile hazardous chemicals (e.g. acrylamide, various mycotoxins)
  Biological agents
  Radioactive agents
  Explosive materials
    Picric acid
    Ethers
Unclaimed chemicals
Asbestos-containing building materials
Lead paint

Items Related to Building Components

Asbestos—insulation, floor tiles, fume hood interiors, work surfaces
Lead paint
PCBs—light ballasts
Plumbing systems—mercury, azides
HVAC systems—perchlorates
Electrical
Radioactive materials

Items Related to Contractor Activities

Construction dust
Organic solvents (paint)
Cleaning chemicals
Welding or cutting

Once packaged, all biological materials must be properly labeled. Labels must include the name of the principle investigator (PI), new location, identification of the agent, Biosafety Level, and a telephone number for assistance in case of breakage.

In general, for moves within a facility, freezers and refrigerators can be moved intact provided that all contents are in sealed unbreakable containers and that the refrigerator/freezer remains closed or, preferably, locked. Since shifting of contents may occur, enclose loose items in boxes, and secure all boxes to avoid breakage and spills when the refrigerator or freezer is reopened. For shipments off-site within the United States, appropriate DOT requirements related to packaging and labeling must be followed.[8]

Moving Radioactive Materials

Radioactive materials require appropriate packaging in accord with the regulations of the NRC, the states, and the DOT.[6] Within your own facility or institution, this packaging and moving must be coordinated with your radiation safety professionals. A complete inventory of radionuclides, amounts, and storage requirements (e.g., room temperature, 0°C, −70°C) must be provided. Portable liquid nitrogen freezers may be necessary in some cases. Any materials that are being sent to another facility require proper packaging and the written permission of the receiving facility along with a copy of appropriate licenses or permits.

During the planning radiation survey, apparatus and equipment that were contaminated with radioactive material must be identified and labeled. If the radioactive contamination is fixed, the laboratory may choose to move the material in an "as-is" condition and not decontaminate to release-level criteria.[4,5,9]

As with biological materials, moves within a facility may be expedited if radioactive materials are packaged in refrigerators/freezers. The same basic principals as described above for biological materials would apply to moving radioactive material in refrigerators/freezers. Prior to the actual move of a refrigerator/freezer packaged with radioactive materials, the radiation safety office should make a final inspection of the packaging and secure the units for moving to their new location. Once secured by radiation safety, the freezers/refrigerators may not be opened until they arrive at their new location.

Gamma and beta counters must be emptied of all samples. The radiation safety office should be contacted to confirm that the reference source is secure. Consideration should also be given to radioactive waste contain-

ers, shielding, and tools such as pipetters and benchtop centrifuges, which may have become contaminated.

Moving should take place only when appropriate safety personnel are readily available. Thus coordination between the movers and safety personnel is essential. If the support of your radiation safety office is available, it is advisable to let it move radioactive sources within a facility.

Moving Chemicals

Unless in-house expertise exists regarding compatibility issues, flammability, and explosiveness, it would be wise to hire an outside company that has the necessary skills and knowledge. Some of the things that must be considered in moving chemicals are:

- Biological and chemical laboratories can have hundreds of chemicals and these must be sorted by compatibility (i.e., acids separated from bases, metals, alcohols), flammability, explosiveness, and reactivity.
- Proper packaging consists of the fairly standard lab pack-fiberboard drum with vermiculite (or similar) packing material. But because vermiculite is not totally unreactive, chemical compatibility with vermiculite must be entered into the equation.
- All chemicals must be labeled with their proper names, not abbreviations or formulas.
- Unlike biological and radioisotopes that can be safely moved in a refrigerator or freezer, chemicals must be removed and properly packaged prior to the move.
- Provisions must be made to keep heat-unstable chemicals at an appropriate temperature to prevent deterioration.
- Principal scientists will have to be consulted regarding whether they are involved in the production of novel chemicals. All properties of novel chemicals may not be known, and these may have to be specially packaged.
- Shock-sensitive chemicals have to be specially handled and packaged to avoid explosions and fires.

Chemical spill kits along with appropriately trained personnel should be on hand in case of accidents.

Moving Equipment

Other equipment, such as fermenters, incubators, water baths, darkroom tanks, fish tanks, and vacuum pumps (oil) must be emptied or drained and, if chemically, radioactively, or biologically contaminated, decontaminated prior to the move. Animal cages, fume hoods, centrifuges, gloveboxes, and biosafety cabinets (BSC) may require special attention depending on the nature of the work. Cages may need to be decontaminated if the animals were excreting oncogenic chemicals, infectious agents, or radioactive markers. Hoods or ventilation systems used for radioactive labeling must be surveyed by the radiation safety office. This survey should include any engineering controls such as filters/filter plenums used to reduce release of radioactive materials. Review the moving instructions that came with the centrifuge or contact your service representative. Decontaminate the chamber and rotor and remove the rotor. It is generally recommended that BSCs and gloveboxes that have been used with potentially infectious materials be gaseously decontaminated prior to moving.[2]

Other details to consider are:

- Return empty compressed-gas tanks.
- Fully back up the computer (and have keys on hand to free the computer from any security device).
- Remove glass thermometers from equipment.

- Disconnect, drain, and decontaminate columns, racks, FPLC, and HPLC.
- Have all necessary keys and lock combinations on hand.
- Some microscopes, such as electron microscopes, may need special preparation.
- Notify service contractors of your new location.
- Any battery-operated sensor alarms should be disabled prior to unplugging the equipment from the wall outlet.

It is advisable to have "Equipment Decontamination Record" stickers available (see Fig. 77.1). Affix these stickers to *all* equipment that will either be moved or that will be discarded or placed in storage. Moving personnel should not move equipment lacking the decontamination sticker. After the move, the PI should complete a laboratory decontamination certification form (see Fig. 77.2) and return it to the safety office.

On moving day, laboratory personnel should be available to oversee the move and train the movers. Coordi-

---

EQUIPMENT DECONTAMINATION RECORD

Principal
Investigator_____

       name   phone   date

This piece of equipment was used with the following:

[] No hazardous materials   [] Biologicals

[] Chemicals   [] PCB's surveyed by IH_____
                                  initials   date

[] Radiation   [] Surveyed by   [] wipe test survey

[] GM detector survey   [] Results report on file at RPO
      RS_____   initials date

[] other hazardous material (specify)_____

Decontaminated with_____

By (Name)_____ Date____

Equipment OK for removal or reuse: YES___ NO___

REMOVE THIS LABEL BEFORE REUSING EQUIPMENT

FIGURE 77.1 Example of an Equipment Decontamination Record label.

1. I certify that the cold/warm rooms and laboratory rooms at the following locations previously used by my laboratory have been emptied of biological materials:

2. The surfaces in these rooms have been decontaminated with: _____

3. All sink traps have been bleached and flushed with water; First, flush with water, then add 1/2 cup of concentrated chlorine bleach, wait 10–20 minutes, then flush thoroughly with water.

4. Please note location(s) of new lab area(s): _____

_____

Signature of Principal Investigator

_____

Print Name _____

New Lab Phone Number _____

Please return completed form to the Safety Office.

FIGURE 77.2 Example of a laboratory decontamination certification form.

nation and minimum awareness training of everyone involved in or "near" the moving activities should be performed so that they have a general idea of what will happen and their specific roles in the process.

The use of proper packaging can minimize spills and injuries; overloaded or overweight boxes can contribute to spills as well as to back injuries. A spill kit consisting of disinfectant, absorbent materials, paper towels, plastic bags, tongs or dustpan and brush, lab coat, gloves, and safety glasses must be readily available during the move. Promptly report all spills to the PI and the appropriate safety office. If a Biosafety Level 2 or higher agent or a volatile toxic chemical spills, evacuate the area first and then call for assistance. For personal injury, or direct contact involving biological materials, wash the affected area with soap and water. For injury or direct contact with a chemical, flood the area with water. In either case, consult with an occupational physician as soon as possible.

After a facility has been cleared by the movers, it may have to be decontaminated. Have standard operating procedures (SOPs) in place so that the radiation safety and necessary lab personnel can proceed with decontamination of remaining equipment and structures.

When the job is done, the radiation safety officer must file a final survey qualifying the facilities and equipment for unrestricted use. A record of the survey must be kept in a permanent file.

Construction and Renovation

Now that the facility has been decontaminated, the construction workers and renovators can begin their work. This is probably the most disruptive and "dirty" portion of the process. The potential hazards primarily apply to the demolition/construction workers, but in some cases they can affect the personnel in surrounding areas and ultimately the entire organization if a catastrophic event occurs or regulations are violated. The fruits of the planning stages will be realized (or suffered) in this phase.

Many of the details of health and safety components during this phase will be the responsibility of the contractor. In some circumstances, if the work is to be done by your institution then other regulatory issues not addressed here must be addressed (see OSHA's General Industry and Construction Standards).[10,11]

During this phase the following control procedures should be implemented:

- General awareness training should be provided to construction personnel, particularly if they have not been associated with previous lab work.
- SOPs must be provided. These should specify what activities construction workers can perform and how they should perform them. It should also delineate things they cannot do, such as handle containers with chemicals, disturb asbestos, and so on. For example, there may be a detailed procedure to follow when removing plumbing systems that might have mercury in them, requiring use of a mercury monitor or calling an on-call health and safety professional.
- SOPs should clearly specify what type of personal protection clothing to use and when to use it.

Reoccupancy

The final phase in this process is reoccupying the renovated space. In this case, the move-preparation phase is done in reverse. Therefore, the same issues of packing must be considered in unpacking. Plans for handling "broken" or "spilled" materials must be made in advance.

It is likely that some construction-related activities will continue after the move, and potential problems must be anticipated. Continuing construction will probably gen-

erate dust, noise, and odors (i.e., from painting, laying floors or rugs). Can these areas be isolated from the reoccupied areas? Are all of the life safety systems operable, particularly fire detectors, alarms, emergency eyewash and showers, emergency egress, and lighting? Are all contaminant control systems working? Are materials available to clean up hazardous spills? Has appropriate training for occupants been provided on any changes unique to the new space?

References

1. Baum, J. Hellmuth, OBATA and Kussabaum, Inc.: St. Louis Missouri, personal communication, January 1996.
2. Phillips, C. C.; Mueller, T. R.; Bader, M.; Haskew, M. W.; Phillips, J. B.; Vick, D. O. *Appl. Occup. Environ. Hyg.* **1994**, *9*, 503–509.
3. Kruse, R. H.; Puckett, W. H.; Richardson, J. H. *Clin. Mircobiol. Rev.* **1991**, *4*, 207–241.
4. *Perchloric Acid Contaminated Hood Decontamination Procedures Manual*; Martin Marretta Energy Systems, Inc., Oak Ridge National Laboratory: Oak Ridge, Tennessee, 1993.
5. *Guidelines for Decontamination of Facilities and Equipment Prior to Release for Unrestricted Use*; U.S. Nuclear Regulatory Commission: Washington, DC, 1993.
6. OSHA, 29 *CFR*, 1910.1030.
7. CDC, 42 *CFR*, 72.
8. DOT Transportation, 49 *CFR*, 1995.
9. NRC, 10 *CFR*, 1995.
10. OSHA, 29 *CFR*, 1910.
11. OSHA, 29 *CFR*, 1926.

78

General Ventilation Design and Control Systems

JOHN P. MARTIN

The principal objective of the laboratory ventilation system is to ensure the health and safety of the laboratory personnel and other building occupants against the hazards posed by chemical fumes and other airborne substances. In order to attain this goal, the ventilation system must be properly designed and controlled. It is therefore extremely important that everyone involved in the design and operation of these systems is

fully aware of exactly what constitutes a proper system.

There are many aspects to what should be considered when designing the laboratory ventilation system. Generally, the following questions must be addressed:

- What is the correct quantity and quality of ventilation air that should be used?

- What are the proper methods for supplying air into a laboratory, extracting exhaust air from the laboratory, and discharging the exhaust air from the building?
- How should the ventilation system be controlled?

Quantity and Quality

To determine the required quantity of ventilation air for a laboratory, it must first be acknowledged that the air system is really made up of three parts:

- the required and actual amount of exhaust air from fume containment devices
- the required and actual amount of room supply air
- the required and actual amount of general room exhaust air

The correct relationship between these quantities is termed proper room pressurization. It is this requirement for pressurization that determines the difference between required and actual amounts of air to be supplied to or exhausted from the laboratory.

The overall quality of the laboratory ventilation air system involves a discussion of the quality of the air supplied to the room and what can be done with the air exhausted from the room, considering its quality.

Exhaust from Fume-Containment Devices

Fume-containment devices generally used in laboratories include fume hoods, canopy hoods, bench-top slot hoods, snorkels, and biosafety cabinets. In all cases, the required volume of exhaust air, measured in terms of cubic feet per minute (cfm), is dependent on the velocity of the air entering the device and the free area of the entrance. Industry standards exist for the proper velocity for each type of fume-containment device. For devices not specifically addressed in this chapter, refer to the latest edition of *ACGIH Industrial Ventilation, A Manual of Recommended Practice.*[2]

Fume Hoods

The industry design standard for fume hood average face velocity is 100 feet per minute (fpm) with a maximum deviation across the entire sash opening area of no more than ±20%. The recommended face velocity for fume hoods by national organizations include the following:

- 29 *CFR* 1910.1450 recommends 60 to 100 fpm.[1]
- ACGIH recommends 60 to 100 fpm.[2]

- ANSI recommends 80 to 120 fpm.[3]
- NFPA recommends 80 to 120 fpm.[4]
- NRC recommends 80 to 120 fpm.[5]

It should be noted that when dealing with fume hood face velocities, more or less is definitely not better. At velocities much lower than 60 to 80 fpm, the velocity is too low for proper fume capture. At velocities approaching or exceeding 150 fpm, turbulence is caused around the periphery of the sash opening that actually reduces the capture efficiency of the fume hood.[5]

Therefore, to determine the required amount of air in cfm to be exhausted from a fume hood, multiply the operational free area in square feet of the fume hood sash opening by 100 fpm. The operational free area of the hood is dependent on the type of sash, the normal operating position of the sash, whether a sill opening is employed on the hood, and whether an air bypass is being used on the hood.

Fume hoods are equipped with one of three different types of sash arrangements: a single-piece vertical-raising sash, multiple horizontal sliding panels, or a combination sash consisting of a vertical-raising sash containing multiple horizontal sliding panels. Both the vertical and combination type of sashes can be fully opened so that the sash opening is nearly equal to the total closed sash area. The horizontal sliding panel type of sash can only be opened to some fraction of the total closed sash area depending on the number of panels and number of sliding panels. For example, take a fume hood that has a sash that is nominally 4.34 ft wide with a height of 36 in. with a sash border of 1.5 in. If this hood were equipped with either a vertical-raising or combination sash, it could be opened to a maximum height of 34.5 in. (36 − 1.5 in.), resulting in a sash opening area of approximately 12.5 square feet. This same hood, if equipped with a two-panel, two-track horizontal sash, would only allow an opening equal to the size of one of the sliding glass panels, which is 2.17 ft wide by 33 in. high or approximately 6 square feet. If a three-panel, three-track horizontal sash were used, the maximum open area would be equal to the area of two of the panels, each of which is 1.45 ft wide by 33 in. high, resulting in a total maximum open area of approximately 8 square feet. Disregarding for the moment all other hood conditions, if the required airflow through the hoods in these examples were only dependent on maximum possible sash opening, these hoods would require 1250 cfm, 600 cfm, and 800 cfm of exhaust, respectively.

For hood operator safety and energy conservation reasons, today's prudent practice is not to design hoods to be normally operated with the sash fully opened or, as in the case of horizontal sliding sashes, with sash panels much wider than 15 in. Consequently, fume hoods with vertical-raising sashes or combination sashes are designed for a normal maximum operating height of 18 in. regardless of the overall height of the sash. Vertical stops are supplied on the hoods to stop and hold the sash at the 18-in. opening level. The sash can still be opened above this level, generally for setup purposes only, but when this occurs a monitor on the hood initiates an audible and visible alarm, indicating that an unsafe condition exists. With this 18-in. opening limitation, the hood in the first example above, which required 1250 cfm of airflow, would now only require approximately 650 cfm.

If a hood is equipped with a sill opening, which is open all of the time regardless of sash height, the area of this sill opening should be added to the operational sash opening area. Likewise, if the hood is also equipped with an airflow bypass louver above the sash, the free area of this louver with the sash at its normal operating position must also be added to the operational sash opening area plus the sill opening area if present. Some hoods equipped with a bypass louver are also equipped with a perforated plate installed behind the louver. In this instance, it is the free area of the plate that should be used in lieu of the louver free area. The resulting total operational free area of the hood is then multiplied by 100 fpm and results in the required total design exhaust airflow for the hood. It should be pointed out that using the sill and bypass free areas in this method of airflow calculation is not entirely correct but the error is small and results in a total airflow that is only slightly higher than that which is actually required. This error is due to the fact that if the average velocity across the sash area is 100 fpm, then, because of different static pressure losses across the sill and bypass, the velocities through these components are in fact less than 100 fpm.

Other Fume Containment Devices

Canopy hood airflow requirements are dependent on the hood face dimensions, the hood face height above the process to be exhausted, and the number of enclosed sides.[5] Letting Q = the required airflow in cfm, W = the hood face width in feet, L = the hood face length in feet, and H = the height of the hood face above the exhausted process in feet, the following equations should be used for determining the required airflow:

- For no enclosed sides (open canopy):
  $Q = 280 \ (W + L)H$
- For two enclosed sides:
  $Q = 100 \ (W + L)H$

- For three enclosed sides:
  $Q = 100 \ WH$ or $100 \ LH$, depending on which side is open

Bench-top slot hood airflow requirements are dependent on slot dimensions and bench-top depth.[5] The equation to use for required airflow (Q) in cfm is $Q = 140 \ LX$, where L is the length of the slot in feet and X is the bench-top depth in feet. The required height (H) of the slot in feet can be calculated as $H = Q/(1000 \ L)$.

Snorkel airflow requirements are dependent on the face area of the snorkel and the distance from the snorkel face to the process to be exhausted.[5] The equation to use for required airflow (Q) in cfm is $Q = 100 \ (10 \ H^2 + A)$, where H is the distance from the snorkel face to the process in feet and A is the face area of the snorkel in square feet.

Biological safety cabinet airflow requirements are dependent on the type and length of cabinet. The airflow requirements for each type and length should be compiled from manufacturers' catalogs. The different types that are in use today are (portion of air not indicated to be exhausted to the outdoors is recirculated within the cabinet or room):

- Class I: For Biological Safety Level 1, 75 fpm face velocity, 100% Exhaust
- Class II Type A: For Biological Safety Level 2, 75 fpm face velocity, 30% Exhaust
- Class II Type B1: For Biological Safety Level 3, 100 fpm face velocity, 70% Exhaust
- Class II Type B2: For Biological Safety Level 3, 100 fpm face velocity, 100% Exhaust
- Class II Type B3: For Biological Safety Level 2, 100 fpm face velocity, 30% Exhaust
- Class III: For Biological Safety Level 4, Fully Enclosed, 100% Exhaust

Total Amount of Exhaust Air from Fume Containment Devices

The total required exhaust air from fume containment devices (RDE) is the sum of the required exhaust air for each containment device in the laboratory. The total actual design exhaust air from fume containment devices (ADE) is equal to the RDE value.

Laboratory Air Supply and General Room Exhaust

The amount of air supply required for a laboratory is that which is required to satisfy the environmental climatic design conditions of temperature and humidity mandated for the laboratory in question. Industry standards generally set indoor conditions for laboratories at 70°F dry bulb and 50% relative humidity. Although the HVAC Applications Handbook of the American Society of Heating, Refrigerating and Air-Conditioning Engineers (ASHRAE)[6] recommends using a dry bulb temperature of 72°F, the peak required air supply is generally dependent on satisfying the heat gain generated in the laboratory during the cooling season. This peak heat gain is dependent on the following factors:

- heat gain through exterior surfaces (fenestration areas, exterior walls, roofs) due to solar radiation and temperature gradients
- heat gain through interior surfaces due to temperature gradients
- heat gain from people in the space
- heat gain due to space lighting
- heat gain from equipment in the space

The tabulation of these simultaneous peak heat gains and the resulting amount of supply air required is a complex process that must be performed in strict accordance with the guidelines outlined by the ASHRAE Handbooks on Fundamentals[7] and Applications.[6] Once the required air supply for the laboratory has been determined, its value must be compared to the exhaust air value, RDE, with consideration being given to room pressurization requirements.

The concept behind the room pressurization for laboratories is that laboratories want to be negative to adjacent spaces so that hazardous fumes generated in a laboratory are exhausted through that space only and can not migrate to any adjacent spaces. In reality what this means is that all of the air exhausted from a given laboratory must be greater than the air directly supplied to that space. The difference between the higher exhaust air quantity and the lower supply air quantity is the air required for negative room pressurization (PA). This air does indeed enter the laboratory, generally from the adjacent corridor that is kept at a positive pressure to all adjacent laboratories as well as to the outdoors. Controlling the amount of PA currently follows two basic methods. One method uses a static differential pressure sensor between the laboratory and the adjacent positive corridor. When this device senses that the laboratory is moving away from being negative to the corridor, it attempts to increase the exhaust air in the laboratory. While in theory this is correct, in practical use, this type of system has failed on numerous occasions due to inaccurate sensors, false inputs, slow response times, and unstable operation. For these reasons, the second method is the preferred approach. This second method is called volumetric or flow tracking control. Under this method, the amount of both the total exhaust air leaving a laboratory and the supply air into the space is continually tracked and one or the other is reset so that the offset between the two always results in more exhaust than supply air, thus maintaining the space at a negative pressure.

To determine the amount of this offset or pressurization air (PA), consideration must be given to the type of overall system (constant or variable air volume) being used and, in the case of variable air volume systems, the accuracy of air volume control devices being used. Both constant and variable volume systems are used in laboratory facilities with equal reliability. The decision to use one over the other is strictly a matter of economic feasibility. A constant volume system can be installed for a lower first cost and will result in less maintenance. However, the constant volume system will result in substantially higher annual energy costs. Generally, the payback period for variable volume systems is quite favorable for most projects.

In either case, the PA is normally set equal to approximately 10% of the total amount of air actually exhausted from the room. In the case of a variable air volume system, this 10% figure assumes the use of air volume control devices having an accuracy of ±5%. These devices would include venturi valves, bladder dampers, and some of the butterfly-type dampers. If the device used has an accuracy higher than ±5%, the PA value of 10% should be increased accordingly.

At this point, for any given laboratory, the required amount of exhaust air from fume-containment devices (RDE), the required amount of supply air (RS), and the required approximate amount of offset or pressurization air (PA) are known. Before finalizing all of the actual design airflows for any laboratory, one last topic must be discussed. This topic has to do with required minimum airflows (MA) in the laboratory.

Most of the national standards referenced in this chapter quote minimum air exchange rates for laboratories. ASHRAE quotes 6 to 10 air changes per hour, NRC states 6 to 12 air changes per hour, and 29 *CFR* 1910.1450 lists 4 to 12 air changes per hour. Obviously there is no consensus on this issue. Also noteworthy is what ACGIH has to say on this subject. Air changes per hour is a poor basis for ventilation criteria where environmental control of hazards, heat, and/or odors is required.[2] The required ventilation depends on the problem, not on the size of the room in which it occurs.[2] Considering all of these references in detail, it is suggested here that the MA for any laboratory be set at approximately 6 air changes per hour. To calculate the actual MA in cfm, the volume of the room in cubic feet is divided by 10.

To determine finally the actual amounts of design exhaust and supply airflows for the laboratory, the following rules should be applied:

- If the RS value is less than 90% of the RDE value and the RDE is equal to or greater than MA: let the actual supply air value (AS) equal 90% of RDE, let the actual exhaust air from containment devices (ADE) equal the RDE value, and let the actual general (AG) equal zero for a constant volume system and equal ADE for a variable volume system. The total actual peak exhaust (TE) from the room will equal ADE.

- If the RS value is less than 90% of the RDE value and the RDE is less than MA: let AS equal 90% of MA, let ADE equal the RDE value, and let AG equal MA – RDE for a constant volume system and equal MA for a variable volume system. The total actual peak exhaust (TE) from the room will equal ADE.

- If the RS value is more than 90% of the RDE value and 1.11 times RS is more than MA: let AS equal RS, let ADE equal the RDE value, and let AG equal (1.11 RS) – RDE for a constant volume system and equal 1.11 RS for a variable volume system. The total actual peak exhaust (TE) from the room will equal ADE.

- If the RS value is more than 90% of the RDE value and 1.11 times RS is less than MA: let AS equal 90% of MA, let ADE equal the RDE value, and let AG equal MA – RDE for a constant volume system and equal MA for a variable volume system. The total actual peak exhaust (TE) from the room will equal MA.

Air Quality

The supply air to all laboratories in which some fume containment devices are employed must be 100% outdoor air. The exhaust air from these laboratories must be discharged to the outdoors and not recirculated within the laboratory itself or any other space within the building. These two rules are supported by all of the national standards referenced in this chapter. The only exemption to these rules comes from ANSI[3] and only applies to general laboratory space exhaust, not exhaust from fume containment devices. ANSI allows recirculation of general room exhaust if one of the following two sets of criteria can be met:

(1) Criteria A
  (a) There are no extremely dangerous or life-threatening materials used in the laboratory.
  (b) The concentration of air contaminants generated by the maximum credible accident will be lower than short-term exposure limits.
  (c) The system serving the exhaust hoods is provided with installed spares, emergency power, and other reliability features as necessary.

(2) Criteria B
  (a) Recirculated air is treated to reduce contaminant concentrations.
  (b) Recirculated air is monitored continuously for contaminant concentrations or provided with a secondary backup air-cleaning device that also serves as a monitor.
  (c) Air cleaning and monitoring equipment is maintained and calibrated under a preventive maintenance program.
  (d) A bypass to divert the recirculated air to atmosphere is provided.

Air Distribution Criteria

In designing a ventilation system for a laboratory, the following special air distribution criteria must be followed:

- Air supplied to the space must be through diffusers that are selected and located such that the diffused supplied air has a terminal velocity directly in front of the air intake of any fume-containment device in the room of not more than 50 fpm and preferably as low as 30 fpm.

- All fume exhaust ducts within the building should be sized for transport velocities of between 1000 and 2000 fpm. Duct velocities below 500 fpm should be avoided to help control the settling of particulates and to minimize vapor condensation. Duct velocities above 3000 fpm should be avoided to reduce objectionable noise and excessive use of fan horsepower.

- Fume exhaust should be discharged to the outdoors using a main stack sized for approximately 2000 fpm in order to contain internal condensation. The stack should contain a drain to collect this condensation and pipe it to the nearest laboratory (acid) waste system. The stack should terminate with a reduction in size so that the discharge velocity to atmosphere is approximately 3000 fpm. The stack should be located at the highest portion of the roof and have its discharge point at a minimum height of 10 feet above the adjacent roof. Also, the stack height above the ground level should be at least 1.3 times the building height for buildings up to three stories in height. Buildings taller than three stories in height should use stack heights above grade that are 0.75 times the building width. Exhaust stacks should be located downwind of and as far from fresh air intakes as possible. Usually, 50 ft is an adequate minimum distance.

• Fume exhaust fans as well as positive-pressure discharge exhaust ducts and stacks should be located outside of the building if at all possible. If parts of these portions of the exhaust system must be located indoors, their number must be kept to an absolute minimum. Fume exhaust fans and stacks must be maintained at a constant volume and velocity even if a variable air volume system is utilized. In this case, an outdoor air bypass controlled by a static pressure controller must be utilized to maintain the constant volume through the fan and stack when fume hood sashes in the system are set below that normal operating level.

• In accordance with NFPA,[4] fire dampers should not be used in laboratory fume exhaust systems so that fume exhaust flow will continue even if a fire is present somewhere in the building. This requirement of NFPA should be followed if possible. However, this does conflict with many local codes and should be resolved with the local code official early in the design process.

• Centralized manifold-type fume exhaust systems are preferred over individual exhaust systems that are subject to a harsh environment because at times exhaust fumes from a single source can be highly concentrated. Central systems have a higher dilution factor and are therefore less likely to be attacked by a single group of chemicals.

Fume Exhaust System Duct Materials

In addition to the above air distribution system criteria, the following duct material guidelines are offered, keeping in mind that no one material can offer complete protection against all possible chemicals or compounds:

• Type 316 stainless steel is one of the most resistant materials that can be used. However, this material should not be used on systems that have high concentrations or amounts of bromine, fluorine, iodine, sulfuric acid, ferric chlorides, stannic chlorides, or mercuric chlorides.

• PVC has excellent resistance to most acids but is unsuitable to acetate compounds, benzene compounds, strong solvents, or temperatures above 120°F. It also must be installed within a fire-rated enclosure inside any building because it emits hydrogen chloride gas if burned.

• Fiberglass reinforced polyester (FRP) has a good overall resistance to chemicals and can withstand vibration.

• Epoxy, teflon, and heresite-coated galvanized steel have excellent resistance to most chemicals. However, because of the difficult process of application, many breaks or junctions can occur in the coating, trapping chemicals and causing serious corrosion.

• Galvanized steel (uncoated) has very limited resistance to many chemicals and is generally not recommended. It may be used if the use of corrosive chemicals is very limited and infrequent.

Fume Exhaust System Duct Construction

Duct systems should be constructed using round ducts that are airtight. Stainless steel ducts must be welded. Use of rectangular ducts and sheet metal-type joints is inappropriate.

Controlling Proper Airflows and Space Temperatures

In the labs, controls for constant volume systems are relatively simple once the system is correctly balanced for the proper design of airflows. Each supply air to the room must be equipped with a reheat coil, which is controlled by a room thermostat or temperature sensor. This is needed to control room temperature, since the supply airflow to the room is constant and based on peak heat gains. Once balanced, the fume hood has no need for face velocity control. Constant volume hoods must be equipped with an air bypass louver in order to aid in maintaining constant volume through the hood as the sash closes. The simplicity of this control system is the reason for this type of system having both a low initial cost and a low maintenance cost. The high use of the supply air reheat coil and the constant volume airflow is the reason for this system's very high annual energy costs.

With a variable volume system, the control system is far more complex, but the energy savings over the constant volume system are high. The complex control system must track all of the supply air and exhaust airflows so that the correct amount of room pressurization air offset is maintained at all times. At the same time, the control system must maintain room temperature control by first attempting to increase or decrease the amount of supply air to the room and then adjusting the general exhaust from the room to maintain room pressurization. Sometimes the supply air cannot be decreased when a decrease is called for by temperature readings. It might have to remain the same or even be increased because of room pressurization demands. For this reason, a reheat coil is also used on the supply air for this system, though its use is low in comparison to the reheat coil used on the constant volume system. Another control that must be used in this system is a face velocity control on the fume hoods. This control is needed to maintain 100 fpm face velocity regardless of sash position and generally is of one of two forms. One method uses a velocity sensor installed in the sidewall of the hood inside the hood. This sensor approximates the sash opening face velocity and adjusts the exhaust airflow through the hood to maintain 100 fpm. This is the least expensive of the two methods employed, but is the least reliable due to continual calibration problems and its ability to transmit false readings if blocked by test equipment placed in the hood. The second method, though somewhat more expensive, is the preferred method of control. With this method, the actual position of the sash is sensed and used by the controller to set the proper airflow rate in order to maintain 100 fpm face velocity for the exact sash area that is open.

References

1. OSHA Laboratory Safety Standard, 29 *CFR* 1910.1450.
2. ACGIH, *Industrial Ventilation, A Manual of Recommended Practice*; American Conference of Governmental Industrial Hygienists: Cincinnati, Ohio, 1995.
3. ANSI/AIHA Z9.5, Laboratory Ventilation; American Industrial Hygiene Association: Fairfax, VA, 1992.
4. NFPA 45, Fire Protection for Laboratories Using Chemicals; National Fire Protection Association: Quincy, MA, 1996.
5. NRC, *Prudent Practices in the Laboratory: Handling and Disposal of Chemicals*; Committee on Prudent Practices for Handling, Storage, and Disposal of Chemicals in Laboratories; National Research Council; National Academy Press: Washington, DC, 1995.
6. ASHRAE, *Handbook of Applications*; American Society of Heating, Refrigerating and Air-Conditioning Engineers: Atlanta, GA, 1995.
7. ASHRAE, *Handbook of Fundamentals*; American Society of Heating, Refrigerating and Air-Conditioning Engineers: Atlanta, GA, 1997.

79

Elevators, Stairs, Ramps, and Step Stools

PHILBERT R. ROMERO

Elevators, stairs, ramps, and step stools are routinely used in everyday operations to facilitate the movement of materials and to provide access to industrial settings as well as to our own homes. Hazards associated with these type of devices include slips, trips, and falls as well as potential crushing injuries resulting from material handling and general usage activities. According to the National Safety Council, an average of approximately 1100 fatalities result annually from falls on or from stairs or steps in the United States.[1] In addition, almost 27% of all injuries can be attributed to accidental falls. Through proper design and maintenance and increasing the safety awareness factor, slip, trip, and fall injuries can be reduced.

Elevators

General Hazards

Elevators pose certain unique hazards that may not be readily apparent. Maintenance and rescue personnel are constantly being exposed to serious injuries while working on or responding to an incident involving elevators. In addition, personnel using the elevator equipment either for accessibility purposes or material movements are exposed to serious hazards.

A principal hazard associated with elevator use is being struck by the elevator door. This particular scenario poses a significant risk to elderly or disabled personnel who are prone to injuries resulting from the physical contact, which can cause further complications. Another major concern about elevators is the hazards related to maintenance and rescue activities. These personnel must ensure that they maintain complete control of the elevator car when they are servicing and/or testing the equipment. In the event that an elevator stalls with passengers in the car, both rescue and trapped personnel must stay out of harm's way. Each year, cases are documented in which trapped personnel attempt to exit a stalled elevator that is not properly secured from unwanted movement, a situation that both rescuers and trapped personnel must acknowledge. Those involved must recognize the potential for serious injuries or even death should an individual be in a precarious position when the elevator car moves. In such a scenario, a car can crush or guillotine a body part between itself and the building's structural supports.

Maintenance personnel are exposed to this same hazard and, in addition, to the hazard of energy sources. Hazard energy sources include physical, electrical, mechanical, and pressure systems that can cause serious injury. Therefore, it is crucial that these energy sources be controlled by adequate means. OSHA[2] requires that any time an energy source is present, it must be controlled by implementation of lockout and tagout procedures. Personnel must be trained in these procedures, including verification and communications to affected parties that the energy source has been isolated and maintained during the servicing of the equipment. In some cases, personnel must move the elevator car under a controlled condition to facilitate extraction, and they must take extreme precautions.

Emergency Response Procedures

It is crucial for any organization that uses elevators to establish emergency response procedures that address steps required to ensure positive control of the elevator car. This will immediately accomplish a safe rescue. In many cases, the first responder to a person trapped in an elevator is the fire department. Therefore, it is important to include personnel from the fire department, as well as other first responders, in training and in the development of referenced procedures.

The following are sample procedural elements that should be considered in the development of a site-specific elevator emergency response procedure.

Element 1

The telephone call originates from a stalled elevator car carrying passengers or originates from another caller who knows of the stalled car situation. The call goes via 911 from a telephone located in the subject car or other nearby telephone to a system where the location of the originating call telephone is displayed on the station emergency call monitor. The display should show the telephone number, site designation, building number, and elevator number. The 911 operator receiving the call can then verify the originating call telephone against the elevator telephone directory retained by the station operators. The receiving operator asks the caller various questions to gain information about the stalled elevator passenger's physical and mental conditions, status of the elevator car, location of the car in the elevator hoistway, and related problems/characteristics of the

stalled car environment (a sign on the elevator telephone cabinet door is available to prompt the caller).

The five questions are:

1. What is your name?
2. Where are you?
3. What is the number of the telephone that you are calling from?
4. How are you feeling?
5. Were you transporting any hazardous or unsecured materials in the car?

The questions are intended to be "nonthreatening" and reassuring to the impaired caller. The answers will provide noncomplicated information for key individuals so they can properly stage the evacuation of personnel from the stalled elevator with the least physical and emotional injury, while minimizing damage to the stalled elevator system.

Element 2

The operator receiving the distress call places two calls to key personnel involved in the proposed evacuation procedure. One call is made to the maintenance department section to activate the mechanic(s) who will stage the elevator car for personnel evacuation. The other call is made to the emergency center to activate the incident command system. The operator provides the received information as answers to the five questions to the maintenance department, who will forward the information to the responding mechanic and to the incident command system operator, who can request additional resources as deemed necessary. Additional communication with the fire department emergency medical service team should also occur simultaneously if needed. The operator should maintain telephone communication with the caller to maintain surveillance with the elevator passenger(s) until the evacuation team has taken charge of the incident site.

Element 3

The maintenance dispatcher, who receives the call, then contacts the maintenance supervisor and directs the supervisor to proceed to the site of the stalled elevator and convey the information received from the dispatcher.

Element 4

The operator contacts the emergency management system that activates the incident command system structure, if needed, which conveys the known information.

Element 5

During normal working hours, the maintenance dispatcher may direct the mechanics or electricians to respond directly to the site as soon as they are available. Facility supervisors should be equipped with a cellular telephone, a pager, and a portable radio that allows direct communications to be maintained. The supervisor should also have keys to the building in which the stalled elevator is located, a key to the elevator fire service key lock box, and key(s) to access the respective elevator equipment rooms.

Element 6

The supervisor, upon arrival at the site of the stalled elevator, continues with the ANSI/ASME Std. 17.4 and other internally referenced evacuation procedures[3] and assesses the physical conditions pertaining to the stalled elevator including the following:

- The supervisor locates the car, if possible, by visual observation or else by direct voice communication with car passengers to determine if the car is stalled at a corridor level or between floors.
- By appropriate means of communication with the car passenger, the supervisor directs rider(s) to push in the elevator emergency "stop" button to secure the car in place.
- As appropriate and within a reasonable amount of time, the supervisor applies yellow caution ribbons across corridor doors and out-of-service tags at each corridor push-button station for the elevator, while proceeding to the elevator equipment room to continue the procedural lockout/tagout of the elevator.
- He/she enters the respective elevator equipment room, unless prevented from doing so by a fire or other life-threatening condition. If the room cannot be entered, he/she proceeds with terminating power to the elevator hydraulic oil pump or traction motor for the elevator at the appropriate electrical disconnect, and installs the correct lock and tag on the respective circuit breakers, disconnect, or other electrical circuit safety components.
- In the company of another individual trained in the evacuation procedure or one available who is appropriately skilled in serving temporarily in this type of emergency, the supervisor continues to stabilize the car preparatory to the passenger evacuation in accordance with applicable standards and procedures.
- He/she determines the location of the stalled car and that the car must be and can be moved in

accordance with the pertinent requirement, then moves the car to the appropriate corridor opening and secures it so passengers may exit. Otherwise, he/she stages the car for personnel evacuation following steps for an ANSI/ASME evacuation procedure under the direction of the emergency management incident commander and fire department emergency rescue teams.

- The incident commander, having received information, mobilizes as appropriate for the incident and directs fire department rescue personnel and other emergency response teams as deemed appropriate for the specific site and circumstances involved.
- Decisions concerning the involvement of any added support teams will fall within the jurisdiction of the incident commander on the scene, using information from the elevator passenger(s), the facility supervisor, and others providing credible information.
- Under the cooperative direction and jurisdiction of the facility supervisor and the incident commander and on completion of the steps necessary to secure the car from unwanted movement, the stalled elevator doors (both corridor and car doors) can be safely opened to permit the safe exit of subject passengers from the secured car.
- The incident commander will have ultimate jurisdiction at *this phase* of the procedure, but will be obligated to exercise cooperative and prudent judgment throughout the rescue effort. The incident commander will direct the involvement of other specialty teams assembled for completion of the elevator car stabilization and personnel evacuation with due regard for the safety of all personnel involved.
- The next steps of the evacuation will be to place appropriate barricades, position ladders at hoistway corridor doors, and position ladders from the appropriate corridor openings to the stalled car hatch and into the elevator as determined to be prudent and appropriate coincident with ANSI/ASME 17.4.
- The supervisor completes the extraction of personnel from the stalled elevator. Emergency rescue personnel will assess the physical and mental well-being of evacuated personnel and release the personnel from the site or direct them to be transported for follow-up health and safety surveillance or medical attention as appropriate. Further, the facility supervisor and accompanying equipment service personnel or other qualified individual(s) will secure the car for repair, operational checks,

and returning the unit to service by certified elevator service personnel.

Stairs

Design

The design of stairs falls into two categories. Stairs being used as working surfaces are typically addressed in the General Industry and Construction Operational Safety and Health Administration standards. For stairs used as a means of egress, the requirements can be found in the National Fire Protection Association standards under the Life Safety Code.[4] In discussing design requirements, both OSHA and NFPA include detailed specifications that are required. Design factors that must be considered in each case include concentrated loads as well as uniform loading conditions. Stairs used as working surfaces are typically designed with a safety factor of at least 5 to 1 with a concentrated load of no less than 1000 pounds. Under the NFPA requirements, design factors must be consistent with construction design factors for the type of building construction. In addition to the strength requirements, both standards require other specific allowable dimensional parameters that must be adhered to. These include minimum width clearances, allowable height of risers, and allowable tread depths. In addition, both standards require that a landing be provided at least every 12 feet in vertical elevation change. The standards also dictate the specifications for handrails and guardrails required for stairways.

Maintenance and Inspection

Procedures to maintain and inspect stairs on a regular basis are critical to achieving a safe configuration. In many cases, stairs can be damaged as a result of being exposed to the elements in an outdoor environment or from normal wear-and-tear use. The following items should be considered in developing a maintenance and inspection program for stairs:

- Inspections should be performed by a competent person.
- Provisions for timely mitigation of any hazard discovered should include posting and barricading. Note that in the event a stair system is temporarily or permanently removed from service, consideration must be given to life safety code requirements to ensure that means of egress from the facility or area are not affected.
- The following checklist can be used as an inspection tool.

1. adequate lighting
2. visible nosing on the edge of the step or tread (highlighted with a distinguishable color and securely fastened)
3. a continuous handrail, cleaned regularly, and painted a light color (handrails should not be loose or contain any sharp edges and should extend beyond the landings to facilitate access onto and off the stairs)
4. stair treads with good traction and a slip-resistant surface
5. worn or defective treads or other parts in good repair
6. stairs clear of obstructions and maintained

Safety

In general, stair use accounts for many injuries as a result of falls and, therefore, personnel should attempt to use elevators in lieu of taking a stairway. It is essential that stairs be properly designed, regularly inspected, and maintained. Devices such as handrails and nosing on treads should be highlighted in contrasting colors that can facilitate safe usage. Based on an inspection program, any unsafe condition must be addressed in a timely fashion to prevent unnecessary injury exposure.

Ramps

Design

Ramps are provided for access to a facility or a workstation. They are also used as an access aid for material-handling operations. Ramps are required for accessibility to a facility by handicapped persons using wheelchairs. Based on this use, ramps must be specifically designed to meet OSHA requirements. In addition, the Americans with Disabilities Act requires specifications that must be met in providing accessibility to handicapped persons. These specifications allow ramp slopes, widths, approach, and discharge areas with respect to maneuverability of a wheelchair on a ramp. Handrail design used with ramps must comply with OSHA stairway specifications. Another design factor that deserves consideration is the need to provide a non–slip-resistant surface on all ramps.

Maintenance and Inspection

A maintenance and inspection program is a critical element in any safety program. Legally, employers are required to provide a means of access to all facilities and workstations, including access for the physically impaired. Ramps are a primary mechanism to provide this function and must be continually maintained and inspected. The ramp and surrounding areas leading to a ramp must never be obstructed. Surfaces must be maintained and adequate lighting should be provided. Any defect discovered on a ramp, such as damaged structural components or handrails, must be corrected in a timely fashion. Also, in terms of maintenance, the ramp must be kept free of snow, ice, and mud, which have the potential to cause problems.

Step Stools

Design

Step stools are primarily used to access an area that cannot normally be reached under standard configurations. They are used extensively throughout industrial environments as well as in residential applications. The major design factor governing step stools is the loading requirement. The stool must be able to carry the anticipated load without deflection or structural failures. Many step stools are labeled with manufacturer warnings that alert the user not to work while standing on the last rung or step. Therefore, the design of any step stool is clearly dependent on a stability factor and load rating, which must be considered.

Maintenance and Inspection

Maintenance and inspection of step stools should be consistent with other access devices such as stairways and ramps. Any defect discovered should be addressed immediately to prevent unnecessary exposure of danger to a person attempting to use the stool. In terms of inspection, items to look for include structural damage, miss-ing or damaged feet covers, which provide non–skid-resistance, and faded warning labels.

Safety

Each year a large number of people fall when using step stools. Step stools are devices to facilitate access to levels normally not reachable from floor levels and must be properly maintained, inspected, and, most importantly, used within manufacturer design applications. Stability is a major factor, as people tend to misuse these items, and the result can cause a serious injury. The top surface, rung, or step should never be used on a step stool because instability causes the stool to topple and basically allows the person's feet to swing away from the stool. Persons using step stools should not attempt to work to one side or the other of a stool plane; step stools should be used to access areas directly perpendicular to the plane, in front of the body, and to reasonable elevations. Persons should not attempt to turn back and reach in different planes while using a step stool. In the event that this type of action is required, changing to a ladder with a platform should be considered for more stability and more working area to access various positions that may be required to complete the task. In addition, step stools must never be used in environments that may cause slippage, such as ponding water areas and oily floors.

References

1. *National Safety Council Accident Facts*, 1995 Edition; National Safety Council: Itasca, IL, 1995.
2. Occupational Safety & Health Administration, *Code of Federal Regulations*, Title 29, Parts 1900–1910, Washington, DC, 1999.
3. ASME A17.1-1990, *Safety Code of Elevators and Escalators*, The American Society of Mechanical Engineers: New York, 1990.
4. *NFPA Life Safety Code Handbook*, National Fire Protection Association: Quincy, MA, 1997.

SECTION VIII

ENVIRONMENTAL MANAGEMENT

RUSSELL PHIFER

SECTION EDITOR

80

The Disposal of Chemical Wastes

RUSSELL W. PHIFER

Hazardous Wastes

Definitions

The Resource Conservation and Recovery Act (RCRA) defines a hazardous waste as a chemical material that is discarded or intended to be discarded and that exhibits one or more hazardous characteristics (flammability, corrosivity, reactivity, or toxicity) or appears on one or more of a series of chemical lists. The purpose of regulating the disposal of these materials was to hold generators accountable for their wastes from cradle to grave. Since the inception of RCRA and its resulting regulations, there have been significant changes in how hazardous wastes are managed. Traditional land disposal methods have been virtually eliminated, and the focus from both a regulatory and a logical viewpoint has been to encourage the minimization of hazardous wastes. Nonetheless, millions of tons of hazardous wastes are generated each year in the United States by hundreds of thousands of industrial, government, medical, commercial, and academic facilities.

Waste Determinations

The first step in managing chemical wastes is a determination of whether or not they are regulated under RCRA. This is done by first evaluating the waste material to see if it meets any of the criteria for characteristic wastes. The waste determination can be made either through knowledge of the process by which the waste was generated or through laboratory analysis. Four lists of hazardous wastes must also be consulted as part of the process. Table 80.1 provides definitions for the various RCRA characteristics.

In addition to the RCRA characteristics, there are four lists of compounds that are regulated as hazardous. These are the F, K, P, and U lists. The F list, "Hazardous wastes from nonspecific sources," consists of a number of spent halogenated and nonhalogenated solvents, wastes generated from various production processes, wastewater treatment operations, and other process wastes involving toxic and/or environmental pollutants. The K list is similar in nature, except that it refers to "Hazardous wastes from specific sources." For example, distillation bottoms from the production of aniline are listed with the K063 hazard code. The U and P lists are both specific to discarded commercial products, not processes. It typically applies to chemicals that are off-spec or surplus and intended to be discarded. The P list is for acutely toxic chemicals only; the U list is for more common compounds, such as ethyl acetate, which is listed as U112.

Waste Accumulation

Wastes are regulated from the time they are initially accumulated until their ultimate disposal is completed. While there is an exception for the storage of waste at or near the point of generation (called "satellite" storage, limited to 55 gal), hazardous waste generators have set time limits and therefore can store these wastes without a permit. Table 80.2 summarizes the three generator classes along with time limits for storage. It should be noted that many states do not recognize the conditionally exempt small-quantity generator designation included in the federal regulations.

Segregation

A key to effective management of hazardous waste is the proper segregation of incompatible wastes. From an economic standpoint, it is also advisable to segregate halogenated from nonhalogenated solvents. It is also important never to mix hazardous wastes with nonhazardous wastes, as the resulting mixture will be regulated as hazardous.

Consolidation

Just as segregation is important, consolidation of compatible wastes can lower disposal costs and improve the efficiency of a hazardous waste management program. For example, nonhalogenated solvents with sufficient BTU value should be consolidated for on- or off-site fuel

TABLE 80.1 Hazardous Waste Characteristics

Characteristic	Definition	Example
Ignitability (flammability)	Liquid with a flashpoint of less than 140°F or a solid that readily sustains combustion	Xylene
Corrosivity	Liquid with a pH of less than or equal to 2 or greater than or equal to 12.5	Hydrochloric acid
Reactivity	Normally unstable and readily undergoes violent change, reacts violently with air or water, or is a cyanide- or sulfide-bearing waste that, when exposed to pH conditions between 2 and 12.5, can generate toxic vapors or gases	Sodium metal
Toxicity	A liquid with specific concentrations of chemicals listed on the Toxicity Characteristic Leaching Procedure (TCLP) list or a solid that, when extracted using the TCLP, has a resulting concentration above the listed levels	Aqueous waste with concentrations of lead above 5 mg/l

recovery. Aqueous wastes can generally be mixed as long as compatibility is considered, and the consolidation of other wastes may make sense from an economic standpoint. Since off-site waste disposal is usually billed on a cost-per-unit basis, for example, a 55-gal drum, consolidating compatible wastes can result in fewer containers being shipped and lower costs.

Disposition Technologies

There are four factors involved in the selection of disposition technologies: suitability of the waste, disposal cost, transportation cost, and future liability. The decision can often be made by assigning a value to each of these factors; often an "RTS" model is suitable. In other words, the first priority should be to consider *reclamation or recycling*, the second option should be *treatment (thermal or chemical)*, and the last option should be *stabilization/ secure landfilling*. From a cost/benefit standpoint, being able to reuse, reclaim, or resell a waste material will present the best combination of low cost and low future liability. For a number of wastes, however, there is only one approved technology—thermal treatment, or incineration. There are also many aqueous wastes that are perfectly suitable for chemical treatment.

Recycling/Reuse/Reclamation

All wastes should be carefully evaluated for recycling, reuse, or reclamation. The type of waste most often recycled is spent solvents. Solvents such as acetone, xylene, and methanol might be redistilled and reused on site. Other flammable or combustible liquids are generally suitable for fuel recovery; it may even be possible to obtain a permit for on-site use as a fuel supplement. There are also permitted commercial facilities that recover solvents, both chlorinated (such as methylene chloride) and nonchlorinated (aromatic hydrocarbons), that meet specific criteria. Careful consolidation and segregation of solvent wastes is advisable, however, as mixing solvents can result in a waste with a much higher disposal cost than one that is reasonably homogeneous. In particular, wastes that contain both halogenated and nonhalogenated solvents are expensive to dispose, since these are virtually impossible to reclaim, and high chlorine content adds significantly to the cost. Technology limits on scrubber design combined with regulatory controls mean that incinerators must dilute the chlorine content with other fuels before burning highly chlorinated wastes. Typically, there is an exponential rise in disposal cost when the chlorine content of a waste is over 2%.

TABLE 80.2 Hazardous Waste Generator Classifications

Generator Status	Time Limit for Accumulation	Generation Rate
Conditionally exempt small quantity generator	No time limit, but must never accumulate more than 6000 kg of hazardous waste on site	Less than 100 kg of hazardous waste per month and less than 1 kg of acutely toxic waste
Small quantity generator	180 days, or 270 days if the designated treatment, storage, or disposal facility is more than 200 miles away	Between 100 kg and 1000 kg per month, but total accumulation must not exceed 6000 kg
Large generator	90 days	More than 1000 kg of hazardous waste per month

Incineration/Thermal Decomposition

Regulatory controls resulting from the RCRA reauthorization in 1984 (formally known as the Hazardous and Solid Waste Amendments, or HSWA) require incineration of many hazardous wastes that were previously landfilled. Unfortunately, the high cost of permitting an incineration facility and public opposition to having these facilities in their neighborhoods has resulted in few facilities being permitted and high costs. Most of these facilities are located in rural areas far from where the wastes are generated. Despite many of the existing plants running at or near capacity, there have been no recent successful attempts to permit new facilities. The primary advantage of incineration, from a generator's standpoint, is the reduced long-term liability associated with the "ultimate" destruction of the waste.

Both liquids and solids may be suitable for incineration in different types of facilities. Incineration of liquids almost always costs less, since less ash is produced and there is a lower retention time. As previously mentioned, many liquid wastes may be suitable for fuel recovery. Cement kilns are among the most popular facilities, since large quantities of fuel are required for the process, and costs are generally low. Solid materials, on the other hand, must be burned in rotary kilns. These facilities require that clean fuel be added to provide the necessary temperatures for thermal destruction. In addition, the ash may need additional treatment or stabilization prior to landfilling. These factors add considerably to the cost.

Chemical Treatment

Aqueous wastes are not generally suitable for incineration, and many can be treated successfully. Typically, compatible wastes are combined in tanks or reaction vessels, undergo one or more treatment processes, then are discharged through a permit issued under the National Pollution Discharge Elimination System, or NPDES. Sludges generated by the process are then dewatered, stabilized, and landfilled. Chemical processes include neutralization, precipitation, sedimentation, flotation, and filtration. There are also a number of biological treatment processes, such as microbial treatment, that are suitable for treating organic wastes in water.

Labpacks as Disposal Units

Labpacks are outer containers used to hold small containers of compatible wastes for both transportation and disposal. An adsorbent material is added between containers for cushioning and as an additional barrier in case of releases inside the drum. Unlike bulk wastes, which provide only one barrier, labpacks have three—the outer container, the adsorbent, and the chemical bottle or container itself. Both fiber and steel drums are used for filling labpacks, so named because they are used primarily for the disposal of laboratory chemicals. The type and drum is usually dependent on the quantity of waste in a particular disposal class. Disposal facilities establish "packaging protocols" suitable for the particular disposal technology to be utilized. For example, fiber drums are frequently used for incineration, since the entire container can be placed in a rotary kiln without additional handling or unpacking. Drums for direct incineration are generally limited to 30 gal in size, since the diameter is suitable for most incinerator "ports." Labpacks, however, can come in 5-, 10-, 12-, 15-, 20-, 30-, and 55-gal sizes. Unfortunately, the same handling applies regardless of size, since drums must be labeled, loaded, transported, unloaded, and logged separately. As a result, the cost difference is not as significant for these different size containers as might be expected.

Labpacks, regardless of the disposal technology, have several key advantages over individual packaging of wastes. First of all, multiple wastes can be disposed of together without the requirement of opening and mixing materials. In addition, disposal facilities rarely require any analysis prior to disposal (unlike bulk wastes); often only the common name or even tradename of the material is suitable for listing on shipping documents.

Disposal Strategies

As noted, there are many strategies to consider in the selection of a disposal method. This also applies to selecting TSD (treatment, storage, and disposal) facilities, transporters, and other contractors that assist in the process (such as waste brokers). Generators should always verify appropriate permitting, insurance, and regulatory compliance. This may be done through the use of independent auditors or by the generators themselves.

Selection of a TSD Facility

It is sometimes difficult to separate the selection of a disposal technology from the selection of a disposal facility. This may be due to the location of the generator relative to possible TSDs, RCRA-mandated disposal technologies, contractual obligations, or the whim of whoever is making the arrangements. Many generators do not have the personnel or expertise to obtain waste approvals; for this reason, transporters, brokers, and/or the facilities themselves actively market the service of "soup-to-nuts" disposal. This includes selecting the disposal facility and transporter, arranging any necessary analysis, completing a waste profile for approval of the waste, and scheduling the shipment.

Nonetheless, generators are advised to take an active part in the selection process. Factors to be considered include disposal cost, financial resources of the TSD, lead time for approval, facility permit limitations, transportation cost (usually based on distance), analytical requirements for approvals, applicable taxes and surcharges, and compliance history.

Use of Brokers

Brokers can provide a useful service in many instances, saving the generator considerable time and effort. For example, a broker may be able to evaluate a number of potential disposal facilities, based on the type of waste offered for disposal. They may also know which transporters have the right equipment and permits, make cost comparisons, and arrange the disposal project from start to finish. Brokers are generally not tied to a specific disposal facility and/or transporter, meaning they can "shop around" for the best price. However, brokers are in business to make a profit, which results in a markup over the actual costs. In addition, a broker may not be as concerned as the generator about long-term liability, resulting in the shipment of wastes to facilities with poor compliance records or poor financial resources. Brokers assume little or no liability under RCRA for illegal or improper waste disposal; the generator maintains liability, which can extend into the future indefinitely.

Transportation

Hazardous waste transporters are required to maintain resources, including liability insurance, specially trained drivers, and fleets in compliance with strict standards. There are also permit fees that vary considerably from state to state. The selection of a transporter is an important consideration, since the generator maintains liability throughout the transportation phase. This includes responsibility for proper completion of the shipping documents (usually a hazardous waste manifest and attachments), packaging, labeling, loading, and placarding. Nonetheless, costs for transportation can vary considerably. Generators should check for appropriate pollution liability insurance (generally $5 million per occurrence, $10 million aggregate), permitting, and experience.

Disposal of Nonregulated Chemical Wastes

Many industrial and laboratory wastes are not regulated under RCRA. While in most cases these wastes can still be shipped to permitted TSD facilities, other options may be available for disposition. These include municipal waste facilities (landfills, incinerators, and treatment plants), recycling/reclamation options (metal recovery), and even a public sewer.

Disposal Technologies

Municipal Waste Facilities

Solid wastes generated by an industrial or laboratory process may be regulated under a reduced hazard classification by the particular state in which the generator is located. This may result in an approval process prior to shipment of a non–RCRA-regulated waste into a facility. It is advisable to consider the potential long-term liability associated with municipal landfills; many have become Superfund sites. Municipal incinerators may be an extremely attractive disposal option for many nonregulated wastes. For example, discarded pharmaceutical products are frequently incinerated, at a much lower cost than with RCRA-regulated waste.

Public Sewer

The use of a public sewer for the disposal of an industrial or laboratory waste frequently requires a permit. The POTW has its own discharge limits to deal with; this results in many otherwise nonregulated wastes being refused. Ironically, NPDES discharge limits are now so restrictive that drinking water in some regions is not suitable for disposal due to relatively small concentrations of such metals as copper, zinc, and nickel. Despite these potential problems, there are numerous waste streams that are perfectly suitable for sewer disposal. Prior to discharge, however, it is recommended that the permitting requirements and discharge limits be reviewed. Submitting a written request to discharge an industrial or laboratory waste is advisable, regardless of quantity or frequency of the generating process.

Recycling

As for RCRA wastes, recycling is frequently the best option for nonhazardous wastes. Industrial and commercial wastes such as paper, cardboard, plastic containers, and many metals should be evaluated for potential recycling. As with other disposition technologies, the generator should determine the ultimate disposition or use of the waste material prior to shipment.

APPENDIX Hazardous Waste Lists (from *Code of Federal Regulations*, Title 40, Part 261).

Sec. 261.31   Hazardous wastes from non-specific sources.

The following solid wastes are listed hazardous wastes from non-specific Sources unless they are excluded under sections 260.20 and 260.22 and Listed in Appendix IX.

Industry and EPA hazardous waste no.	Hazardous waste	Hazard Code
Generic:		
F001	The following spent halogenated solvents used in degreasing: Tetrachloroethylene, trichloroethylene, methylene chloride, 1,1-trichloroethane, carbon tetrachloride, and chlorinated fluorocarbons; all spent solvent mixtures/blends used in degreasing containing, before use, a total of ten percent or more (by volume) of one or more of the above halogenated solvents or those solvents listed in F002, F004, and F005; and still bottoms from the recovery of these spent solvents and spent solvent mixtures.	(T)
F002	The following spent halogenated solvents: Tetrachloroethylene, methylene chloride, trichloroethylene, 1,1,1-trichloroethane, chlorobenzene, 1,1,2-trichloro-1,2,2-trifluoroethane, ortho-dichlorobenzene, trichlorofluoromethane, and 1,1,2-trichloroethane; all spent solvent mixtures/blends containing, before use, a total of ten percent or more (by volume) of one or more of the above halogenated solvents or those listed in F001, F004, or F005; and still bottoms from the recovery of these spent solvents and spent solvent mixtures.	(T)
F003	The following spent non-halogenated-solvents: Xylene, acetone, ethyl acetate, ethyl benzene, ethyl ether, methyl isobutyl ketone, n-butyl alcohol, cyclohexanone, and methanol; all spent solvent mixtures/blends containing, before use, only the above spent non-halogenated solvents; and all spent solvent mixtures/blends containing, before use, one or more of the above non-halogenated solvents, and, a total of ten percent or more (by volume) of one or more of those solvents listed in F001, F002, F004, and F005; and still bottoms from the recovery of these spent solvents and spent solvent mixtures.	(I)

(continued)

APPENDIX *Continued*

F004	The following spent non-halogenated solvents: Cresols and cresylic acid, and nitrobenzene; all spent solvent mixtures/ blends containing, before use, a total of ten percent or more (by volume) of one or more of the above non-halogenated solvents or those solvents listed in F001, F002, and F005; and still bottoms from the recovery of these spent solvents and spent solvent mixtures.	(T)
F005	The following spent non-halogen-ated solvents: Toluene, methyl ethyl ketone, carbon disulfide, isobutanol, pyridine, benzene, 2-ethoxyethanol, and 2-nitropropane; all spent solvent mixtures/blends containing, before use, a total of ten percent or more (by volume) of one or more of the above non-halogenated solvents or those sol-vents listed in F001, F002, or F004; and still bottoms from the recovery of these spent solvents and spent solvent mixtures.	(I,T)
F006	Wastewater treatment sludges from Electroplating operations except from the following processes: (1) Sulfuric acid anodizing of aluminum; (2) tin plating on carbon steel; (3) zinc plating (segregated basis) on carbon steel; (4) aluminum or zinc-aluminum plating on carbon steel; (5) cleaning/ stripping associated with tin, zinc and aluminum plating on carbon steel; and (6) chemical etching and milling of aluminum.	(T)
F007	Spent cyanide plating bath solutions From electroplating operations.	(R,T)
F008	Plating bath residues from the bottom of plating baths from electroplating operations where cyanides are used in the process.	(R,T)
F009	Spent stripping and cleaning bath solutions from electroplating ope-rations where cyanides are used in the process.	(R,T)
F010	Quenching bath residues from oil baths From metal heat treating operations where cyanides are used in the process.	(R,T)
F011	Spent cyanide solutions from salt bath pot cleaning from metal heat treating operations.	(R,T)
F012	Quenching waste water treatment sludges from metal heat treating operations where cyanides are used in the process.	(T)
F019	Wastewater treatment sludges from the chemical conversion coating of aluminum except from zirconium phosphating in	(T)

	aluminum can washing when such phosphating is an exclusive conversion coating process.	
F020...................	Wastes (except wastewater and spent carbon from hydrogen chloride purification) from the production or manufacturing use (as a reactant, chemical intermediate, or component in a formulating process) of tri- or tetrachlorophenol, or of intermediates used to produce their pesticide derivatives. (This listing does not include wastes from the production of Hexachlorophene from highly purified 2,4,5-trichlorophenol.)	(H)
F021...................	Wastes (except wastewater and spent carbon from hydrogen chloride purification) from the production or manufacturing use (as a reactant, chemical intermediate, or component in a formulating process) of pentachlorophenol, or of intermediates used to produce its derivatives.	(H)
F022...................	Wastes (except wastewater and spent carbon from hydrogen chloride purification) from the manufacturing use (as a reactant, chemical intermediate, or component in a formulating process) of tetra-, penta-, or hexachlorobenzenes under alkaline conditions.	(H)
F023...................	Wastes (except wastewater and spent carbon from hydrogen chloride purification) from the production of materials on equipment previously used for the production or manufacturing use (as a reactant, chemical intermediate, or component in a formulating process) of tri- and tetrachlorophenols. (This listing does not include wastes from equipment used only for the production or use of Hexachlorophene from highly purified 2,4,5-trichlorophenol.)	(H)
F024...................	Process wastes, including but not limited to, distillation residues, heavy ends, tars, and reactor cleanout wastes, from the production of certain chlorinated aliphatic hydrocarbons by free radical catalyzed processes. These chlorinated aliphatic hydrocarbons are those having carbon chain lengths ranging from one to and including five, with varying amounts and positions of chlorine substitution. (This listing does not include wastewaters, wastewater treatment sludges, spent catalysts, and wastes listed in Sec. 261.31 or Sec. 261.32.).	(T)

(continued)

F025.....................	Condensed light ends, spent filters and filter aids, and spent desiccant wastes from the production of certain chlorinated aliphatic hydrocarbons, by free radical catalyzed processes. These chlorinated aliphatic hydrocarbons are those having carbon chain lengths ranging from one to and including five, with varying amounts and positions of chlorine substitution.	(T)
F026.....................	Wastes (except wastewater and spent carbon from hydrogen chloride purification) from the production of materials on equipment previously used for the manufacturing use (as a reactant, chemical intermediate, or component in a formulating process) of tetra-, penta-, or hexachlorobenzene under alkaline conditions.	(H)
F027.....................	Discarded unused formulations containing tri-, tetra-, or pentachlorophenol or discarded unused formulations containing compounds derived from these chlorophenols. (This listing does not include formulations containing Hexachlorophene sythesized from prepurified 2,4,5-trichlorophenol as the sole component.)	(H)
F028.....................	Residues resulting from the incineration or thermal treatment of soil contaminated with EPA Hazardous Waste Nos. F020, F021, F022, F023, F026, and F027.	(T)

Note: I is ignitable waste, H is acute hazardous waste, R is reactive waste, and T is toxic waste.

APPENDIX *Continued*

Sec. 261.33 Discarded commercial chemical products, off-specification species, container residues, and spill residues thereof.

Hazardous waste No.	Chemical abstracts No.	Substance
P023	107-20-0	Acetaldehyde, chloro-
P002	591-08-2	Acetamide, N-(aminothioxomethyl)-
P057	640-19-7	Acetamide, 2-fluoro-
P058	62-74-8	Acetic acid, fluoro-, sodium salt
P002	591-08-2	1-Acetyl-2-thiourea
P003	107-02-8	Acrolein
P070	116-06-3	Aldicarb
P203	1646-88-4	Aldicarb sulfone.
P004	309-00-2	Aldrin
P005	107-18-6	Allyl alcohol
P006	20859-73-8	Aluminum phosphide (R,T)
P007	2763-96-4	5-(Aminomethyl)-3-isoxazolol
P008	504-24-5	4-Aminopyridine
P009	131-74-8	Ammonium picrate (R)
P119	7803-55-6	Ammonium vanadate
P099	506-61-6	Argentate(1-), bis(cyano-C)-, potassium
P010	7778-39-4	Arsenic acid H_3 AsO_4
P012	1327-53-3	Arsenic oxide As_2 O_3
P011	1303-28-2	Arsenic oxide As_2 O_5
P011	1303-28-2	Arsenic pentoxide
P012	1327-53-3	Arsenic trioxide
P038	692-42-2	Arsine, diethyl-
P036	696-28-6	Arsonous dichloride, phenyl-
P054	151-56-4	Aziridine
P067	75-55-8	Aziridine, 2-methyl-
P013	542-62-1	Barium cyanide
P024	106-47-8	Benzenamine, 4-chloro-
P077	100-01-6	Benzenamine, 4-nitro-
P028	100-44-7	Benzene, (chloromethyl)-
P042	51-43-4	1,2-Benzenediol, 4-[1-hydroxy-2-(methylamino)ethyl]-, (R)-
P046	122-09-8	Benzeneethanamine, alpha,alpha-dimethyl-
P014	108-98-5	Benzenethiol
P127	1563-66-2	7-Benzofuranol, 2,3-dihydro-2,2-dimethyl-, methylcarbamate.
P188	57-64-7	Benzoic acid, 2-hydroxy-, compd. with (3aS-cis)-1,2,3,3a,8,8a-hexahydro-1,3a,8-trimethylpyrrolo[2,3-b]indol-5-yl methylcarbamate ester (1:1).
P001	\1\ 81-81-2	2H-1-Benzopyran-2-one, 4-hydroxy-3-(3-oxo-1-phenylbutyl)-, & salts, when present at concentrations greater than 0.3%
P028	100-44-7	Benzyl chloride
P015	7440-41-7	Beryllium powder
P017	598-31-2	Bromoacetone

(continued)

APPENDIX *Continued*

P018	357-57-3	Brucine
P045	39196-18-4	2-Butanone, 3,3-dimethyl-1-(methylthio)-, O-[methylamino)carbonyl] oxime
P021	592-01-8	Calcium cyanide
P021	592-01-8	Calcium cyanide Ca(CN)$_2$
P189	55285-14-8	Carbamic acid, [(dibutylamino)-thio]methyl-, 2,3-dihydro-2,2-dimethyl- 7-benzofuranyl ester.
P191	644-64-4	Carbamic acid, dimethyl-, 1-[(dimethyl-amino)carbonyl]- 5-methyl-1H- pyrazol-3-yl ester.
P192	119-38-0	Carbamic acid, dimethyl-, 3-methyl-1-(1-methylethyl)-1H- pyrazol-5-yl ester.
P190	1129-41-5	Carbamic acid, methyl-, 3-methylphenyl ester.
P127	1563-66-2	Carbofuran.
P022	75-15-0	Carbon disulfide
P095	75-44-5	Carbonic dichloride
P189	55285-14-8	Carbosulfan.
P023	107-20-0	Chloroacetaldehyde
P024	106-47-8	p-Chloroaniline
P026	5344-82-1	1-(o-Chlorophenyl)thiourea
P027	542-76-7	3-Chloropropionitrile
P029	544-92-3	Copper cyanide
P029	544-92-3	Copper cyanide Cu(CN)
P202	64-00-6	m-Cumenyl methylcarbamate.
P030	....	Cyanides (soluble cyanide salts), not otherwise specified
P031	460-19-5	Cyanogen
P033	506-77-4	Cyanogen chloride
P033	506-77-4	Cyanogen chloride (CN)Cl
P034	131-89-5	2-Cyclohexyl-4,6-dinitrophenol
P016	542-88-1	Dichloromethyl ether
P036	696-28-6	Dichlorophenylarsine
P037	60-57-1	Dieldrin
P038	692-42-2	Diethylarsine
P041	311-45-5	Diethyl-p-nitrophenyl phosphate
P040	297-97-2	O,O-Diethyl O-pyrazinyl phosphorothioate
P043	55-91-4	Diisopropylfluorophosphate (DFP)
P004	309-00-2	1,4,5,8-Dimethanonaphthalene, 1,2,3,4,10,10-hexa- chloro-1,4,4a,5,8,8a,-hexahydro-, (1alpha,4alpha,4abeta,5alpha,8alpha,8abeta)-
P060	465-73-6	1,4,5,8-Dimethanonaphthalene, 1,2,3,4,10,10-hexa- chloro-1,4,4a,5,8,8a-hexahydro-, (1alpha,4alpha,4abeta,5beta,8beta,8abeta)-
P037	60-57-1	2,7:3,6-Dimethanonaphth[2,3-b]oxirene, 3,4,5,6,9,9-hexachloro-1a,2,2a,3,6,6a,7,7a-octahydro-,

APPENDIX *Continued*

		((1aalpha,2beta,2aalpha,3beta,6beta,6a alpha,7beta, 7aalpha)-
P051	\1\ 72-20-8	2,7:3,6-Dimethanonaphth [2,3-b]oxirene, 3,4,5,6,9,9-hexachloro-1a,2,2a,3,6,6a,7,7a-octahydro-, (1aalpha,2beta,2abeta,3alpha,6alpha,6 abeta,7beta, 7aalpha)-, & metabolites
P044	60-51-5	Dimethoate
P046	122-09-8	alpha,alpha-Dimethylphenethylamine
P191	644-64-4	Dimetilan.
P047	\1\ 534-52-1	4,6-Dinitro-o-cresol, & salts
P048	51-28-5	2,4-Dinitrophenol
P020	88-85-7	Dinoseb
P085	152-16-9	Diphosphoramide, octamethyl-
P111	107-49-3	Diphosphoric acid, tetraethyl ester
P039	298-04-4	Disulfoton
P049	541-53-7	Dithiobiuret
P185	26419-73-8	1,3-Dithiolane-2-carboxaldehyde, 2,4-dimethyl-, O- [(methylamino)-carbonyl]oxime.
P050	115-29-7	Endosulfan
P088	145-73-3	Endothall
P051	72-20-8	Endrin
P051	72-20-8	Endrin, & metabolites
P042	51-43-4	Epinephrine
P031	460-19-5	Ethanedinitrile
P194	23135-22-0	Ethanimidothioc acid, 2-(dimethylamino)-N-[[(methylamino) carbonyl]oxy]-2-oxo-, methyl ester.
P066	16752-77-5	Ethanimidothioic acid, N-[[(methylamino)carbonyl]oxy]-, methyl ester
P101	107-12-0	Ethyl cyanide
P054	151-56-4	Ethyleneimine
P097	52-85-7	Famphur
P056	7782-41-4	Fluorine
P057	640-19-7	Fluoroacetamide
P058	62-74-8	Fluoroacetic acid, sodium salt
P198	23422-53-9	Formetanate hydrochloride.
P197	17702-57-7	Formparanate.
P065	628-86-4	Fulminic acid, mercury(2+) salt (R,T)
P059	76-44-8	Heptachlor
P062	757-58-4	Hexaethyl tetraphosphate
P116	79-19-6	Hydrazinecarbothioamide
P068	60-34-4	Hydrazine, methyl-
P063	74-90-8	Hydrocyanic acid
P063	74-90-8	Hydrogen cyanide
P096	7803-51-2	Hydrogen phosphide
P060	465-73-6	Isodrin
P192	119-38-0	Isolan.
P202	64-00-6	3-Isopropylphenyl N-methylcarbamate.
P007	2763-96-4	3(2H)-Isoxazolone, 5-(aminomethyl)-
P196	15339-36-3	Manganese, bis(dimethylcarbamodithioato-S,S')-,
P196	15339-36-3	Manganese dimethyldithiocarbamate.
P092	62-38-4	Mercury, (acetato-O)phenyl-
P065	628-86-4	Mercury fulminate (R,T)

(continued)

APPENDIX *Continued*

P082	62-75-9	Methanamine, N-methyl-N-nitroso-
P064	624-83-9	Methane, isocyanato-
P016	542-88-1	Methane, oxybis[chloro-
P112	509-14-8	Methane, tetranitro- (R)
P118	75-70-7	Methanethiol, trichloro-
P198	23422-53-9	Methanimidamide, N,N-dimethyl-N'-[3- [[(methylamino)-carbonyl]oxy]phenyl]- , monohydrochloride.
P197	17702-57-7	Methanimidamide, N,N-dimethyl-N'-[2- methyl-4- [[(methylamino)carbonyl]oxy]phenyl]-
P050	115-29-7	6,9-Methano-2,4,3-benzodioxathiepin, 6,7,8,9,10,10- hexachloro-1,5,5a,6,9,9a-hexahydro-, 3-oxide
P059	76-44-8	4,7-Methano-1H-indene, 1,4,5,6,7,8,8- heptachloro- 3a,4,7,7a-tetrahydro-
P199	2032-65-7	Methiocarb.
P066	16752-77-5	Methomyl
P068	60-34-4	Methyl hydrazine
P064	624-83-9	Methyl isocyanate
P069	75-86-5	2-Methyllactonitrile
P071	298-00-0	Methyl parathion
P190	1129-41-5	Metolcarb.
P128	315-8-4	Mexacarbate.
P072	86-88-4	alpha-Naphthylthiourea
P073	13463-39-3	Nickel carbonyl
P073	13463-39-3	Nickel carbonyl Ni(CO)<INF>4</INF>, (T-4)-
P074	557-19-7	Nickel cyanide
P074	557-19-7	Nickel cynaide Ni(CN)<INF>2</INF>
P075	\1\ 54-11-5	Nicotine, & salts
P076	10102-43-9	Nitric oxide
P077	100-01-6	p-Nitroaniline
P078	10102-44-0	Nitrogen dioxide
P076	10102-43-9	Nitrogen oxide NO
P078	10102-44-0	Nitrogen oxide NO<INF>2</INF>
P081	55-63-0	Nitroglycerine (R)
P082	62-75-9	N-Nitrosodimethylamine
P084	4549-40-0	N-Nitrosomethylvinylamine
P085	152-16-9	Octamethylpyrophosphoramide
P087	20816-12-0	Osmium oxide OsO<INF>4</INF>, (T-4)-
P087	20816-12-0	Osmium tetroxide
P088	145-73-3	7-Oxabicyclo[2.2.1]heptane-2,3- dicarboxylic acid
P194	23135-22-0	Oxamyl.
P089	56-38-2	Parathion
P034	131-89-5	Phenol, 2-cyclohexyl-4,6-dinitro-
P048	51-28-5	Phenol, 2,4-dinitro-
P047	\1\ 534-52-1	Phenol, 2-methyl-4,6-dinitro-, & salts
P020	88-85-7	Phenol, 2-(1-methylpropyl)-4,6-dinitro-
P009	131-74-8	Phenol, 2,4,6-trinitro-, ammonium salt (R)
P128	315-18-4	Phenol, 4-(dimethylamino)-3,5-dimethyl- , methylcarbamate (ester).
P199	2032-65-7	Phenol, (3,5-dimethyl-4-(methylthio)-,

APPENDIX *Continued*

		methylcarbamate
P202	64-00-6	Phenol, 3-(1-methylethyl)-, methyl carbamate.
P201	2631-37-0	Phenol, 3-methyl-5-(1-methylethyl)-, methyl carbamate.
P092	62-38-4	Phenylmercury acetate
P093	103-85-5	Phenylthiourea
P094	298-02-2	Phorate
P095	75-44-5	Phosgene
P096	7803-51-2	Phosphine
P041	311-45-5	Phosphoric acid, diethyl 4-nitrophenyl ester
P039	298-04-4	Phosphorodithioic acid, O,O-diethyl S-[2-(ethylthio)ethyl] ester
P094	298-02-2	Phosphorodithioic acid, O,O-diethyl S-[(ethylthio)methyl] ester
P044	60-51-5	Phosphorodithioic acid, O,O-dimethyl S-[2-(methylamino)-2-oxoethyl] ester
P043	55-91-4	Phosphorofluoridic acid, bis(1-methylethyl) ester
P089	56-38-2	Phosphorothioic acid, O,O-diethyl O-(4-nitrophenyl) ester
P040	297-97-2	Phosphorothioic acid, O,O-diethyl O-pyrazinyl ester
P097	52-85-7	Phosphorothioic acid, O-[4-[(dimethylamino)sulfonyl]phenyl] O,O-dimethyl ester
P071	298-00-0	Phosphorothioic acid, O,O,-dimethyl O-(4-nitrophenyl) ester
P204	57-47-6	Physostigmine.
P188	57-64-7	Physostigmine salicylate.
P110	78-00-2	Plumbane, tetraethyl-
P098	151-50-8	Potassium cyanide
P098	151-50-8	Potassium cyanide K(CN)
P099	506-61-6	Potassium silver cyanide
P201	2631-37-0	Promecarb
P070	116-06-3	Propanal, 2-methyl-2-(methylthio)-, O-[(methylamino)carbonyl]oxime
P203	1646-88-4	Propanal, 2-methyl-2-(methyl-sulfonyl)-, O-[(methylamino)carbonyl] oxime.
P101	107-12-0	Propanenitrile
P027	542-76-7	Propanenitrile, 3-chloro-
P069	75-86-5	Propanenitrile, 2-hydroxy-2-methyl-
P081	55-63-0	1,2,3-Propanetriol, trinitrate (R)
P017	598-31-2	2-Propanone, 1-bromo-
P102	107-19-7	Propargyl alcohol
P003	107-02-8	2-Propenal
P005	107-18-6	2-Propen-1-ol
P067	75-55-8	1,2-Propylenimine
P102	107-19-7	2-Propyn-1-ol
P008	504-24-5	4-Pyridinamine
P075	\1\ 54-11-5	Pyridine, 3-(1-methyl-2-pyrrolidinyl)-, (S)-, & salts
P204	57-47-6	Pyrrolo[2,3-b]indol-5-ol, 1,2,3,3a,8,8a-hexahydro-1,3a,8-trimethyl-,

(continued)

APPENDIX *Continued*

		methylcarbamate (ester), (3aS-cis)-.
P114	12039-52-0	Selenious acid, dithallium(1+) salt
P103	630-10-4	Selenourea
P104	506-64-9	Silver cyanide
P104	506-64-9	Silver cyanide Ag(CN)
P105	26628-22-8	Sodium azide
P106	143-33-9	Sodium cyanide
P106	143-33-9	Sodium cyanide Na(CN)
P108	\1\ 57-24-9	Strychnidin-10-one, & salts
P018	357-57-3	Strychnidin-10-one, 2,3-dimethoxy-
P108	\1\ 57-24-9	Strychnine, & salts
P115	7446-18-6	Sulfuric acid, dithallium(1+) salt
P109	3689-24-5	Tetraethyldithiopyrophosphate
P110	78-00-2	Tetraethyl lead
P111	107-49-3	Tetraethyl pyrophosphate
P112	509-14-8	Tetranitromethane (R)
P062	757-58-4	Tetraphosphoric acid, hexaethyl ester
P113	1314-32-5	Thallic oxide
P113	1314-32-5	Thallium oxide $Tl_2 O_3$
P114	12039-52-0	Thallium(I) selenite
P115	7446-18-6	Thallium(I) sulfate
P109	3689-24-5	Thiodiphosphoric acid, tetraethyl ester
P045	39196-18-4	Thiofanox
P049	541-53-7	Thioimidodicarbonic diamide [(H$_2$ N)C(S)]$_2$ NH
P014	108-98-5	Thiophenol
P116	79-19-6	Thiosemicarbazide
P026	5344-82-1	Thiourea, (2-chlorophenyl)-
P072	86-88-4	Thiourea, 1-naphthalenyl-
P093	103-85-5	Thiourea, phenyl-
P185	26419-73-8	Tirpate.
P123	8001-35-2	Toxaphene
P118	75-70-7	Trichloromethanethiol
P119	7803-55-6	Vanadic acid, ammonium salt
P120	1314-62-1	Vanadium oxide $V_2 O_5$
P120	1314-62-1	Vanadium pentoxide
P084	4549-40-0	Vinylamine, N-methyl-N-nitroso-
P001	\1\ 81-81-2	Warfarin, & salts, when present at concentrations greater than 0.3%
P205	137-30-4	Zinc, bis(dimethylcarbamodithioato-S,S')-,
P121	557-21-1	Zinc cyanide
P121	557-21-1	Zinc cyanide Zn(CN)$_2$
P122	1314-84-7	Zinc phosphide Zn$_3$ P$_2$, when present at concentrations greater than 10% (R,T)
P205	137-30-4	Ziram.

\1\ CAS Number given for parent compound only.

(f) The commercial chemical products, manfacturing chemical intermediates, or off-specification commercial chemical products referred to in paragraphs (a) through (d) of this section, are identified as toxic wastes (T), unless otherwise designated and are subject to the small quantity generator exclusion defined in Sec. 261.5 (a) and (g).

APPENDIX *Continued*

[Comment: For the convenience of the regulated community, the primary
hazardous properties of these materials have been indicated by the
letters T (Toxicity), R (Reactivity), I (Ignitability) and C
(Corrosivity). Absence of a letter indicates that the compound is only
listed for toxicity.]

These wastes and their corresponding EPA Hazardous Waste Numbers are:

Hazardous waste No.	Chemical abstracts No.	Substance
U394	30558-43-1	A2213.
U001	75-07-0	Acetaldehyde (I)
U034	75-87-6	Acetaldehyde, trichloro-
U187	62-44-2	Acetamide, N-(4-ethoxyphenyl)-
U005	53-96-3	Acetamide, N-9H-fluoren-2-yl-
U240	\1\ 94-75-7	Acetic acid, (2,4-dichlorophenoxy)-, salts & esters
U112	141-78-6	Acetic acid ethyl ester (I)
U144	301-04-2	Acetic acid, lead(2+) salt
U214	563-68-8	Acetic acid, thallium(1+) salt
see F027	93-76-5	Acetic acid, (2,4,5-trichlorophenoxy)-
U002	67-64-1	Acetone (I)
U003	75-05-8	Acetonitrile (I,T)
U004	98-86-2	Acetophenone
U005	53-96-3	2-Acetylaminofluorene
U006	75-36-5	Acetyl chloride (C,R,T)
U007	79-06-1	Acrylamide
U008	79-10-7	Acrylic acid (I)
U009	107-13-1	Acrylonitrile
U011	61-82-5	Amitrole
U012	62-53-3	Aniline (I,T)
U136	75-60-5	Arsinic acid, dimethyl-
U014	492-80-8	Auramine
U015	115-02-6	Azaserine
U010	50-07-7	Azirino[2,3<ls-thn-eq>3,4]pyrrolo[1,2-a]indole-4,7-dione, 6-amino-8-[[(aminocarbonyl)oxy]methyl]-1,1a,2,8,8a,8b-hexahydro-8a-methoxy-5-methyl-, [1aS-(1aalpha,8beta,8aalpha,8balpha)]-
U280	101-27-9	Barban.
U278	22781-23-3	Bendiocarb.
U364	22961-82-6	Bendiocarb phenol.
U271	17804-35-2	Benomyl.
U157	56-49-5	Benz[j]aceanthrylene, 1,2-dihydro-3-methyl-
U016	225-51-4	Benz[c]acridine
U017	98-87-3	Benzal chloride
U192	23950-58-5	Benzamide, 3,5-dichloro-N-(1,1-dimethyl-2-propynyl)-
U018	56-55-3	Benz[a]anthracene
U094	57-97-6	Benz[a]anthracene, 7,12-dimethyl-
U012	62-53-3	Benzenamine (I,T)
U014	492-80-8	Benzenamine, 4,4-carbonimidoylbis[N,N-dimethyl-

(continued)

APPENDIX *Continued*

U049	3165-93-3	Benzenamine, 4-chloro-2-methyl-, hydrochloride
U093	60-11-7	Benzenamine, N,N-dimethyl-4-(phenylazo)-
U328	95-53-4	Benzenamine, 2-methyl-
U353	106-49-0	Benzenamine, 4-methyl-
U158	101-14-4	Benzenamine, 4,4-methylenebis[2-chloro-
U222	636-21-5	Benzenamine, 2-methyl-, hydrochloride
U181	99-55-8	Benzenamine, 2-methyl-5-nitro-
U019	71-43-2	Benzene (I,T)
U038	510-15-6	Benzeneacetic acid, 4-chloro-alpha-(4-chlorophenyl)-alpha-hydroxy-, ethyl ester
U030	101-55-3	Benzene, 1-bromo-4-phenoxy-
U035	305-03-3	Benzenebutanoic acid, 4-[bis(2-chloroethyl)amino]-
U037	108-90-7	Benzene, chloro-
U221	25376-45-8	Benzenediamine, ar-methyl-
U028	117-81-7	1,2-Benzenedicarboxylic acid, bis(2-ethylhexyl) ester
U069	84-74-2	1,2-Benzenedicarboxylic acid, dibutyl ester
U088	84-66-2	1,2-Benzenedicarboxylic acid, diethyl ester
U102	131-11-3	1,2-Benzenedicarboxylic acid, dimethyl ester
U107	117-84-0	1,2-Benzenedicarboxylic acid, dioctyl ester
U070	95-50-1	Benzene, 1,2-dichloro-
U071	541-73-1	Benzene, 1,3-dichloro-
U072	106-46-7	Benzene, 1,4-dichloro-
U060	72-54-8	Benzene, 1,1-(2,2-dichloroethylidene)bis[4-chloro-
U017	98-87-3	Benzene, (dichloromethyl)-
U223	26471-62-5	Benzene, 1,3-diisocyanatomethyl- (R,T)
U239	1330-20-7	Benzene, dimethyl- (I,T)
U201	108-46-3	1,3-Benzenediol
U127	118-74-1	Benzene, hexachloro-
U056	110-82-7	Benzene, hexahydro- (I)
U220	108-88-3	Benzene, methyl-
U105	121-14-2	Benzene, 1-methyl-2,4-dinitro-
U106	606-20-2	Benzene, 2-methyl-1,3-dinitro-
U055	98-82-8	Benzene, (1-methylethyl)- (I)
U169	98-95-3	Benzene, nitro-
U183	608-93-5	Benzene, pentachloro-
U185	82-68-8	Benzene, pentachloronitro-
U020	98-09-9	Benzenesulfonic acid chloride (C,R)
U020	98-09-9	Benzenesulfonyl chloride (C,R)
U207	95-94-3	Benzene, 1,2,4,5-tetrachloro-
U061	50-29-3	Benzene, 1,1-(2,2,2-trichloroethylidene)bis[4-chloro-
U247	72-43-5	Benzene, 1,1-(2,2,2-trichloroethylidene)bis[4-methoxy-
U023	98-07-7	Benzene, (trichloromethyl)-
U234	99-35-4	Benzene, 1,3,5-trinitro-
U021	92-87-5	Benzidine

APPENDIX *Continued*

U202	\1\ 81-07-2	1,2-Benzisothiazol-3(2H)-one, 1,1-dioxide, & salts
U278	22781-23-3	1,3-Benzodioxol-4-ol, 2,2-dimethyl-, methyl carbamate.
U364	22961-82-6	1,3-Benzodioxol-4-ol, 2,2-dimethyl-,
U203	94-59-7	1,3-Benzodioxole, 5-(2-propenyl)-
U141	120-58-1	1,3-Benzodioxole, 5-(1-propenyl)-
U367	1563-38-8	7-Benzofuranol, 2,3-dihydro-2,2-dimethyl-
U090	94-58-6	1,3-Benzodioxole, 5-propyl-
U064	189-55-9	Benzo[rst]pentaphene
U248	\1\81-81-2	2H-1-Benzopyran-2-one, 4-hydroxy-3-(3-oxo-1-phenyl-butyl)-, & salts, when present at concentrations of 0.3% or less
U022	50-32-8	Benzo[a]pyrene
U197	106-51-4	p-Benzoquinone
U023	98-07-7	Benzotrichloride (C,R,T)
U085	1464-53-5	2,2-Bioxirane
U021	92-87-5	[1,1-Biphenyl]-4,4-diamine
U073	91-94-1	[1,1'-Biphenyl]-4,4'-diamine, 3,3'-dichloro-
U091	119-90-4	[1,1'-Biphenyl]-4,4'-diamine, 3,3'-dimethoxy-
U095	119-93-7	[1,1'-Biphenyl]-4,4'-diamine, 3,3'-dimethyl-
U225	75-25-2	Bromoform
U030	101-55-3	4-Bromophenyl phenyl ether
U128	87-68-3	1,3-Butadiene, 1,1,2,3,4,4-hexachloro-
U172	924-16-3	1-Butanamine, N-butyl-N-nitroso-
U031	71-36-3	1-Butanol (I)
U159	78-93-3	2-Butanone (I,T)
U160	1338-23-4	2-Butanone, peroxide (R,T)
U053	4170-30-3	2-Butenal
U074	764-41-0	2-Butene, 1,4-dichloro- (I,T)
U143	303-34-4	2-Butenoic acid, 2-methyl-, 7-[[2,3-dihydroxy-2-(1-methoxyethyl)-3-methyl-1-oxobutoxy]methyl]-2,3,5,7a-tetrahydro-1H-pyrrolizin-1-yl ester, [1S-[1alpha(Z),7(2S*,3R*),7aalpha]]-
U031	71-36-3	n-Butyl alcohol (I)
U136	75-60-5	Cacodylic acid
U032	13765-19-0	Calcium chromate
U372	10605-21-7	Carbamic acid, 1H-benzimidazol-2-yl, methyl ester.
U271	17804-35-2	Carbamic acid, [1-[(butylamino)carbonyl]-1H-benzimidazol-2-yl]-, methyl ester.
U280	101-27-9	Carbamic acid, (3-chlorophenyl)-, 4-chloro-2-butynyl ester.
U238	51-79-6	Carbamic acid, ethyl ester
U178	615-53-2	Carbamic acid, methylnitroso-, ethyl ester
U373	122-42-9	Carbamic acid, phenyl-, 1-methylethyl ester.

(continued)

APPENDIX *Continued*

U409	23564-05-8	Carbamic acid, [1,2-phenylenebis (iminocarbonothioyl)]bis-, dimethyl ester.
U097	79-44-7	Carbamic chloride, dimethyl-
U389	2303-17-5	Carbamothioic acid, bis(1-methylethyl)-, S-(2,3,3-trichloro-2-propenyl) ester.
U387	52888-80-9	Carbamothioic acid, dipropyl-, S-(phenylmethyl) ester.
U114	\1\ 111-54-6	Carbamodithioic acid, 1,2-ethanediylbis-, salts & esters
U062	2303-16-4	Carbamothioic acid, bis(1-methylethyl)-, S-(2,3-dichloro-2-propenyl) ester
U279	63-25-2	Carbaryl.
U372	10605-21-7	Carbendazim.
U367	1563-38-8	Carbofuran phenol.
U215	6533-73-9	Carbonic acid, dithallium(1+) salt
U033	353-50-4	Carbonic difluoride
U156	79-22-1	Carbonochloridic acid, methyl ester (I,T)
U033	353-50-4	Carbon oxyfluoride (R,T)
U211	56-23-5	Carbon tetrachloride
U034	75-87-6	Chloral
U035	305-03-3	Chlorambucil
U036	57-74-9	Chlordane, alpha & gamma isomers
U026	494-03-1	Chlornaphazin
U037	108-90-7	Chlorobenzene
U038	510-15-6	Chlorobenzilate
U039	59-50-7	p-Chloro-m-cresol
U042	110-75-8	2-Chloroethyl vinyl ether
U044	67-66-3	Chloroform
U046	107-30-2	Chloromethyl methyl ether
U047	91-58-7	beta-Chloronaphthalene
U048	95-57-8	o-Chlorophenol
U049	3165-93-3	4-Chloro-o-toluidine, hydrochloride
U032	13765-19-0	Chromic acid H$_2$ CrO$_4$, calcium salt
U050	218-01-9	Chrysene
U051	....	Creosote
U052	1319-77-3	Cresol (Cresylic acid)
U053	4170-30-3	Crotonaldehyde
U055	98-82-8	Cumene (I)
U246	506-68-3	Cyanogen bromide (CN)Br
U197	106-51-4	2,5-Cyclohexadiene-1,4-dione
U056	110-82-7	Cyclohexane (I)
U129	58-89-9	Cyclohexane, 1,2,3,4,5,6-hexachloro-, (1alpha,2alpha,3beta,4alpha,5alpha,6 beta)-
U057	108-94-1	Cyclohexanone (I)
U130	77-47-4	1,3-Cyclopentadiene, 1,2,3,4,5,5-hexachloro-
U058	50-18-0	Cyclophosphamide
U240	\1\ 94-75-7	2,4-D, salts & esters
U059	20830-81-3	Daunomycin
U060	72-54-8	DDD
U061	50-29-3	DDT

APPENDIX *Continued*

U062	2303-16-4	Diallate
U063	53-70-3	Dibenz[a,h]anthracene
U064	189-55-9	Dibenzo[a,i]pyrene
U066	96-12-8	1,2-Dibromo-3-chloropropane
U069	84-74-2	Dibutyl phthalate
U070	95-50-1	o-Dichlorobenzene
U071	541-73-1	m-Dichlorobenzene
U072	106-46-7	p-Dichlorobenzene
U073	91-94-1	3,3'-Dichlorobenzidine
U074	764-41-0	1,4-Dichloro-2-butene (I,T)
U075	75-71-8	Dichlorodifluoromethane
U078	75-35-4	1,1-Dichloroethylene
U079	156-60-5	1,2-Dichloroethylene
U025	111-44-4	Dichloroethyl ether
U027	108-60-1	Dichloroisopropyl ether
U024	111-91-1	Dichloromethoxy ethane
U081	120-83-2	2,4-Dichlorophenol
U082	87-65-0	2,6-Dichlorophenol
U084	542-75-6	1,3-Dichloropropene
U085	1464-53-5	1,2:3,4-Diepoxybutane (I,T)
U108	123-91-1	1,4-Diethyleneoxide
U028	117-81-7	Diethylhexyl phthalate
U395	5952-26-1	Diethylene glycol, dicarbamate.
U086	1615-80-1	N,N'-Diethylhydrazine
U087	3288-58-2	O,O-Diethyl S-methyl dithiophosphate
U088	84-66-2	Diethyl phthalate
U089	56-53-1	Diethylstilbesterol
U090	94-58-6	Dihydrosafrole
U091	119-90-4	3,3'-Dimethoxybenzidine
U092	124-40-3	Dimethylamine (I)
U093	60-11-7	p-Dimethylaminoazobenzene
U094	57-97-6	7,12-Dimethylbenz[a]anthracene
U095	119-93-7	3,3'-Dimethylbenzidine
U096	80-15-9	alpha,alpha Dimethylbenzylhydroperoxide (R)
U097	79-44-7	Dimethylcarbamoyl chloride
U098	57-14-7	1,1-Dimethylhydrazine
U099	540-73-8	1,2-Dimethylhydrazine
U101	105-67-9	2,4-Dimethylphenol
U102	131-11-3	Dimethyl phthalate
U103	77-78-1	Dimethyl sulfate
U105	121-14-2	2,4-Dinitrotoluene
U106	606-20-2	2,6-Dinitrotoluene
U107	117-84-0	Di-n-octyl phthalate
U108	123-91-1	1,4-Dioxane
U109	122-66-7	1,2-Diphenylhydrazine
U110	142-84-7	Dipropylamine (I)
U111	621-64-7	Di-n-propylnitrosamine
U041	106-89-8	Epichlorohydrin
U001	75-07-0	Ethanal (I)
U404	121-44-8	Ethanamine, N,N-diethyl-
U174	55-18-5	Ethanamine, N-ethyl-N-nitroso-
U155	91-80-5	1,2-Ethanediamine, N,N-dimethyl-N'-2-pyridinyl-N'-(2-thienylmethyl)-
U067	106-93-4	Ethane, 1,2-dibromo-
U076	75-34-3	Ethane, 1,1-dichloro-
U077	107-06-2	Ethane, 1,2-dichloro-

(continued)

APPENDIX *Continued*

U131	67-72-1	Ethane, hexachloro-
U024	111-91-1	Ethane, 1,1'-[methylenebis(oxy)]bis[2-chloro-
U117	60-29-7	Ethane, 1,1'-oxybis-(I)
U025	111-44-4	Ethane, 1,1'-oxybis[2-chloro-
U184	76-01-7	Ethane, pentachloro-
U208	630-20-6	Ethane, 1,1,1,2-tetrachloro-
U209	79-34-5	Ethane, 1,1,2,2-tetrachloro-
U218	62-55-5	Ethanethioamide
U226	71-55-6	Ethane, 1,1,1-trichloro-
U227	79-00-5	Ethane, 1,1,2-trichloro-
U410	59669-26-0	Ethanimidothioic acid, N,N'-[thiobis[(methylimino)carbonyloxy]]bis-, dimethyl ester
U394	30558-43-1	Ethanimidothioic acid, 2-(dimethylamino)-N-hydroxy-2-oxo-, methyl ester.
U359	110-80-5	Ethanol, 2-ethoxy-
U173	1116-54-7	Ethanol, 2,2'-(nitrosoimino)bis-
U395	5952-26-1	Ethanol, 2,2'-oxybis-, dicarbamate.
U004	98-86-2	Ethanone, 1-phenyl-
U043	75-01-4	Ethene, chloro-
U042	110-75-8	Ethene, (2-chloroethoxy)-
U078	75-35-4	Ethene, 1,1-dichloro-
U079	156-60-5	Ethene, 1,2-dichloro-, (E)-
U210	127-18-4	Ethene, tetrachloro-
U228	79-01-6	Ethene, trichloro-
U112	141-78-6	Ethyl acetate (I)
U113	140-88-5	Ethyl acrylate (I)
U238	51-79-6	Ethyl carbamate (urethane)
U117	60-29-7	Ethyl ether (I)
U114	\1\ 111-54-6	Ethylenebisdithiocarbamic acid, salts & esters
U067	106-93-4	Ethylene dibromide
U077	107-06-2	Ethylene dichloride
U359	110-80-5	Ethylene glycol monoethyl ether
U115	75-21-8	Ethylene oxide (I,T)
U116	96-45-7	Ethylenethiourea
U076	75-34-3	Ethylidene dichloride
U118	97-63-2	Ethyl methacrylate
U119	62-50-0	Ethyl methanesulfonate
U120	206-44-0	Fluoranthene
U122	50-00-0	Formaldehyde
U123	64-18-6	Formic acid (C,T)
U124	110-00-9	Furan (I)
U125	98-01-1	2-Furancarboxaldehyde (I)
U147	108-31-6	2,5-Furandione
U213	109-99-9	Furan, tetrahydro-(I)
U125	98-01-1	Furfural (I)
U124	110-00-9	Furfuran (I)
U206	18883-66-4	Glucopyranose, 2-deoxy-2-(3-methyl-3-nitrosoureido)-, D-
U206	18883-66-4	D-Glucose, 2-deoxy-2-[[(methylnitrosoamino)-carbonyl]amino]-
U126	765-34-4	Glycidylaldehyde
U163	70-25-7	Guanidine, N-methyl-N'-nitro-N-nitroso-

APPENDIX *Continued*

U127	118-74-1	Hexachlorobenzene
U128	87-68-3	Hexachlorobutadiene
U130	77-47-4	Hexachlorocyclopentadiene
U131	67-72-1	Hexachloroethane
U132	70-30-4	Hexachlorophene
U243	1888-71-7	Hexachloropropene
U133	302-01-2	Hydrazine (R,T)
U086	1615-80-1	Hydrazine, 1,2-diethyl-
U098	57-14-7	Hydrazine, 1,1-dimethyl-
U099	540-73-8	Hydrazine, 1,2-dimethyl-
U109	122-66-7	Hydrazine, 1,2-diphenyl-
U134	7664-39-3	Hydrofluoric acid (C,T)
U134	7664-39-3	Hydrogen fluoride (C,T)
U135	7783-06-4	Hydrogen sulfide
U135	7783-06-4	Hydrogen sulfide H_2 S
U096	80-15-9	Hydroperoxide, 1-methyl-1-phenylethyl- (R)
U116	96-45-7	2-Imidazolidinethione
U137	193-39-5	Indeno[1,2,3-cd]pyrene
U190	85-44-9	1,3-Isobenzofurandione
U140	78-83-1	Isobutyl alcohol (I,T)
U141	120-58-1	Isosafrole
U142	143-50-0	Kepone
U143	303-34-4	Lasiocarpine
U144	301-04-2	Lead acetate
U146	1335-32-6	Lead, bis(acetato-O)tetrahydroxytri-
U145	7446-27-7	Lead phosphate
U146	1335-32-6	Lead subacetate
U129	58-89-9	Lindane
U163	70-25-7	MNNG
U147	108-31-6	Maleic anhydride
U148	123-33-1	Maleic hydrazide
U149	109-77-3	Malononitrile
U150	148-82-3	Melphalan
U151	7439-97-6	Mercury
U152	126-98-7	Methacrylonitrile (I, T)
U092	124-40-3	Methanamine, N-methyl- (I)
U029	74-83-9	Methane, bromo-
U045	74-87-3	Methane, chloro- (I, T)
U046	107-30-2	Methane, chloromethoxy-
U068	74-95-3	Methane, dibromo-
U080	75-09-2	Methane, dichloro-
U075	75-71-8	Methane, dichlorodifluoro-
U138	74-88-4	Methane, iodo-
U119	62-50-0	Methanesulfonic acid, ethyl ester
U211	56-23-5	Methane, tetrachloro-
U153	74-93-1	Methanethiol (I, T)
U225	75-25-2	Methane, tribromo-
U044	67-66-3	Methane, trichloro-
U121	75-69-4	Methane, trichlorofluoro-
U036	57-74-9	4,7-Methano-1H-indene, 1,2,4,5,6,7,8,8-octachloro-2,3,3a,4,7,7a-hexahydro-
U154	67-56-1	Methanol (I)
U155	91-80-5	Methapyrilene
U142	143-50-0	1,3,4-Metheno-2H-cyclobuta[cd]pentalen-2-one, 1,1a,3,3a,4,5,5,5a,5b,6-

(*continued*)

APPENDIX *Continued*

		decachlorooctahydro-
U247	72-43-5	Methoxychlor
U154	67-56-1	Methyl alcohol (I)
U029	74-83-9	Methyl bromide
U186	504-60-9	1-Methylbutadiene (I)
U045	74-87-3	Methyl chloride (I,T)
U156	79-22-1	Methyl chlorocarbonate (I,T)
U226	71-55-6	Methyl chloroform
U157	56-49-5	3-Methylcholanthrene
U158	101-14-4	4,4'-Methylenebis(2-chloroaniline)
U068	74-95-3	Methylene bromide
U080	75-09-2	Methylene chloride
U159	78-93-3	Methyl ethyl ketone (MEK) (I,T)
U160	1338-23-4	Methyl ethyl ketone peroxide (R,T)
U138	74-88-4	Methyl iodide
U161	108-10-1	Methyl isobutyl ketone (I)
U162	80-62-6	Methyl methacrylate (I,T)
U161	108-10-1	4-Methyl-2-pentanone (I)
U164	56-04-2	Methylthiouracil
U010	50-07-7	Mitomycin C
U059	20830-81-3	5,12-Naphthacenedione, 8-acetyl-10-[(3-amino-2,3,6-trideoxy)-alpha-L-lyxo-hexopyranosyl)oxy]-7,8,9,10-tetrahydro-6,8,11-trihydroxy-1-methoxy-, (8S-cis)-
U167	134-32-7	1-Naphthalenamine
U168	91-59-8	2-Naphthalenamine
U026	494-03-1	Naphthalenamine, N,N'-bis(2-chloroethyl)-
U165	91-20-3	Naphthalene
U047	91-58-7	Naphthalene, 2-chloro-
U166	130-15-4	1,4-Naphthalenedione
U236	72-57-1	2,7-Naphthalenedisulfonic acid, 3,3'-[(3,3'-dimethyl[1,1'-biphenyl]-4,4'-diyl)bis(azo)bis[5-amino-4-hydroxy]-, tetrasodium salt
U279	63-25-2	1-Naphthalenol, methylcarbamate.
U166	130-15-4	1,4-Naphthoquinone
U167	134-32-7	alpha-Naphthylamine
U168	91-59-8	beta-Naphthylamine
U217	10102-45-1	Nitric acid, thallium(1+) salt
U169	98-95-3	Nitrobenzene (I,T)
U170	100-02-7	p-Nitrophenol
U171	79-46-9	2-Nitropropane (I,T)
U172	924-16-3	N-Nitrosodi-n-butylamine
U173	1116-54-7	N-Nitrosodiethanolamine
U174	55-18-5	N-Nitrosodiethylamine
U176	759-73-9	N-Nitroso-N-ethylurea
U177	684-93-5	N-Nitroso-N-methylurea
U178	615-53-2	N-Nitroso-N-methylurethane
U179	100-75-4	N-Nitrosopiperidine
U180	930-55-2	N-Nitrosopyrrolidine
U181	99-55-8	5-Nitro-o-toluidine
U193	1120-71-4	1,2-Oxathiolane, 2,2-dioxide
U058	50-18-0	2H-1,3,2-Oxazaphosphorin-2-amine, N,N-bis(2-chloroethyl)tetrahydro-, 2-

APPENDIX *Continued*

		oxide
U115	75-21-8	Oxirane (I,T)
U126	765-34-4	Oxiranecarboxyaldehyde
U041	106-89-8	Oxirane, (chloromethyl)-
2	123-63-7	Paraldehyde
U183	608-93-5	Pentachlorobenzene
U184	76-01-7	Pentachloroethane
U185	82-68-8	Pentachloronitrobenzene (PCNB)
See F027	87-86-5	Pentachlorophenol
U161	108-10-1	Pentanol, 4-methyl-
U186	504-60-9	1,3-Pentadiene (I)
U187	62-44-2	Phenacetin
U188	108-95-2	Phenol
U048	95-57-8	Phenol, 2-chloro-
U039	59-50-7	Phenol, 4-chloro-3-methyl-
U081	120-83-2	Phenol, 2,4-dichloro-
U082	87-65-0	Phenol, 2,6-dichloro-
U089	56-53-1	Phenol, 4,4'-(1,2-diethyl-1,2- ethenediyl)bis-, (E)-
U101	105-67-9	Phenol, 2,4-dimethyl-
U052	1319-77-3	Phenol, methyl-
U132	70-30-4	Phenol, 2,2'-methylenebis[3,4,6- trichloro-
U411	114-26-1	Phenol, 2-(1-methylethoxy)-, methylcarbamate.
U170	100-02-7	Phenol, 4-nitro-
See F027	87-86-5	Phenol, pentachloro-
See F027	58-90-2	Phenol, 2,3,4,6-tetrachloro-
See F027	95-95-4	Phenol, 2,4,5-trichloro-
See F027	88-06-2	Phenol, 2,4,6-trichloro-
U150	148-82-3	L-Phenylalanine, 4-[bis(2- chloroethyl)amino]-
U145	7446-27-7	Phosphoric acid, lead(2+) salt (2:3)
U087	3288-58-2	Phosphorodithioic acid, O,O-diethyl S- methyl ester
U189	1314-80-3	Phosphorus sulfide (R)
U190	85-44-9	Phthalic anhydride
U191	109-06-8	2-Picoline
U179	100-75-4	Piperidine, 1-nitroso-
U192	23950-58-5	Pronamide
U194	107-10-8	1-Propanamine (I,T)
U111	621-64-7	1-Propanamine, N-nitroso-N-propyl-
U110	142-84-7	1-Propanamine, N-propyl- (I)
U066	96-12-8	Propane, 1,2-dibromo-3-chloro-
U083	78-87-5	Propane, 1,2-dichloro-
U149	109-77-3	Propanedinitrile
U171	79-46-9	Propane, 2-nitro- (I,T)
U027	108-60-1	Propane, 2,2'-oxybis[2-chloro-
U193	1120-71-4	1,3-Propane sultone
See F027	93-72-1	Propanoic acid, 2-(2,4,5- trichlorophenoxy)-
U235	126-72-7	1-Propanol, 2,3-dibromo-, phosphate (3:1)
U140	78-83-1	1-Propanol, 2-methyl- (I,T)
U002	67-64-1	2-Propanone (I)
U007	79-06-1	2-Propenamide
U084	542-75-6	1-Propene, 1,3-dichloro-

(continued)

APPENDIX *Continued*

U243	1888-71-7	1-Propene, 1,1,2,3,3,3-hexachloro-
U009	107-13-1	2-Propenenitrile
U152	126-98-7	2-Propenenitrile, 2-methyl- (I,T)
U008	79-10-7	2-Propenoic acid (I)
U113	140-88-5	2-Propenoic acid, ethyl ester (I)
U118	97-63-2	2-Propenoic acid, 2-methyl-, ethyl ester
U162	80-62-6	2-Propenoic acid, 2-methyl-, methyl ester (I,T)
U373	122-42-9	Propham.
U411	114-26-1	Propoxur.
U387	52888-80-9	Prosulfocarb.
U194	107-10-8	n-Propylamine (I,T)
U083	78-87-5	Propylene dichloride
U148	123-33-1	3,6-Pyridazinedione, 1,2-dihydro-
U196	110-86-1	Pyridine
U191	109-06-8	Pyridine, 2-methyl-
U237	66-75-1	2,4-(1H,3H)-Pyrimidinedione, 5-[bis(2-chloroethyl)amino]-
U164	56-04-2	4(1H)-Pyrimidinone, 2,3-dihydro-6-methyl-2-thioxo-
U180	930-55-2	Pyrrolidine, 1-nitroso-
U200	50-55-5	Reserpine
U201	108-46-3	Resorcinol
U202	\1\ 81-07-2	Saccharin, & salts
U203	94-59-7	Safrole
U204	7783-00-8	Selenious acid
U204	7783-00-8	Selenium dioxide
U205	7488-56-4	Selenium sulfide
U205	7488-56-4	Selenium sulfide SeS$_2$ (R,T)
U015	115-02-6	L-Serine, diazoacetate (ester)
See F027	93-72-1	Silvex (2,4,5-TP)
U206	18883-66-4	Streptozotocin
U103	77-78-1	Sulfuric acid, dimethyl ester
U189	1314-80-3	Sulfur phosphide (R)
See F027	93-76-5	2,4,5-T
U207	95-94-3	1,2,4,5-Tetrachlorobenzene
U208	630-20-6	1,1,1,2-Tetrachloroethane
U209	79-34-5	1,1,2,2-Tetrachloroethane
U210	127-18-4	Tetrachloroethylene
See F027	58-90-2	2,3,4,6-Tetrachlorophenol
U213	109-99-9	Tetrahydrofuran (I)
U214	563-68-8	Thallium(I) acetate
U215	6533-73-9	Thallium(I) carbonate
U216	7791-12-0	Thallium(I) chloride
U216	7791-12-0	Thallium chloride Tlcl
U217	10102-45-1	Thallium(I) nitrate
U218	62-55-5	Thioacetamide
U410	59669-26-0	Thiodicarb.
U153	74-93-1	Thiomethanol (I,T)
U244	137-26-8	Thioperoxydicarbonic diamide [(H$_2$N)C(S)]$_2$ S$_2$, tetramethyl-
U409	23564-05-8	Thiophanate-methyl.
U219	62-56-6	Thiourea
U244	137-26-8	Thiram

APPENDIX *Continued*

U220	108-88-3	Toluene
U221	25376-45-8	Toluenediamine
U223	26471-62-5	Toluene diisocyanate (R,T)
U328	95-53-4	o-Toluidine
U353	106-49-0	p-Toluidine
U222	636-21-5	o-Toluidine hydrochloride
U389	2303-17-5	Triallate.
U011	61-82-5	1H-1,2,4-Triazol-3-amine
U408	118-79-6	2,4,6-Tribromophenol.
U227	79-00-5	1,1,2-Trichloroethane
U228	79-01-6	Trichloroethylene
U121	75-69-4	Trichloromonofluoromethane
See F027	95-95-4	2,4,5-Trichlorophenol
See F027	88-06-2	2,4,6-Trichlorophenol
U404	121-44-8	Triethylamine.
U234	99-35-4	1,3,5-Trinitrobenzene (R,T)
U182	123-63-7	1,3,5-Trioxane, 2,4,6-trimethyl-
U235	126-72-7	Tris(2,3-dibromopropyl) phosphate
U236	72-57-1	Trypan blue
U237	66-75-1	Uracil mustard
U176	759-73-9	Urea, N-ethyl-N-nitroso-
U177	684-93-5	Urea, N-methyl-N-nitroso-
U043	75-01-4	Vinyl chloride
U248	\1\ 81-81-2	Warfarin, & salts, when present at concentrations of 0.3% or less
U239	1330-20-7	Xylene (I)
U200	50-55-5	Yohimban-16-carboxylic acid, 11,17-dimethoxy-18-[(3,4,5-trimethoxybenzoyl)oxy]-, methyl ester, (3beta,16beta,17alpha,18beta,20alpha)-
U249	1314-84-7	Zinc phosphide Zn_3P_2, when present at concentrations of 10% or less

----

\1\ CAS Number given for parent compound only.

[45 FR 78529, 78541, Nov. 25, 1980]

Editorial Note: For Federal Register citations affecting Sec. 261.**33**, see the List of CFR Sections Affected in the Finding Aids section of this volume.

General References

Code of Federal Regulations, Title 40, Parts 260–265, Office of the Federal Register, National Archives and Records Administration: Washington, DC.

Phifer, R.; McTigue, W. *Hazardous Waste Management for Small Quantity Generators*; CRC Press/Lewis Publishers: Boca Raton, FL, 1988.

ACS Task Force on Laboratory Waste Management, *Laboratory Waste Management, a Guidebook*; American Chemical Society: Washington, DC, 1994.

Higgins, T. *Hazardous Waste Minimization Handbook*; Lewis Publishers: Chelsea, MI, 1989.

Freeman, H. *Industrial Pollution Prevention Handbook*; McGraw-Hill: New York, NY, 1995.

Freeman, H. *Standard Handbook of Hazardous Waste Treatment and Disposal*; McGraw-Hill: New York, NY, 1989.

81

Biological Waste Management

ROBERT EMERY
WAYNE R. THOMANN

Facilities working with known or potentially infectious agents, including those associated with human blood or body fluids, inevitably generate wastes as a byproduct of routine operations. Careful consideration and attention must be given to the generation, handling, treatment, and disposal of these materials to minimize any risk of infection to both laboratory personnel and the organizational community. Although the potential for laboratory and clinic-acquired infections has been recognized for decades, the risk of infection from waste materials was not considered as significant until public awareness was heightened by two events: the emergence and reemergence of infectious diseases in the population at large and the notoriety of improper waste disposal events such as occurred on the beaches of New Jersey in 1988. In response to these events, legislation was initiated on the national, state, and local levels in an attempt to protect the public's health, but inconsistencies abound. Therefore it is highly recommended that facilities generating biological wastes carefully check state and local requirements that may affect operations. Even though different municipalities may have promulgated their own peculiar requirements and definitions, a set of basic premises are considered universally prudent when making any decision regarding biological wastes, including:

1. minimize generation
2. avoid the generation of "mixed" hazardous waste
3. minimize handling
4. render waste noninfectious at a point as close as possible to the point of generation or provide adequate packing to prevent possible exposures or releases
5. verify waste treatment efficacy
6. render waste unrecognizable

Before describing these basic tenets, a set of general definitions and a brief background of the biological waste issue is necessary.

Historical Background

Potentially infectious waste has been generated from biological and medical endeavors for years, but no universally accepted terminology has been used to identify or label the waste. Waste materials have been classified as "infectious," "pathological," "biomedical," "biohazardous," "toxic," or "medical."[1] The enactment of the Resource Conservation and Recovery Act (RCRA) in 1976 charged the U.S. Environmental Protection Agency (EPA) as the authority to define and regulate hazardous wastes.[2] Hazardous wastes were subsequently defined as wastes that exhibited any of the following characteristics: toxicity, corrosivity, reactivity, ignitability, or infectiousness.[3] Although the EPA has not promulgated regulations specifically addressing infectious wastes, the agency did issue a guide to assist in the management of these types of wastes. The document proposed consistent management practices for handling and described 12 categories of infectious waste materials. While the categories have not been uniformly embodied in current regulations, they do serve a useful purpose by adequately describing the various components of infectious waste streams as follows:

1. *Isolation wastes*: wastes such as gloves, dressings, and drapes generated as the result of caring for hospital patients with communicable diseases
2. *Cultures and stocks of infectious materials*: cultured specimens collected from patients for diagnosis, and stock materials maintained for research purposes
3. *Blood and blood products*: typically bulk liquid wastes generated by blood banks, dialysis centers, and pharmaceutical companies
4. *Pathological wastes*: body tissues, organs, and parts that are removed during surgery, biopsy, or autopsy
5. *Contaminated wastes from surgery and autopsy*: tubing, sponges, soiled dressings, and drapes contaminated in the course of medical procedures
6. *Contaminated laboratory waste*: culture dishes, absorbent paper, and devices used to mix, transfer, and inoculate specimens
7. *Sharps*: items such as needles, syringes, and razor blades presenting a double hazard because of

their ability to inflict immediate injury as well as induce disease

8. *Dialysis unit wastes*: filters, towels, tubing, and gloves used in the process of blood transfusions

9. *Animal carcasses, body parts, and bedding*: materials generated as the result of purposefully exposing test animals to pathogens for research

10. *Discarded biological*: materials discarded because of failed quality control or outdating; typical generators might be pharmaceutical and veterinary facilities

11. *Contaminated food*: foods, food additives, and colorings that may be contaminated

12. *Contaminated equipment*: equipment contaminated during patient care or microbiological research with infectious agents

As stated earlier, these categories of waste have not been uniformly accepted as warranting special handling as "infectious" during the development of state regulations, and there are significant differences in which categories the individual states have designated as "regulated medical wastes." In addition, a new designation of regulated medical wastes was recently promulgated in the Occupational Safety and Health Administration's Bloodborne Pathogens standard. While the EPA proposal focused on protecting the public and the environment, the OSHA standard is aimed at protecting workers who may be exposed to human blood and body fluids. The OSHA definition includes contaminated sharps, pathological and microbiological wastes containing blood or other potentially infectious materials, *and* contaminated items that would release blood in a liquid or semiliquid state if compressed. Still other entities, such as state environmental protection programs, regulate this last category of waste if it permits blood to flow freely from the surface without compression.

Therefore, a thorough understanding of the state-specific requirements is necessary to ensure regulatory compliance. Medical waste management systems must address both the "internal" requirement for protecting workers, visitors, and patients and the "external" responsibility for safeguarding the public health and environment.

Medical Waste Management

With the historical lack of a consistent system for the classification and labeling of waste types, an accurate estimate of the total quantity of infectious waste generated annually is currently unavailable. However, Rutula

and Sarubbi estimated that a 1000-patient hospital would generate 740 kg (1,635 lbs) of infectious waste per day.[4] Wallace et al. identified the areas of a hospital that generate the largest amounts of infectious waste,[5] namely, that intensive care units and operating rooms produce twice as much waste as other sections such as support units and general medical care facilities. Both of these studies provide only a glimpse of the total infectious waste stream. Information on the amounts of waste currently produced by other types of health care facilities or research institutions is scant. There have also been very few, if any, known and reported occupational infections as a result of handling wastes defined as "regulated medical waste."

The cost associated with the disposal of biological wastes can be significant. Off-site treatment and disposal charges range from approximately $0.20 to over $1.00 per pound. If the waste is mixed with other hazardous constituents, such as radioactivity, the disposal costs can rise considerably. Hence, consideration should be given to reducing the amount of waste generated, thereby producing a dual benefit: a reduced exposure risk to workers and lower disposal costs.

Waste Minimization

Prior to performing any work that may lead to the generation of regulated medical waste, consideration should be given to the methods available to minimize the volumes generated, including the methods that follow.

Source Substitution

Initial consideration should be given to whether an infectious agent must be used in the first place. In some instances, less infectious or noninfectious agents may be used and if possible should be incorporated into the process to eliminate the generation of biological wastes.

Training and Education

All workers can be educated about the risks and costs associated with regulated medical wastes and clear definitions can be provided to assist workers with proper segregation for treatment or disposal. Heightening employee awareness can greatly reduce the amount of waste improperly designated as potentially infectious.

Microscale Techniques

When possible, experiments and clinical procedures should be conducted with small volumes, thus reducing the amount of waste generated. A variety of such microscale techniques are now available and should be considered if possible.

Reusable Equipment

Consideration should be given to the use of reusable equipment and protective clothing, if the cleaning and disinfection of these items can be accomplished in a safe and cost-effective manner. Special consideration should be given to the avoidance of generating mixed hazardous chemical, radiological, and biological waste.

Segregation

Waste materials should be segregated according to hazard type so that noninfectious wastes are not disposed of with the regulated medical waste. The physical separation of waste container types by distance can aid in ensuring proper segregation. And as mentioned previously, a thorough understanding of the local regulations regarding regulated wastes will assist facility personnel by ensuring that waste materials are appropriately designated.

Handling

Once medical waste is generated, the generator is responsible for the proper handling, packaging, labeling, and ultimate disposal of the material. The same precautions used during procedures involving infectious material should also be employed when handling medical waste.

Appropriate packaging of medical wastes is one of the most important control issues. Appropriate packaging will minimize potential exposure by placing barriers against potential contact or release. It is important that the packaging system be consistent with the treatment or disposal method for the individual waste streams. The physical form of the waste dictates the packaging procedures. Solid waste material, such as disposable gloves and gowns, and absorbent paper, can be placed in closeable and labeled containers that are lined with a durable plastic bag. Solids that can become a vehicle for the introduction of pathogens into the body, commonly referred to as "sharps," must be accumulated in rigid, leakproof, and puncture-resistant containers. Generators should be aware that sharps are not limited to syringes and needles, but can also include other objects such as broken glass, scalpels, and glass pipettes that are contaminated with blood or body fluids.

In some states, liquid wastes that do not contain cultures of pathogens can be discharged directly to the sanitary sewer; however, such discharge may involve personnel exposure as containers are opened and emptied. This practice should be carefully evaluated. When liquid wastes are to be accumulated, the original containers should be placed in a plastic or plastic-lined container that will contain any subsequent spills. As with solid waste, the container should be properly labeled.

Tissue samples, animal carcasses, and other bulk objects should be placed in leakproof plastic bags. Larger objects may require double-bagging to prevent leakage when handling. If the waste contains free liquids, vermiculite or other absorbent material should be placed in the bottom of the bag. Tags bearing the universal biohazard symbol and warning can be affixed to the bag closure if such warnings are not imprinted on the bags themselves.

When labeling containers of medical waste, care should be taken to ensure that the warnings can be understood by the persons encountering the waste, and the potential hazard source should also be identified. If the waste is not being processed in a timely manner, a secured refrigerated accumulation area should be designated to prevent decomposition, generation of odors, theft or vandalism, and inadvertent human or animal exposures. Medical waste storage areas must be maintained in a sanitary condition through regularly scheduled cleaning and disinfection.

Transportation

Most facilities will have to address both on-site and off-site transportation of medical wastes. On-site operations involve moving the materials from the point of generation to on-site storage or treatment areas. Off-site transportation will be required when there are insufficient on-site treatment capabilities. Off-site shipment will typically require a manifest or tracking system to monitor the final disposition of the wastes. Transportation of medical waste requires special attention to prevention of accidental exposures or spills. Transport containers, carts, or vehicles must be designed to control spills and prevent inadvertent exposure to the contents. The transport vehicles should be cleaned and disinfected on a routine schedule. Transporters of regulated medical waste are subject to Department of Transportation (DOT) regulations, which require formal notifications, tracking, vehicle enclosures, labeling, record keeping, and reporting. Finally, contingency plans, which address rapid response to releases of medical waste, should be developed. The waste generator should also be cognizant of the potential liabilities associated with the inappropriate transport or disposal of regulated medical wastes, even when performed by an independent contractor. Careful reviews of such contractor operations can greatly aid in reducing such risks.

Options for Treatment at the Point of Generation

Whenever possible, medical wastes should be rendered noninfectious at the point of generation to reduce the risk of infection to other workers and waste handlers. Two of the more common treatment techniques are steam sterilization and incineration.

Steam sterilization, commonly referred to as "autoclaving," is the traditional method of infectious waste treatment. Waste is enclosed in a sealed chamber and exposed to steam under increased pressure. The steam provides a means to elevate the temperature of the waste and results in the disinfection of the pathogenic agents. Sterilization systems range in size from small bench-top models to high-capacity walk-in systems. Systems that are considered to be large capacity may process up to 91 kg (200 lb) of waste per operating cycle.[6] Typical operating cycles are about 1.5 hr in duration, with peak temperatures of 121°C and pressures of 15 lb/in².

The greatest advantage to steam sterilization is the relatively low operating cost and the simplicity of operation. For small-volume generators of infectious waste, steam sterilization is considered the treatment option of choice for on-site processing. The disadvantages of steam sterilization include limited processing capacity, lack of suitability for certain materials, the possibility for incomplete processing, and the lack of change in physical appearance after processing. Wastes that cannot be adequately treated by steam sterilization include such items as body parts, chemotherapy agents, and tissue samples containing large volumes of preservative agents. The autoclave decontamination process will also be incomplete if the steam does not gain intimate contact with the infectious agent for a sufficient length of time. Although the addition of water to bags can aid in ensuring intimate steam contact during the autoclave process, if the material is packed too tightly, such contact becomes difficult if not impossible.[7] Additionally, a significant portion of autoclaved waste material remains recognizable after processing. Individuals subsequently exposed to the waste, such as landfill operators and the general public, have no way of readily determining if the waste still presents a hazard. To allay concerns, generators should inform waste handlers of the processes used to render the waste nonpathogenic, and mechanisms should be put in place to demonstrate the efficacy of the treatment. For instance, some waste bags also include process indicators that change color or label wording upon autoclaving, providing an easy method of identifying processed waste. The regular documented use of biological indicator ampules, containing such agents as the endospore *Bacillus steareothermophilus*, and cycle performance indicator records that document time and temperature of the cycle serve as indicators of proper autoclave operation and should be included as a regular feature in facilities using this technique for waste treatment.[1]

Individuals using the autoclave method for waste processing should also be aware of two other important considerations. Attention should be given to other contaminants in the infectious waste stream, such as volatile chemicals or radioactivity. In some cases, such agents can be released during autoclaving, and unit contamination can occur. Users should also familiarize themselves with any state or local autoclave registration or certification requirements. In some areas, these units are considered to be a type of industrial boiler unit, and several pressure-release or explosion accidents have indeed been reported. In such municipalities, the units may be subject to a set of reviews or controls.

On-site incineration has been a fairly common treatment option in the past. However, increasingly restrictive toxic air regulations have led to the closure of many on-site units that could not achieve the necessary emission limits. Incineration units are available in a variety of operating configurations: All reduce the waste material to a virtually unrecognizable slag or ash, and then, when operating temperatures are achieved and properly maintained, the residuals are rendered nonpathogenic. Of all of the types of incinerators available, controlled air systems make up 90% of the units installed in biomedical institutions in the last 15 years.[1]

Off-Site Treatment Options

For waste materials that cannot be rendered noninfectious within the facility, there are four main treatment options available for the off-site treatment of infectious waste: bulk steam sterilization, shredding followed by chemical disinfection, shredding followed by microwave treatment, and shredding followed by steam autoclaving and incineration.[6] For the most part, these off-site methods were originally conceived and designed for on-site applications, but operational difficulties and high costs associated with installation and maintenance have made the systems more conducive to off-site use.

Bulk steam sterilization is a large-volume autoclave process, and the basis of operation has been described earlier. As mentioned, the autoclave process may not render all wastes unrecognizable and thus may not be the option of choice for some generators. The remaining common processes do possess the capability of transforming the appearance of the waste and the method selected is usually dictated by the amount of waste that can be processed.[8]

Shredding and disinfection involves pulverization of the waste by a hammermill and washing in a spray of sodium hypochlorite or bleach solution. The contact between the free chlorine and the pathogenic agent produces a disinfected shredded material. The microwave system also employs a hammer mill to shred the waste prior to exposure to heating through exposure to a series of microwave generators. The advantage of these systems is that the waste is reduced to an innocuous and unrecognizable pulp that is suitable for direct landfill disposal. The major disadvantage to the chemical treatment is the production of the disinfection solution that may necessitate special approval prior to discharge to the sanitary sewer. Other disadvantages include the lack of suitability for wastes such as body parts, the inability to release residues to the sanitary sewer in some settings due to negative impacts on water treatment operations, and the hazards associated with the production of chlorine.

The other common off-site treatment process for medical waste is incineration. Releases of particulates and gaseous effluents are controlled by sophisticated scrubber and electrostatic precipitator systems. When operated properly, incinerators can decontaminate and detoxify large volumes of waste and produce significant volume and weight reductions. Incineration units can be obtained in a range of sizes to accommodate the particular waste stream and can be fitted with heat-recovery systems. The disadvantages of incineration are high capital costs, high maintenance and repair requirements, emissions control, and public acceptance.

Disposal

The final step in the medical waste management system is disposal. Local and state requirements regarding disposal of treated and untreated medical waste vary significantly. For example, some states permit the direct discharge of liquid wastes that do not contain infectious cultures to the sanitary sewer, while others may require incineration of such waste. Further, the requirement for "rendering medical waste unrecognizable" varies from state to state. Therefore, disposal options must be carefully reviewed prior to selecting the most appropriate method for each waste category.

References

1. U.S. Congress Office of Technology Assessment, *Issues in Medical Waste Management-Background Paper* OTA-BP-0-49; U.S. Government Printing Office: Washington, DC, October 1988, pp. 1–2, 16.
2. U.S. Environmental Protection Agency, *EPA Guide for Infectious Waste Management*; Report No. EPA/S30-SW-86-014; Washington, DC, 1986, pp. v–vii.
3. U.S. Environmental Protection Agency, "Regulations for Hazardous Waste Management," 40 *CFR* 122–124 and 260–271; U.S. Government Printing Office: Washington, DC, 1985.
4. Rutula, W. A.; Sarubbi, F. A. "Management of Infectious Waste from Hospital," *Association for Practitioners in Infection Control Annual Meeting*, Atlanta, GA, 1981.
5. Wallace, L. P.; Zaltman, R.; Burchinal, J. C. Where Solid Waste Comes From; Where it Should Go. *Modern Hospital*, 118(2).
6. Doucet, L. A Comparison of Infectious Waste Treatment Options. *Management of Medical and Infectious Waste-Practical Considerations*, Dec. 1988, Hazardous Materials Control Research Institute. pp. 5–6.
7. Rutala, W. A.; Stiegal, M.; Sarubbi, F. A. Decontamination of Laboratory Microbiological Waste by Steam Sterilization. *Appl. Environ. Microbiol.* **1982**, *43*, 1311–1316.
8. Hunt, D. L. Infectious Waste Management. In *Pollution Prevention: Waste Management Strategies for Hospitals and Clinical Laboratories*; North Carolina Department of Environment, Health and Natural Resources: Raleigh, NC, 1987, pp. 33–36.

General References

National Committee on Clinical Laboratory Standards, *Clinical Laboratory Waste Management: Approved Guidelines*. NCCLS Doc. GP5-A, Vol. 13, No. 22, NCCLS: Villanova, PA, 1993.

Reinhardt, P. A.; Gordon, J. G. *Infectious and Medical Waste Management*; Lewis Publishers: Chelsea, MI, 1991.

Turnberg, W. *Biohazardous Waste, Risk Assessment, Policy and Management*; Wiley & Sons: New York, 1996.

U.S. Environmental Protection Agency, *Medical Waste Management and Disposal*, Landrum, V. J. et al. Eds. Noyes Data Corporation: Park Ridge, NJ, 1991.

82

Safe Handling of Biohazardous Materials for Transport

JOHN H. KEENE

The safe shipment of infectious agents and diagnostic specimens as well as other biohazardous materials (toxins, allergens, venoms, etc.) is of specific concern to laboratory scientists and to postal, airline, and parcel service personnel. The timely delivery of infectious agents and diagnostic specimens is critical in order to minimize spurious laboratory results and to expedite the diagnosis of disease. Damaged or mislabeled packages may result in an occupational exposure to biohazardous materials either during transport or at the time of reception.

Strict regulatory requirements have been developed in the United States to minimize the potential for mishaps resulting from the unsafe transport of biohazardous materials. However, in our global economy, U.S. regulations alone may not be sufficient to ensure safe transport of these materials, and compliance with U.S. regulations may not ensure compliance internationally. The World Health Organization (WHO) has published guidelines[14] regarding the appropriate packaging and shipment of infectious materials. In addition, some international organizations (the Universal Postal Union [UPU], the International Civil Aviation Organization [ICAO], the International Air Transport Association [IATA], and the United Nations Committee of Experts on the Transport of Dangerous Goods) have also developed guidelines to be followed to ensure the safe and expeditious transport of biohazardous materials.[13] Although numerous examples of broken, damaged, and leaking packages of biohazardous materials have been reported, to date no reports of occupational illness attributable to the handling of such materials have been noted. However, it is possible, indeed probable, that occupational exposure to damaged, improperly packaged biohazardous materials has resulted in occupationally acquired illness. These incidents may not have been reported as occupationally acquired because it is difficult, in most cases, to recognize the potential exposure and thus to correlate any resultant illness to that occupational exposure.

While the information provided in this chapter covers the major points of awareness in the transport of biohazardous materials, the reader should understand that regulations and guidelines are constantly being reviewed and revised. Personnel involved with shipping and/or receiving biohazardous materials should be aware of, and familiar with, all applicable regulations and guidelines concerning the packaging and transport of such materials. Local safety and health personnel as well as commercial shippers can be a source of current information and should be consulted when there is any question regarding the appropriate shipping requirements.

Essential in the following of any regulations or guidelines is the adoption of a system of comparable definitions to allow for a universal understanding of the potential hazard. The following definitions, as found in the WHO document, "Laboratory Biosafety"[14] are used in the international community:

Infectious substances are substances containing viable microorganisms, including bacteria, viruses, rickettsia, parasites, fungi or recombinant, hybrid or mutant organisms, that are known or reasonably believed to cause disease in animals or humans. Toxins that do not contain any infectious substances are not included in this definition.

Diagnostic specimens are any human or animal material including, but not limited to, excreta, secreta, blood and its components, and tissue and tissue fluids, being shipped for purposes of diagnosis. Live infected animals are excluded from this definition.

Biological products are either: (a) finished biological products for human or veterinary use manufactured in accordance with the requirements of national public health authorities and moving under special approval or license from such authorities; (b) finished biological products shipped prior to licensing for development or investigational purposes for use in humans or animals; or (c) products for the experimental treatment of animals, and which are manufactured in compliance with the requirements of national public health authorities.

The term "biological products" also includes unfinished biological products prepared in accordance with the procedures of specialized government agencies, such as the U.S. Food and Drug Administration (FDA). Live animal and human vaccines are considered to be biological products and not infectious substances. Under U.S. regulation, biological products are specifically defined as substances manufactured in accordance with 9 *CFR* Parts 102–104 and 21 *CFR* Parts 312 and 600–680.[4,5]

Some licensed vaccines may also present a biohazard in certain parts of the world. Authorities in those places may require the handling and packaging of these vaccines to comply with the requirements for infectious substances or may impose other restrictions. Personnel involved with packaging and shipment of these materials should consult local regulations prior to initiating shipment. In addition, genetically engineered microorganisms are also subject to packaging requirements dependent on the locality from which and to which the materials are being shipped.[14] The reader is reminded that such requirements may change and should be checked before shipping such materials.

U.S. regulations may have slightly different definitions from those of some other countries. These definitions are nonetheless similar in scope. The reader is urged to identify the specific information pertinent to his/her situation and, whenever possible, to adhere to the most stringent definitions. It should be noted that hand carriage and the use of diplomatic pouches for transport of biohazardous materials is strictly prohibited by all jurisdictions.

Packaging

Diagnostic Specimens

Biohazardous materials should be packaged in such a way as to preclude any potential leakage. The U.S. regulation 42 *CFR* Part 72.2 states that for diagnostic specimens the material must be "packaged to withstand leakage of contents, shock, pressure changes, and other conditions incident to ordinary handling in transportation."[2] These packages should withstand rough handling and passage through cancellation machines, sorters, conveyors, and so forth.

Etiologic Agents

In addition to the requirements for packaging and transporting diagnostic specimens as outlined previously, any material known or reasonably believed by the shipper to contain an etiologic agent from a list included in this regulation[2] must be packaged and shipped in accordance with the requirements of 42 *CFR* Part 72.3. The list in

this regulation contains most of the Class 2, 3, and 4 agents, but any etiologic agent should be handled according to the regulation even if it does not appear on the stated list.

As published in 42 *CFR* Part 72.3, there are two sets of instructions for packaging of etiologic agents:

I. Volume Not Exceeding 50 ml[2]

(a) The material to be shipped shall be placed in a securely closed, watertight tube, vial, ampule, or the like that is referred to as the primary container.

(b) The primary container is then placed in a durable watertight container referred to as the secondary container.

(c) Several primary containers can be placed in a single secondary container, as long as total contents of the primary containers does not exceed 50 ml.

(d) Absorbent material must be placed in the space at the top, bottom, and sides between the primary and secondary containers. There must be enough absorbent material to absorb the entire contents of the primary container(s) in case of breakage or leakage and the absorbent material should be nonparticulate, that is, sawdust, vermiculite, and so forth.

(e) Each set of primary and secondary containers is then to be placed in an outer shipping container constructed of corrugated fiberboard, cardboard, wood, or other material of equivalent strength. This means that most, if not all, bags, envelopes, and the like are not acceptable as outer shipping containers (see Fig. 82.1.)

II. Volume Greater than 50 ml[2]

Packaging of these larger volumes of material must comply with all of the foregoing requirements as well as the following:

(a) Shock-absorbent material, in a volume at least equal to that of the absorbent material between the primary and secondary containers, shall be placed at the top, bottom, and sides between the secondary container and the outer shipping container.

(b) Single primary containers shall not contain more than 1000 ml of material; however, two or more primary containers, whose volumes do not exceed 1000 ml, may be placed in a single secondary container.

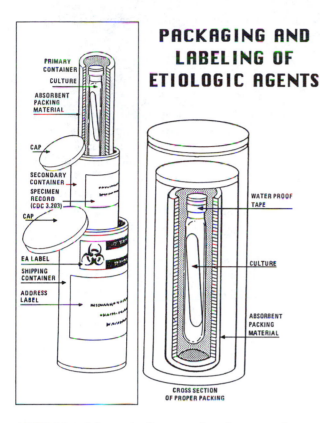

PACKAGING AND LABELING OF ETIOLOGIC AGENTS

PRIMARY CONTAINER
CULTURE
ABSORBENT PACKING MATERIAL
CAP
SECONDARY CONTAINER
SPECIMEN RECORD (CDC 3.203)
CAP
EA LABEL
SHIPPING CONTAINER
ADDRESS LABEL
WATER PROOF TAPE
CULTURE
ABSORBENT PACKING MATERIAL
CROSS SECTION OF PROPER PACKING

FIGURE 82.1 Schematic diagram of packaging and labeling requirements for etiologic agents. (Reproduced from Reference 6.)

ondary container(s) does not become loose within the outer shipping container as the dry ice sublimates. At no time should the dry ice be placed in either the primary or secondary container, both of which are sealed. The WHO recommends that if materials are to be shipped in liquid nitrogen or with other protection from ambient or higher temperatures, "all containers and packaging should be capable of withstanding very low temperatures and both primary and secondary packaging must be able to withstand a pressure differential of at least 95kPA and temperatures in the range of $-40°C$ to $+50°C$."[14] Note that appropriate labeling for the transport of hazardous materials such as dry ice and liquid nitrogen is also required when these materials are present.

The U.S. Department of Health and Human Services (HHS) regulation[2] also contains a list of etiological agents that require special handling in addition to that stated above. These are, by and large, Class 3 and Class 4 agents. The requirement is that they be shipped by "registered mail or an equivalent system which requires or provides for sending notification of receipt to the sender immediately . . . upon delivery." When this notice of receipt is not received "within 5 days following anticipated delivery" the sender must notify the Centers for Disease Control and Prevention (CDC).

Note that, under the U.S. regulation, "interstate" shipping is interpreted to include "intrastate" shipping.

(c) The maximum amount of etiological agent that may be enclosed within a single outer shipping container may not exceed 4000 ml.

Note that the WHO requires three layers for packaging of all biohazardous materials, including infectious substances and diagnostic material[14]:

(a) a primary watertight receptacle containing the specimen
(b) a secondary watertight receptacle enclosing enough absorptive material between it and the primary receptacle to absorb all of the fluid in the specimen in case of leakage
(c) an outer package, which is intended to protect the secondary package from outside influences such as physical damage and water while in transit (see Fig. 82.1)

Miscellaneous Requirements

If dry ice is used, it must be placed between the secondary container(s) and the outer shipping container. The shock-absorbent material must be placed so that the sec-

Contaminated Equipment

Any equipment that has become contaminated with infectious agents or materials that may be considered to contain infectious agents must be decontaminated prior to shipment for repair or replacement.[9] If the equipment cannot, for any reason, be decontaminated, a readily observable biohazard label must be affixed to the equipment. This label must state which areas of the equipment have been contaminated and not decontaminated. It is preferable to include on the labels the nature of the contamination. Such contaminated equipment must also be packaged in such a way that the contaminated areas cannot become accessible to the personnel who must subsequently handle the equipment.

Imports/Exports

Importation of biologically hazardous materials into the United States is strictly regulated by a number of different regulatory bodies. The importation of etiologic agents of human disease and materials contaminated, or potentially contaminated, by these agents is regulated under HHS regulation 42 *CFR* Part 71, "Foreign Quarantine," and Part 71.54, "Etiologic Agents, Hosts, and Vectors."[3] Under these regulations, a person may neither import into the United States, nor distribute after importation, any etiologic agent or any arthropod or other animal host or vector of human disease, or any exotic living arthropod or other animal capable of being a host or vector of human disease unless accompanied by a permit issued by the Director of the CDC. In addition, such imported materials will not be released from custody of the U.S. Customs Service (USCS) until a permit issued by the Director of the CDC has been received. Packaging of such agents and materials must conform with USPHS, U.S. Department of Transportation (DOT), and other applicable regulations and guidelines.[2,8,12,13]

In addition to the requirements stated above for etiologic agents of human disease, the U.S. Department of Agriculture (USDA) Animal and Plant Health Inspection Service (APHIS) is the permitting agency for the importation of infectious agents of plants and animals as well as biological materials containing plant and animal materials. Note that tissue culture materials that may contain growth stimulants of bovine or other livestock origin are controlled by the USDA to reduce the potential risk of introduction of exotic animal diseases into the United States. The importation of certain animals, including live bats, is regulated by the U.S. Department of Interior (USDI). The export of infectious materials may require a license from the Department of Commerce. Export of infectious materials and materials potentially contaminated with such infectious materials requires strict adherence to all applicable regulations regarding packaging and transport.[3,7,10,11]

Communication

Any transfer of infectious or potentially infectious materials requires coordination among the sender, carrier, and recipient. U.S. regulations[2] require notification of receipt in the case of BL 3 and BL 4 agents, while the WHO[14] recommends notification of receipt for any agent. Specific requirements for registration of facilities and notification of responsible facility officials, outlined below, have been amended to the existing U.S. regulations (42 *CFR* 72).[1]

The sender should ensure that the material is appropriately packaged in accordance with the carrier's requirements and applicable regulations and should notify the recipient of the impending transport. The receiver should ensure that all appropriate documentation is made available to the sender and should provide the

sender with acknowledgment of receipt of the shipment in a timely manner.

Emergency Response

The carrier and the recipient should have some type of written emergency response procedure in place to handle transport-associated accidents and spills. Written spill cleanup procedures should be available, and all personnel should be trained in the appropriate mechanism for cleanup and decontamination of spilled biological materials. In addition, appropriate notification of affected personnel as well as the initiator of the shipment should be made by a mutually acceptable method. The WHO Lab Safety Manual[14] provides an example decision tree for "action in transport-associated accidents to infectious materials in shipment." While this is one way of handling the problem, it is important that each organization develop its own methodology to ensure prompt, efficient handling of the situation.

Labels

Infectious agents and other biohazardous materials are classed as "dangerous goods" for purposes of transportation. Personnel handling packages containing these materials have a right to be warned with regard to potential exposure in order to minimize that exposure in the case of accidental breakage and leakage. Therefore, packages containing these substances must bear appropriate labeling. U.S. regulations specify the size, color, and specific wording necessary for inter- and intrastate transport of biohazardous materials. A special label (Fig. 82.2) must be placed on the outer shipping container. This label identifies the package as containing etiologic agents and directs anyone observing damage to the package or leakage of contents to call the CDC.[6] In addition, the U.S.

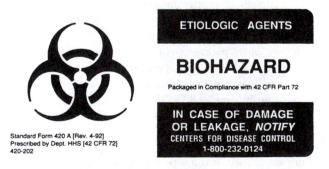

FIGURE 82.2 U.S. Department of Health and Human Services label for shipping etiologic agents.

FIGURE 82.3 World Health Organization label for shipping biohazardous materials.

Occupational Safety and Health Administration (OSHA) requires that biohazard labels be present on shipments of human blood/body fluids and other potentially infectious materials. As of this writing, the CDC has proposed new labeling requirements but has not finalized the requirements. Therefore, the reader is cautioned to determine the current status of the labeling regulations and to be in compliance with the applicable regulation prior to shipping any biohazardous materials.

Bold use of the biohazard sign is necessary, but different jurisdictions require different labeling. For example, the WHO has a "Biohazard Label"[14] (Fig. 82.3) that is appropriate for worldwide transport and is accepted by the IATA. Personnel involved in the packaging and shipment of biohazardous materials must have a working knowledge of the approved warning labels and must ensure their proper use.

Antiterrorism Requirements

Finally, in addition to the above rules, guidelines, and regulations, the U.S. HHS, in response to public and political concern for biological terrorism, and in accordance with the requirements of Section 511 of Public Law 104-132, "The Antiterrorism and Effective Death Penalty Act of 1996," has issued an amendment to 42 *CFR* 72, "Additional Requirements for Facilities Transferring or Receiving Select Infectious Agents (42 *CFR*

72.6, 72.7 and Appendix A)." This regulation requires the registration of facilities transferring or receiving select agents as listed in the appendix of the regulation. A registered facility will be surveyed by HHS and must be appropriately equipped and capable of handling agents at Biosafety Levels 2, 3, and/or 4 depending on the agent to be used. These registered facilities will be inspected and a registration number will be issued to those facilities found to be qualified.[1]

Requests for agents on the list, published as Appendix A of the regulation, as amended, will require completion of a CDC form that will include:

- the name of the requestor and requesting facility
- the name of the transferor and transferring facility
- the names of the responsible facility officials for both the transferor and requestor
- the requesting facility's registration number
- the transferring facility's registration number
- the name of the agent(s) being shipped
- the proposed use of the agent(s)
- the quantity (number of containers and amount per container) of the agent(s) being shipped

The form must be signed by responsible facility officials representing both the transferring and requesting facilities and retained by both parties for a period of 5 years after the agents are consumed or properly disposed, whichever is longer.

The requesting facility must acknowledge receipt of the agent telephonically or otherwise electronically within 36 hours of receipt and provide a paper copy of receipt to the transferor within 3 business days of the receipt of the agent. The transferor must then provide a copy of the transfer form to the registering entity within 24 hours. This form will be filed in a centralized repository.

Penalties for violation of this part of the regulation are severe. Individuals in violation are subject to a fine of no more than $250,000 or one year in jail, or both. Organizations are subject to a fine of no more than $500,000 per event.

Summary

In order to minimize potential occupational and environmental exposure to biohazardous agents during transport, myriad guidelines, rules, and regulations pertaining to the packaging and transport of these materials have been promulgated. It is important that persons responsible for handling such materials be aware of the regulations as well as of specific packaging requirements. Generally, materials must be packaged in such a way as to preclude leakage to the outside during transport, must be appropriately labeled, and must be shipped in accordance with all applicable regulations. Communication between sender and receiver must be in accordance with applicable regulations and must ensure the notification of safe receipt of the materials. Emergency response protocols should be developed by transporters and receiving facilities to ensure proper handling of damaged and spilled materials. New rules regarding registry of facilities for the handling, shipping, and receiving of organisms considered to be particularly dangerous have been promulgated to minimize the potential for terrorist activity. Personnel involved with the packaging and shipping of etiologic agents and other biohazardous materials should contact their shipping agents for specific requirements applicable to their situation.

References

1. DHHS, Interstate Shipment of Etiologic Agents, Amended, 42 *CFR* 72.6 & 72.7.
2. DHHS, Interstate Shipment of Etiologic Agents, 42 *CFR* 72.
3. DHHS, Foreign Quarantine, 42 *CFR* 71.
4. DHHS, FDA, Biological Products, 9 *CFR* 102, 103, and 104.
5. DHHS, FDA, Biologics, 21 *CFR* 312, 600 to 680.
6. DHHS, *Biosafety in Microbiological and Biomedical Laboratories*, Richmond, J. Y., McKinney, R. W., Eds.; HHS Publication No. (CDC)93-8395, U.S. Government Printing Office: Washington, DC, 1993.
7. DOC, Bureau of Export Administration, Biological Warfare Experts Group Meeting: Implementation of Changes to Export Administration Regulations: ECCNs 1C991, 1C61B, 1B71E, and 1C91F, 15 *CFR* 774 and 799A.
8. DOT, Infectious Substances, Final Rule, 49 *CFR* 171 et al.
9. OSHA, Bloodborne Pathogen Standard, 29 *CFR* 1910.1030.
10. USDA, Animal Biologics, Exports, Imports, Reporting and Recordkeeping Requirements, 9 *CFR* 101 and 112.
11. USDA, Exports, Plant Diseases and Pests, Reporting and Recordkeeping Requirements, 7 *CFR* 353 and 354.
12. U.S. Postal Service Rules and Regulations, 39 *CFR* 111, 124.381 and 124.382a.
13. United Nations, *Recommendations on the Transport of Dangerous Goods*, 7th ed. revised; New York, 1989.
14. World Health Organization, *Laboratory Biosafety Manual*, 2nd ed.; Geneva, 1993.

83

Radioactive Waste Management and Transportation

ROBERT E. UHORCHAK

Using radioactive materials is a "nuclear" decision. If top management support is not available to deal with the direct costs of supporting a radiation safety program and the indirect costs of dealing with public relations issues, the contingent liability of wastes disposed, and other similar costs, then it would be prudent to investigate other alternatives. These include use of nonradioactive techniques or total transfer of this work component to another company. Once the company has decided to use radioactive materials, a radiation safety/radioactive waste program needs to be established.

Goals

The major goals for a radioactive waste management program are:

- to derive maximum benefit from the use of radioactive materials
- to minimize unnecessary employee, public, and environmental exposure to radiation
- to eliminate or minimize the production of radioactive waste overall, waste that cannot be disposed of, and costs, both direct and indirect

To accomplish these goals, several philosophical principles that are universally adopted by all staff handling radioactive material are helpful. The EPA has issued guidance[1] for establishing a hazardous waste minimization program, which can be adopted into a radioactive waste management program. The six points in the program are:

1. Obtain top management support
2. Characterize waste generation and costs

3. Conduct periodic waste minimization assessments
4. Allocate costs to the department producing the waste
5. Encourage technology transfer
6. A comprehensive program needs to be implemented and regularly evaluated

Implementing these points into an effective program requires day-to-day users of radioactive material to have an operational mindset that works toward these goals. Adopting traditional loss-control principles[2] provides such a means. The principles are avoidance, prevention, reduction, segregation, and transfer. All of these are captured in Covey's[3] second principle: "Start with the end in mind."

Radioactive Waste Program

Avoidance

As the term "avoidance" suggests, avoiding production of waste is the simplest, most cost-effective means of waste control. The three methods that follow are readily available.

Alternative Techniques

A good review of nonradioactive substitutes for use in biomedical research has been prepared by F. Party and E. L. Gershey.[4] These include colorimetric and chemiluminescent assays, amplification techniques, and bioluminescent assays. In some applications, colored microspheres can replace radioactive microspheres. The use of stable isotopes, such as deuterium or carbon-13, should be considered. Finally, radioactive substitution, for example, ^{33}P or ^{35}S, shortens decay times and allows for shorter waste storage time until the material is no longer radioactive.

Training Staff

Traditionally, the radiation safety support staff accepted radioactive waste "as presented" from laboratory staff. An ongoing education program regarding waste-handling problems, segregation techniques, waste regulations, and waste costs, can give the laboratory staff the knowledge to help solve these problems in tandem with the support staff. A written manual, periodically updated with new techniques for elimination/reduction, is a starting point. Periodic group meetings of radioactive users can foster ideas and give radiation safety staff a forum to reiterate such fundamentals as basic segregation.

Planning/Experimental Design

Planning and design may be done in several forms. First, a written plan of the experiment using radioactive materials can be reviewed by knowledgeable, independent peers and by staff with specialized regulatory knowledge. Both can make suggestions on methods to eliminate/minimize waste production. Second, the experiment can be done "cold" to find both technical problems and mechanical problems. Then a "tracer" experiment is done, and finally the "hot" work. Third, a "waste flowchart" for the entire experiment should be developed as part of the experimental design to help plan appropriate disposal (see generalized example under "Segregation" below). Fourth, record keeping and detailed container labeling needs to be practiced (see #1 under "Contamination Control" below).

Prevention

Contamination Control

Prevention means eliminating to the extent possible the causes of waste production. This includes waste from production and from analytical processes as well as inadvertent contamination from handling, spills, and aerosols. Prevention includes using proper facilities and equipment (see Section VII of this volume on laboratory design), trained staff, and proper materials. Some contamination control techniques include the following:

1. Always label containers of radioactive material clearly, indicating the nuclide, total activity, compound, solvent(s), date, chemist, and level of radiation on the surface of the container. The research notebook number and page reference will allow access to further details if needed at a later date. Labeling prevents excess handling and future waste generation from reanalyzing unknown samples and assists in segregation, reduction, and packaging by others beyond the point of generation.
2. Use radioisotopes only in designated areas to prevent spread of contamination.
3. Gloves and lab coats should be worn when handling radioisotopes. Remove these when leaving the area to prevent spread of contamination.
4. Hands should be washed frequently to minimize potential exposures and contamination transfer.
5. All operations with radioisotopes should be carried out in a tray of sufficient capacity to contain the material if spilled. The tray should be lined with absorbent paper.

6. Benches should be covered with absorbent paper with plastic backing. The absorbent side should be facing up. Cover any porous surfaces such as wood with absorbent paper. Only contaminated sections need to be removed as radioactive waste. Use a survey meter and/or wipe tests to identify the sections.

7. Operations that could lead to dust formation or aerosols should be carried out in a fume hood or glovebox.

8. Pouring solutions that contain radioisotopes creates aerosols with contamination potential. Use a transfer pipette.

9. Never leave items with radioisotopes projecting over the edge of the bench. They may drop or fall, spreading contamination to the floor.

10. Transport any appreciable amount of radioactive material within two containers. The outer container should be impervious and unbreakable. If the radioactive material is in a liquid form, the outer container should contain enough absorptive material to absorb all of the radioactive material should it be spilled from breakage of the inner container.

11. To avoid contamination of your scintillation cocktail and to prevent splashing, add the cocktail to the counting vial first and then add the radioactive sample.

12. Monitor the work area regularly to avoid cross-contamination of experiments.

13. When concentrating a solution with a stream of air or nitrogen, adjust the air or nitrogen flow first and then place your vial under it. Do these operations in a fume hood.

14. A plastic bag should be readily available to dispose of solid radioactive waste. When radioactive work is done in a fume hood, reduce contamination risk by placing the bag in the hood to minimize movement in or out of the hood.

Additional Precautions When Handling Radioiodine

1. To minimize contamination from volatilization, vials containing radioiodine should be opened for as brief a time as possible and capped tightly when not in use.

2. Dispose of or clean contaminated apparatus and glassware as quickly as possible to minimize release of volatile iodine into the hood or room air. Contaminated items, such as vials, syringes, and pipette tips, should be wrapped in a double poly-

ethylene bag, taped securely, and placed immediately in radioactive waste disposal.

3. Use a solution consisting of 0.1 M NaI, 0.1 M NaOH, and 0.1 M Na$_2$S$_2$O$_3$ in liquid waste containers to stabilize any iodine waste.

4. Do not add acids to radioiodine solutions. The volatility of ^{125}I is significantly enhanced at low pH.

Mixed Waste Prevention

Mixed waste is a combination of radioactive material and an RCRA-regulated hazardous waste. The RCRA rules defining hazardous waste are given in 40 *CFR* 261. Scintillation fluid containing flammable liquid scintillation cocktails is a common example. Great effort should be used to substitute nonflammable scintillation cocktails to eliminate this source of mixed waste. Similar efforts to avoid production of mixed waste in synthesis and other analytical techniques is warranted because disposal is expensive, if available. If unavailable, indefinite storage on site will be required. This may involve procurement of a "treatment, storage, disposal" (TSD) permit from the EPA.

Other radioisotope-specific techniques may apply and the radioisotope manufacturer should be contacted for additional guidance. Many references cover general laboratory handling of chemicals to minimize staff exposure and waste production: "Prudent Practices in the Laboratory: Handling and Disposal of Chemicals"[5] is one of the most comprehensive.

Reduction

Reduction means mitigating the waste (e.g., volume, radioactive content) once it is produced. Common reduction techniques include:

- use the minimum amount of radioactivity compatible with the objectives of the experiment
- use tissues to control drips or sprays, such as drips off separatory funnels and pipettes
- use minivials for liquid scintillation counting
- scraping of thin-layer chromatography plates should be done in a glove bag in a hood
- compaction of waste containing space voids
- incineration of waste at an approved incinerator, then disposal of the radioactive ash

Segregation

General Considerations

1. In order to efficiently manage radioactive waste, a flowchart describing what kind and quantity

of waste and how it is to be treated/disposed of should be made before beginning an experiment.

2. DO NOT mix radioactive waste with nonradioactive waste.

3. DO NOT mix organic solvent waste with aqueous waste.

4. Determine your facilities' radioactive waste storage capacity. Store for decay as much waste as possible. As a rule of thumb, ten half-lives are needed but longer storage may be necessary based on rad meter surveys or concentrations. Separate short versus long (e.g., tritium and carbon-14) half-life isotopes.

5. If a solution (aqueous or organic) does not contain volatile radioactive material, an attempt should be made to separate the radioactive material from the solvent by evaporation or distillation.

6. If a solution contains volatile radioactivity, an attempt should be made to convert the radioactive components to solids. For example, salts can be made of acids or bases; basic hydrolysis of a carboxylic ester will produce a solid salt if the radioactivity is not in the alcoholic portion of the ester; solid derivatives such as oximes, hydrazones, and bisulfite complexes can be made of ketones and aldehydes; alcohols react with phenylisocyanate to give solid derivatives; quaternary salts can be made of alkylhalides.

7. Trace amounts of volatile carbon-14, tritium, and iodine may be exhausted through a hood. Calculate these amounts in advance using the rules in 10 *CFR* Part 20, Appendix B.

8. Make sure that equipment such as stills and evaporators are free of contamination in order to avoid contaminating a solvent that otherwise would be free of radioactivity. Contaminated solutions cannot be disposed of with the nonradioactive waste even if they contain less than the *de minimis* amount (0.05 µCi of ^{3}H or ^{14}C/ml) of radioactivity. The *de minimis* ruling ONLY applies to scintillation fluids and animal tissue [10 *CFR* 20.2005(a)]. Also, some states include ^{125}I in this category.

9. Liquid radioactive waste produced by high-performance liquid chromatography (HPLC), containing a scintillant, has been classified by some state agencies as scintillation fluid. This waste, if it meets the *de minimis* amount for ^{3}H and ^{14}C, may be disposed of following scintillation fluid rules [10 *CFR* 20.2005(a)].

10. Aqueous waste containing any organic solvents (except ethanol) should be classified as organic waste.

11. Liquid storage containers should be marked with radioactive tape and labeled with the required information

   - isotope(s) present
   - number of mCi of each
   - solvent(s) and volume
   - source (HPLC, synthesis, cleaning, etc.)
   - chemist, date, notebook and page numbers

Radioactive Waste Flowchart

An example flowchart with major decision/segregation points is shown in Figure 83.1. The flowchart also suggests actions that can be taken based on half-lives, physical form, and specific activities of isotopes/chemicals used. Along with a general facility radioactive waste flowchart, a similar flowchart for each project will help focus review of the project on the most difficult waste produced.

Decision/Segregation Points

1. $t_{1/2}$—the physical half-life of each isotope. Selection of short-lived radioisotopes is one of the most significant methods of reducing radioactive waste and warrants extensive study. From an exposure standpoint, the general rule is to "hold-for-decay" ten half lives followed by an external survey to verify no radiation above background is present. Two critical factors must always be considered when "hold-for-decay" reduction is used. First, the Nuclear Regulatory Commission (NRC) does not want licensees to hold radioactive waste for indefinite periods of time and has established a 5-year review cycle.[6] A licensee may petition the NRC or agreement state authority for a radioisotope half-life decay-in-storage time to exceed the typical 60–90 day half-life granted. Second, if the radioisotope is a constituent of, or mixed in with, an EPA-regulated waste, it is "mixed waste," and the EPA hazardous waste storage rules apply. The maximum storage time varies by your company generator status (e.g., large quantity—90 days and small quantity—180 days). Exceptions may be sought from the EPA or state authority. Alternatively, a TSD permit may be obtained. However, significant potential legal liabilities and administrative burden accompanies possession of this permit. Legal counsel should be sought prior to obtaining the TSD permit.

Flow Chart I: Radioactive Waste Flow Chart

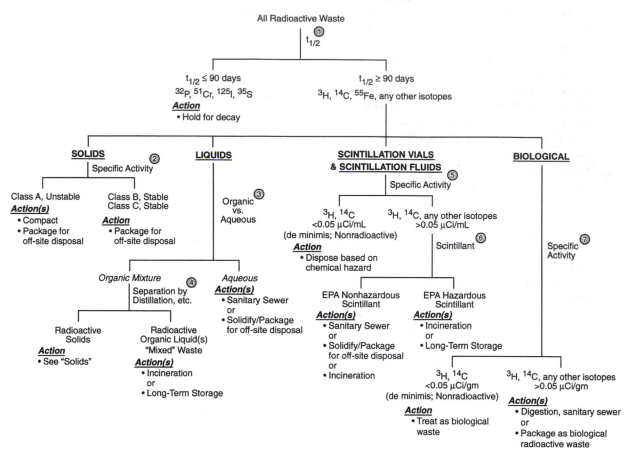

FIGURE 83.1 Radioactive waste flowchart.

2. Specific Activity—the radioactive content per unit volume. Separation can be made by disposal class. Three references must be used to properly establish the governing conditions:
   - 10 *CFR* Part 61
   - Branch Technical Position on Concentration Averaging and Encapsulation, NRC, January 17, 1995
   - waste disposal site license

   Part 61 establishes the overall specific activity parameters, by isotope, and establishes classification. The technical position paper provides further guidance to generators with various types and forms of waste. The disposal site license may have more restrictive conditions to be met.
3. Aqueous means containing either or both water and ethanol.
4. Separation may include distillation, molecular sieve, and similar techniques.[5,7]
5. 10 *CFR* 20.2005(a)(1); some states also include ^{125}I in the *de minimis* rule.

6. 40 *CFR* Part 261.
7. 10 *CFR* 20.2005(a)(2); some states also include ^{125}I in the *de minimis* rule.

Transfer (Disposal)

Receipt of Material

Radioactive waste begins with the radioactive material package receipt. Prior to receipt, the sender must verify that you have a license authorizing possession of the isotope, the quantity, and chemical/physical form. Likewise, prior to sending isotopes to another party, including disposal sites, a copy of the recipient's license should be in your file.

Within 3 hours of receipt of a labeled radioactive package or one with physical damage, a "leak" check is required. If radioactive contamination outside the package has occurred, immediate notification to the shipper and regulatory agency may be required (10 *CFR* 20. 1906). Cleanup should be done as soon as possible to minimize the spread of contamination. A package survey is conducted prior to transfer to the user. For low-energy beta emitters, such as tritium and carbon-14, a wipe test is necessary. One wipe over each package and each inner package is normally sufficient. If at background, all material except the container with radioisotope may be discarded as ordinary trash once any radioactive material labels, tape, and similar identifiers have been defaced.

Waste Disposal

With the exception of liquid scintillation fluid and animal tissue, the NRC has not adopted *de minimis* amounts of radioactive material. However, a given state may have adopted exceptions, and individual licensees may request them. Contact your state authority for clarification. Thus, all radioactive waste must be either stored on site or transferred to a new location.

Extended storage on site requires physical protection in a secure area, good containment, and efficient use of available space. Designation as radioactive waste does not remove the radioactive amount from the facilities' licensed possession limit. The NRC has issued two information notices that address licensed possession authority for waste in storage and extended storage guidance.[6,8] A problem with on-site storage for decay is finding sufficient space to handle the volume of material to be decayed. Creative use of available space, such as vertical shelving, may be necessary. If mixed-waste storage is involved, EPA storage rules must also be followed.

On-Site Disposal

Sanitary Sewer

Aqueous liquids and digested tissues that are biodegradable may be amenable to disposal via the sanitary sewer system. Three information sources are needed:

- 10 *CFR* 20.2003, Appendix B, Table 3
- facility sanitary sewer permit conditions
- monthly facility sanitary sewage discharge volume

Using the isotope-specific monthly average concentrations, facility average monthly discharge volume, and any additional permit limits, an upper limit of radioisotope discharge can be computed. Annual limits for tritium and carbon-14 may not exceed 5.0 curies and 1.0 curie, respectively, and all other isotopes combined may not exceed 1.0 curie (10 *CFR* 20.2003).

Dilution to Air (Incineration)

Incineration on site is authorized under 10 *CFR* 20. 2004. Research and development facilities desiring to incinerate radioactive waste beyond *de minimis* waste must obtain approval as noted in 10 *CFR* 20.2002. Discharge into the air must meet 10 *CFR* 20, Appendix B limits. Incineration technology and regulation are complex. A good overview of this technique should be done prior to selection of this disposal method.[9] More commonly, waste is sent to an approved incinerator.

Off-Site Disposal

Off-site disposal consists of transfer for secure burial. Required information includes:

- 10 *CFR* Subpart K, Waste Disposal, 20.2001–20.2007
- 10 *CFR* Subpart D, 61.55–61.58
- 49 *CFR* Subpart I, Radioactive materials, 173.401–173.478
- waste disposal site license (with disposal conditions)
- disposal state notification forms
- waste transporter company requirements/guidelines and manifest
- compact requirements

Each licensee must package radioactive waste to satisfy all of the conditions of all regulations. Since disposal site requirements may be more restrictive than NRC requirements are, the most direct approach is to start with the disposal site license conditions to determine acceptable packaging. Similarly, the transporter should be contacted to see if they have requirements beyond DOT regulations, and the disposal state may require transport notification. If these requirements are not more restrictive than NRC or DOT, their rules apply. Additionally, each state belongs to a radioactive waste compact. Licensees cannot ship waste out-of-compact for disposal unless both the host compact and out-of-compact authorities have approved the transfer.

With the multiple regulations to follow, preparation of a detailed checklist for each waste stream is suggested. DOT requirements for transporting a radioactive package include:

1. Shipper's Declaration for Dangerous Goods, 49 *CFR* Subpart C
2. Materials' Classification, 49 *CFR* 173 Subpart I
   a. low specific activity (LSA)
   b. limited quantities
   c. Type A quantities

3. Package Labels, 49 *CFR* 172.400–403
   a. white-I; yellow-II; yellow-III
   b. other chemical hazards
4. Package Markings 49, *CFR* 172.300–338
   a. material's proper shipping names and ID numbers
   b. package specification
   c. physical or chemical warnings
5. Emergency Response Information, 49 *CFR* 172 Subpart G
6. Training, 29 *CFR* 172 Subpart H

Transportation of radioactive waste and current advice on proper packaging and transport rules are normally available through licensed broker/transporter companies. Site-disposal criteria and assistance are available from the companies operating the sites. A list of some companies providing these services is given in Table 83.1.

Summary

The management of radioactive waste is a complex task involving detailed familiarity with multiple regulations. While once the domain of the radiation safety administrative staff, the avoidance, prevention, reduction, and segregation of the waste can best be accomplished by knowledgeable radioactive material users. They need to become active participants in the radioactive waste management process. The disposal and transfer of the waste incurs not only direct costs but can give rise to needs for emergency action or contingent liabilities. Top management support and guidance is similarly based on their understanding of the problems and costs involved. A successful radioactive waste management program is dynamic, technically in response to changing techniques and regulations, and administratively in terms of selection of which risk-control techniques should receive the monies available.

References

1. EPA Guidance to Hazardous Waste Generators on the Elements of a Hazardous Waste Minimization Program. *Fed. Regist.* **1993** (May 28).
2. *Essentials of Risk Control*, 2nd ed.; Head, G. L., Ed.; Insurance Institute of America: Malvern, PA, 1989.

TABLE 83.1 List of Treatment/Disposal and Transportation/Services Facilities

Treatment/Disposal	Transportation/Services
Allied Technology Group, Inc. 99A Midway Lane Oak Ridge, TN 37830 (423) 482-3275	Adco Services, Inc. 17650 Duvan Drive Tinley Park, IL 60477 (708) 429-1660
Chem-Nuclear Systems, LLC 140 Stoneridge Drive Columbia, SC 29210 (803) 256-0450	Bionomics, Inc. P.O. Box 817 Kingston, TN 37763 (615) 376-0053
DSSI P.O. Box 863 Kingston, TN 37763 (615) 376-0084	Norvell Protective Clothing Mfgs. 164 Edgewood Street Alexandria, TN 37012 (615) 529-2855
Envirocare of Utah, Inc. 46 W. Broadway, Suite 240 Salt Lake City, UT 84101 (801) 532-1330	RSO, Inc. P.O. Box 1526 Laurel, MD 27025-1526 (301) 953-2482
NSSI, Inc. 5711 Etheridge St. Houston, TX 77078 (713) 641-0391	Teledyne Brown Engineering 50 Van Buren Avenue Westwood, NJ 07675-1235 (201) 664-7070
Perma-Fix Environmental Services 1940 N.W. 67th Place Gainsville, FL 32653-1692 (352) 373-6066	US Ecology, Inc. 109 Flint Road Oak Ridge, TN 37830 (615) 482-5532
SEG Oak Ridge, TN 37831 (615) 376-8169	

3. Covey, Stephen R. *The Seven Habits of Highly Effective People, Restoring the Character Ethic*; Simon & Schuster: New York, 1989.
4. Party, E.; Gershey, E. L. *Health Physics* **1995**, 69, 1–5.
5. *Prudent Practices in the Laboratory: Handling and Disposal of Chemicals*; Committee on Prudent Practices for Handling, Storage and Disposal of Chemicals in Laboratories; National Academy Press: Washington, DC, 1995.
6. NRC Information Notice 90-09. *Extended Interim Storage of Low-Level Radioactive Waste by Fuel Cycle and Materials Licensees*; U.S. Nuclear Regulatory Commission: Washington, DC, 1990.
7. Gordon, A. J.; Ford, R. A. *The Chemists Companion: A Handbook of Practical Data, Techniques, and References*; Wiley & Sons: New York, 1973.
8. NRC Information Notice 89-13. *Alternative Waste Management Procedures in Case of Denial of Access to Low-Level Waste Disposal Sites*; U.S. Nuclear Regulatory Commission: Washington, DC, 1989.
9. *Radiation and Mixed Waste Incineration*, EPA Research Series, Government Institutes, Inc.: Rockville, MD, 1994.

84

Waste Minimization: General

CYNTHIA KLEIN-BANAI

Waste minimization or pollution prevention is the ultimate approach to waste management. Just as solid waste recycling programs have significantly reduced the amount of solid waste reaching the landfills, hazardous waste minimization should significantly reduce the amount of waste reaching hazardous waste treatment facilities and subsequently the environment.

Since waste minimization is the most recent approach to hazardous waste management, there have been increasing numbers of publications related to this subject. This chapter is intended to provide an introduction to the field, define the terms and the assessment process, and illustrate some waste minimization processes. Should you wish to pursue this topic further, the references provide a good beginning.

Why Minimize Waste?

In order to make the effort toward waste minimization, there must be the motivation to do so. Although many pollution-prevention programs offer financial incentives or recognition for waste minimization, there are many inherent incentives.

Safety

Although it may not be immediately apparent, there is good reason to include a discussion of waste minimization in a book whose topic is safety. One of the main incentives for waste minimization is safety. Waste minimization inevitably involves the reduction of the amount and toxicity of waste products, thus reducing the hazards involved with even temporary storage of such waste. Waste minimization also involves a reduction in the amount and hazards of the raw materials used. Both of these processes make for a safer work environment. This means less personal protection equipment used, improved air quality, reduction of spill hazards, fewer material-handling concerns, and fewer waste-management concerns.

Economics

Probably the prime motivators for waste minimization in the business world are the economic incentives. Reduction of raw materials, particularly hazardous materials, means reduction of purchase and operating costs. Often, there is some recycling and/or reuse of materials. Of course, waste-disposal costs will be significantly lower when quantities of generated waste are reduced.

Pollution prevention can improve plant productivity by using raw materials more efficiently, updating technologies, and improving process control and efficiency.

Environmental Liability

Reduction in waste generation decreases the possibility of creating future environmental liabilities. Although certain disposal practices may have been legal in the past, they have often caused unanticipated environmental damage that has been costly to repair. Pollution prevention can reduce long-term liability by reducing the amount and hazard of waste generated. This further protects the environment, which, after all, is the purpose of hazardous waste management. Waste minimization can also give an institution a more positive and greener image within the community.

Regulatory Incentives

If the above incentives are not enough to stimulate waste reduction, then there are a few regulatory reasons to do so. All generators of hazardous waste certify on each hazardous waste manifest that goes with each hazardous waste shipment that there is "a program in place" to minimize waste. Second, there is the biennial waste minimization program reporting. Most recently, the U.S. EPA issued a "Guidance to Hazardous Waste Generators on the Elements of a Waste Minimization Program"[1] that, while not actually mandating waste minimization, makes it clear that the EPA wants voluntary waste minimization. Other related regulatory incentives include land disposal restrictions and bans as well as increasing permitting requirements for waste handling and treatment.

Definitions

Waste Minimization versus Pollution Prevention

Although these terms are often used interchangeably, there is a difference between them. *Waste minimization,* as defined by the EPA,[2] is "the reduction, to the extent feasible, of hazardous waste that is generated or subsequently treated, stored, or disposed of." This can be achieved in one of two ways—either by the reduction of the volume or quantity of hazardous waste, by the reduction of the toxicity of the waste, or both—as long as this reduction minimizes the current and future threats to human health and the environment.

Pollution prevention, on the other hand, is defined as only the reduction of the volume or quantity of pollutants and/or the toxicity of the pollutants that would have entered any waste stream or have been released into the environment prior to management techniques such as recycling, treatment, or disposal. Pollution prevention does include in-process recycling, which means using or reclaiming the material within a process. It also refers to "pollutants," in general, not just RCRA hazardous waste. The stress is on source reduction, which is defined next.

Source Reduction

The EPA also defines the preferred order of exploration of waste minimization possibilities. The most desirable level is that of *source reduction.* This is any activity that reduces or eliminates the generation of "any hazardous substance, pollutant or contaminant entering any waste stream or otherwise released into the environment (including fugitive emissions)" at the source. This is usually achieved by making changes within a process.[3]

Recycling and Reclamation

The next area of exploration is recycling. A material is considered recycled if it is used, reused, or reclaimed. In order to use or reuse a product, it must be used as an ingredient to make another product. The material may also be used in a particular function as an effective substitute for a commercial product. A material is considered reclaimed if it is processed to recover a useful product or if it is regenerated.

Treatment

The last area of exploration is treatment. Although this is not technically waste minimization, this may be considered after exploring all the other alternatives. Some examples of these processes are discussed later in this chapter.

The Waste Minimization Assessment Process

How is waste minimization implemented? One way is to just get some ideas by reading, talking to colleagues, and from presentations at professional meetings, and then to try them out. Depending on a company's organizational structure, this may or may not work. An individual probably will not have all the authority needed to make the purchasing, processing, or waste management changes that are called for. Even if this person is able to implement some changes, he/she may have difficulty showing the results, or may not have the data to prove that the work has been effective.

Therefore, the other way to approach this problem is through a waste minimization assessment process. The EPA has a manual,[2] guidance,[1] and many "Guides to Pollution Prevention"[4] that can be used to address your specific needs. This section describes the main aspects of that process and provides some tools to get started.

Figure 84.1 gives an overview of the waste minimization assessment procedure. The very first step is to recognize the need to minimize waste. Once this has been done the assessment procedure can start.

Planning and Organization

Waste minimization affects many parts of an organization, such as management, purchasing, production, and environmental. The program must encompass all these different groups in order to reduce waste. Depending on your company's size and complexity, the formality of the program will vary.

Management Commitment

Once management or administration has been convinced that waste minimization is a worthwhile endeavor (that benefits outweigh costs), this support should be translated into a policy statement or management directive. Upper management is responsible for obtaining a formal commitment throughout all areas of the organization. Mean-

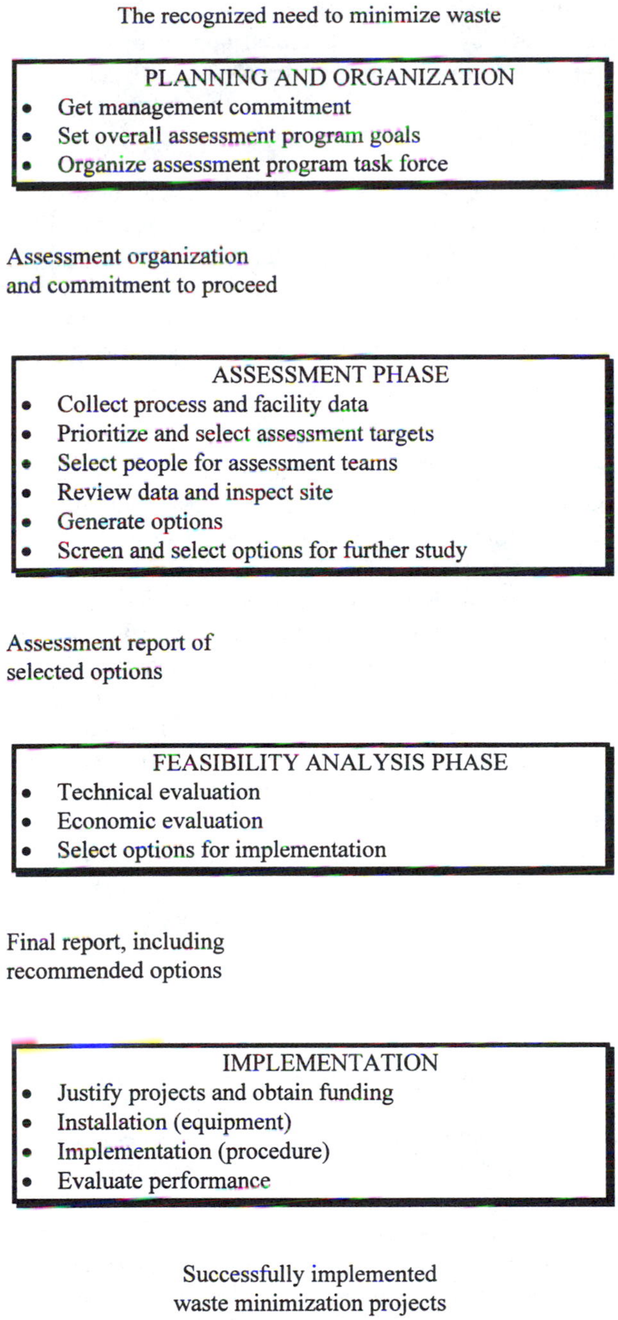

The recognized need to minimize waste

```
┌─────────────────────────────────────────┐
│        PLANNING AND ORGANIZATION         │
│  • Get management commitment             │
│  • Set overall assessment program goals  │
│  • Organize assessment program task force│
└─────────────────────────────────────────┘
```

Assessment organization
and commitment to proceed

```
┌─────────────────────────────────────────┐
│            ASSESSMENT PHASE              │
│  • Collect process and facility data     │
│  • Prioritize and select assessment targets│
│  • Select people for assessment teams    │
│  • Review data and inspect site          │
│  • Generate options                      │
│  • Screen and select options for further study│
└─────────────────────────────────────────┘
```

Assessment report of
selected options

```
┌─────────────────────────────────────────┐
│        FEASIBILITY ANALYSIS PHASE        │
│  • Technical evaluation                  │
│  • Economic evaluation                   │
│  • Select options for implementation     │
└─────────────────────────────────────────┘
```

Final report, including
recommended options

```
┌─────────────────────────────────────────┐
│             IMPLEMENTATION               │
│  • Justify projects and obtain funding   │
│  • Installation (equipment)              │
│  • Implementation (procedure)            │
│  • Evaluate performance                  │
└─────────────────────────────────────────┘
```

Successfully implemented
waste minimization projects

FIGURE 84.1 The waste minimization assessment procedure. (Reproduced from Reference 2.)

while, environmental affairs personnel are obligated to inform management of the importance of waste minimization and the need for their commitment. Waste minimization can be incorporated within a total quality management program (TQM) or into a health or environmental safety program.

Organizing the Program Task Force

The program task force should be made up of employees from any group that will be involved with waste minimization. This would include people with responsibility for production, facilities, maintenance, quality control, and waste management. The size of the team will depend on the size and structure of the facility. However, there should be two people involved at a minimum.

The responsibilities of the task force should include implementing this assessment process. They should get commitment and policy from management, set waste minimization goals, prioritize the waste streams, and select assessment teams. Assessment teams will be concerned with a particular waste stream or production area, while the task force is concerned with the entire institution.

A waste minimization or pollution-prevention coordinator should be appointed to be responsible for facilitating the implementation, monitoring, and evaluation of the program. This person should be an advocate of the program and motivate people.

Goals

The first priority for the task force is to establish goals that meet the policy that management has adopted. These goals should be explicit for reducing the volume and toxicity of waste streams. They should be achievable within a reasonable period of time. They do not, however, have to be strictly quantitative—they can be qualitative as well. They can be as comprehensive as an overall 5% waste reduction per year, or they can target a particular waste stream for elimination.

Since waste minimization assessments are not one-time projects, periodic review of goals is recommended. There will be changes in available technology, raw materials, environmental regulations, and economics.

Employee Involvement

You can see immediate results from your efforts if you include an employee awareness program. Training should include defining the terms, asking for input of ideas from employees, identifying goals, encouraging employee participation, and explaining the benefits to the company and individual employees.

Not everyone in a company can be on a task force or assessment team; however, for waste minimization to really work the involvement of most, if not all, of the employees is critical. This can be done by recognizing individual and collective accomplishments. While the company as a whole benefits from the inherent incentives toward waste minimization, the individual employee may not. Reward employees who pinpoint cost-effective opportunities. Also publicize success stories, not only to recognize those who are involved but also to inspire others to become involved.

Assessment Phase

During the assessment phase, the processes, operations, and waste management practices will be examined. This information will be used to develop a set of waste minimization options. These options will then be reviewed to identify the more attractive options that require further examination.

Collecting Data

Information is needed as to the following:

- what waste streams are generated and in what quantity
- the processes or operations generating these wastes
- classification of the waste streams as hazardous and their hazardous components
- the raw materials used to generate the waste streams and their quantities
- efficiency of the process
- separation and mixing of waste streams
- housekeeping practices used to limit quantity of waste generated
- process controls to improve efficiency

Environmental Records In order to identify and characterize the facility waste streams, records will need to be examined. The most obvious place to find out about waste quantities is to look at hazardous waste manifests, which include the description and quantity of waste shipped for disposal or treatment. However, the manifests may not provide enough detail regarding the chemical content of the waste, the source of the waste, and when the waste was generated. Annual and biennial reports to the EPA for hazardous waste, NPDES (National Pollutant Discharge Elimination System), toxic-substance release inventories, and so forth, can provide additional information on waste generation. Also look at waste stream lab analyses for information about specifically targeted chemicals.

Flowcharts Prepare flowcharts that show the waste-generating systems, if processes are amenable to this. These charts should start from the raw materials and end with the final products and wastes generated. When working

on a small scale, with many processes, such as a laboratory, this may not be a practical approach. However, even listing the starting materials and the chemical products, including waste, along with the type of procedure will give an overall picture.

Mass Balances Mass or material balances can be used to calculate losses or emissions that were unaccounted for or that you may not even be aware of. They also provide a baseline for tracking waste minimization success, the means to estimate the size and cost of new equipment or changes, and data to analyze economic performance. Using the equation

$$\text{Mass in} = \text{Mass out} + \text{Mass accumulated}$$

calculate material balances for all components that enter and leave the process. A significant mass accumulation indicates that there are unaccounted-for losses.

While this is more difficult in a laboratory situation than in an industrial situation, it was done during a study on laboratory waste minimization performed at the University of Illinois at Urbana–Champaign.[5] The total solvents purchased and disposed of as hazardous waste by one organic chemistry research group were measured. This was done by examining the purchase records for solvents and comparing them to hazardous waste management pickup records for solvent wastes for this group and the inventory on hand. About 50% of the solvents were unaccounted for. It was speculated that these solvents were lost through evaporation or down the drain.

Waste Accounting System The EPA recommends maintaining a system to track the types and amounts of wastes and their hazardous constituents. The rates and dates they are generated are important too. Subsequent calculations and waste minimization assessments will be much easier to perform if these data are readily available.

Figure out the total costs involved with waste management and cleanup. These include regulatory oversight compliance; paperwork and reporting requirements; loss of production potential; costs of materials found in the waste stream; transportation, treatment, storage, and disposal costs; employee exposure and health care; liability insurance; and possible future corrective action costs. The volume and toxicity of the waste generated should be taken into account. The best way to do this depends on each organization.

Prioritizing Waste Streams

The waste minimization assessment should identify opportunities at all points in a process where materials can be prevented from becoming waste. This can be done by using less material, recycling the material in the process, finding less hazardous substitutes, or making equipment/process changes. Analyze these opportunities based on the true costs associated with waste management and cleanup.

Waste minimization assessments should start with the most important waste problems first and then go on to the less important problems. This means the most effective and obvious waste-reduction process will be visible initially. When management sees results, they will be pleased and support a continuation of the process.

Assessment Teams

Since the program task force is concerned with the whole plant, if the operation is large enough, assessment teams will be needed. The focus of these teams will be on a particular waste stream or area of the plant. The teams might include a department manager, environmental manager, an outside consultant with expertise in that particular area of production, and an engineer involved in that process. Production and line employees can also provide waste minimization suggestions.

Site Inspection

If all the assessment team members are not familiar with the process site, the team should visit it. They should follow the process from the point where raw materials enter to where wastes leave. This may also include maintenance operations and raw material and finished product storage areas. Housekeeping aspects of the operation should be observed. Storage of chemicals, signs of spills or leaks, and overall cleanliness of site and odors should be noted.

Generating Options

After collecting all the pertinent data, viewing the site and understanding the processes, creative waste minimization options should be generated. These options can be generated from team members' own education and experience. Additionally, there are now many publications and resources for material through many organizations. Some of these are:

- trade associations—information about environmental regulations and various available techniques for complying with these regulations
- other similar companies/facilities—seek and exchange technical information on waste minimization

- published literature—technical magazines, trade journals, government reports, and research briefs
- university or government technical assistance programs—many states and local agencies have EPA and/or state funded programs that include technical assistance and information on industry-specific waste minimization techniques
- equipment vendors—vendors can provide information on equipment that can reduce waste generation
- consultants—a waste minimization consultant with experience in a particular industry is most desirable

Much information can be obtained through the Internet as well. A good web site to start with is http://www.epa.gov. They have a listing of EPA publications and their numbers. These publications provide industry-specific waste minimization suggestions. They also have links to other areas of interest. Some general suggestions are also discussed below.

Options for Further Study

Options will need to be screened to determine which of them warrant further study. The ones that appear marginal, impractical, or inferior should be eliminated. This review can be informal or quantitative. Common group decision-making techniques can be utilized.

Some questions that need to be considered are:

- What benefits are gained by implementing this option?
- Does the appropriate technology exist to develop the option? What is its cost? Is it cost-effective?
- Can the option be implemented in a reasonable time period with no disruption of production?
- Does the option have a good chance of success?

There are options that require little capital investment and can be implemented quickly. These do not require further evaluation. If an option is desirable and has a potential cost savings, it should be implemented right away.

Feasibility Stage

This stage determines if the options generated and weeded out in the previous stage are technically and economically feasible.

Technical and Economic Evaluation

The technical evaluation will probably involve some actual testing of materials, if this does not involve major equipment installation or modifications. If equipment changes are needed, try to visit existing installations, arranging these visits through equipment vendors and industry contacts. Sometimes vendors will install equipment on a trial basis or for bench-scale or pilot-scale experiments. This evaluation should also consider facility constraints and product requirements.

The economic evaluation uses standard measures of profitability. These include payback period, return on investment, and net present value. Depending on the size of the project, a calculation of capital costs and operating costs and savings, profitability analysis, and adjustments for risks and liability will all need to be made. Some useful guidance for this evaluation can be found in "A Primer for Financial Analysis of Pollution Prevention Projects."[6]

Selecting Options

The results of these evaluations as well as recommendations to implement the feasible options should be included in a report. It is helpful to consult with the affected departments and obtain their support. This will improve the probability that the projects will be implemented.

Implementation

The waste minimization report will now be used as the basis for getting funding for waste minimization projects. The tangible and intangible benefits should have been described there. The waste minimization coordinator should be flexible enough to develop alternatives or modifications to the proposed projects. This stage involves obtaining funding, installation of equipment, implementation of procedural and materials changes, and follow-up.

Measuring Waste Reduction

Not only is it important to demonstrate that cost savings can be achieved through waste minimization; if possible, a means of measuring waste reduction should be employed. The simplest way to measure waste reduction is to compare quantities of waste generated before and after a project has been implemented. However, this measurement disregards other factors that affect the quantity of waste generated, such as production rate and changes in production lines. Thus, the ratio of waste generation to production rate is a useful way of measuring waste reduction.

Some types of organizations do not lend themselves to this type of calculation. Research and education institu-tions are hard put to quantify production. Would the number of students, graduates, or papers published be considered production? Even comparing waste generation to past years can be difficult, since there are constant changes in research activity as well as researcher awareness of waste management. A discussion of this topic can be found in "Alternatives for Measuring Hazardous Waste Reduction."[7]

Periodic Reviews

In any case, a way will be found to evaluate the success of waste minimization projects. Conduct periodic reviews of program effectiveness. Feedback should be utilized to identify areas for improvement. Additionally, periodically go back to the assessment phase and look for new options in a cycle of continual improvement.

The final goal of a waste minimization program should be to reduce waste generation as much as possible. Ideally, waste minimization will become an integral part of the company's operations and be considered in all planning processes.

General Types of Approaches

Waste minimization options that reduce waste at the source should be explored initially. Then consider the options that involve in-process recycling. Some general waste minimization approaches and techniques can be found in Figure 84.2.

Source Reduction Techniques

When looking at source reduction, probably the approach that is most easy to implement is good operating practices. These are procedural, administrative, or institutional measures. They involve little cost and have a high return on investment.

Good Operating Practices

Some good operating practices that are simple to identify and incorporate are:

- Ship/receive materials in bulk to eliminate drum disposal when large quantities are used.
- Reuse containers wherever possible.
- Order materials "just in time" to avoid overstocking and expiration.
- Create a central stockroom/inventory control system—prevents duplication of ordering and encourages sharing of materials when feasible.

Inventory Management & Improved Operations

- Inventory and trace all raw materials
- Purchase fewer toxic and more nontoxic production materials
- Implement employee training and management feedback
- Improve material receiving, storage, and handling practices

Modification of Equipment

- Install equipment that produces minimal or no waste
- Modify equipment to enhance recovery or recycling options
- Redesign equipment or production lines to produce less waste
- Improve operating efficiency of equipment
- Maintain strict preventive maintenance program

Production Process Changes

- Substitute non-hazardous for hazardous raw materials
- Segregate wastes by type for recovery
- Eliminate sources of leaks and spills
- Separate hazardous from non-hazardous wastes
- Redesign or reformulate end products to less hazardous
- Optimize reactions and raw material use

Recycling and Reuse

- Install closed-loop systems
- Recycle onsite for reuse
- Recycle offsite for reuse
- Exchange wastes

FIGURE 84.2 Waste minimization approaches and techniques. (Reproduced from Reference 8.)

- Segregate waste streams—some waste streams may be more difficult to manage or minimize than others—for example, individual solvents are easier to redistill than mixtures.

Some other ideas include management and personnel practices (employee training, incentives, and bonuses), loss prevention (avoiding leaks from equipment and spills), cost-accounting practices (allocate waste treatment and disposal costs directly to departments that generate waste rather than to overhead), and production scheduling (that reduces the amount of equipment cleaning).

Technology Changes

These changes involve process and equipment modifications to reduce waste, primarily in a production setting. They can include changes in the production process, equipment changes, use of automation, and changes in operating conditions. This could be as simple as getting a cover for solvent baths to reduce losses through evaporation.

Input Material Changes

Material substitution reduces or eliminates the hazardous materials that enter the production process. Material purification changes the nature of the input material to avoid generation of hazardous wastes within the process. One area where many substitutes are available is solvent/cleaner applications. Additionally, the numbers of different solvents used can be reduced.

Product Changes

The manufacturer of a product can change the product in order to reduce waste resulting from a product's use. These changes can include product substitution, product conservation, and changes in product composition. An excellent example of this can be seen in the paint manufacturing industry where water-based coatings (latex) are being used in many applications where solvent-based paints were used before. Therefore, when paints are disposed of they do not need to be treated as hazardous wastes or cleaned with solvents, and they create a healthier working environment.

Recycling Techniques

Use and Reuse

Recycling involves the return of a waste material either to the originating process as a substitute for an input material or to another process as an input material. Reuse of solvents is an example of this. For example, in an operation involving parts cleaning where no satisfactory substitute is available, move solvents from cleaner parts cleaners to dirtier ones. Only dispose of the solvent left in the dirtiest parts cleaner as waste.

Reclamation

While considered waste minimization, reclamation is not pollution prevention. It is the recovery of a valuable material from a hazardous waste. This differs from use and reuse techniques, since the recovered material is not used in the facility but is sold to another company. An example of this is the use of an electrolytic deposition cell in photoprocessing to recover silver from the rinsewater found in the film-processing equipment. The silver is then sold to a recycler. On the other hand, gold regenerated from a cyanide solution used in gold plating could be put back into the system for plating purposes (use/reuse).

Waste Exchanges

One person's waste is another person's fortune. At least that is the philosophy of the many waste exchanges operated throughout the United States. These publicly funded organizations periodically publish lists of solid and hazardous materials that companies no longer have a use for. This could be due to overstocking or unneeded waste product from a production process. They also publish a wish list of materials that companies and institutions are seeking. Companies can usually publicize anonymously at no charge. The only costs involved are those related to the shipping and transport of material. Contact your state environmental agency to obtain the address of the waste exchange in your area.

References

1. EPA Guidance to Hazardous Waste Generators on the Elements of a Waste Minimization Program; *Fed. Regist.*, Part VII, Environmental Protection Agency, May 28, 1993.
2. EPA, *The EPA Manual for Waste Minimization Opportunity Assessments*; Publication No. EPA/600/2-88/025; Hazardous Waste Engineering Research Laboratory: Cincinnati, OH, 1988.
3. *Pollution Prevention and Waste Minimization in Laboratories*; Reinhardt, P. A., Leonard, K. L., Ashbrook, P. C., Eds.; CRC Press/Lewis Publishers: Boca Raton, FL, 1995.
4. EPA, *Guides to Pollution Prevention Research and Educational Institutions*; Publication No. EPA/625/7-90/010; Risk Reduction Engineering Laboratory: Cincinnati, OH, 1990.
5. Ashbrook, P. C.; Klein-Banai, C.; Maier, C. Determination, Implementation and Evaluation of Laboratory Waste Minimization Opportunities; ENR Contract No. HWR 91085, Illinois Dept. Of Energy and Natural Resources, Hazardous Waste Research & Information Center, draft report, July 1, 1992.
6. EPA, *A Primer for Financial Analysis of Pollution Prevention Projects*; Publication no. EPA/6000/R-93/059; U.S. Environmental Protection Agency and American Institute for Pollution Prevention: Cincinnati, OH, April 1993.
7. HWRIC, *Alternatives for Measuring Hazardous Waste Reduction*; Publication No. HWRIC RR-056, Illinois Hazardous Waste Research & Information Center: Champaign, IL, 1991.
8. EPA, *Waste Minimization Environmental Quality with Economic Benefits*; Publication no. EPA/530-SW-90-044; U.S. Environmental Protection Agency: Washington D.C., 1990.

85

Waste Minimization: Laboratories

CYNTHIA KLEIN-BANAI

Many industrial operations and businesses can find particular applications to the concepts of waste minimization described in the preceding chapter. However, research and teaching institutions, as well as analytical and industrial laboratories, may find it hard to apply these ideas. Since laboratories typically generate a large number of diverse, yet small, waste streams, calculations made to determine whether one waste minimization approach or another is viable for each waste stream may not be a practical approach.

Application to Laboratories

With some creativity, applications of the techniques described in the preceding chapter to laboratories can be found; however, this section lays out some more specific examples. Many of these ideas are described in detail in *Pollution Prevention and Waste Minimization in Laboratories*[1] and in a column published in *Chemical Health and Safety*.[2] Figure 85.1 gives some specific examples of source reduction, recycling, and treatment as they apply to laboratory work.

Good Material Management

What is considered good operating practice in industry can be called good material management in laboratories. Many of the waste minimization options applicable to laboratories fall into this category. These steps take few financial resources to implement and can show results quickly.

Good Housekeeping

It all starts with good housekeeping. If laboratory chemicals, supplies, and equipment are scattered all over the place, you cannot make prudent use of your resources. Some suggestions for cleaning up a laboratory are:

- Properly label all chemical containers, including solutions, waste, and water.
- Organize chemical storage according to hazardous characteristics.
- Clean laboratory equipment right after use, and store it properly when not in use.
- Minimize clutter.
- Dispose of wastes, surplus chemicals, and surplus equipment in a timely manner.

SOURCE REDUCTION

Good Material Management

- Good housekeeping
- Regular chemical inventories
- Centralized purchasing of chemicals
- Purchase chemicals in quantities needed
- Centralized waste management
- Appropriate waste segregation

Good Laboratory Practices

- Increased instrumentation and automation
- Scaling down experiments
- Microscale and videodiscs for teaching labs
- Solid-phase extractions
- Review techniques for waste generated
- Spill prevention

Substitution

- Citrus based solvents for xylene
- Resins for DNA preparation
- Ozone treatment or soap & water for degreasing applications
- Detergent or nochromix for chromerge
- Red liquid thermometers for mercury
- Biodegradable scintillation cocktail

RECYCLING

Use and Reuse

- Chemical redistribution program
- Reusing solvents after rotary evaporation
- Using spent solvents for initial degreasing
- Redistilling mercury
- Using a solvent recycler for HPLC unit

Reclamation

- Distillation of solvents
- Off-site reclamation (e.g. for oil and mercury)

TREATMENT

Biological

- Phenol degradation with sludge

Chemical

- Neutralization of corrosive wastes with no heavy metals
- Oxidation/reduction
- Precipitation of metals

FIGURE 85.1 Examples of source reduction, recycling, and treatment for laboratories.

- Keep an inventory of chemicals and their locations in the lab.
- Do an annual spring cleaning.

Source Separation

Since laboratories generate such diverse types of waste, there is a tendency to combine waste. While this can be useful at times, it is wise to separate waste into optimum categories to increase the waste minimization options and lower disposal costs.

Solvents Usually halogenated solvents are separated from nonhalogenated solvents for cost control purposes. However, solvent redistillation may be a viable option if solvents are segregated to avoid forming azeotropes.

Oil Oil can be shipped to a recycler for rerefining. However, if it contains more than 1000 ppm of chlorine, it must be treated as a hazardous waste. Flammable solvents and other wastes often found in used oil can render it unsuitable for recycling.

Acids and Bases Those corrosive wastes that do not contain heavy metals may be suitable for neutralization. Separate them from solutions that do contain heavy metals.

Other Wastes Mercury-bearing wastes are extremely expensive to dispose of and some may be amenable to recycling. Multihazardous wastes—those that contain chemical, radioactive, and/or biohazardous wastes combined—are difficult to manage. Avoid using these types of materials in combination.

Spill Prevention and Preparedness

Spill debris and cleanup materials produce a much larger volume of waste than typical laboratory work produces. Perhaps the most common spill in a laboratory is mercury. It is worthwhile to take steps to prevent, and yet be prepared for, spills. Prevention involves evaluating the equipment, chemicals, and procedures before you begin working in order to anticipate problems. Preparedness means having secondary containment for mercury-containing equipment and other hazardous chemicals that will prevent their spread in the event of spillage. Have an emergency plan (see Chapter 24). Use appropriate personal protective equipment.

Good Laboratory Practices

Good laboratory practices generally reduce the amount of chemicals being used during laboratory procedures.

Scaling Down Experiments

Sometimes it is possible when doing research to reduce the scale of the experiment, thereby reducing the amount of chemical waste generated. This field has been extensively explored in teaching laboratories and utilizes microscale techniques. These techniques reduce chemical usage by two to three orders of magnitude from traditional teaching methods. There are several books on the market that specify microscale techniques for a wide variety of teaching laboratories.

Another way to achieve waste reduction in teaching laboratories is by eliminating chemical use altogether. A series of videodiscs were developed at the University of Illinois at Urbana–Champaign that simulate experiments. These discs are most suitable for students who are not planning to work with chemicals in a laboratory.[3]

Changing Experimental or Analytical Techniques

During the assessment phase of the waste minimization process, experiments and analytical techniques should be analyzed for waste-reduction options. Currently used procedures should also be reviewed periodically and new ones should be examined, taking into account the wastes generated. New techniques and equipment are constantly being marketed to help the chemist achieve better, faster, more accurate results. The increasing use of instrumentation and automated equipment leads to smaller sample size, less waste due to human error, and the ability to run larger batches at one time. All of these techniques also generate less hazardous waste.

Solid-Phase Extractions

Traditionally, labs have used liquid/liquid extraction techniques, such as liquid chromatography and Soxhlet extractions. Solid-phase extraction (SPE) is analogous to this kind of extraction, but it extracts the compounds from the sample into a column or membrane. Solid-phase microextraction (SPME) is a similar technique, involving two processes: partitioning analytes between the coating and the sample followed by desorption of the concentrated analytes into an analytical instrument such as a gas chromatograph.

SPE reduces the use of solvent by up to 90% and SPME can eliminate the use of solvents altogether. This means fewer fugitive emissions to the air and wastewater and lower exposure of workers to solvents.

DNA Preparations

An example of process-specific techniques that reduce the generation of hazardous waste can be found in recombinant DNA research. DNA has typically been isolated and purified using phenol, chloroform, and other organic solvents. New techniques have been developed in recent years that almost eliminate the need for these solvents. These kits incorporate purification resins and are produced by several molecular biology products companies.

Substitution

There are many opportunities for substitution of less toxic or nontoxic materials in the laboratory. Engineering laboratories will find that many of the products used in industry for cleaning high-tech equipment can also be used in a laboratory. Again, all procedures should be reviewed and assessed to determine what substitutions can be made.

Cleaning Materials

Almost all laboratories utilize some kind of cleaning agent. Often it is a chromic acid/sulfuric acid or potassium permanganate/sulfuric acid solution. Consider replacing these solutions with nonhazardous detergents produced for specific cleaning purposes. If these are not adequate for your cleaning needs, try a solution that can be neutralized and subsequently sewered, such as potassium hydroxide/ethanol, dilute hydrochloric acid, aqua regia, or oxidizing agents that do not contain chromium or other metals.

Mercury

Mercury is found in virtually every laboratory, in reagents, thermometers, bubblers, and other laboratory equipment. Mercury-filled thermometers can be substituted with red alcohol- or mineral spirit-filled thermometers. Thermocouples may be used to replace thermometers in physics labs. Manometers can be replaced by pressure transducers in mechanical engineering work.

Histological Solvents

The histology labs have been targeted by chemical manufacturers as a good outlet for solvent substitutes. There are numerous citrus-based xylene substitutes as well as other solvent substitutes that are considered less toxic than xylene. Substitutes have also been manufactured for other commonly used histological chemicals.

Recycling

Since laboratories generate many small waste streams, they may not appear to be amenable to recycling, which often requires a minimum quantity of a material to be viable. In-process recycling in a laboratory would involve use of a reaction product or conversion of the waste product as a starting material for the next batch.

Chemical Redistribution Program

Another form of recycling that has known much success in a laboratory environment is chemical redistribution. These programs take surplus chemicals no longer needed by one lab and find another user for them. Some tips for creating a successful program are given in Figure 85.2.

Reclamation

Distillation of solvents generated from histology laboratories and from certain chromatography processes is an example of reclamation. These solvents may be reused in the same processes or recycled to another user, who perhaps does not need them at the same purity as they were originally.

Aside: Treatment

Although technically not waste minimization, if no other waste minimization options were found, treatment of laboratory wastes can be considered. Often, chemical products can be rendered nonhazardous as a final step

---

GETTING STARTED

- Appoint a chemical recycling coordinator
- Segregate excess chemicals of good quality from the waste stream and stored in a central area
- Publish a list of available chemicals several times a year and distribute it throughout the institution
- Distribute requested chemicals at no charge on a first-come, first-serve basis
- Keep records of the redistributed chemicals

PROGRAM IMPROVEMENTS

- Improve publicity through modern technology - send list via e-mail, publicize list on the local area network
- Identify new markets and uses for chemicals
- Put chemicals on the list, even if you don't see an immediate use for them. Someone else may.
- Issue savings reports as to how much money "customers" have saved by using these chemicals (include saved purchase and avoided disposal costs)
- Evaluate the program regularly

FIGURE 85.2 Elements of a chemical redistribution program.

to a procedure. Elementary neutralization of acids and bases that do not contain heavy metals can significantly reduce corrosive waste streams.

A Continuous Cycle

Waste minimization is part of a continuous improvement cycle. Processes must be regularly reevaluated and new options assessed for waste reduction. Much of waste minimization is just common sense and frugality. Employees need to receive training in waste minimization, perhaps as part of their regular safety training. Ulti-

mately, pollution prevention means creating a safer work environment, a healthier planet to live on, and a better quality of life.

References

1. *Pollution Prevention and Waste Minimization in Laboratories*; Reinhardt, P. A., Leonard, K. L., Ashbrook, P. C., Eds.; CRC Press/Lewis Publishers: Boca Raton, FL, 1995.
2. Ashbrook, P. C.; Klein-Banai, C. *Chem. Health Saf.* **1995–1996**, Column on "Laboratory Waste Minimization."
3. Smith, S. G.; Jones, L. L. *J. Chem. Educ.* **1989**, *66(1)*, 8–11.

86

Environmental Controls and Liabilities

RUSSELL W. PHIFER
JAMES HARLESS

Overview

Environmental legislation and regulation in the United States was initially enacted to control industrial effluents, such as air pollution and wastewater, whose negative impacts on the environment were easiest to observe. Later regulation of industrial chemicals and wastes arose after highly publicized incidents such as that at Love Canal and the noted impact of certain pesticides on wildlife. More recently, legislation and regulations have focused on worker and community right to know, pollution prevention, and expansion and refinement of earlier programs. Most environmental laws are enacted for set periods of time, such as 5 years, and must be periodically reauthorized. Changes are often enacted at the time of reauthorization.

The U.S. Environmental Protection Agency (EPA) was established to oversee the federal environmental programs created by Congress. The EPA has many responsibilities, including supporting research into a wide variety of environmental impact, monitoring, and control issues, developing and promulgating regulations to carry out legislative mandates, and enforcement of environmental laws and regulations. In the early years of each environmental initiative, (e.g., air pollution control, water pollution control, and waste management) the EPA usually had sole responsibility for promulgating and en-

forcing regulations. Later, as individual states enacted similar environmental laws and regulations, the EPA began sharing enforcement (and permitting) responsibility, with stringent oversight. Today, most federal environmental legislation is structured such that the EPA establishes baseline regulatory criteria for each environmental program. States that then pass laws and regulations at least as stringent as the federal standards are delegated the authority to run those programs within their borders.

Environmental regulation in the United States has profoundly affected the way hazardous substances are managed and business is conducted. Some of the most significant changes have occurred in the area of liability. Environmental legislation created liability schemes, repeatedly upheld by the Supreme Court, which were previously unknown in this country. Generators of hazardous waste are liable for those wastes and any resulting contamination associated with releases from "cradle to grave." This liability remains with the generator even if someone else, such as a transporter or disposal site, mishandles the material and causes contamination. Persons that currently, or ever, owned a contaminated site are liable for the cleanup of that site simply through the act of their ownership (known as status liability and liability without fault or causation), even if they did not cause the contamination. All parties responsible for contamination at a site are jointly liable for the cleanup, and any one party can be held individually (severally) responsible for the entire cleanup if the other responsible parties cannot be located or if those parties do not have the resources to contribute their fair share toward the cleanup.

The potentially massive costs of harm to human health, environmental cleanup, effluent management, regulatory reporting, lawsuits for damages, and other environmental compliance activities have significantly changed American manufacturing and chemical management practices. Some products, such as PCBs and chlorinated pesticides, have disappeared completely. Others, such as chlorinated solvents (used as degreasers), aromatic hydrocarbons, and Freon are produced and currently used in much lower quantities than before. Manufacturing processes and chemicals have been changed to reduce the need for costly emission controls or waste disposal. These changes also have been encouraged by the public availability of mandatory annual Toxic Release Inventories documenting the volumes of hazardous substances used and discharged by manufacturing facilities across the United States.

Based on the potential costs of cleaning up contaminated sites, liability is a major issue in all transfers of commercial and industrial property. Environmental assessment prior to accepting ownership of a site often determines whether a sale is completed. In addition, ownership is a major issue at abandoned or financially unstable disposal sites. This has resulted in much greater care in how and where hazardous wastes are disposed of, with far greater emphasis placed on recycling and thermal destruction technologies. These handling methods result in a loss of a waste's identity more readily than with other techniques such as stabilization and landfilling.

The Nature of Environmental Controls

Environmental laws passed by the U.S. Congress generally require the U.S. Environmental Protection Agency to develop and promulgate regulations. There are six steps in this process:

1. research into appropriate and fair regulatory structures and limits
2. publish notice of proposed rulemaking
3. review comments from the regulated community and interested parties
4. publish proposed rules
5. receive and review public comments
6. promulgate final rules

Federal environmental laws have generally allowed states to promulgate regulations and issue appropriate permits. A nearly universal thread among environmental laws and controls is the shift of regulation from the federal government to individual states. This has been accomplished without creating nightmares of consistency through a provision common to nearly all federal environmental laws—the requirement that a state's regulations be "at least as strict as" the federal model. Specific requirements of the federal laws cannot be compromised by state programs, though many states, due to specific regional circumstances, have chosen to tighten regulations well beyond federal law. Examples include stricter air controls to address California's severe air pollution problems, restrictive water management requirements in Florida, with its shallow water table, and additional chemical waste listings in New Jersey, with its large volume of chemical businesses. The general political awareness of a state's citizens also has a bearing on how environmental controls are implemented and enforced. The issue of jurisdiction is not always clear; in some cases, for instance, EPA regional offices may take enforcement responsibilities in addition to or instead of state environmental agencies. Clarifications (written addendums to regulatory language in response to specific questions from the regulated community) can also change enforcement policies. Interpretations, either by an individual regulator (inspector), an agency, or the facility being regulated, may differ considerably; there is, for instance, no guarantee that today's inspector will interpret a regulation the same as one who inspects the same facility tomorrow. Despite these difficulties, however, the prudent facility manager or responsible party will take a common sense approach to each issue and will request a written clarification when it appears that there are significant differences of opinion.

Regulating Agencies/Authority

The U.S. Environmental Protection Agency is responsible for issuing and enforcing regulations on the environment at the federal level. Other agencies, however, may be involved when issues cross over into their jurisdiction. The U.S. Department of Transportation, for instance, has nearly total control over the packaging, labeling, marking, placarding, transportation, and paperwork associated with the movement of hazardous materials and wastes. The Occupational Safety and Health Administration (OSHA) has jurisdiction over all worker protection issues. These include hazard communication (the employee's right to know about hazardous materials in the workplace), workplace exposures to air contaminants and other hazardous materials, and worker training in environmental areas such as waste management and response to chemical emergencies.

Ironically enough, academic students are not protected under these laws and regulations, since they are not considered employees.

In the majority of states, a state agency has been designated to issue regulations and enforce state laws. A list of these agencies is included at the end of this chapter. Some states have multiple agencies to address different issues, such as air or water quality. Most states also have an OSHA-equivalent agency.

National Environmental Policy Act

Federal facilities have traditionally been exempt from most environmental laws. Without the hammer of regulations to force funding, environmental programs at facilities such as military testing and defense research operations were given little attention. This resulted in some of the most polluted sites in the United States being located at defense facilities. To address this problem, Congress passed the National Environmental Policy Act, which essentially requires federal facilities to comply with public laws and regulations, and further directs all federal agencies to consider the environmental impact of any decision making early in the process. An environmental impact statement (EIS) is required for any project that may have a significant effect on the environment or that may be controversial due to environmental concerns.

Air Controls

Clean Air Act and Amendments

The Clean Air Act of 1990 is the basis for nearly all regulation of air emissions involving hazardous pollutants. Of primary concern are such contaminants as volatile organic compounds, nitric oxides, and particulates. The law regulates both mobil and stationary sources; permits are required for significant emissions. Emissions calculations are frequently based on a facility's potential to emit (PTE), which assumes around the clock operation of the emission source. While there are exemptions for *de minimis* emissions and several specific industrial/commercial categories, the PTE method of emissions accounting results in many "batch" operations with small actual emission levels having to obtain operating permits. The Act puts an onus on states to meet certain attainment levels (national ambient air quality standards) and to protect areas that already meet these levels. The EPA sets minimum standards that individual states must meet and then works with state agencies on such issues as procedures, permitting, monitoring, and recordkeeping. The Clean Air Act's amendments allow for tighter controls than are currently in place, but only if a health benefit can be shown.

Indoor Air Quality/Personnel Exposure

Ironically, there is no single federal agency with control over air quality issues. OSHA essentially regulates indoor air quality, while the EPA regulates outdoor air quality. An additional irony is that *only* workers are normally protected by OSHA regulation. For instance, in an academic setting, faculty, staff, and maintenance workers are provided with rights under the Hazard Communication Standard; students are not covered. OSHA regulates indoor air quality by setting personnel exposure limits to various air contaminants such as formaldehyde, glutaraldehyde, carbon monoxide, cadmium, lead, and volatile organic compounds. There are also training requirements associated with procedures and personal protective equipment designed to prevent exposures.

Water Controls

Clean Water Act

The Clean Water Act of 1972 is the principal authority for federal water pollution control regulations and policies. Designed essentially to eliminate the discharge of hazardous substances to streams, rivers, and other bodies of water, the Act mandates the permitting program that controls these discharges—the National Pollution Discharge Elimination System (NPDES). The Act also sets pretreatment standards for wastewater treatment operations, and includes grants for construction and improvements to these such facilities.

In addition to surface discharges, the Clean Water Act is designed to protect groundwater supplies. This effectively means that discharges to the ground that have the potential to migrate and affect groundwater quality are also regulated. This interpretation explains the application of the permitting program to industrial stormwater discharges. With few exceptions, facilities that store hazardous materials where they might be contacted by stormwater runoff are required to obtain a permit and perform stormwater monitoring.

One major limitation of the Clean Water Act is its neglible impact on agricultural discharges, which account for significant water quality problems. Of particular concern is the high nitrogen content of stormwater runoff from livestock grazing areas.

National Pollution Discharge Elimination System

This permitting program, mandated by the Clean Water Act, is designed to eliminate the discharge of industrial wastes to bodies of water. The basic means of meeting goals for improving water quality is through the setting of effluent limits on a variety of toxic pollutants. All industrial discharges require a permit, though both general (industry-specific or state-generic) and individual permits are allowed. State programs also have significant leeway in the setting of monitoring requirements, and many states have set up streamlined requirements for general permitting.

Safe Drinking Water Act

The basis for the Safe Drinking Water Act (SDWA) of 1974 is that much of the United States obtains its drinking water from groundwater supplies. An additional concern was the aging condition of many of the public water systems; treatment technologies were also inadequate or poorly designed. There was also concern over the level of training of operational personnel and the resulting substandard monitoring of water quality. The Act sets primary and secondary standards for various contaminants, sets treatment standards, and provides mechanisms for programs designed to protect water sources. Under the Act, states must submit compliance plans to the EPA that meet or exceed federal standards for drinking water quality. The Act also placed controls on underground injection of hazardous materials.

Waste Controls

Resource Conservation and Recovery Act of 1976

The Resource Conservation and Recovery Act (RCRA), which was last reauthorized in 1984, mandates "cradle-to-grave" management of hazardous wastes. It regulates both materials that are specifically listed and those that meet certain characteristics. The "U" and "P" lists refer to specific chemical compounds that are hazardous when discarded as off-spec or surplus. The "K" list refers to byproducts of specific industrial processes, and the "F" list refers to wastes generated by nonspecific processes such as cleaning or degreasing. The four characteristics, which cover the largest range of waste material, are ignitability, corrosivity, reactivity, and toxicity. Toxicity calculations are based on laboratory simulation of how the waste will react under landfill conditions; the list of contaminants covered under the toxic category includes a number of heavy metals (As, Ba, Cd, Cr, Hg, Pb, Se, Ag) as well as numerous common solvents and pesticides.

RCRA, and its resulting regulations at the state and federal levels, place requirements on generators to properly label, store, and document the accumulation of hazardous wastes. Other requirements include training of employees, recordkeeping, and documentation of all shipments through a cradle-to-grave manifest tracking system. While recycling and reclamation of hazardous wastes is encouraged, on-site treatment and/or disposal of wastes typically requires extensive permitting. As a result, most small companies and educational institutions accumulate wastes on-site and arrange for off-site handling of wastes. There are significant restrictions on the amount of time that wastes can be accumulated on-site without a storage permit. Large generators, those that produce over 1000 kg of hazardous waste per month, are allowed to accumulate for only 90 days before wastes must be shipped off site. Smaller generators, those that generate between 100 and 1000 kg per month, may store for up to 180 days, or under some circumstances, for up to 270 days. Generators that produce less than 100 kg of hazardous waste per month are exempt from accumulation time limits, but are restricted as to the total volume that may be stored on site. Generating acutely toxic wastes in quantities greater than 1 kg per month will result in classification as a large generator.

RCRA provides significant civil and criminal penalties for improper management of wastes, and the Environmental Protection Agency (EPA) continues to actively enforce the provisions of the Act. The total amount of penalties against individuals and companies in 1996 was $173 million; of this total, $76.6 million was for criminal fines, the highest total ever recorded.[1] It should be noted that individuals within a company may have criminal and civil liability for willful acts that result in environmental impairment. Such cases may also result in state violations and fines.

Comprehensive Environmental Response, Compensation and Liability Act

Few other environmental laws have had the financial impact of the Comprehensive Environmental Response, Compensation and Liability Act (CERCLA) of 1980. CERCLA, also known as Superfund, allows the EPA to determine cleanup costs for abandoned and closed waste sites and to then assign financial liability to those it deems to be responsible parties. Many generators have had to pay twice for the disposal of the same hazardous wastes as the result of CERCLA actions.

The legal issue of joint and several liability raised by

CERCLA has raged in federal courts almost since the Act was first passed by Congress. Many questions concerning CERCLA liability still need to be addressed; most efforts to amend the Act have been aimed at making the assignment of financial liability more equitable. The EPA also has the right "to act when there is a release or *threat of a release* of a pollutant from a site which may endanger public health." The decision of when a threat of a release exists is left up to the EPA, which has broadly interpreted this clause in many instances.

The advent of CERCLA and similar laws and regulations addressing liability have spurred industries to institute environmental management systems with significant auditing of programs. Regulatory compliance audits may be done either by internal staff or by outside consultants, and frequently cover not only the industries themselves, but also their waste management contractors and disposal facilities.

Toxic Substances Control Act

The Toxic Substances Control Act (TSCA) of 1976 primarily addresses the manufacture of toxic substances that might present a future unreasonable risk. The most significant provision of the Act was giving EPA the right to prohibit the manufacture, sale, use, or disposal of a new or existing chemical. Each new chemical must undergo toxicity testing to determine its impact on public health, and existing compounds are selected each year for testing as well.

Of particular interest within the Act are the specific controls on the use and disposal of polychlorinated biphenyls (PCBs), since their disposal is controlled not by RCRA, but specifically by TSCA. This was apparently due to the immediacy (at the time of TSCA passage) of public concerns over their toxicity. Their manufacture was discontinued in the 1970s, and while the cleanup of PCB-contaminated sites continues, most PCB transformers and capacitors have been either disposed of or retrofilled with safer heat-transfer fluids.

The Act also addresses public concerns over asbestos through amendments in 1986 (the Asbestos Hazard Emergency Response Act, or AHERA). This amendment precipitated the inspection for, and removal of, asbestos in all public schools and thousands of other public buildings.

Superfund Amendments and Reauthorization Act (SARA) of 1986

While the name of this act implies a primary objective of reauthorizing CERCLA, its more sweeping impact is the result of Title III of the act, known as the Emergency Planning and Community Right to Know Act. SARA Title III has arguably had more impact on the reduction of hazardous chemical emissions and releases than any other single piece of legislation. This has been accomplished by requiring industry to report all emissions of a substantial list of chemicals. It has been particularly effective both because of a thorough review of reports and because of the public availability of emission figures. Those companies with the highest emission figures are encouraged by public pressure to make reductions. Ironically, most government efforts to require reductions have been through voluntary programs. Specific reductions are not required by the act.

The act has also had a significant impact on the handling of chemical emergencies, largely through the development of a hierarchy of local and state emergency response planning committees. State Emergency Response Commissions (SERCs) and Local Emergency Planning Committees (LEPCs) are charged with developing plans for handling emergencies that threaten human health or the environment, and they must develop local emergency response (Haz Mat) teams to deal with hazardous material emergencies. Industries that store chemicals above threshold planning quantities must report to state and local agencies, and many states assess fees based on the number of reportable chemicals at individual facilities. The Act also requires specific notifications in the event of a release.

SARA has also had a significant impact on a less obvious factor in emergency management-training. OSHA regulations require significant training of those personnel involved in responses to hazardous chemical emergencies at SARA sites.

Environmental laws and regulations will continue to have a significant impact on industry, both in the United States and abroad. While political conditions dictate the extent and direction of changes in standards, environmental controls and liabilities are here to stay. It is largely through industry recognition of the benefits of maintaining effective environmental management programs that pollution-prevention efforts remain strong. As long as there are serious liabilities associated with noncompliance, environmental laws and regulations will continue to have a major impact.

Reference

1. *Chemical & Engineering News*, American Chemical Society, March 3, 1997.

Index

absenteeism, illness and, 218
absorption. *See also* specific absorption ratio
  dermal, 84
access control, 22, 23, 497, 498, 500
accident/incident investigation, 205–8
accident investigators, 205–8
accident-prevention program, 206
accident reporting, 205–6
accidents. *See also* injuries
  anticipating, 209
  defined, 205
  laser, 464
  window of opportunity for, 255
  and working alone, 9–10
accountability, 98
acids and bases, separation of, 624
Action Level, 112. *See also* Permissible Exposure Limits
activity (radiation), 437
acylation, 151
Administration-Controlled Substances Code Number, 16,
  17
Administrative Controls, 40
adverse reactions
  reporting of significant, 56
  threshold limit values and, 76
aerosols, 355, 480
age
  carcinogens and, 142
  and toxicity, 150
air condenser, 240
air contaminants, toxic
  in confined spaces, 169
Air Contaminants Standard, 76
air distribution criteria, 564–65
air handling equipment and accessories servicing tasks,
  218
air hoods. *See* hoods
air pressurization, 562–64
air quality, 564, 628
air-reactive chemicals, 340–41
airflow monitors, 304
airflow(s), 547
  controlling proper, 565–66
aldrin, 131
aliphatic hydrazines, 144
aliphatics, halogenated, 144
alkali metals, 404–5

allergic reactions, 320
allergies, 12
alumina, activated, 369
American Conference of Governmental Industrial
    Hygienists (ACGIH), 75
  ACGIH-AIHA Task Group on Occupational Exposure
    Databases, 98
  Threshold Limit Values for Chemicals and Physical
    Agents, 75–77, 89, 414
American Industrial Hygiene Association (AIHA), 88, 94,
  96
  Exposure Assessment Strategies Committee, 98
American Industrial Hygiene Association Workplace
    Environmental Exposure Level (AIHA-WEELs), 77
American National Standards Institute (ANSI) standards
  on Emergency Eyewash and Shower Equipment,
    186–93
  on laboratory ventilation, 294–95
American Society of Heating, Refrigerating and Air-
    Conditioning Engineers (ASHRAE) *Guide* and
    Standards, 535–36
Americans with Disabilities Act of 1990 (ADA), 193, 533,
    539, 551–52
amino acid conjugation, 151
amino azo dyes, 143
analytical lab requirement and access, 498
analytical studies, 137
  characteristics of, 137
analytical techniques, changing, 624–25
analytical toxicology, 154
animal carcasses, body parts, and bedding, 599
animal-research protocols involving hazardous agents,
    health and safety considerations in, 472–73
animals, research
  biosafety, 471, 474–75
    guidelines and standards, 471, 472
    hazard recognition, 471–72
    methods of dose administration and potential for
      exposure, 474
    risk reduction approaches, 473–75
  biosafety levels, 473, 475
annual limit on intake (ALI), 453
anthropometry, 514
antiseptic, 479–80
aromatic amines, 143
aromatic hydrocarbons, 143
arsenic, 131

aryl hydrazines, 144

as low as reasonably achievable (ALARA), 442, 443, 457

asbestos, 630

aseptic filling isolators, 491, 492

aseptic processing, 491–94
    special design considerations for, 491–94

asphyxiating gases, 381–82. *See also* oxygen deficiency

asphyxiation hazards, cryogenics and, 403–4

assets, protection of, 22

assets protection program, 21

Assistant Plant Manager, 198

at-risk work practices and behavior, 25, 28–31
    defining, 28–29

atmospheres
    ovens, furnaces, and toxic, 288

atmospheric testing, 169–70

atomic mass, defined, 436

atomic number, 436

attendants, authorized, 170, 171

attenuation, defined, 254

attitudes, addressing, 26–28

audit surveys, 95

auto ignition temperature, defined, 319

autoclaving, 601

autocorrelation, 103

autopsy, contaminated wastes from, 598

auxiliary air hoods, 301–2

average(s), 106, 108, 109

"avoid contact," 320

awareness-level training, 185

azo, 143, 144

azoxy, 143, 144

back posture and back problems, 526–27

bacterial tests, 144

barricades, 380

barrier isolators, 488

barriers. *See also* glove boxes; high-pressure test cells and
        barriers
    primary and secondary, 466

baseline sampling surveys, 94

batch polymerization process, 161

batch purging, 228–29, 231

batch reactors and reactions, 383–85

behavior-based interventions, 26–28, 30–31
    guidelines for designing, 30

behavioral feedback, providing workers with, 30, 31

bench clamps, 202

bench locations, 546–48

bench tops, radioactive materials and, 499

benches, laminar flow clean, 310

benchmark dose (BD), 65

benzene, 63, 131

bias
    defined, 139
    types of, 139

bioassay program, in radiosynthesis labs, 500

"Biohazard Label," 606, 607

biohazard spill kit, 476, 480

biohazardous materials, safe shipment of, 603–4, 607–8.
        See also biological/medical waste management
    antiterrorism requirements, 607
    communication and, 606
    imports/exports, 605–6
    packaging, 604–5

biological containment, methods of, 465–66

Biological Exposure Indices (BEIs), 84

biological markers of exposure, 138

biological materials
    discarded, 599
    moving, 556–57

biological/medical waste, 598. *See also* biohazardous
        materials; waste
    components of infectious waste streams and, 598–99
    handling of, 600
    transportation of, 600–1

biological/medical waste management, 599
    disposal, 602
    historical background of, 598–99
    off-site treatment, 601–2
    treatment at point of generation, 601

biological monitoring, 84, 133

biological products, defined, 603–4

biological safety. *See also* biohazardous materials; biosafety
    decontamination, 477–82
        cleaning surfaces and equipment, 483
        PPE and, 483
    emergency response considerations and, 476–77

biological safety cabinets (BSCs), 293, 296, 307, 313,
        562
    characteristics and applications, comparison of, 311
    classification of, 307–10
    detcontamination of, 483
    installation and certification of, 312
    lab design and, 547
    principles of containment in, 307
    selection of, 310–12

biological safety programs, 464–70
    risk assessment, 467

biomarkers. *See* biological markers

biomechanics, 514

biosafety. *See* animals, research; biohazardous materials;
        biological safety

biosafety levels (BSL-1, 2, 3, and 4), 467–68
    assignment of, 466–68

birth defects, 132

bleach, 479

blending. *See* mixing and blending

blood
    and blood products, 598
    toxicity and, 153

blood-cell-count diluting pipets, 257, 275

Bloodborne Pathogen Standard, OSHA, 469, 470

bloodborne pathogens, 468–70

body hygiene. *See* personal hygiene

body piercing, 12

boiler servicing tasks, 220
Boiling Liquid Expanding Vapor Explosion (BLEVE), 349, 353
boiling point, 319
    measurement of, 244
boiling point methods, micro and ultramicro, 244, 245
bookshelves, 204, 205
boots, 326, 335
brain cancer, 423
breathing air, compressed, 383
"Brief and Scala model," 98
brokers, use of, 576
brush discharges, 258
building codes, 508–10
building materials-related sources, health impact from various, 216
Building-Related Illness (BRI), 215
buildings and facilities. *See also specific topics*
    preventive maintenance for, 236, 569, 570
bump caps, 324
Bunsen burners, 273
burets, 275. *See also* glassware
burning. *See also* incineration
    definition of, 347
burns
    chemical, first aid for, 210
    thermal, 286
bypass hoods, 300–1

cabinet door restraints, 203
cabinets. *See also* biological safety cabinets
    earthquake safety and, 203–5
    wall
        countertop clearance for, 544
        secured, 544
cadmium, 131
cancer. *See also* carcinogenesis
    radio-frequency radiation, microwaves, and, 423
canopy hoods, 293, 297
capture hoods, 293–94
carbon disulfide, 131
carbon monoxide poisoning, 215
carbon-supported catalysts, 346
carcinogenesis, 141
    historical development, 141
    use of bioassay data and, 145–46
carcinogenic hazards, and dose-response relationship, 64–65
carcinogenicity, testing for, 144–45
carcinogens, 146, 320
    factors influencing, 142–43
    probable, 146
    types/categories of, 76, 141–42
cardiovascular system, toxicity and, 153
case-control studies, 137
casework, adjustable, 553
catalyst baskets, 385, 387

catalysts, 345
    handling procedures for, 346
    Raney and activated metal, 346
    supported precious metal, 345–46
cataracts, 421–22
causal interpretation, features strengthening a, 137–38
causality, criteria for, 137–38
ceiling coverings in radioisotope labs, 498–99
cellular phones, 423
CEN (Comité Européen de Normalisation), 88, 94, 96, 97, 103, 104
censored data, 111
centrifuge installation and maintenance, 266
centrifuge safety considerations, 266–71
    chemical, radioactive, and biological hazards, 268–71
    mechanical hazards, 266–68
    safe use practices, 271
centrifuges, 265–66, 271
    electrostatic hazards associated with, 262–63
    rotor maintenance, 271
chairs. *See also* seating
    knee spaces for, 553
charge. *See* electrostatic charge accumulation
Checklist analysis, 162
Chemical Abstracts Service (CAS) registry number, 39, 54, 319, 393, 394
Chemical Diversion & Trafficking Act (CDTA), 15, 20
Chemical Hygiene Officer (CHO), 41, 43–44
Chemical Hygiene Plan (CHP), 4, 38, 40–44
    required elements of any, 41
chemical identification, 393–94
chemical information, reporting of, 55–56
chemical inventory, 41, 42
    TSCA requirements regarding, 54–55
chemical manufacturing process, life cycle of, 158
chemical reaction conditions, changed, 241
chemical reactors, 355–56
chemical sensitivity, multiple, 216
"chemical substance," definition and scope of the term, 53, 54
Chemical Use Inventory, 59
chemicals. *See also specific topics*
    classification of, 42
    global commerce in, 57
    measurement of, 240
    moving, 557
    new, 54–55
    ordering, storage, and disposal of, 42–43. *See also* disposal; storage
    structure of, 143–44
    transferring, 241
chemistry of hazardous reactions, understanding, 249
children, 14
chlorinated compounds, 131. *See also* PCBs
chloroform, 131
choice, perceiving, 28
chronic disease agents, 88
civil disturbances, 24–25

Clean Air Act of 1990, 628

clean rooms. *See also under* semiconductors
   defined, 508
   laminar flow, 509

Clean Water Act of 1972, 628–29

cleaning materials, 625

clinical toxicology, 154

closed circuit video transmission (CCVT), 22, 23

clothing, 13, 14. *See also specific items*
   permeation testing of, 333–34
   protective, 332–36, 600
      hazard assessments for selection of, 336
      research animals and, 473
      types of, 334–36

clothing barriers, chemical permeation of, 333–34

clothing programs, protective, 336

co-op students, 7–8

coating, 356

cocarcinogens, 142

code 39, 393

Code of Federal Regulations (CFR), 15–21

cohort studies, 137

cold, common, 12

cold rooms, walk-in, 289, 290

combination unit (eyewash/shower), 187

combustible gas, nitrogen required per mole of, 225, 226

combustible liquid properties, 251

combustible liquids, terminology and classification of, 348–49

combustible solids/dusts
   experimental determination of flammable limits of, 222
   flammability characteristics of, 222

combustibles
   in laboratories, 355
   special products and facilities, 355–57

combustion, 167, 221. *See also* flammable gases and vapors
   heat of, 349

combustion triangle, 221

communication, risk, 71–72, 606
   defined, 71
   follow-up, 71
   preparation for, 70
   seven cardinal rules for, 70
   steps in effective, 68–71

communication principles, risk, 67–68

communications objectives, setting clear, 69

compliance, 89, 94

compliance statistics, 106, 107, 109–10, 117–19

Comprehensive Environmental Response, Compensation and Liability Act (CERCLA), 629–30

compressed air, 383

Compressed Gas Association (CGA), 379

compressed gas checklist, 377–83

compressed gases, 377, 511

computer operators, ergonomic guidelines for, 517–18

computer workstations, ergonomic design and, 519–20

concrete barriers, reinforced, 505, 506

conductors. *See also* semiconductors
   charge accumulation in isolated, 257

cone discharges, 258–59

confidential business information, 58

confidential paper files, 22

confined space, 172
   dangers of, 167, 168
      biological hazards, 169
      chemical/atmospheric hazards, 167–68
      physical hazards, 168–69
   OSHA definition of, 166
   permit-required, 167, 168

confined-space entry, 169–70
   personnel, 170–71
   PPE and, 171
   rescue, 171–72

confined-space permits, 169

confounding, 139

conical vial, 240

consolidation of compatible wastes, 573–74

construction contractors, 33

contact electrification, 256

contact lenses, 11–12, 344

containers
   breakage of, 203, 204
   labeling, 392–93
   size restrictions on, 398
   storage of, 203, 204, 351

containment. *See also* hoods, containment-control; ventilation, local exhaust/containment control
   biological, 307, 465–66
   fume-containment devices, 561–62
   of spills, 482, 497

containment categories, performance-based, 79

containment testing, 295, 305

contamination control, 536–38, 609–10

continuous-flow reactors, 385–86

"continuous improvement" concept, 84

continuous purging, 231–32

contractor health and safety process, 37
   challenges in implementing a global standard, 37
   components for a successful, 34–35
   developing and implementing, 35–36
   process and outcome measures, 36–37

contractors, 32–34
   training for, 8

control, ease of
   defined, 254

control areas, 510

control chart techniques, 114–18

control systems, 253–54

controlled substance analogs, 16–17

controlled substances
   disposal of, 19
   laboratory use of, 18
   ordering, 18
   record keeping obligations for, 19–20
   reporting loss, theft, or breakage involving, 18

schedules of, 16–17
security requirements for, 17–18
corona discharges, 258
correlational/"ecologic" studies, 137
corrosion, 251, 252, 267–68, 320
corrosives and irritants, 320, 343–44
first aid for exposure to, 344
identification of, 344
precautionary measures regarding, 344
corrosivity, defined, 574
cosmetics, 11
countertops
ergonomic design of, 553
materials used for, 548–51
coupling multiplier (CM) table, 516
cover glass, 276
credibility, 67, 69
CPM (counts per minute), 448
critical exposure group, 91
cross-contamination, 250
cross-sectional studies, 137
cryogen frostbite first aid, 401
cryogenic gases, 401
cryogenic liquid tanks, lab design and handling of, 543
cryogenic liquids, 418
cryogenics safety checklist, 401–4
cryogens, physical properties of common, 403
crystal growth, 510–11
crystallization, purification by, 247
culture bottles and tubes, 275. *See also* glassware
cultures and stocks of infectious materials, 598
cumulative trauma disorders (CTDs), 514, 515, 522–23
factors that contribute to, 523, 524
cupboards, 545
custodial staff, training for, 8

daily operations audit, 24
"danger," 320
data analysis, 83, 102–6, 111. *See also specific topics*
descriptive and compliance statistics, 106–11
data collection, 83, 92–97
data interpretation, 83, 111–20. *See also* statistical interpretation; *specific topics*
data management, 83, 97–98
databases, formal, 48
DBCP, 131, 133–34
dealkylation, 151
decision logic, 83
decontaminants
common, 478
levels of, 479
types of, 477–80
decontamination, 556–59. *See also under* biological safety
equipment, 558
radioactivity, 450, 451
decontamination certification form, laboratory, 558, 559
degenerative disease, 522
demonstration materials, 395

density, measurement of, 244
dermal absorption, 84
descriptive statistics, 106–7, 109, 115–18
descriptive studies, 137
designated area, 44
developmental effects of radio-frequency radiation, 422
Dewar leaks and tip-overs, 403
Dewars, preventing injuries when moving, 401
diagnostic specimens, 604
defined, 603
diagnostic surveys, 95
dialkyl peroxides, methods of removing, 370
dialysis unit wastes, 599
dichloroethylene, 131
dieldrin, 131
diet
carcinogens and, 143
and toxicity, 149, 150
diffusers, 564
dip strips, 367, 368
dipping, 356
"direct control" model, 87, 88
disease, degenerative, 522
disease agents, chronic, 88
dishes, 275. *See also* glassware
disinfectant, 478
disinfection of waste, 602
disposal, 42–43, 602. *See also* radioactive waste disposal; waste disposal
of controlled substances, 19
goals of, 374
of peroxide forming chemicals, 407–8
disposal and treatment options, federal (EPA) definitions and classifications of, 371–73
disposal operations, 372–76, 483
chemical definitions related to, 371, 372
consequences of error in, 374
inventory and, 397
iterative planning process for, 374
organizational assets, 374–75
regulatory definitions related to, 371
disposal units, labpacks as, 575
distillation, 245–46
spinning-band, 246
DNA preparations, 624–25
DNA synthesis/repair, 145
DO IT process, 28–30
Domestic Chemical Diversion Control Act of 1993, 20
dose-equivalent (radiation), 437
dose-response assessment, 64–65
defined, 64
dosimeters (radiation), personnel, 454–57
characteristics of ideal, 454
pocket, 454, 455
dosimetry (radiation), 452
personnel monitoring program, 457–58
double-layer charging, 256–57
DPM (disintegrations per minute), 448

drench hose, hand-held, 187

drinking, 12

Drug Enforcement Agency (DEA), 15
Form 222, 18, 19
inspection by, 19–20
List I and List II chemicals, 20–21
regulations and registrations, 15–16
security requirements, 17–18

dry ice, 605

drying, electrostatic hazards associated with, 264

dual limits, 115

dust deflagration index (K_{St}), 227, 231

dust deflagration pressure-time characteristics, 230

dust explosions, ignition sources for, 232

dust sphere, 229

e-mail, 46

e-mail lists/discussion groups, 46, 49–50. *See also* SAFETY

eardrum rupture due to blast pressure, 504

earrings, 12

earthquake emergency planning, 400

"earthquake-safe" buildings, 201–2

earthquake safety, 201–5

earthquake zone shelving, 544, 545

ease of control, 254

eating, 12

"ecologic" studies, 137

education, 25–27. *See also* training

effect level (ED_{10}), 65

egress, 546

egress corridors, 534

elastomer gloves, 334, 335

elbows, "winged" out to side of body, 524, 526

electric and magnetic fields, 416–18
sub radio-frequency, 418–20

electrical classification of laboratories, 251

electrical equipment, and flammability, 352

electrical hazards of centrifuges, 266

electrical power
emergency, 540
normal, 539–40

electrical system design, 539–40

electromagnetic radiation. *See also* radiation
defined, 436

electrostatic charge accumulation, 257

electrostatic charge development, 256–57

electrostatic discharges, 257–59

electrostatic hazards associated with equipment and operations, controlling, 259–65

electrostatic ignition sources, 259–64

elevators
emergency response procedures, 567–69
general hazards posed by, 566–67

emergencies, and pressure/vacuum-containing systems and equipment, 281

emergency equipment, 186, 187
selection of, 192–93
six types of, 186–88

emergency exits, 497

Emergency Eyewash and Shower Group (Industrial Safety Equipment Association), 192

Emergency Planning and Community Right-to-Know (SARA Title III), 183

emergency power, 540

emergency procedures, 43. *See also* evacuation

emergency public information, 200

emergency response, 497
biological safety and, 476–77, 606

emergency response planning
history of, 181
which to do first, 181

Emergency Response Plans (ERPs), 182–83
exercises to measure effectiveness and viability of, 184
legal requirements, 181–82
v. SOPs, 182
types of, 182–83

emergency response training, 3, 191
for chemical emergencies, 181, 184–86
legal training requirements, 184–85

emergency response training providers, selection of, 185–86

emergency safety equipment terminology, 186

emergency systems, maintenance for, 237

employee information and training, 41

employees. *See also* personnel; workers
affected, 174
authorized, 174
maximum-risk, 90n, 91, 93, 95–96, 112, 120

employer-employee responsibilities, 41

empowerment, for health and safety, 25–26

enclosure hoods, 293

enclosures, specialty, 293

endothermic compounds and reactions, 338

endurance (physiological), 514

energy
hazardous
strategies used to control, 173
types of, and methods for protecting personnel from their accidental flow, 173, 174
stored, 176

energy control procedures, 175–78

energy releases in confined spaces, 169

energy sources, 172–73

engineering, 25

Engineering Controls, 40

engulfment, 169

entry. *See also* confined-space entry
primary route of, 319

entry area, of radioisotope lab design, 497

environment impact statement (EIS), 628

environmental controls and liabilities, 615, 626–27
  nature of, 627
  regulating agencies/authority, 627–28. *See also specific agencies*
Environmental Protection Agency (EPA), 53–60, 65, 626, 627
  carcinogens, bioassay data, and, 145–46
  Master Testing List, 57
  rules and regulations
    chemical-specific, 56
    for ERPs, 182
    notification requirement, 56
    on waste management, 598, 615
environmental records, 618
enzyme inducers, carcinogens and, 143
enzyme induction and inhibition, 153
enzymes, 151, 153
epidemiologic approach, strengths and limitations of, 138–39
epidemiology, 136, 140–41
  methodology, 136–38
  observational, 136–37
  v. toxicology, 139–40
equipment. *See also* personal protective equipment; *specific pieces of equipment*
  contaminated, 599
  and hygiene, 14
  modification of, and waste minimization, 621
  moving, 557–58
  removal of, from service, 177
  requirements that it be functioning properly, 43
  reusable, 600
  unattended operation of, 10
Equipment Decontamination Record label, 558
equipment shut down, 176
ergonomic checklists, 517–21
ergonomic concerns in laboratories, 522–23
ergonomic design, 521
  high-performance labs and, 521–22
  purpose of, 552
  steps for implementing, 527–28
ergonomic guidelines
  for computer operators, 517–18
  for laboratory employees, 517
ergonomic hand tools, selection of, 518
ergonomic intervention strategies, 523–24, 526–27
ergonomic interventions, 525–26
ergonomics, 513, 515, 516, 522
ergonomics issues, laboratory-specific, 524, 526–27
ergonomics program approach, 514
  matching task requirements with worker capability, 514–15
Escherichia coli (*E. coli*), 144
"essential data elements" in exposure database, 98
ethane, air, and nitrogen mixtures
  flammability envelope of, 222, 224
ethylene oxide, 134

etiologic agents, packaging/shipping of, 604–6
evacuation, 43, 194–96, 209
  attitudes regarding, 194
  impediments to decision making regarding, 194–95
  and personnel accountability, 199–200
  planning considerations for, 195–96
exceedance fraction, 107–10, 119
"excluded" items (inventory), 395
exhaust, 536–37
  fume, 536–37
  general, 299
  laboratory air supply and general room, 562–64
  "total," 309
exhaust fans, 537–38, 565
exothermic compounds and reactions, 338
experimental/analytical techniques, changing, 624–25
experiments/experimental studies, 136
  scaling down, 624
explosion hazard data, 319, 320
explosion hazards. *See also* high-pressure test cells and barriers; lower explosive limit; reactive chemicals
  unusual, 319
explosion safe refrigerators/freezers, 289
explosion venting, 287
explosive concentration, 167
exports, reporting of, 57
exposure assessment(s), 65–66
  defined, 64
  in epidemiologic studies, 138
  predictive value of, 102
  quantitative *v.* semiquantitative *v.* qualitative, 64, 83, 102
  suggested readings on, 99
exposure category rating scheme, 88, 89
exposure control, rating the degree of, 88–89
exposure data, independence of, 103
exposure distribution, stationary, 103
exposure group-based strategies, 93–94
exposure groups, 82, 93
  homogeneity of, 105–6. *See also* homogeneous exposure group (HEG) concept
  prioritization of, 91
  unique, 91
exposure hazards, chemical, 250. *See also specific topics*
exposure histories, 92
exposure limits, 65. *See also* occupational exposure limits
  permissible, 40, 76, 82, 89, 112, 319
exposure management. *See also* occupational exposure management; risk management (RM) process
  as long-term responsibility, 84
exposure measurement(s), 85–86, 90
  defined, 85
  interpretation of a single, 114–15
  "sufficient" number of, 90
  suggested readings on, 99
  "valid and representative," 90
exposure monitoring programs, 90–92, 102, 103

exposure profile parameters, 106
exposure profiles, 86
    acceptable *v.* unacceptable, 83, 86–88, 106
exposure rating, 91
exposure routes, 149–50
exposure sampling, periodic, 90. *See also* resampling
exposure sampling strategies and data analysis, suggested
      readings on, 99
exposure zone. *See* exposure groups
exposure(s), 208. *See also specific topics*
    acceptable *v.* unacceptable, 91, 92, 114
    documentation of the absence of, 84
    measures to reduce, 242
      criteria for implementing, 43
    time since first, 138–39
extinguishing media, recommended, 319
Extremely Hazardous Substances (EHS), 183, 196
extremely low frequency (ELF) magnetic and electric fields,
      417–20
eye/face wash, plumbed, 187–88, 191
eye protection, 12, 42, 324–26, 344
eyeglasses, 324–25
    shaded, 325
eyes
    first aid for chemical burns to, 210
    toxicity and, 153–54
eyewash, 191–92, 534
    personal, 187
    plumbed, 187

face shields, 325
facepiece respirators, 328, 329
Failure Modes and Effects Analysis (FMEA), 163, 164
fans, exhaust, 537–38, 565
fastening, 487
fatigue, 522
Federal Emergency Management Agency (FEMA), 184
Federal Register, 60
feedback, providing workers with, 30, 31
feet
    hygiene of, 14
    protection of, 326
female reproductive health, 132
ferromagnetic objects, 418
ferrous salt, 369–70
ferrous thiocyanate method of peroxide detection, 367
fetal exposure to radiation, 440–41, 453–54
fetal protection policies, 134. *See also* reproductive hazards
fiber drums, 351
fiberglass reinforced polyester (FRP), 565
Fillibren's test, 104, 120
film, photographic, 442
film badges, 456
    advantages and disadvantages of, 456–57
filter cake (centrifuges), 262–63
fingernails, 13
fire chemistry, 347–49
fire codes, 508–10, 530, 532

fire dampers, 565
fire emergency planning, 400
    biological safety and, 476–77
fire fighting devices, manual, 352
fire hazard data, 319
fire hazards, unusual, 319
fire protection systems, 539
fire safety
    centrifuges and, 268
    ovens, furnaces, and, 287
fire safety cabinets, 205
fire spread mechanisms, 349
first aid
    chemical, 208. *See also* safety stations
      for exposure to corrosives and irritants, 344
      processes/components of, 208–10
    cryogen frostbite, 401
first aid treatment of chemical exposure, 210–11
flammability, 320
    defined, 574
flammability characteristics of liquids and gases, 222,
      223
flammability limits, 222–25
    of common chemicals, 348
    defined, 319
    estimating, 223–24
flammable atmospheres, 259–62
flammable gas cylinders, precautions for, 382
flammable gases and vapors, 169, 170, 353, 355
flammable limits apparatus, 225, 227
    high-pressure, 225, 228
flammable liquid cabinet, 354
flammable liquid drums, 351
    bonding and grounding of, 352, 353
flammable liquid properties, 251
flammable liquids
    defined, 397–98
    fire control and, 349–51
    labeling of, 349, 350
    storage of, 351–52
    terminology and classification of, 348–49, 352
    use and handling of, 352–53
flammable material storage refrigerators/freezers,
      289
flammable solids, 355
flammable solvents and plastic containers, 204
flammables
    in laboratories, 355
    scale-up process and, 251
    special products and facilities, 355
flash point
    of common chemicals, 348
    defined, 319
flexible intermediate bulk containers (FIBCs), 263
flood contingency plans, 198–200, 400
floor coverings in radioisotope labs, 498–99
floor drains, 191
floor size and number of floors, 532

floors, 351–52

flowcharts, 618

flue gases, 232

food, contaminated, 599

foot hygiene, 14

foot protection, 326

freezers, 289, 291

   safe practices, 290–92

   selection criteria, 289–90

freezing (hazard), 401

freezing (packaging), 605

frequency multiplier (FM) table, 516

frequency re-sampling, 96–97

frostbite, 401

fume-containment devices, 561–62

fume exhaust, 561–62. *See also* exhaust

fume exhaust ducts, 564

fume exhaust system duct construction, 565

fume exhaust system duct materials, 565

fume exhaust systems, centralized manifold-type *v.*
    individual, 565

fume hood setup area, 541

fume hoods, 300–301, 488, 561–62

furnaces (laboratory), 282

   box, 282–85

   hazards associated with, 286–88

   high-temperature inert gas and vacuum, 285–86

   tubular, 282, 283

furniture, radioactive materials and, 499

galvinized steel, 565

garments, 335–36

gas cabinets, 511

gas cylinder leaks, preventing/troubleshooting, 378–79

gas cylinder safety, 202–3

gas cylinders, 232, 378. *See also* compressed gases

   lab design and handling of, 543

   moving, 378

   ordering, 377

   point-of-use, 378

   storage of, 377–78

   of toxic, corrosive, and pyrophoric gases, 382

gas deflagration index (K_G), 225

gas piping, 511–12

gas system, safety manifold for high-purity, 280

genetic effects of radiation, 439–41

genetic influences, carcinogens and, 143

genotoxic *v.* nongenotoxic carcinogens, 142

geometric mean (GM), 106, 115–16

geometric standard deviation (GSD), 106–7

germicidal lights, 311–12

glass containers, 351

glass terminology, 276–77

glassware, 12

   chemicals and, 272

   cleaning and storage of, 274–76

   connecting, 241

   handling, 272–73

heating, 273

   selection of, 240–41

gloveboxes, 293, 294, 488

   advantages and disadvantages of, 491

   aseptic processing, 491–94

   checklist of field performance tests for, 495

   with controlled environment, 491, 492

   design suggestions for, 491

   lab design and, 547–48

   performance criteria for, 494–95

   simple, 489, 490

   specialized, 490–94

   trends in, 489–90

   trends that are driving the move to, 490

   types of, based on application, 490

gloves, 12, 334–35

glucoside and glucuronide conjugation, 151

glutathione conjugation, 151

goggles, safety, 12, 42, 325

Good Laboratory Practice (GLP) precepts, 148

goodness-of-fit, 115

goodness-of-fit tests/testing, 103–5, 117

government planning for emergencies

   federal, 184

   local, 183

   state, 183–84

grease, 273, 274

green chemistry, 243

grinding, electrostatic hazards associated with,
    264–65

guide words, 162

hair hygiene, 11–12

hairnets and caps, 324

half-life, defined, 436

hand hygiene, 12–13

hand lotions, 13

hand shields, 325

hand washing, 12

handkerchiefs, 12

"hard hats." *See* helmets

Hartmann tube, 231

Haws emergency equipment, 187–90

hay fever. *See* allergies

Hazard and Operability Study (HAZOP), 162–64

hazard assessment/evaluation, 82, 83

   scale-up process and, 249–51

hazard assessment form, 250

Hazard Communication (Haz Comm) Standard, OSHA,
    38–39, 130, 398

hazard control, 82–84

"hazard control" plan, 82–83

hazard identification, 249–51

   defined, 64

hazard identification checklist, 249, 250

hazard information, chemical

   sources of, 39–40

hazard recognition, 81, 82

hazardous materials. *See also specific topics*
  special precautions for work with particularly, 44. *See also* Extremely Hazardous Substances
  Web sites on, 47
hazardous production materials (HPM), 510
hazardous waste. *See* waste
hazards. *See also specific topics*
  checklist of, 250
head hygiene, 11–12
head posture, 525
head protection, 323–24
headphones, 12
health, safety, and environmental (HS&E) aspects of commercial plants, 249
health-effect rating, 91
health hazard data, 319
Health Industry Bar Code (HIBC), 393
hearing damage. *See* eardrum rupture due to blast pressure
heat guns, 273
heat transfer, 253, 349
heating, ventilation, and air conditioning (HVAC) equipment
  inspections and maintenance of, 219, 220
heating, ventilation, and air conditioning (HVAC) systems, 215, 301, 535–38
  problems with, 216
    costs of correcting air quality problems relating to, 216, 217
    health risks from, 217
heating and stirring chemical mixtures, 241
heating mantles, 273
helmets, 323–26
HEPA filters. *See* high-efficiency particulate air (HEPA) filters
hepatitis B virus (HBV) research laboratories and production facilities, 470
hexachlorocyclohexane, 131
hierarchy, as impediment to decision making, 194
high-efficiency particulate air (HEPA) filters, 307–9, 328, 500
high hazard
  defined, 488
  equipment alternatives for chemicals of, 488–95
high-pressure test cells and barriers, 501, 507
  design of, 502–3
    donor systems, 503–4
    protective systems, 505–6
    receiver/acceptor systems, 504–5
  use of, 506
    safety rules, 507
    space allotment, 506–7
histological solvents, 625
homogeneous exposure group (HEG), 102–3, 105
homogeneous exposure group (HEG) concept, 93–94. *See also* exposure groups
hood exhaust monitoring (radioactive materials), 497–98

hood face/slot velocity, 295, 303
hood performance, measures of, 295–96, 305–6
hood users
  communication to, 296, 306
  work practices followed by, 296–97, 299, 306
hoods, 43. *See also* fume hoods
  containment-control, 295–99
    guidelines regarding, 294–95
    main purpose of, 295
    regulations regarding, 294
    types of, 293–94, 297, 298
  laboratory chemical, 300, 306
    guidelines regarding, 305
    lab design and, 547
    regulations regarding, 304
    types of, 299–304
  radioisotope, 496, 499–500
horizontal sliding sash hoods, 303
hoses
  drench, 187
  electrostatic hazard associated with, 261–62
hot plates, 273
human-equipment interface, 254
human immunodeficiency virus (HIV), 467, 470
human immunodeficiency virus (HIV) research laboratories and production facilities, 470
humidification and dehumidification equipment tasks to reduce safety risk, 219
hydraulic fluids, 356
hydrochlorinated biphenyls, 131
hydrogen, 341
  properties of, 386–87
hydrogenation reactions, stoichiometry of, 388
hydrogenation techniques, 383
  batch reactions, 383–85
  continuous-flow reactors, 385–86
  hazard identification and, 387–88
    chemistry and, 388
    pressure and, 388–89
    temperature and, 389–90
hydrogenations, 383
hydroperoxides, 362–63, 369
hypersensitivity pneumonitis, 215
hypothermia, 189

ignitability. *See* flammability
ignitable atmospheres and explosions
  ovens, furnaces, and, 286–87
ignition sources, 221, 232
  electrostatic, 259–64
immune system, toxicity and, 154
imports, certification of, 57
impurities, scale-up process and, 249–50
incapacitations, working alone and, 9–10
incentive/reward programs, 30
incident commanders, 185

incident control tactics. *See also* Emergency Response
    Plans
    Offensive and Defensive, 185
incineration and thermal decomposition of waste, 575,
    601, 602
incompatibles. *See also* segregation of incompatibles
    defined, 338
    uncontrolled, 342
inconsistent data points, 111
"indirect control" model, 86, 88
indoor air quality (IAQ), 215
    in microscale chemistry lab, 242
indoor air quality (IAQ) problems, 215–16, 220
    financial consequences of, 216–19
    inspections and maintenance tasks and, 220
    preventative strategies, 219–20
indoor air quality (IAQ) regulations, 218, 628
induction charging, 257
Industrial Safety Equipment Association, 192
industrial toxicology, 154
inert gas sources, 232
inert gases, 381–82
inerting, 221–23
    defined, 221
    when to inert, 232–33
inerting procedures, 228–29, 231–32
infectious substances. *See also* biohazardous materials;
    biological/medical waste
    defined, 603
information, protection of, 22
information processing (cognition), 515
information transfer, insufficient, 254
Infrared radiation (IR), 413–16
ingestion, chemical
    first aid for, 210–11
inhalation, chemical, 149–50, 210
    first aid for, 210
inherent safety, 254
initial-exposure sampling surveys. *See* baseline sampling
    surveys
initiation-promotion-progression, 142
injection exposures, chemical
    first aid for, 210–11
injuries. *See also* accidents
    microscale chemistry laboratories and reduction of,
    242
    prevented by PPE, 25
injury-free workplaces, 32
inspection (safety), 191, 278, 569–70, 619
Institute of Electrical and Electronics Engineers, Inc. (IEEE),
    423, 424
instrumentation, specialized, 215–20
integrated circuit (IC) fabrication, 510, 511
Integrated Emergency Management System (IEMS), 182,
    183
intensification, defined, 254
interlocks, 415
interlocks systems, 253–54

International Agency for Research on Cancer (IARC),
    318
International Commission on Illumination (CIE), 413
International Non-Ionizing Radiation Committee of the
    International Radiation Protection Association
    (INIRC/IRPA), 414–15
Internet, 45
    chemical safety information on, 45–52
        providing, 50–51
    networking on, 49
    as research tool, 47–49
Internet information, 46
    using, 45–46
Internet information tools, 46–47
Internet search strategy, 47–48
Interruption Matrix, 160, 161
inventory
    chemical, 391
        conducting the initial, 394–96
        safety issues and, 394, 396
    periodic physical, 397
inventory bar codes, 393, 396
inventory management, 621
inventory programs/projects, chemical
    implementing, 391–92
inventory records, chemical
    building, 392–94
    data needed for complete, 392
    maintaining, 396–97
inventory tools, chemical, 396
Inventory Update Rule, 59
Iodide tests, 367, 68
ionizing radiation. *See also* radiation
    defined, 436
iron. *See* ferromagnetic objects
irradiance and radiant exposure (lasers), 462–63
irritants. *See* corrosives and irritants
isolation facility, PPE in, 488–89
isolation space
    entrance and exit from, 487
    ergonomics of, 487
isolation technology, 487, 488
    definitions in, 487–88
    principles of, 487
isolation wastes, 598
isolators
    aseptic filling, 491, 492
    defined, 488
isotopes, defined, 436–37

jewelry, 13
Joule-Thompson inversion points of various gases,
    387

keywords in searching for chemical safety Web sites,
    48
kidneys, toxicity and, 154
kinesiology, 514–15

labels/labeling, 39, 349, 350, 366, 392–93, 558, 606–7

labile compounds, 338–39

labile structures and energies, 339

laboratories. *See also specific topics*

  flexibility in, 535, 542–43

  moving, renovating, and decommissioning, 498, 500, 554, 556

    construction and renovation, 559–60

    decontamination and moving, 556–59

    hazards associated with decommissioning, 556

    planning phase, 554–56

    reoccupancy, 560

laboratory activities, requirements for prior approval of, 43

Laboratory Administration, 6

laboratory aisles, 546–47

  minimum recommended widths of, 543

laboratory design criteria, 532

  bench and major equipment locations, 546–48

  building and fire code factors, 530, 532

  casework, 553

    ADA compliance, 551–52

    materials, properties, and standard systems, 548–51

  for engineering systems, 535–40

  ergonomic design, 552–53

  materials handling, 543–46

  modular design approach, 532–33

  personnel and materials circulation, 533–34

  room design performance criteria, 531

  safety and performance factors, 535

  safety systems and distribution, 534–35

  structural strategies, 533

laboratory design process, programming, 528–30

  project cost components, 530, 532

laboratory employees

  with experience, training for, 6–7

  inexperienced, 6

  new, 6–7

  part-time and temporary, 8

laboratory fitout, design criteria for, 540–43

laboratory module layout, double, 529

laboratory practices. *See also specific topics*

  good, 623, 624

laboratory security, 17–18, 21–25, 450, 497, 500

Laboratory Standard (OSHA), 38, 40, 44, 72, 79

  Appendix to, 44

Laboratory Ventilation Management Program, 295

labpacks, as disposal units, 575

laced reinforcing, 505, 506

lachrymators, 320

laminar flow clean benches, 310

laminar flow clean room, 509

laser accidents, common causes of, 464

laser wavelengths, 459–62

lasers, 458–59

  best safe practices, 464–65

biological effects of, and exposure limits, 459–62

  hazard classification, 462

  hazard evaluation, 462–64

laundry, 13

LC50 (lethal concentration 50), 319

LD50 (lethal dose 50), 148, 319

lead, 131–33

leather, 14

Legionnaire's disease, 215

liability, 629–30

lifting and lowering tasks, ergonomic design and, 518–19

lifting index (LI), 515

light bulbs, 352

lighting design, 540

lightning-like discharges, 259

limit of detection (LOD), 111

limitation of effects, defined, 254

limiting oxidant concentration (LOC), 221–23, 229, 231, 259

  estimating/calculating, 224–25

"liquid-proof" protective clothing barriers, 333

liquids

  charge accumulation in, 257

  low- *v.* high-conductivity, 257

  in vessels/pipes, 259–60

liver, toxicity and the, 154

Local Emergency Planning Committees (LEPCs), 183–84

lockers, storage, 356

lockout (LO)

  contractor, 178

  v. tagout (TO), 176

lockout/tagout device, affixing, 176

lockout/tagout inspection/audit checklist, 177, 178

Lockout/Tagout Standard (LOTO), 172

  compliance requirements, 174–78

  devices/hardware, 176

  group, 178

  release from, 176–78

  when it does not apply, 173

  when to use, 173

Lockout/Tagout Standard (LOTO) procedures and use, 175–78

  inspection/audit of, 178

log-probability plot, 104

log-probability plotting techniques, 104–5

log-probit curves, 116, 119

log-probit plot, 104

lognormal distribution, 123–25

lognormal distribution assumption, 103–5, 108

lognormally distributed data, 121–22, 126

lower confidence limit (LCL), 102

lower explosive limit (LEL), 167, 169, 170, 348

lower flammable limit (LFL), 222–25

lowest observed adverse effect level (LOAEL), 65, 149

lubricants, 273

lungs. *See* respiratory system

machine-equipment-specific energy control procedures, 175

machine isolation, 176

magnetic field(s)
  earth's, 416–17
  static, 416
    biological effects of, 417
    common quantities and conversions, 417
    deleterious effects from, 418
    exposure standards, 417, 418
    safety considerations, 417–18
    sources of, 416–17

magnetic resonance imaging (MRI), 416–17

magnetic-sensitive materials, 418

magnitude of effect, 138

maintenance contractors, 33

maintenance items in buildings and facilities, 234

maintenance needs, priorities, codes, and standards, 238

maintenance (preventive), 234, 569, 570
  assessing needs for, 235
  frequency of, 237
  for laboratories and pilot plants, 235–37
  and management of change, 238–39
  planning for, 234–35
  providing/assigning, 235
  safety and training considerations, 235–39

maintenance process, preventive, 234

male reproductive health, 132

man-down transmitter, 10, 11

manufacture, processing, and distribution (MPD), 58

masks, 328

mass balances, 618

material management, 623–24

material releases in confined spaces, 169

material safety data sheet (MSDS), 39, 80, 130, 209, 317, 318, 391
  employee training and, 318–19, 321
  format of, 318
  origin and history of, 317–18
  PPS, biological safety, and, 483
  reliability of, 318
  safety information and, 394
  terms used in, 319
  using, 360

material safety data sheet (MSDS) phrases and recommended precautions, 320–21

maximum allowable concentrations (MACs), 75

maximum allowable water level (MAWL), 389

maximum allowable water pressure (MAWP), 389

maximum permissible exposure (MPE), 459

maximum-risk employees (MREs), 90n, 91, 93, 95–96, 112, 120

mean, arithmetic, 106, 108, 109

measurement averaging time, 96

median, 109, 110

medical consultations and examinations, 43

medical surveillance, 474

medical waste. See biological/medical waste

medications, 14

mercury, 131, 624, 625

mercury oxycyanide, 375

metabolism
  phase I and II, 151, 153
  and toxicity, 151, 153

metal containers, 351. See also corrosion
  electrostatic hazard posed by small, 260, 261

metal fatigue, 267

metal reactors, tanks, and vessels
  electrostatic hazards associated with, 259–60
  vessels with internal linings or coatings, 260

metatarsal guards, 326

methane, 222, 223, 225

methyl ethyl ketone peroxide (MEKP), 376

micro-capillary bell method, 244

microanalysis, 248

microbiological practices, standard safe, 466

micropycnometer, construction of a, 244, 245

microscale center, national, 247

microscale chemistry, 239
  elimination of chemical wastes in, 243, 599–600
  health and safety benefits from, 242
  history of, 247–48
  nature of, 239–41

microscale conversion, 248

microscale glassware, 241

microscale laboratory techniques, 244–47

microwaves, 423
  biological effects of, 421–23
  exposure guidelines, 423–25

Mild Threat, 27–28

milling and grinding, electrostatic hazards associated with, 264–65

minimization, and emergency preparation, 194–95

minimum variance unbiased estimator (MVUE), 107, 117

mixing and blending, electrostatic hazards associated with, 263–64

mixing processes, scale-up of, 253

modifying factor (MF), 65

molecular sieves, 342

monitors, specialized, 215–20

monomers, 340

motivation, developing internal, 27–28

mounting methods (casework), 548, 549

movement/posture (kinesiology), 514–15

moving. See under laboratories

multiple chemical sensitivity (MCS), 216

mutagenicity, 144–45

mutagens, 320

N-nitroso compounds, 144

nails, 13

name tags, visitor, 9

National Academy of Sciences (NAS) report, 63

National Consensus Codes, 277, 278

National Environmental Policy Act, 628

National Fire Protection Association (NFPA) Standards, 251, 530, 539, 565
  checklist for, 347

National Institute of Occupational Safety and Health (NIOSH), 76, 113
  Lifting Equation, Revised, 515
  "Proposed National Strategy for the Prevention of Disorders of Reproduction," 135
  sampling strategy, 92, 93, 112

National Sanitation Foundation (NSF) standards, 311

natural disasters. *See also* flood contingency plans
  biological safety and, 477

naturally occurring radioactive material (NORM), 439

neck posture, 526

nervous system, toxicity and, 154

nitric acid, 341

nitro compounds, 408
  contamination and desiccation of, 408–9
  precursor, 143

nitrogen mixtures, 222–24

nitrogen required per mole of combustible gas, 225, 226

no-observed-adverse-effect level (NOAEL), 65, 77, 78, 85, 149

no-observed-effect level (NOEL), 76

noise, 535

nominal hazard zone (NHZ), 463–64

nonparametric decision logics, 112–13

nonparametric statistical analysis, 112–13, 127–28

nonparametric statistics, 109–11, 119

normal distribution, 123–25

normal distribution assumption, 108

normally distributed data, 126

Northridge earthquake, 201

nose, runny, 12

Notice of Commencement of Manufacture (NOC), 55

notifying EPA of substantial risk information, 56

nuclear magnetic resonance (NMR), 417

nuclear radiation, defined, 437

Nuclear Regulatory Commission (NRC), 428–30, 497

O-rings, 241, 269

observation bias, 139

observation intervals, 91

"observational approach," 105

observing at-risk work behavior, 29–30

occupant complaints and investigation, indoor air quality and, 218–19

occupational exposure databases, 83
  categories for essential data elements in, 98

occupational exposure limits (OELs), 75, 85–90, 111, 114
  components of, 85
  current, 75–77
  documentation and communication of, 80
  formula for calculating, 77
  long-term average (LTA), 87, 88, 96, 106n, 115
  methods for establishing, 77–80
  performance-based, 78–79

TWA, 85, 87–89, 96, 113–15, 119
  working statistical definitions for, 87
  worldwide, 77

occupational exposure management, 81–85. *See also* exposure management; risk management (RM) process
  program performance characteristics, 91–92
  role of judgment and experience in, 84–85

"Occupational Exposure to Hazardous Chemicals in Laboratories." *See* Laboratory Standard

Occupational Safety and Health Administration (OSHA), 627
  regulations, 3–4, 38, 63. *See also* Hazard Communication (Haz Comm) Standard; Laboratory Standard; Permissible Exposure Limits; *specific topics*
    DBCP Standard, 133, 134
    Ethylene Oxide Standard, 134
    General Industry Standard for ERPs, 182, 184, 185
    Lead Standard, 132, 133
    Worker's RTK/Haz Comm, 38–40

Occupational Safety and Health Administration (OSHA) agencies, state, 182, 628

off shift audits, 24

Office of Technical Assistance, 59, 60

oil, 624

ongoing review, 252

Operating Characteristic curve, 91

operations-level responders, 185

oral route of exposure, 149

Organization for Economic Cooperation and Development (OECD), 58

oven gloves, 286

ovens, laboratory, 282, 286–88, 356

oxidant concentration. *See* limiting oxidant concentration

oxidant-reactive chemicals, 341

oxidation, 151, 341, 347

oxidizers, 320, 356
  defined, 356

oxidizing agents
  insidious hazards pertaining to, 360
  list of common, 358
  reactions with reducing agents, 358–60

oxygen
  precautions for liquid, 403
  proportion of, in compounds, 338

oxygen, nitrogen, and methane mixtures
  flammability of, 222, 223

oxygen analyzers, process, 232

oxygen concentration, measurement of, 232

oxygen content of confined spaces, 167–68

oxygen cylinders, 380–81

oxygen deficiency, 167, 381–82. *See also* asphyxiating gases

oxygen-deficient environments, 327

oxygen-deficient hazard (ODH), 403, 404

oxygen-enriched atmosphere (OEA), 380

oxygen enrichment, 167

ozone protection, 415

P450 genes, 151, 152
pacemaker implants, 417
paper review, 251–52
parametric decision logics, 113–14
parametric statistics, 106–9
particulate respirators, 328
PCBs (polychlorinated biphenyls), 400, 630
pegboards, materials used for, 550
perchloric acid, 409–10
perchloric acid hoods, 303–4
performance-based category enrollment criteria, 79
performance-based containment categories, 79
performance-based standard(s), 40, 41
Performance Characteristic curve, 91
periodic table, 358–60
permissible exposure limits (PELs), 40, 76, 82, 89, 112, 319
permit-required confined space, 167, 168
permits, confined-space, 169
peroxidation, 405–6
  rate of, 361–62
peroxidation chemistry, 361
peroxidation inhibitors, 365
peroxide concentration, hazardous, 363, 365
peroxide detection methods, 367–69
peroxide-forming chemicals, 320, 407
  classes of, 363, 364
  label for, 366
  open detonation of, 408
  safe storage period for, 365–66
  treatment and disposal options for, 407–8
  use of, 367
peroxide test strips, 406
peroxides, 405
  explosive hazard for, 406
    hazardous levels of, 363
  laboratory testing for presence of, 406–7
  removal of, 369–70
  stabilization/reduction of, 408
peroxidizable chemicals
  organic, 361
    control of, 365
    secondary reaction products, 362–63
  purchase and storage of, 365–66
  surveillance and, 366–67
peroxidizable organic moieties, 362
person-based inverventions, 26–28
personal hygiene, 11–14, 42
personal protective equipment (PPE). *See also specific items of equipment*
  bloodborne pathogens and, 470
  deciding when to use, 322–23
  human behavior and the use of, 27
  injuries prevented by, 25
  payment for, 37
  training employees in how to use, 3
  worker education and training regarding, 336

personal protective equipment (PPE) rules and requirements, 40, 42
  OSHA requirements, 322–23
personnel. *See also* employees; workers
  designation of responsible, 43–44
personnel changes, during lockout/tagout, 178
pesticides, 153
pets, 14
photosensitivity, 414
physical review, 252
picric acid, 409
pilot plants. *See* scale-up of laboratory operations
pipet washers, automatic, 275
pipets, 275. *See also* glassware
  microdelivery, 244
piping and hoses, electrostatic hazard associated with, 261–62
piping and instrumentation diagram (P&ID), 251
Plant Manager, 198, 199
plastic containers, 260–61, 351
plastic items and implements, electrostatic hazard posed by, 260–61
plumbing, 538–39
point estimates, 106, 107
"poison," 321
pollution prevention, 615, 616. *See also* recycling; waste minimization; waste reduction
polychlorinated biphenyls (PCBs), 400, 630
polycyclic aromatic hydrocarbons, 131
polyester, fiberglass reinforced, 565
polymerization, compounds posing hazards due to peroxide initiation of, 408
polymers, 35, 334, 335
Pontiac fever, 215
population parameters, 106
posture. *See* ergonomics
potassium iodide (KI) test, 406–7
potential to emit (PTE), 628
powered air-purifying respirators (PAPRs), 327
pre-start-up safety review, 252
precommissioning review. *See* pre-start-up safety review
predictive value, 102
Premanufacture Notification (PMN), 54–55
pressure differentials, creating, 487, 538
pressure effects, quantitative, 342
pressure excursions, 388
pressure explosions, preventing, 402–3
pressure indicating control (PIC) device, 232
pressure purging, 229
pressure-relief, 287
pressure safety, 379–80
pressure terms, relationships of defined, 380, 381
pressure/vacuum-containing systems and equipment, 277–78, 281
  codes and standards for, 278
  hardware, 279–80
  safe practices with, 280–81
private sector planning, for emergencies, 183

process design, 251, 252, 255
Process Safety Management (PSM) program, 157
process safety review, procedure for, 157
process safety review methodologies, 160–65
Process Safety Review report, contents of, 165
process safety review team, 158
    forming a, 159–60
process safety review team members, areas of expertise for,
    160
process safety reviews
    background information, 159
    data that should be considered for, 159
    documentation and follow-up, 165
    objectives of, 157–58
    scope definition, 158–59
    selecting methodology, 159
    when to do, 158
product generation, compared with product need, 243
Production Manager, 198
propagating brush discharges, 258
protective equipment. *See* personal protective equipment
purchases, chemical, 365–66, 396
purging, 228–31
purification, methods of, 245–47
pushing and pulling tasks, ergonomic design and, 519
pycnometer, 244, 245

quantitative risk assessment (QRA), 64, 83, 102

radar development staff, 423
radiation, 436–38, 446. *See also* radioisotope lab
    best safe practices, 415, 416, 419–20
    biological effects of, 414, 416, 417, 419, 421–23,
        439–42
    defined, 436
    detection of, 442
    dose, 437, 443, 450. *See also* dosimetry
    dose limits, 453
        to members of general public, 454
        occupational, 453–54
    exposure guidelines and standards, 414–20
    external exposure, 453
        factors that influence the hazard of, 443–44
    internal exposure, 452–53
        hazard of, 444, 446
    non-ionizing. *See also* microwaves; radio-frequency (RF)
        radiation
        defined, 436
    optical, 413. *See also* ultraviolet (UV) radiation
    protection principles, 442–43
    sources of, 413–14, 416–17, 419, 438
    terminology, 436–37
    visible and infrared, 415–16
radiation emergencies
    closure, 450–51
    defined, 446–47
    precautions and, 447

response planning and, 447–49
    scope of, 447
radiation emergency procedures, 448–52
    PPE and, 451
radiation emergency response equipment and supplies,
    448–49
radiation exposure contributions to U.S. populations,
    average annual, 438–39
radiation models, 439–40
radiation producing machines, guidelines for using, 443
radiation safety
    audits, 433–35
    inspections, 431–33
radiation safety manual topics, 432
radiation safety manuals, 431
radiation safety program elements, 428, 431
    environmental concerns, 433
    licensing and registration, 430–31
    records, 433
    regulations and guidelines, 428–30
radiation source security procedures, 450
radio-frequency (RF) radiation
    biological effects of, 421–23
    control measures regarding, 426
    exposure guidelines, 423–25
    frequency and field strength of, 423–25
    hazard evaluation of, 425–26
    sources of, 421, 422
radioactive material contamination, defined, 495–96
radioactive material contamination zones, 496
radioactive materials (RAM)
    guidelines for using, 444–45
    moving, 557
    nonradioactive substitutes for, 609
    storage for, 496, 498
radioactive substitution, 609
radioactive waste, 497, 500
    avoidance of, 609
    prevention of, 609–10
    receipt of, 612–13
    reduction of, 610
    segregation of, 610–12
radioactive waste disposal, 612–13
    off-site, 613–14
    on-site, 612–13
radioactive waste flowchart, 611–12
radioactive waste management and transportation, 608, 614
radioactive waste management program, 609–13
    goals of, 608–9
radioactive waste treatment/disposal and transportation/
    services facilities, 614
radioactivity, defined, 436
radioiodine, precautions when handling, 610
radioisotope hoods, 499–500
radioisotope lab design, 495–98
    construction materials, 498–99
    decommissioning considerations, 498, 500

entry area, 497
separate services and systems, 496
user considerations in, 500–501
radioisotope lab(s)
diagram of a, 497
layout of, 496–97
radiosynthesis labs, licensing concerns for, 500
ramps, 569–70
reaction conditions. See chemical reaction conditions
reactive chemicals, common, 404–10
reactivity
defined, 574
mutual, 340–42
reactivity hazard data, 321
reactivity hazards, chemical, 250
receipt (inventory), 396–97
reclamation of waste. See recycling, reuse, and reclamation of waste
recommended exposure limits (RELs), 76
Recommended Weight Limit (RWL), 515
recovery team (emergencies), 200
recycling, reuse, and reclamation of waste, 243, 574, 576, 616, 621–23, 625
redistribution program, chemical, 625
redox compounds, 339–40
"reduce, recover, and recycle" (three Rs), 243
reduccant-reactive chemicals, 341–42
reducing agents
insidious hazards pertaining to, 360
list of common, 358
reflective v. nonreflective materials, 415
refrigerators, 289, 291
safe practices regarding, 290–92
selection criteria for, 289–90
regulators (compressed gas), installation of, 379
regulatory issues, 250. See also specific issues
in scale-up of operations, 254
regulatory toxicology, 154
renovation. See laboratories, moving, renovating, and decommissioning
repeat measurements, analysis of, 111
repetitive hand tasks, ergonomics and, 519, 521
"representative employee." See maximum-risk employees
reproductive effects of radio-frequency radiation, 422
reproductive hazards
radiation, 440–41, 453–54
in workplace, 130–31, 134
employer guidelines, 134–35
reproductive outcomes, adverse
environmental toxicants and, 131
reproductive toxicity
evaluating, 132
future of, 135
reproductive toxins, OSHA-regulated, 132–34
REPROTOX, 131
resampling, 90, 92, 97
Research and Development (R&D), 252

residue management, 375
Resource Conservation and Recovery Act of 1976 (RCRA), 182, 573, 575, 598, 610, 629
respirator classes, protection factors for common, 330
respirator facepieces, 328, 329
respirators
selection and fit-testing of, 330–31
types of, 327–29
air-purifying, 328–31
respiratory hazards, types of, 327
respiratory protection, 326–27
standards for, 327
respiratory protection programs, 331–32
respiratory system, toxicity and, 154
respiratory tract infections, upper, 218
reuse of waste. See recycling, reuse, and reclamation of waste
reward/incentive programs, 30
rewarding safe behavior, 28, 30, 31
rings, wearing, 12–13
risk, perception of, 67, 68
risk assessment, 63, 66
risk assessment process
definition of, 63–64
model of, 63, 64
risk assessment research, 95
risk characterization, 66
defined, 64
risk information, substantial
notifying the EPA of, 56
risk management (RM) process, 58. See also occupational exposure management
risk standard, unreasonable, 59
rocket reactors, 384–85
rotors
maintenance, 270, 271
stress on, 266–67
roundup (inventory), 395
route(s) of entry, primary, 319
rubber, 14
rule-based decision logics, 112

Safe Drinking Water Act (SDWA) of 1974, 629
safety. See also specific topics
three Es of, 25
safety and health
addressing the human dynamics of, 25–32
behavior- v. person-based approaches to, 26–28
safety and health studies, reporting them to EPA, 56–57
safety can cutaway, 353, 355
safety data sheet. See material safety data sheet (MSDS)
SAFETY e-mail list, 39, 48, 51–52
safety equipment, 3
safety interlocks, 253–54
Safety Manager, 198
Safety Permits, 176

safety professional, 206
safety review
  ongoing, 252
  pre-start-up, 252
safety stations, 534
safety systems, maintenance for, 237
Safety Video Evaluation checklist, 5
Salmonella, 144
sample location, 95
sample materials, 395
sample size, 96, 138
sample statistics. *See* point estimates
sampling, reduction of, 97
sampling and analytical error, *v.* environmental
      variability, 98–99
sampling strategies, 83, 92–94
  regulatory compliance, 97
sampling surveys, 94
scale-up of laboratory operations, 249. *See also*
      laboratories, moving, renovating, and
      decommissioning
  design issues in, 252–54
  design review process, 251, 252
  hazard identification and assessment, 249–51
  human factors and, 254–55
  inherent safety and, 254
scintillation detectors, 442
sealing, 487
search engines, 47
search (radiation emergencies), 450
seasonal differences, 96
seating, laboratory. *See also* chairs
  ergonomic design and, 553
security audit checklist, 24
security audits, 22, 24
Security Contact/Manager, 21
Security Contacts, 22
security equipment, 22
A Security Outline of the Controlled Substances Act of 1970,
      15, 16
security personnel, 22
security procedures, radiation source, 450
security program, purpose/goals of, 21
security requirements for controlled substances,
      17–18
security work station, 23
segregation of incompatibles, 342, 573, 600, 610–11. *See
      also* source separation
seismic safety, 201–5
selection bias, 139
self-contained breathing apparatus (SCBA), 328–30
self-reactivity, factors for, 338–40
Semiconductor Equipment and Materials International
      (SEMI), 508, 510
semiconductor loss history, 508, 509
semiconductor manufacturing, materials processing in,
      510–11
  control technology, 511–12

semiconductors, 442
  clean room design and construction, 508
      building and fire codes, 508–9
  future trends regarding, 512–13
sensitization, 78
separation, types/methods of, 245–47
serological tubes, 275–76. *See also* glassware
service providers, 34
sewer, public, 576
sex differences
  in carcinogenesis, 143
  and toxicity, 150
shaker reactors, 384
sharp items, 598–99
shelf restraints, wire, 202, 203. *See also* shelves
shelter-in-place plans, 196–97
  impediments to decision making regarding,
      194–95
sheltering-in-place, 194, 196–97
shelves, 544–45. *See also* shelf restraints, wire
  materials used for, 545–46, 550
  reagent, 545, 546
shielding, 415
shielding effectiveness (SE), 426
shields, 380
shift changes, during lockout/tagout, 178
shift selection, 96
shoes, 14, 326
short-term exposure level (STEL), 75
short-term exposure limit (STEL), 87
shoulder posture, 525, 526
showers, emergency, 188, 534, 535
shredding and disinfection of waste, 602
shutdown, preparation for, 176
sick building syndrome (SBS), 215–16
signage, 474, 540
silicone sealant, clear, 203
silver, 342
silver fulminate, 375–76
similar exposure group. *See* exposure groups
simplification, defined, 254
single-shift excursion limit, 88
sinks, 541, 542
  materials used for, 550
siphon purging, 229
skin
  first aid for chemical burns to, 210
  hazards affecting, three types of, 332–33
  toxicity and, 153
skin exposure, 150
slides, 276
slot hoods, 293–94, 297, 298
"smelling" chemicals, 12
smoke visualization, 295, 305
solid-phase extraction (SPE), 624
solid-phase microextraction (SPME), 624
solvent storage cabinets, specification for,
      398

solvents, 624
  flammable, 204, 353
  histological, 625
  safety, 353
source reduction, 243, 623
  defined, 616
source reduction techniques, 620–21
source separation, 624. *See also* segregation
source substitution. *See* substitution
space issues, and scale-up operations, 252
spark discharges, 258
specialists, 185
species
  carcinogens and, 142
  and toxicity, 150
specific absorption ratio (SAR), 424, 426
spectacles. *See* eyeglasses
spill prevention and preparedness, 624
spills. *See also* incompatibles, uncontrolled
  biological safety and, 476, 480–82
  cleaning up, 480–82
  containing, 482, 497
  emergency planning for, 400, 476
  of flammable liquids, in storage rooms, 351–52
  involving biological safety cabinet, 482
  involving equipment, 482
  liquid biohazard, 480, 481
  multiple hazards, 482
  radioactive, 497
  solid biohazard, 482
spray finishing, 357
sprinkler systems, 352, 539
stagnant spots, 253
stainless steel, 565
stairs, 569
standard deviation, 106–7
standard operating procedure (SOP) drafts
  preparing the final draft, 73–74
  writing and revising, 73
standard operating procedures (SOPs), 42–43, 72,
      176–77
  defined, 72
  revising accepted, 74
standard (safe) work practices, 466
State Emergency Response Commission (SERC),
      183–84
static electricity, 256–59
static muscle loading and work postures, 527
statistical interpretation, 85, 86, 88–89. *See also* data
      interpretation; *specific topics*
statistics
  compliance, 106, 107, 109–10, 117–19
  descriptive, 106–7, 109, 115–18
  nonparametric, 109–13, 119, 127–28
  parametric, 106–9
  suggested reading on, 99
steam. *See* water vapor
steam sterilization, 601, 602

steel, 565
  stress-strain curves for, 507, 508
steel barriers, 505, 506
step stools, 570
sterilant, 477–78
stirred reactors, 385, 386
stockroom, microscale chemistry laboratories and
      reduction of risks in, 242
storage, 42–43, 397
  emergency planning and, 400
  ergonomic design and, 553
  lab design and, 543–46, 553
  of nonhazardous supplies, 544–45
  for radioactive materials, 496
  rules for
    general, 399–400
    OSHA, 397–99
storage containers. *See also* containers
  indoor *v.* outdoor, 351
storage locations, 351
  inventory and, 394
  laboratory *v.* storeroom, 399
  overhead, 399
storage lockers, hazardous materials and, 356
storage room specifications, inside, 398–99
storage rooms, 351–52, 356
strain differences
  carcinogens and, 142–43
  in toxicity, 150
strain hardening, 506
stratification, 253
"stress concentrator" and stress corrosion, 267
students, 6–8
substitution, source/material, 599, 623, 625
  defined, 254
  radioactive, 609
suits. *See* garments
sulfate conjugation, 151
sunlight, chronic exposure to, 414
Superfund Amendments and Reauthorization Act of 1986
      (SARA), 182, 183, 196, 630
supervisors, 170–71
  and accident investigation, 206
supply air hoods, 301–2
support staff, nontechnical
  training for, 8
surgery, contaminated wastes from, 598
surveillance sampling, 94–95
survey periods. *See* observation intervals
sweep-through purging, 231
system design, defined, 522

tagout (TO). *See* lockout/tagout
tank farms, 357
tanks
  cryogenic liquid, 543
  electrostatic hazards associated with, 259–60
  portable, maximum allowable size of, 398

task complexity and unfamiliarity, 255
teaching assistants (TAs), 7
technician-level responders, 185
telecommunications, 540
temperatures, controlling space, 565–66
temporary services, 33–34
tensile strength, 506
tepid (tempered) water, 188
tepid (tempered) water systems, 189, 190
teratogenic effects of radiation, 439
teratogens, 321
terrorism, biological, 607
testing of chemical substances, 57–58
testing programs, 57
thermoluminescence, 442
thermoluminescent dosimeter (TLD) badge, 455
thermoluminescent dosimeters (TLDs), 454–55
    advantages and disadvantages of, 454–56
thermostatic tempering/mixing valves, 189. *See also*
        temperatures, controlling space
threshold dose (no-effect level), 77
threshold limit value ceiling concentration (TLV-C), 75–76
threshold limit value short-term exposure level (TLV-
        STEL), 75
threshold limit value time-weighted average (TLV-TWA),
        75
*Threshold Limit Values for Chemical Substances and Physical
        Agents and Biological Exposure Indices* (*TLV Booklet*),
        75, 84
threshold limit values (TLVs), 75–76, 89, 321, 414
time-weighted average (TWA), 75. *See also under*
        occupational exposure limits
tissue, facial, 12
titanium sulfate, 368–69
titration, 244
TNT (trinitrotoluene) equivalency ratings, 501–3, 506–7
toe caps, 326
tolerance, defined, 254
"toxic," 321
    origin of the word, 148
toxic products. *See also specific products*
    unexpected, 342
Toxic Substances Control Act (TSCA), 53, 630
    purpose of, and reporting policy, 53–54
    regulation of hazardous chemical substances and
        mixtures, 58
    requirements, 54–58
        expertise and organization required for compliance
            with, 59
        issues that may shape future, 58–59
    stakeholder participation in public policy and, 59–60
toxicity. *See also specific topics*
    acute, 148
    chronic, 149
    defined, 574
    determinants of, 150–51
    subchronic, 149

Toxicity Characteristic Leaching Procedure (TCLP), 574
toxicology, 82
    applications of, 154
    v. epidemiology, 139–40
    future directions in, 154–55
    history and background, 148
    resources in, 155
    subspecialties in, 153–54
Toxics Release Inventory (TRI), 59
TPA, 142
trade secret information, 58
training (chemical lab safety), 3, 6–9, 474. *See also*
        contractors; visitors
    awareness-level, 185
    methods of, 5–6
    specialized, 4
    on waste management, 500, 609
training requirements, LOTO, 174–75
training topics, 3–5
training videos, 5
transferees, 6
transferring chemicals, 241
trash storage, ordinary, 546
treatment, storage, and disposal (TSD) facilities (TSDFs),
        374
    selection of, 575–76
treatment of waste, 616, 623, 625. *See also* waste
trichloroethylene, 131
trinitrotoluene (TNT) equivalency ratings, 501–3, 506–7
twisting one's body, 525

ultra-micro method, 244
ultraviolet (UV) laser radiation, exposure to, 461–62
ultraviolet (UV) lights, 311–12
ultraviolet (UV) radiation, 413
    best safe practices, 415
    bioeffects of, 414
    exposure guidelines, 414–15
    sources of, 413–14
unattended operations, 10
uncertainty/uncertainty factor (UF), 77, 84–85
Uniform Building and Fire Codes (UBC/UFC), 509–10
uniform exposure group. *See* exposure groups
uninterruptable power supply (UPS), 540
"untagged" items (inventory), 395, 397
upper confidence limit (UCL), 102
upper explosive limit (UEL), 167, 169, 348
upper flammable limit (UFL), 222–25
upper respiratory tract infections (URI), 218
utility distribution strategies, 535
utility service, maintenance for, 236–37

vaccines, 604
vacuum pumps, 512
vacuum purging, 228–29
vacuum systems, central, 538–39
vaporization, 347

vapors
  density, 348–49
  flammability characteristics of, 222
vapors and gases
  in air, estimating flammability limits of, 223–24
  effects of temperature and pressure on flammability limits of, 225
  estimating the LOC for, 224–25
  experimental determination of flammable limits of, 225
  LFL and UFL of mixtures of, 225
variable-air-volume hoods, 304
vented balance safety enclosure, 298
ventilation, 170
  comfort, 292, 299
  in first aid, 209
  local exhaust/containment control, 292, 299. See also exhaust
    types of, 292–93, 299–300
  purpose of, 292, 299
  scale-up of, 253
ventilation control systems, 292, 496, 560–61
  air distribution criteria, 564–65
  constant volume (CV) v. variable air volume systems (VAV), 536–38
  controlling proper airflows and space temperatures, 565–66
  quantity of ventilation and quality of, 561–64
  in radioisotope labs, 496
ventilation criteria, 536
vessels
  containing flammable liquids, addition of solids to, 263
  electrostatic hazards associated with, 259–60, 263
  heating thick-walled, 273
vibrations, 524–26, 533
video transmission, closed circuit, 22, 23
violence, workplace, 24
visitor/business invitee health and safety process
  challenges in implementing a global standard, 37
  components for a successful, 36
  process and outcome measures, 36–37
visitors, 36
  training for, 8–9

wall coverings, in radioisotope labs, 498–99
wall cupboards, 545
walls, 351
washing hands, 12
washrag, 524, 525
waste. See also biological/medical waste management
  avoidance and prevention of, 609–10
  characteristics of hazardous, 573, 574
  contaminated laboratory, 598
  definitions of hazardous, 573
  mixed, 610
  pathological, 598
  transportation of, 576
waste accounting system, 618–19

waste accumulation, 573–74
waste controls, 629–30
waste determinations, 573
waste disposal. See also disposal
  of nonregulated chemical, 576
waste-disposal goals, 373–74
waste disposal strategies, 575
  selection of a TSD facility, 575–76
waste disposition technologies, 574–76. See also disposal
waste exchanges, 622
waste facilities, municipal, 576
waste generator classifications, hazardous, 573, 574
waste identification, 372–73
waste lists, hazardous, 577–97
waste minimization, 599–600, 615, 616
  application to laboratories, 622–26
  approaches to and techniques of, 620–22
  defined, 616
  as part of continuous improvement cycle, 626
  v. pollution prevention, 616
  reasons for, 615–16
waste minimization assessment process, 616
  assessment phase of, 617–19
  feasibility stage of, 617, 619–20
  planning and organization, 616–18
waste reduction, 243, 610
  measuring, 620
waste storage, 546
waste streams, prioritizing, 618–19
waste water collection and discharge, radioactive, 497
water
  radioactive waste, 497
  use in first aid, 209–10
water controls, 628
  national pollution discharge elimination system, 629
water miscibility, 349
water-reactive chemicals, 341
water vapor, 232
weather. See also flood contingency plans
  biological safety and, 477
Web browsers, 46
Web directories, 46–47
Web searches, conducting, 47
Web site(s)
  major health and safety, 52
  making them effective, 51
  planning a, 50
  selecting, 48–49
welding helmets, 326
What If/Checklist, 162
What If process safety review (analysis), 160–62
work culture, and emergency preparation, 194
work environment(s), 90
  defined, 90
  "reliably classified," 91
work schedules, nontraditional, 98

worker protection, 40

worker selection, 95–96

workers. *See also* employees; personnel strength, 514

workers' compensation claims, 218

Worker's Right To Know (RTK), 38

working alone, 9–11

working conditions, 255. *See also specific conditions*

working environment, 515

workplace characterization, 82

Workplace Environmental Exposure Level (WEEL), 77

workplace exposure limits. *See* occupational exposure limits

workstation design principles
computer, 519–20
general, 517

wrist postures, bent, 524

X-ray machines, analytical, 449